Gene Expression

Gene Expression

Gurbachan S. Miglani

Alpha Science International Ltd.
Oxford, U.K.

Gene Expression
532 pgs. | 317 figs. | 49 tbls.

Gurbachan S. Miglani
Visiting Professor
School of Agricultural Biotechnology
Punjab Agricultural University
Ludhiana, Punjab

ALPHA SCIENCE INTERNATIONAL LTD.
7200 The Quorum, Oxford Business Park North
Garsington Road, Oxford OX4 2JZ, U.K.

www.alphasci.com

Printed from the camera-ready copy provided by the Author.

ISBN 978-1-84265-819-2

Printed in India

In the sweet memory of my
Father-in-law
Jagjit Singh Bajaj
and
Mother-in-law
Harbans Kaur

Foreword

Gene Expression, written by my class-fellow, Dr. Gurbachan Singh Miglani, is a magnificent effort to make available to teachers and students of genetics various concepts related to genes and their function. Gene expression is fundamental to understanding how genotype and phenotype are related and determined.

Dr. Miglani has a knack for explaining difficult concepts in a simple and easy-to-understand manner. He possesses vast knowledge of genetics, which shines through this book. Being a teacher, he endeavors to present the matter in a learner-friendly manner. This is important, as teaching is becoming more and more learner-centered around the globe.

Gene Expression has 14 chapters. Each chapter is thoroughly illustrated with diagrams or drawings to make complex concepts easy to understand. This clearly shows Dr. Miglani's in-depth knowledge of the subject matter and his desire to facilitate learning. In my opinion, a good teacher is one who explains every little concept to build a strong foundation for the student. Dr. Miglani truly falls in this category.

The numerous references listed at the end of each chapter reveal the amount of published works that have been consulted in preparing this book. The reference lists contain not only the latest citable, scientific works but also some of the classic, landmark papers on genetics. These references should be valuable for those who wish to explore a particular area more extensively and to expand their knowledge further.

The glossary at the end of the book is a valuable feature of the book. I am sure both teachers and students will appreciate this facility.

I must congratulate Dr. Miglani on producing yet another excellent teaching tool for us. This book contains encyclopedic knowledge of the author on genetics, and I trust it will be adopted for not only undergraduate courses but also for graduate level courses. Dr. Miglani deserves our thanks for meticulously writing this book.

Manjit S. Kang
Adjunct Professor
Kansas State University
Manhattan, Kansas, USA

Former Vice-Chancellor
Punjab Agricultural University
Ludhiana, India

Preface

To become proficient in teaching, research, and applications of genetics, the underlying concepts, phenomenon, hypotheses, theories, laws and principles must be thoroughly understood at the molecular level. The blueprint or genetic material responsible for the physical appearance of an organism resides inside the cell. This genetic material stores all the information for the development and survival of the individual and transmission of biological properties from one generation to the next. The first step for studying molecular biology/genetics/biotechnology is to gain an in-depth understanding of the nature, structure, molecular forms, location, biosynthesis, organization, analysis, sequencing, organo-chemical synthesis, packaging, replication, recombination, damage, repair, protection and evolution of genetic material in viruses, prokaryotes and eukaryotes. Genetic material performs its function by organizing itself in the form of genes, popularly called units of heredity. Knowing above-mentioned aspects of genetic material sets the stage for knowing about gene. In order to understand how a gene expresses itself to perform its function, its various aspects ş gene structure, gene organization, genome evolution, gene function, transcription, RNA processing, genetic code, and translation, and fate of finished proteins, ribosomes and used messenger RNAs ş need to be understood. This book *Gene Expression* has been written with the objective of providing comprehensive knowledge on these aspects of a gene. However, it is assumed that the reader has knowledge about the fundamental principles of genetics, and structure, organization, evolution and properties of genetic material.

Gene Expression is primarily designed as a text book. However, teachers and research scholars can very conveniently use it as a reference book. Undergraduate and graduate students, teachers and researchers in any discipline of life sciences, agricultural sciences, medicine, and biotechnology in all the conventional and agricultural universities, research institutes, and molecular biotechnology companies/colleges/institutes/schools across the world should find it useful.

This book provides a brief historical background in introductory paragraph(s) of every chapter. Recent progress is also discussed. Nobel Prize winning works find a special mention. Various hypotheses, principles, concepts, phenomenon of molecular genetics have been dealt with in a simple and lucid language. The text is supported by sufficient number of original references, tables, figures and flow diagrams where necessary. Special attention has been paid to the precise definitions of various terms in the subject in light of the present day knowledge. For this, there is a separate section on glossary of important terms used in the book. Recent literature relevant to the subject matter has been cited in the text and all such references have been listed at the end of every chapter. Where there are more than six authors, only first three have been mentioned, followed by et al. Throughout efforts have been made to pin-point applications/implications of different discoveries in the area of molecular genetics and biotechnology. Extensive cross-referencing has been done. *Gene Expression* should be extremely useful to those who are preparing for national and international level competitive examinations, entrance tests, and interviews for jobs/fellowships. I trust the book will be an enjoyable reading experience. Strenuous efforts have been made to include all relevant information; however, if something is missing, the readers are urged to inform the author so that deficiencies can be corrected in a subsequent printing/edition. Style of presentation and selection of examples are purely my choice. I am responsible for any omissions and commissions. Readers' feedback will be appreciated.

Students have always been a guiding force to me in an indirect way in deciding order of different chapters in the book and organization of the contents in a chapter. Thus contribution of my present and past students in my books has been tremendous. During the writing of this book I have consulted a large number of original research papers, review articles, books, monographs, and web sites. My head bows with respect before all these great authors for their landmark works. I thank various copyright holders for granting me permission for use of their works in this book.

I am grateful to Dr. Kuldeep Singh, Director, School of Agricultural Biotechnology, Punjab Agricultural University, Ludhiana, India for his moral support and for providing facilities for my work.

Dr. S.S. Gosal, Director of Research, Punjab Agricultural University, Ludhiana, India, who himself is an experienced author, made very useful suggestions in the preparation of this book.

Stewardship provided by Dr. Manjit Singh Kang, former Vice-Chancellor, Punjab Agricultural University, Ludhiana, India, who himself has edited and authored several books in the area of genetics, is gratefully acknowledged. I also thank him for writing the Foreword of this book.

I must make mention of my immediate family, relatives and friends without whose involvement and support, completion of this book would not have been possible. No one can do any creative work such as writing of a book without the cooperation of his/her spouse. My wife Harjit showed utmost patience and cooperation during the preparation of this book. My son Jemmy and daughter-in-law Parvi made sure that I was free of tedious household duties like depositing bills and making purchases from the market. My daughter Simmi and son-in-law Shally f motivated me by checking up on the progress of the book. My grandchildren, Prabhasis, Harmehar and Bhavneet were more of an encouragement rather than a disturbance to me. Because of his experience in writing manuals and books, I often consulted my dear nephew, Dr. Sandeep Singh, Assistant Entomologist, Punjab Agricultural University, Ludhiana. He also assisted me in compiling the subject index.

Mr. Gagandeep Singh typed the manuscript and drew figures meticulously. Mr. Gurdeep Singh of ALPS Educational Services, Ludhiana formatted the book for camera-ready printouts. I am very happy the way the book *Gene Expression* has come up with respect to format and design. For this, I whole heartedly thank Mr. N.K. Mehra, Managing Director, Narosa Publishing House Pvt. Ltd., New Delhi.

Gurbachan S. Miglani
Visiting Professor
Former Adjunct Professor
School of Agricultural Biotechnology

Professor of Genetics (Retired)
Department of Plant Breeding, Genetics and Biotechnology
Punjab Agricultural University, Ludhiana
Email: miglanigs1945@gmail.com

Contents

1

Gene Structure and Colinearity

GENE STRUCTURE

Structure of gene can be understood in terms of its two functions — recombination and mutation. The concept of gene structure has been changing with the developments in the subject. This changing scenario can be discussed under three phases — classical, transitional and modern.

Classical Phase

In the beginning, gene was regarded as a hypothetical particle, performing a specific function and recombining with other genes. A gene was considered a unit of function and the unit of recombination was regarded no larger than a gene. After the independent discovery of Sutton (1903) and Boveri (1904), gene was considered a concrete particle — a small part of the chromosome. There are several genes carried by a chromosome. Genes located in the same chromosome less than 50 map units apart do not show independent assortment. Genes showing their location on the same chromosome can also show recombination by crossing-over. Studies of Morgan (1926) showed that multiple allelism arises from alternative modification of the same gene (e.g., *white* locus in *Drosophila melanogaster*). Multiple alleles (w and w^e) do not show complementation in F_1 (F_1 is not wild-type) and do not show recombination in F_2. Alleles determined on the bases of complementation test were termed functional alleles while those determined on the bases of recombination test were called structural alleles. At this stage unit of recombination was regarded no smaller than a gene. A gene thus was considered a unit of recombination. T.H. Morgan was awarded Nobel Prize in 1933 for discovering functions of chromosomes during hereditary transmission.

Sturtevant and Morgan (1923) studied *Bar* locus in *D. melanogaster*. A *Bar* mutation consists of insertion of a *Bar* gene (duplication) at that locus (16A). When this repeated segment was lost and the mutant reverted back to wild-type. Ultrabar had two repeats. Duplication of 16A segment reduced number of facets in the eye: B^+/B^+, 779; B^+/B, 358, B/B, 68, Ultrabar, 45 (Figure 1.1). Ultrabar arose due to unequal pairing and crossing-over (Sturtevant 1925). These observations lead to the conclusion that a gene does not necessarily represent a constant amount of genetic material and its alleles may differ from each other in this regard. When such recombined sections appear in *cis* position, the phenotype produced may be different than when they are in *trans* position. Muller (1927) observed that mutation could be produced after being hit by particles, which clearly indicated particulate nature of the gene. At this stage, gene was regarded as a unit of mutation. Muller (1940) defined gene as simultaneously a unit of function and a unit of recombination. This concept also implied that gene is differentiated internally.

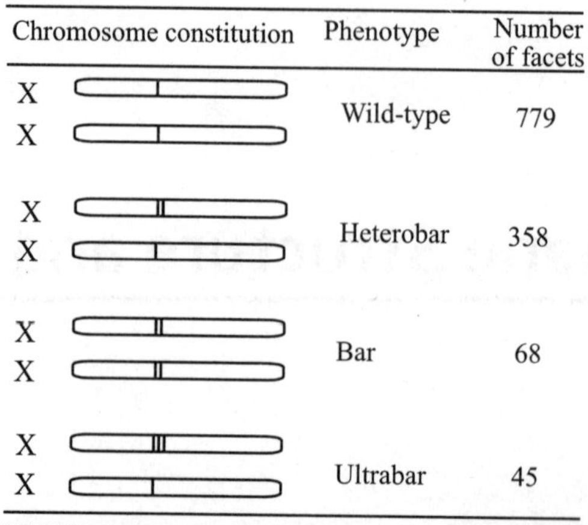

Chromosome constitution	Phenotype	Number of facets
X X	Wild-type	779
X X	Heterobar	358
X X	Bar	68
X X	Ultrabar	45

Figure 1.1 Number of facets in different phenotypes of *Bar* locus of *D. melanogaster*

Transitional Phase

This phase deals with pseudoalleles. Two mutations *Star* (*S*) and *asteriod* (*ast*) showed no complementation and *S* and *ast* mutations were concluded as allelic. Lewis (1941, 1951), however, showed that *S* and *ast* show 0.02 per cent recombination. Such mutations which are allelic in complementation test and non-allelic on recombination test were termed pseudoalleles by McClintock (1945). Chovnick (1961) made similar observations at garnet locus of *D. melanogaster*. It was suggested that such alleles (pseudoalleles) seem to occupy complex locus which can be divided into subloci between which recombination can occur.

Modern Phase

Pontecorvo (1952) claimed that if progeny obtained for detecting recombination was studied in thousands (10^5), recombination can be observed between alleles of multiple series. He suggested that the term pseudoallelism should be discarded and our concept of gene should be suitably modified. A gene may be regarded as a unit of function but not as a unit of recombination. Now we agree that recombination can take place within a gene; the phenomenon known as intragenic recombination. Thus two mutations are allelic if they do not show complementation. Recombination between different alleles implies that the two alleles had mutated at different places in the gene. A gene, therefore, may not be regarded as a unit of mutation. At this stage it appeared that a gene was a unit of function but not a unit of recombination or that of mutation. Gene can no longer be considered a structureless molecule but one with a well-defined internal structure.

Complementation mapping

Lozenge (*lz*) in *D. melanogaster* is a complex locus (Green and Green 1949; Green 1961) which consists of four groups, each containing one or more different *lozenge* alleles (Figure 1.2). Alleles of different groups recombine. All these mutants are recessive. When two mutations at two subloci are present in *cis* position, they have their wild-type alleles on the other homolog. However, when two mutations are placed in *trans* position many heterozygous combinations of different *lozenge* alleles do show mutant effects. From this observation, Lewis hypothesized that this effect arose from sequential nature of gene product synthesis. The discovery of *cis-trans* effect enabled simple functional test for

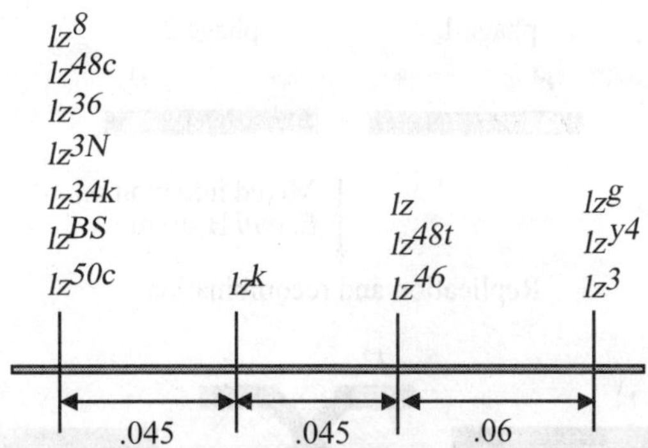

Figure 1.2 Some alleles of *lozenge* occupying different subloci within the *lozenge* locus of *D. melanogaster*

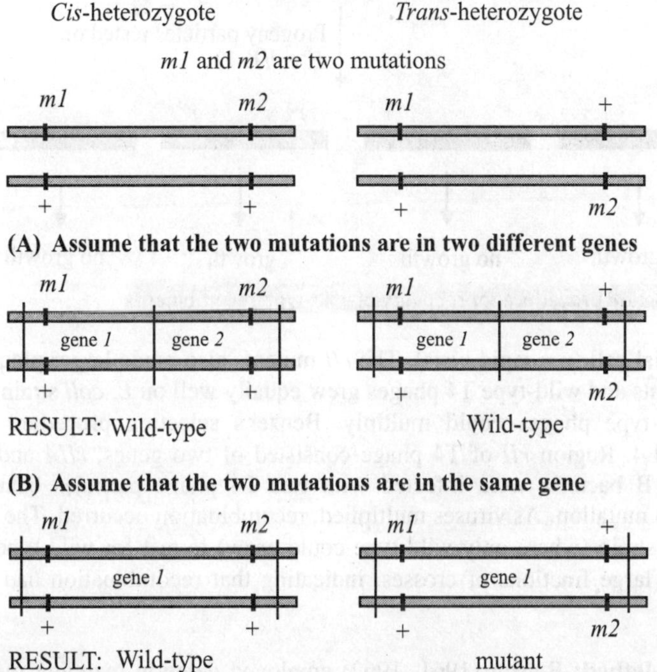

Figure 1.3 *Cis-trans* complementation test for knowing whether two mutations are located in two different genes or in the same gene

allelism with similar effect. Two mutations in *trans*-heterozygote are allelic if they produce a mutant phenotype and non-allelic if they produce a wild-type phenotype. The ability of two recessive mutations to restore wild-type phenotype (partially or completely) has been termed as complementation (Figure 1.3).

The *rII* locus

Complementation test determines if two mutations are in the same gene. Work on *rII* locus of bacteriophage T4 yielded interesting results. Certain mutations in *rII* region influenced length of life

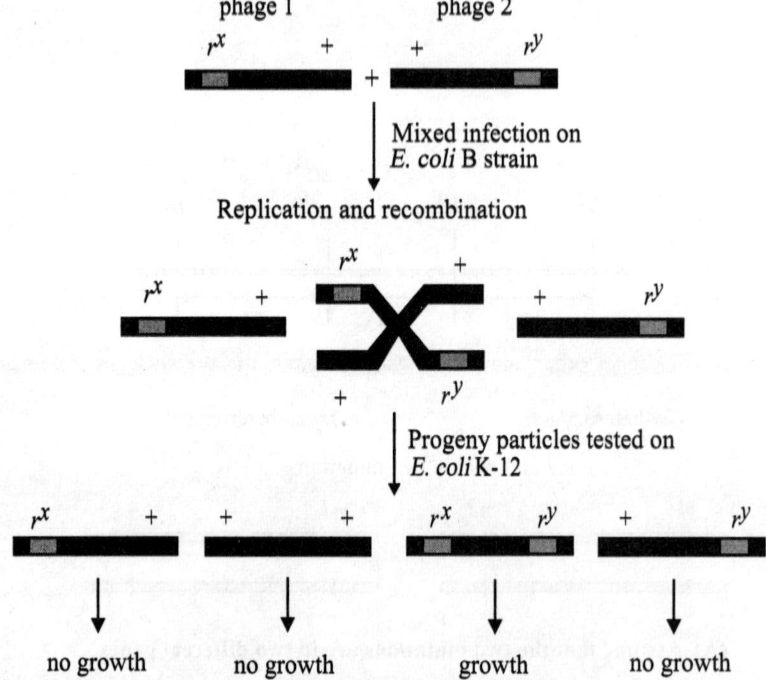

Figure 1.4 Benzer's selection procedure for recovery of wild-type recombinants

cycle of T4 in bacterial cell (r = rapid lysis). The *rII* mutants also made larger plaques than wild-type phages. The *rII* mutants and wild-type T4 phages grew equally well on *E. coli* strain B but on *E. coli* K (λ) strain only wild-type phages could multiply. Benzer's selection procedure (Benzer 1956) is described in Figure 1.4. Region *rII* of T4 phage consisted of two genes, *rIIA* and *rIIB*. In a typical cross, *E. coli* strain B bacteria were infected with two T4 phage particles bearing independently isolated *rIIA* (or *rIIB*) mutation. As viruses multiplied, recombination occurred. The progeny were then grown on *E. coli* (λ) strain (where only wild-type could grow) to test for wild-type. Normal particles were found in many large fractions of crosses, indicating that recombination had occurred within a gene (Figure 1.5).

Deletion Mapping Method: Benzer (1961, 1962) employed deletion mapping method to determine whether or not deletion included site of a mutation (Figure 1.6). Benzer found that through recombination he could distinguish between "point" mutations affecting only one mutation site and longer "multisite" mutations. The distinction was made by observing whether r^+ recombinants were produced from a cross between an unknown r mutant and r mutants previously known to recombine with each other to produces r^+ recombinants. Were such r^+ recombinants formed, the mutation could be considered a point mutation. If r^+ recombinants were not formed, the mutation was of multisite nature. For example, r mutants 155 and 274, known to recombine with each other to produce wild-type phages, did not produce r^+ recombinants when either of them was crossed to mutant *r164*. Mutant *r164* was, therefore, considered to be a multisite mutation that covered both the 155 and 274 regions. The fact that the presence of such multisite mutation could be shown to decrease recombination frequencies for mutations on either side of them (e.g., 288 and 196) indicates that they were probably deletions. Also, as expected of deletions, such mutations never showed spontaneous reversion to wild-type,

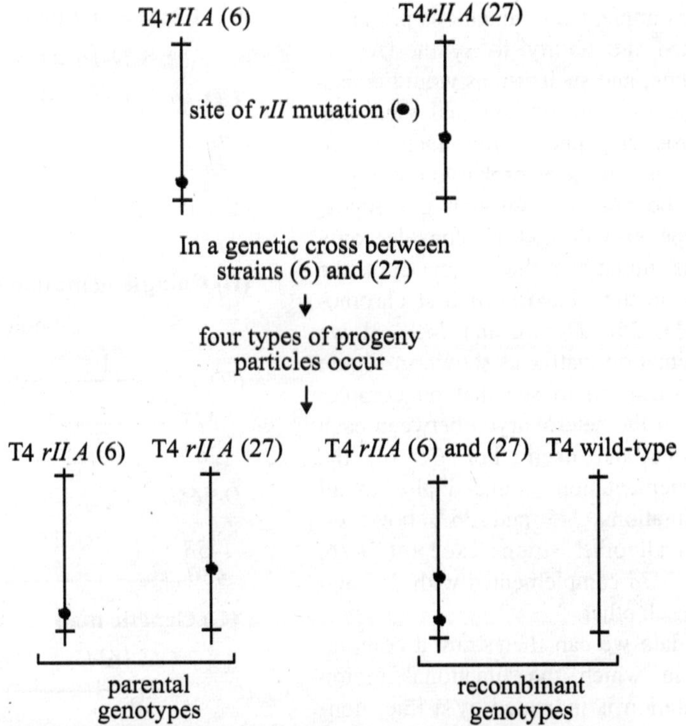

Figure 1.5 The use of T4 *rII* mutations in demonstration of crossing-over within the gene

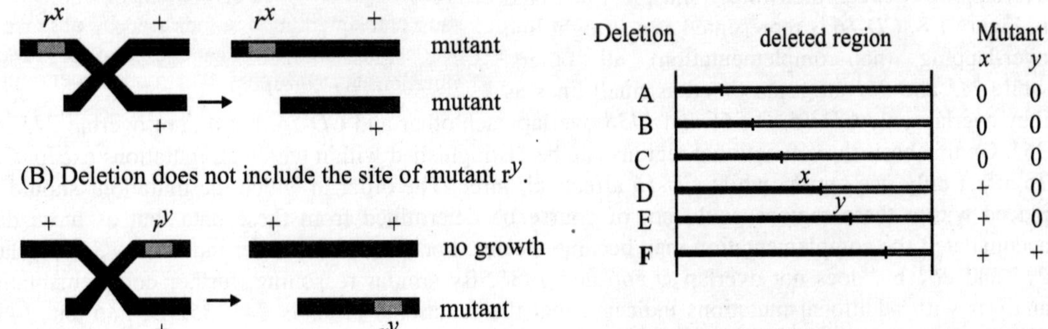

Figure 1.6 Principle of deletion mapping

Figure 1.7 Procedure for locating the position of unknown mutations r^x and r^y

whereas the point mutations occasionally did revert. Benzer used 47 large and small overlapping deletions to locate *rII* mutations within cistrons *A* and *B*. Procedure used for locating position of unknown mutations is described in Figure 1.7.

Complementation Matrix: The abundance of nutritional mutations in microorganisms and the ease with which they could be detected made them especially suitable for complementation studies. In

Neurospora, for example, many mutations produced strains that lacked the ability to synthesize the amino acid histidine, and such strains would consequently be unable to grow on minimal medi-um. However, by introducing nuclei from each of two such different strains into a heterokaryon, complementation might be observed through the appearance of wild-type growth. Let us consider only fivesuch histidine mutations that mapped in the same *his-3* locus on the *Neurospora* first chromosome, *CD-16*, *245*, *261*, *D-566*, and *1438*. If we draw a complementation matrix as shown in Figure 1.8, we are not surprised to see that no complementation occurs in the heterokaryon between each of these mutations and itself. For *CD-16*, the absence of complementation extended also to all other strains. Mutations *245* and *261*, however, complement with all other strains except *CD-16*, while *D-566* and *1438* complemented with *245* and *261* but not with each other.

Using these data we can then draw a complementation map in which the functional region affected by a mutation is indicated by a line; non-complementation between mutations is indicated by over-lapping of their lines; and complementation between mutations is indicated by the absence of overlapping between their lines. Thus, as illustrated in Figure 1.8, *CD-16* is represented as a straight line over-lapping (non-complementation) all others, while *245* and *261* are represented as small lines as they overlap only *CD-16*, *D-566* and *1438* overlap each other and *CD-16*, but do not overlap *245* and *261*. On this basis three functional regions can be distinguished within which all mutations except *CD-16* affect only one region, while *CD-16* affects all three. The order in which the mutations should be placed within these regions could not of course be determined from these data, but as more data accumulated the complementation map became clearer. For example, another mutation, *430*, overlaps *245* and *261* but does not overlap *D-566* and *1438*. By similar reasoning, further complementation analysis with additional mutations indicated that the functional order is *245*, *261-D*, *566* and *1438*. Since *CD-16* affected all three functional regions, its complementation position could not be determined from these data alone, but other data indicated that it extended to the left of *245*. A comparison between the complementation map and the genetic map determined by recombinational analysis is of great interest. As shown in Figure 1.8, the genetic map for these closely linked mutations preserve the same order as the complementation map, i.e., they are colinear. However, colinearity between the genetic and complementation map is not always true, and the allele that maps in one position genetically, may affect another functionally. The *CD-16* allele on the left of the genetic map, for example, affects distant functional regions at the extreme right. Mapping by overlapping deletions is illustrated through a matrix given in Figure 1.9.

(A) Complementation matrix

	CD-16	*245*	*261*	*D-566*	*1438*
CD-16	0	0	0	0	0
245		0	+	+	+
261			0	+	+
D-566				0	0
1438					0

(B) Complementation map

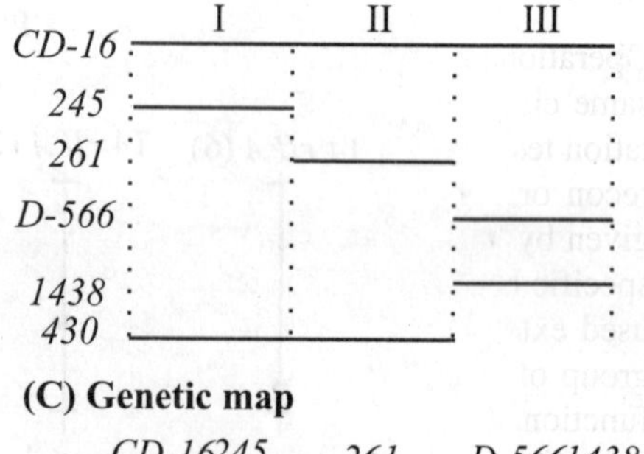

(C) Genetic map

Figure 1.8 (A) Complementation matrix, (B) complementation map (region affected) is indicated by a line; non-complementation between mutations is indicated by overlapping of their lines and complementation is indicated by absence of overlapping lines, (C) genetic map of various mutations in *Neurospora*

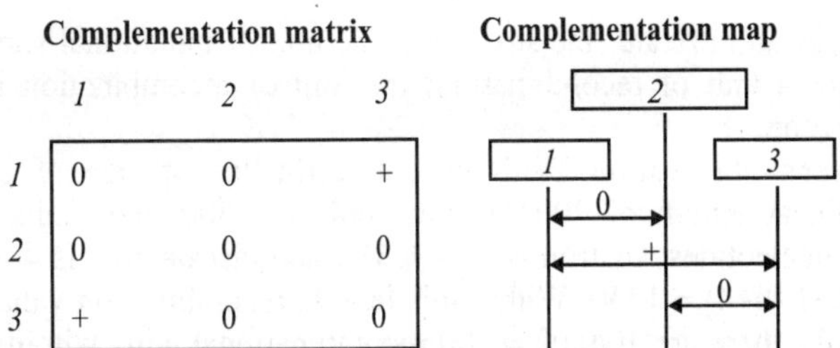

Figure 1.9 Complementation matrix and complementation map of three mutations

Fine structure of *rII* locus

Operationally, a cistron is defined by *cis-trans* complementation tests. Two mutations belong to the same cistron if they do not show complementation in *cis-trans* test or in a simple *trans* complementation test. Both muton and recon are defined by recombination test. Two mutations belong to the same recon or muton if they do not show recombination. In light of the molecular structure of DNA (as given by Watson and Crick 1953) gene can be considered as a part of the DNA molecule possessing a specific base pair sequence and a mutation as a change in the base pair sequence. Benzer (1955, 1956) used extensively *cis-trans* complementation test in his study of *rII* locus in T4 phage. He observed a group of non-complementary mutants falling within a short map segment, corresponding to a unitary function. Since *cis-trans* complementation test was used by Benzer, he proposed the term cistron. The term cistron and gene are now used synonymously. Benzer located 2,400 independent mutants at over 300 sites within *rII* locus (Figure 1.10). The existence of many mutational sites in the *rII* locus implies

(1) Simultaneous infection of *E. coli* K 12 (λ) with two phage particles each containing a separate *rII* A mutation.

rIIA *rIIB*

→ No phage multiplication

(2) Simultaneous infection of *E. coli* K12 (λ) with two phage particles each containing a separate *rII B* mutation.

rIIA *rIIB*

→ No phage multiplication

(3) Simultaneous infection of *E. coli* K12 (λ) with two phage particles one an *rII A* and the other an *rII B* mutation.

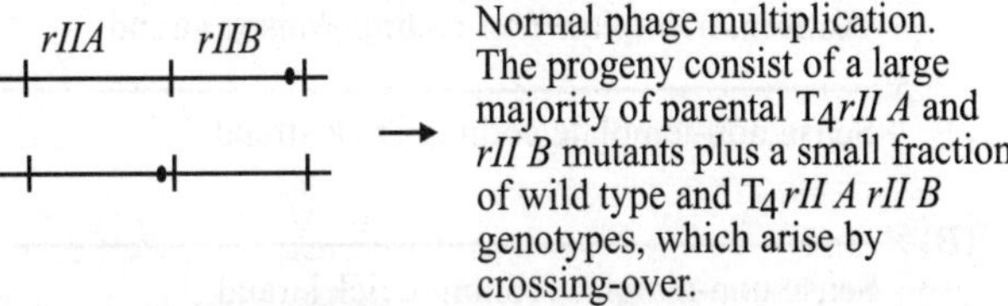

Figure 1.10 Demonstration that *rII* region consists of two distinct genes which can complement each other during simultaneous infection

that this locus possesses an intricate fine structure. The unit of mutational sites shows recombination. Gene is, therefore, not a unit of recombination; the unit of recombination is a much smaller unit, termed by Benzer as recon.

Comparing Benzer's observations with general estimates of size of phage chromosome, T4 chromosome has a linear length of 200,000 nucleotides, 1,500 map units (MU). Thus, 1 MU = 200,000/1,500 = 133 nucleotides. Cistron *rIIA* is 6 MU and cistron *rIIB* is 4 MU. Number of nucleotides in *rII* region = 133 (6+4) = 1330. Within this length, recombination values range from 10 to0.02 per cent. In other words, there are 10/0.02 = 500 recombinational units within *rII* locus. In all, Benzer estimated 400-500 mutational sites in rII region. Number of nucleotides per mutational sites = 1330/500 ≈ 3. It was suggested to be the size of the recombinational unit. These estimates were not considered as precise. With the advances in detection and estimation of recombination frequency and mutational activity, the present view is that the smallest recombinational unit probably is two neighboring nucleotides and probably the smallest mutational unit is a single nucleotide. In light of insights into the structure of the gene, from a geneticist's point of view a gene is a discrete chromosomal region which is responsible for synthesis of a specific cellular product (mRNA, tRNA, rRNA), and it consists of a linear collection of potentially mutable sites each of which can exist in several alternative forms and between which crossing-over can occur.

TYPES OF GENES

From structure point of view genes are classified into three types — simple gene, compound gene, and complex gene. These three different types of genes are described below:

Simple Gene

A continuous sequence in a nucleic acid that specified a particular polypeptide or functional RNA is defined as a simple gene. This condition exists in the simple RNA viruses. The functional portion of the gene obviously is the sequence of nucleotides that encodes a particular macromolecule which plays a role in the structure or metabolic processes of the cell. With respect to a particular gene product, the strand which is transcribed into RNA is called antisense strand, template strand, non-coding strand, or Watson strand. The strand complementary to the template strand is called sense strand, non-template strand, coding strand, or Watson strand.

Because first product of gene is RNA whose sequence is complementary to the template strand of DNA, it is customary to represent the gene by non-template strand sequence of duplex DNA. A gene in a double-stranded DNA molecule is located in antisense strand but in literature it is usually represented by the sense strand (Figure 1.11) because its sequence is same as that of its RNA product except that

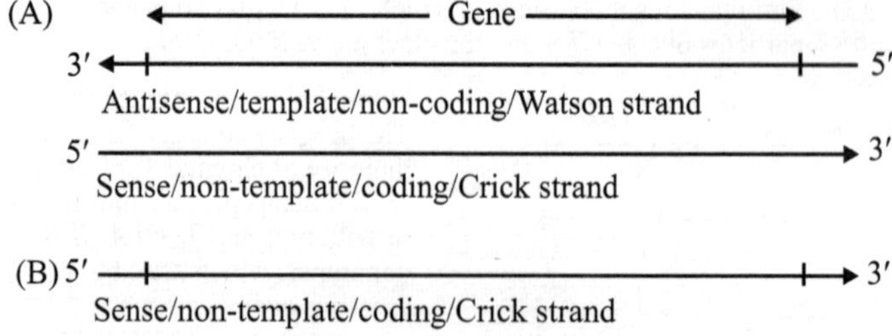

Figure 1.11 (A) Double-stranded DNA molecule showing location of a gene on the sense strand. (B) Antisense strand of DNA is actually used to denote a gene in literature

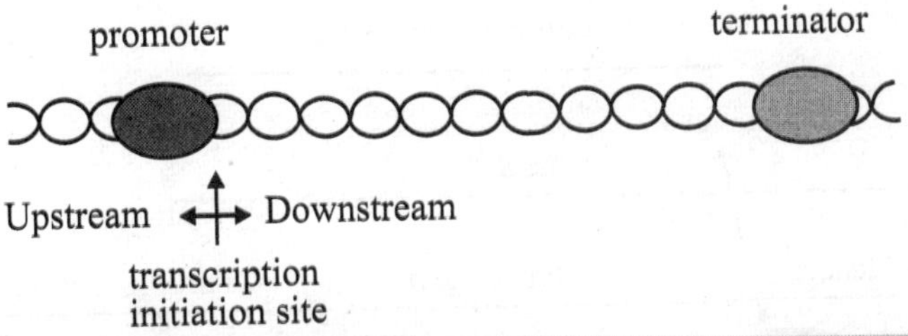

Figure 1.12 Transcriptional unit showing position of promoter and terminator. Transcription initiation site, and upstream and downstream sequences are indicated by arrows

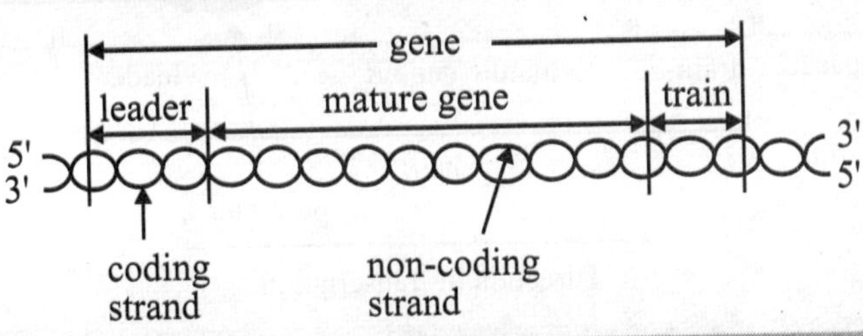

Figure 1.13 Gene structure showing leader and train as part of gene coding and non-coding strands are also indicated

where there is thymine in DNA, there is uracil in RNA. A transcriptional unit is a sequence of DNA transcribed into a single RNA, starting at the promoter and ending at the terminator (Figure 1.12). Intergenic spacers are the genomic regions that are not directly involved in the final product and are found between actual coding sectors of two adjacent genes. These spacers vary in length from a single base pair to many thousands base pairs. These are frequently referred to as flanking regions. Certain parts of these sequences exhibit specialization of function. The sector of the non-coding sequence preceding 5'-end of gene proper (mature gene, antisense strand) is called leader while that following the 3'-end is the train (Figure 1.13).

Leader

Leader is actually located before the 3'-end of the functional region of the coding strand and the train follows the 5'-end. One given strand of DNA does not necessarily bear the coding sequences of all the genes located in any single sector of the genome; sometimes some genes are located on one strand while others are located on the opposite one. Since cistrons are always read in 3'←5' direction on the coding strand, those on opposite strands also differ in polarity as shown in Figure 1.14. Thus train of one gene may be located under a mature gene of opposite polarity or even below its leader. The promoters and enhancers that regulate transcription of protein-encoding genes in mammalian cells are composed of multiple genetic elements, or modules (Dynan 1989).

The cellular transcriptional machinery is able to gather and integrate the regulatory information conveyed by each module, allowing different genes to evolve distinct, often complex, patterns of transcriptional regulation.

Promoter: To assist in locating, reading and processing the gene and its product, a number of signaling devices appear to exist. Promoters are specific sites of a given gene that serve in recognition of the gene and subsequent attachment with the transcribing enzyme. Promoter is part of the leader region. Promoter contains Pribnow box (term used for prokaryotic genes) (Pribnow et al. 1981) or

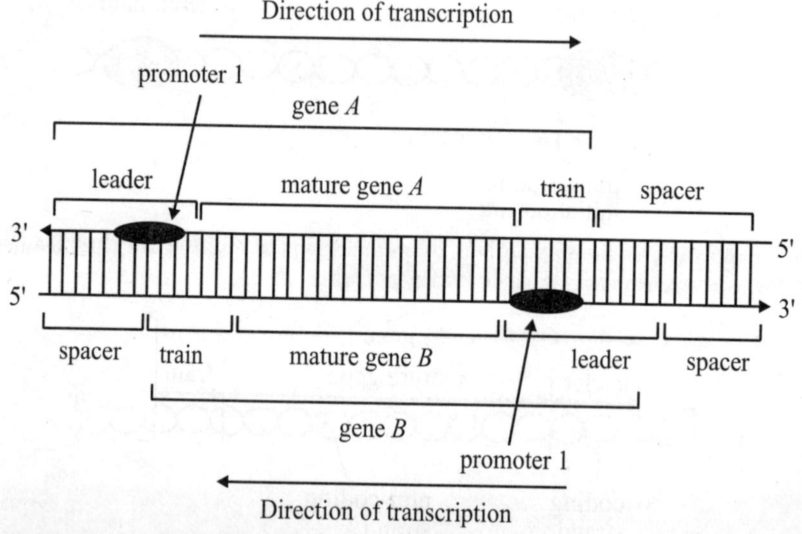

Figure 1.14 Two genes, *A* and *B*, showing that train of one gene may be located under the mature gene of opposite polarity or even below its leader

Goldberg-Hogness box (term used for eukaryotic genes) (Lifton et al. 1978). In certain genes coding for small species of RNA, e.g., tRNAs and 5S rRNA, promoter sequences are located within the mature gene itself. The sequence of DNA needed for initiation reaction for transcription defines the promoter. Transcription enzyme, RNA polymerase binds at this site. The site at which the first nucleotide is added during transcription is called startise or startpoint. RNA polymerase finds target promoter rapidly on DNA by random diffusion. Promoters belong to two general classes. Some promoters can be recognized by RNA polymerase holoenzyme alone. Other promoters are not by themselves adequate to support transcription. Ancillary protein factors are needed for transcription initiation to occur. These proteins recognize the sequence on DNA close to, or overlap with, the sequence bound by RNA polymerase itself.

Minimum size of promoter is 12 bp. Consensus sequence is defined by aligning all known examples so as to maximize their homology. Some generalizations in this regard are known in *E. coli.* The start point in greater than 90 per cent cases is a purine. Just upstream of the start point, a 6-bp region is recognized in almost all the promoters. The centre of the hexamer generally is close to 10 bp upstream of the start point. This hexamer is, therefore, often called the –10 sequence. This –10 sequence is TATAAT; conservation of these bases varies from 45 to 96 per cent (T_{80} A_{95} T_{45} A_{60} A_{50} T_{96}). At –35 sequence of promoter, consensus sequence is T_{82} T_{84} G_{78} A_{65} C_{54} A_{45}. Distance between –10 and –35 sequence in 90 per cent of cases is between 16 and 18 bp. Actually, sequence of these 16-18 bp may not be important, the distance between two sites may be important in appropriate separation for the geometry of RNA polymerase. Thus an optimal promoter is defined as a sequence consisting of the –35 hexamer, separated by 17 bp from the –10 hexamer, lying 7 bp upstream of the start point (Figure 1.15).

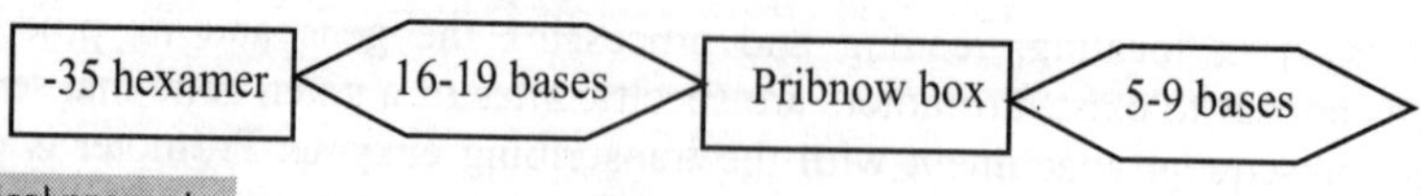

Figure 1.15 A typical promoter

Typical promoter can use the –35 and –10 sequences to be recognized by RNA polymerase. Down mutations in the promoter decrease the transcription rate whereas up mutations in the promoter increase the transcription rate compared to transcription rate of the normal promoter. Study of down and up mutations of the promoter suggest that the function of the –35 sequence is to provide the signal for recognition by RNA polymerase while the –10 sequence allows the complex to convert from "closed" to "open" form. RNA polymerase may initially "touchdown" at the –35 sequence and then extends its contact over the –10 region. The enzyme covers ≈60 bp of DNA. The initial "closed" binary complex is converted by melting of a sequence of 12 bp that extends from the –10 region to the start point. The A-T rich base pair composition of the –10 sequence may be important for melting reaction. In certain genes, coding for small species of RNA (tRNAs, 5S rRNA, etc.), promoter sequences are located within the mature gene itself.

Regarding the simplest functional promoter, transcription of the lymphocyte-specific terminal *deoxyribonucleotidyl transferase* gene begins at a single nucleotide, but no TATA box is present (Smale and Batlimore 1989). A 17-bp element is sufficient for accurate basal transcription of this gene. This initiator (*inr*) contains within itself the transcription start site. Thus *inr* constitutes the simplest functional promoter. There are downstream promoters for RNA polymerase III. For example, (a) genes coding for 5S RNA in *Xenopus* have been shown to have promoters downstream the initiation site and within the transcription unit +55 and +80 bp, within the gene; (b) adenovirus gene *VA1* promoter lies between +9 and +72; (c) gene $tRNA_f^{Met}$ promoter of *Xenopus* lies in two parts, between +8 and +30 and +51 and + 72. Several other tRNA genes also have similar promoter regions.

Synthetic promoter as a model to predict gene expression: Transcription binding sites are being discovered at a rapid pace. It is now necessary to turn attention towards understanding how these sites work in combination to influence gene expression. Quantitative models that accurately predict gene expression from promoter sequence will be a crucial part of solving this problem. Gertz et al. (2009) present such a model, based on the analysis of synthetic promoter libraries in budding yeast. The information encoded by combinations of *cis*-regulatory sites is interpreted primarily through simple protein-DNA and protein-protein interactions, with nucleosome modifications being downstream events.

Ancillary Sites: These sequences are located between –30 and –40 position, also referred to as –35 sequence. In bacterial genes, it is known as CAP (cAMP-activated protein) site. The term ancillary site applies to both prokaryotic and eukaryotic genes.

Enhancers: First enhancer found was 72-bp tandemly repeated sequence located 100 nucleotides upstream of the ancillary site in DNA of SV40 (Figure 1.16). A common feature of the series is GGTGTGG AAAG. Enhancers have been detected in eukaryotic genes also. Enhancers mostly are *cis*-acting but *trans*-acting ones are also known. Enhancers are effective whether lying upstream or downstream from the promoter. They are active whether they lie in same or opposite polarity as the mature gene. They are equally effective regardless of the organism from which the gene is derived when attached to foreign DNA.

SV40 enhancer region upto several kb from the promoter is given in Figure 1.17 (Ondek et al. 1987; Atchison 1988). The SV40 enhancer is composed of 15-20 bp. The enhancer elements cooperate with one another or duplicates of themselves to enhance transcription (Ondek et al. 1988; Dynan 1989). These elements are bipartite, being composed of subunits called enhansons, which can be duplicated or interchanged to create new enhancer elements. Enhansons differ from the enhancer elements because they are very sensitive to changes in spacing. This prototypic enhancer therefore contains two distinct levels of organization, each of which requires redundancy to be effective.

Complete sequence of leader of *trp* gene is shown in Figure 1.18.

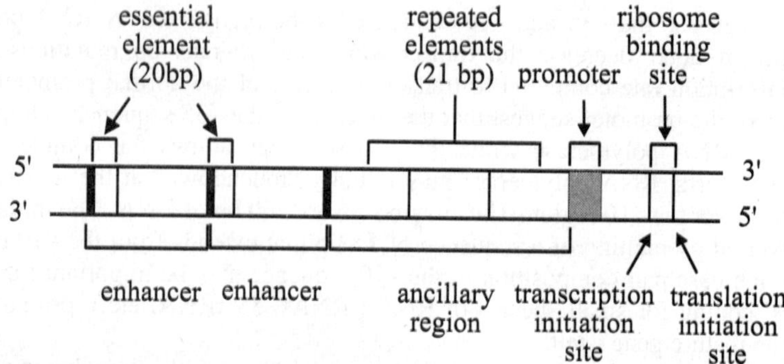

Figure 1.16 Sites upstream of translation initiation site of gene

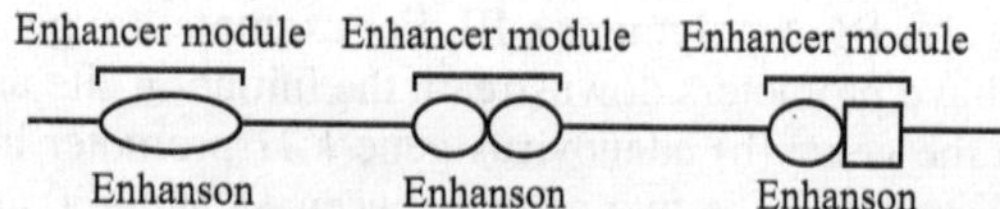

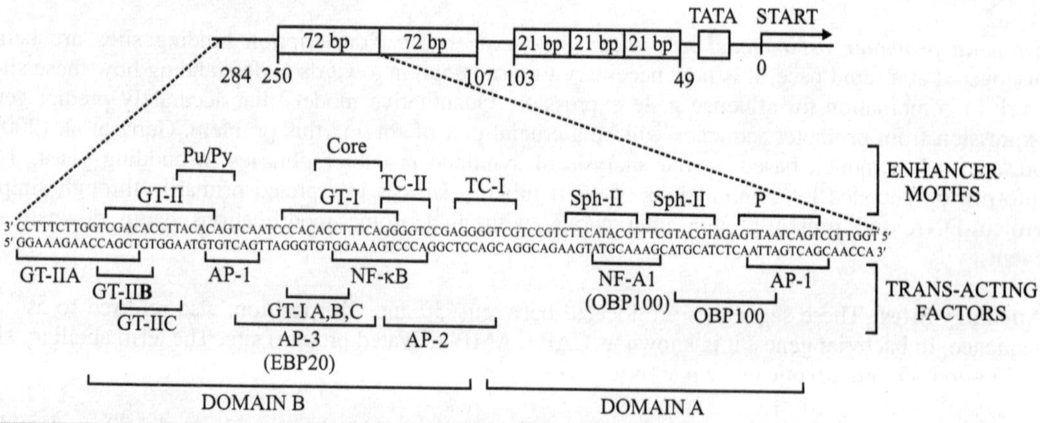

Figure 1.17 SV40 enhancer region upto several kilobases from the promoter

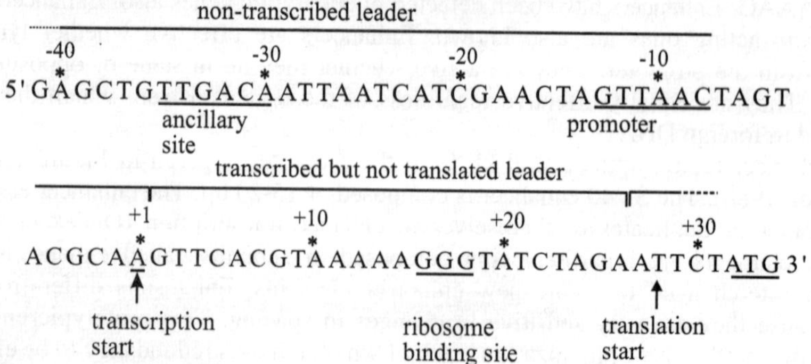

Figure 1.18 Structure of *trp* gene leader

Train

Train contains signals concerned with termination of transcription. Typically, there exists a series of T-A bp, in prokaryotes usually six or more in number, while in metazoans four times seems to suffice. A stem-and-loop structure could conceivably form in the transcript.

Terminators: RNA polymerase continues transcription until it meets a terminator sequence. At this point, the enzyme stops adding nucleotides to the growing RNA chain, releases the transcript and dissociates from the DNA template. Identification of terminators is provided by systems in which RNA polymerase terminates in vitro. Many prokaryotic and eukaryotic terminators require a hairpin to form in the secondary structure of the RNA being transcribed. This indicates that termination depends on the RNA product and is not determined simply by scrutiny of the DNA sequence during transcription. At some terminators, termination event can be prevented by specific ancillary sequences (factors) that interact with RNA polymerase. Responsibility for termination lies with the sequences already transcribed by RNA polymerase.

Antitermination causes the enzyme to continue transcription past the terminator sequence, an event called readthrough. Antitermination is used as a control mechanism in both bacterial operons and phage regulatory circuits. Antitermination is used as a control mechanism in operons controlled by attenuation to provide a link between translation and transcription. In this case, termination of transcription occurs when the ribosome is unable to move along a leader segment of the mRNA. During phage infection, different ancillary proteins (antitermination factors) allow RNA polymerase to bypass specific terminator sequences. Termination (as well as initiation) of transcription requires breaking of hydrogen bonds and additional proteins to interact with core enzyme.

Attenuation: Attenuation is a mechanism that links the supply of an aminoacyl~tRNA to the ability of RNA polymerase to read through a termination site. The terminator is located at the beginning of the cluster of structural genes coding for the enzymes that synthesize amino acid carried by tRNA. In this way, synthesis of amino acid responds to the level of aminoacyl~tRNA; if aminoacyl~tRNA is available, synthesis is inhibited, but if aminoacyl~tRNA runs out, more amino acid is synthesized. Attenuation was discovered in the tryptophan (*trp*) operon whose five structural genes are arranged in a contiguous series (Figure 1.19). Transcription starts at a promoter at the left end of the cluster. Transcription of the structural genes is partially terminated at a rho-independent site, *trpt*, 36 bp beyond the end of the last coding region. About 250 bp later there is a rho-dependent terminator *trpt'*.

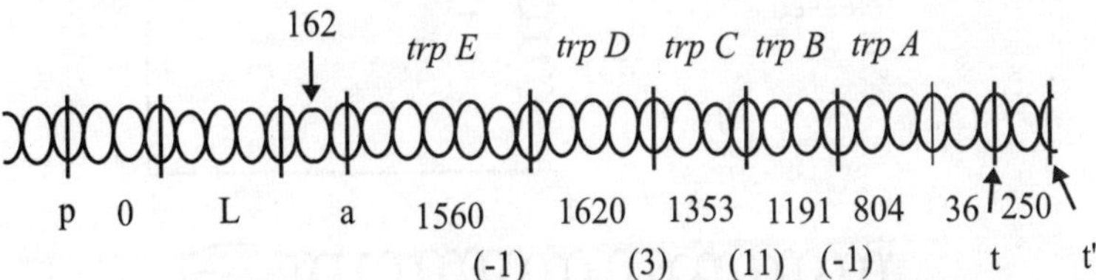

Figure 1.19 Tryptophan (trp) operon showing five genes arranged contiguously. Promoter (*p*) is at the left of the cluster. Rho-independent terminator (*trp t*) is located 36 bp beyond the end of the last coding region. Rho-dependent terminator (*trp t'*) is located about 250 bp later. L denotes a 162-bp attenuator

In addition to the promoter-operator complex, a sequence between the operator and the *trpE* coding region (denoted as L) which is 162-nucleotide long leader that precedes the initiation codon for the *trp E* gene (Figure 1.20). This regulator site is known as attenuator which serves as a barrier to transcription. The termination event at this site responds to the level of transcription. Figure 1.21

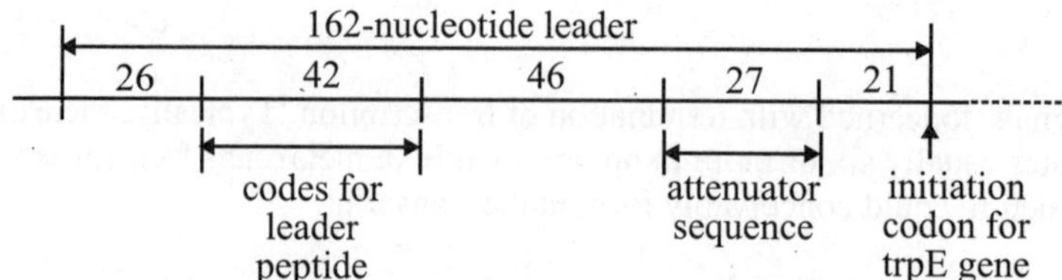

Figure 1.20 Structure of 162-bp leader of *trp* gene comprising of code for leader peptide and attenuator sequence

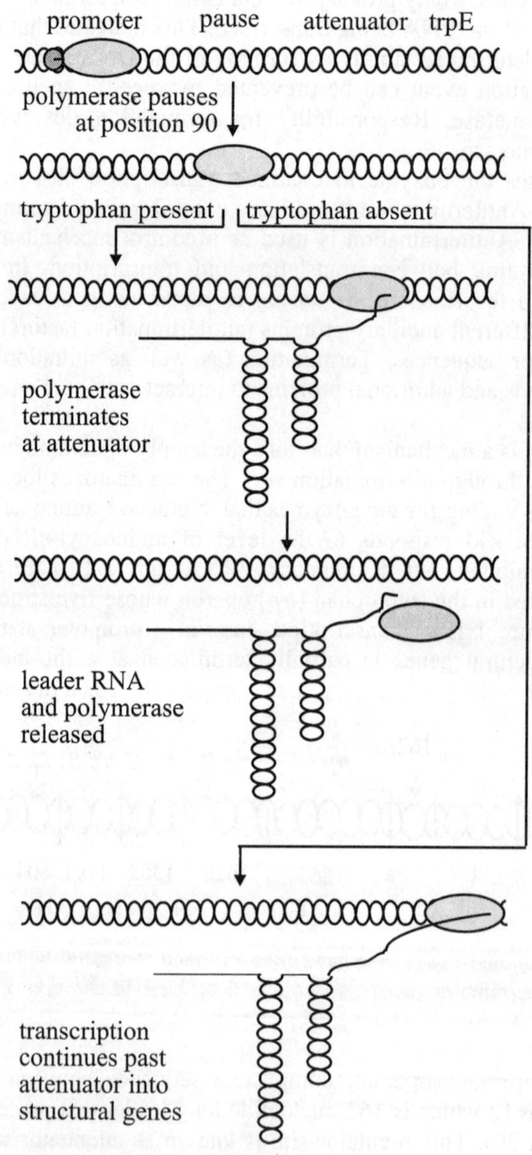

Figure 1.21 Termination events at the leader of *trp* operon

shows that RNA polymerase terminates at attenuator and leader RNA and polymerase is released in presence of tryptophan whereas in absence of tryptophan transcription continues past attenuator into structural genes.

The *trp* leader region can exist in alternative base-paired conformations as shown in Figure 1.22. Figure 1.22A shows the four regions (1-4) that can pair. Region 1 is complementary to region 2, which is complementary to region 3, which in turn is complementary to region 4. Conformation shown in Figure 1.22B is produced when region 1 pairs with 2 and 3 pairs with 4. Alternate conformation (Figure 1.22C is obtained when region 2 pairs with 3, leaving regions 1 and 4 unpaired. These signals are also known as attenuators and serve in regulation of gene's activity. Attenuator also marks the site for the attachment of poly(A) tail.

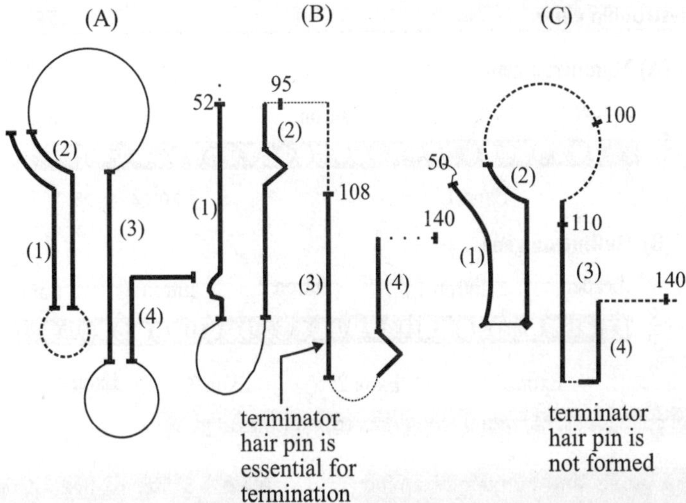

Figure 1.22 Alternate base-pair conformations in *trp* leader region: (A) Four regions (1)-(4) where region (1) is complementary to region (2) which is complementary to region (3) which in turn is complementary to region (4). (B) Conformation produced when region (1) pairs with region (2) and region (3) pairs with region (4). (C) Conformation produced when region (2) pairs with region (3) leaving regions (1) and (4) unpaired

Compound Gene

Compound genes are also known as split genes. Discovery of split genes was made independently by two groups led by Phillp A. Sharp (Berget et al. 1977a,b) and Richard J. Roberts (Chow et al. 1977) in the genes of adenovirus 2. Phillip A. Sharp and Richard J Roberts were awarded 1993 Nobel Prize for discovery of split genes and splicing of messenger RNA. Split genes were also reported by Wellauer and David (1977), David and Wellauer (1978), and Glover and Davidson (1977) in genes for 28S rRNA in *Drosophila* and by Chambon (1977, 1979, 1981) in *β-globin* genes, ovalbumin genes and tRNA genes.

In split genes, coding sequences (exons) are separated by non-coding sequences (introns). Numbers of introns in different genes are given in Table 1.1. Based on the number of introns in a split gene, the genes are classified into two types. First are the monintron genes as they have only one intron (Figure 1.23A). Such genes have been reported in case of $tRNA^{Tyr}$ gene in *Xenopus* and $tRNA^{Leu}$ gene in *Drosophila*. The transcripts of such genes are not used for translation and they as such perform a biological function.

Table 1.1 Number of introns in various genes

Gene	*Number of introns*
Actin gene of yeast	1
α and β globin gene	2
Chick ovalbumin gene	7
Xenopus leavis vitellogenin gene	3
Chicken α-2 collagen gene	50
Chicken ovomucoid gene	3
Chicken lysozyme gene	7
Human blood clotting factor VIII gene	25
Chicken collagen gene (one type)	51
Human dystrophin gene	78

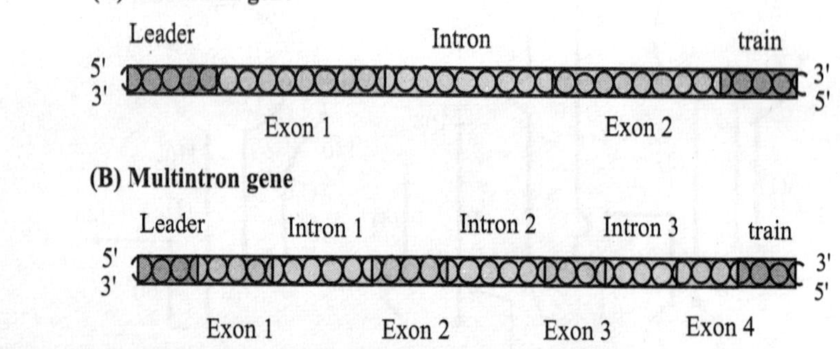

Figure 1.23 Types of split genes: (A) monintron gene; (B) multintron gene

Table 1.2 Size of six exons and number of amino acids coded by each exon of rat muscle α-actin gene

Exon number	*Size (nucleotides)*	*No. of amino acids coded*
1	93	31
2	273	91
3	282	94
4	121	40+
5	24	8
6	12	4

Table 1.3 Size of five introns of rat muscle α-actin gene

Intron number	*Size (nucleotides)*
1	1349
2	1439
3	307
4	395
5	612

The importance of this intron is indicated by the fact that removal of this intron leads to lack of processing and loss of function of the transcript. The second are the multintron genes as they have several introns (Figure 1.23B). Transcripts of multintron genes are translated into proteins, e.g., rat muscle α-actin gene has six exons. Size of each intron and number of amino acids coded by each exon is given in Table 1.2. Size of introns is usually much larger than the exons. Introns are spliced out during processing of the pre-mRNA (Table 1.3). The discovery of introns burst upon an unsuspecting world in 1977. Since then much information has accumulated about the ubiquity of introns in the genes of higher eukaryotes and their absence from those of prokaryotes and lower eukaryotes, but it is still not really clear why they are there or what they do. There is still argument on whether they were present in the primordial genes and were eliminated in the interest of efficiency by prokaryotes, or whether, instead, they were introduced after the eukaryotes separated from the prokaryotes.

Discovery of split genes

Breathnach et al. (1977) and Chambon (1977) discovered intervening sequences in ovalbumin gene of chicken by hybridization technique and found that this gene was split into three coding sequences interrupted by two intervening sequences. However, using restriction mapping and electron microscopy of mRNA-cDNA hybrids, Kourilsky and Chambon (1978) revealed the presence of seven intervening sequences separating 8 coding sequences. O'Farrell et al. (1978) discovered presence of intervening sequences in the anticodon loop of yeast tRNA gene. Such intervening sequences have also been observed in immunoglobulin gene of mouse, *28S rRNA* gene of *D. melanogaster* and SV40 virus. Intervening sequences have been given different names like spacers, inserts, silent DNA, non-transcribed region, intron, etc. Gilbert (1978, 1981, 1985) preferred to call coding sequences as exons and non-coding sequences as introns.

A large portion of human genome is intronic DNA. 1.5 per cent of human genome is exonic DNA (i.e., some 25,000 genes). One-half of exonic DNA is "traditional" single gene unique sequence DNA. 24.0 per cent is intronic multigene family DNA. All DNA sequences present in a split gene are not represented in finished mRNA. The sequences which are present in both DNA and its finished mRNA transcript are termed as exons while the sequences that are present in DNA but are lacking in finished mRNA are termed as introns.

Chambon (1977) suggests three possibilities by which split genes can produce a mature mRNA – (a) looping mechanism, (b) splicing mechanism, and (c) independent transcription. Evidences available support splicing mechanism. In case of yeast $tRNA^{Tyr}$ gene splicing occurs with the help of excision-ligase enzyme, which has the property of first breaking and then rejoining the regions. Intervening sequences were suggested to be involved in gene regulation. Doolittle (1978) suggested that split gene type of organization might have legacy from the last common ancestor of eukaryotes and prokaryotes. Gilbert (1978) suggested that split gene type organization would speed evolution by providing mechanism for the generation of novel proteins from old parts of the genes. Split gene organization seems to be a later evolutionary innovation.

The amount of DNA per cell increases approximately 890-fold from *E. coli* to mammals. Not all eukaryotic genes are split. Human and rat cytochrome C genes contain introns whereas yeast cytochrome C gene lacks introns. *Drosophila Adh* gene has introns whereas its yeast counterpart does not. Vast majority of metazoan genes are split genes. Genes without introns are exceptions. In yeast, only certain genes contain introns. Several classes of introns given by Danchin (1987) are: (1) short or very short introns are found in many tRNA genes. (2) Much longer introns are found in nuclear RNA and many mitochondrial mRNA genes. (3) The longest introns (a few hundred to tens of kb long) are found in genes specifying proteins. Intron encoded proteins act as helpers in RNA splicing.

It is scarcely tenable to propose that all introns are functionless, opportunistic pieces of selfish DNA which have invaded eukaryotic genes by taking advantage of the biochemical machinery that normally excises them. If so why then the excision machinery evolved in the first place? At least some introns, therefore, must have had a function, at some stage in the evolution of the gene. The most plausible suggestion is that the existence of introns makes it much easier for large and complex proteins to evolve, as they can be assembled from the small functional units or domains into which introns split genes. Moreover, new functions could in principle be produced by rearranging such domains. Once introns evolved, however, it became possible for parasitic DNA to take advantage of the excision machinery. Thus, although some, perhaps most, introns must have a functional origin, it is not necessary to assume this for every intron.

Split genes in chloroplasts

Split genes for ribosomal RNA (rRNA), transfer RNAs and some proteins have also been reported in the chloroplast genomes of several plants including *Chlamydomonas* and *Nicotiana*. Introns found in chloroplast genes are classified into three groups on the basis of intron boundary sequences. Group I introns (e.g., in *trnL*) can be folded in a secondary structure similar to self-splicing rRNA procedure of *Tetrahymena*. These can be removed either by self-splicing or by a 'maturase' (as in cytochrome b and cytochrome oxidase mRNA precursors). Group II introns (e.g., majority of genes including *trnA* and *trnl*) can be folded into a complex secondary structure (as in introns of mitochondrial genes for cytochrome oxidase in maize and yeast). Group III introns (e.g., *trnG*, *trnK*, *trnV*, *rpl2*, *rps12*, *rps16*, etc.) have conserved sequence at their borders (GTGCGNY at 5'-end, and ATCHR YY(N)YYAY at 3'-end), similar to those in the eukaryotic nuclear genes.

Split genes in mitochondria

Split genes are also found in mitochondria. Introns of these split genes in fungal mitochondria are classified into two groups on the basis of their internal organization. Group I introns, are found in majority of the fungal mitochondrial split genes, do not carry any conserved sequence at intron-exon junctions, but carry internally a short conserved sequence called 'internal guide sequence'. By internal pairing this guide sequence brings the two intron-exon junctions together and helps in splicing out the introns. These group I introns are also found in the nuclear genes coding for rRNA in *Tetrahymena* (a ciliate) and *Physarum* (a slime mold). Features similar to those of group I introns are also found in introns of phage T4 genes. Group II introns resemble nuclear genes and have consensus sequences (GT and APy) and a branch sequence that resembles the TACTAAC box. These introns are excised as lariats. In fungal mitochondria, for some genes it has been observed that introns may code for RNA maturases or endonucleases, which take part in RNA processing and DNA recombination, respectively.

Eukaryotic Promoters

In eukaryotic promoters, Hogness box, comparable to Pribnow box of prokaryotes and hence also called as TATA box, is 7-bp long located between position –20 and –60 bp of the transcription start site. Sequence of the Hogness box is $T_{82}A_{97}T_{93}A_{35}A_{43}/T_{27}A_{83}A_{50}/T_{37}$. Hogness box is surrounded by G.C-rich sequence. Further upstream is another sequence, known as CAAT box, which is necessary in transcription initiation and is conserved in some promoters (e.g., *β' globin* gene) while not in others (e.g., *tk* gene of herpes virus). CAAT box, having sequence GGT/CAATCT, lies between –70 and –80. Eukaryotic promoters may also possess elements, called enhancers, located 100 or 200 bp upstream which interact with proteins other than RNA polymerases and thus regulate activity of promoters. These elements may lead upto 200-fold increase in transcriptional activity.

The events leading to transcription of eukaryotic protein-coding genes culminate in the positioning of RNA polymerase II at the correct initiation site. The core promoter, which can extend ~35 bp upstream and/or downstream of this site, plays a central role in regulating initiation (Smale and Kadonaga 2003). Specific DNA elements within the core promoter bind the factors that nucleate the assembly of a functional preinitiation complex and integrate stimulatory and repressive signals from factors bound at distal ends. Although core promoter structure was originally thought to be invariant, a remarkable degree of diversity has become apparent. Figure 1.24 shows core promoter motifs. Some of the sequence elements can contribute basal transcription from a core promoter.

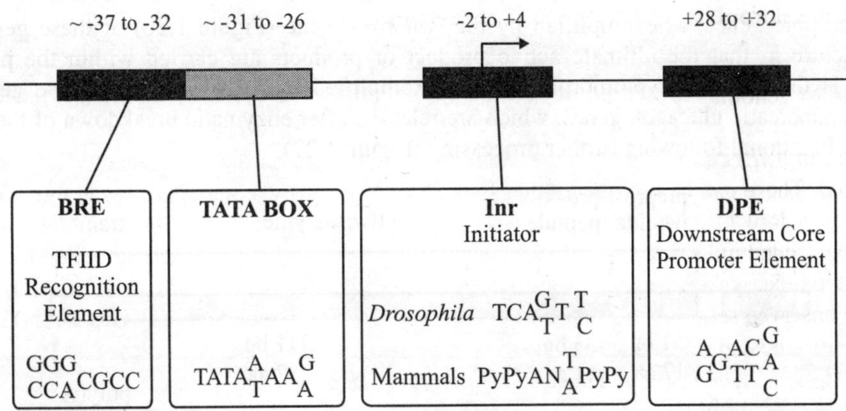

Figure 1.24 Core promoter motifs

A particular promoter may contain some, all, or none of these elements. The TATA box can function in the absence of BRE (TFIIB recognition element), *Inr* (initiator), and DPE (downstream core promoter element). In contrast, the DPE motif requires the presence of an *Inr*. The BRE is located immediately upstream of a subset of TATA box motifs. The DPE consensus was determined with *Drosophila* core prompters. The *Inr* consensus is shown for both mammals and *Drosophila*. Internal promoters are present in *5S rRNA*, *VA1*, $tRNA_f^{Met}$ genes. RNA polymerase III transcribes all these genes.

Complex Gene

Complex genes are so called because either there is rearrangement of DNA at different levels before 3'←5' strand of DNA is available for transcription (as is case of immunoglobulin gene) (Figure 1.25) or pre-mRNA is cleaved through a series of steps before mRNA in finished form is available (as is the

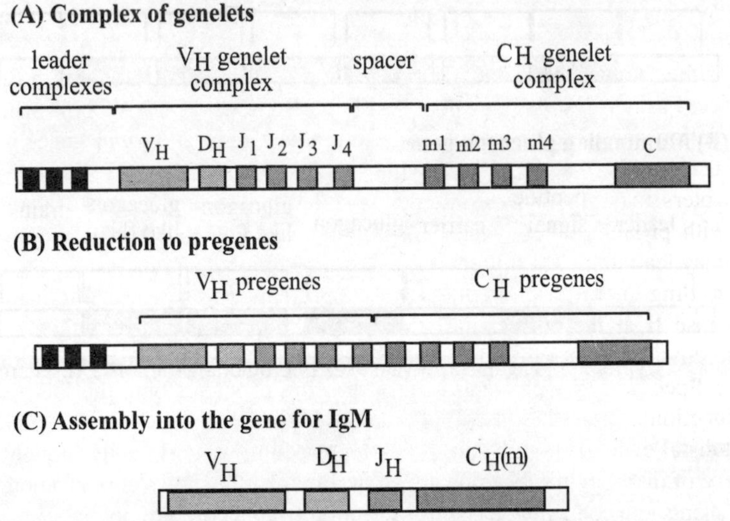

Figure 1.25 Rearrangement of DNA at various levels for assembly into gene for immunoglobulin M (IgM)

case of dimorphic genes, as exemplified by the *kallikrein* gene (Figure 1.26) or these genes have a cryptic structure in that the ultimate active product or products are carried within the precursorial protein (as is the case of cryptomorphic genes, exemplified by yeast sex pheromone gene (*MF1*), mammalian pancreatic glucagon genes, which are released after enzymatic breakdown of the precursor and become functional following further processing (Figure 1.27).

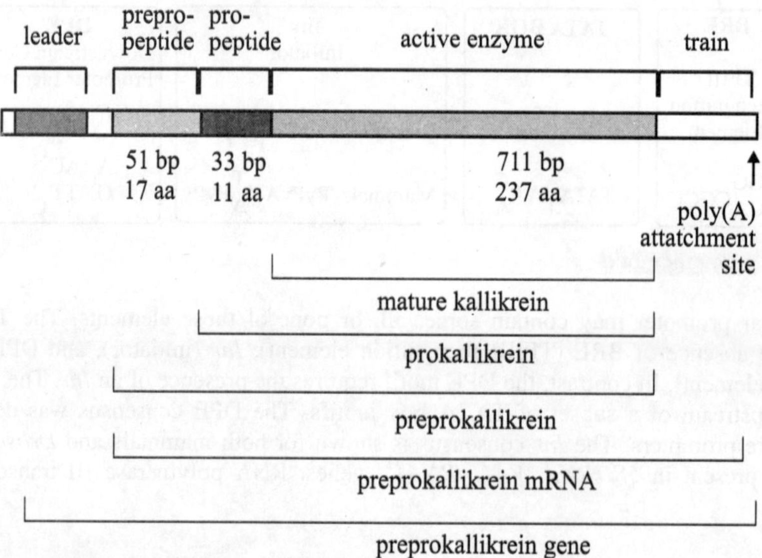

Figure 1.26 *Kallikrein* gene structure — a dimorphic gene

(A) Yeast sex pheromone gene

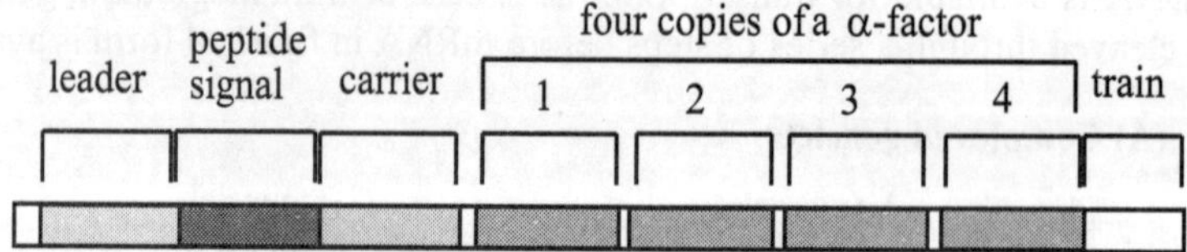

(B) Mammalian glucagon gene

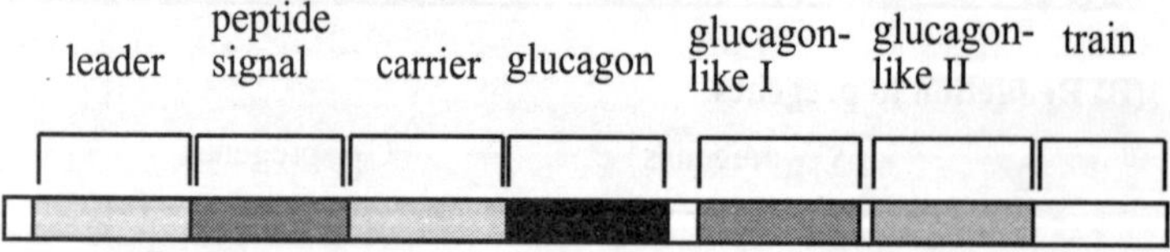

Figure 1.27 Structure of cryptomorphic genes: (A) yeast sex pheromone gene, *MF 1* (B) mammalian pancreatic glucagon gene

WHAT IS A GENE?

Mendel's "particles" are unit of heredity responsible for phenotype. Morgan's "loci" are genes in a chromosome, i.e., it is a cellular entity that is part of chromosome and is mapable. According to

Watson and Crick, gene is a sequence of specific nucleotides along the length of a double helical DNA molecule. One nucleotide (n) has length 0.34nm. Thus tRNA is 81n x 0.34 = 27.5 nm long. Mass of one nucleotide is 340 amu. Thus tRNA gene is 81n x 340 = 27,540 amu. Modern functional definition of gene is: a DNA sequence coding for a specific polypeptide but split gene DNA sequence must also include introns (non-coding segments) and exons (coding segments that make proteins) and. others DNA pieces. Any definition of gene must also include promoters, enhancers, regulator genes, operators, and also segments that code for rRNA, tRNA, and snRNP's. The gene is a sequence in a nucleic acid (DNA, except in RNA viruses) that provides a code for a protein, polypeptide, or RNA of direct value to cellular metabolic processes, plus those parts, adjacent or internal, important to its being located, identified, transcribed, often translated, and processed. Internal transcribed regions, such as introns, are also a part of the gene.

Thus, the sequence of a typical prokaryotic gene can be classified into the following regions: recognition region (~50 bp), transcription initiation site, 5' untranslated region, translation initiation, coding region, translation stop site, 3' untranslated region and translation stop site. For eukaryotes the gene structure is similar to prokaryotes except that the coding region' for prokaryotic genes is replaced by an alternating sequence of exons and introns separated by splice sites, and there is a polyadenylation signal to account for. Thus eukaryotic gene has following regions: recognition region (~50 kb), transcription initiation site, 5' untranslated region, translation initiation, alternating exon/ intron, splice donor and acceptor sites, translation stop site, 3' untranslated region, polyadenylation signal, and transcription stop site.

INTRONS

Consensus sequences at the boundaries of intron-exon junctions have shown that GT was always found at the 5′-side of the intron (left splice junction) and AG at 3′-side (right splice junction). This is known as Chambon's rule. Introns are DNA junk or sophisticated genetic control elements. Introns are present in abundance in eukaryotes. They are absent in prokaryotes: they have few non-coding DNA sequences. As eukaryotic complexity grows so does non-coding DNA which makes up greater than 95 per cent of the DNA. Less than 1.5 per cent of human genome encodes proteins, but all of DNA is transcribed. 40 per cent of human genome is transposons and repeat genetic elements.

Intron sequences are recognized by short sequence elements within exons, known as exonic sequence/splicing enhancers (ESEs). But the mechanism by which they function is poorly understood. The fact that the splicing process occurs contemporaneously with RNA polymerase transcription elongation allows for the immediate splicing upon the recognition of an exon-intron junction sequence, and may account for the high accuracy of this system. Additionally, some evidence suggests that the C-terminal domain of RNAP II itself helps to tether the exon and splicing machinery together.

The relative uniformity of exon length in protein coding genes is not, as claimed by Straus and Gilbert (1985), a good evidence against an insertional origin. The periodic structure of chromatin may prevent totally random insertion. The correspondence between the three major peaks in exon size distribution and the DNA lengths in nucleosome core particles, linkers, and whole suggests that the junctions between linkers and core particles have been hot spots for intron insertion and therefore that introns in protein-encoding genes originated after the origin of eukaryotic chromatin. By contrast, exon length in mitochondria and chloroplasts, which lack histones, is much less uniform and much random than in nuclei. Ovalbumin gene has 8 exons (E_1-E_8). Size of these introns is: E_1 = 47 bp; E_2 = 185 bp; E_3 = 51 bp; E_4 = 129 bp; E_5 = 118 bp; E_6 = 143 bp; E_7 = 156 bp; and E_8 = 1,043 bp. These exons are interrupted by 7 introns (A-G) (Chambon 1981) (Figure 1.28). Entire chicken ovalbumin gene (all exons + introns) is 7,700 bp in length. Removal of introns is in two steps. In step 1, five introns are removed and in step 2, two introns are removed. Mature mRNA is 1,872 nucleotides long.

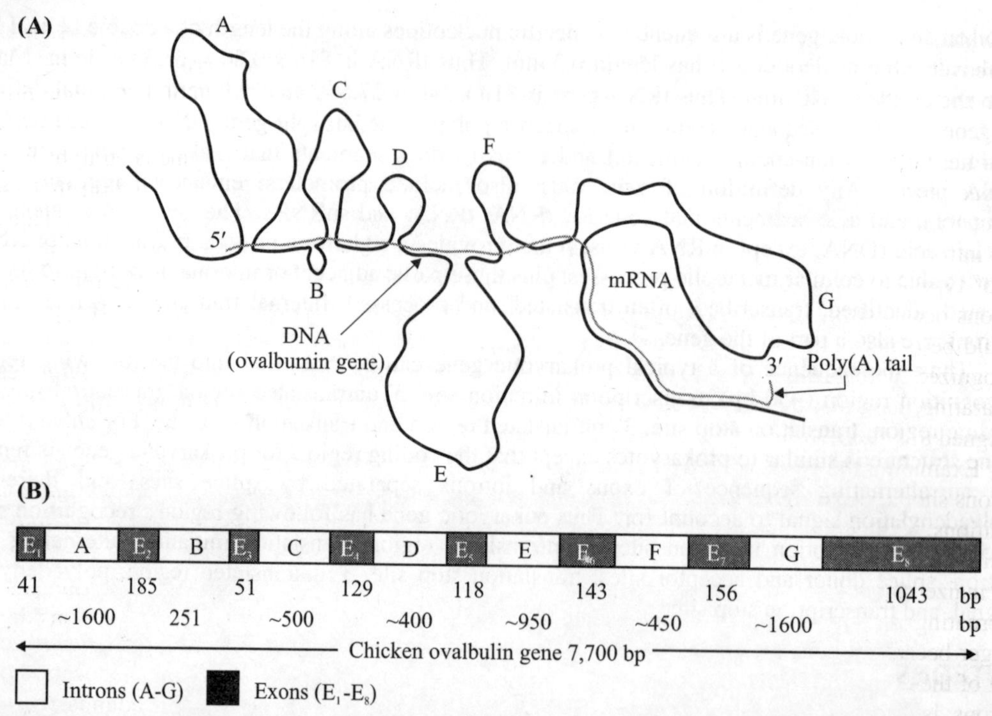

Figure 1.28 (A) The map of electron micrograph prepared after hybridizing mRNA (thin line) of the ovalbumin gene with its single-stranded (thickline). The loops represent introns. (Redrawn, with permission, from Gupta, P.K. 2008. *Molecular Biology and Genetic Engineering*. Meerut: Rastogi Publications) (B) Structure of chicken ovalbumin split gene. Exons are numbered as E_1-E_8. Introns are numbered as A-G

Classification of Introns

There are actually four different types of introns: Type 1 introns are self splicing and are employed in nuclear, mitochondrial and chloroplast rRNA, tRNA and mRNA. In Type 1 introns, the 3′ OH group of a free guanosine acts a nucleophile, attacking the 5′ phosphate at the splice site, displacing it from the exon. The 3′ OH at the exon, usually a U, acts as a nucleophile, attacking the phosphodiester bond, removing the other end of the intron, and reforming the phosphodiester bond with itself. Type 2 introns are also self-splicing, and are used in genes in mitochondria and chloroplasts of fungi algae and plants. In Type 2 introns, a similar mechanism is used, except that an internal 2′ OH of an adenosine is used as the nucleophile, attacking the 5′ splice site of the exon to produce a lariat structure. In the final step, the 3′ OH of the 5′ exon attacks the phosphodiester bond of 3′ uracil, to remove the lariat. Type 3 introns require a spliceosome and small nuclear ribonucleoproteins (snRNPs) and are responsible for solely eukaryotic splicing. In type III introns, a lariat structure is formed, but the snRNPs are used to form the secondary structures and to mediate the reaction, though it still occurs without free energy input. There are 6 snRNP subunits that make up the splicesome. Type 4 introns require ATP and an endonuclease.

The case of glyceraldehyde-3-phosphate dehydrogenase gene

Stone et al. (1985) classify introns in glyceraldehyde-3-phosphate dehydrogenase (GAPDH) gene of the chicken as type A which result from the original assembly of the gene from smaller units and type

B, which have never had any functions. They suggest that both kinds of introns are recognizable. Moreover, they believe that type A introns can be further classified according to whether they derive from the creation of a domain from smaller units (A1), or from the duplication of a complete domain (A2), or from the joining of two dissimilar domains (A3). Classification of the three introns that occur at or near the boundaries of the four recognized structural domains of the enzyme is straight forward — one is type A2, two are type A3. In addition, the catalytic domain contains a region that makes about 85 per cent of hydrogen bonds within itself and which also corresponds to an exon. Thus, the two introns which flank this exon can be regarded as resulting from the joining of protein domains and so of type A3. Apart from one intron in the non-coding region of mRNA, all of the remaining six introns occur within domains, and are more difficult to classify, because both A1 and B type introns would be expected in similar positions. Nevertheless, R.J. Schwartz believes that subdomains can be recognized within the domains and that several of the remaining introns can be interpreted on separating them that they are of type A1 (Stone et al. 1985). Only one intron in the coding region is assigned to type B.

Examining the genes of comparable proteins can test this type of classification. Types A2 and A3 introns should occur in other genes at domain boundaries; type A1 introns should occur at homologous positions within homologous domains; but type B introns should occur at arbitrary and unrelated positions. Structure of the gene for chicken pyruvate kinase has been compared with that of the gene for maize alcohol dehydrogenase. All these proteins contain a mononucleotide-binding domain with alternating regions of β-pleated sheet and α-helix though this domain in GAPDH is considerably longer because it contains an extra piece of β-sheet (which does not interfere with the structure of the rest of the domain). The comparison is shown in Figure 1.29 with arrows indicating the location of the introns, is disappointing; intron II of GAPDH, classified as the type A1 occur at the boundary between and regions, but there is no intron at this boundary in either of the other two genes; intron III, on the other hand, suggested to be of type B, corresponds almost exactly to an intron of the alcohol dehydrogenase (*Adh*) gene (Cornish-Buwden 1985).

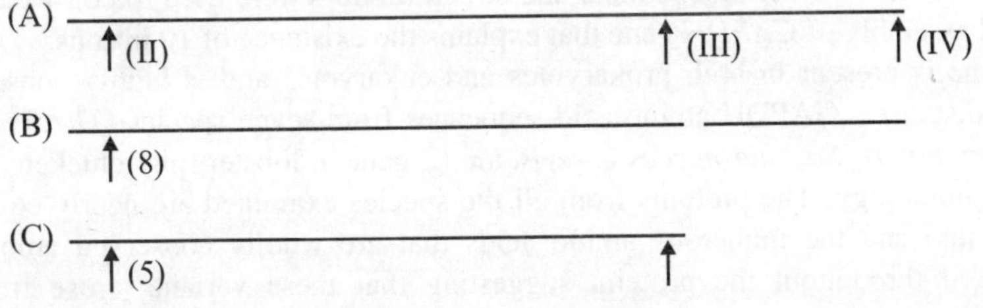

(D) β β β β α α α α α α α α α β β β β β α α α α α α [β β β β β β β β]. β β β β

Figure 1.29 Intron arrangement in the mononucleotide binding-site regions of pig glyceraldehyde 3-phosphate dehydrogenase (A) chicken pyruvate kinase (B) and horse alcohol dehydrogenase (C), together with the consensus secondary structure (D) in terms of α-helices β-pleated sheets

The case of pyruvate kinase gene

Examining the pyruvate kinase gene as a whole, it is clear that introns are not placed at random in the sequence, and at least two systematic characteristics are evident. The ten exons in this gene are of much more uniform length than one would expect if the coding regions were interrupted at random, and the introns tend to fall between discrete regions of secondary structure in the protein. Surprisingly, there is little tendency for introns to occur at domain boundaries, most noticeably, the boundary between domains B and A2 occurs near the middle of exon 5, the largest exon in the gene. It would be appealing to suggest that an intron at this boundary has been lost, because if it were present it would not only separate the two domains but would also make the exon size even more uniform. Although there is now much to suggest that introns are an ancient relic of primordial genes, convincing proof must await the discovery of clearly corresponding intron arrangements in genes that arose by duplication before the separation of prokaryotes and eukaryotes.

Functions of Introns

Introns are not junk, but rather genetic control elements. Are microRNAs derived from introns? They are found to occur in plants, animals, and fungi. MicroRNAs help control timing of developmental processes as cell proliferation, apoptosis, and stem cell maintenance. They also help tag chromatin with methyl and acetyl groups and may also help in alternative splicing mechanisms. The functions of introns in the evolution of genes can be explained in at least two ways: either introns appeared late in evolution and therefore could not have participated in the construction of primordial genes, or RNA splicing and introns existed in the earliest organisms but were lost during the evolution of the modern prokaryotes. The latter alternative allows the possibility of intron participation in the formation of primordial genes before the divergence of modern prokaryotes and eukaryotes. Blake (1979, 1983) suggested that evidence for intron-facilitated evolution of a gene might be found by comparing the borders of functional protein domains with the placement of introns. For this, glyceraldehyde phosphate dehydrogenase (GAPDH), a glycolytic enzyme, was examined. X-ray crystallographic studies demonstrated structurally independent protein 'domains' which were highly conserved during the divergence of prokaryotes and eukaryotes. A study of genomic organization mapped introns in the gene. Sequencing of the chicken *GAPDH* gene revealed 11 introns. The sites of three of the introns (IV, VI and XI) correspond closely with the borders of the NAD-binding, catalytic and helical tail domains of the enzyme, supporting the hypothesis that introns did have a role in the evolution of primordial genes. In addition, other biochemical and structural data were used to construct a model of the intron-mediated assembly of *GAPDH* gene that explains the existence of 10 introns.

GAPDH enzyme is present in both prokaryotes and eukaryotes and is highly conserved across great evolutionary distances. GAPDH amino acid sequences from seven species (*Thermus aquaticus*, *Bacillus stereothermophilus*, *Saccharomyces cerevisciae* (2 genes), lobster, pig, chicken, and human) were compared for homology. The proteins from all the species examined are nearly equal in length (332-334 amino acids) and the numerous amino acids that are wholly conserved among all eight variants are scattered throughout the protein, suggesting that these variants arose from a single ancestral gene and that no exon addition, deletion, or rearrangement occurred after the divergence of the modern prokaryotes and eukaryotes. Therefore, the proteins compared in Figure 1.30 are similar not only to each other but also to the primordial GAPDH protein (Stone et al. 1985). Thus, many structural clues to the ontogeny of the ancestral *GAPDH* gene may remain relatively undisturbed in an organism that has intervening sequences in its DNA.

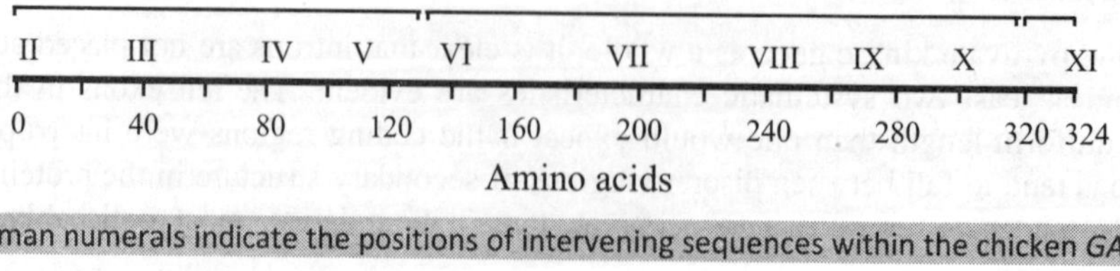

NAD-binding domain | Catalytic domain | Tail

II III IV V VI VII VIII IX X XI

0 40 80 120 160 200 240 280 320 324

Amino acids

Figure 1.30 Roman numerals indicate the positions of intervening sequences within the chicken *GAPDH* gene

Introns fulfill a broad spectrum of functions in cell and are virtually involved in every step of mRNA splicing. The life span of the intron can be divided into five phases: genomic intron (which is the DNA sequence of the intron), transcribed intron (the phase in which the intron is under active transcription), spliced intron (spliceosome is assembled on the intron and is actively excising it), excised intron (intronic RNA sequence released upon the completion of splicing reaction) and EJC harboring transcript. Separate functions are associated with each phase and these functions are mediated by different intronic properties. Sequence- dependent functions are mediated by sequence elements within the intron; length-dependent functions are mediated by the length of the intron, regardless of its nucleotide content; position-dependent functions are mediated by the position of the intron with respect to the exons; and splicing-dependent functions are mediated by the mere fact that splicing had occurred during the maturation of the mRNA (Chorev and Carmel 2012).

Presence of introns despite of the high cost of maintaining them and requirement of elaborate splicing machinery needed to remove them, suggests that introns play more fundamental and evolutionary conserved role and have evolved various functions such as alternative splicing (Morello and Breviario 2009), intron-mediated enhancement (Moabbi et al. 2012), intron-dependent spatial expression (Jeong et al. 2006), gene regulation (Chen et al. 2012) and markers (Breviario et al. 2007).

Associative and Divisive Introns

Conceptually, intervening sequences could have had two broad roles in the formation of modern genes: associative (type A) and divisive (type B). An associative intron would come into being when two unrelated exons were joined by an unequal recombination between two genes. The introns would render this gene fusion event insensitive to a shift in the reading frame and allow a novel unequal recombination to succeed within a target site of thousands of nucleotides instead of one or just a few. Hence a type A intron would function evolutionarily by increasing the frequency of successful shifting of gene segments. If the early organisms contained such introns (and the necessary RNA splicing machinery) they also might have tolerated random insertions of DNA (bearing rudimentary splice signals) into previously intact genes. Such type B introns could arise by aberrant recombination, retrovirus or transposon integration, or retrovirus-mediated insertion of cellular mRNA sequences.

If introns arose in this manner during the evolutionary assembly of a gene, some evidence for these events might still be found in the structures of the protein and gene: for example, if an associative intron facilitated the original assembly of a crystallographically defined domain, one might expect to find that two or more amino acid residues which are essential to the function of the domain, are encoded by gene segments split by an intron (type A1). If an intron mediated the duplication of a complete domain (type A2) the portion of the gene encoding the junction of the duplicated domains should contain the intervening sequence. Such A2 introns probably mediated the assembly of the three domains of ovomucoid, the four domains of the immunoglobulin heavy chain constant region, the numerous 18-animo acid domains of collagen, and the three domains of α-fetoprotein. If an intron participated in the novel association of two different functional domains (type A3), one would expect to find the same concordance of domain border and intron placement as seen in the A2 case. The A3 intron is probably responsible for the joining of a very similar NAD-binding domain to the different catalytic domains of the various dehydrogenases. Finally, if an intron arose as a divisive element (type B), it would separate a previously contiguous exon into two halves and would closely mimics the A1 case. However, one could still differentiate between the two cases whenever the divisive event occurred in a previously duplicated domain, that is, an A1 intron would be duplicated when the gene segment encoding the domain was duplicated, while a late-arriving B intron would have no homolog in the duplicated segment.

The chicken *GAPDH* gene lies within a 4.65-kb DNA fragment that was cloned and sequenced (Stone et al. 1985) (Figure 1.31). Comparison of this genomic DNA sequence with the previously published GAPDH mRNA sequence revealed the existence of 11 introns; the amino acid codons split by these introns are given in this figure.

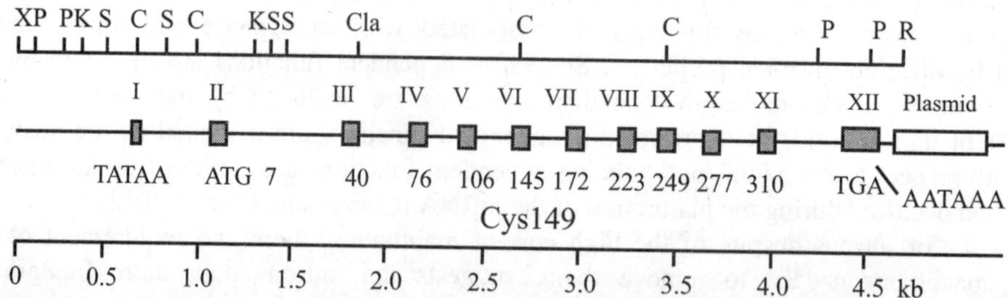

Figure 1.31 Structural map of the chicken *GAPDH* gene. A detailed restriction map of the 4.65 kb XhoI/EcoRI fragment was obtained from a plasmid subclone (PGAD-17) of chicken genome DNA, and the entire *GAPDH* gene was sequenced as reported by E.M. Stone (Stone, E.M. et al. 1985. Nature 313: 498-500) Restriction sites: Cla, ClaI; K, KpnI; S, SmaI; and P, PstI. Solid bars indicate GAPDH mRNA sequences; solid lines represent introns and flanking DNA. Small numbers indicate the terminal codon of each exon. The polymerase entry site (TATAA), initiation codon (ATG), catalylic cysteine (Cys 149) and poly(A) addition signal (AATAAA) are also indicated

Table 1.4 Concordance of protein domain borders with exon borders

Domains	*Domain border (amino acid number)*	*Exon border (amino acid number)*	*Intron*
Nucleotide-binding site 1 Nucleotide-binding site 2	75-90	76-77	IV
Catalytic site	148-149	144-146	VI
Helical element	315-316	309-311	XI

Intron I falls in the 5' non-coding region of GAPDH mRNA and thus could not have participated in the evolutionary association of functional protein domains or their component subdomains, as suggested for all but one of the remaining introns. The location of three of the intervening sequences corresponds closely with the boundaries of four protein domains detected by X-ray crystallography (Table 1.4). Specifically, intron IV divides the two mononucleotide-binding domains (type A2); intron VI separates the catalytic domain from the dinucleotide-binding domain (type A3); and intron XI separates the helical element (shared by several different dehydrogenases) from the remainder of the protein (type A3). This concordance strongly suggests that introns participated in the construction of the *GAPDH* gene and therefore supports the hypothesis that RNA splicing occurred in very early organisms.

Examination of the hydrogen-binding of the peptides encoded by the various exons should indicate the degree of structural independence of each polypeptide from the rest of the protein. The degree of autonomy of a protein segment may reflect the likelihood of it retaining its function if transferred into a new protein by a gene fusion event. Table 1.5 gives an analysis of GAPDH hydrogen- binding, determined by X-ray crystallography (Stone et al. 1985) (Figure 1.32). The domains given in Table 1.4 share less than or equal to 15 per cent of their hydrogen bonds with residues of other domains. The polypeptide encoded by exon X is independent by this criterion as is the second mononucleotide-binding domain encoded by exons V and VI. Thus, it has been suggested that the introns flanking exon X are of the A3 type.

Table 1.5 Quantitation of half-hydrogen bonds within and outside protein segments encoded by chicken GAPDH exons (Calculated from the main-chain hydrogen bonding scheme of Moras, D.M. et al. 1975. J. Biol. Chem. 250: 9137-62)

Exon	*In*	*Out*	*% out*
II	0	9	100
III	24	12	33
IV	24	8	25
V	8	25	75
VI	18	12	40
VII	26	10	27
VIII	24	16	40
IX	2	27	93
X	22	4	15
XI	14	11	44
XII	38	2	5

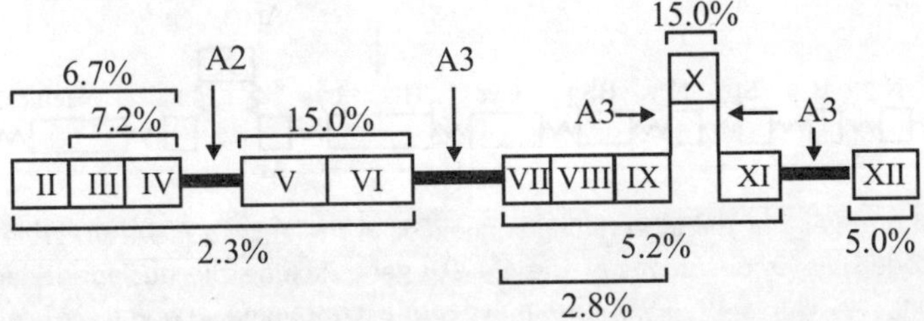

Figure 1.32 Representation of hydrogen-bonding dependence of several regions of GAPDH protein. Percentages were calculated from the main chain hydrogen bonding scheme of D.M. Moras (Moras et al. 1975. J. Biol. Chem. 250: 9137-62). Roman numerals denote portions of the protein encoded by the various exons of the chicken gene. A2 and A3 are intron types as described in the text. Calculation of per cent hydrogen bonding not including exon X is indicated by an asterisk. The number of an intron is the same as the exon that it follows

The exons separated by introns II, III V, VII and VIII are highly dependent on others for the structure of the peptides they encode, and thus encode subdomains. Therefore, these introns must be of the A1 or B type. Introns VII and VIII separate three exons that each contains at least one essential component of the catalytic site. Exon VII contains the catalytic cysteine 149, exon VIII the destabilizing histidine 176 as well as the phosphate-binding lysine 191, and exon IX the other phosphate-binding residue, arginine 231. Thus introns VI and VIII probably participated in the initial A1 type assembly of the catalytic domain. The NAD-binding domain was formed by the duplication of a mononucleotide-binding unit and an A1 type intron that existed in the mononucleotide unit must have been duplicated. Projected into the protein, introns II and V exist in structurally equivalent sites in the NAD-binding domain (they each fall between the two β-sheets of a mononucleotide unit) and are thus most probably duplicated A1 intron. Conversely, intron III does not have a homolog in the second mononucleotide-binding unit and, therefore, arose by B type DNA insertion into a previously intact exon.

Figure 1.33 summarizes the hypothetical evolutionary steps in the construction of the *GAPDH* gene (Stone et al. 1985). This scheme predicts that when the genes for other dehydrogenases are

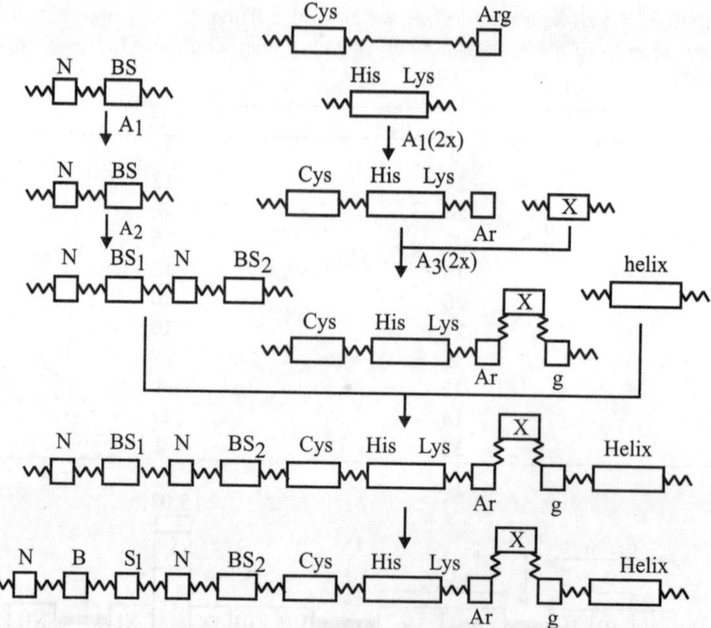

Figure 1.33 Proposed ontogeny of the intervening sequences in the ancestral *GAPDH* gene. Functional domains and subdomains encoded by various exons of the *GAPDH* gene include the nucleotide-binding site (NBS), the polypeptide containing cysteine 149 (Cys), the polypeptide containing arginine 231 Arg), the polypeptide containing histidine 176 and lysine 191 (His-lys), the polypeptide encoded by exon X, and the α-helical element shared by several dehydrogenases. A1, A2, A3 and B indicate the class of exotic connection mediated by the various introns

sequenced, introns will be found dividing the mononucleotide-binding domain as well as the NAD-binding and catalytic domains. Similarly, it predicts that *GAPDH* gene sequences from other species will share the A type introns but may differ in the number and position of the B type introns. Sequencing of the genes for other glycolytic enzymes of higher eukaryotes should provide additional information about the evolution of primordial genes.

WHY ARE THE EUKARYOTIC GENES SPLIT?

Even since the discovery of split genes, there has been a debate about why they are split? This can be resolved into three separate problems — the origin of the introns, the role of introns in evolutions and their present functions, if any. Ideas on these matters can be tested against the serine proteases (proteins that have particularly clear patterns of domains) now that more than a dozen of these genes have been sequenced. In agreement with one hypothesis, the different domains are encoded in separate exons, some of them apparently transferred from unrelated genes. But it is also clear that new introns have been inserted at various times in evolution (Rogers 1985).

Exon Shuffling

Exon shuffling can lead to significant genome rearrangement (evolutionarily significant) exons are mixed and matched due to meiotic division. Errors occur via unequal crossing-over at non-sister transposable element sites. Proteins often have a modular architecture made up of discrete structural and functional regions called domains. Different exons code for the different domains of a protein. For example, tissue plasmogen activator (tPA), an extracellular protein that prevents blood clotting protein, has four domains – three each coded by different exons and one duplicate exon. Each exon is also found in other proteins. Origin of such genes may have been by exon shuffling.

Exon shuffling is believed to be responsible for evolution of split genes in nuclear genes. Perlman and Buton (1989) have stated that mobility of a class of related introns elements was another source of split genes. Some of the group I introns are mobile. There are certain multifunctional intron-encoded proteins. Synergism among introns is possible. Independent mobility of intron and intron like reading frames has been noted.

During the splicing stage, different coding sequences (segments) could splice in a large number of different ways, thus generating a large number of different genes. Gilbert (1978) has called this process as "exon shuffling". This process can take place even today, resulting in the production of new genes. Split genes, therefore, retain evolutionary flexibility. Low density lipoproteins (LDL) receptor gene has one region homologous to the precursor of a peptide hormone epidermal growth factor (EGF) and another region homologous to the complement component Co. Evolution of LDL receptor gene can be explained as follows: the present day long coding sequences of the prokaryotes may have evolved from the split gene processing mechanism. Doolittle (1978) calls this process of discarding the intervening sequences (IVS) as a streamlining process. Such genomes gained efficiency at the cost of evolutionary flexibility. Let us consider the following questions: Why the IVS that are not expressed in proteins but are nevertheless quite long and are maintained in the eukaryotic genome? Why an IVS sequence for one mRNA may be included at another time in another mRNA? All IVS may not exist for the same reason. For instance, those at the 5'-end in particular are involved in regulating transcription or translation. But certainly there are IVS in certain genes that may never be used as parts of an mRNA. Perhaps many eukaryotic cells simply do not have facility for deleting "excess" sequences so that non-functional "intervening" sequences would be maintained for a longer time in eukaryotes. A genome that is build by recruitment of ligatable pieces might have been successful only if its system for deleting unnecessary DNA was inefficient.

Discussions of the role of introns in evolution have centered round idea of "exon shuffling", in which exons are supposed to be units in the evolutionary rearrangement of genes. One should bear in mind that exon shuffling, however, advantageous in the long run of evolution, cannot explain the origin of introns, nor does it amount to a present day function for them. Nevertheless there has been considerable interest in the idea that some exons encode protein domains, or at least coherent structural motifs, which would be good material for evolution by exon shuffling. Thus, they could be subject to duplications within a gene, as has happened in several genes such as that for fibronectin and the serine proteins. In addition, exons could be transferred between unrelated genes to create new proteins out of successful old domains. The nucleotide-binding fold of metabolic enzymes is not encoded separately, and the amino-terminal signal peptide of secreted proteins, which is often more or less contained in a single exon, is too variable in sequence for its ancestry to be reliably traced. But good evidences of exon shuffling are now available from the exon/intron organization of the serine protease genes.

The serine protease superfamily includes digestive enzymes (such as trypsin), enzymes of the blood clotting and complement cascades and various other polypeptide processing enzymes such as the kallikreins. New serine proteases have been arising by gene duplications for most of biological history. Those just mentioned probably diverged more than 500 million years ago, and there are even more divergent homologs in bacteria. The protease domain itself is activated by cleavage from a larger precursor. The non-enzymatic portion often includes various structurally distinct domains with particular physiological functions.

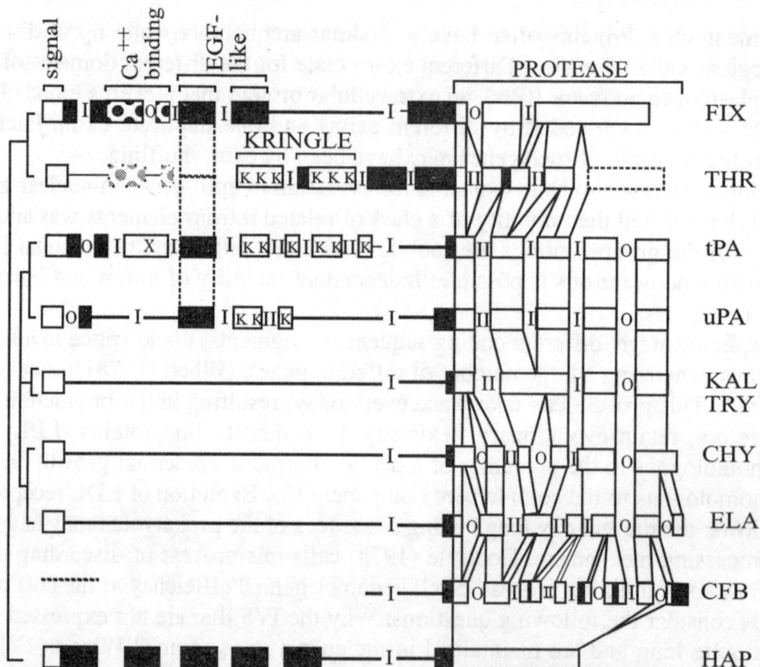

Figure 1.34 Structure of the coding parts of serine protease genes. The family tree, as deduced from protein sequencing; is summarized schematically at left, the dashed branches being tentative. FIX is human clotting factor IX; human thrombin (gene structure not reported for regions shown dashed); tPA, human tissue plasminogen activator; uPA, pig urokinase; KAL, mouse Kallikrein; TRY, rat trypsin (gene structures are identical for two trypsin genes and four kallikrein genes, including serine growth factor subunits; CHY, rat chymotrypsin b; ELA, rat elastase two similar genes); CFB, Human complement factor B (the exons shown are preceded by other non-homologous exons); HAP, human haptoglobin. Exons are shown as boxes shaded according to their homologies. Sequences with no discernible homology to each other are in black. Introns are not to scale, but their phase relative to the reading frame is indicated, according to whether they follow the first (I), second (II), third (0) nucleotide of a codon

Exon shuffling is best illustrated in the 5'-halves of the dozen sequenced genes (Figure 1.34). In each case, the coding region outside the protease domain is split by at least one intron in phase I of the reading frame and various extra domains within the region are encoded by separate exons or groups of exons, also separated by introns in phase I. It is presumably this coincidence of phasing that has permitted the assembly of diverse exons and the tandem duplications of some of them. An example is "kingle domain" (so named because of its topological resemblance to a traditional type of Scandinavian cake). The region of the gene that encodes the kingle flanked by phase I introns has been duplicated independently in the genes for thrombin, tissue plasminogen activator and plasmin. This has happened even though the domain has different internal splicing patterns in these three genes (intron insertion?) and even though the donor splice site following the kingle seems to have been replaced by a splice site that is 9 codons further downstream in the gene for uPA, another serine protease (intron sliding).

The *tPA* gene contains another two exons which provide first example of exon transfer between otherwise unrelated genes (Ny et al. (1984). The third coding exon, unique to *tPA*, encodes a disulfide linked finger that is homologous to the tandemly repeated fingers of the fibronectin molecule. In fibronectin, each finger is encoded by a separate exon, and although their boundaries have been mapped only approximately, the one sequenced does have a potential splice site in a position precisely equivalent to that in *tPA*. Thus, the *tPA* exon seems to have come from a *fibronectin* gene. The fourth coding exon, the *tPA* encodes a region of homology to epidermal growth factor (*EGF*) gene; a homologous exon is duplicated in the gene for clotting factor IX. This exon may have come from a gene like the *EGP*, which contains several *EGF* like segments in separate exons; but any firm conclusion must await the sequencing of the *EGF* gene. Even further back, the *fibronectin*-related and *EGF*-related exons of all these genes may have had a common origin, as all three exons encode sequences approximating, to cys-X-cys-X-X-Gly/Asn-X-X-Gly-X-X-cystine near their 3'-ends.

The domain that encodes the protease activity itself sheds light on the origin of introns. As in several other superfamilies, the positions of introns are different in different genes, and the question is whether this results from insertions or excisions. Clearly introns can be excised, at least in mammals and that they can also be inserted, at least in invertebrates, is suggested by the apparently random positioning of introns in invertebrate gene families such as actin and myosin, but this idea has never been proved. The different serine proteases, which are also pre-vertebrate in origin, show an unambiguous history of intron insertions (Figure 1.34).

This is the clearest within the trypsin family (middle five lines in the Figure 1.34), the family tree of which, as deduced from protein sequences, is shown at the left. All these genes have three introns in common. As one follows the branches of the family tree, new unique introns appear in different genes, one extra in *tPA* and *uPA*, two extra in *chymotrypsin* and three extra in *elastase*. It seems that the shared introns were also inserted, as the introns in the trypsin family are entirely different from those in complement factor B. Such unrelated patterns could hardly have arisen by random deletions of introns. Therefore, most or all of the introns in the protease domain have been inserted. The only apparent exceptions are the two introns of the factor IX protease domain, which match introns in *elastase* and *factor B*, respectively. But according to the amino acid homologies, the first of these is in a segment that may have come from a *chymotrypsin* or *elastase* like gene by recombination, while the second is in a region of weak and uncertain relationship.

Given that these introns have been inserted, do their positions show us anything about how or why this happened? Some regularities have been observed: (1) Figure 1.34 suggests that new insertions tend to occur near the middle of pre-existing exons. Thus, a gene evolves towards a uniformity of exon size. This has been noted independently for the *elastase* gene and for genes in general. (2) Another regularity observed is that many of the introns map to surface loops, and particularly to variable-length loops, in the protein structure. Craik et al. (1982, 1983, 1984) suggested ways in which this length variation could be caused by the presence of the introns, for example, by substitution of alternative splice sites (intron sliding). Another possibility is that introns generate length variation because they are inserted inaccurately. As introns do not resemble any known transposable element, one can speculate that a new intron might be included as part of a longer insertion or might be induced when an insertion activates a cryptic splice site that is nearby.

Some introns are in conserved coding regions, and these include some of the most striking cases of independent insertions into the same region. Thus, five genes in Figure 1.34 contain an intron that is 14-21 codons into the protease domain, but at least three of these are in different positions relative to codons for conserved amino acids. Even more remarkable, the penultimate introns in complement *factor B* and *elastase* fall in the same conserved glycine codon, but one nucleotide apart. These coincidences cannot be reasonably explained by intron sliding. This would require vanishingly unprobable coincidences of mutations in order to shift an intron within a conserved coding region, or to shift the phase of an intron. The non-random positions of these introns seem to imply either sequence specificity at the insertion, or selection following it, for reasons that are still mysterious. The methodological lesson, however, is clear. Correlation of introns with protein structure does not, per se, tell us whether the introns are original or recent additions.

A final puzzle is presented by the gene for haptoglobin, a member of the superfamily but not known to have protease activity. Although this gene has plenty of introns in its 5'-half, dividing duplicated phase I exons as in the other genes, there are no introns in the 3'-half, which encodes the protease like domain. This could represent the condition of the ancestral serine protease gene, intronless like genes of bacteria. Alternatively, it could represent a much more recent processed gene, derived from the mRNA of a split serine protease gene via reverse transcription, which became attached to a different split gene. A similar history has been postulated for part of the (unrelated) clotting factor VIII gene; this mechanism has also been proposed as a general way of evolving new proteins. Processed genes are common only in mammals, so if one is incorporated either in *haptoglobin* or in *factor VIII*, the gene from which it was derived may still be present and detectable by nucleic acid hybridization.

Origin of Introns

Introns may have been self-splicing mobile genetic elements that inserted themselves into host genomes. Advent of spliceosomes (catalytic RNA/protein complexes) seems to be intimately involved with the evolution of introns. Spliceosomes snip pieces of RNAs (introns) out of pre-mRNAs. This would encourage introns to proliferate, mutate and evolve.

Transposon hypothesis

The selfish DNA or transposon theory of the origin of introns is dismissed by Cornish-Bowden (1985). As introns are very often not at domain boundaries, the hypothesis of evolutionary exon shuffling to recombine protein domains is hardly the most plausible suggestion for their function. It does not adequately explain either the origin or the evolutionary maintenance of RNA splicing. Although fusion of two genes by recombination between existing introns probably has been evolutionarily important, it does not explain the origin of the introns in the first place. De novo intron formation by fusing two genes lacking introns is far more complex than fusion to produce an unsplit gene. It could hardly have occurred in a single step in an organism previously lacking RNA splicing? Split genes and RNA splicing could not have evolved merely because they might, millions of year later, help a new protein evolve. Evolution lacks foresight. The transposon hypothesis can explain the origin of introns and also of RNA splicing itself. The presence of introns in mitochondria and chloroplasts, in particular, is best explained by transposition from the nucleus. The different positions of the introns of the chloroplasts *psbA* genes of *Chlamydomonas* and *Euglena*, coupled with their absence from the ancestral cyanobacteria, supports insertional origin for chloroplast introns, rather than one from primordial gene fusions that created the *psbA* gene. The coding of proteins involved in splicing by certain introns in fungal mitochondria, and in a chloroplast gene, supports the thesis that such introns originated as defective transposons in which the RNA splicing enzymes evolved from pre-existing DNA splicing enzymes specific for their termini.

This can also explain the origin of introns in general. Suppose a transposon capable of precise excision is inserted into a cellular gene, thus inactivating its transcript. This intron would be exactly spliced out of the RNA if the DNA excision of the transposon is mutated so as to be able to cut RNA instead of DNA but retain its specificity for the sequence at either end of the transposon. This would simultaneously prevent its self-transposition and reserve the cell from its harmful insertion mutation. This mutation would protect the cell from insertion mutations by any transposons generating similar terminal sequences upon insertion and, therefore, have allowed the rapid intragenomic spread of kindred non-defective transposons. Moreover, these would have provided DNA excision activity allowing the rapid intragenomic spread of the defective RNA splicing transposon itself. So many genes would quickly acquire introns that RNA splicing could never be lost. All descendants of this cell would be permanently injected with useless introns. RNA splicing could later acquire real functions for at least some genes, much as differential splicing to make different proteins from one and the same transcriptional unit. This rescue hypothesis provides an immediate and powerful mode of spreading for introns and RNA splicing. At one and the same time it involves intragenomic kin selection at the cellular level for protection from the harmful effects of certain selfish transposons. Alternatively, RNA splicing might have originated in a virus before being taken over by the cell.

The transposon theory of intron origin best fits the nuclear mRNA and rRNA splicing mechanism, both of which depend on specific sequences within as well as at the end of introns. It was suggested that they originated in an early eukaryote and were never present in prokaryotes. The small nuclear RNAs involved in nuclear pre-mRNA splicing might have been synthesized from transposon promoters lying on the DNA strand opposite the transcribed strand of the cellular gene into which it had inserted. But nuclear tRNA splicing does depend on specific intron sequences and may have originated independently. The presence of similar rRNA introns in the archaebacterium sulpholobus suggests that it originated before archaebacteria and eukaryotes separated. Chloroplast rRNA introns seem unrelated to nuclear and archaebacterial tRNA introns. They resemble mitochondrial and nuclear rRNA introns and have features suggestive of defective transposons. The improbability of losing every single intron by accidental deletion and the extra complexity that splicing poses for the origin of protein coding, make it improbable that RNA splicing was the general rule in the primordial cell, and favor the view that archaebacteria and eukaryotes both evolved from a eubacterium lacking split genes, Once introns evolved, their number and total length per genome would increase or decrease in response to generalized selection of larger or smaller total genome size and transcript size, as well as to selection for specific functions for at least some introns.

Intron of one gene may contain exon of another gene. A few examples of this phenomenon are amylase gene in mouse, adenovirus late expression, *Troponin T* gene for rat muscles, and immunoglobulin gene. Exon sequences are conserved but intron sequences vary. Two hypotheses, namely "late intron" and "early intron" have been proposed. Spliceosomal introns are present in the proteins encoded by nuclear genes but they have not been detected in eubacteria, archaebacteria and organelles.

"Early intron" theory

"Early intron" theory, also known as "exon theory" of origin of introns, explains that exons are descents of ancient minigenes and the introns are descendents of the spacers between them. Genes large enough to encode proteins were first assembled from sets of exons. The machinery of splicing originated in an ancient RNA world. Introns are completely lost from both kingdoms of bacteria as well as several protist groups. Exons existed as independent genetic units or minigenes and their association with introns helped in generation of diversity in genes through *trans*-splicing and exon shuffling. These are examples where new genes were generated through exon shuffling through intron recombination events.

"Late intron" view

"Late intron" view holds that split genes arise from uninterrupted genes by insertion of introns. Genes large enough to encode proteins first arose (presumably from smaller genes) without participation of introns. The machinery of spliceosomal splicing arose from fragmented self-splicing introns.

Spliceosomal introns were never observed in the ancestors of those organisms that now lack them. Existence of five spliceosomal introns in identical positions in nuclear genes encoding glyceraldehyde-3-phosphate dehydrogenases (GAPDHs) of eubacterial (chloroplast Gap A/B) and uncertain (cytosolic Gap C) ancestry provides strong evidence in favor of the "early intron" hypothesis.

Logsdon and Palmer (1994) suggested that either the introns are mobile elements that can insert at matching position in different copies of a gene or they are ancient relics lost from nearly all bacterial genes. If later position is correct then why tiny fractions of bacterial-derived genes have retained introns that are identical with tiny fractions of bacterial genes. Therefore, they suggested that identical positions of 5 introns across chloroplast and cytosolic genes (*GapA/B* and *GapC*, respectively) are best explained by parallel insertions at common target sites rather than by common ancestry. The "early intron" view of Logsdon and Palmer (1994) is based on a large extent on negative evidence. Lack of introns has been observed in many *GAPGH* genes. For example, in *Arabidopsis GapC* lacks three introns, which are present in maize and pea; three losses in *Arabidopsis* are more likely to occur than six independent gains at identical positions in maize and pea.

There are other such examples which clearly contradict "late intron" view. Stoltzfus et al. (1994) opine that if "late intron" view is accepted then a question arises how genes and long contiguous open reading frames were assembled in early evolution if concept of intron-mediated exon shuffling for primordial gene evolution was not accepted. According to the "late intron" view, introns invaded the genome of eukaryotic ancestors as endosymbionts and evolved in mitochondria and chloroplasts as group II introns. It is postulated that invasion took place at mRNA level, which was followed by reverse transcription and recombination with DNA producing split genes. These introns gave rise to other types of introns later. Thus it was attempted to explain why introns were absent in eubacteria, archaebacteria and lower eukaryotes. To explain the origin of introns, "early intron" and "late intron" hypotheses may have to be combined. Evidence strongly favors the view that split genes have not arisen from unsplit genes but from primordial DNA sequences. Introns have been discovered in ribosomal RNA genes of archaebacteria (Kaine et al. 1983); they may represent third cell type that branched from common ancestor earlier than prokaryote and eukaryotes. Although Senapathy (1986) has provided strong evidence in favor of origin of introns from primordial DNA yet suggestion of Kaine et al. (1983) and Garret (1985) caused a shadow of doubt regarding this view.

Periodic introns

The introns-early view has been challenged for several genes; prominent instances are *triose phosphate isomerase* (*TPI*), *aldolase*, *pyruvate kinase* (*PK*), *alcohol dehydrogenase* (*ADH*), *glyceraldehyde-3-phosphate dehydrogenase* (*GAPDH*) and *myosin heavy chain*. While some of their introns appear to be phylogenetically ancient and/or to delineate exons corresponding to protein modules, a considerable number seemingly do not. But it is argued here that many of these anomalous introns are periodic, that is, relics of internal sequence repetitions within the ancestral gene. Some of these periodic-intron patterns are shared between related genes, as in the αβ-barrels of *TPI*, *aldolase* and *PK*, or the Rossmann nucleotide-binding domain common to *PK*, *ADH* and *GAPDH*. This is further evidence for the ancestral status of these introns (Elder 2000). The myosin heavy chain C-terminal rod region is paradoxical in that its sequence is clearly periodic but its intron placements are not; however, they exhibit a remarkable coherence of intron translational phases, suggesting that these introns may also have originally had a periodic arrangement now obscured by intron slipping.

In prokaryotes, genes for related enzymes of a pathway are linked in operons. In eukaryotes, genes for related enzymes of a pathway are scattered in different chromosomes. If eukaryote genes evolved from prokaryote genes, why did eukaryotes give up the evolutionary advantage of an efficient mechanism for the coordinate regulation of a whole metabolic pathway? Most globular protein polypeptide chains are split into two or more structural/functional domains, for example, GAL4 regulatory protein. Many introns divide genes into exons, each of which codes for a different functional domain. Such a non-random distribution of introns is not consistent with random insertion of nucleotide sequence into contiguous prokaryotic genes. This is, however consistent with the hypothesis of the evolution of split genes by splicing of short oligonucleotides each of which coded for an oligopeptide performing a distinctive function.

There are two hypotheses regarding evolution of split genes. First, the split genes (eukaryotic genes) have evolved from non-split genes (prokaryotic genes). In response to the need for more sophisticated control, IVS evolved by first the development of RNA-RNA splicing ability with subsequent insertions of nucleotide sequences into the structural genes. And this insertion must be random. Second view is that split genes of eukaryotes did not evolve from un-split genes of prokaryotes. The most primitive genes were split genes which evolved into the present day contiguous prokaryotic genes as well as the split genes of present day eukaryotes.

Reading frame lengths in computer generated random sequences

Senapathy (1986) analyzed nucleotide sequence data bank, listing 2.1×10^6 bases in 1,320 sequences and a random sequence of length 600,000 nucleotides was generated using the computer, for the distance between successive stop codons, i.e., reading frame lengths (RFLs). It was found that in the computer simulated random sequences there existed an upper limit of about 200 codons (600 nucleotides) for reading frame lengths (RFLs). In the present day eukaryotic DNA sequences also, RFLs had an upper limit of 200 codons. In the present day prokaryotic DNA sequences, however, RFLs significantly exceed the upper limit. It is also important to note that the length of exons has an upper limit of 600 nucleotides in most of the eukaryotic genes.

Progenote Stage: Based on these data it has been proposed that DNA sequences of the most primitive unicellular eukaryotes may have originated from primordial DNA sequences which had been randomly generated. Before a self-replicating cell could come into existence but after biological energy-generating system had evolved, DNA molecules were synthesized in the primordial soup by random addition of the four nucleotides without the help of templates. The nucleotide sequences that coded for proteins were selected from the randomly generated nucleotide sequences in the primordial soup by natural selection. This represents the “progenote” stage in the terminology of Woese and Fox (1977).

The long coding sequences of present day split genes of eukaryotes must have arisen by the splicing of the short coding sequences and excising the sequences containing interfering stop codons and the adjacent non-coding sequences, if any (Senapathy 1995). The coding sequences became the exons and non-coding sequences alongwith those that contained interfering stop codons, became the intervening sequences. For evolutionary reasons, these sequences were conserved at the DNA level and were removed at the RNA level.

Sharing of Exons in Gene Evolution: One likely mechanism for sharing of exons coding for different proteins is through the duplication and migration of exons during evolution. This view is supported by the following examples.

Human Lipoprotein Receptor: Exon organization of gene for human low density lipoprotein (LDL) receptor is more than 45 kb in length and contains 18 exons thirteen of which encode protein sequences that are homologous to sequences in other proteins; five of the exons encode a sequence, similar to the one in the complement component Co, the terminal component of complement cascade.

Immunoglobulin Heavy Chain: Immunoglobulin heavy chain consists of 6 domains: (i) N-terminal signal peptide which helps in the secretion of the antibody. (ii) V_H forms the antigen binding site. (iii) CH_1 is the attachment point for the heavy chain. (iv) Hinge probably transfers information from Fab (CH_1 paired with C_L together V_H and V_L domains, form the Fab part of the antibody molecule) to the F2 fragment (formed by pairing of the CH_L and CH_3 domains of one heavy chain with equivalent domains of the second heavy chain in the antibody molecule). (v) The paired CH_2 domains are involved in complement fixation. (vi) The paired CH_3 domains are involved in interacting with the surface of the cell. In the immunoglobulin heavy chains, CH_3, CH_2, CH_1 and hinge domains are encoded by independent exons.

Hemoglobin Chains: A globin chain (α and β) of hemoglobin consists of 3 domains: heme-binding domain binds heme, N-terminal domain involved in heme-heme interaction, and C-terminal domain.

Glyceraldehyde-3-Phosphate Dehydrogenase in Chicken: Comparison of genomic DNA sequence with mRNA of glyceraldehyde-3-phosphate dehydrogenase (*GAPDH*) gene revealed existence of 11 introns. Sites of three of the introns (IV, VI and XI) correspond closely with the borders of the NAD-binding, catalytic and helical tail domains of the enzymes, supporting the hypothesis that introns did have a role in the evolution of primitive genes.

Globin Genes: In the globin genes, each domain is coded by an exon and the three exons are separated by two intervening sequences.

OVERLAPPING AND NESTED GENES

Khorana et al. (1967) while deciphering the genetic code as words of triplets implied that genes are the definite sets of non-overlapping codons so genes were also thought to be non-overlapping. Most genes are discrete non-overlapping units, i.e. they do not share information with other genes. In strictest sense, overlapping genes are defined as a single nucleotide sequence coding for more than one polypeptide. Thus overlapping genes share some of the same sequence. Such an arrangement of genes was discovered by Barrell et al. (1976) and Sanger et al. (1977) in phage ϕX174, Shaw et al. (1978) in G4, and Contreas et al. (1977) and Fiers et al. (1978) in the animal virus SV40. Later these genes were also reported in the genomes of other bacteriophages, animal viruses, bacteria, and mitochondria (Normark et al. 1983). Overlapping genes have also been found in eukaryotic multicellular organisms such as *Drosophila melanogaster*. Overlapping genes can be located on the same DNA strand or on opposite strands. A particular case of overlapping genes is the so-called hidden genes, hidden frames or mini-cistrons found in the poliovirus (Pierangeli et al. 1998). These are short open reading frames, i.e., sequences initiated by the alternative translation initiation codons ACG, AUA, and GUG in the 50-terminal extra-cistronic region of poliovirus RNA. Neither the classical nor the neoclassical view of the gene encompassed the possibility of overlapping genes, since genes were believed to reside on the chromosome always in tandem.

Sanger et al. (1977) sequenced ϕX174 DNA and reported a sequence of 5,375 nucleotides. This much DNA could code for maximum of 200,000 Da proteins but observed amount was 250,000 Da. It was thought that one gene could code for more than one polypeptide chain. This discrepancy was solved by Sanger et al. (1978) when they delineated the genes on the basis of amino acid sequences. For this work Fredrick Sanger was awarded his second Nobel Prize in 1980. In ϕX174, genetic material is single-stranded DNA and is distributed among nine genes (*A, B, C, D, E, J, F, G* and *H*). Genetic map of ϕX174, as worked out by Sanger et al. (1981), is given in Figure 1.35. Amino acid sequences of major proteins coded by all the nine genes of ϕX174 have also been worked out. The

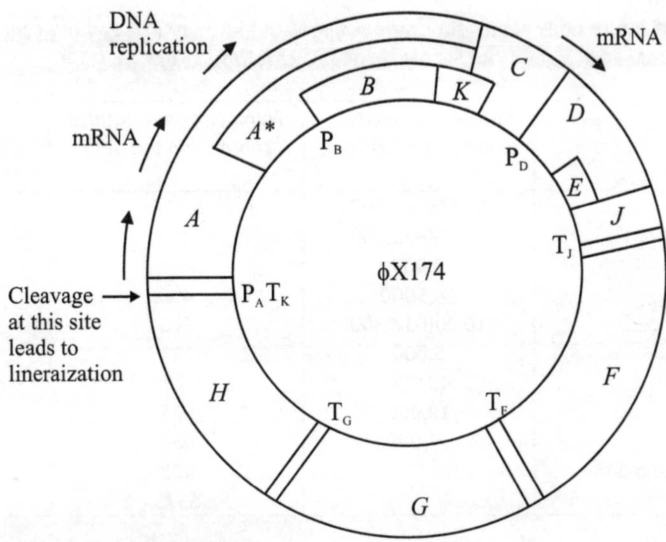

Figure 1.35 Gene map of ϕX174 bacteriophage

estimated number of nucleotides required for synthesis of these proteins exceeded 6,000. It was suggested that sequences in the same segment could be utilized by two different cistrons for different proteins. Barrell et al. (1976) reported nucleotide sequence of DNA containing this segment and found that cistron *E* is present between *D* and *J* and the cistron *E* overlaps cistron *D*. It could be shown that amber mutations in cistron *E* lie within the cistron *D* and these amber mutations do not influence the translation of cistron *D*. Similarly, some other non-sense mutations in cistron *E* also lie in the cistron *D* suggesting that the cistrons *D* and *E* overlap in the DNA sequences. A 133-amino acid-long polypeptide chain is coded by the nucleotide sequence extending from 3,973→5,375+1. Another 48-amino acid-long polypeptide chain is coded by the nucleotide sequence extending from 5,064→5,375+1. Thus gene *B* is contained entirely within gene *A*. Reading frame of gene *A* is different from that of gene *B*. Upto the first nucleotide of gene *B*, we have 1,092 nucleotides (3,973→5,064) which form 364 complete triplets. Thus, the first nucleotide of the first codon of gene *B* is the third nucleotide of the previous codon of gene A. Thus, the reading frames of the two genes are out of phase with respect to each other thus synthesizing unrelated polypeptide chains.

The sequence for gene *E* is entirely contained within the sequence of gene *D* but is translated in different reading frame. In the same manner, gene *B* is entirely contained within the sequence of gene *A* and a newly discovered gene *K* (Pollock et al. 1978) overlaps the terminating codon of gene *C* at 3′-end of gene *A* and terminates within gene *C*. However, with the overlapping genes, the genome has more coding capacity than had been originally supposed to be on the assumption that each gene was physically separate. Most striking feature of the ϕX174 DNA sequence is the way in which various functions of genome are compressed within its genome. Number of nucleotides comprising various genes of ϕX174 phage and molecular weight of proteins encoded by these genes are given in Table 1.6. An interesting feature of genetic code in ϕX174 DNA is very low occurrence of codons starting with AG, particularly in overlapping genes. Base composition of ϕX174 DNA is A, 23.9; C, 21.5; G, 23.3; and T, 31.2 per cent.

In the G4 phage, the situation was even more complicated. In this case, several genes overlapped, encoding different proteins read from the same DNA strand but in different reading frames. The same DNA strand encoded as many as three different proteins, the messenger RNAs of which were transcri-

Table 1.6 Number of nucleotides comprising various genes of phage ϕX174 and molecular weight of proteins encoded by these genes (Extracted from Sanger, F. et al. 1977. Nature 265: 687-95)

Genes	*Protein molecular weight from SDS gels*	*Number of nucleotides comprising the gene*	*Protein molecular weight from sequence information*
A	55,000-67,000	1,536	56,000
B	19,000-25,000	360	13, 845
C	7,000		
D	14,5000	456	16,811
E	10,000-17,500	273	9,940
J	5,000	114	4097
F	48,000	1,275	46,000
G	19,000	525	19,053
H	37,000	984	35,800
Non-coding and C		485	
Total		5375	

bed overlappingly in all three possible reading frames. Further, multiple gene products *A1*, *A2* from *A*; *A1'*, *A2'* from *A'*; and *G1*, *G2*, *G3* and *G4* from *G* have been assigned with multiple initiation and termination sites. A possible explanation for the functions of these gene products is that they provide the organism with selective advantages. The capacity to initiate translation from more than one site to read through a terminal TGA codon or to modify the progenitor protein differently might be a frugal way to mimic aspects of heterozygosity and could confer an advantage by creating for a single gene a heterogeneity that would aid in adaptation to variable growth conditions and may serve different functions. The large multiplicity of gene products signifies an extraordinary potential for variability, adaptation and evolution.

Overlapping sequences have also been discovered in tryptophan mRNA of *E. coli*. In *E. coli*, the 3' 886-nucleotide long portion of *frd-D* gene includes two parts. First part comprises of 29 nucleotides which codes for the ten C-terminal amino acids of the frd-D polypeptide and second part contains 857 nucleotides which is a 3' untranslated region. The 5' 35-nucleotide long portion of the sequence is the promoter for the *AmpCi* gene. The 851-nucleotide long 3'-part is the 5'-part of the *AmpCi* gene, it is translated. Thus there is partial overlap between *frdD* and *ampC* genes. The part of the *frdD* gene coding for the 10 C-terminal amino acids of fumarate reductase D protein and the 3' untranslated part of gene also constitutes the *ampC* promoter and a part of the untranslated leader sequence of *ampCi*. The numbering of the nucleotides is with respect to the mRNA start point of the *ampC* gene. Thus, nucleotide sequence 36 to 851 is part of both *frdD* and *ampC* genes.

In principle, genes can overlap at two levels. In bacterial systems and other situations where space conservation is necessary (e.g., in RNA virus genomes and in animal mitochondrial DNA), genes may overlap at the level of reading frame, so that the same information is used to generate two or more unrelated proteins. The open reading frame may be transcribed from opposite strands, may be translated in different directions, and may be out of frame with respect to each other, i.e., the genes have nothing in common except that they share some of the same space. An example is the lysis protein gene of the levivirus (which includes bacteriophage MS2); this overlaps with the replicase and coat protein genes but is translated in the opposite direction and in a different reading frame. In some species, the lysis protein gene is completely inset within the replicase gene. In IS4, there are two open reading frames (ORFs). ORF-1 extends from 85 to 1413 nucleotides and codes for 442 amino acids. ORF-2 extends from 609 to 217 nucleotides and codes for 131 amino acids. In eukaryotic systems, genes can overlap at the level of transcription unit, but exons can remain discrete. Thus the same information is never expressed in the protein products of both genes because DNA regarded as exon material in one gene is part of the intron of the overlapping gene (e.g. *human TCRA* and *TCRD T-cell* receptor genes overlap at the exon level). Occasionally complete genes may be embedded within the intron of a larger gene; small open reading frames encoding proteins concerned with intron metabolism are often found in self-splicing introns. Three small genes are also found in intron 26 of the large human gene *NF-1*. Overlapping genes may also reflect a mechanism of regulation. In plasmids, genes encoding antisense RNA tend to overlap with the gene they regulate.

A nested gene is a single gene which produces two or more nested products by modulating the end point of protein synthesis. This can occur by leaky read through of a termination codon (e.g. in the case of the Q virus coat protein gene) or by co-translational frameshifting (e.g., in the case of the *E. coli dmaX* gene and the F-plasmid *traX* gene). Similar strategies are used by eukaryotic RNA viruses (e.g., retroviruses), and nested products can also be produced from eukaryotic genes by alternative splicing or use of alternative polyadenylation sites.

Nested genes serve as an example of a situation in which one gene resides within an intron of another gene, was first demonstrated in the *Gart* locus of *D. melanogaster* by Henikoff et al. (1986). In this particular case the nested genes were on opposite strands of DNA. Chen et al. (1987) demonstrated that in the large intron of the *dunce* locus of *D. melanogaster* there were actually two other genes residing, of which one was the known *Sgs-4* gene. In this case the nested genes were on the same strand of DNA. Levinson et al. (1990) were the first to demonstrate nested genes in man. The 22nd intron of the blood *coagulation factor VIII* gene included another gene in its opposite strand. The large intron of the human neurofibromatosis gene includes a total of three other transcription units in two opposed orientations. The existence of nested genes is in contradiction to the central hypothesis adopted by both the classical and neoclassical gene concept, that genes are located in linear order on the chromosome.

THE SMALLEST KNOWN GENE

Micron C7 (Mcc C7), is a modified linear heptapeptide that inhibits protein synthesis in Gram negative Enterobacteriaceae, consists of Acetyl-Met-Arg-Thr-Gly-Asn-Ala-Asp. It is synthesized from 21- bp open reading frame gene called *mcc*. This is one of the smallest genes known so far (González-Pastor et al. 1994). In the translation inhibitor microcin C7 (MccC7), N terminus has been replaced by an N-formyl group and whose C terminus has been replaced by the phosphodiester of 5'-adenylic acid and n-aminopropanol. MccC7 production and immunity determinants lie on a 6.2-kb region of the *E. coli* plasmid *pMccC7*. This region was entirely sequenced. It contains six open reading frames, which were shown to be true genes by different complementary approaches. Five genes, *mccABCDE*, which are transcribed in the same direction, are required to produce mature extracellular microcin. The sixth gene, *mccF*, adjacent to *mccE*, is transcribed in the opposite direction and encodes specific self-immunity. Genes mccA to mccE constitute an operon transcribed from a promoter (*mccp*) located upstream of *mccA* (Gonzalez-Pastor et al. 1995).

PSEUDOGENES

Defective copies of genes present in an organism's genome are called pseudogenes. In human, mouse and *Drosophila*, pseudogenes have been found for *α-globin* gene and *β-globin* gene. Both these pseudogenes are non-translatable because they may have mutations in initiation codons. One of the pseudogenes in mouse has no intron. This pseudogene seems to arise from reverse transcription of mRNA. U snRNA series of pseudogenes in human beings includes U1, U2 and U3 snRNAs. In *Drosophila*, histone pseudogenes have been discovered. Two important characteristics of pseudogenes are: (a) Most of the pseudogenes outnumber their genes and are therefore repetitive sequences. (b) They are flanked by 6-21 bases long direct repeats.

Two hypotheses have been proposed for origin of pseudogenes. (a) mRNA or snRNA itself becomes incorporated into DNA, or into retrovirus RNA followed by incorporation of retrovirus. (b) Reverse transcript of mRNA or snRNA was synthesized as cDNA, which subsequently became integrated. The second hypothesis is considered to be more probable for origin of pseudogenes.

COLINEARITY

Gene-Protein Colinearity

Tryptophan synthetase A polypeptide of *E. coli*

A prokaryotic gene and its protein product have a point-for-point correspondence. In other words, a prokaryotic gene and its protein product were colinear. Sequence of nucleotides in a prokaryotic gene determines sequence of amino acids in a protein/polypeptide. This conclusion was drawn by a study carried out by Yanofsky et al. (1964a,b) and Yanofsky (1967) on many mutants of tryptophan synthetase A polypeptide. Different mutants were crossed together and a genetic map was obtained, showing the order of the mutated sites along the DNA of the gene. Tryptophan synthetase polypeptide A was purified from each mutant strain, and the sequence of amino acids compared to that of wild-type (Figure 1.36). The order of the changes in the amino acid sequence corresponds exactly to the order of the genetic changes in DNA.

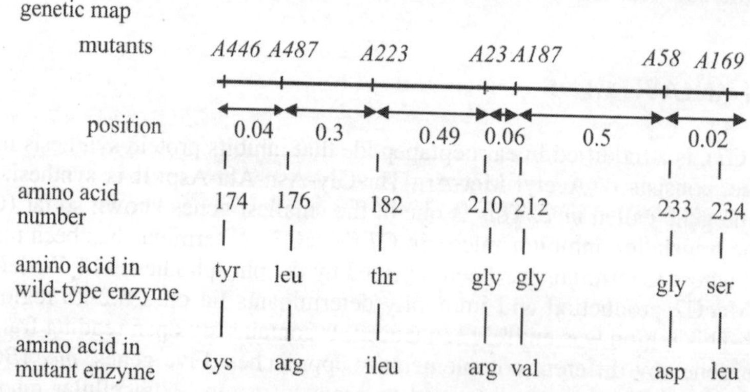

Figure 1.36 Colinearity shown by *trp* gene in *E. coli* and its product (Redrawn, with permission, from Yanofsky, C. et al. 1967. Proc. Natl. Acad. Sci. USA 57: 296-8)

Bacteriophage T4 gene for the major structural protein of the phage head

Sarabhai et al. (1964) demonstrated a similar colinearity between the position of mutations in gene of bacteriophage T4 which codes for the major structural protein of the phage head, and the positions in the polypeptide affected by mutations. They studied *amber* (UAG-chain-termination) mutations and demonstrated a direct correlation between the length of the polypeptide fragment produced and the position of the mutation within the gene. Formation of an amber (UAG) chain-termination mutation and its effect is shown in Figure 1.37. Suppressor tRNA for tyrosine has anticodon AUC which incor-

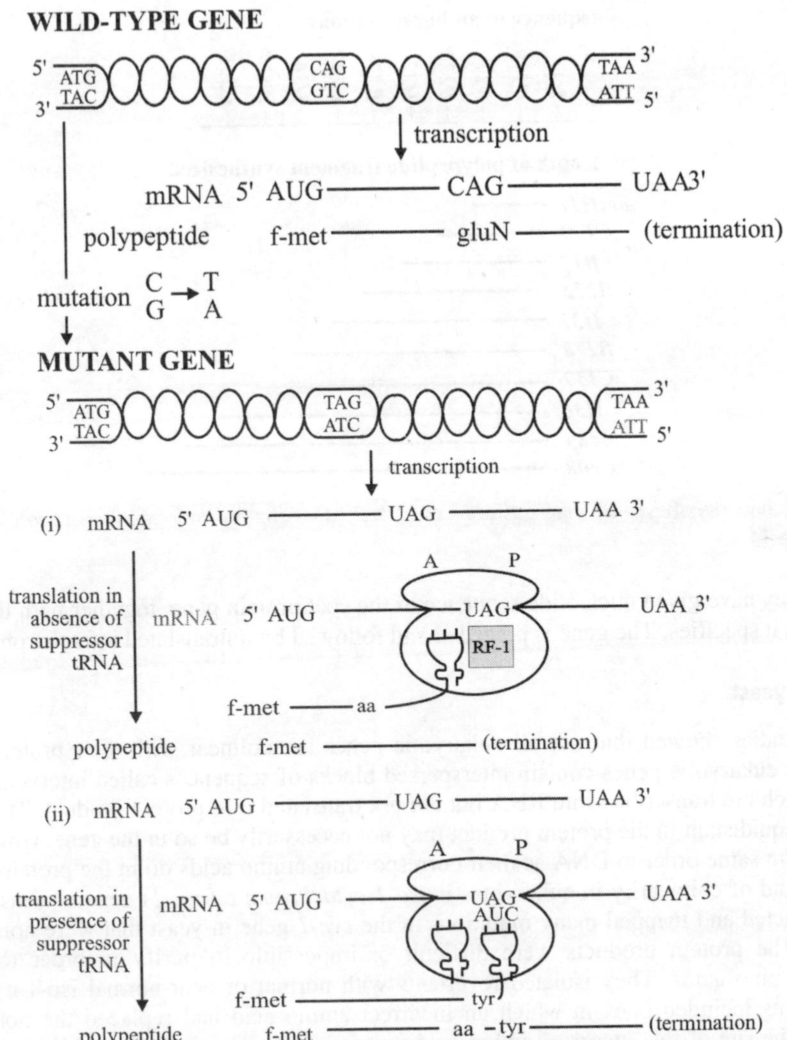

Figure 1.37 Procedure used by Sarabhai and co-workers for detection of amber (UAG) mutations by using suppressor tRNA which incorporates tyrosine at the site of amber mutation. The polypeptide containing tyrosine may or may not be functional

porates tyrosine at the amber termination codon. Polypeptide containing tyrosine may or may not be functional. Suppression will, of course, be observed only if the tyrosine-containing polypeptide is active. Colinearity between map positions of amber (UAG chain-termination) mutations in gene *23* and its protein product of bacteriophage T4 is shown by the genetic map and size of peptides synthesized by each mutant (Figure 1.38).

Coat protein gene of RNA bacteriophage MS2

Definitive evidence for colinearity has also been provided by correlated nucleic acid sequencing of *coat protein* gene of RNA bacteriophage MS2 and amino acid sequence of coat protein by Min Jou et

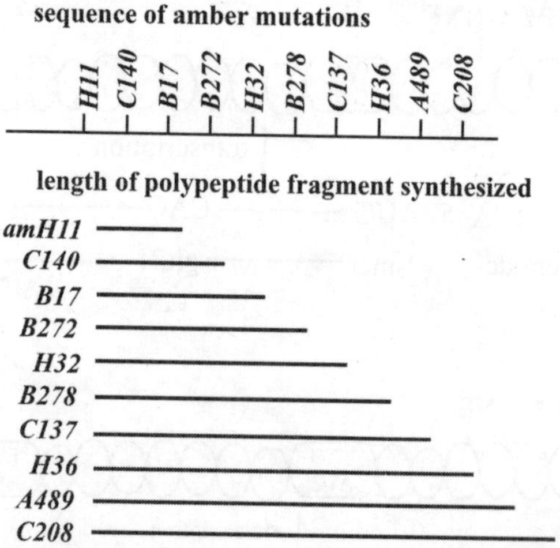

Figure 1.38 Colinearity shown by map positions of amber (UAG-chain-termination) mutations in gene *23* of bacteriophage T4

al. (1972). They have given nucleotide sequence of the coat protein gene, together with the 129 amino-acid sequence it specifies. The gene is preceded and followed by untranslated intercistronic regions.

cyc-1 gene in yeast

Subsequent studies showed that not all eukaryotic genes are colinear with their proteins. This is so because many eukaryotic genes contain interspersed blocks of sequences called intervening sequences or introns which are transcribed into RNA but are not translated into protein product. The two pairs of sites that are equidistant in the protein product may not necessarily be so in the gene. However, codons would appear in same order in DNA as their corresponding amino acids do in the protein. The clearest case of this kind of colinearity in eukaryotes is *iso-1-cytochrome c* (*cyc-1*) gene of yeast. Sherman et al. (1975) selected and mapped many mutations in the *cyc-1* gene in yeast that were non-sense chain-terminating. The protein products were difficult or impossible to purify in order to analyze the sequence of amino acids. They isolated revertants with normal or near-normal iso-1-cytochrome C. These revertants included ones in which an incorrect amino acid had replaced the non-sense chain termination. The site of this incorrect amino acid was inferred to be the site of the original mutation. The amino acid analysis and genetic map are shown in Table 1.7 and Figure 1.39, respectively. These results have corroborated the linear correspondence of the fine structure map of cyc-1 and inferred locations of the non-sense triplets in the *cyc-1* gene product.

Table 1.7 Position of amino acids and the amino acids found in revertants derived from mutants of cyc-1 gene in yeast (Extracted, with permission, from data of Sherman, R. et al. 1975. Genetics 81: 51-73)

Mutant	*Position of amino acid*	*Amino acid found in revertant*
Cyc1-31	3	Phe
-239	5	Ala
-6	12	Ala
-166	64	Trp
-72	66	Glu
-76	71	Glu
-140	93	Glu

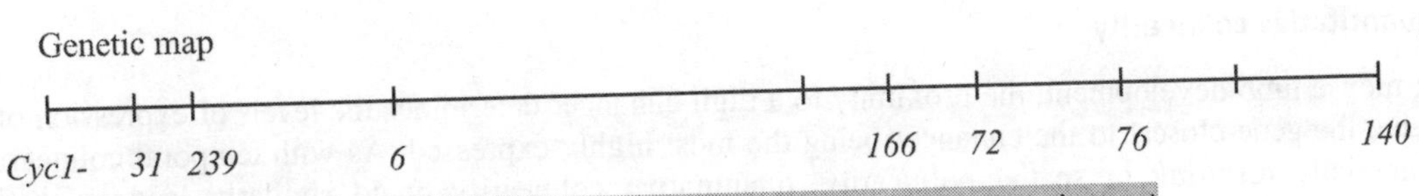

Figure 1.39 Genetic map representing position of different mutants of *cyc-1* gene in yeast

Temporal, Spatial and Quantitative Colinearity of *Hox* Genes

Spatial colinearity

The various *Hox* genes are situated very close to one another on the chromosome in groups or clusters. Interestingly, the order of the genes on the chromosome is the same as the expression of the genes in the developing embryo, with the first gene being expressed in the anterior end of the developing organism. The reason for this colinearity is not yet completely understood.

The deuterostomes are the clade of animals for which we have the most detailed understanding of *Hox* cluster organization. With the *Hox* cluster of amphioxus (*Branchiostoma floridae*) we have the best prototypical, least derived *Hox* cluster for the group, whilst the urochordates present us with some of the most highly derived and disintegrated clusters. Combined with the detailed mechanistic understanding of vertebrate *Hox* regulation, the deuterostomes provide much of the most useful data for understanding *Hox* cluster evolution (Monteiro and Ferrier 2006). Considering both the prototypical and derived deuterostome *Hox* clusters leads us to hypothesize that temporal colinearity is the main constraining force on *Hox* cluster organization, but until we have a much deeper understanding of the mechanistic basis for this phenomenon, and know how widespread across the Bilateria the mechanism(s) is/are, then we cannot know how the *Hox* cluster of the last common bilaterian operated and what have been the major evolutionary forces operating upon the *Hox* gene cluster.

The *Hox* genes are a set of transcription factor-encoding genes that pattern the anterior-posterior body axis of animals. They tend to have a very distinctive organization in the genome, being arranged in gene clusters in which the order of the genes within the cluster corresponds to or is colinear with some aspect of the gene expression. Spatial colinearity was first recognized in *D. melanogaster*, whereby the domains of action of each homeotic locus of the *Bithorax complex* (*BX-C*) are staggered along the anterior-posterior axis in an order that matches their position along the chromosome. Such spatial colinearity is also observed in the *Antennapedia Complex* (*ANT-C*), which together with the *BX-C* constitutes the *Hox* cluster of *D. melanogaster*. The phenomenon was also extended to the *Hox* clusters of vertebrates soon after the discovery of the homeobox; the homeobox being the major motif of the *Hox* genes, that is 180-bp long and encodes a sequence-specific DNA-binding domain.

Temporal colinearity

In addition to spatial colinearity, temporal colinearity is also observed in some *Hox* clusters; principally those of vertebrates in which the genes at the 3'-anterior end of the cluster become transcriptionally active first, and then progressively genes towards the 5'-posterior end of the cluster are initiated later. This temporal control of the mammalian *Hox* clusters is correlated with modulation of their chromatin conformation. Temporal and spatial colinearity can be mechanistically linked, with the time of gene initiation determining the axial limits of expression. However the two phenomena are also separable. *Drosophila* exhibits spatial colinearity without any obvious temporal colinearity, as does the larvacean urochordate *Oikopleura dioica*, and distinct regulatory regions that drive temporal and spatial colinearity have recently been discovered in mice.

Quantitative colinearity

In mouse limb development, the proximity to a digit enhancer determines the levels of expression of a gene, the gene closest to the enhancer being the most highly expressed. As with temporal colinearity potentially resulting in spatial colinearity, quantitative colinearity could similarly lead to spatial colinearity as well. The organization of the *Hox* cluster and the position of the genes within it are thus intimately linked to the transcriptional regulation of the genes.

Hox gene colinearity in some shape or form seems to be widespread, and so is usually assumed to be the major reason that the genes remain clustered. *Hox* genes from flies to humans clearly are homologous and even orthologous in most instances, and undoubtedly they were clustered in the last common ancestor of the bilaterian animals (and probably even deeper in animal evolution - before the origin of the Cnidaria). If we can discern what the homologous features of *Hox* cluster organization, function and regulation are across the animal kingdom, then we can discover how these genes were arranged and were operating in an organism that lived more than 525 mya, from which all subsequent animal lineages diverged.

With such an approach it has been proposed that the mechanism(s) producing temporal colinearity is/are potentially the principal constraining force on *Hox* cluster organization, because the various derived animal lineages that have broken up their *Hox* clusters almost invariably possess a mode of development in which there is a lack of opportunity for temporal colinearity to exist. The disintegrating clusters of Drosophilids, nematodes and urochordates correlate with the rapid embryogenesis of these animals, which can still however retain elements of *Hox* spatial colinearity, thus showing that the mechanisms generating spatial colinearity do not absolutely require *Hox* gene clustering. The mechanisms generating temporal colinearity may well thus be the principal reason that *Hox* gene clusters are maintained as clusters. However a much closer examination of these mechanisms is required in order to discover whether these similarities are truly homologous, and conversely how some lineages 'escape' from these 'fundamental' constraints during their divergence and diversification.

REFERENCES

Atchison, M.L. 1988. Enhancers: mechanisms of action and cell specificity. Annu. Rev. Cell. Biol. 4:127-53.

Barrell, B.G., G.M. Air, and C.A. Hutchison III. 1976. Overlapping genes in bacteriophage ϕX174. Nature 264: 34-41.

Benzer, S. 1955. On the topography of the genetic fine structure. Proc. Natl. Acad. Sci. USA 47: 403-415.

Benzer, S. 1956. Genetic fine structure and its relation to DNA molecule. Brookhaven Symp. Biol. 8: 3-5.

Benzer, S. 1961. Genetic fine structure. In: *Harvey Lectures 56*. New York: Academic Press.

Benzer, S. 1962. The fine structure of the gene. Scient. Am. 206(1): 70-64.

Berget, S.M., C. Moore, and P.A. Sharp. 1977a. Spliced segments at 5' terminus of adenovirus 2 late messenger-RNA. Proc. Natl. Acad. Sci. USA 74: 3171-5.

Berget, S.M., A.J. Berk, T. Harrison, P.A. Sharp. 1977b. Spliced segments at 5′ termini of adenovirus-2 late messenger-RNA - role for heterogeneous nuclear-RNA in mammalian-cells. Cold Sp. Harb. Symp Quant. Biol. 42: 523-9.

Blake, C. 1979. Exons code protein functional units. Nature 277: 598.

Blake, C. 1983. Exons - present from the beginning? Nature 306: 535-7.

Boveri, T. 1904. *Erebnnisse über die Konstitution der chromatischen substanz des Zellkerns*. G. Fischer, Jena.

Breathnach, R., J.L. Mandel, and P. Chambon. 1977. Ovalbumin gene is split in chicken DNA. Nature 270: 314-9.

Brevario, D.M., W.M. Baird, K. Hilu, P. Blumetti, and S. Giani. 2007. High polymorphism and resolution in targeted fingerprinting with combined β-tubulin introns. *Mol. Breeding* **20:** 249-59.

Chambon, P. 1977. Molecular biology of eukaryotic genome is coming of age. Cold Sp. Harb. Symp. Quant. Biol. 42: 1209-34.

Chambon, P. 1979. Determination of the organization of coding and intervening sequences in the chicken ovalbumin split gene. Differentiation 13(1): 43-4.

Chambon, P. 1981. Split genes. Scient. Am. 244(5): 48-59.

Chen, D., Y. Meng, and C. Yuan. 2012. Plant siRNAs from introns mediate DNA methylation of host genes. *RNA* doi: 10.1261/rna.2589011.

Chorev, M., and L. Carmel. 2012. The function of introns. *Front. Genet.* doi: 10.3389/fgene.2012.00055.

Chovnick, A. 1961. The garnet locus in *Drosophila melanogaster*. I. Pseudoallelism. Genetics 46: 493-507.

Chow, L.T., R.E. Gelinas, T. R. Broker, and R.J. Roberts. 1977. An amazing sequence arrangement at the 5′ ends of adenovirus 2 mRNA. Cell 12: 1-8.

Clark, D.P. 2005. *Molecular Biology*. London: Elsever Academic Press.

Contreras, R., R. Rogiers, V.A. Van, and W. Fiers. 1977. Overlapping of the *CP2-VP3* gene and the *VP1* gene in the SV40 genome. Cell 12: 529-38.

Cornish-Bowden, A. 1985. Are introns structural elements or evolutionary debris? Nature 313: 434-5.

Craik, C.S., Q.L. Choo, G.H. Swift, C. Quinto, R.J. MacDonald, and W.J. Rutter. 1984. Structure of two related rat pancreatic trypsin genes. J. Biol. Chem. 259: 14255-64.

Craik, C.S., S. Sprang, R. Fletterick, and W.J. Rutter. 1982. Intron-exon splice junctions map at protein surfaces. Nature 299: 180-182.

Craik, C.S., W.J. Rutter, and R. Fletterick. 1983. Splice junctions: Association with variation in protein structure. Science 220: 1125-9.

Danchin, A.Q. 1987. Significance of split genes to developmental genetics. Adv. Genet. 24: 243-84.

David, I.B., and P.K. Wellauer. 1978. Ribosomal DNA and related sequences in *Drosophila melanogaster*. Cold Sp. Harb. Symp. Quant. Biol. 42: 1185-94.

Doolittle, W.F. 1978. Origins of split gene organization. Nature 272: 581-2.

Dynan, W.S. 1989. Modularity in promoters and enhancers. Cell 58: 1-4.

Elder, D. 2000. Split gene origin and periodic introns. J. Theor. Biol. 207: 455-72.

Fiers, W., R. Contreras, G. Haegeman, et al. 1978. Complete nucleotide sequence of SV40 DNA. Nature 273: 113-210.

Garret, R.A. 1985. The uniqueness of Archaebacterium. Nature 318: 233-5.

Gertz, J., E.D. Siggia, and B.A. Cohen. 2009. Analysis of combinatorial cis-regulation in synthetic and genomic promoters. Nature 457: 215-8.

Gilbert, W. 1978. Why genes in pieces? Nature 271: 401.

Gilbert, W. 1981. DNA sequencing and gene structure. Science 214: 1305-12.

Gilbert, W. 1985. Genes-in-pieces revisited. Science 228: 823-4.

Glover, D. and N. Davidson. 1975. A novel arrangement of the 18S and 28S sequences in the repeating unit of *Drosophila melanogaster*. Cell 10: 67-176.

Gonzalez-Pastor, J.E., J.L. San Millan, M.A. Castilla, and F. Moreno. 1995. Structure and organization of plasmid genes required to produce the translation inhibitor microcin C7. J. Bacteriol. 177: 7131-40.

González-Pastor, J.E., San Millán, J.L. and Moreno, F. 1994. The smallest known gene. Nature 369: 281.

Green, M.M. 1961. Phenogenetics of the *lozenge* loci in *Drosophila melanogaster*. II. Genetics of *lozenge-Krivshenko* (*lz*^{k}). Genetics 46:1169-76.

Green, M.M. and Green, K.C. 1949. Crossing over between alleles at the *lozenge* locus in *Drosophila melanogaster*. Proc. Nat. Acad. Sci. USA 35: 586-91.

Henikoff, S., M.A. Keene, K. Fetchel, J.W. Fristrom. 1986. Gene within a gene: Nested *Drosophila* genes encode unrelated proteins on opposite DNA strands. Cell 44: 33-42.

Kaine, B.P., R. Gupta, and C.R. Woese. 1983. Putative introns in tRNA genes of prokaryotes. Proc. Natl. Acad. Sci. USA 80: 3309-12.

Khorana, H.G., H. Buchi, H. Ghosh, et al. 1967. Polynucleotide synthesis and the genetic code. Cold Sp. Harb. Symp. Quant. Biol. 31: 39-49.

Kourilsky, P., and Chambon, P. 1978. The ovalbumin gene: an amazing gene in eight pieces. Trends. Biochem. Sci. 3: 244-7.

Levinson, B., S. Kenwrick, D. Lakich, G. Hammonds Jr, and J. Gitschier. 1990. A transcribed gene in an intron of the human factor VIII gene. Genomics 7: 1-11

Lewis, E.B. 1941. The *Star* and *asteroid* loci in *Drosophila melanogaster*. Genetics 27: 153-4.

Lewis, E.B. 1951. Pseudoallelism and gene evolution. Cold Sp. Harb. Symp. Quant. Biol. 16: 159-72.

Lifton, R.F., M. Goldberg, R.W. Karp, and D.S. Hogness. 1978. The organization of the histone genes in *Drosophila melanogaster*: functional and evolutionary implicartions. Cold Sp. Harb. Symp. Quant. Biol. 42: 1042-51.

Logadon, J.M. and J.D. Palmer. 1994. Origin of introns – early or late. Nature 369: 526-528.
McClintock, B. 1945. *Neurospora.* I. Preliminary observations of the chromosomes of *Neurospora crassa*. Am. J. Bot. 32: 671-8.
Min Jou, W., G. Haegeman, M. Ysebaert, and W. Fiers. 1972. Nucleotide sequence of the gene coding for the bacteriophage MS2 coat protein. Nature 237: 82-8.
Moabbi, A.M., N. Agarwal, B.E. Kaderi, and A. Ansari. 2012. Role for gene looping in intron-mediated enhancement of transcription. *Proc. Natl. Acad. Sci. USA* 109: 8505-10.
Monteiro, A.S., and D.E.K. Ferrier, 2006. Hox genes are not always colinear. Int. J. Biol. Sci. 2: 95-103.
Moras, D.M., K.W. Olsen, M.L. Sabsen, M. Buehner, G.C. Ford, and M.G. Rossmann. 1975. Studies of assymetry in the three-dimensional structure of lobster D-gyceraldehyde-3-phosphate dehydrogenase. J. Biol. Chem. 250: 9137-62.
Morello, L., and D. Breviario. 2008. Plant spliceosomal introns: not only cut and paste. *Curr. Genomics* 9: 227-38.
Morgan, T.H. 1926. *The Theory of the Gene*. New Haven: Yale University Press.
Muller, H.J. 1927. Artificial transmutation of the gene. Science 66: 84-7.
Muller, H.J. 1940. An analysis of process of structural changes in chromosomes of *Drosophila*. J. Genet. 40: 1-66.
Normark, S., S. Bergström, B. Jaurin, F.P. Lindberg, and O. Olsson. 1983. Overlapping genes. Annu. Rev. Genet. 17: 499-525.
Ny, T., F. Elgh, and B. Lund. 1984. The structure of human tissue-type plasmogen activator gene: Correlation of intron and exon structures to functional and structural domains. Proc. Natl. Acad. Sci. USA 81: 5355-9.
O'Farrell, F.Z., B. Cordell, P. Valenzuela, W.J. Rutter, and H.M. Goodman. 1978. Structure and processing of yeast precursor tRNAs containing intervening sequences. Nature 274: 438-45.
Ondek, B., L. Gloss, and W. Herr. 1988. The SV40 enhancer contains 2 distinct levels of organization. Nature 333: 40-45.
Perlman, P.S., and R.A. Buton. 1989. Mobile introns and intron-induced proteins. Science 246: 1106-9.
Pierangeli, A., M. Bucci, M. Forzan, P. Pagnotti, M. Equestre, and B.R. Perez. 1998. Identification of an alternative open reading frame ("Hidden Gene"?) stringently required for infectivity of poliovirus cDNA clones. Microbiologica 21: 309-20.
Pollock, J.T., I. Tessman, and E.S. Tessman. 1978. Potential for variability through multiple gene products of bacteriophage ϕX174. Nature 274: 34-7.
Pontecorvo, G. 1952. The genetical formulation of gene structure and action. Adv. Enz. Mol. 13: 121-40.
Pribnow, D., Sigurdson, C., Gold, L., Singer, B.S., Napoli, C., Brosius, J., Dull, T.J. and Noller, H.E. 1981. rII cistrons of bacteriophage T4 DNA sequence around the intercistronic divide and positions of genetic landmarks. J. Mol. Biol. 149: 337-76.
Rogers, J. 1985. Split gene evolution; exon shifting and intron insertion in serine protease genes. Nature 315: 458-9.
Sanger, F. 1981. Determination of nucleotide sequence in DNA. Science 214: 1205-10.
Sanger, F., A.R. Coulson, T. Friedmann, et al. 1978. The nucleotide sequence of bacteriophage ϕX174. J. Mol. Biol. 125: 225-46.
Sanger, F., G.M. Air, B.G. Barrell, N.L. et al. 1977. Nuleotide sequence of bacteriophage ϕX174 DNA. Nature 265: 687-95.
Sarabhai, A., A.D.W. Stretton, S. Brenner, and A. Bolle. 1964. Colinearity of the gene with polypeptide chain. Nature 201: 13-37.
Senapathy, P. 1986. Origin of eukaryotic introns: a hyothesis based on codon distribution statistics in genes, and its implications. Proc. Natl. Acad. Sci. USA 83: 2133-7.
Senapathy, P. 1995. Introns and the origin of protein-encoding genes. Science 268: 1366-7.
Shaw, D.C., J.E. Walker, F.D. Northrop, B.G. Barrell, G.N. Godson, and J.C. Fiddes. 1978. Gene K, a new overlapping gene in bacteriophage G4. Nature 272: 510-4.
Sherman, R., M. Jackson, S.W. Liebman, A.M. Schweinyruber, and J.W. Stewart. 1975. *Cyc-1* mutants and its correspondence to mutationally altered iso-1-cytochrome C of yeast. Genetics 81: 51-73.
Smale, S.T., and D. Batlimore. 1989. The initiator as a transcription control element. Cell 57: 103-13.
Smale, S.T., and J.T. Kadonaga, 2003. The RNA polymerase II core promoter. Annu. Rev. Biochem. 72: 449-79.
Stoltzfus, A., M. Spencer, M. Zuker, J.M. Logsdon, and W.F. Doolittle. 1994. Testing the exon theory of genes. The evidence from protein structure. Science 265: 202-7.
Stone, E.M., K.N. Tothblum, and R.J. Schwartz. 1985. Intron dependent evolution of chickem glyceraldehydes phosphate dehydrogenase. Nature 313: 498-500.

Straus, D. and W. Gilbert. 1985. Genetic engineering in the Precambrian: structure of the chicken triosephosphate isomerase gene. Mol Cell Biol. 5: 3497-506.

Sturtevant A.H. 1925. The Effects of unequal crossing over at the *Bar* locus in *Drosophila*. Genetics 10:117-47.

Sturtevant, A.H., and T.H. Morgan. 1923. Reverse mutation of *Bar* gene correlated with crossing over. Science 57: 746-7.

Sutton, W.S. 1903. The chromosomes in heredity. Biol. Bull. 4: 213-51.

Watson, J.D., and F.H.C. Crick. 1953. A structure for deoxyribose nucleic acids. Nature 171: 737-8.

Wellauer, P.K., and I.B. David. 1977. The structural organization of ribosomal DNA in *Drosophila melanogaster*. Cell 10: 193-212.

Woese, C.R., and G.E. Fox. 1977. Phylogenetic structure of prokaryotic domain: The primary kingdoms. Proc. Natl. Acad. Sci. USA 74: 5088-90.

Yanofsky, C., B.C. Carlton, J.R. Guest, D.R. Helinski, and U. Henning. 1964a. On the colinearity of gene structure and protein structure. Proc. Nat. Acad. Sci. USA 51: 266-72.

Yanofsky, C., G.R. Drapeau, J.R. Guest, and B.C. Carlton. 1967. The complete amino-acid sequence of the tryptophan synthetase A protein (α subunit) and its colinear relationship with the genetic map of the A gene. Proc. Nat. Acad. Sci. USA 57: 296-8.

Yanofsky, C., V. Horn, and D. Thorpe. 1964b. Protein structure relationships revealed by mutational analysis. Science 146: 1593-4.

2

Gene Organization and Genome Evolution

Organization of the genes may be considered from different points of view. For example, one may consider whether individual genes are independently controlled or a group of them are under the control of one and the same operator, whether the message is contiguous or split, genes are overlapping or non-overlapping, genes are functional or non-functional, genes are repeated or not, whether the genes occupy a fixed position or they shift from one place to another. Present gene organization in different organisms is net result of concerted efforts of genome evolution. So it will be appropriate to have a look at genome evolution alongwith gene organization.

GENE ORGANIZATION

Multigene Families

A multigene family is a group of nucleotide sequences or genes that exhibits four properties, namely, multiplicity, close linkage, sequence homology, and related or overlapping phenotypic functions. Model of a multigene family is given in Figure 2.1. A multigene family may well be a fundamental unit of chromosome organization encompassing a broad spectrum of gene families encoding both simple and complex phenotypic traits.

Multigene families are classified into three types – simple-sequence family, multiplicational family and informational family. The simple sequence family encompasses segments of DNA derived from 10^2-10^7 repetitions of a short fundamental sequence, generally 6-15 nucleotides in length. Degree of homology amongst the repeat units is often 80-100 per cent. The multiplicational family consists of 10^1 to 10^4 copies of a gene, 80-100 nucleotides long. Repeat units are essentially identical. It has individual members of the informational family may differ markedly in sequence from one another although all are homologous and obviously show an ancient ancestry.

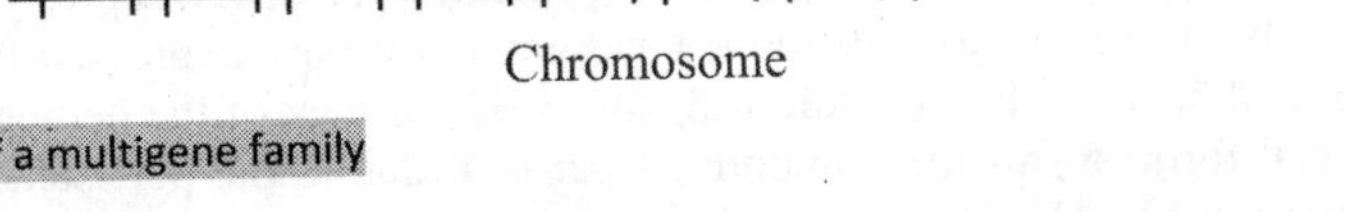

G_1 G_2 G_3 G_4 G_5 G_6 G_7 ... G_m G_n

Chromosome

Figure 2.1 Model of a multigene family

Repetitions of genes are of two types – dosage repetition and variant repetition. In case of dosage repetition there are many copies per nucleus. The rRNA, 5SRNA, tRNAs and histone genes show dosage repetition. For variant repetitions, variant repeated genes exist and their products are not identical but related. The different members of a variant repetitive gene family evolve by gene duplication and divergence of an ancestral gene. A repeated gene family consists of many identical or nearly identical genes that cohabit a single haploid genome. In some cases the genes are contained within tandemly repeated units.

Multigene families may comprise of divergent or identical genes.

Divergent Multigene Families

In case of divergent multigene family, different members of the family have resulted from concerted evolution to the formation of divergent sets of duplicates. Some well-known examples of divergent multigene families are immunoglobulin gene family, globin gene family, and actin gene family.

Immunoglobulin genes

The C regions of heavy chain define various classes of serum immunoglobulins. IgG, IgH, and IgA are three major classes and IgD and IgE are two minor classes in man. The mammalian germ cell appears to contain at least three families of antibody genes – λ, k, and H – each located on a different chromosome. Thus, antibody genes constitute a multigene system which comprises of three unlinked multigene families. A model of the genes encoding the three antibody families of man are shown in Figure 2.2.

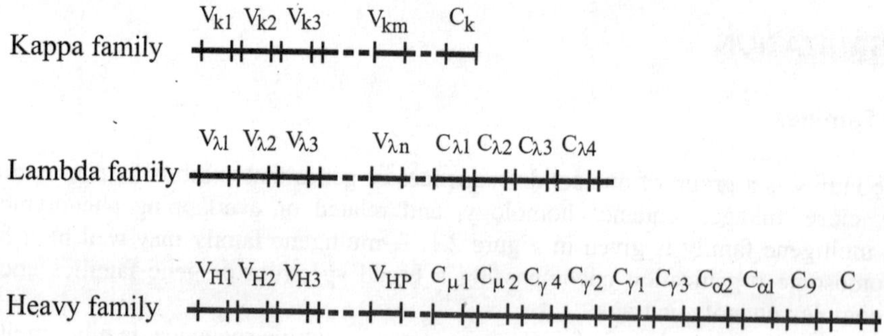

V_{k1} V_{k2} V_{k3} V_{km} C_k

Kappa family

$V_{\lambda 1}$ $V_{\lambda 2}$ $V_{\lambda 3}$ $V_{\lambda n}$ $C_{\lambda 1}$ $C_{\lambda 2}$ $C_{\lambda 3}$ $C_{\lambda 4}$

Lambda family

V_{H1} V_{H2} V_{H3} V_{HP} $C_{\mu 1}$ $C_{\mu 2}$ $C_{\gamma 4}$ $C_{\gamma 2}$ $C_{\gamma 1}$ $C_{\gamma 3}$ $C_{\alpha 2}$ $C_{\alpha 1}$ C_9 C

Heavy family

Figure 2.2 A model of genes coding for antibody families of man

Globin gene family in human genome

Although in adult humans only single genes, each specifying alpha (α) and beta (β) polypeptide chains of hemoglobin are expressed, this is not enough during development, such that a whole family of related α globin genes and another family of β globin genes function. These genes are turned off and on in a coordinated and sequential manner during embryo development till the fetus is born, depending upon the intracellular milieu. The constitution of hemoglobins found in different stages of development is shown in Table 2.1 and Figure 2.3. During development, zeta (ξ) is expressed first and is later replaced by α. Similarly, in β pathway, epsilon (ε) and gamma (γ) are expressed first, which are later replaced by delta (δ) and beta (β). In the adult, $\alpha_2\beta_2$ forms 97 per cent of the hemoglobin, $\alpha_2\delta_2$ forms 2 per cent and $\alpha_2\gamma_2$ (fetal form) forms the remaining 1 per cent due to the persistence of the fetal form (Figure 2.3). In humans, α-like globin genes are located in chromosome 16 and β-like globin genes are located in chromosome 11 (Figure 2.4). Based on similarity in amino acid sequences of globin of different mammals, evolutionary trees have been prepared (Figure 2.5 and Figure 2.6). Efstratiadis et al. (1980) have described the structure and evolution of the human β-globin gene family.

Table 2.1 Tetramer-dimers of natural human hemoglobins

Hemoglobin	*Subunit in common*	*Classification*
A ($\xi_2\beta_2$)	β	Adult
Portland-2 ($\xi_2\beta_2$)	β	Embryonic
F ($\alpha_2\gamma_2$)	γ	Fetal
Portland-1 ($\xi_2\gamma_2$)	γ	Embryonic
Gower-2 ($\alpha_2\,\varepsilon_2$)	ε	Embryonic
Gower-1 ($\xi_2\,\varepsilon_2$)	ε	Embryonic
A2 ($\alpha_2\,\delta_2$)		Adult

Hemoglobin $\xi_2\delta_2$ has not been isolated and has not been named.

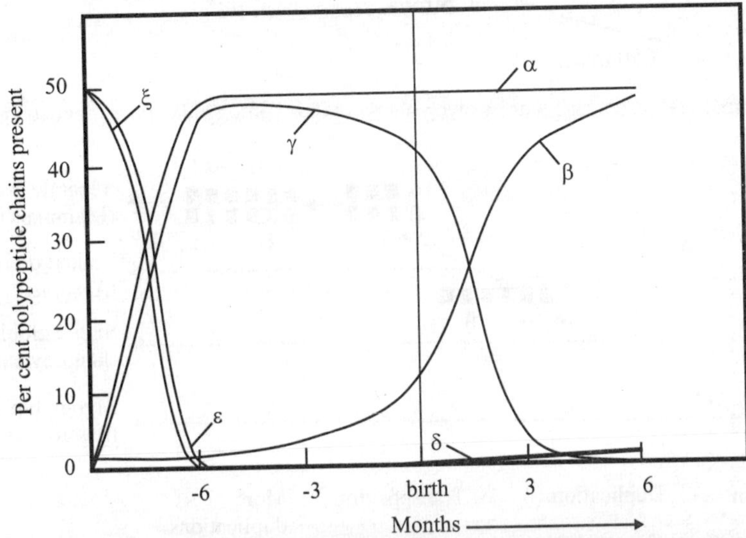

Figure 2.3 Temporal regulation of the synthesis of different polypeptides representing α and β polypeptide chains of hemoglobin molecules during pregnancy and after birth in human beings (Redrawn, with permission, from Gupta, P.K. 2008. *Molecular Biology and Genetic Engineering*. Meerut: Rastogi Publications)

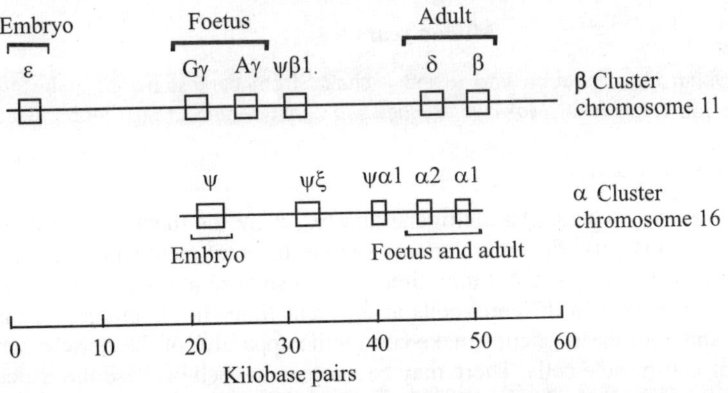

Figure 2.4 Arrangement of human α and β globin gene clusters on chromosomes 11 and 16 (Redrawn, with permission, from Gupta, P.K. 2008. *Molecular Biology and Genetic Engineering*. Meerut: Rastogi Publications)

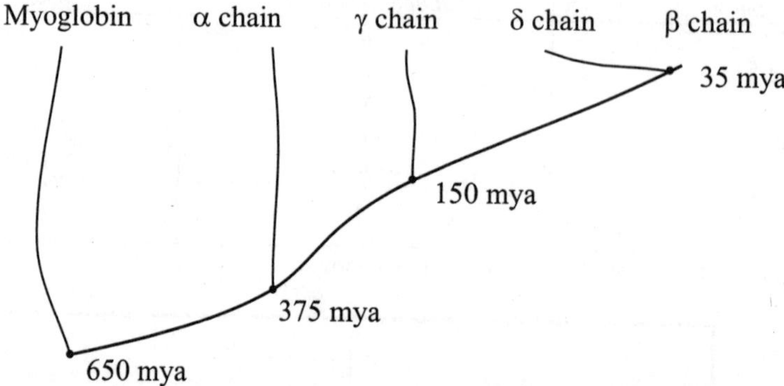

Figure 2.5 Phylogenetic relationships among myoglobin and four hemoglobin chains deduced from amino acid substitutions

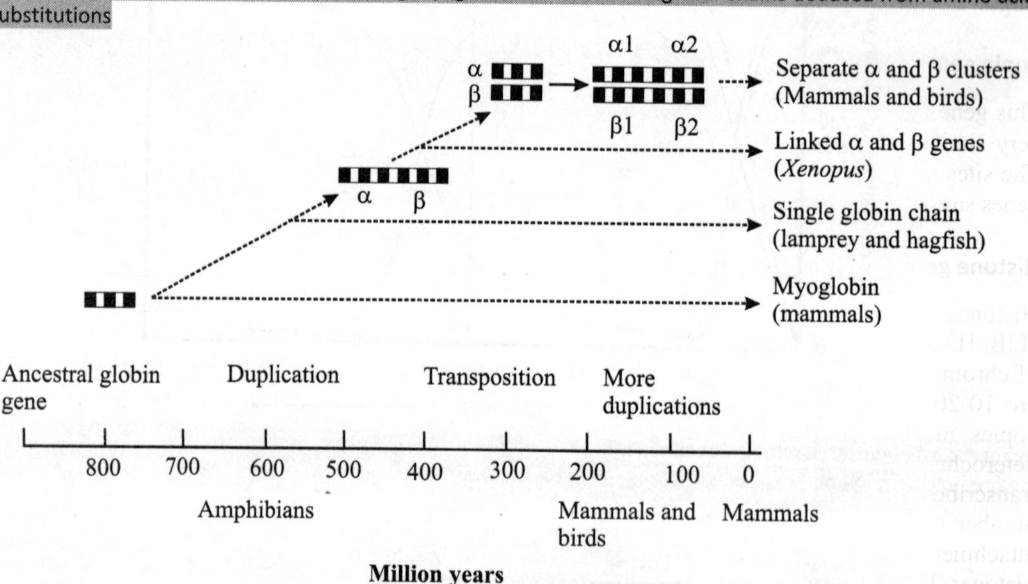

Figure 2.6 Myoglobin differentiation into α and β chains from earliest period in history (Redrawn, with permission, from Gupta, P.K. 2008. *Molecular Biology and Genetic Engineering*. Meerut: Rastogi Publications)

Actin gene family

Actin genes in eukaryotes represent a multigene family, where the members are homologous but non-identical, giving rise to slightly different variants so that different members function either at different times or in different tissues at the same time. For instance, a minimum of six closely related polypeptides are synthesized in different cells at different times in mammals. Actins may be broadly classified as α, β and γ; of these γ actins make contractile apparatus of the muscle, while β and γ actins are constituents of non-muscle cells. There may be variants in each of these three classes, making upto 20-30 actin related sequences in human genome. In a number of plant systems (e.g., tobacco, potato, etc.) also, more than 20 related but variant actin sequences have been reported. In contrast to this, single actin gene is known in yeast.

Some other examples of multigene families with divergent members are albumin and α-fetoprotein (major components of blood plasma), serine proteases, the interferons (for defense against viral infection) genes, and chorion proteins making the major component of egg shell in insects.

Identical Multigene Families

In case of identical multigene family, different members of the family are multiple copies of the same gene. Some examples of multigene families with identical genes are 412 gene family, copia gene family, histone gene family, ribosomal RNA gene family, 5S RNA gene family, tRNA gene family, small nuclear RNA gene family and storage protein gene family. Tartoff (1975) reviewed the then status of knowledge about redundant genes.

412 gene family

The 412 gene family at each of the 30 or so sites contains a complete *412* gene. Each is 7.0 kb in length. It is terminally redundant.

Copia gene family

This gene family is 4-5 kb long. The copia family, like 412 gene family appears to consist of genes of very similar structure located at multiple sites that are characterized by different flanking sequences. The sites are scattered throughout the genome. Direct repeats also exist at or near the termini of copia genes suggesting that terminal redundancy may be a general characteristic of dispersed families.

Histone gene family

Histones are involved in packaging of DNA into chromatin. There are five types of histones H1, H2A, H2B, H3 and H4. During S-phase of cell cycle, the requirement of histones increases due to replication of chromosomes. Genes synthesizing these histones are present in multiple copies. For example, there are 10-20 copies of histone genes in birds and mammals, ~100 copies in *Drosophila* and 600-800 copies in Amphibians. These genes are present in form of heteroclusters. Structure of histone heterocluster is shown in Figure 2.7. This cluster differs from rRNA heterocluster as each gene is transcribed separately. Where ever histone gene number is higher, cell division rate is rapid than where number of histone genes is less. The non-transcribed spacers have repeat units. Repeat units have attachment sites called nuclear attachment cites, which are attached to nucleus for rapid transfer of histones. Histone genes lack introns and the transcripts lack poly(A) tails.

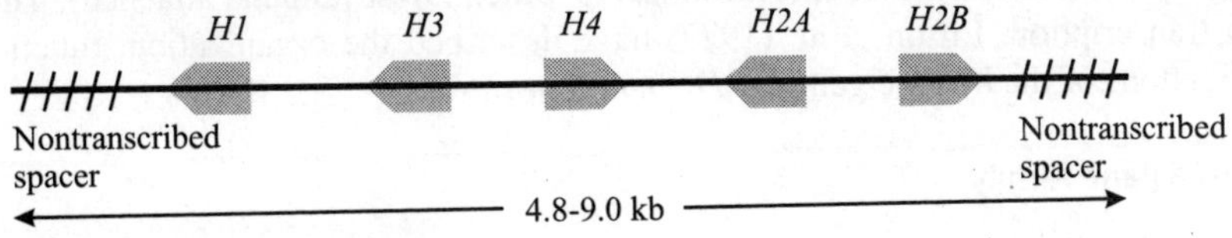

Figure 2.7 Structure of histone gene clusters

The five histone genes coding for these five types of histones form a basic unit that is repeated, although different genes within a repeat unit may differ in orientation (Figure 2.8). Transcripts of histone genes are devoid of introns and poly(A) tails. Arrangement of the histone genes in *D. melanogaster* is in form of repeat units. In two classes of repeat units, one differs from another by an insertion (Figure 2.9).

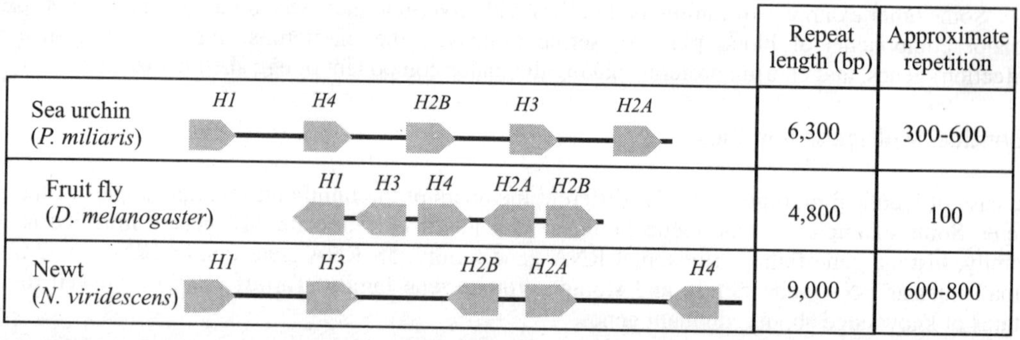

		Repeat length (bp)	Approximate repetition
Sea urchin (*P. miliaris*)	*H1* *H4* *H2B* *H3* *H2A*	6,300	300-600
Fruit fly (*D. melanogaster*)	*H1* *H3* *H4* *H2A* *H2B*	4,800	100
Newt (*N. viridescens*)	*H1* *H3* *H2B* *H2A* *H4*	9,000	600-800

Figure 2.8 Organization of highly repeated histone gene clusters in several organisms; genes are separated by non-transcribed spacer (NTS) region; the direction of transcription is shown by arrows; the clusters are tandemly repeated in sea urchin and fruit fly, but not in the newt

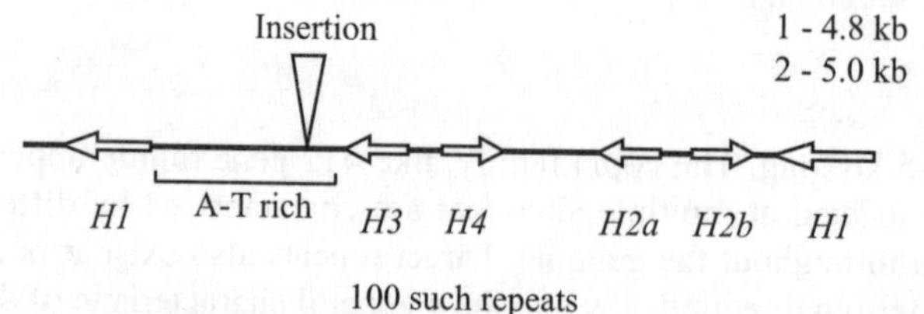

Figure 2.9 Arrangement of the histone genes in *D. melanogaster*

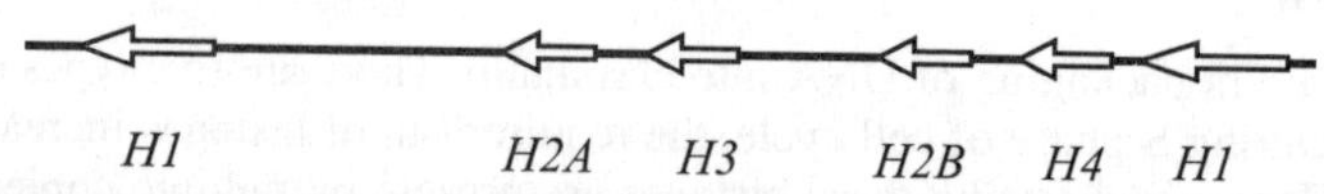

Figure 2.10 Arrangement of the histone genes in *S. purpratus*

Hundred copies of *D. melanogaster* histone genes appear to span most of the 12 chromomeres in 39DE region of chromosome 2. Arrangement of the histone genes in *Strongylocentrotus purpuratus* (sea urchin) is shown in Figure 2.10. In *Drosophila* multigene family of histone genes, there is present an attachment site per repeat unit (Figure 2.11). One hundred such repeat units are present in each haploid genome. With the help of attachment site, the genes remain attached to nuclear scaffold (scaffold is the protein backbone of chromosomes to which DNA remains attached). These attachment sites help in transcription. Lifton et al. (1977) have described the organization, functions and evolutionary implications of the histone genes in *D. melanogaster*.

Ribosomal RNA gene family

Williams et al. (1990) have explained in detail the superstructure of the *Drosophila* ribosomal gene family. Ribosomes have two subunits – large subunit and small subunit. Large subunit has three rRNAs of size 28S, 5.8S and 5S. Small unit has 18S rRNA. There are 4 RNA genes. *18S rRNA*, *5.8S RNA* and *28S rRNA* genes are present in form of clusters, 18S rRNA-5.8S rRNA-28SrRNA, known as heteroclusters. 5S forms a separate cluster. In *E. coli*, rRNA gene family has 3 genes with 7 copies each. Prokaryotes have less number of ribosomes per cell (≈1,000). In eukaryotes, number of copies of rRNA genes ranges from 150-15,000. Plants have 600-8,500 copies of this heterocluster. Eukaryotes

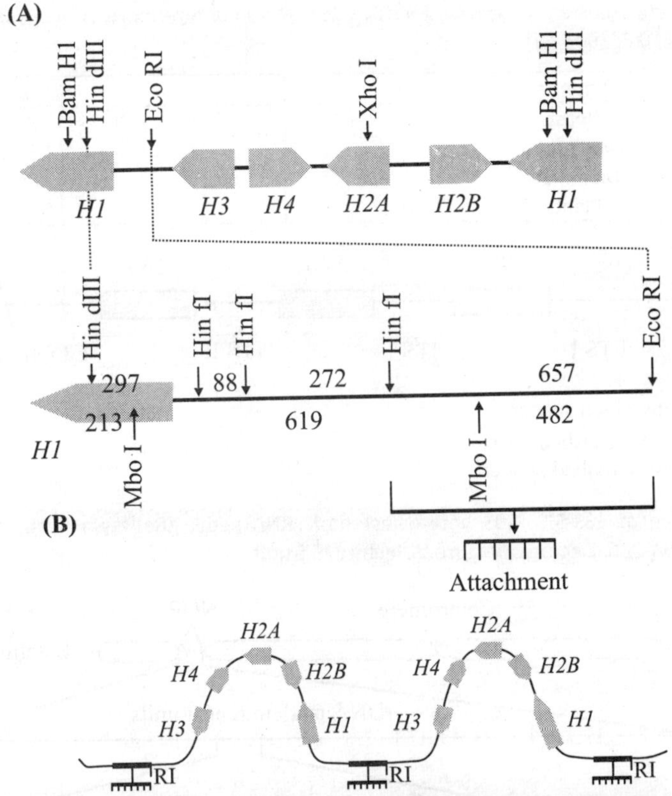

Figure 2.11 (A) A restriction map of major histone gene cluster in *Drosophila*; each repeat unit has one specific attachment site to nuclear scaffold, this attachment site is HinI/EcoRI DNA segment 657 bp long. (B) Schematic representation showing how *Drosophila* histone repeats might be attached to the nuclear scaffold (Redrawn, with permission, from Gupta, P.K. 2008. *Molecular Biology and Genetic Engineering*. Meerut: Rastogi Publications)

have more number of ribosomes per cell ($\approx 10^6$). Structure of *18S rRNA-5.8S rRNA-28SrRNA* heterocluster is given in Figure 2.12. Size of heterocluster is variable in different organisms (Table 2.2). Non-transcribed spacers are of different lengths in different individuals and are also inverted repeats of repeat units of different sizes, viz., 60, 80, and 90 bp long. The rRNA heteroclusters are present in nucleolar organizer region (NOR). For the production of about 10 million ribosomes per cell in eukaryotes, rRNA genes are present in multiple copies in tandem array at nuclear organizer regions (NOR) of specific satellite chromosomes. The number of these genes may vary from 50-30,000 in a cell and this number may be equally distributed on NORs. The DNA comprising these genes is called ribosomal DNA (rDNA), which is repetitive in nature. Each repeat unit has a coding region with genes that specify 18S, 5.8S and 28S rRNA molecules, a spacer region called intergenic spacers (IGS) and internal transcribed sequences (ITS) one each between *18S rRNA* and *5.8S rRNA* genes. Non-transcribed spacer (NTS) makes only a part of the IGS (Figure 2.13), the remaining part of IGS being external transcribed sequences (ETS). The IGS has a region consisting of tandem array of variable number of subrepeats ranging from 100-300 bp in length. The variation in number and size of the subrepeats in IG is responsible for variation in length of the repeat unit (IGS + coding region).

Table 2.2 Size of heterocluster 18S rRNA-5.8S rRNA-28SrRNA in different organisms (*Source*: Manning, L.R. et al. 2007. Protein Sci. 16: 1641-58)

Organism	*Size (kb)*
Yeast	8
Drosophila	8
Mammals	13
Plants	8-18

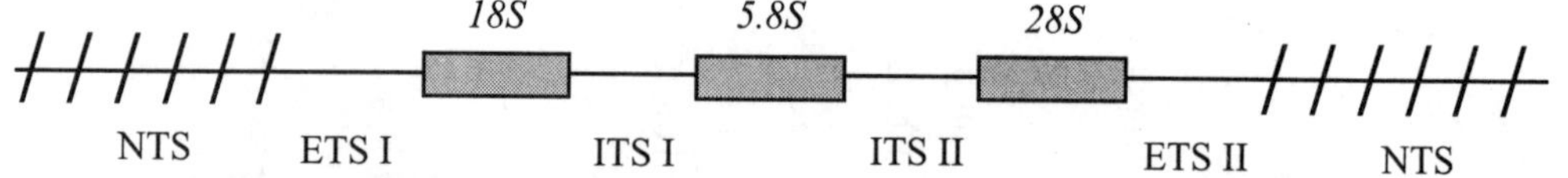

NTS = Nontranscribed spacer
ETS = External transcribed spacer
ITS = Internal transcribed spacer

Figure 2.12 Structure of 18S-5.8S-28S heterocluster of rRNA genes (Redrawn from http://statistics.arizona.edu/courses/EEB600A-2003/lectures/lecture25/lecture25.html)

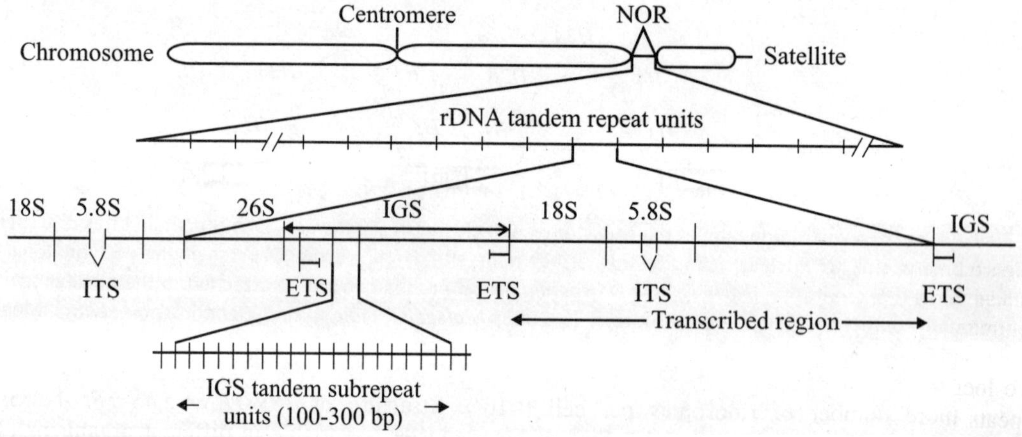

Figure 2.13 Organization of ribosomal DNA (rDNA) in nucleolar organizer region (NOR) of a eukaryote chromosome (IGS = intergenic spacer); NTS = non-transcribed spacer ; ETS = external transcribed spacer; ITS = internal transcribed spacer; 18S, 5.8S, 26S = different genes responsible for 18S rRNA, 5.8S rRNA and 26S rRNA) (Redrawn, with permission, from Gupta, P.K. 2008. *Molecular Biology and Genetic Engineering*. Meerut: Rastogi Publications)

5S RNA genes

5S RNA genes are present independently of the rDNA, which are localized in the NOR in eukaryotes. However, in prokaryotes and yeast, *5S rRNA* genes are present in close vicinity of rDNA (Figure 2.14). The *5S RNA* genes are also organized in tandem repeats, each repeat consisting of a gene 120 bp long and a spacer region. These repeat units may sometimes carry pseudogenes instead of genes, and may not synthesize any RNA. The *5S rRNA* genes are present on different chromosomes and are present as different tandem repeats. The length of the complete repeat is 375 bp in *Drosophila*. In wheat, there are

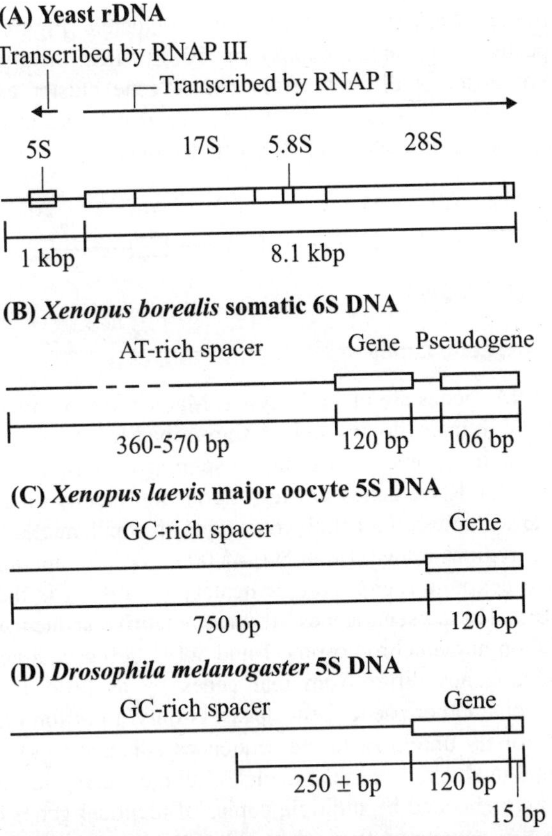

Figure 2.14 Organization of 5S rDNA (produces 5S RNA for ribosomes) in several organisms; only in yeast it is linked with repeat unit of rDNA; in other cases it is independent on a separate chromosome (Redrawn, with permission, from Gupta, P.K. 2008. *Molecular Biology and Genetic Engineering*. Meerut: Rastogi Publications)

two loci for 5S rRNA having repeats of different lengths; 480 bp and 500 bp. There are 200-50,000 repeats in different animals and 200-9,000 in different plants. Lengths of non-transcribed sequences (NTSs) in *5S rRNA* gene cluster range from 250-750 bp. In case of *Xenopus*, there are 20,000 copies of *5S rRNA* gene. These clusters in *Xenopus* are of two types – (a) cluster of 400 genes is expressed in all somatic cells and (b) cluster of 19,600 genes is expressed only at oocyte stage. In *Xenopus*, these clusters are located at telomeres of all the chromosomes only whereas in other eukaryotes, these are located in one or two chromosomes only. *5S rRNA* genes have internal promoters. Heterocluster units are transcribed by RNA polymerase I while *5S rRNA* genes are transcribed by RNA polymerase II. In oocytes, rRNA genes are amplified thousand times which allows an oocyte to accumulate about 10^{12} ribosomes. The process is called gene amplification. After performing their function, the amplified rRNA genes are destroyed.

tRNA gene family

E. coli has 60 tRNA genes whereas *Xenopus* has 4,000 tRNA genes. Each type of tRNA gene has 10 to several hundred copies per haploid genome. Humans have more than 1,300 tRNA genes per haploid genome. The tRNA genes are also present in form of heteroclusters (Figure 2.15). These heteroclusters may be tandemly repeated (e.g., in *Xenopus*) or 50-60 sites in different chromosomes (e.g., *Drosophila*). This heterocluster is different from rRNA gene cluster as each gene is transcribed separately. The genes for each of the different tRNAs are also found in multiple copies to meet the heavy demand of the cell for the production of tRNA molecules.

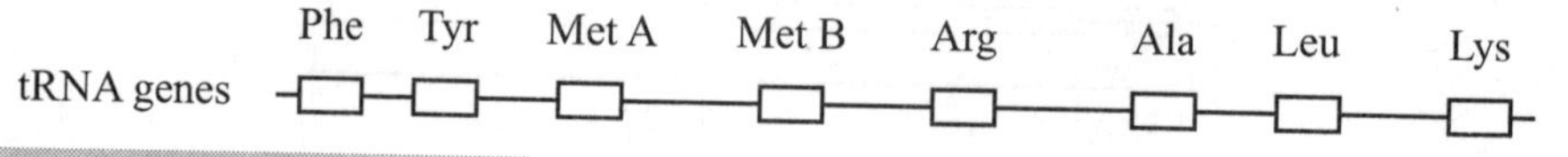

Figure 2.15 Heterocluster of tRNA genes

Small nuclear RNA (snRNA) gene family

Small nuclear RNA (snRNA) genes are of 11-13 types. Major 5-6 types of snRNA genes (U1, U2, U3, U4, U5 and U6) are present in many copies and are transcribed frequently. These small nuclear RNAs, called U RNAs, constitute spliceosome formed during splicing of introns from precursors of messenger RNA. The rest small nuclear RNA genes are less frequent. Usually small nuclear RNA genes are organized into tandem array and contain only one type of small nuclear RNA. SnRNA genes are variable in size. Spacer regions range from 800-45,000 kbp. In humans and mammals, a most unexpected feature of the genomic region complementary to snRNA is that pseudogenes outnumber real genes by at least 10:1. These pseudogenes are thus repetitive sequences. There are some 30 true U1 snRNA genes located on human chromosome 1 and 500-1,000 pseudogenes distributed throughout the genome.. Some pseudogenes differ from real genes by as little as one nucleotide and have homologous flanking regions. Other pseudogenes possess only a portion of snRNA sequence and are flanked by DNA that is totally unrelated to the sequences adjacent to the real genes. Small nuclear RNAs (snRNAs) are found in abundance in the nuclei of all the eukaryotes and represent neither tRNA nor rRNA, The snRNAs are encoded by multiple copies of identical genes organized in tandem array of repeat units, each snRNA gene being flanked on either side by spacer DNA ranging from 800 bp to 45,000 bp in length in different organisms.

Storage protein gene family

Storage proteins in crop plants are encoded by multiple gene families and are represented mainly as prolamins (major storage proteins in cereals – soluble in aqueous alcohol) or globulins (major storage proteins in legumes – soluble in salt). Three to ten genes for each of the ten prolamin families are reported in maize; more than a dozen genes for prolamin represent several families in wheat; and 18 genes belonging to three families of genes are known in pea. Homologies in genes coding for prolamins and globulins in different crop plants has also been shown. Their multiplicity meets the demand for rapid synthesis of storage proteins in the developing seeds.

FOLD-BACK ELEMENTS OF DROSOPHILA

When DNA from any eukaryote is denatured and allowed to reanneal at low concentration, the first sequences to become double-stranded are those that occur as inverted repeats. The repeats may be immediately adjacent or separated by upto several thousand base pairs, and as the reaction is intramolecular, the structures formed are called snap-back (SB) or fold-back (FB) structures. There are 2,000-4,000 pairs of inverted repeats or potential fold-back structures in *D. melanogaster* which comprise about 3 per cent of the total genome.

EVOLUTION OF GENOME SIZE

DNA content per nucleus is constant in various cells of an organism. There are of course some exceptions to this generalization. Sperm and egg cells contain about half as much DNA as ordinary somatic cells. Amount of DNA in cell nucleus of various groups of organisms (invertebrates, fish, amphibians. reptiles, birds and mammals) was compared which suggested that amount of DNA per cell in invertebrates increased from primitive to more highly developed ones. This relationship did not hold for vertebrates. Hinegardner (1976) reviewed DNA content of 2,000 different species. The information is summarized in Figure 2.16. These species were divided into four classes based on the amount of DNA present per cell. These classes are: (a) bacteria, (b) fungi, (c) almost all animals and some plants, and (d) many plants, Salamender, and some non-teleost fishes.

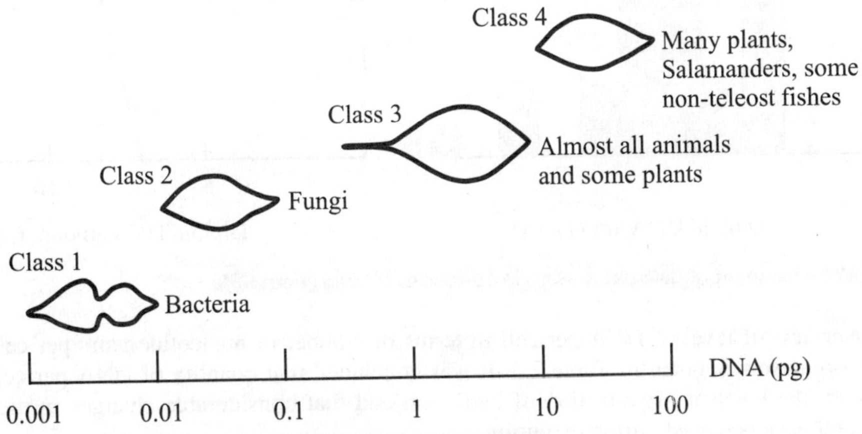

Figure 2.16 Organisms classified according to their amount of DNA

Bachmann et al. (1972) reviewed amount of DNA per nucleus in 19 species of toad genus *Bufo* (Figure 2.17). The amount of DNA per nucleus ranges from about $7.2x10^9$ to about $15.4x10^9$ nucleotide pairs (or from 8 to 7 pg), with a mode about 9.5×10^9 nucleotide pairs. In eukaryotes, fungi have least amount of DNA. In vertebrates, as we go from simpler to complex, DNA content increases with exception to *D. melanogaster* which has very low amount of DNA. In vertebrates, both increases and decreases of DNA have occurred. For example, reptiles have more DNA than birds, reptiles have less DNA than birds and amphibians have 20x more DNA than a typical mammalian cell. *Chaos chaos*, a single-celled protozoan, has $1x10^{12}$ nucleotide pairs per cell. This is the highest amount of DNA per cell. Organisms within each of the four broad classes mentioned above have similar amount of DNA; in each group the amount of DNA in most of the organisms varies over only one order of magnitude, i.e., differs by a factor from 1 to 10. From all organisms, from bacteria to plants and animals, the amounts of DNA vary over five orders of magnitude.

When genome sizes of many teleosts, anuran, or placental species were arranged in a frequency diagram, they conform to a logarithmic normal distribution around a single mode. The cumulative frequencies of species plotted against diploid DNA amount, on a logarithmic scale falls very nearly in a straight line. It was thus concluded that changes in DNA content have been numerous and small that are approximately proportional to pre-existing amount of DNA.

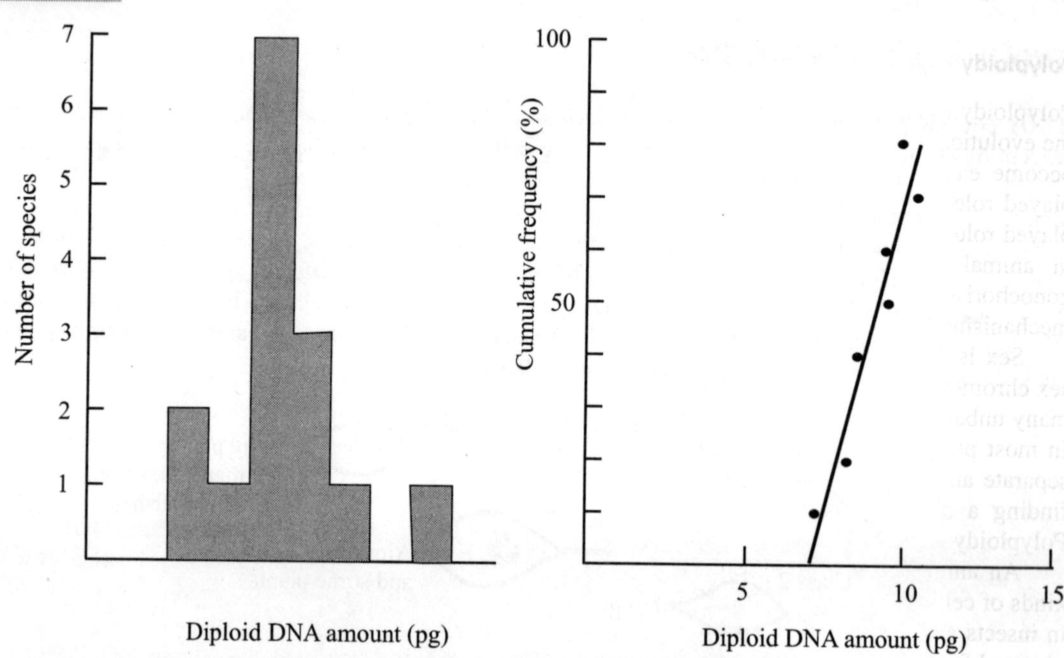

Figure 2.17 The amount of DNA per nucleus in 19 species of toad genus *Bufo*

An overview of level of DNA per cell in terms of number of nucleotide pairs per cell in various groups of organisms is given in Table 2.3. It was concluded that quantity of DNA per cell cannot be equated with the level of organization of evolution and that considerable changes in the amount of DNA per cell have occurred during evolution.

Table 2.3 Number of nucleotide pairs per cell in various groups of organisms (Extracted from Dobzhansky, Th. et al. 1977. *Evolution*. San Francisco: W. H. Freeman and Co.)

Group	*Range (nucleotide pairs/cell*	
	From	*To*
Algae	1.2×10^7	4.0×10^{11}
Gymnosperms	8.4×10^9	1.4×10^{11}
Angiosperms	2.0×10^9	1.2×10^{11}
Insects	1.7×10^8	1.0×10^{10}
Amphibians	2.1×10^9	1.0×10^{11}
Teleosts	8.0×10^8	8.8×10^9
Reptiles	4.5×10^9	Narrow range
Birds	2.0×10^9	3.0×10^9
Mammals	5.7×10^9	Narrow range

Evolutionary Changes in Genome Size

There are five mechanisms that are responsible for bringing out evolutionary changes in genome size – polyploidy, polyteny, deletions or deficiencies, duplications and transposons.

Polyploidy

Polyploidy refers to multiple copies of a complement in a single cell. It is a common phenomenon in the evolution of most species of plants but much rarer phenomenon in animals since polyploidy cannot become easily established in species with separate sexes and regular outcrossing. Polyploidy has played role in evolution by having several copies of the same genome but it does not seem to have played role in causing evolutionary changes in genome size. Why has polyploidy played a minor role in animal evolution? Four reasons are attributed to this – imbalance of sex determination, gonochorism, developmental difficulties and effect of polyploidy on size of the cells. These mechanisms are described here briefly.

Sex is determined in most animals by a balance of sex factors, usually distributed over both the sex chromosomes and autosomes. The progeny of a tetraploid female and a tetraploid male will contain many unbalanced and hence more or less sterile individuals. Selection will thus act against polyploids. In most plants, same flower produces male and female gametes. In animals, however, the sexes are separate and reproduction is biparental. A newly arisen polyploid animal will thus have difficulty in finding a mate and reproduce. Incidentally, polyploidy is also rare in self-incompatible annuals. Polyploidy is, however, possible where there is hermaphroditism as in the planarian genus *Dueesia*.

An animal is in general a much more complex organism than a plant. The number of different kinds of cells varies from 47-52 in a flowering plant, whereas this number is 66 in earthworm, 100-150 in insects and 200-250 in a mammal. Moreover, the integration and delicate balance between organs required for the mobility and sense reception in animals are far greater than anything existing in plants is. Thus, the probability that elements from two different patterns can be combined to make a functional intermediate is much greater in plants than in animals. An increase in the size of the individual cells, perhaps, the most widespread effect of polyploidy, is often accompanied by a decrease in the number of cells. For instance, increased cell size reduces the number of neurons that can exist in the brain and other nerve centers, thereby greatly diminishing the ability of artificially induced autopolyploids of amphibia to react to stimuli.

Polyteny

It is the multiplication of the number of DNA strands within a chromosome. This occurs in certain animal tissues, e.g., the salivary glands, the gut and the Malphighian tubules of Diptera. Polyteny can increase amount of DNA manifold in somatic cells but not in germ cells. Thus polyteny is not a general mechanism for increasing amount of DNA in evolution.

Since changes in DNA content have been numerous and small, duplications and deficiencies, rather than polyploidy or polyteny, may cause evolution of genome size.

Deletions or deficiencies

Various staining techniques are known to detect deletions in the chromosomes. In homozygous form even small deletions are often lethal, as would be expected if gene(s) responsible for essential function(s) is (are) deleted, because in such cases certain important functions in physiology and development of the organism may be missing, e.g., *notch* mutation in *D. melanogaster*. Large deletions are often lethal or at least seriously deleterious, even in heterozygous form presumably owing to developmental imbalance. Certain deletions may not be lethal and might perhaps be favored by natural selection in the evolution of sporophytic or parasitic way of life when the functions of groups of genes may not be needed. It is well established that deletions have occurred in the evolution of the genome (Hinegardner 1976).

Duplications

Visible evidence of different DNA duplications in the phylogeny has been obtained in the study of giant polytene chromosomes of various dipteran species. Homology of the repeat segments is manifested by not only the similarity of the bands but also by their ability to pair each other in polytene nuclei. Duplications may be found contiguous to each other in different parts of the chromosomes. One of the effects of duplications may be visible effect on phenotypes. For example, duplication of 16A segment of X-chromosome of *D. melanogaster* leads to reduction in eye size. Flies homozygous for wild-type allele (B^+/B^+) have 779 facets in one compound eye, those heterozygous for *Bar* allele (B^+/B) have 358 while those homozygous for *Bar* allele (*B*/*B*) have 68 facets.

Some duplication may not have any visible effect on the phenotype. For example, *Clarkia unguicalata* may have whole chromosome present three times, rather than normal two, without exhibiting a detectable change in its phenotype or vigor. Duplications in DNA have played role in evolution of new genes.

Redundant Genes: Presence of several copies of a single gene allows the organism to obtain large amounts of the gene product in short time interval. These are known as redundant genes. Some of the examples are:

- rRNA genes are of two type – type I rRNA genes produce 5S rRNA while type II rRNA genes produce 18S and 28S rRNA genes. In *E. coli*, there are 5-10, in *D. melanogaster*, there are 1300 and in *Xenopus*, there are 400 copies of each rRNA gene.
- There are 30-40 different types of tRNA genes in *E. coli* and there is only one copy of each tRNA genes. In yeast, there are 60 different types of tRNA genes and each type of genes has 5-7 copies. In *D. melanogaster*, each type of tRNA genes has 13 copies.
- Histone genes are present in multiple copies in eukaryotes. During cell division, histones are needed in large amounts.
- Antibody genes are also present in eukaryotes in multiple copies. For providing defense to the body, antibodies are needed in very large amounts in a very short interval.

Transposons

Insertion sequences are identified in a group of many prokaryotes (bacteria) and eukaryotes *Drosophila*, maize, etc.), which have the ability to get inserted at various different places of the genome. These insertion sequences, if carrying any other gene(s), have the ability to insert them also in the genome alongwith itself. This system is thus capable of incorporating genes from unrelated species into another species. This leads to an increase in the genome size. This is also a mode for generating genetic variability.

EVOLUTIONARY CHANGES IN DNA SEQUENCES

Nucleotide sequencing techniques has made it possible to know primary structure of genes and compare pairs of homologous nucleotide sequences of a wide variety of genes in different organisms. This enables us to evaluate the extent of sequence divergence in coding regions (exons) and non-coding regions (introns) of a gene. From the comparison of DNA sequences in coding regions it becomes possible to evaluate the extent of sequence divergence for two types of substitutions – (a) a nucleotide replacement leading to an amino acid change (mis-sense mutations) and (b) a nucleotide replacement leading to a synonymous codon change (same-sense mutations). In molecular terms, at DNA level, evolutionary rate is expressed as difference in nucleotide per site per year.

Extent of Nucleotide Sequence Divergence

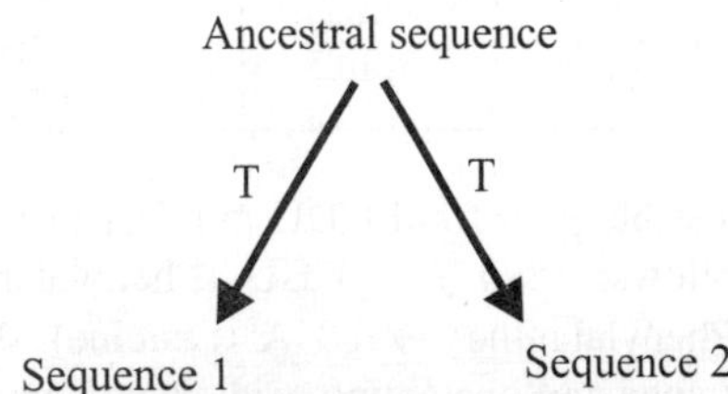

Figure 2.18 How the homologous sites differentiated from each other in the course of evolution (Extracted from Kimura, M. 1980. J. Mol. Evol. 16: 110-20)

The sequence difference, K, is defined as the number of nucleotide substitutions per site in a pair of aligned sequences. Using RNA code, one can calculate how the homologous sites differentiated from. The sequence difference, K, is defined as the number of nucleotide substitutions per site in a pair of aligned sequences. Using RNA code, one can calculate how the homologous sites differentiated from each other in the course of evolution starting from the common ancestor T years back (Figure 2.18). At individual sites bases are substituted one after another during evolution. This is called model of evolutionary base substitution. In this model, rate of transition type substitution may not be equal to the transversion type substitution. Total rate of substitution per site per year (K) = Rate of transition type substitution (A) + Rate of transversion type substitution (B). Here 'A' and 'B' refer to evolutionary rates of mutant substitutions in a species rather than the ordinary mutation rates at the level of individuals. The total number of base substitutions which separate two sequences and therefore involve two branches each with length T is given by 2TK.

Efstratiadis et al. (1977) compared nucleotide sequence of rabbit *β globin* gene with that of chicken *β globin* gene. As many as 438 nucleotide sites were compared corresponding to 146 amino acid sites. For 58 sites, the two sequences showed transition type differences and for 63 sites, transversion type differences were observed. Thus frequency of transitions (P) = 58/438 = 0.132 and frequency of transversions (Q) = 63/438 = 0.144. $K = -\frac{1}{2}\log_e(1-2P-Q) = 0.348$.

Evolutionary Rates for Three Codon Positions

It is more interesting to estimate separately the evolutionary rates for the three codon positions (Table 2.4). The relationship depicted by above comparison is $K_2 < K_1 < K_3$, which holds in general. Thus evolutionary mutant substitutions are most rapid at position 3, less rapid at position at position 1 and least rapid at position 2. This observation is interpreted by neutral theory of evolution as follows: Base substitutions at position 2 tend to be more drastic in the physiological properties of amino acid than those at position 1. Thus mutational changes in position 1 have a higher chance of not being harmful than those at position 2, and they therefore have a higher chance of being fixed in the species by random drift. This reasoning is valid more effectively to position 3 since majority of mutational changes at this position does not cause amino acid changes. Neutral theory regards protein and DNA polymorphism as a transient phase of molecular evolution.

Minimum Substitution Number

The reader is referred to genetic code dictionary for understanding the discussion below. Consider a codon change from UUU (Phenylalanine) to GUA (Valine) or vice-versa. The codon UUU reaches GUA through a minimum of two successive one-step changes. For this case, minimum substitution (MSN) is 2. Assuming that no more substitutions than MSN have occurred during evolution, the

Table 2.4 Evolutionary rates for the three codon positions (Extracted from data of Kimura, M. 1980. J. Mol. Evol. 16: 111-20)

Codon position	*Probability*		*Evolutionary rate (K)*
	P	*Q*	
1	15/146	21/146	$K_1 = 0.300$
2	7/146	18/146	$K_2 = 0.195$
3	36/146	24/146	$K_3 = 0.635$

possible path from UUU to GUA or vice-versa can be traced through successive one step changes as follows: *Path 1* – UUU (Phenylalanine) → GUU (Valine) → GUA (Valine); *Path 2* – UUU (Phenylalanine) → UUA (Leucine) → GUA (Valine). (9) Since path 1 contains one synonymous change and one amino acid change and path 2 contains no synonymous change, weight factor for each path (w1 and w2) show that path 1 is more likely.

Evolutionary Changes in Coding Regions

Histone is known to be the most slowly evolving protein. Even in histone genes synonymous substitutions as compared to amino acid substitutions occurred at a very high rate. This gene has undergone synonymous codon changes at the highest known rate whereas Immunoglobulin London type variable region genes in mouse have undergone mis-sense changes in DNA at a very high rate. The high and approximately uniform rate of synonymous substitutions as compared to amino acid substitutions implies that majority of the mutations leading to synonymous codon changes are much less subjective to selective constraints than amino acid altering change. Use of synonymous substitutions as evolutionary clock might be adequate because its rate is approximately constant regardless of the type of gene.

Evolutionary Changes in Non-Coding Regions

Sequence divergence of various non-coding sequences (KN) of mRNA for β-globulin, insulin and growth hormone genes revealed that each functional block of non-coding region as well as the synonymous sites of coding region was approximately uniform regardless of the type of gene. Comparison of a mouse α-globin pseudogene sequence with its functional counterpart revealed that the pseudogene evolved at a very high rate, which is higher than rate of change between synonymous codon in functional genes. Recombination can change rate of evolution. Recombination may play important role in evolution of duplicator genes. Recombination within introns leads to more dynamic and flexible evolution of genes.

DNA Sequence Polymorphism in Eukaryotes

Less than 10 per cent nuclear DNA of eukaryotes is translated into proteins. The techniques of DNA cloning and sequencing have revealed the following facts: (a) Genes are separated from each other by long DNA sequences that are not transcribed into RNA. (b) Genes themselves have a complex organization at both ends as they have relatively short sequences that are present in mature mRNA transcripts but do not code for amino acids (untranslatable exons). (c) most genes, in addition, contain introns that separate the segments that code for amino acid translatable codons. The introns are transcribed in the nucleus but they are spliced out before mRNA migrates to the cytoplasm.

A number of genes have been sequenced in two or more related species. It has become apparent that different segments of genome evolve at different rates. This suggests that different kinds of segments may have different levels of polymorphism. This hypothesis has been supported by direct evidence. Slightom et al. (1980) sequenced two alleles of the *Aγ* gene which codes for one of the polypeptides (IgG2a) of the fetal hemoglobin. The two alleles are from a single individual, one paternal and one maternal. The results depicted in Fig. 2.19 revealed that three exons had identical sequences but nucleotide substitutions occurred in the 5′ flanking sequence and both the introns. Most of the substitutions occurred in the 5′ region of the larger intron. The two alleles also differed in this region by 2,4-nucleotide gaps. The length of the DNA sequenced is 1,468 nucleotide pairs and the total number of substitutions is 24. The nucleotide difference thus is 2.2 per cent counting the two gaps as differences.

NUCLEOTIDE POLYMORPHISM

Rapid DNA sequencing allows allelic differences to be identified at nucleotide level. It not only solves the problem of detecting all amino acid substitutions between proteins, it also identifies nucleotide variation that is not translated into protein differences. Sequence variation in non-coding regions, such as introns or between synonymous codons, which is presumably not subject to natural selection acting through phenotype of the encoded protein, is the best available source for estimating the frequency of polymorphism due to neutral base substitutions in DNA. Direct DNA comparison is essential for understanding the role of neutral selection and genetic drift in the maintenance of genetic variation. Here we discuss the case of nucleotide polymorphism at alcohol dehydgrogenase (*Adh*) locus.

Physical Organization of *Adh* Gene

Kreitman (1983) studied nucleotide polymorphism at alcohol dehydrogenase (*Adh*) locus of *D. melanogaster*. Table 2.5 gives size (in nucleotide pairs) of different components of *Adh* gene in adult and larvae of *D. melanogaster*.

Table 2.5 Size (in nucleotide pairs) of different components of *Adh* gene in adult and larvae of *Drosophila melanogaster* (Extracted from data of Kreitman, M. 1983. J. Mol. Evol. 16: 111-20)

Region	*Adult Adh gene*	*Region*	*Larval Adh gene*
5′ flanking sequence	−63 to −1	5′ flanking sequence	708-777
Exon 1	1 to 87		
Intron 1	88 to 741		
Exon 2	742 to 876 (36 leader + 99 coding)	Exon 1	70 leader + 99 coding
Intron 2	877 to 941	Intron 1	877 to 941
Exon 3	942 to 1346	Exon 2	942 to 1346
Intron 3	1347 to 1416	Intron 2	1347 to 1416
Exon 4	1417 to 1861	Exon 3	1417 to 1861
3′ flanking sequence	1862 to 2628	3′ flanking sequence	1862 to 2628

Figure 2.19 represents map of *Adh* gene. Capping sites of adult and larval *Adh* genes are also indicated. Sequence comparison of eleven independently cloned *Adh* genes from 5 geographically distinct populations of *D. melanogaster* was done. A consensus DNA sequence based on the six *Adh-S*

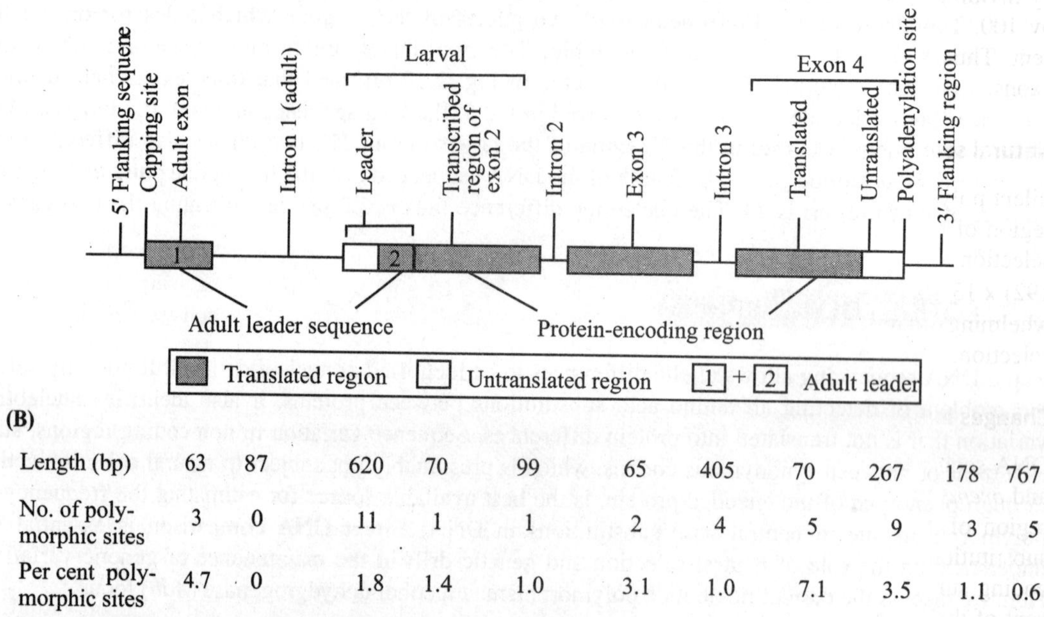

Figure 2.19 Map of adult and larval *Adh* gene of *D. melanogaster*

Table 2.6 Polymorphism in different regions of *Adh* gene of *Drosophila melanogaster* (Extracted from data of Kreitman, M. 1983. J. Mol. Evol. 16: 111-20)

Region	*Length (nucleotide pairs)*	*Polymorphic sites*	
		Number	*Per cent*
5′ flanking sequence	63	3	4.7
Exon 1	87	0	0.0
Intron 1	620 + (34 + 36)	11 + (1)	1.8 (1.4)
Exon 2	99	1	1.0
Intron 2	65	2	3.1
Exon 3	405	4	1.0
Intron 3	70	5	7.1
Exon 4	267	9	3.5
3′ flanking sequence	178	2	1.1
Further downstream	767	5	0.6

genes contains the complex the *Adh* untranslated leader is made up of residues 1-87 and 742-777 and that of larval mRNA of residues 708-777. The 3′-end of both mRNAs is at position 1,858.

The whole structure is 2,721 bp long. Larval leader sequence is from nucleotide position 708-777. In 2,721-bp stretch of *Adh* gene studied, a total of 43 nucleotide positions are polymorphic (Table 2.6). Of the 14 polymorphic sites in the three coding regions, only one leads to an amino acid replacement and it accounts for two allelomorphs *Adh-F* and *Adh-S* differing by a single amino acid (Threonine versus Lysine) at position 192. Remaining exon polymorphisms represent silent changes. Out of 43 observed polymorphisms, 18 are transitions while 25 are transversions, which does not differ significantly from 1:1 ratio. The per cent polymorphism in coding regions of *Adh* gene was calculated by dividing observed number of silent substitutions by effective number of silent sites and multiplied by 100. This value comes out to be 6.7 per cent. Similarly, polymorphic sites in introns are 2.4 per cent. Thus, there is a significant difference in the per cent silent polymorphism between introns and exons.

Natural selection against replacement substitutions

Silent polymorphisms in this analysis out number amino acid replacement polymorphism in the coding region of the *Adh* gene. The observed numbers seem to reflect differences in the intensity of natural selection acting on two classes of mutations. Estimated number of replacement polymorphisms (765–192) x 13 ÷ 192 = 39, which is different from the observed number. It clearly indicates that the overwhelming majority of amino acids in the ADH polypeptide sequence are constrained by natural selection.

Changes in DNA sequences in sibling species

DNA sequences were determined for *Adh* gene in *Drosophila* in sibling species *simulans*, *mauritiana* and *orena* and compared with *melanogaster* species (Bodmer and Ashburner 1984). Within the coding region of 768 bp, there are 192 effectively silent (synonymous) sites at which a single base pair substitution cannot result in a coding change. There are nearly 10 times as many synonymous as coding substitutions (Table 2.7). Overall, of the 20.8 per cent of the synonymous sites, only 2.4 per cent of the non-synonymous sites differ. In the present comparison, 25 of the 182 sequence variations in four sibling species alter length. Rate of nucleotide substitution is higher in non-coding than in coding regions. This implies selective constraints against changes in coding regions. In other words, natural selection operates against coding mutations most of which are likely to be deleterious. Among the exons, exon 2 is the most highly conserved and exon 1 is the least conserved.

Rate of change in the untranslated leader sequences (8.4%) is similar to that in the coding DNA (6.9%) and about one-half of that seen in the intron sequences (14.1%). This suggests that relative constraints are exerted on non-coding regions of mRNAs. This may be due to functional requirements for transcription and modification of mRNAs or to the demands for the interaction of other molecules with mRNA. There is clearly a hierarchy of constraints on DNA changes in different parts of the *Adh* sequence, the per cent changes descending in order: exon silent sites (20.8%), introns (14.1%), untranslated leader sequences (8.4%) and amino acid replacement sits (2.4%). Intron boundaries are very highly conserved. It is remarkable that silent sites within exons are so much more variable than the intron sequences.

ANCIENT DNA

Ancient DNA refers to the organisms which lived long ago and some of which have become extinct. Such DNA is a genetic treasure, which is useful in studying the molecular changes that might have occurred in a species during evolution. Higuchi et al. (1984) were the first to isolate ancient DNA from soft tissues of guagga (*Equus guagga*), an extinct zebra-like animal. Since then DNA has been extracted from various preserved remains.

Paabo (1989) studied properties of DNA extracted from dry remains of soft tissues of ages between 4 to 13,000 years. The DNA obtained was invariably of low molecular size, which was damaged presumably by endogenous hydrolysis and oxidative processes. The reduction in size was not

Table 2.7 Summary of alcohol dehydrogenase (*Adh*) gene sequence changes in four sibling species of *Drosophila*

Length	*(bp)*	*No. of changes*					
		mel	*sim*	*mau*	*ore*	*Total*	*Per cent*
A. Non-coding regions							
1. Leader							
Adult	134	3	1	0	9	13	9.7
Larval	70	1	0	0	3	4	5.7
Sub-total	204	4	1	0	12	17	8.4
2. Introns							
Intron 1	648	16	8	11	44	79	12.2
Intron 2	67	2	2	1	6	11	16.4
Intron 3	70	4	2	3	12	21	30.0
Subtotal	785	22	12	15	62	111	14.1
Total non-coding	989	26	13	15	74	128	12.9
B. Coding regions							
1. Exon 1	96						
Synonymous	24	0	2	2	6	10	41.7
Replacement	72	1	0	1	1	3	4.2
2. Exon 2	405						
Synonymous	101*	1	1	3	10	15	14.8
Replacement	304*	1	1	1	2	5	1.6
3. Exon 3	267						
Synonymous	67*	4	0	2	9	15	22.4
Replacement	200*	0	0	1	5	6	3.0
4. Subtotal							
Synonymous	192*	5	3	7	25	40	20.8
Replacement	576*	2	1	3	8	14	2.4
Total coding	768	7	4	10	33	54	6.9
C. Grand total	1757	33 (1.9)	17 (1.9)	25 (1.4)	107 (6.1)	182 (10.3)	

*Mutational sites

correlated with age of the specimen. The molecular cloning of such degraded DNA was difficult. However, use of recent techniques, like polymerase chain reaction, has been used to amplify such DNA and examine it in greater detail (Paabo 1993).

Comparison of DNA sequences of existing and extinct populations is useful to understand their phylogenetic relationship, evolution of parasites, inter- and intraspecific gene transfers and in solving old forensic cases (Paabo 1986; Cooper et al. 1992; DeSalle et al. 1992). Isolating a gene from an extinct species and its introduction into existing species may create organisms that mimic an aspect of the extinct species.

REFERENCES

Bachmann, K. O.B. Goin, and C.J. Goin. 1972. The nuclear DNA amounts in vertebrates. In: Evolution of Genetic Systems. Ed. Smith, H.H. Brookhaven Symp. Biol. 23: 419-50.

Bodmer, M., and M. Ashburner. 1984. Conservation and change in the DNA sequences coding for alcohol dehydrogenase in sibling species of *Drosophila*. Nature 309: 425-30.

Cooper, A, C. Mourer-Chauviré, G.K. Chambers, et al. 1992. Independent origins of the New Zealand moas and kiwis. Proc. Natl. Acad. Sci. USA 89: 8741-4.

Deselle, R., J. Gatesy, W. Wheeler, and D. Grimlady, 1992. DNA sequences from a fossil termite in Oligo-Miocene amber and their phylogenetic implications. Science 257: 1933-36.

Dobzhansky, Th., F.J. Ayala, G.L. Stebbins, and J.W. Valentine. 1977. *Evolution.* San Francisco: W. H. Freeman and Company.

Efstratiadis, A., F.C. Kafatos, and T. Maniatis. 1977. The primary structure of rabbit beta globin mRNA as determined from cloned DNA. Cell 10: 571-85.

Efstratiadis, A., J.W. Posakony, T. Maniatis, et al. 1980. The structure and evolution of the human β-globin gene family. Cell 21: 653-68.

Higuchi, R., B. Barbara, M. Freiberger, O. Ryder, and A.C. Wilson. 1984. DNA sequences from the quagga, an extinct member of the horse family. Nature 312: 282-4.

Hinegardner, R. 1976. Evolution of genome size. In: *Molecular Evolution.* Ed. Ayala, F.J. Sunderland, MA: Sinauer Associates, Inc. pp. 179-199.

Kimura, M. 1980. A simple method for estimating evolutionary rates of base substitutions through comparative studies of nucleotide sequences. J. Mol. Evol. 16: 111-20.

Kreitman, M. 1983. Nucleotide polymorphism at the *alcohol dehydrogenase* locus of *Drosophila melanogaster.* Nature 304: 412-7.

Lifton, R.P., M.L. Goldberg, R.W Karp, and D.S. Hogness. 1977. The organization of the histone genes in *Drosophila melanogaster.* Functions and Evolutionary implications. Cold Sp. Harb. Symp. Quant. Biol. 42: 1047-51.

Pääbo, S. 1986. Molecular genetic investigations of ancient human remains. Cold Sp. Harb. Symp. Quant. Biol. 51: 441-6.

Pääbo, S. 1989. Ancient DNA: extraction, characterization, molecular cloning, and enzymatic amplification. Proc. Natl. Acad, Sci. USA 86: 1939-43.

Pääbo, S. 1993. Ancient DNA. Scient. Am. 269 (5): 60-6.

Slightom, J. L., A. E. Blechl, and O. Smithies. 1980. Human fetal *Go*- and *ky*-globin genes: complete nucleotide sequences suggest that DNA can be exchanged between these duplicated genes. Cell 21: 627-38.

Tartoff, K.D. 1975. Redundant genes. Annu. Rev. Genet. 1: 355-85.

Vold, B.S. 1985. Structure and organization of genes for transfer ribonucleic acid in *Bacillus subtilis.* Microbiol. Rev. 49: 71-80

Williams, S.M., L.G. Robbins, P.D. Cluster, R.W Allard, and C. Stroback. 1990. Superstructure of the *Drosophila* ribosomal gene family. Proc. Natl. Acad. Sci. USA 87: 3156-3160.

Zuckerkandl, E., and W.A. Schroeder. 1961. Amino-acid composition of the polypeptide chains of gorilla haemoglobin. Nature 192: 984-5.

3

Defense Genes and Vaccines

A brief description of otherwise very complex phenomenon of antibody genes and some other important aspects of immunoglobulins, like generation of antibody diversity, is provided here. Another important aspect discussed in this chapter is human leukocyte antigen (HLA) complex. Human leukocyte antigen (HLA) complex comprises of genes that are responsible for synthesis of antigens that determine acceptance or rejection of a tissue graft. Finally, vaccines will be discussed.

IMMUNOGLOBULINS

Immunoglobulins (Ig) and T-cell receptors (TCRs) are antigen-binding proteins synthesized by B-lymphocytes and T-lymphocytes, respectively. F.M. Burnet and P.B. Medawar were awarded Nobel Prize in 1960 for giving clonal selection theory of antibody formation. Immunoglobulins may be displayed on the cell surface as B-cell receptors or secrete antibodies (Burnet 1957; Billingham et al. 1953). Both types of molecules have a fundamental heterodimeric structure.

R.R. Porter and G.M. Edelman were awarded Nobel Prize in 1972 for discovering chemical structure of antibodies (Porter 1958; Porter 1959; Edelman 1959). The structural diversity of antibodies is reflected by the diversity of their antigenic determinants associated with antigen-combining properties. The antigenically-related proteins are named as immunoglobulins (Ig). The combining ability of antibodies depends solely on the primary structure of their polypeptide chains. Differences in amino acid sequences between immunoglobulin determine antigenic specificities. There are five types of human immunoglobulins which differ in their composition and function — IgG, IgM, IgA, IgD and IgE. Structure of IgG is given in Figure 3.1. IgG molecule has four polypeptide chains of which two are heavy (H) and two are light (L) chains. Heavy and light chains have intra- and interchain disulphide bonds. Within the L and H chains, disulfide bonds lead to the formation of loops. L and H chains are held together by interchain disulfide bonds. The two H chains are also joined by disulfide bonds. L chain has constant and variable domains, respectively, referred to as C_L and V_L domains. Similarly, H chain has constant and variable domains, respectively, referred to as C_H and V_H domains. Antigen specificity of a particular antibody is conferred by amino acid sequence of V_L and V_H domains. Particularly important in specificity are hypervariable regions. H chain contains four different members of gene families, namely, V_H, D, J_H and C_H. L chain contains three different members of gene sequences, namely, V, J_L and C. Amino and carboxyl ends of polypeptide chains are represented by $NH3^+$ and COO^-, respectively. One carbohydrate group (CHO) is attached with each one of the C_H regions. A fork-shaped structure, known as hinge, is formed at the point where two V_H chains separate from each other. A model of generalized Ig molecule illustrating locations of loops is shown in Figure 3.2. Features of the amino acid sequence of light and heavy chains are given in Figure 3.3.

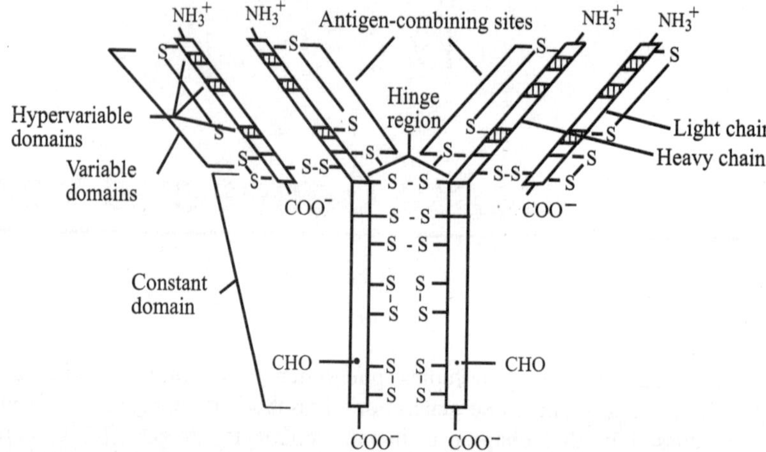

Figure 3.1 Schematic representation of structure of mammalian immunoglobulin G (IgG)

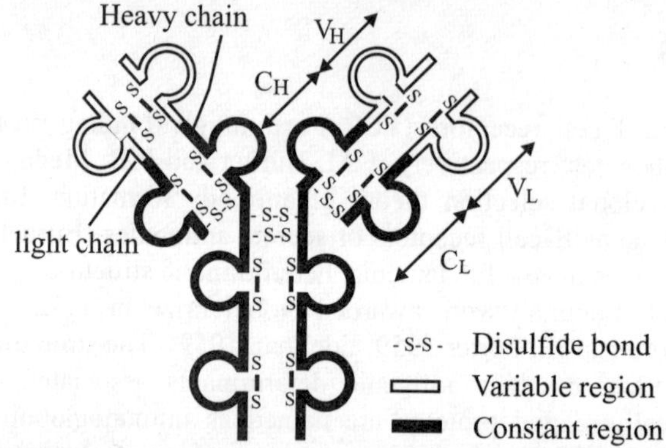

Figure 3.2 Structure of immunoglobulin G (IgG) molecule showing formation of loop by disulfide bonds

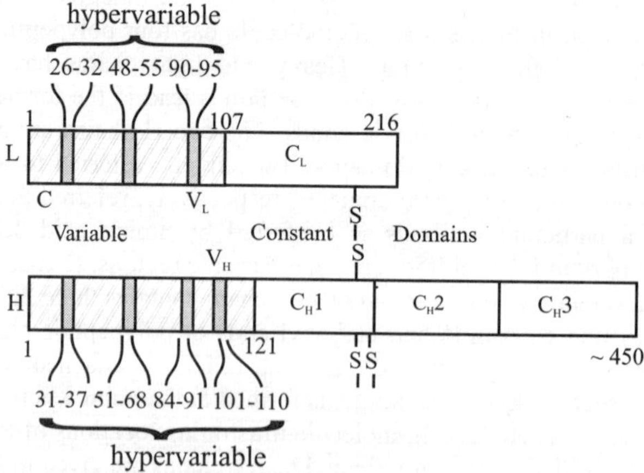

Figure 3.3 A model of generalized immunoglobulin molecule illustrating features of amino acid sequences of light and heavy chains

Specific binding of antigen with the antibody is executed by the two combining sites that are formed cooperatively by the hypervariable regions of heavy and light chains. Complement binding which involves interaction with cellular receptors, and control of catabolic rate is brought about by constant domains.

IMMUNOGLOBULIN GENES

If convention that one polypeptide is coded for by one gene is accepted, each of the final DNA segment coding a complete heavy or light chain should be considered as a single gene. Various alternative sequences within each set (V, J, D or C) should according to this definition be termed simply as DNA sequences since they are all parts of a single gene coding for one polypeptide chain. With respect to immunoglobulin genes, the genome of somatic cells is not constant.

Only two types of light chains are known, k and λ, each differing in amino acid sequence of C_L segment and includes L chains with many different VL sequences. There are five types of heavy chains of constant region (C_H) known: μ, δ, γ, ε and α which define five types of immunoglobulins, IgM, IgD, IgG, IgE and IgA, respectively. Each of the five heavy chain classes may be associated with either k or λ light chains. The most interesting problem in antibody diversity is how a virtually infinite number of DNA sequences that encode different V regions but the same C region is generated. To understand this we need to know the structure of immunoglobulin genes.

Light Chain Genes

λ type light chain gene structure seems to be the simplest of all the immunoglobulin genes. Rearrangement of murine germline λ chain genes during differentiation from a stem cell is shown in Figure 3.4. The arrangement is followed by transcription and processing that produces the mature mRNA. The family of genes determining k chain consists of at least 300 tandemly arranged V_λ sequences, 5 J_λ sequences and 1 C_λ segment in chromosome 2 in humans. V_λ array is 23 kbp away from the J_λ sequences. *L* and *V* are separated by an intervening sequence, I_1, 93 bp long. J_λ and C_λ are separated by an intervening sequence, I_2, 1,250 bp long.

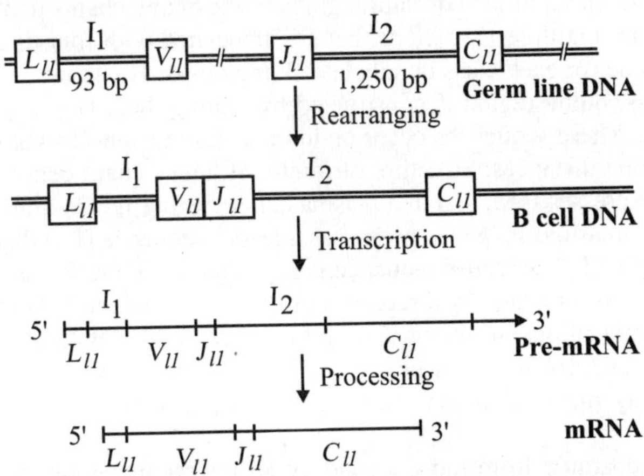

Figure 3.4 Rearrangement of murine germ line λ chain gene during differentiation from a stem cell to a B cell

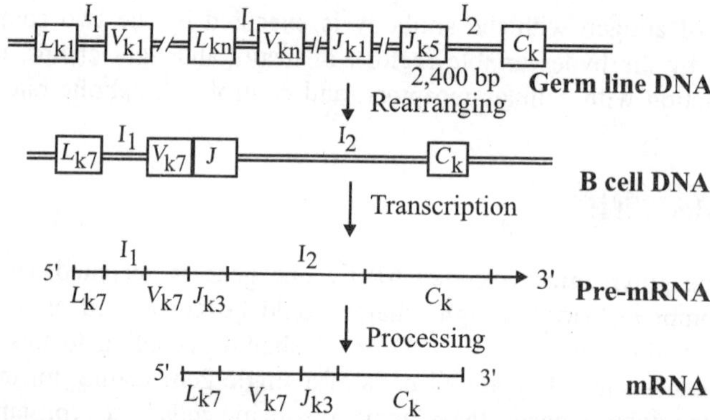

Figure 3.5 Rearrangement of murine germ line k chain gene during differentiation from a stem cell to a B cell

Structure of *k* type of light chain gene is shown in Figure 3.5. Rearrangement of murine germline *k* chain genes during differentiation from a stem cell to a B-cell is outlined in this figure. Functional γ or *k* L chain genes are formed in much the same way and the processes are similar in most of the substrates. Each V_k sequence has three functional parts, a promoter followed by two exons L_k and V_k. L_k encodes a signal sequence that directs the nascent polypeptide to endoplasmic reticulum. Family of genes determining the k chain consists of at least 100-200 V_k, 4 or more J_k gene sequences, and a single C_k gene in the germline DNA. The rearrangement at DNA level is followed by transcription and processing that yields mature mRNA. Arbitrarily, in Figure 3.5, the selected gene is shown to consist of *L*, *V*, and *J* sequences. Nucleotide sequence analysis of the cloned V_k genes demonstrates that each gene consists of a L_k portion encoding the hydrophobic leader peptide, which is separated from the V_k gene by an intervening sequence, I_1. The J_k sequence closest to the C_k gene is separated by intervening sequence, I_2, about 2,400 bp long.

Heavy Chain Genes

At least eight classes and subclasses of immunoglobulins can be identified in mice on the basis of the different amino acid sequences of the constant regions of the heavy chains μ, δ, γ1, γ2a, γ2b, γ3, α, and ε. Similar to V_k, there are multiple V_H regions that are associated with four distinct J_H sequences. Each cell contains one C_H gene for each class or subclass. *V* regions of H chains are more diverse than those of L chains because V_H coding region is constructed by joining three (V_H, D_H and J_H) rather than two (V_H, and J_H) sequences. These sequences occur on human chromosome 14 but cluster of V_H's, D_H's and J_H's are separated by long distances. Structure of family of heavy chain genes in germline DNA, in an immature B-cell that expresses IgM, and in a plasma cell secreting IgG2a immunoglobulin is shown in Figure 3.6. Each of the multiple V_H genes possesses a leader sequence (L_H) that codes for hydrophobic amino acids –19 through –5. The leader sequence is separated from the V_H sequence by an intervening sequence, I_1. V_H and J_H are separated by a region for diversity, *D*, which in germline DNA is present in 10 copies. Random combinations of one of several hundreds of V_H's, with one of 30 *D*'s, and one of 5 J_H's yield between 10^4 and 10^5 different heavy chains. Each contains one C_H gene for each class or subclass. DNA rearrangement of heavy chain genes requires two steps and generates a *V-D-J* sequence.

The J_H region is separated from the C_H gene by an intervening sequence, I_2. The C_H genes are arranged on the chromosome in the following order: $C\mu$, $C\delta$, $C\gamma 3$, $C\gamma 1$, $C\gamma 2b$, $C\gamma 2a$, $C\varepsilon$ and $C\alpha$. The

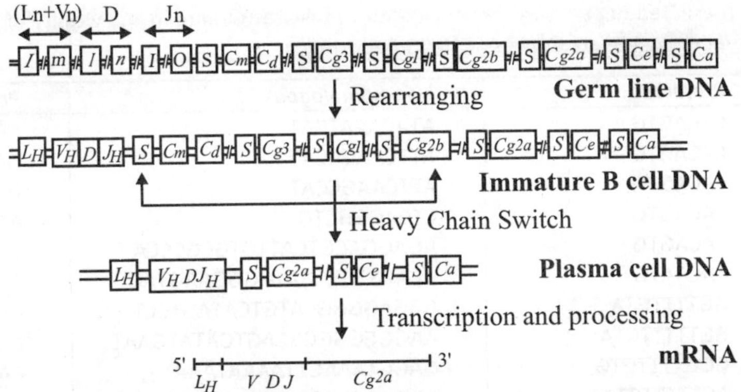

Figure 3.6 Family of heavy chain genes in germ line DNA in an immature B cell that expresses IgM and in plasma cell secreting IgG2 immunoglobulin

structure of these genes is similar, but it may be noteworthy that the hinge regions of *Cδ*, *Cγ* and *Cα* chains are encoded by separate genes. The heavy and light chains thus provide diversity of 18 million only due to somatic recombination. Additional diversity is generated due to recombination among V_H, *D* and J_k sequences of heavy chain gene and between V_k and J_k sequences of the light chain gene.

RECOGNITION SYSTEM FOR RECOMBINATION

V(D)J Recombination

In late 1970s S. Tonigawa and colleagues showed that the genes that encode immunoglobulin heavy and light chains have a different structure in embryonic cells from that found in tumors derived from B cells (Tonigawa et al. 1976). S. Tonegawa was awarded Nobel Prize in 1987 for discovery that DNA rearrangements are responsible for antibody diversity. Developing B lymphocytes assemble immunoglobulin genes from widely scattered gene segments, using a somatic DNA rearrangement process known as V(D)J recombination, and that developing T cells play the same trick to assemble functional T-cell receptor genes. The V(D)J recombination is catalyzed by a "recombinase" enzymatic machinery.

Analysis of the DNA sequence in the vicinity of several V_L and V_H, and the J_L and J_H sequences shows a remarkable conservation of hepta- and decanucleotides (Table 3.1). These conserved sequences are separated by a 10-11 or 21-23 bp non-homologous stretch of nucleotides. Presumably similar sequences are found in the vicinity of *D* regions of the heavy chain genes. An immunoglobulin heavy chain variable region of heavy chain gene is generated from three sequences of DNA, V_H, *D* and J_H. These conserved hepta- and decanucleotides in vicinity of *V*, *D*, and *J* genes form recognition system required for recombination.

Diagrammatic view of the recognition system that is believed to be involved in the recombination between the *V-J* sequences in the light chains and the *V-D-J* sequences in the heavy chains are presented in Figure 3.7. For sake of simplicity, only one strand of DNA double helix is shown. It is suggested that recombination enzymes recognize the seven-nucleotide (heptamer) and nine-nucleotide (nonamer) sequences. The double-stranded heptamer sequence would appear palindromic, with an axis of symmetry at the central nucleotide; i.e., it reads similarly from opposite ends of each strand, CACA GTG/GTGTCAC. Between each heptamer and nonamer sequence is a "spacer" that may be either

Table 3.1 Highly preserved hepta- and decanucleotides (italicized) found in the vicinity of V and J genes and separated by non-homologous sequences of 10-23 nucleotides

Gene	*Preserved*	*Non-homologous*	*Preserved*
V	CACAGTG	ATACAAATCAT	AACATAAACC
	CACAGTG	ATTCAAGCCAT	GACATAAACC
	CACAGTG	ATTCAAGCCAT	GACATAAACC
	CACAGTG	CTCAGGGCTG	AACAAAAAACC
	CACAGTG	AGAGGACGTCATTGTGCGCCCA	GACACAAACC
	CACAATG	ACATGTGTAGATGGGGAAGTAG	ATCAAGAACA
	GGTTTTGTA	GAGAGGGGCATGTCATAGTCCT	CACTGTG
	GGTTTTTGTA	AAGGGGGGCGCAGTGATATGAAT	CACTGTG
	GGGTTTTGTG	GAGGTAAAGTTAAAATAAAT	CACTGTA
	AGTTTTTGTA	TGGGGGTTGAGTGAAGGACAC	CAGTGTG
	GGTTTTTGTA	CAGCCAGACAGTGGAGTACTAC	CACTGTG
	AGTTTTAGTA	TAGGAACAGAGGCAGAACAGA	GACTGTG
	GGTTTTTGTA	CACCCACTAAAGGGGTCTATGA	TAGTGTG
	GGTTTTTGCA	TGAGTCTATAT	CACAGTG

12±1 or 23±1 nucleotides long. A recombination enzyme complex can presumably pair two recognition sequences with spacers of different lengths, e.g., 23/12 or 12/22, but cannot pair sequences of similar lengths, i.e., 12/12 or 23/23. Each V sequence in the *kappa* gene family has a 12-spacer recognition sequences on its "downstream" side and each *J* sequence has a 23-spacer recognition sequence on its "upstream" side. This arrangement is reversed in the lambda gene family. As a result 12/23 pairing can take place between any of the V_k and J_k sequences (designated by subscript i), and a recombinational event then deletes the intervening nucleotide sequence between the paired V_k and J_k sequences. The excised intervening sequence may cover thousands of nucleotides, and include other V_k and J_k sequences that are thus discarded.

Note also that the paired recognition sequences are "inverted repeats" and the ability for these repeats to come together followed by excision of the intervening sequence is reminiscent of phenomena observed in transposons. In the heavy chain, since a 23-spacer sequence lies downstream from each V sequence and 22-spacer sequence lies upstream from each J sequence, direct V_H-J_H pairing does not occur. Instead, the D sequences of the heavy chain, which possess 12-spacer sequences of each side, allow 23/12 and 12/22 pairing to take place between V_H-D-J_H.

Histone H3 Trimethylation at Lysine 4 is involved in V(D)J Recombination

Nuclear processes such as transcription, DNA replication and recombination are dynamically regulated by chromatin structure. Eukaryotic transcription is known to be regulated by chromatin-associated proteins containing conserved protein domains that specifically recognize distinct covalent post-translational modifications on histones. However, it has been unclear whether similar mechanisms are involved in mammalian DNA recombination. Matthews et al. (2007) show that RAG2 – an essential component of the RAG1/2 V(D)J recombinase, which mediates antigen-receptor gene assembly – contains a plant homeodomain (PHD) finger that specifically recognizes histone H3 trimethylated at lysine 4 (H3K4me3). The high-resolution crystal structure of the mouse RAG2 PHD finger bound to H3K4me3 reveals the molecular basis of H3K4me3 recognition by RAG2. Mutations that abrogate RAG2's recognition of H3K4me3 severely impair *V(D)J* recombination in vivo. Reducing the level of H3K4me3 similarly leads to a decrease in *V(D)J* recombination in vivo. Notably, a conserved trypto-

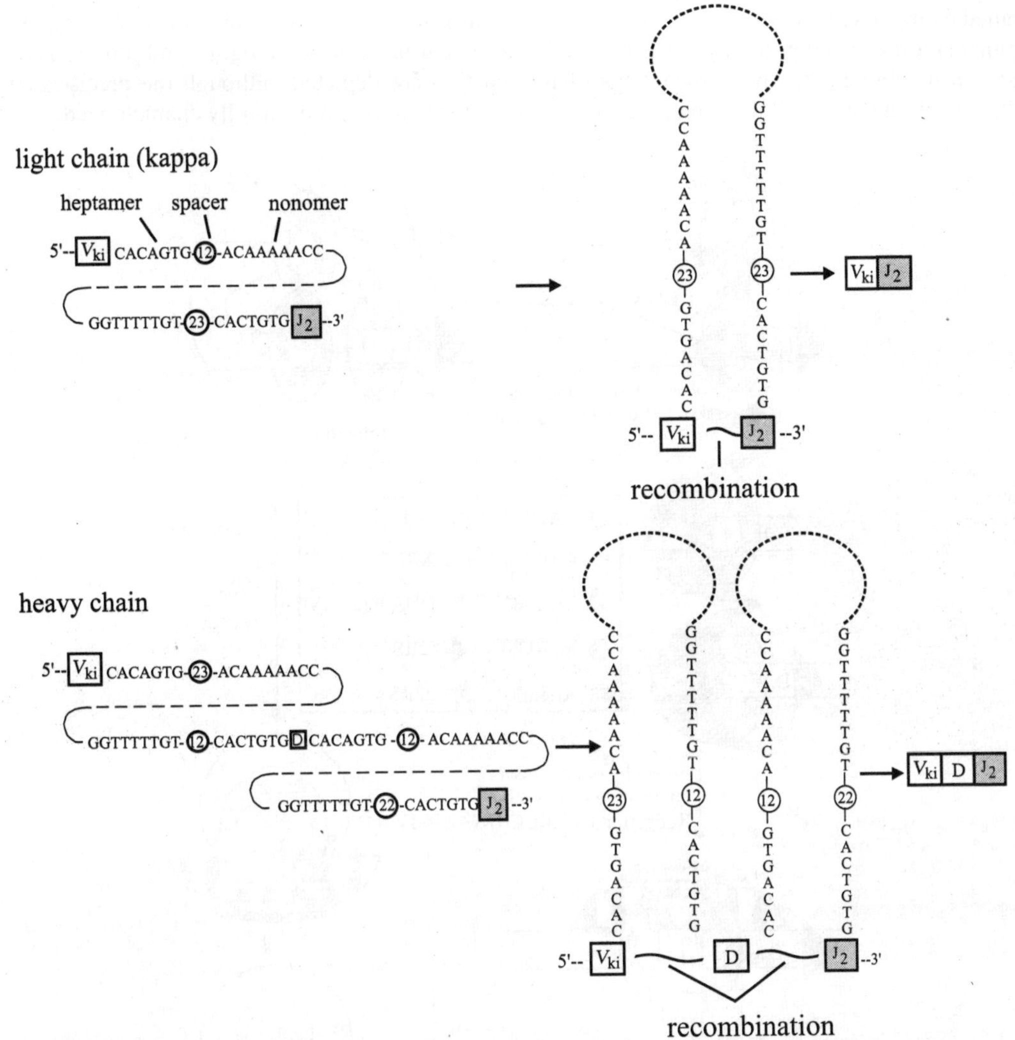

Figure 3.7 Recognition system believed to be involved in recombination between V-J segments in light chain and V-D-J segments in heavy chain

phan residue (W453) that constitutes a key structural component of the K4me3-binding surface and is essential for RAG2's recognition of H3K4me3 is mutated in patients with immunodeficiency syndrome; together, these results identify a new function for histone methylation in mammalian DNA recombination. Furthermore, these results provide the first evidence indicating that disrupting the read-out of histone modifications can cause an inherited human disease.

V(D)J Recombination assembles Antigen Receptor Genes

V(D)J recombination assemble antigen receptor genes from component gene segments. Jung and Alt (2004) review findings that have shaped our current understanding of this remarkable mechanism – the detailed comparison of germline and rearranged antigen receptor loci and the discovery of the recombination activating gene-1. *V(D)J* recombination and the non-homologous end-joining pathway is shown in Figure 3.8. The known steps of the reaction are depicted, although the precise identity and/or stoichiometry of all the proteins involved at each step have not been fully characterized.

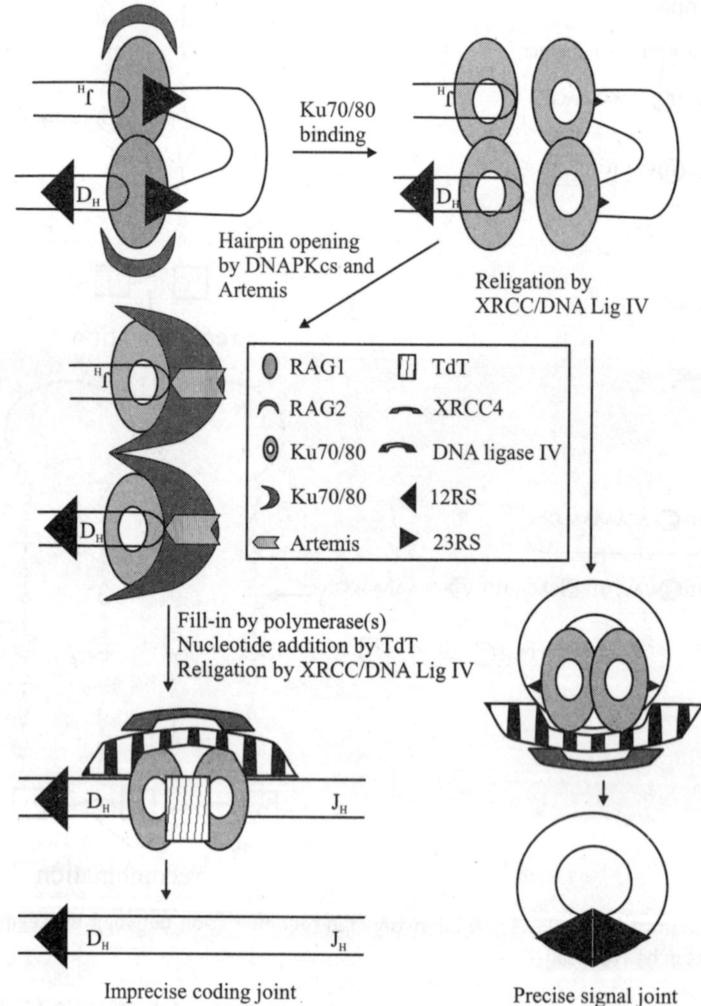

Figure 3.8 V(D)J recombination and the non-homologous end-joining pathway

ATM functions in the Repair of DNA DSBs Generated during Lymphocyte Antigen Receptor Gene Assembly

The ATM (Ataxia-telangiectasia mutated) protein kinase mediates early cellular responses to DNA double-strand breaks (DSBs) generated during metabolic processes or by DNA-damaging agents. ATM deficiency leads to ataxia-telangiectasia, a disease marked by lymphopenia, genomic instability and an increased predisposition to malignancies with chromosomal translocations involving lymphocyte antigen receptor loci. ATM activates cell cycle checkpoints and can induce apoptosis in response to DNA DSBs. However, defects in these pathways of DNA damage response cannot fully account for the phenotypes of ATM deficiency. Bredemeyer et al. (2006) show that ATM also functions directly in the repair of chromosomal DNA DSBs by maintaining DNA ends in repair complexes generated during lymphocyte antigen receptor gene assembly. When coupled with the cell cycle checkpoint and pro-apoptotic activities of ATM, these findings provide a molecular explanation for the increase in lymphoid tumors involving antigen receptor loci associated with ataxia-telangiectasia.

Class-Switch Recombination

Variable diversity and joining [*V(D)J*] recombination and class switch recombination use overlapping but distinct non-homologous end-joining pathways to repair DNA double-strand break intermediates. 53BP1 is a DNA damage response protein that is rapidly recruited to the sites of chromosomal double-strand breaks, where it seems to function in a subset of ataxia telangiectasia mutated (ATM) kinase-, H2A histone family member X (H2AX, also known as H2AFX)- and mediator of DNA damage checkpoint 1 (MDC1)-dependent events. A 53BP1-dependent and end-joining pathway has been described that is dispensable for *V(D)J* recombination but essential for class-switch recombination. Dofilippantonio et al. (2008) reported a previously unrecognized defect in the joining phase of V(D)J recombination in 53BP1-deficient lymphocytes that is distinct from that found in classical non-homologous end-joining-, *H2ax*-, *Mdc1*- and *Atm*-deficient mice. Absence of 53BP1 leads to impairment of distal *V-D-J* joining with extensive degradation of unrepaired coding ends and episomal signal joint reintegration at *V(D)J* junctions. This results in apoptosis, loss of T-cell receptor α locus integrity and lymphopenia. Further impairment of the apoptotic checkpoint causes propagation of lymphocytes that have antigen receptor breaks. These data suggest a more general role for 53BP1 in maintaining genomic stability and during long-range joining of DNA breaks. A model for the role of 53BP1 in promoting and/or maintaining synapsis during *V(D)J* recombination is given in Figure 3.9 (Dofilippantonio et al. 2008). Before recombination, 53BP1 is loosely associated with chromatin (step 1). RSSs that are at a close physical distance (a) spend more time near the RAG1/2-bound recombination signal sequence (RSS) than distant RSSs and (b) have a high degree of synapsis whether or not 53BP1 is present. 53BP1 homo-oligomerization increases the effective local concentration of distant RSSs, thereby promoting their pairing. When RAG1/2 associates with and cleaves a single RSS, 53BP1 accumulates at the lesion and binds tightly to flanking chromatin in a manner dependent on H2AX and MDC1 (steps 2 and 3). After RAG1/2-mediated DSB formation (step 4), 53BP1 stabilizes the post-cleavage complex, thereby increasing the efficiency of non-homologous end-joining (NHEJ).

DNA cleavage step

Activation induced cytidine deaminase (AID) is required for the DNA cleavage step in immunoglobulin class-switch recombination (CSR) (Begum et al. 2004). AID is proposed to deaminate cytosine to generate uracil (U) in either mRNA or DNA. In the second instance, DNA cleavage depends on uracil DNA glycosylase (UNG) for removal of U. Using phosphorylated histone γ-H2AX focus formation as a marker of DNA cleavage, Begum et al. (2004) found that the UNG inhibitor Ugi did not inhibit DNA cleavage in immunoglobulin heavy chain (Ig_H) locus using CSR, even though Ugi blocked UNG binding to DNA and strongly inhibited CSR in $UNG^{-/-}$ B cells. These results indicate that UNG is involved in repair step of CSR yet by an unknown mechanism. The dispensability of U removal in the DNA cleavage step of CSR requires a reconsideration of the model of DNA deamination by AID. A clarification to this is given by Begum and Hongo (2004).

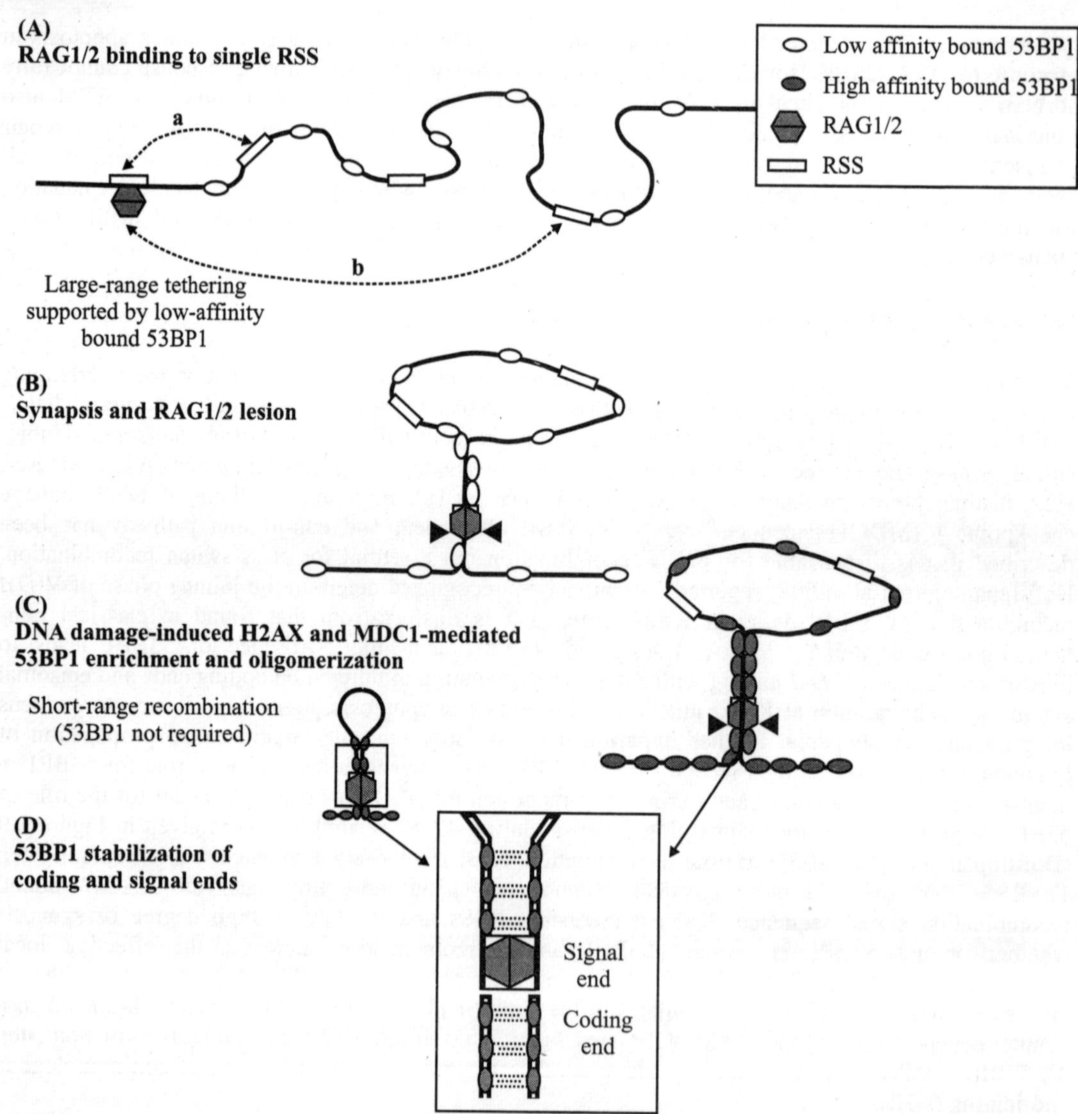

Figure 3.9 Model for the role of 53BP1 in promoting and/or maintaining synapsis during *V(D)J* recombination

Uracil removal introduces immunoglobulin class-switch recombination

The residual uracil removal activities of UNG single mutants may be sufficient to introduce double-strand breaks and induce class-switch recombination (CSR) (Stivers 2004).

VSG Switching is Result of DSB by Break-induced Replication

Trypanosome brucei is the causative agent of African sleeping sickness in humans and one of the causes of nagana in cattle. This protozoan parasite evades the host immune system by antigenic variation, a periodic switching of its variant surface glycoprotein (VSG) coat. VSG switching is spontaneous and is thought to occur predominantly through gene conversion, a form of homologous recombination initiated by a DNA lesion that is used by other pathogens to generate surface protein diversity, and by B lymphocytes of the vertebrate immune system to generate antibody diversity. Introduction of a DNA double-strand break (DSB) adjacent to the ~70 bp repeats upstream of the transcribed *VSG* gene increases switching in vitro ~250-fold, producing switched clones with a frequency and features similar to those generated early in an infection. Boothroid et al. (2009) detected spontaneous DSBs within the 70-bp repeats upstream of the actively transcribed *VSG* gene, indicating that a DSB is a natural intermediate of VSG gene conversion and that *VSG* switching is the result of this DSB by break-induced replication. Antigenic switching is induced by a single I-sceI-generated DSB (Boothroid et al. 2009) (Figure 3.10).

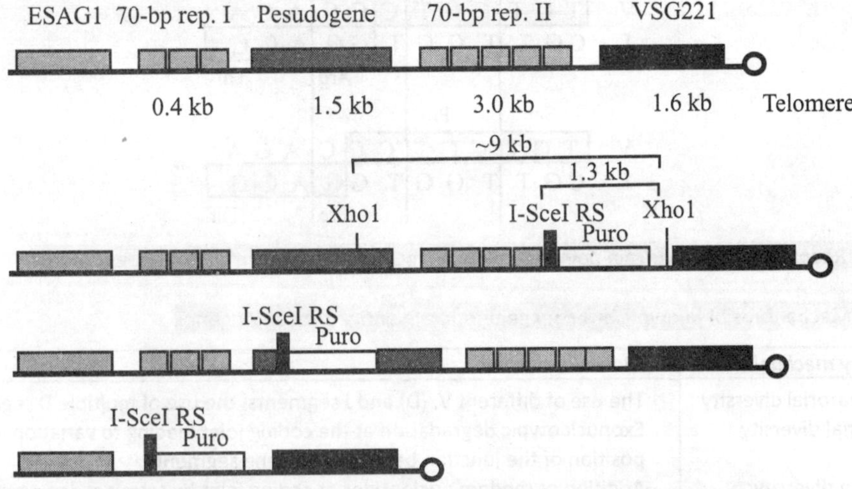

Figure 3.10 Schematic representation of the telomeric region of the VSG 221 expression site

NEW CODON GENERATION

Recombination at certain points between V and J sequences can generate new codons in the hypervariable regions (Figure 3.11). Recombination between different triplets can generate yet other new triplets. For example, amino acids 95 and 96 of an active light chain are at the position of exchanges between V and J sequences. Exchanges in codons for amino acid 96 can lead to arginine or proline instead of tryptophan, depending on site of recombination. This provides a means of generating a vast array of different gene products. Various mechanisms of idiotypic generation in antigen-receptor genes and the molecular basis involved are given in Table 3.2. Further diversity is generated at the protein level by combination of different heavy and light chains or α, β, γ subunits (TCRs).

ALLELIC EXCLUSION

In immunoglobulin-producing cells or plasma cells, only one member of each pair of alleles, is involved in the synthesis of immunoglobulin being expressed. Correct rearrangement of only one chromosome is needed for the generation of each expressed light or heavy chain genes. Since DNA rearrangement of one chromosome takes place, expression of the alternate allele is excluded. This phenomenon is known as allelic exclusion.

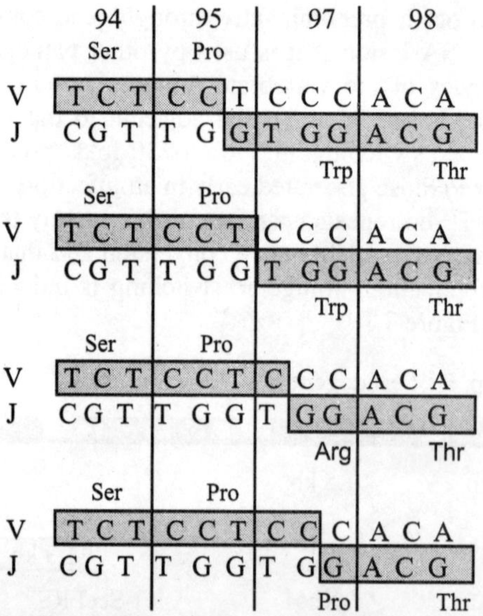

94 95 97 98
Ser Pro
V TCT CC T CCC ACA
J CGT T G GT GG ACG
Trp Thr

Ser Pro
V TCT CCT CCC ACA
J CGT T GG T GG ACG
Trp Thr

Ser Pro
V TCT CCT C CC ACA
J CGT T GGT GG ACG
Arg Thr

Ser Pro
V TCT CCT CC C ACA
J CGT T GGT G G ACG
Pro Thr

Figure 3.11 Recombination at certain points between V and J sequences can generate new codons

Table 3.2 Mechanisms of idiotypic diversity generation in antigen-receptor genes

Diversity mechanism	*Molecular basis*
Combinatorial diversity	The use of different V, (D) and J segments, the use of multiple D segments
Junctional diversity	Exonucleotypic degradation at the coding joint leading to variation in the position of the junction between the gene segments
N-region diversity	Addition of random nucleotides at coding joint by terminal deoxynucleotidyl transferase
V-gene replacement	Further recombination between an assembled but unproductive V(D)J exon, and remaining V segments
Somatic hypermutation	Point mutation of residues in a variable region to facilitate affinity maturation; occurs in immunoglobulin loci only

T-CELL RECEPTOR GENES

There are four types of TCR monomer (α, β, γ and δ) and these form two types of heterodimer – the αβ-TCR and γδ-TCR). Like immunoglobulins, TCR chains also comprise of an N-terminal variable region which is concerned with antigen binding and a C-terminal constant region, which is concerned with efffector functions. Alternative constant regions are also found in TCRβ and TCRγ molecules. The structure of γδ-TCR, indicating the position of constant and variable regions, is shown in Figure 3.12 (Twyman 1998).

T-cell receptors genes are more complex in structure than immunoglobulin genes. The immunoglobulin genes show a relatively simple organization with various genes segments lying in the same order in which they are assembled. The *TCRD* gene lies embedded within the *TCRA* gene and the *V* segments are mixed. Additionally, *V* segments lie both upstream and downstream of the corresponding *C*-gene segments, and rearrangements involving downstream *V*-gene segments cause large inversions.

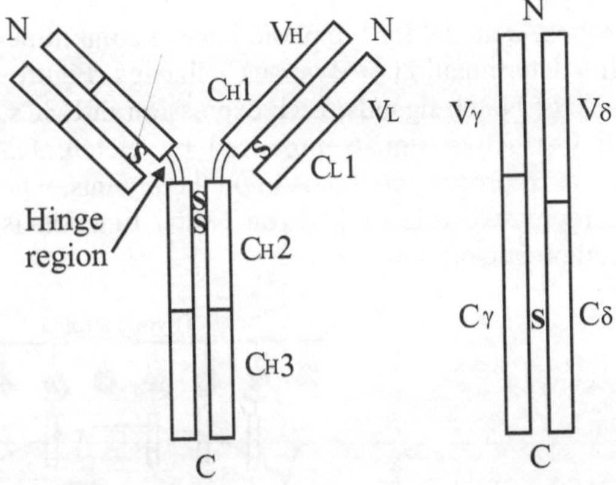

Figure 3.12 The structure of γδ-TCR, indicating the position of constant and variable regions

T-Cell Development

A signaling pathway in bone marrow progenitor cells leads to T-cell development unless a cellular factor intervenes and turns them towards a B-cell fate (Maillard and Pear 2007). The proto-oncogene *LRF* is encoded by *Zbtb7a* gene, formerly known as *Pokemon*. Notch signaling is reduced by the cellular factor LRF, thus allowing B-cell development in the bone marrow. In the absence of LFR-inhibitory activity, progenitor cells develop into T cells in thymus (Figure 3.13).

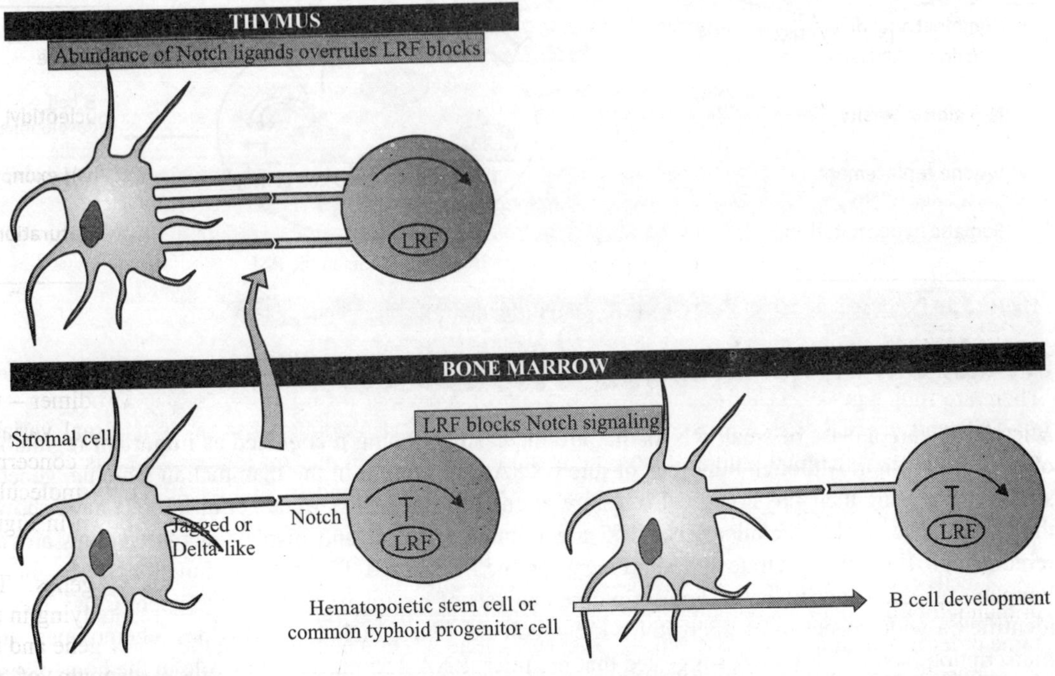

Figure 3.13 Notch block

Maeda et al. (2007) propose a model for lymphoid lineage commitment, in which, LRF acts as a master regulator of the cell's determination of B versus T lineage (Figure 3.14). Bone marrow (BM) cells express moderate levels of Notch ligands, LRF expression in HSCs and lymphoid progenitors function to repress T-cell instructive signals produced by Notch. ICN represents intracellular domain of Notch. However, once progenitors home in on the thymus, where Notch ligands are more abundantly expressed this repressive role of LRF on Notch function is overruled, which allows efficient production of T-cell precursors.

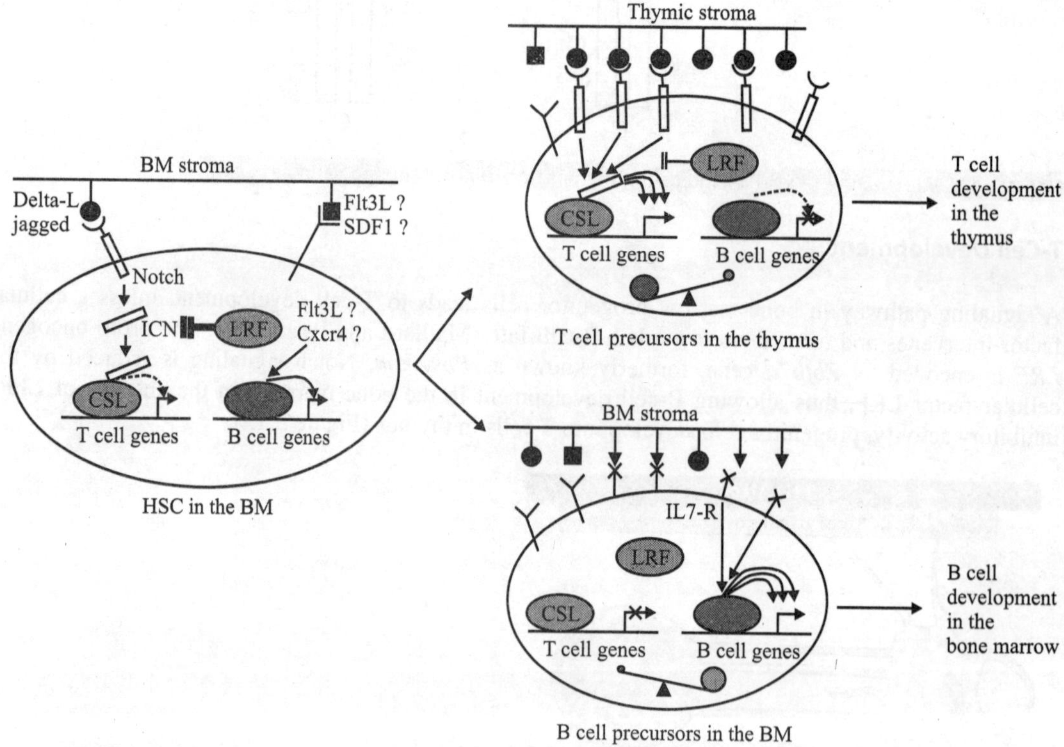

Figure 3.14 Proposed model for the role of LRF in determining B versus T lineage fate

MicroRNAs IN THE HOMEOSTASIS AND FUNCTION OF THE IMMUNE SYSTEM

MicroRNAs are a class of small RNAs that are increasingly being recognized as important regulators of gene expression. Although hundreds of microRNAs are present in the mammalian genome, genetic studies addressing their physiological roles are at an early stage. Rodriguez et al. (2007) have shown that mice deficient for bic/microRNA-155 are immunodeficient and display increased lung airway remodeling. They demonstrate a requirement of bic/microRNA-155 for the function of B and T lymphocytes and dendritic cells. Transcriptome analysis of bic/microRNA-155-deficient CD4$^+$ T cells identified a wide spectrum of microRNA-155-regulated genes, including cytokines, chemokines, and transcription factors. This work suggested that bic/microRNA-155 played a key role in the homeostasis and function of the immune system.

THEORIES OF ANTIBODY DIVERSITY

Germline Theory

The genome contains thousands of related but unique DNA sequences, each of which specifies one of the *V'* regions synthesized by an organism. Thus a large percentage of the mammalian genome is devoted to carrying variable sequence families and each person will inherit and transmit all V regions through his or her germline. Each lymphocyte will selectively express only one of these regions (genes) for the *V* portion of a light chain and only one of the *V* portions of a heavy chain. Because most lymphocytes will happen to select different combinations of sequences to express, the organism will come to express its vast diversity of antibody producing cells (Figure 3.15A).

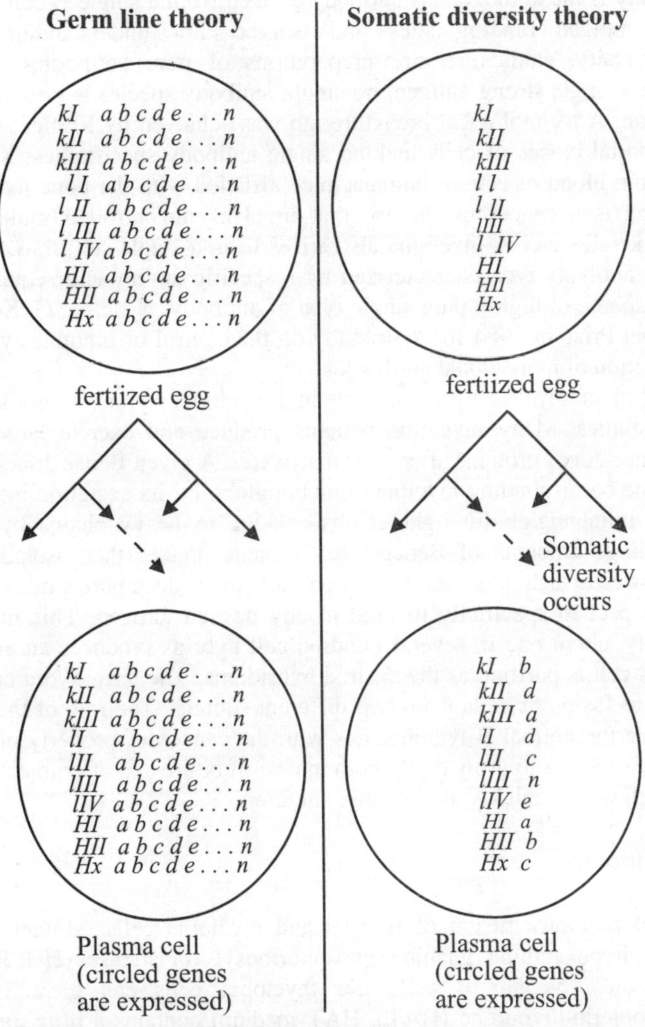

Figure 3.15 The germplasm versus somatic diversity theory of antibody diversity; only *V* sequences are shown

Somatic Diversity Theory

Only a small number of V sequences are transmitted from parents to offspring, perhaps as few as 3 V_k sequences, 5 V_γ sequences and a few V_H sequences. These sequences then experience thousands of different base-pair changes in individual lymphocyte lines creating the observed diversity. One lymphocyte clone will for example carry sequences $V_{\delta 1a}$ and $V_{\gamma 11c}$, another $V_{\delta 1b}$ and $V_{\delta 11e}$, and so on. Two of these sequences are then selected for expression in a given lymphocyte clone (Figure 3.15B).

Neither of these two theories of origin of antibody diversity is proven or disapproved.

HYBRIDOMAS AND MONOCLONAL ANTIBODIES

Monoclonal antibody is the homogenous antibody derived from a single B-cell clone and therefore all bearing identical antigen binding sites and isotype. For understanding molecular basis of immunological diversity, studies require preparations of pure antibodies. Even after continued immunization with a single strong antigen, no single antibody species is present in pure enough form to allow its isolation. A technological breakthrough was achieved by Kohler and Milstein (1975) for production of immortal clones of cells making single antibody specificities. Single antibody species can be obtained from blood of certain humans/mice afflicted with the bone marrow disease, multiple myeloma. Myeloma is a cancerous disease that involves uncontrolled multiplication of antibody producing cells. Like all cancers, myeloma also arises from a single cell; thus, a given specific tumor produces only one antibody type characterized by a specific amino acid sequence. Thus a myeloma patient becomes a source of highly pure single type of antibody. N. Jerne, G. Koehler and C. Milstein were awarded Nobel Prize in 1984 for their study on the control of immune system and discovery of principle for production of monoclonal antibodies.

The amount of given myeloma protein made in a myeloma patient is very large, easily permitting protein sequence studies. Many myeloma patients produce and excrete large amounts of specific proteins, called Bence-Jones proteins after their discoverer. A given Bence-Jones protein is identical to the light chain of the corresponding myeloma immunoglobulin. Its excretion into urine is the result of overproduction of immunoglobulin light chains relative to heavy chains by myeloma cells. The presence of very large amounts of Bence-Jones proteins makes their isolation extremely simple. Recently, a very powerful technique has made it possible to produce pure antibodies called monoclonal antibodies, with respect to specificity to bind to any desired antigen. This method is illustrated in Figure 3.16. Usually, about one in several hundred cell hybrids produces an antibody of the desired specificity, and this cell is purified as the desired hybridoma. The term hybridoma is applied to fused cells resulting due to fusion of cells from two different sources. The role of the myeloma cell in this process is to provide the normal B lymphocytes with the cancerous property of uncontrolled growth, thereby allowing hybridomas to grow easily both in tissue culture and after injection into mice. Normal non-cancerous B-cell would quickly die in either situation.

Selection of Hybridomas

Polyethylene glycol promotes fusion of B cells and myeloma cells. Mutant HGPRT IG does not synthesize enzyme hypoxanthine guanine phosphoribosyl transferase (HGPRT) and has stopped synthesizing antibodies. Neither B cells nor myeloma cells can grow in selective medium, hypoxanthine-aminopterin-thymidine (HAT). HAT medium contains a drug aminopterin that blocks one pathway for nucleotide synthesis, making the cell dependent on a salvage pathway that uses the HGPRT enzyme. Only hybridomas survive which inherit *HGPRT* gene from normal B-cell and

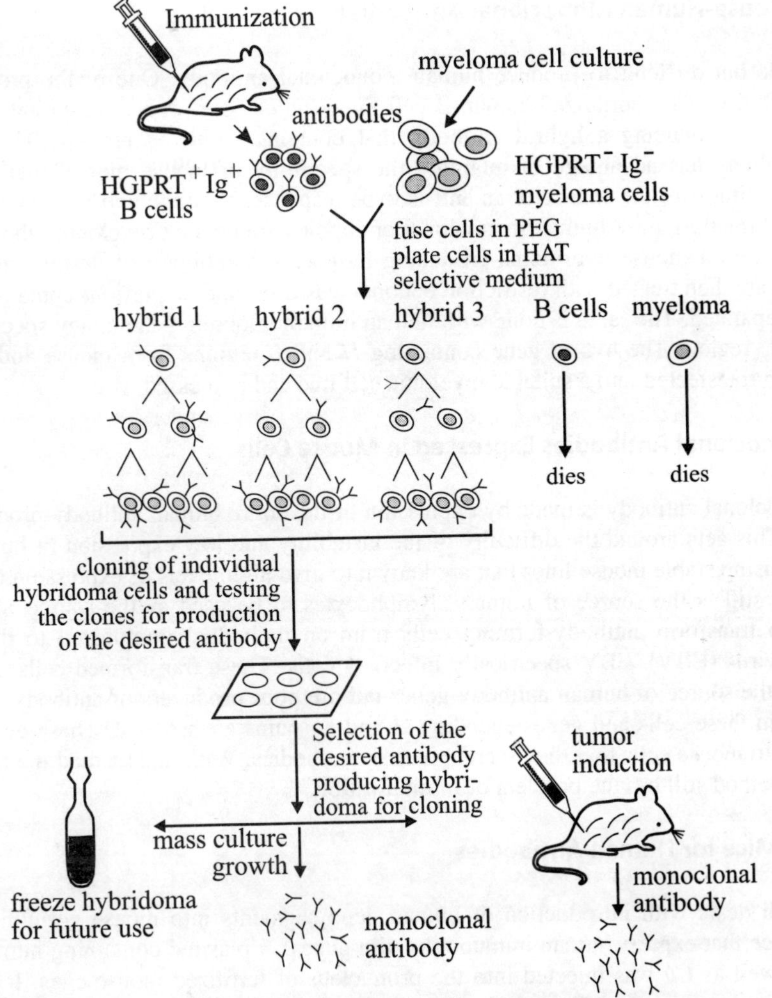

Figure 3.16 Procedure for making and selecting hybridoma for production of monoclonal antibodies

cancerous property of immortal growth from myeloma cell. Hybridomas that secret monoclonal antibody that can bind to the immunizing antigen are screened. Purification of antibodies is done by ion-exchange chromatography or antigen-affinity chromatography.

Alternatives to Hybridoma for Production of Monoclonal Antibodies

As an alternative to hybridoma, monoclonal antibodies can be produced by use of gene technology. Antibody genes are isolated from lymphocytes of immunized animals and then cloned and expressed in bacteria. The antibodies produced in bacteria under the control of cloned genes are screened for binding to specific antigens. Thus, while hybrid technology can immortalize antibody producing cells, gene technology immortalizes antigen producing genes. Computer graphic techniques are also being used to build specific antigen binding sites in antibodies.

Chimeric Mouse-Human Monoclonal Antibodies

It is desirable but difficult to produce human monoclonal antibody. One of the problems has been relative instability of immortalized human B-cell lines and/or poor production of antibodies. One can compromise by producing a hybrid antibody that contains mouse *V* regions. This mouse-human chimeric antibody has advantage of retaining the specificity of mouse monoclonal but is a human isotype. This eliminates the chance of an anti-isotype response, although anti-idiotypic response is still a problem. Chimeric mouse-human monoclonal antibodies are created by cloning the H-chain and L-chain genes from a mouse hybridoma-producing monoclonal antibody of desired specificity. These cloned genes are then treated with restriction endonucleases so that the portions containing the V and C regions are separated. The same is done with human immunoglobulin gene of any specificity; it will be a source of *C* region. The hybrid gene containing *H* and *L* regions from mouse and *C* region from human is then transfected into a suitable myeloma cell line and expressed.

Human Monoclonal Antibodies Expressed in Mouse Cells

Human monoclonal antibody is made by expression of the entire human antibody-producing gene in a mouse cell. This gets around the difficulty of the instability and low expression of human hybridoma cell lines by using stable mouse lines that are known to give high levels of expression of the molecule. The problem still is the source of immune lymphocytes to be used as the source of antibody. The strategy is to transform antibody-forming cells from an individual immunized to the antigen with Epstein-Bar virus (EBV). EBV specifically infects B cells. These transformed cells grow in culture, and serve as the source of human antibody genes but are poor producers of antibody. cDNA libraries are made from these cells and genes encoding H and L chains are isolated. These genes can then be transfected into mouse cells that then secrete human antibodies. With this method the antibody is fully human, but method still has the problem of immunization.

Transgenic Mice for Human Antibodies

This approach deals with introduction of human gene segments into mouse germline, i.e., produce transgenic mice that express human immunoglobulin genes. A plasmid containing human V_H, *D* and *J* segments as well as $C\mu$ was injected into the pronucleus of fertilized mouse eggs. It was found that when mice were born, i.e., 'minilocus' was rearranged in B cells and these B cells produced human Ig.

Phage Display Technology for Human Antibodies

This technology allows production of monoclonal antibodies (Mabs) for chosen target antigen, without the use of any animal or hybridoma. DNA sequences coding for antibody V region are fused with DNA sequence coding for amino-terminal of the minor coat protein pIII of the phage. This fused DNA sequence, inserted in phage genome, will produce a fused protein (carrying the antibody) that will display on the surface of the phage (due to coat protein sequence). A large repertoire of sequences coding for diverse antibodies can be created by polymerase chain reaction amplification and used for gene fusions that will be inserted into phage genome. The phage particles having a variety of fused gene constructs will produce a variety of antibodies on their surface. From this mixture, phage particles producing antigen-specific antibodies can be selected. When antigen immobilized on a surface is exposed to antibody producing phage, only the phage producing antigen-specific antibody will be retained on the surface coated with antigen. These phage particles are recovered from the solid surface and used for re-infecting the bacteria for the production of antigen-specific antibodies. The use of this phage display technology is expected to gradually replace hybridoma technology in all its applications in research and diagnosis.

Uses of Monoclonal Antibodies

Diagnosis, screening and therapy

Monoclonal antibodies are now an essential tool of much bio-medical research and are of great commercial and medicinal value. Monoclonal antibodies are used in ELISA test for detection of viruses, immunopurification, imaging and therapy. In diagnosis, pregnancy can be detected by assaying of hormones with monoclonals. Similarly, pathogens can be detected in a few hours sparing several days of cell culturing earlier needed. Immunopurification involves separation of one substance from a mixture of very similar molecules. For instance, individual interferons could be purified using monoclonal antibodies and could be used for inactivating T-lymphocytes responsible for rejection of organ transplant. Removal of tumor cells from bone marrow is another therapeutic use of monoclonal antibodies.

Vaccine production

Antibodies have been used to immunize against certain diseases in humans and cattle. The most promising outcome in this area is the prospect of developing antimalarial vaccines in future. Monoclonal antibodies that inhibit the in vitro multiplication of *Plasmodium* and the anti-gametocyte antibodies that inactivate male gametes have been developed. Monoclonal antibodies that destroy merozoite infected red blood cells have also been developed now. Such antibodies may prove useful as vaccines.

Enzymes (abzymes)

The antibodies may often bind specific ligands (haptens), but may not carry out chemical reactions. By modifying these ligands, antibodies may be generated that will catalyze specific reactions just like enzymes. Production of these enzymes is based on the following two principles: (a) Enzymes work by binding the transition state of a reactant better than ground state and (b) antibodies which bind to specific small molecules to a protein carrier and using this protein for immunizing experimental animals. If this small molecule is transition state analog (molecule that will mimic the shape and electronic configuration of transition state) then that antibodies that are produced to bind to this molecule will function as enzyme towards the substrate of this reaction. Considerable success in the production of abzymes catalyzing acyl transfer reaction, carbon-carbon bond formation and carbon-carbon bond cleavage can be achieved.

Immunosuppressive agents

The first monoclonal antibody to be approved for use in human was developed by Kung et al. (1980). The antibody called OKT3 was directed against an epitope on one of the chains of the CD3 molecule associated with the T-cell receptor and was found virtually on all T cells. It has been possible to prevent rejection of kidney grafts by injection of OKT3 because T cells are responsible for graft rejection. OKT3 reacts with cells in vivo and prevents them from functioning. Because CD3 is present on all T cells, the use of OKT3 causes a general immunosuppressor by eliminating the function of all T cells.

Immunotoxins

Immunotoxins are produced by plants that inhibit protein synthesis in any cell that they can get into. These consist of toxic moiety, the A-chain, and a specific moiety, the B-chain. The A-chain is inactive while it is disulfide bonded to the B-chain. Interaction with cell occurs through the B-chain's reaction with cell-surface receptors. In an immunotoxin, the A-chain is linked to an immunoglobulin via a disulfide bond. The specific reaction at the cell surface is through the antigen binding site of the antibody. When the molecule is internalized, the disulfide bond is broken and the A-chain exerts its toxic actin by inactivating 60S ribosomal subunit. If the antibody is specific for an antigen on a tumor cell, this becomes a powerful tool for delivering toxins to specifically eliminate tumor cells.

Purification and quantitation of other molecules

Antibodies have been used in the purification and quantitation of certain molecules present in trace amounts such as hormones, enzymes, antigens. This assay is extremely sensitive and quantities as low as nanopicomolar concentrations can be detected in small volumes (1 ml) of body fluids such as plasma, urine, cerebrospinal fluid. Following steps are involved in the purification and quantitation of enzymes and other proteinaceous molecules:

Raising the Antibodies: Monoclonal antibodies for a specific enzyme or protein are raised using traditional hybridoma technology. The enzyme is first extracted and partially purified using most of the conventional steps such as extraction, gel/column filtration. The preparation is diluted and emulsified with 'Freunds complete adjuvant'. Then it is injected intramuscular into mice to raise antibodies. The spleen cells from mice are collected and fused with myeloma (tumor) cell lines derived from a rat. The resultant hybridoma cells are screened for their capacity to produce antibodies and secrete them out. The hybridoma cells are thawed and recultured for the production of antibodies.

Purification and Quantitation: The supernatant from cultured hybridoma cells is used for detection and purification of an enzyme. The most common methods for enzyme purification/detection are affinity chromatography, radioammunoassay and immunoprecipitation.

Affinity Chromatography: The monoclonal antibody is immobilized on an insoluble matrix, such a cross-linked dextran or agarose beads, by a covalent linked agent. The immobilized antibody is generally packed in a column where its average concentration is 5-10 mg antibody per ml of bed volume. Protein solution containing the specific enzyme is applied over these immobilized antibodies. The specific enzyme binds with the antibody while other components do not and thus the specific enzyme is separated. The enzyme can be recovered later by using suitable reagents. Antigens can also be purified using affinity chromatography with monoclonal antibodies. The antigen binds with the immobilized antibody and is eluted and recovered from the immobilized support by suitable washing procedures.

Radioimmunoassay: It is an extremely sensitive method for the detection and quantitation of any substance that is antigenic and that can be labeled with a radioactive isotope. Basically, the method depends upon the competition between the labeled (known) and unlabelled (unknown) antigen for the same antibody. A known amount of labeled antigen (enzyme), a known amount of specific antibody and an unknown amount of unlabeled antigen are allowed to reach equilibrium. By measuring amount of labeled antigen bound to the antibody, a measure of the antigen in unknown sample can be obtained. By suitably adjusting the amount of antibody and labeled antigen, even small amount of antigen in an unknown material can be detected in this way.

Immunoprecipitation: If normal conditions are employed, antigen and antibody react to from a precipitate. This precipitation phenomenon is used to separate and purify enzymes and other antigenic substances in a variety of ways. One application of immunoprecipitation is immunoelectrophoresis. This is the most sensitive method for the detection of enzymes and antigens in a mixture. It involves combination of electrophoresis and el diffusion. In this procedure, the mixture containing the enzyme (or antigen) is placed in a small well, cut in a layer of agar placed on a glass sheet or a microscope slide. It is then subjected to electrophoresis by application of an electric current. The enzyme and other components of the mixture migrate through the agar at variable rates. Following electrophoresis, a trough is cut in the gel parallel to te elctrophoretic migration. An antiserum (containing the antibody of that specific enzyme) or monoclonal antibody preparation is then placed in the trough. The reactants diffuse towards one another and form separate arcs of precipitate for each antigen-antibody complex. If only one monoclonal antibody preparation is used, then only one precipitation arc is formed.

HUMAN LEUKOCYTE ANTIGENS

Histocompatibility antigens are those antigens that determine the acceptance or rejection of a tissue graft. These antigens are produced by histocompatibility genes. Histocompatibility genes are the genes that code for histocompatibility antigens. Human leukocyte antigen (HLA) complex comprises of genes that are responsible for synthesis of antigens that determine acceptance or rejection of a tissue graft. Histocompatibility is also known as tissue compatibility. Tissues can be transplanted between genetically identical animals without concern for immunologic rejection, whereas transplants between genetically non-identical animals are usually rejected with time. The timing and intensity of the rejection reaction are functions of the genetic differences between donor and recipient. Each of these genes has a series of alleles. This allelic series provides an excellent example of multiple allelism in humans. The HLA type of an individual is determined by testing his or her lymphocytes with a battery of standard antibodies from donors of known HLA type. In a transplant, the recipient's immune system will recognize the graft as "foreign" and rejects it if an allele is present in the donor tissue that is not present in the recipient. In the typing test, this rejection is observed as lymphocyte death. The immune system will not reject tissue that lacks some alleles present in the recipient. It only rejects the "strange" alleles. Table 3.3 shows examples using different alleles of A and B genes assuming that all alleles of other genes between graft and recipient are identical.

Table 3.3 Strange alleles of the donor are rejected by the recipient (Source: Zeleski, M.B. et al. 1987. *Immunogenetics*. John Wiley & Sons Inc.)

Transplant No.	*Recipient genotype*	*Donor genotype*	*Result*
1	A1A2, B5B5	A1A1, B5B7	Rejected (B7)
2	A2A3, B7B12	A1A2, B7B7	Rejected (A1)
3	A1A2, B7B5	A1A2, B7B7	Accepted
4	A2A3, B7B5	A3A3, B5B5	Accepted

HUMAN LEUKOCYTE ANTIGEN COMPLEX

G. Snell, A. Benacerraf and J. Dausset were awarded Nobel Prize in 1980 for their work on histocompatibility gene complex (Snell et al. 1953; Snell 1953; Benacerraf and McDevitt 1972; Benacerraf and Katz 1975; Dausset 1958; Dausset 1971). Genes involved in the production of cell-surface antigens that are recognized by the rejection or tolerance of tissue transplants are collectively called histocompatibility genes or loci. The action of antigen-producing alleles at many of these loci seems to be codominant. Individuals will reject the tissue of donors that carry alleles which they themselves do not carry. HLA system is based on three classes of HLA genes on chromosome number 6 which determine immunological acceptability (Figure 3.17).

Class	Genes
Class I	*B, C, A*
Class II	SB (α, β) DR (α, β) DC (α, β) BR (α, β)
Class III	*C2, BF, C4A, C4B*

Note: SB, DR, DC, and BR are regions, each containing α and β genes.

Figure 3.17 Three classes (I-III) of genes in human leukocyte antigen (HLA) complex

Table 3.4 Multiple alleles of HLA complex in man

Gene	*Number of alleles*
HLA-A	17
HLA-B	32
HLA-C	8
HLA-D	12
HLA-DR	10
HLA-C2	5
HLA-C4A	6
HLA-B4	4
HLA-BF	11

Each gene may possess 8 to 40 alleles (Table 3.4), each allele specifying a particular antigen. A particular number 6 chromosome may have one of 75,000 theoretically possible combinations of HLA alleles; each particular combination is called a haplotype. An individual usually possesses two different haplotypes, one from each parent. This provides millions of different possible HLA diploid genotypes. This enormous populational variability helps explain the difficulties in successful transplantation of tissues and organs, since most unrelated people will differ markedly in their HLA genotypes. Tests to obtain as close an HLA match as possible between transplant recipient and donor are necessary. Hypothetical pedigree for the inheritance of HLA-A and HLA-B antigen alleles are given in Figure 3.18. Each parent carries two haplotypes and each haplotype in commonly transmitted unchanged. Rare crossovers do occur and occasional siblings may show similar antigens. HLA tests can also be used to help decide question of genetic relatedness (such as paternity problems). There is a correlation between particular HLA antigens and the incidence of particular diseases. In autoimmune disease, individual's own tissues and organs are rejected.

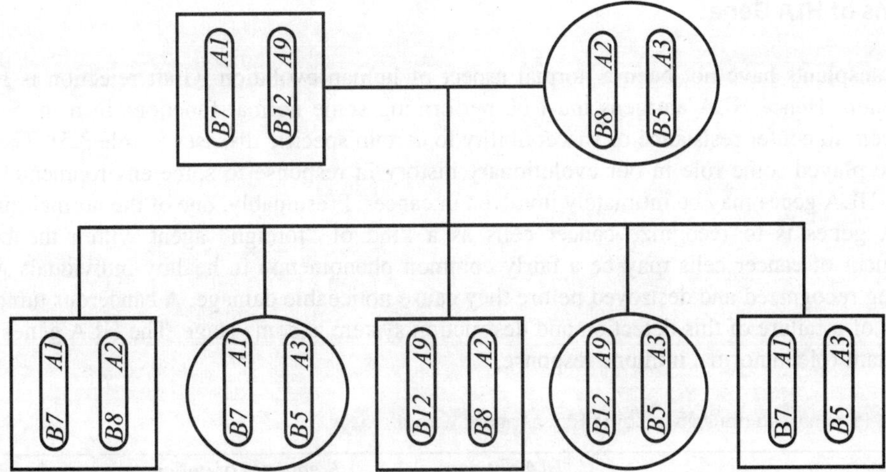

Figure 3.18 A hypothetical pedigree for inheritance of HLA antigen alleles

DNA Mismanagement leads to Immune System Oversight

Trex1, a major 3' DNA exonuclease in mammalian cells has been thought to act primarily in DNA replication or repair. Surprisingly, the major phenotype resulting from Trex1 deficiency in humans and mice is a chronic inflammatory disease (Coscoy and Rauler 2007). Multiple mechanisms could account for the inflammatory disease in mice caused by Trex1 deficiency (Figure 3.19). Yang et al. (2007) report that Trex1 deficiency causes chronic activation of ATM-dependent DNA damage checkpoint and accumulation of a discrete single-stranded DNA species in the cytoplasm, either of which could contribute to chronic inflammation.

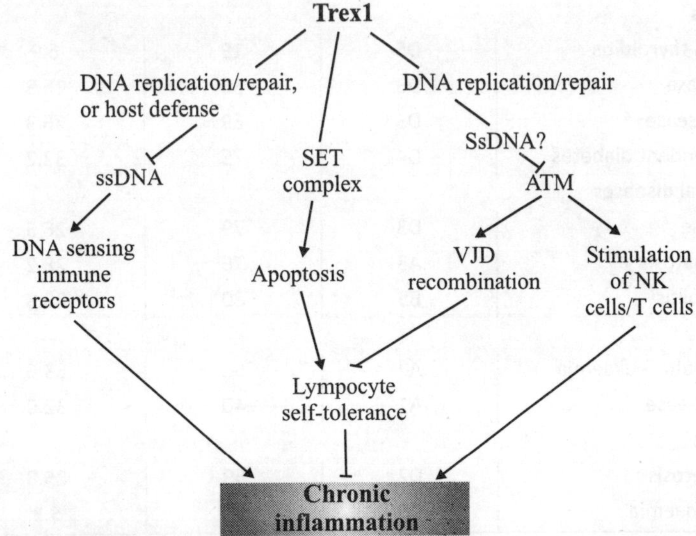

Figure 3.19 Inflammatory disease and Trex1 deficiency

Functions of HLA Genes

Tissue transplants have not been a normal aspect of human evolution. Graft rejection is man-made phenomenon. Hence HLA antigens must be performing some normal functions in man. Some HLA alleles seem to confer resistance or susceptibility to certain specific diseases (Table 3.5). These alleles may have played some role in our evolutionary history in response to some environmental selection pressure. HLA genes may be intimately involved in cancer. Presumably, one of the normal functions of the HLA genes is to recognize cancer cells as a kind of "foreign" agent within the body. The development of cancer cells may be a fairly common phenomenon in healthy individuals with these cells being recognized and destroyed before they cause noticeable damage. A cancerous tumor may be the result of a failure of this detection and destruction system at some stage. The HLA genes also play an important role in normal immune response.

Table 3.5 Possible relationship of some HLA antigens with diseases

Disease	*HLA antigen*	*Frequency of antigen in*		*Relative risk*
		Patients (with disease)	*Controls (without disease)*	
Joint diseases				
Ankylosing spondylitis	B27	90	9.4	87.0
Reiter's disease	B27	79	9.4	37.0
Rheumatoid arthritis	D4	50	19.4	4.2
Skin diseases				
Psoriasis	Cw6	87	33.1	13.3
Pemphigus	D4	87	32.1	14.4
Systemic lumpus erythematosus	D3	70	28.2	5.2
Gland diseases				
Hashimoto's thyroiditis	D5	19	6.9	3.2
Grave's disease	D3	56	26.3	3.7
Addison's disease	D3	69	26.3	6.3
Insulin-dependent diabetes	D4	75	32.2	6.4
Gastrointestinal diseases				
Celiac disease	D3	79	26.3	10.8
Hemochromatosis	A3	76	28.2	8.2
Ulcerative colitis	B5	80	30.8	9.3
Malignancies				
Acute lymphatic leukaemia	A2	60	53.6	1.3
Hogdkin's disease	A1	40	32.0	1.4
Other diseases				
Multiple sclerosis	D2	59	25.8	4.1
Pernicious anaemia	D5	25	5.8	5.4

ESCAPING THE IMMUNE DETECTION

Human cytomegalovirus (HCMV) prevents the display of class I major histocompatibility complex (MHC) peptide complexes at the surface of infected cells as a means of escaping the immune detection. Two HCMV-encoded immunoevasins, US2 and US11, induce the dislocation of class I MHC heavy chains from the endoplasmic reticulum membrane and target them for proteasomal degradation on the cytosol. Although the outcome of the dislocation reactions catalyzed is similar, US2 and US11 operate differently: Derlin-1 is a key component of the US11 but not the US2 pathway. So far, proteins essential for US2-dependent dislocation have not yet been identified. Loureiro et al. (2006) compare interacting partners of wild-type US2 with those of a dislocation-incompetent US2 mutant, and identify signal peptide peptidase (SPP) as a partner of the active form of US2. They show that decrease in SSP levels by RNA-mediated interference inhibits heavy chain dislocation by US2 but not by US11. These data implicate SPP in the US2 pathway and indicate the possibility of a previously unknown function for this intramembrane-cleaving aspartic protease in dislocation from the endoplasmic reticulum.

VACCINES

Vaccines are defined as antigens that do not reproduce but elicit production of antibodies. Vaccination is one of man's most significant inventions. It has eliminated smallpox and brought under control such diseases as diphtheria, poliomyelitis and measle. Yet, the process remains imperfect; the safety of vaccines cannot be absolutely ensured. Vaccination is any preparation which is used to confer immunity to a disease by inoculation.

Attenuated or Killed Viruses as Vaccines

Immunization against viral infection is achieved by injecting a virus that elicits antibodies capable of neutralizing the agent of a disease. The injected virus is not itself likely to cause infection because it has either been attenuated or killed. An attenuated virus is one whose virulence in human beings has been reduced by mutation in the course of passage through some different animal host, so that the virus becomes adapted to survival in that animal rather than in human beings. The cowpox virus, which confers immunity to smallpox is, in effect, a naturally attenuated virus. An attenuated virus is, however, a living organism. It can mutate further, possibly increasing in virulence. This possibility is eliminated if the virus is killed and thereby inactivated, but there have been instances where an incompletely inactivated virus has caused disease. Certain cells of the immune system, called lymphocytes, after encountering the virus, are triggered for an immune response, one result of which is the proliferation of plasma cells that secrete antibodies against sites on the surface of the virus. The antibodies bind to those sites and coat the surface of the virus. The latter is thus neutralized and unable to infect the cells. The site that is recognized by a specific lymphocyte and the site to which the antibody binds is called an antigenic determinant.

Engineered Vaccines

In order to avoid dangers of incomplete inactivation, vaccines can be prepared by recombinant DNA technology. Research has been initiated to engineer safe vaccines against not only viral diseases but bacterial and protozoan diseases also. Such vaccines as are prepared by use of recombinant DNA technology are called engineered vaccines. There are four steps involved in the preparation of vaccines by recombinant DNA technology. The first step is to identify the surface antigens of the parasite. Second step is to clone the gene for the antigen. Thirdly, gene is got expressed in *E. coli* or some other system. Finally, purified antigen is used as a vaccine. The recombinant DNA technology has been exploited to prepare a vaccine against malaria. This task is difficult as there are four different species of *Plasmodium* which cause malaria in humans and each species has different strains that carry on its surface different sets of antigens in successive developmental stages (sporozoites, merozoites, gametocytes). Gene for the major sporozoite antigen of *P. falciparum* has been expressed in *E. coli*. The expressed antigen has proved immunogenicity. Monoclonal antibodies against this protein when injected into chimpanzees gave a measure of passive immunity against malaria. Similarly, one of the surface antigens has expressed in *E. coli*. Monkeys have been immunized with partial success. A major antigen of the male gamete of *P. yoelli* has been described. A realistic malarial vaccine shall be a cocktail of antigens or a multivalent vaccines. These techniques are also being used to prepare large amounts of the sporozoite molecule for possible protection against East Cost Fever caused by the protozoan parasite, *Theileria parva*. Research is also going on for preparing a vaccine against leprosy.

Immunopurification of Antigens

Immunopurification involves separation of a specific antigen from a mixture of very similar antigens. This purified antigen can then be used for developing vaccine against a pathogen. The purification can be very effectively achieved using monoclonal antibodies, which are very specific in their reaction against an antigen. Individual interferons, which are proteins having a property of inhibiting viral infection and cell proliferation, have also been purified using monoclonal antibodies. After such purification, interferons were used for clinical trials before these were released recently for commercial use.

Synthesis of Antigens through Cloned Genes

Hundreds of genes in eukaryotes have been cloned either from genomic DNA or from cDNA. These cloned genes included a number of genes for specific antigens, and in some cases have been used for the synthesis of antigens leading to the preparation of vaccines. Following two examples can be used to illustrate the use of cloned genes for vaccine preparation.

Cloning of hepatitis B virus (HBV) genome

The HBV genome has been cloned in the plasmid pBR322 and propagated in *E. coli*. From this clone, antigen could be produced in good quantity which reacted with hepatitis B core antibody (HBAb). This has, therefore, been used to produce hepatitis B vaccine, which was later approved for mass vaccination in several countries.

Cloning of human malarial gene

Despite the great menace and threat to human health due to malarial parasite *P. falciparum*, no antimalaria vaccine, could be developed so far. Recently, with cloning of a gene coding for surface protein of the sporozoite of *P. falciparum*, there is a hope for developing a vaccine.

In human host, malarial parasite passes, through several antigenically distinct phases: (i) sporozoite, the form in which the parasite is injected with mosquito bite; sporozoites enter the liver and multiply and develop into (ii) merozoites, which in turn invade and multiply in red blood cells; small fraction of these merozoites in red blood cells form (iii) gametocytes, which may be picked up by a mosquito to start another cycle. Therefore, vaccinated individual may also block its spread; antimerozoite vaccine which will protect or ameliorate the patient but will not check the spread of the disease and antigametocyte vaccine, which would prevent the spread without helping the patient. Of these, antisporozoite (CS) protein leading to the development of vaccine which was used for clinical trials could not reach the commercial stage till the end of the year 2001. This gene was obtained directly from DNA of erythrocytic form of parasite, rather than as cDNA from mRNA. This cloned gene may, in course of time, lead to the synthesis of vaccine by synthesizing CS protein by cloned gene.

Synthetic Vaccines

These vaccines do not contain intact viruses or complete polypeptides but merely small peptides that have been synthesized in laboratory to mimic a very small region of the outer coat of the virus (Cox 1985). These peptides elicit antibodies capable of neutralizing the virus. A synthetic vaccine against malaria consists of a combination of three synthetic peptides, corresponding to partial sequences of one of the major proteins on the surface of malaria red blood cells (RBCs). A contraceptive vaccine for humans based on the carboxyl terminal region of human chorionic gonadotrophin (HCG) is being tested. This vaccine is prepared by a US-Australian group. Another contraceptive vaccine is being developed by G.P. Talwar's group in India (Jayaraman 1986). This vaccine is a cocktail of antigens. There are two formulations — one formulation is a mixture of ovine luteinizing hormone (OLH) and HCG-TT (tetanus toxoid) and OLH-HCG-CHB (cholera toxin B) and the second formulation is a mixture of OLH-TT-CHB and HCG-TT-CHB. Both these formulation have been extensively tested in baboons. Experimental vaccines against foot disease virus, mouth disease virus, influenza virus, hepatitis B virus, AIDS virus, etc. have been prepared.

Synthetic peptides as vaccines

Vaccines can also be prepared through short synthetic peptide chains, which have, therefore, become a subject of considerable research activity. In order to synthesize peptides to be used as vaccines, structure and function of proteins involved would be studied. Since, it is the three dimensional structure (TDS) and not the amino acid sequence, which is responsible for immunogenic response, it may be necessary to find out the protein region involved in immunogenic response. For instance, in Foot and Mouth Disease Virus (FMDV), it is the amino acid 114-160 of virus polypeptide, which can produce antibodies neutralizing FMDV and thus provide protection. Neutralization of FMDV was also possible through the region of 201-213 amino acids of the same protein. It has thus been shown that small synthetic peptides representing these regions of proteins can show immunogenic response and can, therefore, be used for development of a vaccine.

An alternative approach to find out the immunogenic region of protein is through the study of gene coding the protein. Recently, it has been shown that a cloned gene of an immunogenic protein of a pathogen feline leukemia virus (FLV) can be cut into fragments by DNAase I and these fragments can be cloned in lambda phage where they may express. Phage colonies (plaques) having different cloned fragments are screened with a specific monoclonal antibody that neutralizes the pathogen. The fragments which react with antibody must be synthesizing the immunogenic peptide fragment. This cloned DNA fragment can then be sequenced. It was possible, in this manner to identify a 14-amino acid immunogen of the envelope protein of FLV. The corresponding synthetic peptide was also found to compete with the virus for antibody. When injected in guinea pigs, such synthetic peptides also elicited a partial immunogenic response. Therefore, there is a great promise for the use of such synthetic peptides to be used as vaccines. Actually, a vaccine for malaria in the form of a synthetic peptide has already been prepared and is being tested for its suitability. This is the first example of a vaccine developed in the form of a synthetic peptide.

Recently, it has been shown that immunogenic region of protein of a pathogen can also be identified from purified MHC molecules (MHC = major histocompatibility complex). Different MHC allelic variants are available in cells for binding of different proteins and they can be purified using specific T cells. Peptides can be eluted from these purified MHC molecules and sequences of such peptides can be determined and used for manufacturing synthetic peptides to be used as vaccines. Using these approaches, vaccines against several pathogens, including the following pathogens have either been produced in recent years or are expected to be produced in the near future: Rabies virus, Foot and Mouth Disease virus, and *Salmonella typhimurium*.

Live Vaccines

Vaccines against multiple pathogens

Recent advances in molecular genetics have led to the possibility of using large DNA viruses, such as Vaccinia virus as a biological delivery causing agents. Vaccinia virus is a member of the poxvirus family. It has a linear double-stranded DNA genome of approximately 1,85,000 bp and replicates in the cytoplasm of infected cells. The relative inoccuvity of Vaccinia virus which has been used extensively to control and eradicate smallpox in man, has stimulated its development as a cloning and expression vector. A mutant of Vaccinia virus which has a spontaneous 9-Kb deletion located for insertion and expression foreign DNA. The Vaccinia virus *tk* locus was chosen as a site of insertion for foreign DNA, primarily because it provided an efficient method of selecting infectious recombinants (Buller et al. 1985).

Insertion of multiple foreign genes into Vaccinia virus

Perkus et al. (1985) succeeded in constructing Vaccinia virus recombinants expressing multiple foreign genes. The modification of a spontaneously occurring viable deletion mutant of Vaccinia virus to express 1,780 bp cDNA of the RNA segment encoding the Influenza virus hemagglutinin (InfHA) was done by Panicalli et al. (1983). This recombinant virus, VP53 containing the *InfHA* coding sequence, was used as a substrate for insertion of 1,330 bp herpes simplex virus glycoprotein D (*HSVgD*) coding sequence (Perkus et al. 1985). The resulting Vaccinia virus, VP124, containing both the *InFHA* and *HSVgD* coding sequences was used in turn as substrate for insertion of the 1,090-bp hepatitis B virus surface antigen (*HBsAG*) coding sequence. Thus, a triple Vaccinia virus recombinant VP168 was generated. All the three foreign coding sequences were inserted at non-essential sites in the Vaccinia genome, and transcription of the inserted genes was right to left relative to the Vaccinia genome. In addition to Vaccinia virus, the other vaccine vectors that are being tested are: Avipozvirus, Adenovirus, Poliovirus, *S. typhimurium*, Herpes Simplex virus, Cytomegalovirus, Varicella virus.

Detecting presence and expression of foreign genes in Vaccinia virus

To detect the presence of three foreign genes, *InfHA*, *HSVgD*, and *HBsAg*, in the recombinant Vaccinia virus, the strain VP168 was plaque-purified and the DNA was extracted and cleaved with HindIII or BamHI restriction endonuclease. DNA fragments were separated by Southern method and probed with ^{32}P-labeled nick translated sequences homologous to the foreign genes. The expression in vitro of all the three genes by Vaccinia virus recombinant VP168 was detected by means of serological tests.

Immunological response to vaccination with recombinant Vaccinia virus

The rabbits inoculated either intravenously or intradermally for immunization against unrelated disease with Vaccinia virus recombinant VP168 responded by making antibodies to all three of the foreign antigens synthesized under Vaccinia virus regulation. The intravenous route of inoculation also provided a criterion for the absence of toxicity in the preparation of the Vaccinia recombinant.

Revaccination with novel recombinant vaccinia virus

To determine whether a primary vaccination precludes revaccination with Vaccinia virus recombinants engineered to express novel foreign genes, Perkus et al. (1985) used two approaches: (1) A rabbit inoculated with Vaccinia virus recombinant expression, *HBsAg* retained high titers of antibody to HBsAg for more than one year. Revacciniation with a Vaccinia recombinant expressing the same foreign gene led to increased levels of antibodies reactive with HBsAg. (2) To obtain direct evidence for the induction of antibodies reactive to a foreign antigen or revaccination to a previously immune animal, a rabbit was inoculated initially with a Vaccinia virus recombinant expressing the HBsAg. Antibodies to HBsAg persisted at high titer for a year beyond the initial inoculation at which time the same rabbit was revaccinated with a Vaccinia virus recombinant expressing the *InfHA* instead of *HBsAg*. This revaccination has essentially no effect on the titer of the antibodies to HBsAg. However, InfHA antibodies were detectable in the serum. Genetically-engineered Vaccinia virus containing multiple foreign genes can be used through a single vaccination to immunize an individual against heterologous pathogens. Vaccinia virus has the ability to accommodate 20-25 kbp of foreign DNA. Additional space for insertion of foreign genes can be obtained by using viable deletion mutants of the virus. Vaccinia virus can be genetically engineered to express several vaccines for immunization against multiple pathogens would also increase benefit-risk ratio.

Nucleic Acid Vaccines

Injection into muscle of plasmid DNA encoding an antigen of interest has been shown to result in sustained expression of the antigen and generation of an immune response. This approach, termed as "nucleic acid vaccines," is receiving much attention for several reasons. First, in animal models, these vaccines appear to stimulate persistent humeral and cell-mediated immune responses, with integration of plasmid into chromosomal DNA. The animals are protected from lethal virus challenge. This is potentially useful in protecting against viral infections in which the antibody response alone is not protective, or where there is antigenic diversity of surface proteins among strains. Second, several routes of vaccine administration are possible — parenteral, mucosal, or via a gene "gun" that delivers tiny amounts of DNA-coated gold beads. Finally, this strategy results in relevant antigen production in primates without the use of infectious agents. Thus, this approach to vaccine development may be pertinent to several diseases, including AIDS, and will undoubtedly continue to receive intense scrutiny. Hypothetical safety concerns, including the potential integration plasmid DNA into the host genome or the generation anti-DNA antibodies, will need to be addressed.

Regulation of Immune Responses

Forkhead transcription factors play key roles in regulation of immune responses. Lin et al. (2004) identify a role for one member of this family, Foxj1, in the regulation of T-cell activation and autoreactivity. Foxj1 deficiency resulted in multiorgan systemic inflammation, exaggerated Th1 cytokine production and T-cell proliferation in autologous mixed lymphocyte reactions. Foxj1 suppressed NF-κB transcription activity in vitro and Foxj1-deficient T cells possessed increases NF-κB activity in vivo, correlating with the ability of Foxj1 to regulate 1κB proteins, particularly 1κBβ. Thus, Foxj1 likely modulates inflammatory reactions and prevents autoimmunity by antagonizing proinflammatory transcriptional activities. These results suggest a potentially general role for forkhead genes in the enforcement of lymphocyte quiescence.

Protective Immunity

Vaccination with irradiated sporozoites can provide protective immunity. Protective immunity can be confirmed by immunization with a recombinant *Salmonella* expressing only the circum sporozoite protein that normally covers the sporozoites. A new commitment is urged to the far greater challenges ahead, especially that of vaccine development against HIV (Fauci 2008). Françoise Barré-Sinoussi and Luc Montagnier were awarded Nobel Prize in 2008 for their discovery of human immunodeficiency virus (HIV) that causes AIDS (Barré-Sinoussi et al. 1983; Montagnier 2002).

REFERENCES

Barré-Sinoussi, F., J.C. Chermann F. Rey, et al. 1983. Isolation of a T-lymphotropic retrovirus from a patient at risk for acquired immune deficiency syndrome (AIDS). Science 220: 868-71.

Begum, N.A., and T. Honjo. 2004. Response to comment on "Uracil DNA glycosylase activity is dispensible for immunoglobulin class switch". Science 306: 2042.

Begum, N.A., K. Kinoshita, N. Kakazu, et al. 2004. Uracil DNA glycosylase activity is dispensible for immunoglobulin class switch. Science 305: 1160-3.

Benacerraf, B. and D.H. Katz. 1975. The histocompatibility-linked immune response genes. Adv Cancer Res. 21: 121-73.

Benacerraf, B., and H.O. McDevitt. 1972. Histocompatibility-linked immune response genes. Science 175: 273-9.

Billingham, R.E., L. Brent, and P.B. Medawar. 1953. 'Actively acquired tolerance' of foreign cells. Nature 172: 603-6.

Boothroid, C.E., O. Dreesen, T. Leonova, K.I. Ly, L.M. Figueiredo, G.A.M. Crose, and L.M. Papavasiliou. 2009. A yeast-endonuclease-generated DNA break induces antigenic switching in *Trypanosoma brucei*. Nature 459: 278-81.

Bredemeyer, A.L., G.G. Sharma, C.-Y. Huang, et al. 2006. ATM stabilizes DNA double-strand-break complexes during V(D)J recombination. Nature 442: 466-70.

Buller, R.M.L., G.L Smith, K. Cremer, A.L. Notkins, and B. Moss. 1985. Decreased virulence of recombinant Vaccinia virus expression vectors is associated with a thymidine kinase-negative phenotype. Nature 317: 813-5.

Burnet, F.M. 1957. A modification of Jerne's theory of antibody production using the concept of clonal selection. Aust. J. Sci. 20: 67-9.

Çoscoy, L. ,and D.H. Rauler. 2007. DNA mismanagement leads to immune system oversight. Cell 131: 836-8.

Cox, E.F.G. 1985. Malaria vaccine: components of a cocktail. Nature 318: 212-3.

Dausset, J. 1958. Iso-leuco-anticorps. Acta Haematol. 20: 156-66.

Dausset, J. 1971. The polymorphism of the HL-A system. Transplant. Proc. 3: 1139-46.

Dofilippantonio, S., E. Gapud, M. Wong, et al. 2008. 53BP1 facilitates long-range DNA end-joining during V(D)J recombination. Nature 456: 529-33.

Edelman, G.M. 1959, Dissociation of γ-globulin. J. Am. Chem. Soc. 81: 3155-6.

Fauci, A.C. 2008. 25 years of HIV. Nature 453: 289-90.

Jayaraman, K.S. 1986. Contraceptive vaccines. Trials in India launched. Nature 323: 661.

Jung, D., and F.W. Alt. 2004. Unravelling V(D)J recombination: Insights into gene regulation. Cell 116: 299-311.

Kohler, G., and C. Milstein. 1975. Continuous cultures of fused cells secreting antibody of predefined specificity. Nature 256: 495-7.

Kung, P.C., M.A. Talle, M.E. DeMaria, M.S. Butler, J. Lifter, G. Goldstein. 1980. Strategies for generating monoclonal antibodies defining human t-lymphocyte differentiation antigens. Transplant Proc. 12(3 Suppl 1): 141-6.

Lin, L., M.S. Spoor, A.J. Gerth, S.L Brody. and S.L Peng. 2004. Modulation of Th1 activation and inflammation by the NF-κB repressor Foxj1. Science 303: 1017-20.

Loureiro, J., B.N. Lilley, E. Spooner, V. Noriega, and D. Tortorella. 2006. Signal peptide peptidase is required for dislocation from the endoplasmic reticulum. Nature 441: 894-7.

Maeda, T., T. Merghoub, R.M. Hobbs, et al. 2007. Regulation of B versus T lymphoid lineage fate decision by the proto-oncogene LRF. Science 316: 860-6.

Maillard, I. and W.S. Pear. 2007. Keeping a tight leach on notch. Science 316: 840-2.

Matthews, A.G.W., A.J. Kuo, S. Ramon-Maiques, et al. 2007. RAG2 PHD finger couples histone H3 lysine 4 trimethylation with V(D)J recombination. Nature 450: 1106-10.

Montagnier L. 2002. Historical essay. A History of HIV Discovery. Science 298: 1727-8.

Perkus, M.E., A. Piccini, B.R. Lipinskas, and E. Paoletti. 1985. Recombinant Vaccinia virus: Immunization against multiple pathogens. Science 229: 981-4.

Porter, R.R. 1958. Separation and isolation of fractions of rabbit gamma-globulin containing the antibody and antigenic combining sites. Nature 182: 670-1.

Porter, R.R. 1959. The hydrolysis of rabbit γ-globulin and antibodies with crystalline papain. Biochem. J. 73: 119-27.

Rodriguez, A., E. Vigorito, S. Clare, et al. 2007. Requirement of bic/microRNA-155 for normal immune function. Science 316: 608-11.

Snell G. D., P. Smith, and F. Gabrielson. 1953. Analysis of the histocompatibility-2 locus in the mouse. J. Natl. Cancer. Inst. 14: 457-80.

Snell, G.D. 1953. The genetics of transplantation. J. Natl. Cancer Inst. 14: 691-704.

Stivers, J.T. 2004. Comment on "Uracil DNA glycosylase activity is dispensible for immunoglobulin class switch". Science 306: 2042.

Tonegawa, S., N. Hozumi, G. Matthyssens, and R. Schuller. 1976. Somatic changes in the content and context of immunoglobulin genes. Cold Sp. Harb. Symp. Quant. Biol. 41: 877-89.

Twyman, R.M. 1998. *Advanced Molecular Biology*. Oxford: Bios Scientific Publishers Limited.

Yang, Y.-G., T. Lindahl, and D.E. Barnes. 2007. Trex1 exonuclease degrades ssDNA to prevent chronic checkpoint activation and autoimmune disease. Cell 131: 873-86.

Maillard, I. and W.S. Pear. 2007. Keeping B cells in check: [illegible] notch. Science 316: 840-2.

4

Oncogenes and Antioncogenes

Genes that confer the ability to convert cells to a tumorigenic state are called oncogenes. All human and subhuman species contain a complement of oncogenes that in their non-activated state produce substances necessary for normal cell proliferation and cell surface properties. There are four main classes of oncogenes. Class I oncogenes (e.g., *src*, *yes*, *neu*, *abl*, *fps*, *fms*, *erb*, *ros*, *mos*, *fgr*) are related to synthesis of cell surface receptor proteins. The cells surface receptor proteins are activated by growth factors produced by other cell types. An activated cell will proliferate as long as the second cell type produces its growth factor in response to some external stimulus. Class I oncogene products have kinase activity and are found in cytoplasm. Class II oncogenes (e.g., *HA-ras*, *k1-ras*, *N-ras*) are commonly found in human tumors. Their protein products have a common characteristic of regulating cellular metabolism. Class III oncogenes (e.g., *myc*, *myb*, *fos*, *ski*, *p53*) regulate nuclear activities, possibly cell cycling. Class IV oncogenes (e.g., *cis*, *rel*, *B-lym*, *erb-A*, *ets*, *met*) are relatively unrelated oncogenes, some of which produce cell growth factors (proteins).

Tumors are aggregates of cells derived from an initial aberrant founder cell that although surrounded by the normal tissue is no longer integrated into the environment. Cancer cells have a number of properties — rapid division, invasion of new cellular territories, high metabolic rate, new membrane antigens, altered shapes, and so on. In cancerous cells, the factors regulating normal cellular differentiation have been altered. These cells obey no community rules and multiply without any regard to tissue organization. These cells are termed as anti-social, malignant, or simply cancer cells. While dealing with oncogenes, one tries to answer the following questions. What is the primary cause of alterations? Is the multiplicity of changes a reflection of several defects within the tumor cells or a pleiotropic result of a single lesion? The observation that radiation and certain chemicals are carcinogenic suggested one or more targets of such agents in cells. The known mutagenicity of many of the carcinogens also suggested that DNA might be one of the targets. About 80 per cent of all tumors are induced by environmental factors. Shih and Weinberg (1982) found that DNA of tumor viruses is oncogenic. This suggested that multiple phenotypic changes signaling cell transformation are pleiotropic effects of a small number (perhaps even one) of lesions (Weinberg 1988).

L. Hartwell, R.T. Hunt and P.M. Nurse were awarded Nobel Prize in 2001 for their discoveries on the nature of cancer cell development (Evans et al. 1983; Hartwell and Weinert 1989; Hartwell et al. 1970; Hartwell et al. 1973; Hartwell et al. 1974; Hartwell 1971a,b; Hartwell 1973; Murray and Hunt 1993; Nebreda et al. 1995; Nurse and Thuriaux 1977; Nurse and Thuriaux 1980; Nurse et al. 1998; Nurse 1975; Nurse 1990; Nurse 2000; Weinert and Hartwell 1988).

RETROVIRUSES CONTAIN ONCOGENES

The Rous Sarcoma Virus (RSV) is a retrovirus (having RNA as genetic material). Many of the oncogenic retroviral genomes carry coding segments other than the usual viral genes. These segments are modified cellular genes that account for oncogenicity; they are called viral oncogenes. The RNA is reverse transcribed into DNA. Transformation of cells by RSV is found to result from the action of a single gene, *src*, located on one end of the RNA molecule. Transformation here means cancerous growth of cells. A single gene that causes cancer is known as an oncogene.

NORMAL CELLS CONTAIN PROTO-ONCOGENES

Normal cellular genes from which the viral oncogenes are derived are called proto-oncogenes. Schematic diagram showing a typical "wild-type" retroviral genome and retroviral genomes carrying viral oncogenes (v-onc) is given in Figure 4.1.

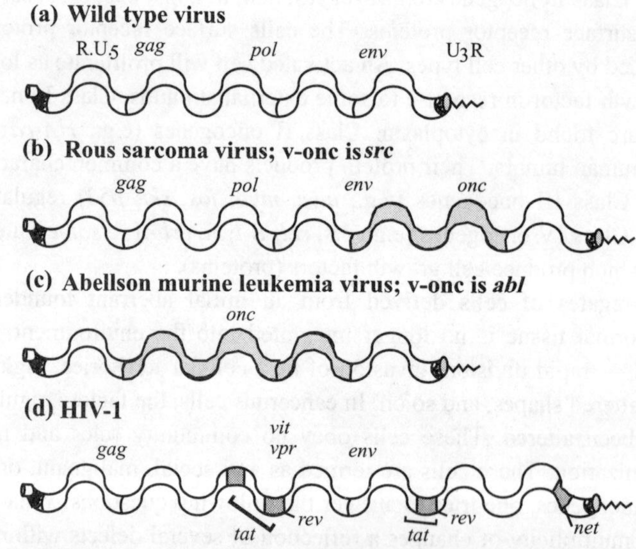

Figure 4.1 A wild type retroviral genome (a), a nondefective retroviral genome carrying the v-onc gene, *src* (b), a defective retroviral genome carrying the v-onc gene *abl* but cannot replicate independently because it lacks *pol* and *env* genes (c), and HIV-1 virus which causes AIDS showing interrupted or overlapping reading frames. *gag*, *pol* and *env* are the genes for nucleocapsid, reverse transcriptase, and envelope transcripts, respectively

Proto-oncogenes do not produce cancer cells under normal circumstances but can do so if they are modified or changed by their incorporation into viral genomes. Identification of the proteins encoded by some of the 50 or so known proto-oncogenes indicates that several of them may be involved in normal cell growth. Method used to demonstrate that proto-oncogenes exist in normal host cells is illustrated in Figure 4.2. There are about 40 proto-oncogenes in our genome which are normal cellular genes performing some vital functions in relation to cell proliferation. The oncogenic virus RNA is used to make DNA, which is hybridized to mutant viral RNA lacking the oncogene. The viral oncogene can be separated by column chromatography as unhybridized material, which in turn is used

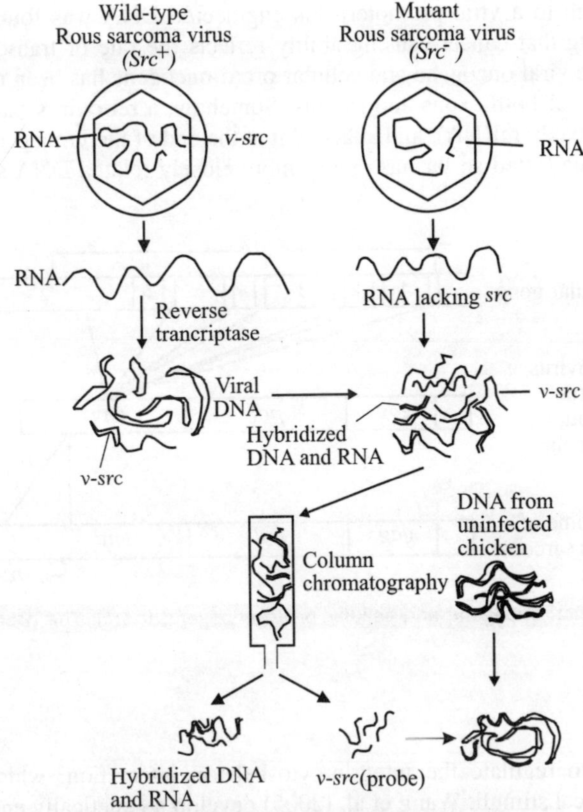

Figure 4.2 Demonstration that proto-oncogenes exist in normal chicken cells

as a probe to hybridize with complementary sequences in normal chicken DNA. Cellular proto-oncogenes have been found in mammals in chicken and even in *Drosophila*, thereby showing that they must have existed more than 600 million years ago before these diverse groups separated. J.M. Bishop and H.E. Varmus were awarded Nobel Prize in 1989 for giving cellular theory of oncogenesis (Hughes et al. 1979; Spector et al. 1978a; Spector et al. 1978b; Varmus et al. 1971; Varmus et al. 1972; Varmus et al. 1974). Genes can cause cancer if they malfunction. Viral oncogenes are altered normal cellular gene.

src ONCOGENE

Virus can multiply even when the *src* locus is deleted, so this oncogene is not necessary for survival. src gene encodes an enzyme called protein kinase that phosphorylates tyrosine. Phosphorylation affects activity of a protein. This function of *src* gene is regulatory. Using radioactive *src* RNA, it was found out that in uninfected chicken cells, complementary DNA existed. This shows that *src* DNA is normal part of the chicken genome. The chicken *src* DNA is typical of eukaryotes with a number of introns whereas virally made DNA lacks introns. This suggests that the virus had somehow picked up processed mRNA (from which the introns were cleaved). Almost every retrovirus oncogene tested so far has one or more closely related DNA sequences in uninfected host cells. When isolated cellular oncogenes were attached to a viral promoter, this engineered DNA was found to transform cells in vitro, thereby suggesting that cancer-causing ability reflects the rate of transcription of the DNA. A comparison between the viral oncogene and cellular proto-oncogene has been made in Figure 4.3. The proto-oncogene consists of both exons and introns. Somehow, a retrovirus managed to pick up RNA without the introns (probably mRNA), and attached it at the end of the virus genome. Almost every (15 out of 16) viral oncogene tested so far has one or more closely related DNA sequences in uninfected host cells.

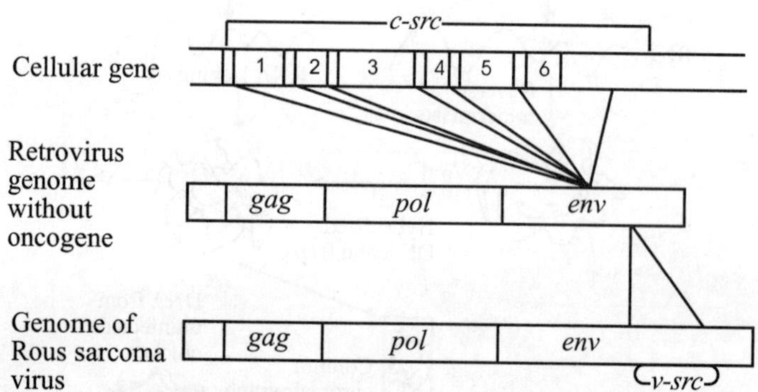

Figure 4.3 Structure of viral oncogene and cellular proto-oncogene (*c-src*). The proto-oncogene consists of both exons and introns

Functions of *src* Gene

Src protein is known to regulate the integrin-cytoskeleton interaction, which is essential for the transduction of mechanical stimuli. Wang et al. (2005) develop a genetically encoded Src reporter that enables the imaging and quantification of Src in liver cells. Using the Src reporter, they observed a rapid distal Src activation along the plasma membrane. This force-induced directional and long-range activation of Src was abolished by the disruption of actin filaments or microtubules. Src reporter has thus made it possible to monitor mechano-transduction in living cells with spatio-temporal characterization. Transmission of mechanically-induced Src activation is a dynamic process that directs signals via the cytoskeleton to spatial destinations. A proposed model depicting the mechanism, by which local mechanical forces induce directional and long-range Src activation, as given by Wang et al. (2005), is shown in Figure 4.4.

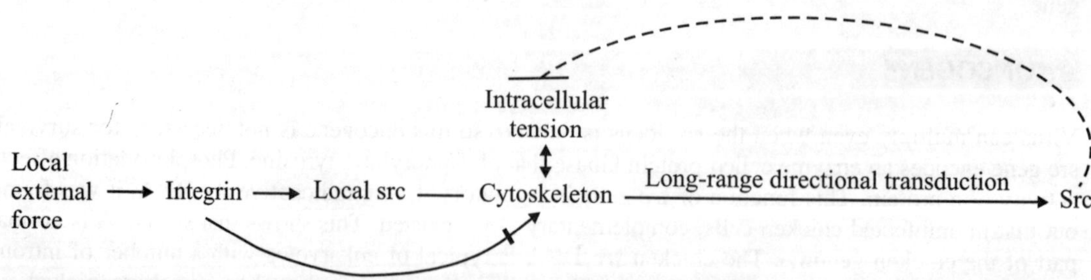

Figure 4.4 Model depicting the mechanism by which local mechanical forces induce directional and long-range *Src* activation

ORIGIN OF ONCOGENES

A hypothetical scheme for the formation of a viral oncogene and a cellular oncogene from a cellular proto-oncogene is given in Figure 4.5. Different oncogenes encode different kinds of functions and their tumorigenicity follows different mechanisms. Some of the proto-oncogenes encode growth factors and others plasma membrane receptors for growth factors. Two *H-ras* proto-oncogenes are found in yeast. Yeast cells cannot grow if the two *H-ras* genes are eliminated, but a human H-ras gene inserted into the yeast DNA corrects the defect. In yeast, ras like proteins interact with the system that regulates the levels of cyclic AMP, the second messenger that mediates responses of all eukaryotic cells to altered extracellular environments (i.e., hormones). It is hoped that oncogenic viruses would be the primary cause of cancer. For full development of oncogenic potential, at least two mutational events are required. In most of the cases, these take place in different genes but in *c-Ha-ras* gene, these occur in the same gene, i.e., one at position 12 and other at position 2,719.

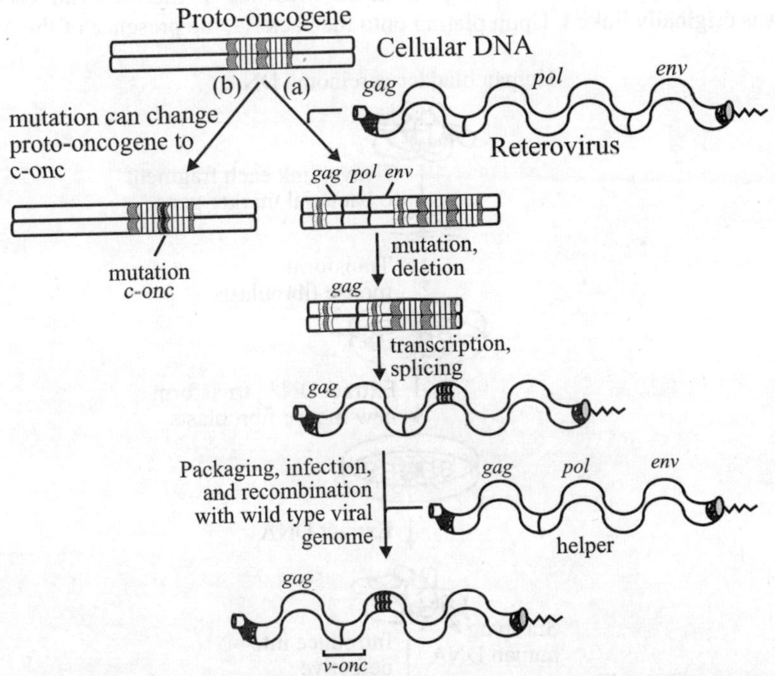

Figure 4.5 A hypothetical scheme for the formation of a viral oncogene (a) and a cellular oncogene; (b) from a cellular proto-oncogene

Unregulated Ras Activation Associated with Cancer

To investigate the unregulated Ras activation associated with cancer, Stites et al. (2007) developed and validated a mathematical model of Ras signaling. The model-based predictions and associated experiments help explain why only one of the two classes of activating *Ras* point mutations with in vitro transformation potential is commonly found in cancers. Model-based analysis of these mutants uncovered a system-level process that contributes to total Ras activation in cells. This predicted behavior was supported by experimental observations. They also used the model to identify a strategy in which a drug could cause a stronger inhibition on the cancerous Ras network than on the wild-type network. This system-level analysis of the oncogenic Ras network provides new insights and potential therapeutic strategies.

R. Furchgott, P. Murad and L. Ignarro were awarded Nobel Prize in 1998 for their discovery leading to the use of Viagra as an anti-impotency drug (Furchgott et al. 1980; Ignarro et al. 1987; Murad et al. 1978). This discovery has applications in treatment of cardio-vascular diseases, shock and possibly cancer as well as impotency.

IDENTIFICATION OF AN ONCOGENE

Method for identifying the DNA clone carrying a human oncogene is illustrated in Figure 4.6. DNA from human bladder tumor was extracted and cleaved. Each fragment was attached to a bacterial marker gene and then used to transform mouse cells. DNA from transformed cells is then cleaved and cloned in a defective phage that can only grow in the presence of the bacterial gene to which the human DNA was originally linked. Upon plating onto the bacteria, the presence of the bacterial gene is

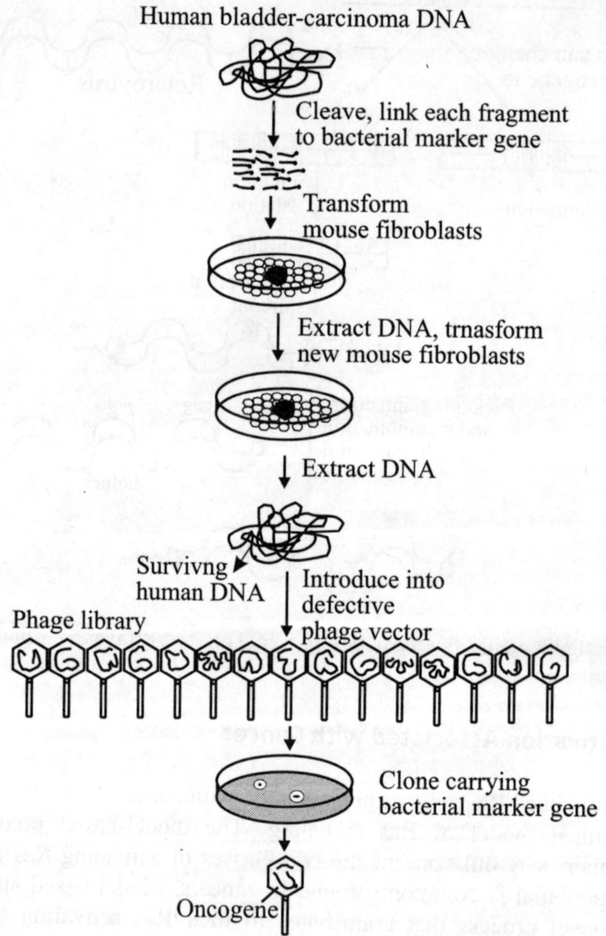

Figure 4.6 A method for identifying the DNA clone carrying a human oncogene

indicated by phage plaques, which are then tested for the oncogene. The identical oncogenes show that tumor formation results from a change in a small segment of the total genome.

'Sleeping Beauty' System for identifying Cancer Genes

Ancient jumping DNA found napping in fish has been revived and is being used to identify cancer genes in mice. But the benefits of this aptly named 'Sleeping Beauty' system could reach far beyond cancer (Weiser and Justice 2005). It can be used in human cells, or in any organism that is a model for human disease.

Annotating the Entire Cancer Genome

Studies by Greenman et al. (2007) provide first unbiased, large-scale analysis of DNA mutations across an array of cancers. They report more than 1,000 somatic mutations found in 274 Mb of DNA corresponding to the coding exons of 518 protein kinase genes in 210 diverse human cancers. There was evidence for 'driver' mutations contributing to the development of the cancers studied in approximately 120 genes. This will help in identifying the molecular mechanisms responsible for tumor initiation and progression (Haber and Settleman 2007).

Tumor-Progression Genes and Metastatic Genes

Some genes are involved in the development of new tumors; others specifically promote the dissemination of its cancerous cells to other organs (Christofori 2007). These genes are classified as 'tumor-progression genes' and 'metastatic genes' (Figure 4.7). A set of four genes seems to be required for both processes (Gupta et al. 2007). The epidermal growth factor ligand epiregulin (EGFR), the cyclo-oxygenase 2 (COX2), and the matrix metalloproteinases 1 and 2 (MMP1 and MMP2) genes when expressed in human breast cancer cells, collectively facilitate the assembly of new tumor blood cells, the release of tumor cells into the circulation, and the breaching of lung capillaries by circulating tumor cells to seed pulmonary metastasis. Epiregulin is a ligand for the EGFR and is essential for the growth, survival and progression of several types of cancer. The MMP-1 and MMP-2 participate in the formation of new blood vessels to supply tumors (angiogenesis), as well as in tumor-cell migration and invasion. The COX2 enzyme mediates wound healing and inflammatory responses.

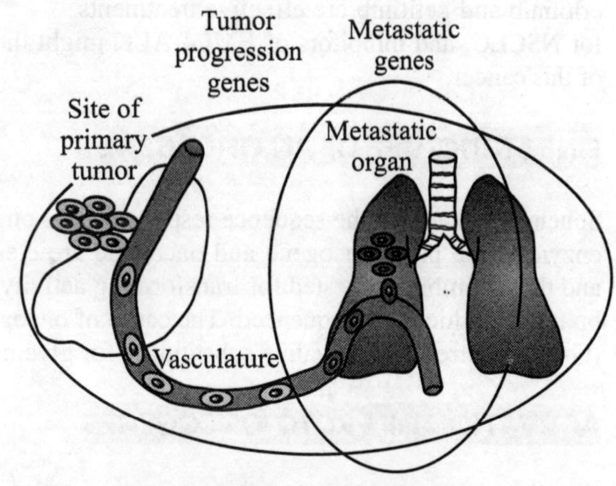

Figure 4.7 Role of different genes in cancer

DNA Sequence of Cancer Genomes

All cancers arise as a result of changes that have occurred in the DNA sequence of the genomes of cancer cells. Much has learnt about these mutations and the abnormal genes that operate in human cancers. We are, however, moving in an era in which it will be possible to obtain the complete DNA sequence of large number of cancer genomes (Stratton 2009). These studies will provide us with a detailed and comprehensive perspective on how individual cancers have developed.

Hybrid Gene Formation is Frequent in Blood-related Cancers

Mutations that cause potions of two genes to fuse together and form a hybrid gene are frequent in blood-related cancers. New findings implicate one such fusion gene in the most common type of lung cancer (Meyerson 2007). In case of non-small-cell lung cancer (NSCLC), a chromosomal rearrangement results in a fusion gene (EML4-ALK), a product of which is echinoderm microtubule-associated protein-like 4 (EML4)-anaplastic lymphoma kinase (ALK) (Soda et al. 2007) (Figure 4.8). This fusion protein functions as an activated tyrosine kinase, and thus might stimulate the EGFR-mediated signaling-pathway. Tyrosine kinase inhibitors such as erlotinib and gefitinib are effective treatments for NSCLC, and inhibitors of EML4-ALK might therefore be equally promising drugs in the therapy of this cancer.

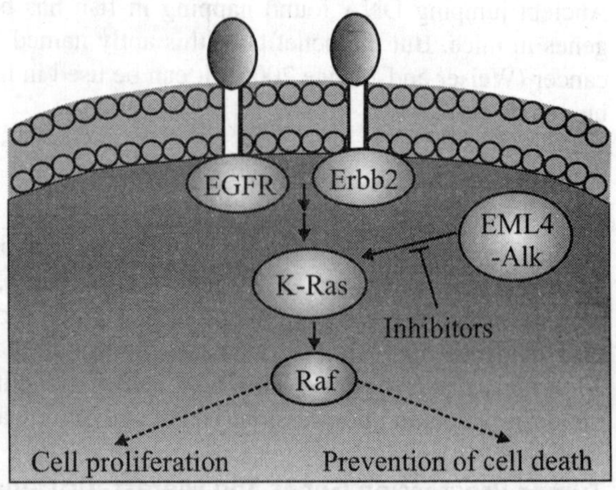

Figure 4.8 The EML4-ALK fusion protein and lung cancer

FINE STRUCTURE OF AN ONCOGENE

Scheme for locating the sequence responsible for oncogenicity is shown in Figure 4.9. With restriction enzymes, the proto-oncogene and oncogene are cleaved at the same site, the fragments recombined, and the recombinants tested for transforming activity. In this way, exact position was narrowed to 350-bp region, which was sequenced. The cause of oncogenic activity was found to be a single base change (G→A) that results in a valine substitution for glycine.

ACTIVATION OF PROTO-ONCOGENES

The current model of how a proto-oncogene may be activated to produce a tumor is given in Figure 4.10. A point mutation induced by a chemical or radiation carcinogen can cause a change in the protein, thereby initiating cancerous growth. Another possibility is the induction of a chromosomal rearrangement that places the proto-oncogene next to the regulatory region of an immunoglobulin gene. This will lead to the expression of the proto-oncogene inappropriately. If the proto-oncogene is amplified either as repeat segments within the chromosome or extrachromosomally, the gene product will be over produced. The final possibility is the retroviral transfer of a proto-oncogene picked up in another animal cell. Thus, transcription of proviral regulatory signals, not by the signals associated with the proto-oncogenes in the cell's genome. Abnormal regulation and/or abnormal product of v-onc expression yield the tumor phenotype.

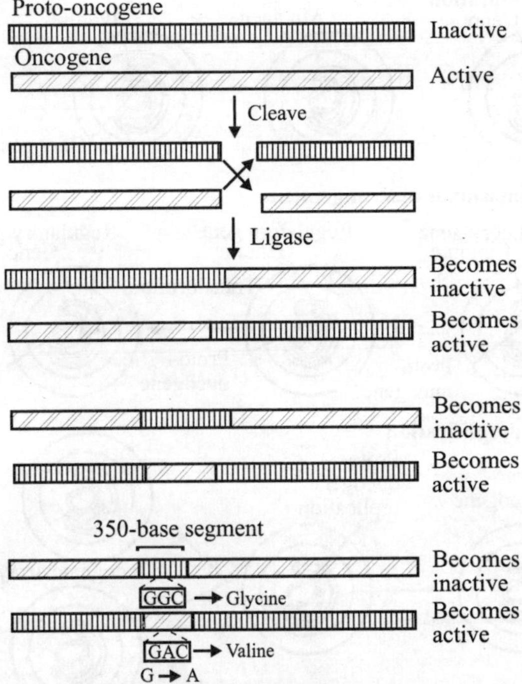

Figure 4.9 Location of a sequence responsible for oncogenicity

GENETIC ABNORMALITIES AND ABERRANT FEATURES OF CANCER CELLS

Pinpointing the genes involved in cancer will help chart a new course across the complex landscape of human malignancies. Collins and Barker (2007) present the following chronology of connection between genetic abnormalities and the aberrant features of cancer cells.

1890-1914 : Abnormal chromosome distribution during cell division suggests a role in malignancy.

1950s-1960s : Tumor viruses cause cancer by injecting their genes into cells; 1960: First genetic defect associated with a specific cancer. Philadelphia chromosome is discovered in chronic myelogenous leukemia (CML) cells.

1966 : P. Rous was awarded Nobel Prize in for discovering tumor viruses (Rous 1911).

1976 : Discovery that *src*, a non-viral gene, found in animal cells, can cause cancer.

1979 : *p53* gene, later found to be the most frequently mutated gene in human cancer, discovered.

1981 : *H-RAS* is the first human oncogene discovered. Alteration in this gene is cancer-promoting.

1983 : Altered methylation of DNA, suspected to affect gene activation, found in cancer cells.

1986 : First tumor-suppressor gene, *RB1*, is identified.

1987 : Fused gene *BRC-ABL* in Philadelphia chromosome is found to cause chronic myelogenous leukemia.

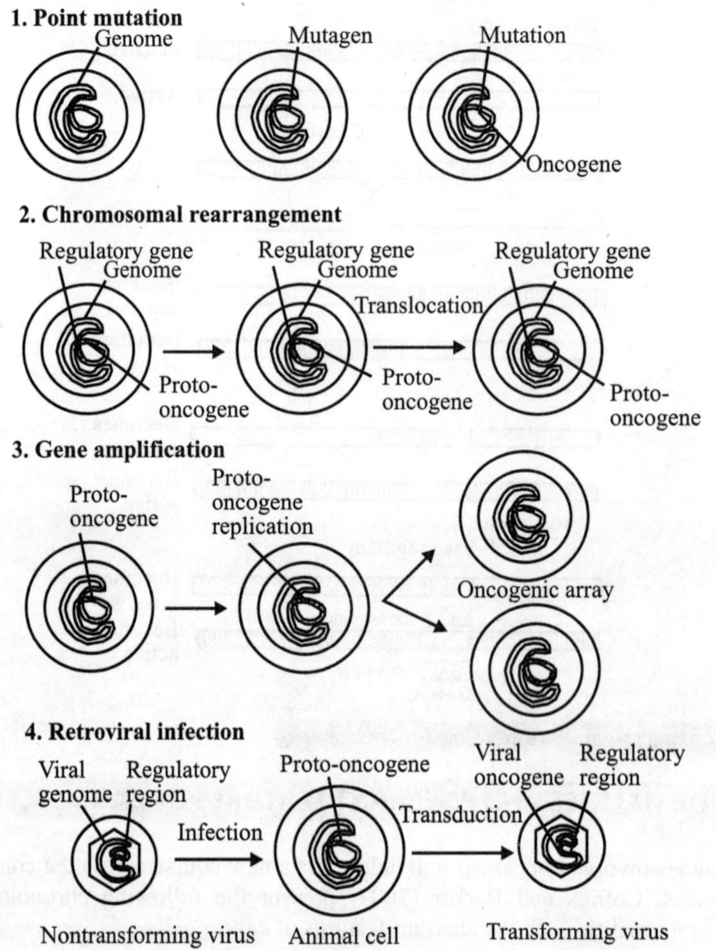

Figure 4.10 Proposed mechanisms of activation of proto-oncogenes

1990 : Model of multistep tumor genesis clarifies the role of accumulated gene changes in cellular transformation to malignancy.

1990 : Human Genome Project begins.

1993 : The first therapy developed to target a known gene-based cause of cancer.

1999 : Gene activity profiles are first shown to distinguish between cancer types and to predict chemotherapy response.

2002 : Tumor genome survey discovers a mutation in *B-RAF* gene common to 70 per cent of melanomas.

2003 : Human Genome Project is completed;

2005 : The Cancer Genome Atlas (TCGA) project is announced by the National Institute of Health; TCGA names three cell types for sequencing and genetic analysis.

2007-2010 : TCGA will collect and analyze tumor samples and data will be made available to the wider research community.

p53 IS THE MOST FREQUENTLY MUTATED GENE IN HUMAN CANCERS

Tumor-Suppressor Protein is Nuclear Transcription Factor

The principal tumor-suppressor protein, p53, accumulates in cells in response to DNA damage, oncogene activation and other stresses (Green and Kroemer 2009). It acts as a nuclear transcription factor that transactivates genes involved in apoptosis, cell cycle regulation and numerous other processes. An emerging area of research unravels additional activities of p53 in the cytoplasm, where it triggers apoptosis and inhibits autography. These previously unknown functions contribute to the mission of p53 as a tumor suppressor.

Methylation of p53 Regulates itself

p53 is a tumor suppressor that regulates the cellular response to genotoxic stresses. p53 is a short-lived protein and its activity is regulated by stabilization via different post-translational modifications. Churkov et al. (2004) reported a novel mechanism of p53 regulation through lysine methylation by Set9 methyltransferase. Set9 specifically methylates p53 at one residue within the carboxyl-terminus regulatory region. Methylated p53 is restricted to the nucleus and the modification positively affects its stability. Set9 regulates the expression of p53 target genes in a manner dependent on the p53-methylation site. The crystal structure of a ternary complex of Set9 with a p53 peptide and the cofactor product S-adenosyl-L-homocysteine (AdoHcy) provides the molecular basis for recognition of p53 by this lysine methyltransferase.

p53 and Ras Cooperate to Transform Normal Cells into Cancer Cells

Transformation of normal cells into cancer cells entails concerted changes in the expression of many genes. Identifying which of those genes are crucial will provide insight into the mechanism underlying malignancy (Luo and Elledge 2008). Oncogenic mutations in the transcription factor p53 and in the small GTPase protein RAS – which individually have limited effects on promoting cancer – cooperate to transform normal cells into cancer cells (Figure 4.11).

p53 is Involved in MicroRNA Processing

MicroRNAs (miRNAs) have emerged as key post-transcriptional regulators of gene expression, involved in diverse physiological and pathological processes. Although miRNAs can function as both tumor suppressors and oncogenes in tumor development, a widespread downregulation of miRNAs is commonly observed in human cancers and promotes cellular transformation and tumorigenesis. This indicates an inherent significance of small RNAs in tumor suppression. However, the connection between tumor suppressor networks and miRNA biogenesis machineries has not been investigated in depth. Suzuki et al. (2009) show that a central tumor suppressor, p53, enhances the post-transcriptional maturation of several miRNAs with growth-suppressive function, including miR-16-1, miR-143 and miR-145, in response to DNA damage. In HCT116 cells and human diploid fibroblasts, p53 interacts with the Dorsha processing complex through the association with DEAD-box RNA helicase p68 (also known as DDX5) and facilitates the processing of primary miRNAs to precursor miRNAs. They also found that transcriptionally inactive p53 mutants interfere with a functional assembly between Dorsha complex and p68, leading to attenuation of miRNA processing activity. These findings suggest that

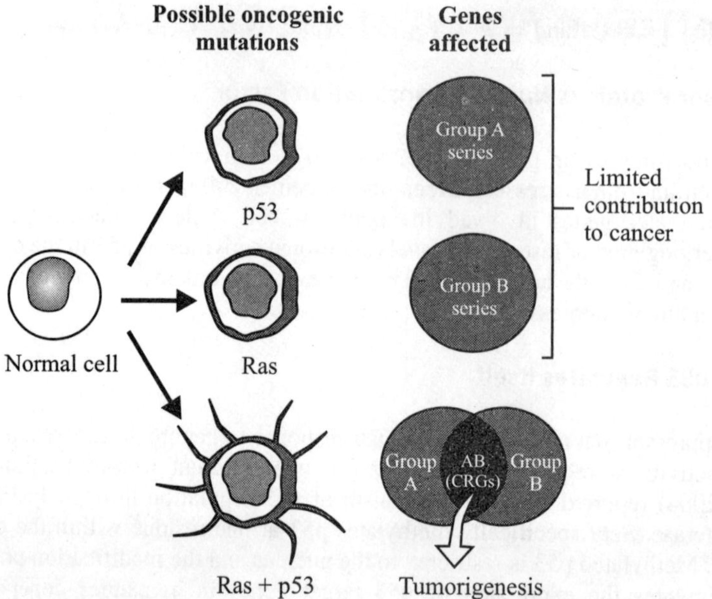

Figure 4.11 Cooperation response genes

transcription-independent modulation of miRNA biogenesis is intrinsically embedded in a tumor suppressive program governed by p53. This study reveals a previously unrecognized function of p53 in miRNA processing, which may underlie key aspects of cancer biology.

p53 is Regulated by Methylation on Lysine 370

Specific sites of lysine methylation on histone correlate either activation or repression of transcription. The tumor suppressor p53 is one of only a few non-histone proteins known to be regulated by lysine methylation. Huang et al. (2006) report a lysine methyltransferase, Smyd2, which methylates a previously unidentified site, Lys 370, in p53. This methylation site in contrast to the known site Lys372, is repressing to p53-mediated transcriptional regulation. Smyd2 helps to maintain low concentrations of promoter-associated p53. They show that reducing Smyd2 concentration by short interfering RNA enhances p53-mediated apoptosis. They find that Set9-mediated methylation of lysine 372 inhibits Smyd2-mediated methylation of Lys 370, providing regulatory cross-talk between post-translational modifications. In addition, they show that inhibitory effect of Lys 370 methylation on Lys 370 methylation is caused, in part, by blocking the interaction between p53 and Smyd2. Thus similar to histones, p53 is subject to both activating and repressing lysine methylation. These results also predict that Smyd2 may function as a putative oncogene by methylating p53 and repressing its tumor suppressive function.

Restoration of p53 Activity Results in Tumor Regression

Most malignant tumors disrupt the p53 signaling pathway in order to grow and survive. Although many genes in addition to p53 are mutated in tumors, recent studies by Ventura et al. (2007) and Xues et al. (2007) suggest that restoring p53 function alone is sufficient to cause regression of several different tumor types in mice and thus might represent a potent therapeutic strategy to treat certain human cancers. Restoration of p53 activity results in tumor regression but add the sobering caveat that tumors may be able to quickly generate resistance by finding other ways to disrupt the p53 pathway (Kastan 2007). Pathways to p53 activation are shown in Figure 4.12.

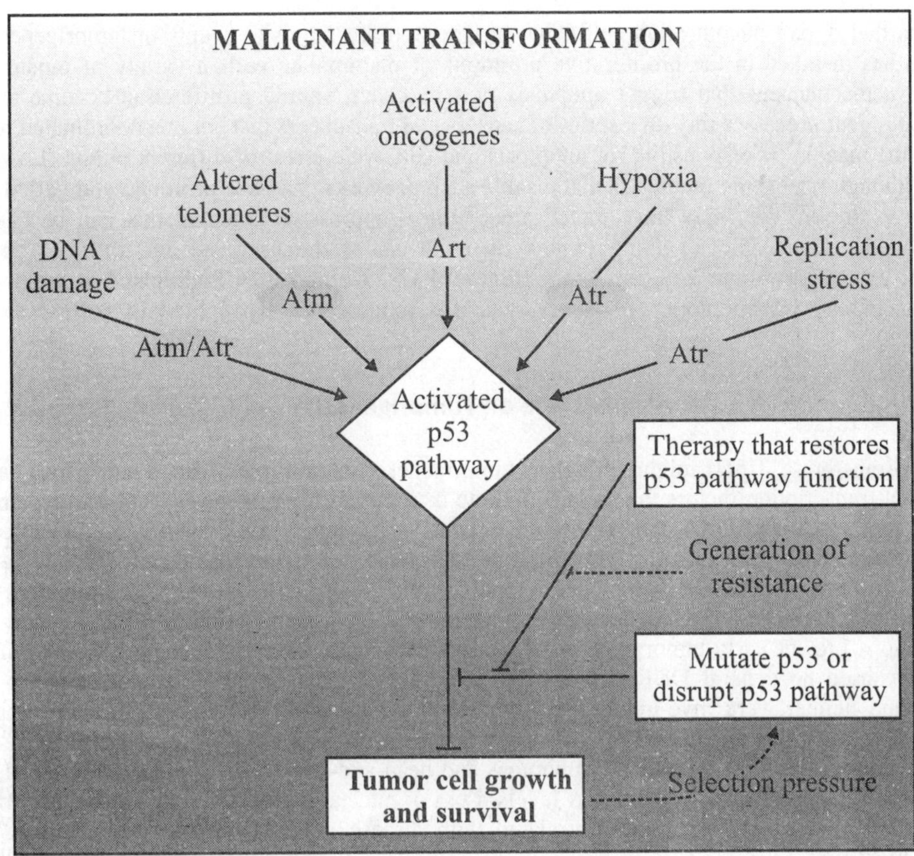

Figure 4.12 Pathways to p53 activation

EVOLUTION OF TUMOR SUPPRESSOR MECHANISMS

Meiosis in the female germline of mammals is distinguished by a prolonged arrest in prophase of meiosis I between chromosome recombination and ovulation. DNA damage detection is thought to involve p53, a central tumor suppressor in mammals. Suh et al. (2006) show that p53 homolog p63, and specifically the TAp63 isoform, is constitutively expressed in female germ cells during meiotic arrest and is essential in the process of DNA damage-induced oocyte death not involving p53. They also show that DNA damage induces both the phosphorylation of p63 and its binding to p53 cognate DNA sites and that these vents are linked to oocyte death. Their data support a model whereby p63 is the primordial member of the p53 family and it is in the conserved process of monitoring the integrity of the female germline, whereas the functions of p53 are restricted to vertebrate somatic cells for tumor suppression. These findings have implications for understanding female germline fidelity, the regulation of fertility and the evolution of tumor suppressor mechanisms.

Tumor-Suppressive Mechanisms may Trigger Apoptosis

Mutations that derive uncontrolled cell cycle progression are requisite events in tumorigenesis. But evolution has installed in the proliferative programs of mammalian cells a variety of innate tumor-suppressive mechanisms that trigger apoptosis or senescence, should proliferation become aberrant. These contingent processes rely on a series of sensors and transducers that act in a coordinated network to target the machinery responsible for apoptosis and cell cycle arrest of different points (Lowe et al. 2004). Although oncogenic mutations that disable such networks can have profound and varied effects on tumor evolution, they may have intact latent tumor-suppressive potential that can be harnessed therapeutically. Oncogenic signaling targets many levels of the extrinsic and intrinsic apoptotic programs, as well as some key regulators (Figure 4.13). Components highlighted in bold can be downregulated by pro-apoptotic oncogenes, whereas components highlighted in normal are often upregulated.

Multiple Influences of an Acyltransferase on Tumorigenesis

The acetyltransferase Tip60 might influence tumorigenesis in multiple ways. First, Tip61 is a co-regulator of transcription factors that either promote or suppress tumorigenesis, such as myc and p53. Second, Tip60 modulates DNA damage response (DDR) signaling, and a DDR triggered by oncogenes can counteract tumor progression. Using *Eμ-myc* transgenic mice that are heterozygous for a *Tip60* gene (*Htatip*) knockout allele (denoted as Tip60$^{+/-}$ mice), Gorini et al. (2007) show that Tip60 counteracts myc-induced lymphomagenesis in a haploinsufficient manner and in a time window that is restricted to a pre- or early-tumoral stage. *Tip60* heterozygosity severely impaired the myc-induced DDR but caused no general DDR defect in B cells. Myc- and p53-dependent transcription is not affected, and neither were myc-induced proliferation, activation of the ARF-p53 tumor suppressor pathway or the resulting apaot、otic response. They found that human *Tip60* gene (*Htatip*) is a frequent target for mono-allelic loss in human lymphomas and head-and-neck and mammary carcinomas, with concomitant reduction in messenger RNA levels. Loss of nuclear TIP60 staining was demonstrated in mammary carcinomas. These events correlated with disease grade and frequently concurred with mutation of p53.

Transcriptional Repressor Pokemon is a Critical Factor in Oncogenesis

Aberrant transcriptional repression through chromatin remodeling and histone deacetylation has been postulated to represent a driving force underlying tumorigenesis because histone deacetylation inhibitors have been found to be effective in cancer treatment. However, the molecular mechanisms by which transcriptional derepression would be linked to tumor suppression are poorly understood. Maeda et al. (2005) identify the transcriptional repressor Pokemon (encoded by the *Zbtb7* gene) is a critical factor in oncogenesis. Mouse embryonic fibroblasts lacking Zbtb7 are completely refractory to oncogene-mediated cellular transformation. Conversely, *Pokemon* overexpression leads to overt oncogenic transformation both in vitro and in vivo in transgenic mice. Pokemon can specifically repress the transcription of the tumor suppressor gene ARF through direct binding. They find that Pokemon is aberrantly overexpressed in human cancers and that its expression levels predict biological

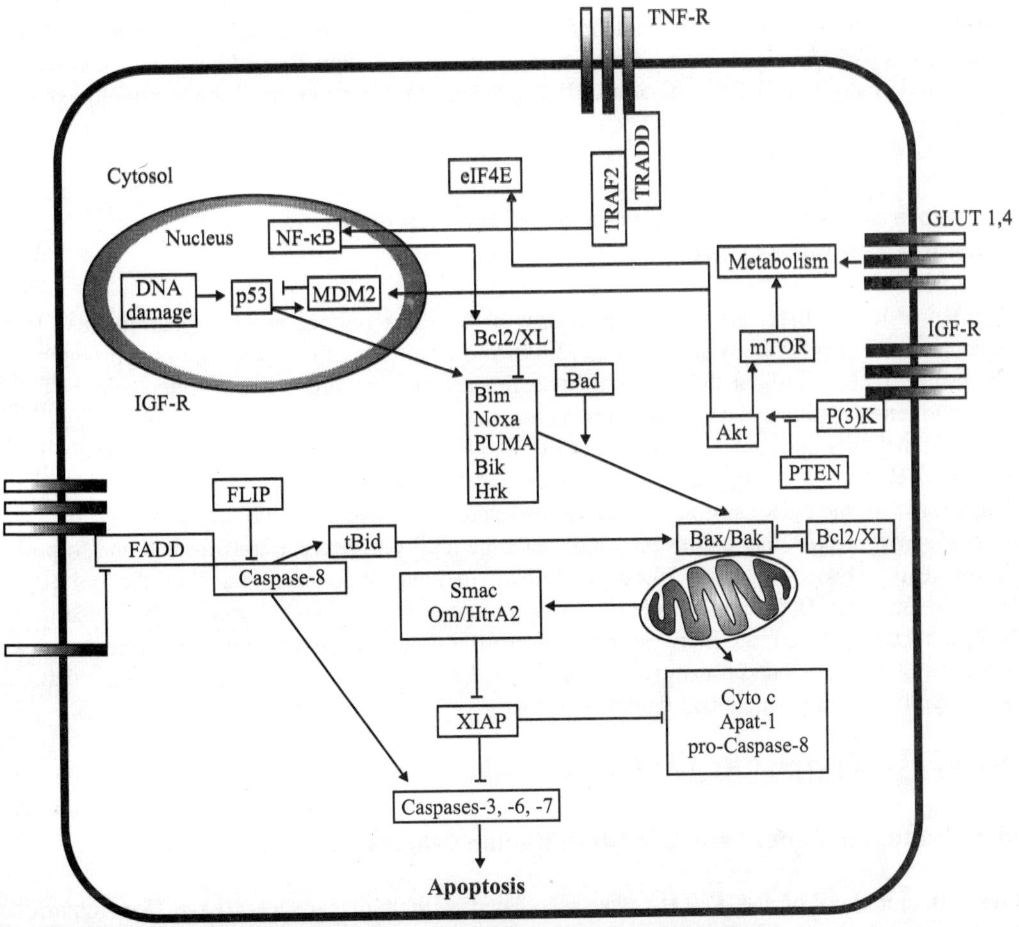

Figure 4.13 Oncogenic signaling targets many levels of the apoptotic machinery

behavior and clinical outcome. Pokemon's critical role in cellular transformation makes it an attractive target for therapeutic intervention. A model proposed for the role of Pokemon in oncogenesis, as suggested by Maeda et al. (2005), is given in Figure 4.14. Abrogation of Pokemon function leads to cell cycle arrest, cellular senescence and apoptosis, and hence could be exploited for cancer therapy.

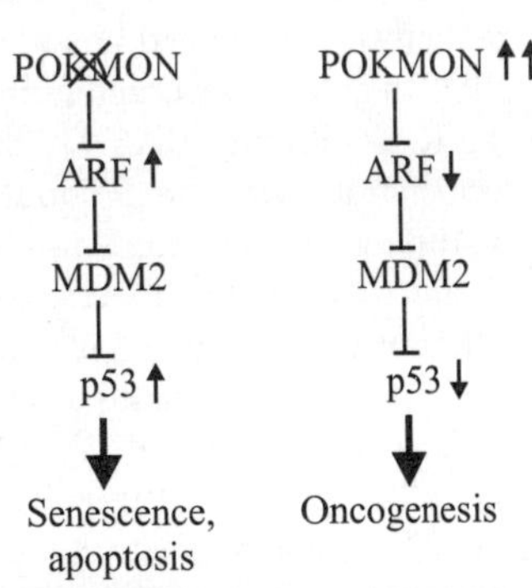

Figure 4.14 Model proposed for the role of Pokemon in oncogenesis

RNA POLYMERASES AND CANCER

Oncogenic Stimuli and Tumor Suppressors Control RNA Polymerase III-Dependent Transcription

Regulating the production of RNAs involved in protein synthesis can induce the transformation of cells to an oncogenic state (Johnson and Johnson 2008). Oncogenic stimuli and tumor suppressors control

TFIIB, and RNA polymerase III-dependent transcription (Figure 4.15). Positive and negative regulators of TFIIB may alter the expression or function of TBP or Brf1. Alterations in TFIIB produce differential changes in RNA polymerase III-dependent transcription, resulting in altered expression of specific products. This results in selective changes in protein translocation, contributing to oncogenic transformation of a cell.

Overexpression of a Transcription Factor of the RNA Polymerase III Apparatus Transforms Cells

Overexpression of Brf1, a transcription factor of the RNA polymerase III apparatus, can transform cells in vivo (Berns 2008). Marshall et al. (2008) show that one of the transcriptional products of RNA polymerase III, the initiator $tRNA^{Met}$, mediates this effect, revealing an unexpected role for this tRNA in tumorigenesis. Figure 4.16 shows translation initiation as a control node in tumorigenesis (Berns 2008). Translation initiation factors such as eIF2 and eIF4, the structure of mRNAs, and the translation initiator $tRNA_i^{Met}$ are important determinants of the translational control of protein synthesis. Alterations in their expression or activation state can promote tumorigenesis by the selective translation of mRNAs that confer on tumor cells the ability to become transformed and to proliferate. Components shown with thick ovals are known to have oncogenic potential; the pol III-specific transcription factor Brf1 and the pol III transcriptional product $tRNA_i^{Met}$ now can be added to this list. Brf1 induction caused an increase in cell proliferation and oncogenic transformation, whereas depletion of Brf1 impeded transformation (Marshall et al. 2008). The data also indicate that elevated tRNA synthesis can promote cellular transformation.

GENES ASSOCIATED WITH CANCERS

Identification of Genes Associated with Human Cancers

Over 30 mutations of the *B-RAF* gene associated with human cancers have been identified, the majority of which are located within the kinase domain. Wan et al. (2004) show that of 22 *B-RAF* mutants analyzed, 18 have elevated kinase activity and signal to ERK in vivo. Surprisingly, three mutants have reduced kinase activity towards MEK in vitro, but by activating C-RAF in vivo, signal to ERK in cells. The structure of wild-type and oncogenic V599EB-RAF kinase domains in complex with the RAF inhibitor BAY43-9006 show that the activation segment is held in an active conformation by association with the P loop. The clustering of most mutations to these two regions suggests that disruption of this interaction converts B-RAF into its active conformation. The high activity mutants signal to ERK by directly phosphorylating MEK, whereas the impaired activity mutants stimulate MEK by activating endogenous C-RAF, possibly via an allosteric or transphosphorylation mechanism. Activating and impaired activity B-RAF mutants stimulate ERK signaling through distinctive mechanisms (Figure 4.17). (A) The activated mutants can directly phosphorylate and activate MEK and although they also activate C-RAF, their ability to signal to ERK does not require this pathway. (B) The impaired activity *B-RAF* mutants cannot stimulate efficient activation of MEK, but can stimulate C-RAF activity, which then activates MEK. Their ability to signal to ERK is therefore dependent on C-RAF protein. Ngo et al. (2006) describe an approach for identifying genes that become essential for the survival of cancer cells.

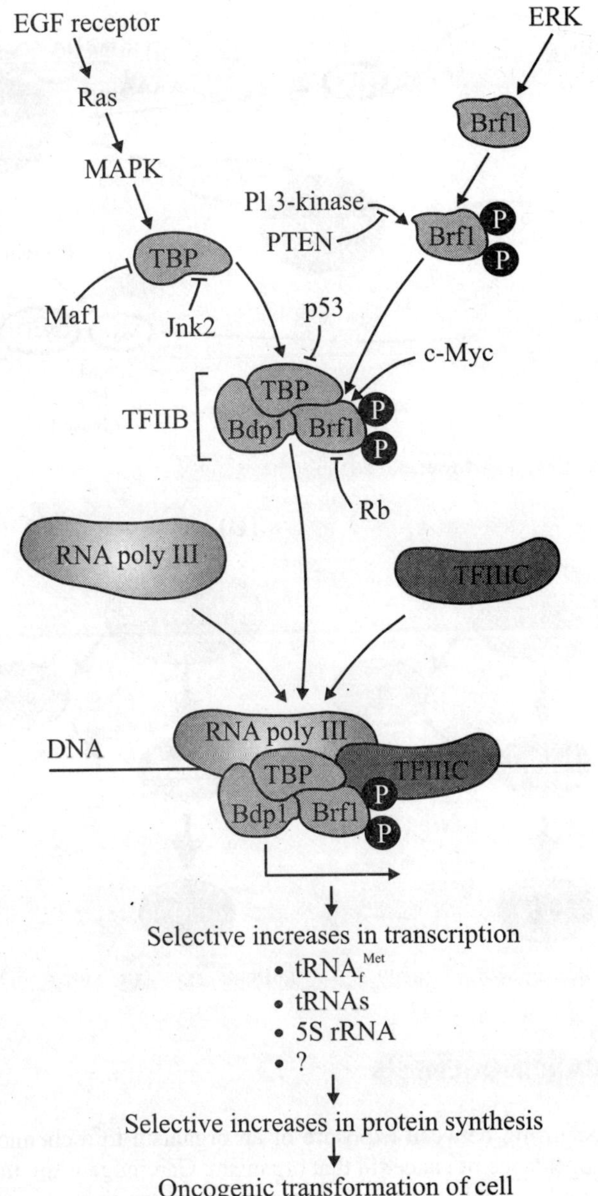

Figure 4.15 Metabolic consequences

Tumor-Related Gene Mutations Reduce Fitness

Tumor cells tend to carry many gene mutations, but at a potential cost to their overall fitness. Studying the interactions between genes on a large scale could be a way of identifying the chinks in the tumor cell armor (Kaelin 2006).

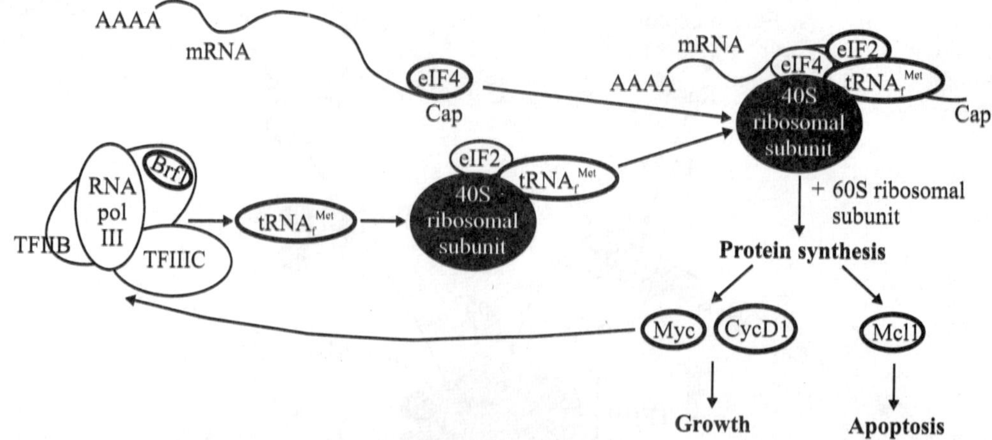

Figure 4.16 Translation initiation as a control node in tumorigenesis

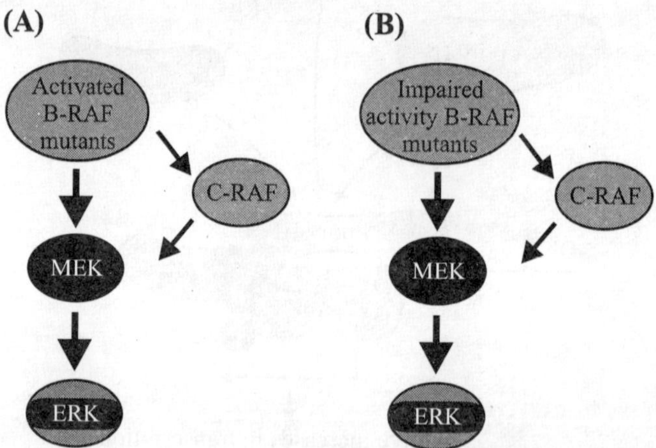

Figure 4.17 Activating and impaired activity B-RAF mutants stimulate ERK signaling through distinctive mechanisms

PALL'S MODEL OF CARCINOGENESIS

There is long latency occurring between exposure of an organism to a chemical carcinogen or to an ionizing radiation and appearance of cancer in that organism. Carcinogens are thought to act as somatic cell mutagens and no long lag would be expected between mutation induction and expression. Cancer-causing viruses are assumed to carry cancer-causing oncogenes, the gene product of which is required for neoplastic transformation. The non-transformed host cell carries a similar or identical gene, a proto-oncogene, which normally fails to produce neoplastic transformation in that case. Evidence to this comes from sarcoma virus. Amount of gene product produced by a proto-oncogene is insufficient to produce transformation but the retroviral oncogene produces much larger amount of product that it is sufficient to transform the host cell. Pall (1981) proposed a model that induced carcinogenesis is due to the production of tandem duplications of a proto-oncogene in the cell genome, producing 5-10 fold increase of the proto-oncogene product. The initial event in carcinogenesis is proposed to be a mutation that produces a single tandem duplication of a proto-oncogene. This should produce a 50 per cent increase in the gene product in a diploid cell, insufficient to transform that cell. However, after DNA replication has occurred, two sister chromatids will be formed each of which carries a tandem duplication (Figure 4.18). This cell or one of its progeny is supposed to undergo unequal sister chromatid crossing-over, generating a chromatid with three tandem genes and one with one. After subsequent rounds of cell division and DNA replication, additional unequal sister chromatid exchanges may occasionally occur generating still higher duplications, until the minimum amount of gene product is synthesized to transform the cell. This model of gene amplification of proto-oncogene requires a series of unequal crossing-over that would be relatively rare and each must occur in a different cycle. Thus considerable time would always elapse between exposure to carcinogen and the appearance of the transformed cell.

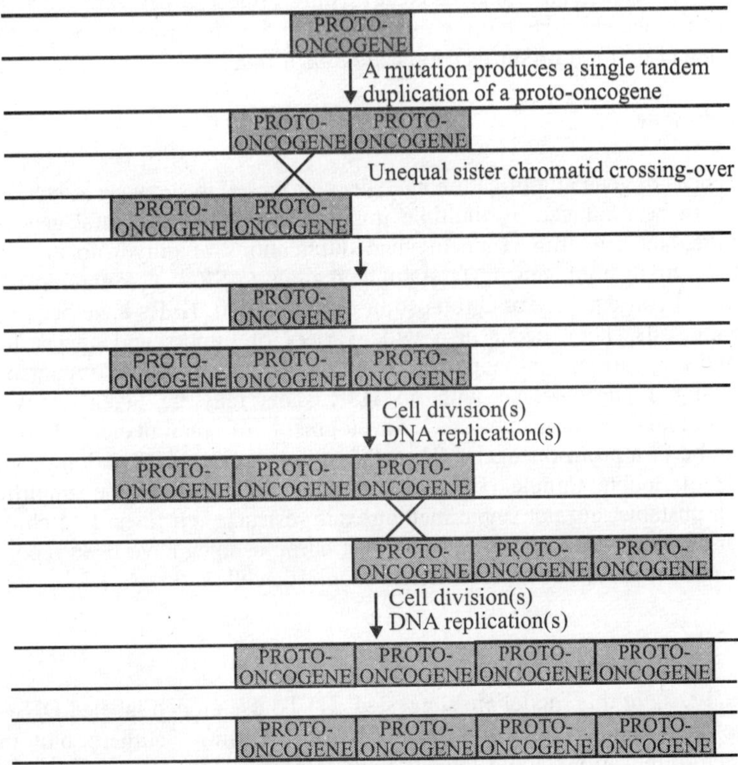

Figure 4.18 Unequal sister chromatid exchange leading to gene amplification of proto-oncogenes (Redrawn, with permission, from Pall 1981. Proc. Natl. Acad. Sci. USA 78: 2465-8)

Oncogene Evidence

There exist properties of oncogenes in the vertebrate genome. Several of the viral oncogenes have been studied by nucleic acid hybridization and found to be similar or identical to DNA sequences in the host genome. Most of these virulent transforming retroviruses are defective, having replaced a portion of the required genes for virus replication by oncogenic sequences. Both hybridization studies and the isolation of new defective virulent transforming viruses on growth of non-defective retroviruses led to the proposal that each defective virus was formed by replacement of some viral genetic information by an oncogenic DNA sequence from the host cell genome. Several of the proto-oncogenes have been shown to be transcribed in non-transformed cells. Example of proto-oncogenes is the best in the case of avian sarcoma virus oncogene, *src*, and its homolog proto-oncogene, *proto-src*. *Proto-src* is present, transcribed and translated in the non-transformed cell. The translation product of *src* gene is very similar in structure, immunological activity and enzymatic activity to *src* gene product in the avian sarcoma virus-transformed cell. The *proto-src* gene can be transferred to the avian sarcoma virus genome by recombination into a mutant deleted for *src* and, when so transferred, appears to be capable of high-level transcription and translation of transforming the host cell. All except possibly extreme 5'-end of the *src* gene in some of these recombinants is derived from the *proto-src* sequence of the host cell. Thus it appears that all or almost all of the genetic information in this oncogene is derived from the host cell genome. Non-transformed cells have only almost 1 per cent as much *src* type gene product as the transformed cells, supporting the suggestion that the level of this gene product is crucial in determining whether or not it produces neoplastic transformation.

Cytogenetic Evidences

Cytogenetic evidences of gene amplification have been proposed in stepwise induction of methotrexate which is believed to be produced by multiple implications of the structural gene for dihydrofolate reductase. Methotrexate is a drug. Tandem gene duplications of dihydrofolate reductase gene are usually found to occur in homogeneously staining regions (HSRs) of a chromosome. This may be viewed as cytological consequence of chromosome amplification. HSRs have been widely reported to occur in malignant cells. They occur in a wide variety of tumors and appear to be common in neoplastic cells and very rare in non-neoplastic cells. In one recent study, 5/16 human tumors had most or all cells showing a chromosome with an HSR. This may be taken as evidence that gene amplification commonly accompanies the development of malignancy. It is suggested that amplification may be of a proto-oncogene. The other cytogenetic manifestation of gene amplification is the occurrence of double minute (DM) chromosomes. They carry the amplified dihydrofolate reductase genes in unstable, but not stable methotrexate resistant cell lines. DM chromosomes appear to be derived from HSRs in at least some cells. DM chromosomes have been reported to occur in a wide variety of neoplastic cells but are rare in non-neoplastic cells.

Experimental Tests

Three experimental tests of this model are suggested: (1) To use cloned labeled DNA probes to detect changes in proto-oncogene DNA in transformed cells by using Southern blot method. Changes produced by tandem duplication should include both an increase in intensity of restriction fragments included in such duplications and possible production of new fragments showing DNA contained in two adjacent duplications. (2) To use above labeled probes to perform in situ hybridization of neoplastic cells possessing HSRs. It is predicted in this model that homogeneously staining regions will be labeled by probes. (3) To isolate DM chromosomes from tumors and to clone DNA segments from them. Such cloned segments could then be used as probes to determine if other similar tumors have experienced amplification for the same section of the genome. Promoters of carcinogenesis in "gene amplification" model are predicted to act by stimulating both cell division and sister chromatid exchange. A known promoter (phorbol ester) has been shown to stimulate sister chromatid exchange. Certain antipromoters inhibit this simulation. Many complete carcinogens that act as initiators and promoters stimulate sister chromatid exchange (SCE). Bloom syndrome individuals, who are predisposed to cancer have high frequencies of SCE.

Applicability of the Model

Three specific categories of malignancy must be considered to be unlikely candidates for generation by this mechanism. Malignancies characterized by specific deletions or translocations. In these cases, this model will be applicable only if it is assumed that the processes leading to deletion or translocation might also generate tandem duplication next to the point of chromosome break. Virus-induced transformation also does not fit the simple gene amplification model. Tumors can be of non-genetic origin also. For example, Murine teratocarcinoma cells are genetically normal but their neoplastic transformation is due to developmental aberrations of gene expressions. Tandem duplications can revert back at high frequencies. The extra gene copies are readily deleted by crossing-over, or loss of DM chromosomes or both. Thus it is possible that teratocarcinoma was generated by gene duplication and that genetically normal progeny cells were produced from the tumor cells by loss of extra gene copies. If the model can be confirmed by this or other tests, it should lead to important practical approaches to cancer treatment. Initiation could be detected by developing mutagen screening systems for specifically measuring rates of induction of large tandem duplications. Promoters should be screened via their effects on SCEs. Oncoproteins perform several normal cellular functions. For example, *myc*, *myb*, *fos*, *jun*, *rel* and *erbA* are located in the nucleus. src is a membrane-associated protein. Proteins mos, abl and fps are located in the cytoplasm while erb, neu, fms, ras and mas are transmembrane proteins.

MORE ABOUT CARCINOGENESIS

Carcinogenesis Proceeds by Misappropriating Homeostatic Mechanism

Cancer is increasingly being viewed as a stem cell disease, both in its propagation by a minority of cells with stem-cell-like properties and in its possible derivation from normal tissue stem cells. But stem cell activity is tightly controlled, raising the question of how normal regulation might be subverted in carcinogenesis (Beachy et al. 2004). The long-known association between cancer and chronic tissue injury, and more recently appreciated roles of Hedgehog and Wnt signaling pathway in tissue regeneration, stem cell renewal and cancer growth together suggest that carcinogenesis proceeds by misappropriating homeostatic mechanisms that govern tissue repair and stem cell self-renewal.

Self-Renewal of Leukemia Stem Cells

Rare cells with the properties of stem cells are integral to the development and perpetuation of leukemias. A defining characteristic of stem cells is their capacity to self-renew, which is markedly extended in leukemia stem cells. The underlying molecular mechanisms, however, are largely unknown. Viale et al. (2009) demonstrate that expression of the cell cycle inhibitor p21 is indispensable for maintain self-renewal of leukemia stem cells. Expression of leukemia-associated oncogenes in mouse hematopoietic stem cells (HSCs) induce DNA damage and activates a p21-dependent cellular response, which leads to reversible cell cycle arrest and DNA repair. Activated p21 is critical in preventing excess DNA damage accumulation and functional exhaustion of leukemia stem cells. These data unravel the oncogenic potential of p21 and suggest that inhibition of DNA repair mechanisms might function as potent strategy for the eradication of the slowly proliferating leukemia stem cells.

Stem Cells have Potential to Turn Malignant

A dark side of stem cells – their potential to turn malignant – is at the root of a handful of cancers and may be the cause of many more. Eliminating the disease could depend on tracking down and destroying these elusive killer cells (Clarke and Becker 2006).

Cooperation Response Genes Contribute to Tumor Formation

Malignant cell transformation requires the cooperation of a few oncogenic mutations that cause substantial reorganization of many cell features and induce complex changes in gene expression patterns. This cooperation depends on synergistic modulation of downstream signaling circuitry. McMurray et al. (2008) show that a large proportion of genes controlled synergistically by loss-of-function. p53 and Ras activation are critical to the malignant state of murine and human colon cells. Notably, 14 out of 24 cooperation response genes were found to contribute to tumor formation.

RP80 Targets BRCA1 to Specific Ubiquitin Structures at DNA Damage

Mutations affecting the BRCT domains of the breast cancer-associated tumor suppressor BRCA1 disrupt the recruitment of this protein to DNA double-strand breaks (DSBs). The molecular structures at DSB recognized by BRCA1 are presently unknown. Sobhian et al. (2007) report the interaction of the BRCA1 BRCT domain with RAP80, a ubiquitin-binding protein. RAP80 targets a complex containing the BRCA1-BARD1 (BRCA1-associated ring domain protein 1) E3 ligase and the deubiquitinating enzyme (DUB) BRCC36 to MDC1-γH2AX-dependent lysine- and lysine-linked ubiquitin polymers at DSbs. These events are required for cell cycle checkpoint and repair responses to ionizing radiation, implicating ubiquitin chain recognition and turnover in the BRCA1-mediated repair of DSBs (Sobhian et al. 2007). Model for recruitment of a BRCA1-BARD1-BRCC36-RAP80 complex to sites of DNA damage by binding to non-K48-linked ubiquitin structures, as reported by Sobhian et al. (2007), is shown in Figure 4.19.

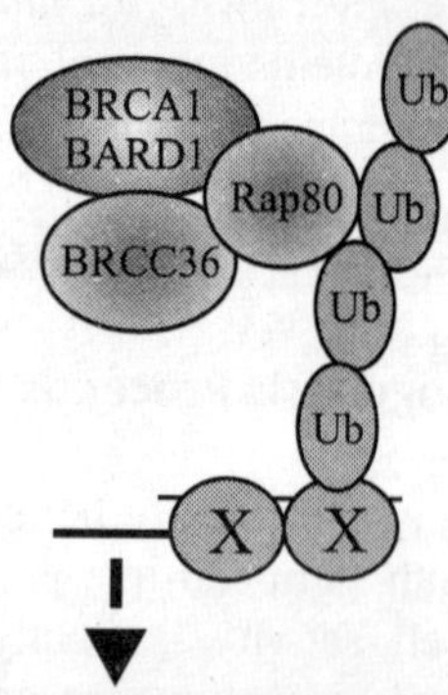

Figure 4.19 Model for recruitment of a BRCA1-BARD1-BRCC6-RAP80 complex to sites of DNA damage by binding to non-K48-linked ubiquitin structures

DNA Damage Checkpoint Response Activation and Oncogene-Induced Senescence

Early tumorigenesis is associated with the engagement of the DNA damage checkpoint response (DDR). Cell proliferation and transformation induced by oncogene activation are restrained by cellular senescence. It is unclear whether DDR activation and oncogene-induced senescence (OIS) are causally linked. Di Micco et al. (2006) show that senescence, triggered by the expression of an activated oncogene (*H-RasV12*) in normal human cells, is a consequence of activation of a robust DDR. Experimental activation of DDR abrogates OIS and promotes cell transformation. DDR and OIS are established after a hyper-replicative phase occurring immediately after oncogene expression. Senescent cells arrest with partly replicated DNA and with DNA replication origins having fired multiple times. In vivo DNA labeling and molecular DNA combing reveal that oncogene activation leads to augmented numbers of active replicons and to alterations in DNA replication fork progression. Oncogene expression does not trigger a DDR in the absence of DNA replication. Oncogene activation is associated with DDR activation in a mouse model in vivo. OIS results from the enforcement of a DDR triggered by oncogene-induced DNA hyper-replication. Events that occur after oncogene expression, as suggested by Di Micco et al. (2006), are shown in Figure 4.20.

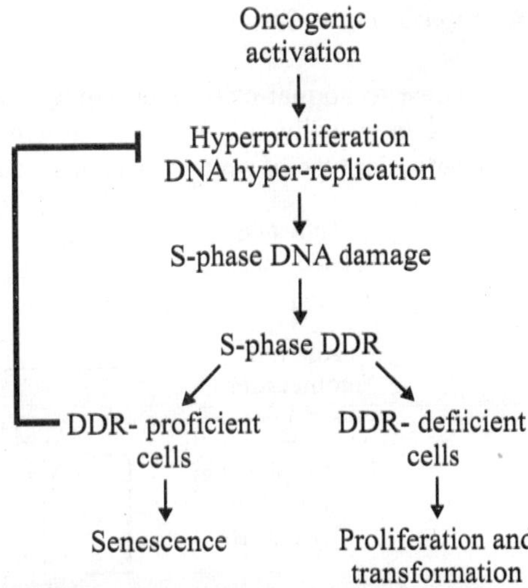

Figure 4.20 Proposed model for the events after oncogene expression

Oncogene-Associated Senescence and DNA Replication Stress

Recent studies have indicated the existence of tumorigenesis barriers that slow or inhibit the progression of preneoplasitc lesions to neoplasia. One such barrier involves DNA replication stress, which leads to activation of the DNA damage checkpoint and thereby to apoptosis or cell cycle arrest, whereas a second barrier is mediated by oncogene-induced senescence. The relationship between two barriers, if any, has not been elucidated. Bartkova et al. (2006) show that oncogene-associated senescence is associated with signs of DNA replication stress, including prematurely terminated DNA replication forks and DNA double-strand breaks, inhibiting the DNA double-strand break response kinase ataxia telangiectasia mutated (ATM) suppressed the induction of senescence and in a mouse model led to an increased tumor size and invasiveness. Analysis of human precancerous lesions further indicated that DNA damage and senescence markers cosegregate closely. Thus, senescence in human preneoplastic lesions is a manifestation of oncogene-induced DNA replication stress and, together with apoptosis, provides a barrier to malignant progression.

Oncogene-Induced DNA Replication Stress

Of all types of DNA damage, double-strand breaks (DSBs) pose the greatest challenge to cells. One might have, therefore, anticipated that a sizable number of DSBs would be incompatible with cell proliferation. Yet, recent experimental findings suggest that, in both precancerous lesions and cancers, activated oncogenes induce stalling and collapse of DNA replication forks, which in turn leads to formation of DSBs. This continuous formation of DNA DSBs contributes to the genomic instability that characterizes the vast majority of human cancers. In addition, pre-cancerous DSBs activate p53, which by inducing apoptosis or senescence raises a barrier to tumor progression (Halazonetis et al. 2008). Breach of this barrier by various mechanisms, most notably by p53 mutations that impair the DNA damage response pathway allows cancer to develop. Thus, oncogene-induced DNA damage may explain two key feature of cancer: genomic instability and the high frequency of p53 mutations. Oncogene-induced model for cancer development and progression is given in Figure 4.21. Genomic instability and tumor suppression are direct outcomes of oncogene-induced DNA replication stress that are both present from the beginning of cancer development, before the transition from precancerous lesion to cancer.

Aneuploidy and Cancer

In contrast to normal cells, aneuploidy – alterations in the number of chromosomes – is consistently observed in virtually all cancers. A growing body of evidence suggests that aneuploidy is often caused by a particular type of genetic instability, called chromosomal instability, which may reflect defects in mitotic segregation in cancer cells (Rajagopalan and Lengauer 2004). A better understanding of the molecular mechanisms leading to aneuploidy holds promise for the development of cancer drugs that target the process. Molecular mechanisms of chromosomal instability are depicted in Figure 4.22. There are multiple roads to aneuploidy. Several pathways within the cell cycle (indicated in bold letters) can be disrupted. Genes indicated in normal letters associated with these processes and structures have been found to be mutated or functionally altered in aneuploid cancers.

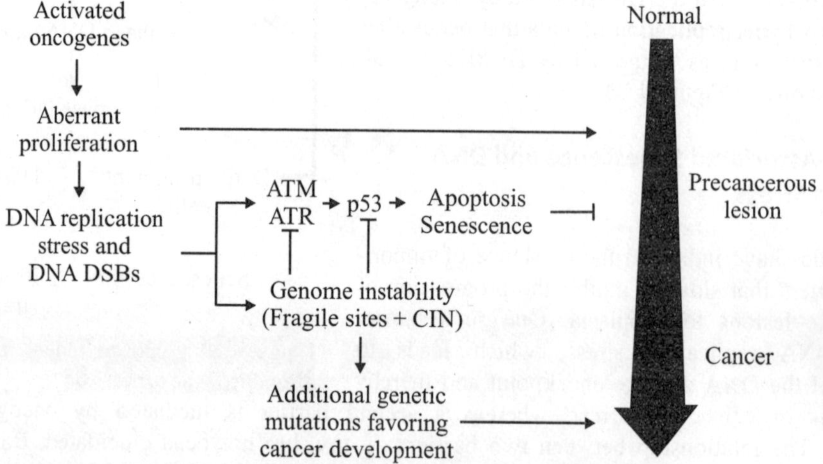

Figure 4.21 Oncogene-induced model for cancer development and progression

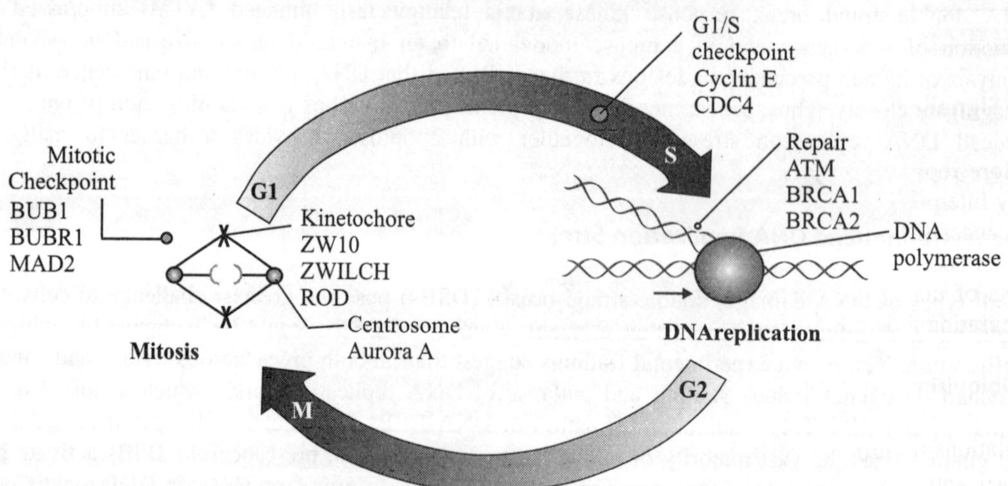

Figure 4.22 Multiple roads to aneuploidy

Current wisdom on the role of genes in malignancy may not explain some features of cancer, but stepping back to look at the bigger picture into cells reveals a view that just might by looking at exceptions to the current rule (Duesberg 2007). Aneuploidy could cause cancer. How aneuploidy could cause cancer is shown in Figure 4.23. Abnormal chromosome numbers in a cell create conditions that lead to further chromosome damage and disarray. With each new generation, resulting cells grow increasingly unstable and develop ever more malignant traits.

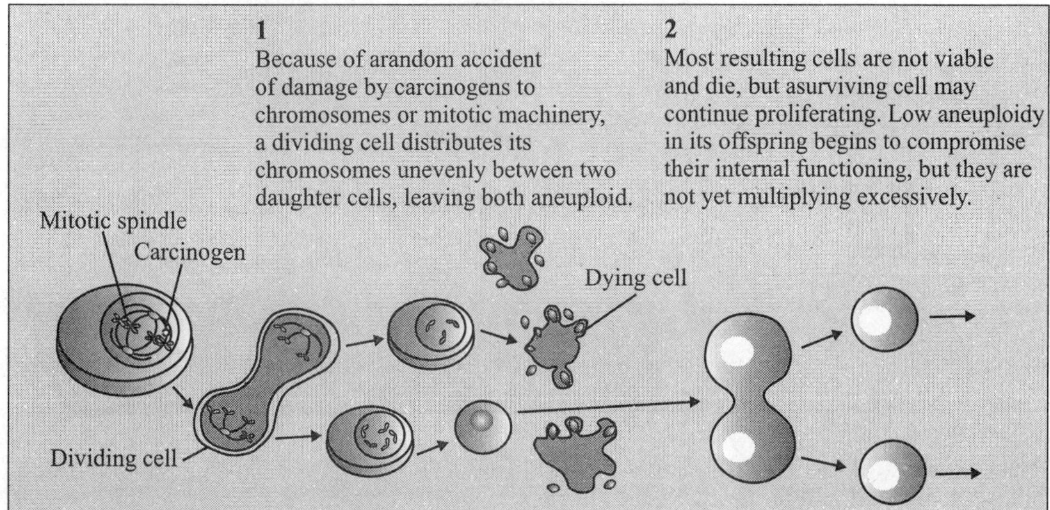

Figure 4.23 How aneuploidy could cause cancer

G1 Signaling Networks Define the Roots of Malignancies

Before replicating DNA during their reproductive cycle, cells enter a phase called G1 during which they interpret a flood of signals that influence cell division and cell fate. Mistakes in this process lead to cancer (Massague 2004). An increasingly complex and coherent view of G1 signaling networks, which coordinate cell growth, proliferation, stress management and survival, is helping to define the roots of malignancies and shows promise for the development of better cancer therapies. Networks integrating growth, survival and proliferation signals are depicted in Figure 4.24.

A Ubiquitination-Dependent Signaling Pathway in DNA Damage Response

Mutations in the breast cancer susceptibility gene 1 (BRCA1) are associated with an increased risk of breast and ovarian cancers. BRCA1 participates in the cellular DNA damage response. Kim et al. (2007) report that identification of receptor-associated protein 80 (RAP80) as a BRCA1-interacting protein in humans. RAP80 contains a tandem ubiquitin-interacting motif domain, which is required for its binding with ubiquitin in vitro and its damage-induced foci formation in vivo. Moreover, RAP80 specifically recruits BRCA1 to DNA damage sites and functions with BRCA1 in G2/M checkpoint control. Together, these results suggest the existence of ubiquitination-dependent signaling pathway involved in the DNA damage response.

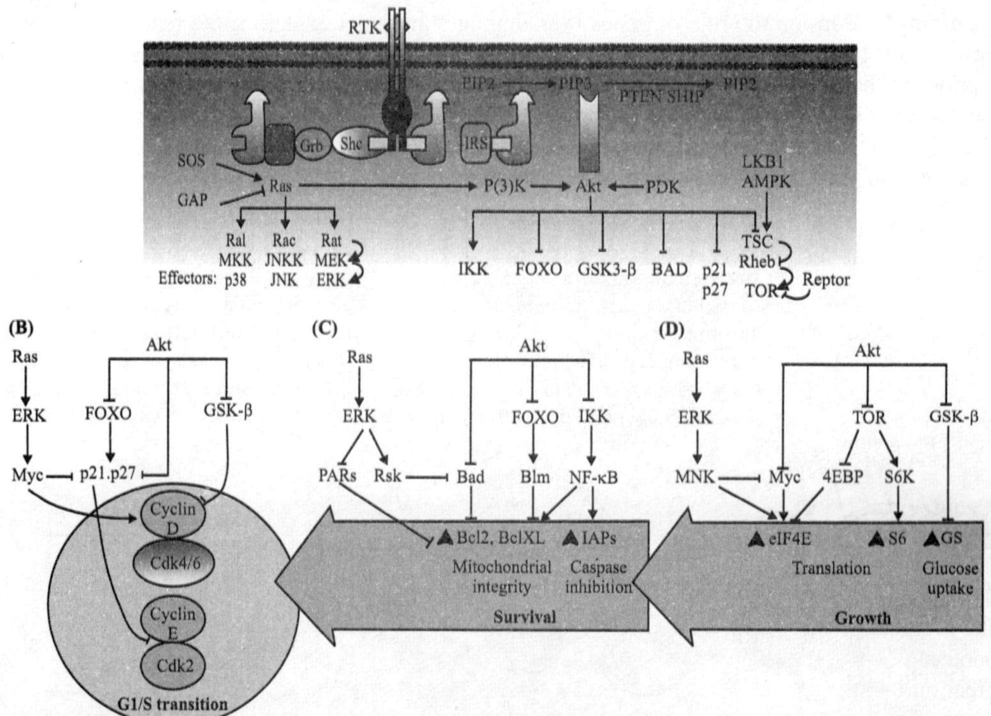

Figure 4.24 Networks integrating growth, survival and proliferation signals

Cancer-Related Responses to DNA Damage

All life on earth must cope with constant exposure to DNA-damaging agents such as the Sun's radiation. Highly conserved DNA-repair and cell cycle checkpoint pathways allow cells to deal with both endogenous and exogenous sources of DNA damage (Kastan and Bertek 2004). How much an individual is exposed to these agents and how their cells respond to DNA damage are critical determinants of whether that individual will develop cancer. These cellular responses are also important for determining toxicities and responses to current cancer therapies, most of which target the DNA. General scheme of responses to DNA damage or replication-fork arrest and impact on cell fate, genomic instability and cancer development, as illustrated by Kastan and Bertek (2004), is shown in Figure 4.25. Replication-fork arrest stimulates the initiation of of cellular ATR activity, whereas DNA damage can directly activate ATM and can lead to replication-fork arrest, thereby also activating cellular ATR kinase. Once active, both the ATM and ATR kinases, functioning in combination with other proteins and substrates, help determine the outcome of the cell. If genome instability ensues, this can contribute to cellular transformation.

Antitumor Drugs Impede DNA Uncoiling by Topoisomerase I

Camptothecins, such as topotecan, induce cell death by poisoning DNA topoisomerase I, an enzyme capable of removing DNA supercoils. Topotecan is thought to stabilize covalent topoisomerase-DNA complex, rendering it an obstacle to DNA replication forks. Koster et al. (2007) used single-molecule nanomanipulation to monitor the dynamics of human topoisomerase I in the presence of topotecan.

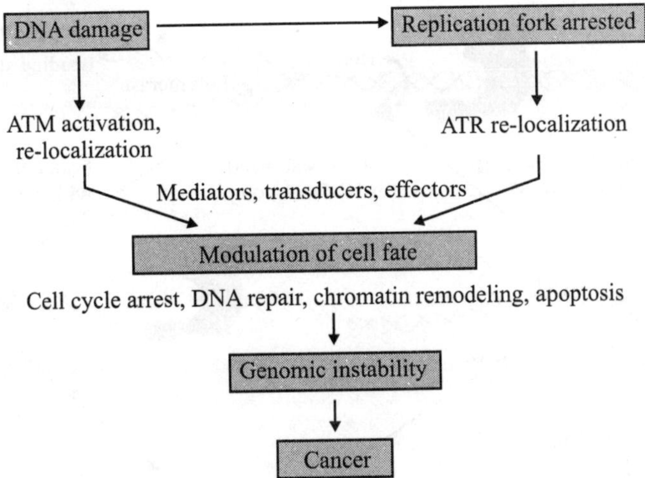

Figure 4.25 A general scheme of responses to DNA damage or replication-fork arrest and impact on cell fate, genomic instability and cancer development

This allows them to detect the binding and unbinding of an individual topotecan molecule in real time and to quantify the drug-induced trapping of topoisomerase on DNA. Their findings show that topotecan significantly hinders topoisomerase-mediated DNA uncoiling, with a more pronounced effect on the removal of positive (overwound) versus negative supercoils. Positive supercoiling would accumulate during transcription and replication as a consequence of camptothecin poisoning of topoisomerase I. Positive supercoils were, however, not induced by drug treatment of cells expressing a catalytically active camptothecin-resistant topoisomerase I mutant. This combination of single-molecule and in vivo data suggests a cytotoxic mechanism for camptothecins, in which the accumulation of positive supercoils ahead of the replication machinery induces potentially lethal DNA lesions. Figure 4.26 depicts mechanisms of camptothecin-induced toxicity derives from TopIB-dependent accumulation of positive supercoils (Koster et al. 2007). A replication fork generates positive supercoils in the DNA, which are removed by TopIB. In the presence of camptothecin, fork stalling and collapse have been predicted to result from the physical collision of the advancing replication complex with the drug-stabilized TopIB-DNA covalent complex (right). The data of Koster et al. (2007) thus suggest a second scenario, in which fork progression and integrity are indirectly impaired by unresolved positive supercoils (left).

MicroRNAs AND DEVELOPMENT OF CANCER

Multiple Roles of MicroRNAs

Small RNAs bound to argonaute proteins recognize partially or fully complementary nucleic acid targets in diverse gene-silencing processes. A subgroup of the Argonaute proteins – known as the Piwi family – is required for germ- and stem-cell development in invertebrates, and two Piwi members – MILI and MIWI – are essential for spermatogenesis in mouse. Small RNAs associate with Argonaute proteins and serve as sequence-specific guides to regulate messenger RNA stability, protein synthesis, chromatin organization and genome structure. In animals, Argonaute proteins segregate into two subfamilies. The Argounate subfamily acts in RNA interference and in microRNA-mediated gene

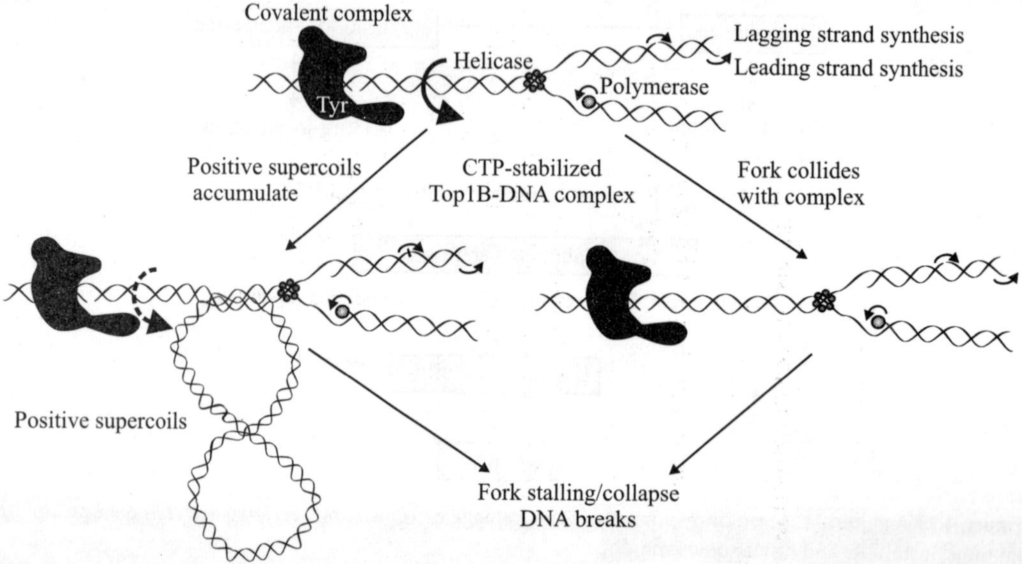

Figure 4.26 Mechanisms of camptothecin-induced toxicity derives from TopIB-dependent accumulation of positive supercoils

regulation using 21-22-nucleotide RNAs as guides. The Piwi subfamily is involved in germline-specific events such as germline stem cell maintenance and meiosis. Girard et al. (2006) show that MIWI, a murine Piwi protein, binds a previously uncharacterized class of ~29-30-nucleotide RNAs that are highly abundant in testis. Girard et al. (2006) have named these Piwi-interacting RNAs (piRNAs). piRNAs have distinctive localization patterns in the genome, being predominantly grouped into 20-90 kb clusters, wherein long stretches of small RNAs are derived from only one strand. Similar piRNAs are also found in human and rat, with major clusters occurring in syntenic locations. Although their function must still be resolved, the abundance of piRNAs in germline cells and the male sterility of Miwi mutants suggested a role in gametogenesis.

Aravin et al. (2006) have described a new class of small RNAs that bind to MILI in male mouse germ cells, where they accumulate at the onset of meiosis. The sequence of the over 1,000 identified unique molecules share a strong preference for a 5′ uridine, but otherwise cannot be readily classified into sequence families. Genomic mapping of these small RNAs reveals a limited number of clusters, suggesting that these RNAs are processed from long primary transcripts. The small RNAs are 26-31 nucleotide (nt) in length – clearly distinct from the 21-23 nt of microRNAs (miRNAs) or short interfering RNAs (siRNAs) – and Aravin et al. (2006) refer them as 'Piwi-interacting RNAs' or piRNAs. Orthologous human chromosomal regions also give rise to small RNAs with the characteristics of piRNAs, but the cloned sequences are distinct. The identification of this new class of small RNAs provides an important starting point to determine the molecular function of Piwi proteins in mammalian spermatogenesis. Temporal expression of piRNAs, MILI and MIWI, during mouse spermatogenesis, following Aravin et al. (2006), is shown in Figure 4.27.

MicroRNAs and Development of Cancer

The microRNAs are now definitely linked to the development of cancer (Meltzer 2005). Process of microRNAs production is illustrated in Figure 4.28. The precursor of a microRNA (pri-miRNA) is

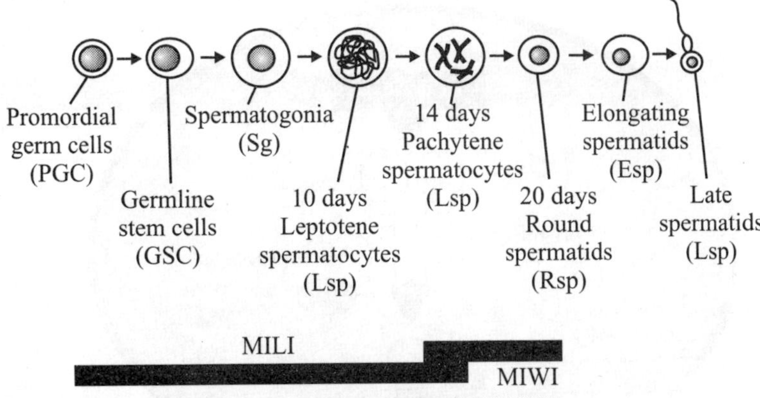

Figure 4.27 Temporal expression of piRNAs during mouse spermatogenesis

transcribed in the nucleus. It forms a stem-loop structure that is processed to form another precursor (pre-miRNA) before being exported to the cytoplasm. Further processing by the Dicer protein creates the mature miRNA, one strand of which is incorporated into the RNA-induced silencing complex (RISC). Base pairing between the miRNA and its target directs RISC to either destroy the mRNA or impede its translation into protein. The initial stem-loop configuration of the primary transcript provides structural clues that have been used to guide searches of genomic sequence for candidate miRNA genes.

Down-Regulation of miRNAs in Tumor Cells

Lu et al. (2005) present a systematic expression analysis of 217 mammalian miRNAs from 334 samples, including multiple human cancers. The miRNA profiles are surprisingly informative, reflecting the developmental lineage and differentiation state of the tumors. They observe a general down-regulation of miRNAs in tumors compared with normal tissues. Furthermore, they were able to successfully classify poorly differentiated tumors using miRNA expression profiles, whereas messenger RNA profiles were highly inaccurate when applied to the same samples. These findings highlight the potential of miRNA in cancer diagnosis.

Genome-Wide RNA Interference Screen in Transformed Cells

The conversion of a normal cell to a cancer cell occurs in several steps and typically involves activation of oncogenes and the inactivation of tumor suppressor and pro-apoptotic genes. In many instances, activation of genes critical for cancer development occurs by epigenetic silencing, often involving hypermethylation of CpG-rich promoter regions. It remains to be determined whether silencing occurs by random acquisition of epigenetic marks that confer a selective growth advantage or through a specific pathway initiated by an oncogene. Gazin et al. (2007) report a genome-wide RNA interference (RNAi) screen in K-ras-transformed NIH3T3 cells and identify 28 genes required for Ras-mediated epigenetic silencing of pro-apoptotic *Fas* gene – six genes (*Kalm*, *Mapk1*, *Map3K9*, *Pdpk1*, *Ptk2b*, *S100z*) are required for signal transduction; seven genes (*Ctcf*, *Eid1*, *E2f1*, *Rcor2*, *Sox14*, *Trim66*, *Zfp354b*) are required for transcription regulation; eight genes (*Bmi1*, *Dnmt1*, *Dot1l*, *Esd*, *Ezh2*, *Hdac9*, *Mrgbp*, *Smyd1*) are required for chromatin modification; three genes (*Asf1a*, *Baz2a*, *Npm2*) are required for chromatin remodeling; and four genes (*Sirt6*, *Sipa1l2*, *Trim37*, *Zcchc4*) genes

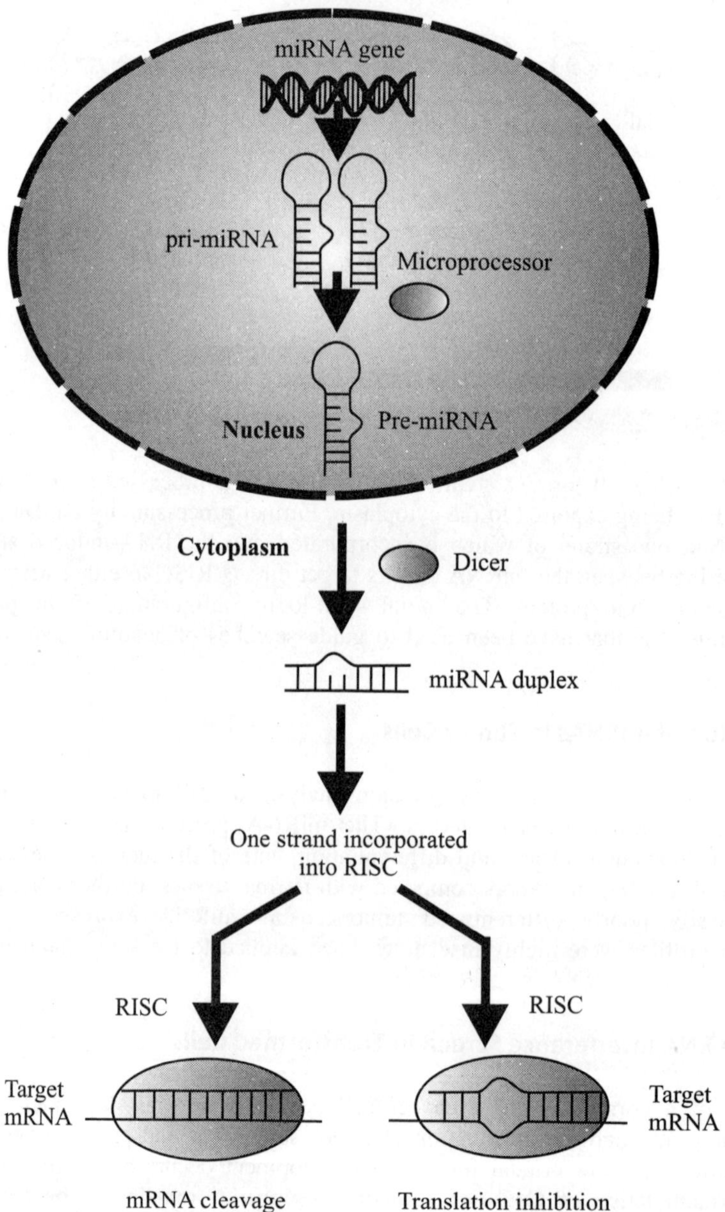

Figure 4.28 Process of microRNA production

are required for genome stability/aging/unknown processes. At least nine of these RESEs (Ras epigenetic silencing effectors), including the DNA methyltransferase DNMT1, are directly associated with specific regions of the *Fas* promoter in K-ras-transformed NIH3T3 cells but not in untransformed NIH3T3 cells. RNAi-mediated knockdown of any of the 28 RESEs results in failure to recruit DNMT1 to the *Fas* promoter, loss of a *Fas* promoter hypermethylation, and derepression of *Fas* expression. *Ras* directs the silencing of multiple unrelated genes through a largely common pathway. They show that nine RESEs are required for anchorage-independent growth and tumorigenicity of K-ras-transformed NIH3T3 cells; these nine genes have not previously been implicated in transformation by *Ras*. These results show that Ras-mediated epigenetic silencing occurs through a specific complex pathway involving components that are required for maintenance of a fully transformed phenotype.

Genomic Loss of a MicroRNA Results in Cancer Progression

Enhancer of zeste homolog 2 (EZH2) is a mammalian histone methyltransferase that contributes to the epigenetic silencing of target genes and regulates the survival and metastasis of cancer cells. EZH2 is overexpressed in aggressive solid tumors. Varambally et al. (2008) show that the expression and function of EZH2 in cancer cell lines are inhibited by microRNA-101 (miR-101). Analysis of human prostrate tumors revealed that miR-101 expression decreases during cancer progression, paralleling an increase in EZH2 expression. One or both of the two genomic loci encoding miR-101 were somatically lost in 37.5 per cent of clinically localized prostrate cancer cells and 66.7 per cent of metastatic disease cells. They propose that the genomic loss of miR-101 in cancer leads to overexpression of EZH2 and concomitant dysregulation of epigenetic pathways, resulting in cancer progression.

Proto-oncogene c-myc has a Direct Role in DNA Replication Control

The *c-myc* proto-oncogene encodes a transcription factor that is essential for cell growth and proliferation and is broadly implicated in tumorigenesis. Dominguez-Sola et al. (2007) show that *c-myc* has a direct role in the control of DNA replication. C-Myc interacts with the pre-replicative complex and localizes to early sites of DNA synthesis. Depletion of *c-myc* from mammalian cells as well as *Xenopus* cell-free extracts, which are devoid of RNA transcription, demonstrates a non-transcriptional role for c-myc in the initiation of DNA replication. Overexpression of *c-myc* gene causes increased replication origin activity with subsequent DNA damage checkpoint activation. These findings identify a critical function of c-myc in DNA replication and suggest a novel mechanism for its normal and oncogenic functions.

MicroRNAs can modulate Tumor Formation

By the year 2005, more than 200 micro RNAs have been described in humans; however, the precise functions of these regulatory non-coding RNAs remains largely obscure. One cluster of microRNAs, the *mir-17-92* polycistron is located in a region of DNA that is amplified in human B-cell lymphomas. He et al. (2005) compared B-cell lymphomas samples and cell lines to normal tissues, and found that the levels of the primary or mature microRNAs derived from the *mir-17-92* locus are often substantially increased in cancers. Enforced expression of the *mir-17-92* cluster acted with *c-myc* expression to accelerate tumor development in a mouse B-cell lymphoma model. Tumors derived from hematopoietic stem cells expressing a subset of the mir-17-92 cluster and c-myc could be distinguished by an absence of apoptosis and that was otherwise prevalent in c-myc-induced lymphomas. Together, these studies indicate that non-coding RNAs, specifically microRNAs, can modulate tumor formation, and implicate the *mir-17-92* cluster as a potential human oncogene.

MicroRNAs Regulate Cell Proliferation

Strict tissue- and developmental-stage-specific expression is critical for appropriate miRNA function. Some mammalian transcription factors regulate miRNAs. The proto-oncogene *c-myc* encodes a transcription factor that regulates cell proliferation, growth and apoptosis. Dysregulated expression or function of c-myc is one of the most common abnormalities in human malignancy. O'Donnell et al. (2005) show that c-myc activates expression of a cluster of six miRNAs on human chromosome 13. c-myc binds directly to this locus. The transcription factor E2F1 is an essential target of c-myc that promotes cell cycle progression. Expression of E2F1 is negatively regulated by two miRNAs in this cluster, miR-17-5p and miR-20a.

Most Cancerous Mutations Abolish All Tumor-Suppressor Functions

The core domain of the p53 protein has been found to affect microRNA processing – its third known antitumor activity. Most cancerous mutations affect this domain and may abolish all tumor-suppressor functions (Toledo and Bardot 2009). Three antitumor functions of p53 are depicted in Figure 4.29. The DNA-binding domain of p53 lies in the core of the protein and has three antitumor functions. It binds to DNA and enables the activation of target genes (including the miR-34 gene family) to induce apoptosis or cell cycle arrest; it stimulates apoptosis through an interaction with proteins of the Bcl2 family at the mitochondrion; and, as Suzuki et al. (2009) show, it interacts with proteins of the Drosha complex to promote processing of a subset of miRNAs, including miR-16-1 and miR-143, which supp-

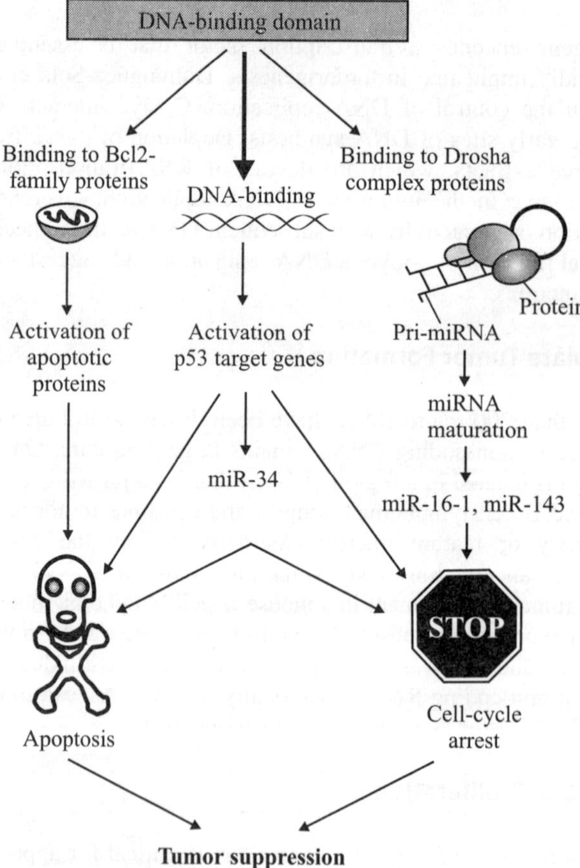

Figure 4.29 Three antitumor functions of p53

sress cell proliferation. Most p53 mutations in human cancers lie in the DNA-binding domain and may affect all three functions.

Some MicroRNAs Function as Oncogenes or Tumor Suppressors

MicroRNAs have been implicated in regulating diverse cellular pathways. There is emerging evidence that some microRNAs can function as oncogenes or tumor suppressors. Ma et al. (2007) show using a combination of mouse and human cells that microRNA-10b (miR-10b) is highly expressed in metastatic breast cancer cells and positively regulates cell migration and invasion. Overexpression of miR-10b in otherwise non-metabolic breast tumors initiates robust invasion and metastasis. Expression of mi-10b is induced by the transcription factor Twist, which binds directly to the putative promoter of mir-10b (MIRN10B). The mi-10b induced by Twist proceeds to inhibit translation of the messenger RNA encoding homeobox D10, resulting in increased expression of a well-characterized pro-metastatic gene, RHOC, Significantly, the level of miR-10b expression in primary breast carcinoma correlates with clinical progression. These findings suggest the workings of an undescribed regulatory pathway, in which a pleiotropic transcription factor induces expression of a specific microRNA, which suppresses its direct target and in turn activates another pro-metastatic gene, leading to tumor cell invasion and metastasis. A model for the regulation and function of miR-10b in cancer metastasis, as proposed by Ma et al. (2007) is given in Figure 4.30. RISC represents RNA-induced silencing complex.

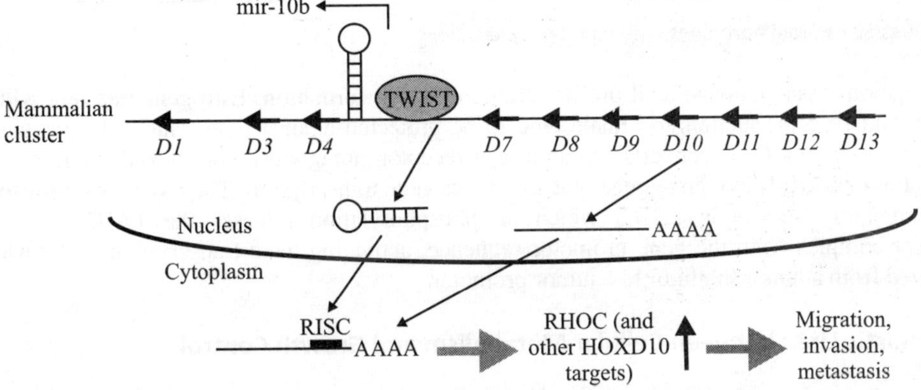

Figure 4.30 miR-10b expression level is associated with the metastasis outcome in breast cancer patients

TUMOR DEVELOPMENT

Interaction between Signals Induced by Sex Hormones and Inflammation, and Tumor Development

Signals induced by sex hormones and inflammation have been viewed as different aspects of tumor development. But a three-way interaction between these two classes of signal and carcinogenesis has emerged (Mantovani 2007). Relationship between sex steroid hormones, inflammation and cancer is shown in Figure 4.31. In the liver, a carcinogen (dimethylnitrosamine, DEN) causes tissue damage that (presumably acting via Toll-like receptors, TLR) activates the MyD88-NF-κB signal pathway in Kupffer cells. These cells, a type of macrophage, produce interleukin-6 (IL-6), which in turn promotes

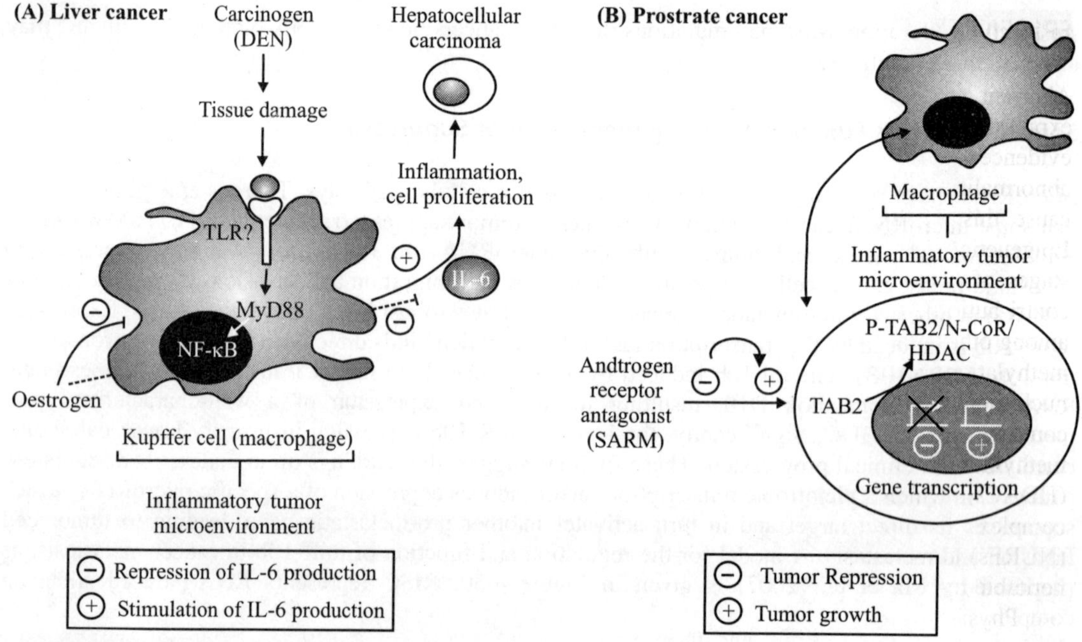

Figure 4.31 Sex steroid hormones, inflammation and cancer

inflammation, tissue damage, cell proliferation and tumor formation. Estrogens interfere with NF-κB activity and IL-6 production so female tend to be protected against liver cancer. In prostate cancer, macrophage-driven IL-1 converts an androgen-receptor antagonosis (a steroid androgen receptor modulator, or SARM) into an agonist that stimulates gene transcription. The modular sensor for the IL-1 inflammatory signals is TAB2, which on phosphorylation releases the TAB2/N-Co-R/HDAC repressor complex from the gene promoter sequence, activating gene transcription. A SARM is thus converted from a tumor inhibitor to a tumor promoter.

Mechanistic Links between Cellular Metabolism and Growth Control

In contrast to normal differentiated cells, which rely primarily on mitochondrial oxidative phosphorylation to generate the energy needed for cellular processes, most cancer cells instead rely on aerobic glycolysis, a phenomenon termed "the Warburg effect". Aerobic glycolysis is an inefficient way to generate adenosine 5′-triphosphate (ATP), however, the advantage it confers to cancer cells has been unclear. Vander Heiden et al. (2009) propose that the metabolism of cancer cells, and indeed all proliferating cells, is adapted to facilitate the uptake and incorporation of nutrients into the biomass (e.g., nucleotides, amino acids, and lipids) needed to produce a new cell. Supporting this idea are recent studies showing that (i) several signaling pathways implicated in cell proliferation also regulate metabolic pathways that incorporate nutrients into biomass; and that (ii) certain cancer-associated mutations enable cancer cells to acquire and metabolize nutrients in a manner conducive to proliferation rather than efficient ATP production. A better understanding of the mechanistic links between cellular metabolism and growth control may ultimately lead to better treatments for human cancer.

EPIGENETIC ALTERATIONS AND NEOPLASIA

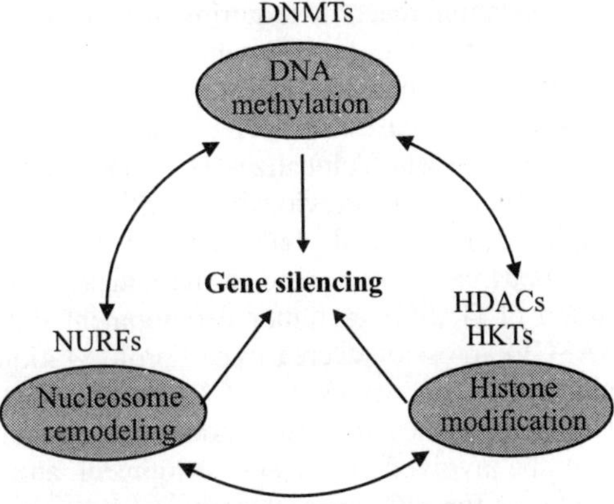

Figure 4.32 Gene silencing in normal cells

Aberrant gene function and altered patterns of gene expression are key features of cancer. Growing evidence shows that acquired epigenetic abnormalities participate with genetic alterations to cause this dysfunction (Jones and Baylin 2007). Epigenetic alterations participate in the earliest stages of neoplasia, including stem/precursor cell contri-butions. Heritable gene silencing involves, among other processes, the interplay between DNA methylation, histone covalent modifi-cations, and nucleosomal remodeling. Some of the enzymes that contribute to these modifications include DNA methyltransferases (DNMTs), histone deacetylases (HDACs), histone methyltransferases (HMTs), and complex nucleosomal remodel-ing factors (NURFs) (Jones and Baylin 2007) (Figure 4.32). The interplay between these processes establishes a heritable repressive state at the start site of gene resulting in gene silencing.

Physiologically, silencing is critical for development and differentiation. Pathologically, silencing leads to disease such as cancer. Recent evidence suggests global changes in all three processes in cancer, perhaps reflecting their interrelationships. Epigenetic silencing genes p16, SFRPs, GATA-4 and -5, and APC in stem/precursor cells of adult cell-renewal systems may serve to abnormally lock these cells into stem-like states that foster abnormal clonal expansion (Jones and Baylin 2007) (Figure 4.33). These genes are termed 'epigenetic gate keepers' because the normal epigenetic pattern of expression should allow them to be activated during stem/precursor cell differentiation as needed to properly control adult cell renewal. The repertoire of abnormal gene silencing then allows abnormal survival of the cells in the setting of chronic stress, such as inflammation (Jones and Baylin 2007). The resulting pre-invasive stem cells become "addicted" to the survival pathways involved so that selection for mutations in genetic gate keeper genes provide an even stronger stimulus for further tumor progression. The bulk of the resulting tumor is composed of a subpopulation of cancer stem cells and neoplastic progeny. Networks of gene-silencing events are shown in Figure 4.34.

Such networks help to foster early and later steps during neoplastic progression. Examples of early gene silencing (X) occur at multiple points in key tumor pathways to allow abnormal cell survival after stress and early clonal expansion. Examples of gene-silencing events that foster subsequent silencing events (arrows linking SIRT1 to silencing of GATA-4 and -5 and SFRPs) are depicted.

Histone H3 Dimethylation of Lysine 9 and Tumor Development

Methylation of lysine and arginine residues on histone tails affects chromatin structure and gene transcription. Tri- and dimethylation of lysine 9 on histone H3 (H3K9me3/me2) is required for the binding of the repressive protein HP1 and is associated with heterochromatin formation and transcriptional repression in a variety of species. H3K8me3 has long been regarded as a "permanent epigenetic mark'. In a search for protein and complexes interacting with H3L9me3, Cloos et al. (2006) identified the protein GASc1 (gene amplified in squamous cell carcinoma 1), which belongs to the JMJD2 (jamnoji domain containing 2) subfamily of the jumonji family, and is also known as JMJD2C.

They show that three members of this subfamily of proteins demethylate H3K9me3/me2 in vitro through a hydroxylation reaction requiring iron and α-ketoglutarate as cofactors. Further, they demonstrate that the ectoptic expression of GASC1 or other JMJD2 members markedly decreases H3K9me3/me2 levels, increases H3K9me1 levels, delocalizes HP1 and reduces heterochromatin in vivo. Previously, GASC1 was found to be amplified in several cells derived from esophageal squamous carcinomas, and in agreement with a contribution of GASC1 to tumor development, inhibition of GASC1 expression decreases cell prolife-ration. Thus, in addition to identifying GASC1 as a histone trimethyl demethylase, they suggest a model for how this enzyme might be involved in cancer development, and propose it as a target for anti-cancer therapy.

HOW IMMUNE SYSTEM COUNTERACTS ONCOGENES?

The most dangerous characteristic feature of cancer cells is their uncontrolled divisions, break off from the original tumor and establishment of tumorous growth (metastasis) in other tissues. Three approaches, namely surgery, radiation and chemotherapy, have been used to fight cancer. Immune system is an ideal weapon against any infectious disease (Boon 1993). Tumor cells display special antigens which do not occur on normal cells. These antigens are called tumor rejection antigens or simply tumor antigens. These antigens form the basis of immunotherapies. The proposed three therapies that use this specificity and power of immune system to eradicate cancer are immunotrixin production, adoptive immunotherapy and tumor antigens. For immunotrixin production, a toxin molecule is conjugated to an antibody directed specifically against cancer cells (Gilliland et al. 1980). Adoptive immunotherapy was made possible with the discovery and knowledge of mode of action of interleukin-2 (IL-2) (Smith 1988). Administration of lymphokine-activated killer (LAK) cells alongwith IL-2 lead to tumor repression both in mice and humans (Vujanovic 1988a). Further, tumor infiltrating lymphocytes (TILs) from tumors were discovered, which when administered alongwith IL-2 were found to be twice as effective as LAK (Vujanovic 1988b). Therapy based on tumor antigens was possible with the work of Boon (1993). He isolated gene encoding a tumor antigen called antigen-E. Cancer patients are tested for the expression of this gene in tumor cells, after the removal of original tumor from the body. If positive, patient is injected with killed cells of a tumor expressing antigen-E. These may stimulate an immune response which eradicates the metastasized cancer from the body.

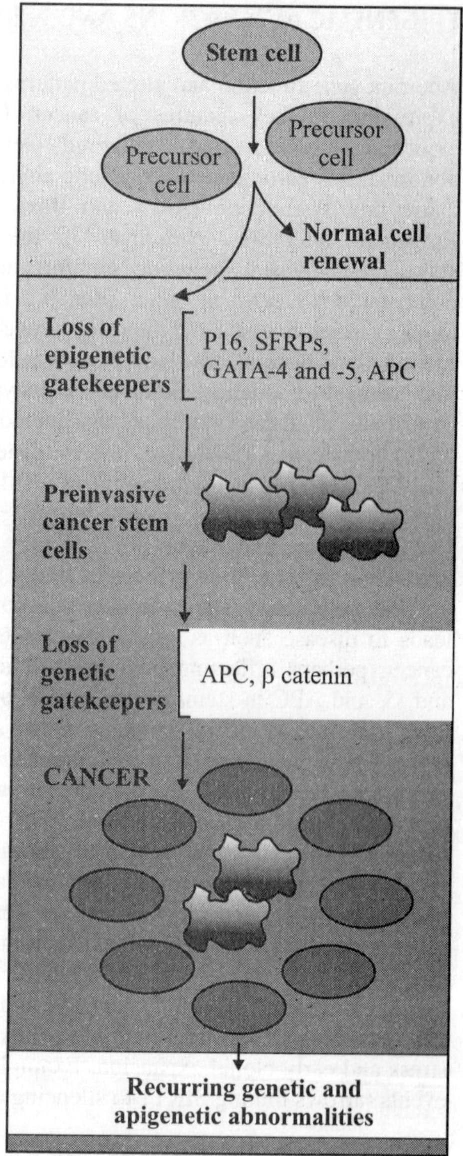

Figure 4.33 "Epigenetic gatekeeper" prevents early tumor progression

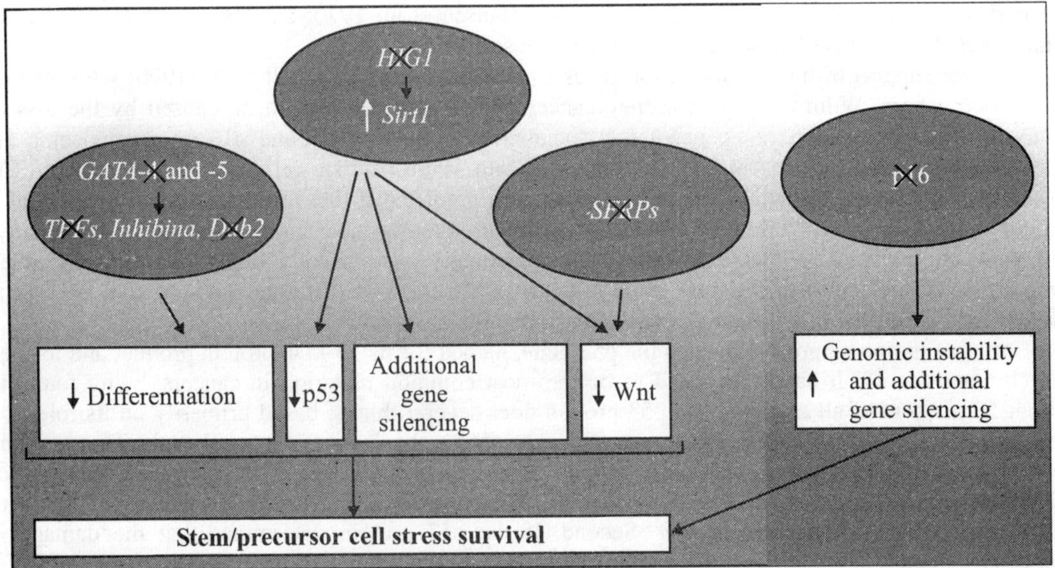

Figure 4.34 Networks of gene-silencing events

ANTI-ONCOGENES

Proteins encoded by anti-oncogenes can prevent tumorigenic transformation of cells. Anti-oncogenes are also known as tumor-suppressor genes (TSGs) or recessive oncogenes. These genes act by suppressing malignant growth. Examples include the genes associated with Wilms' tumor (WT1) of kidney, retinoblastoma (RB1), cell proliferation regulation (TP53), RAS-GTPase activation (NF1), cyclin-dependent kinase inhibition (P21), CDK4 and CDK6-D cyclin complex inhibition (P16), breast cancer development (BRCA1 and BRCA2). The humans who inherit a single mutated *RB* gene (hemizygotes) produce the gene product from the remaining functional allele and are normal. However, if the functional allele mutates in a somatic cell, the cell can initiate a tumor. The RB gene encodes a nuclear, DNA-binding protein. Transfection of the gene into tumor cells in culture reverses the tumor phenotype. Thus, anti-oncogene product appears to function in some manner to suppress growth, in distinction to many oncogene products, which stimulate cell proliferation.

Yunis (1986) examined cells from several retinoblastoma patients. He found that there was frequently a deletion of part of chromosome 13, specifically band q14. The exact points of deletion varied from individual to individual, indicating that the phenomenon was due to loss of gene action rather than enhancement of gene activity due to the new placement of genes previously separated by the deleted material. The *RB* gene has been isolated and cloned. The RB gene is 200,000-bp long and it specifies a 105-kDa protein (p105) found in the nucleus, as would be expected if it were a suppressor of DNA transcription. It binds with at least three known oncogene proteins: the EIA protein of adenovirus, the SV40 (a simian virus) large T antigen, and the 16E7 protein of human papilloma virus, a virus associated with 50 per cent of cervical carcinomas. The implications of these findings are that these three viruses may use a similar mechanism in transformation, and this mechanism may involve inactivation of the retinoblastoma p105 protein (Green 1989). This interaction is the first demonstration of a physical link between an oncogene and an antioncogene. H. zur Hausen was awarded Nobel Prize In 2008 for finding human papilloma viruses that cause cervical cancer, the second most common cancer among women (zur Hausen et al. 1970; zur Hausen et al. 1974a; zur Hausen et al. 1974b; zur Hausen et al. 1975; zur Hausen 1976).

Further support of tumor-suppressor genes came from work by Standbridge (1990) with another childhood cancer, Wilm's tumor, a kidney cancer that is also believed to be caused by the loss of activity of a tumor-suppressor gene. It is associated with the loss of band p13 on chromosome 11. Researchers introduced a normal chromosome 11 into Wilm's tumor cells growing in culture. The result was normal cell growth, exactly what would be predicted if the introduced normal gene was a tumor-suppressor gene. Children with the condition frequently have had a loss of the maternally-inherited chromosome, a fact consistent with a phenomenon, currently being investigated, called imprinting or molecular or parental imprinting in which there is differential expression of a gene depending on whether that gene was maternally or paternally inherited.

Another tumor-suppressor gene is the p53 gene, named for its 53-kDa protein product and located on chromosome 17. It has been found to be the most common mutation in cancers, being found in about 50 per cent of all cancers. The p53 protein does several things, based primarily on its role as a transcription factor. First, the p53 protein seems to control a checkpoint in the cell cycle. That is, if the p53 protein detects extensive genetic damage, it can stop the cell cycle to allow more time for repair of DNA damage. That is, normal cells stop at the G1 phase of the cell cycle when irradiated with ultraviolet light; p53 mutants do not. Second, if the p53 protein determines that the damage is especially extensive, it can cause apoptosis, cell death. Thus mutations of p53 result in increased DNA damage and can result in cancer. Currently, about a dozen tumor suppressor genes are known.

NF1 is a RAS-GTPase activating protein (GAP), i.e., a protein whose function is to antagonize RAS signaling by accelerating the RAS intrinsic GTPase activity. RAS is a guanine nucleotide-binding protein; it is active when bound by GTP but inactive when bound by GDP. Active RAS recruits RAF to the cell membrane, where it is phosphorylated and then able to initiate the MAP kinase-signaling cascade. Loss of NF1 results in the inability of the cell to shut down RAS signaling, and hence constitutive activation of the mitogenic MAP kinase pathway. This occurs even when the ubiquitous GTPase activating protein GAPp120 is present. Germline mutations in NF1 are associated with neurofibromas and chronic myelogenous leukemia.

The *WT1* gene encodes at least four zinc finger proteins by alternative splicing. The main splice variant is a transcriptional repressor of several genes involved in growth regulation, including BCL2, IGFII and myc. Mutations in the *WT1* gene which abolish DNA-binding are often seen in cases of Wilm's tumor of the kidney and the Denys-Drash syndrome. WT1 acts as a dimer and WT1 mutations are, therefore, often dominant negatives. p21 is a general cyclin-dependent kinase inhibitor and is activated by p53 in response to DNA damage. p16 is a specific inhibitor of CDK4 and CDK6-D cyclin complexes. Loss-of-function mutations in the genes encoding both inhibitors prevent normal growth inhibitory signals blocking the cell cycle at G1. D cyclins are responsible for initiating the events which promote into the S phase.

ANTISENSE MOLECULES AS ANTICANCER DRUGS

Recent approach in search of anticancer drug is application of oncogene-targeted antisense DNA and RNA oligonucleotides. Antisense RNAs are small, diffusible transcripts produced from non-coding DNA strand. Hence, antisense RNA and messenger RNA are transcribed from opposite strands of the same DNA template by opposing promoters. Antisense RNA cannot be translated by ribosomes to form proteins. Antisense RNA has some natural biological functions. Tomizawa et al. (1981) and Mizuno et al. (1984) reported that bacteria and viruses regulate at least some genes during their life cycle with antisense RNAs that pair to specific complementary target messenger RNAs and thus selectively turn off or modify the activity of any gene. Antisense RNAs are also known to control plasmid replication and conjugation (Tomizawa 1986). Antisense molecules have now become valuable research tools in understanding the gene function, creation of mutations and target homologous recombination. Viruses can cause cancer by directly introducing an oncogene. Employment of an oncogene-targeted antisense oligonucleotide to inhibit oncogenic viral integration, multiplication and cell transformation offers a novel approach to the problem of oncogenesis. The first evidence to inhibit the replication and cell transformation of a malignant tumor virus, RSV, by a specific antisense DNA of tridecamer was provided by Zamecnick and Stephenson (1978). The tridecamer (AATGGTAAAATGG) sequence of deoxyribonucleotide complementary to thirteen nucleotides of 3′ and 5′ reiterated terminal sequences of RSV 35S RNA was added to chick embryo fibroblast tissue culture, which was previously infected with RSV. The antisense DNA was taken up by the cells and it hybridized with the 3′ and 5′ ends of 35S RNA, resulting in the inhibition of viral production and cell transformation.

The antisense oligonucleotides can either be employed as unmodified form or their modified derivatives can be used. Heikkila et al. (1987) used an unmodified antisense oligonucleotides to block the expression of *c-myc* oncogene by employing antisense DNA of short length, complementary to the 5′-end of human *c-myc* gene's second exon. Holt (1988), Wickstrom et al. (1988) and Sklar et al. (1991) show that an oligomer complementary to *c-myc* messenger RNA inhibits the proliferation of HL-60 promyelocytic cells and induces differentiation by binding with *c-myc* messenger RNA. A large number of potential drawbacks like poor uptake efficiency by cells and their resistance to nuclease attack are associated with the unmodified oligonucleotides. To surmount these problems and to increase the antimessenger activity of such molecules, various chemical modifications were introduced into the oligos. Modifications were either in the sugar phosphodiester backbone of the anti-messenger oligonucleotides to increase their resistance against nuclease attack (Agris et al. 1986; Smith et al. 1986) or the antisense oligos were covalently linked to some active reagents like acridine to increase their affinity for the target sequences. Modified antisense oligonucleotides having a metal complex linked active group like EDTA-Fe or O-phenanthroline have been used to inactivate specific genes (Chu and Orgel 1985; Dreyer and Dervan 1985). Triple helix-forming compounds have also been shown to inhibit the expression of oncogenes (Le Doan et al. 1987; Moser and Dervan 1987; Cooney et al. 1988).

SIGNAL TRANSDUCTION BY ONCOPROTEINS

Signal transduction pathways precede most of the mechanisms of regulation gene expression. In eukaryotes, a cascade of molecules leading to the activation of one or more specific transcription factors is involved. The extracellular signals in the form of ligands are transcribed across the plasma membrane via a variety of molecules called second messengers. Effector molecules which bind to specific receptors to initiate pathways for the production of second messengers are often called first messengers. The binding of this effector to receptor is often mediated by proteins whose activity depends on GTP/GDP binding. These GTP/GDP binding proteins are called G-proteins, so named due to their affinity to bind guanine residues. They are heterotrimeric proteins with three subunits α, β and γ, located in plasma membrane and transduce extracellular signals received by transmembrane receptors to effector proteins, in many eukaryotes. Since G-proteins transduce signals, they are also described as transducers. The intracellular molecules can be diverse, including adenylate cyclase, phospholipases and ion channels.

G-proteins are activated when GTP binds to the α subunit (by displacing GDP) and causes its dissociation from the βγ dimer (Figure 4.35). Such activation is common to a variety of GTP binding proteins. The separated βγ dimer may carry the message from receptor to effector, although in some cases α subunit may do this job. Among oncoproteins, G proteins are represented by Ras proteins,

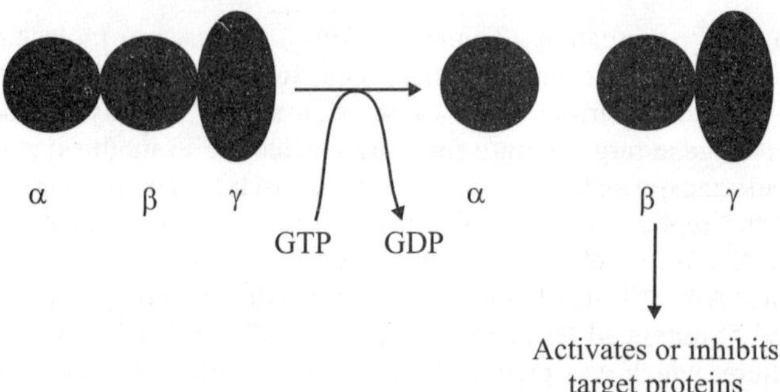

Figure 4.35 A typical trimeric G-protein comprising of α, β and γ subunits, where GTP binding release α subunit from βγ dimer. α subunit or βγ dimer become free to act upon target proteins (Redrawn, with permission, from Gupta, P.K. 2009. *Genetics*. 4th edition. Meerut: Rastogi Publications)

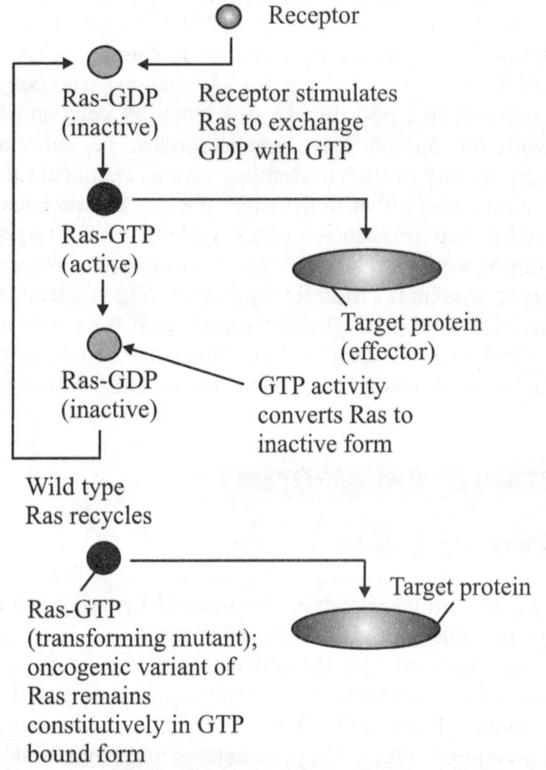

Figure 4.36 Activation of Ras, which stimulates replacement of GDP by GTP; the active protein recognizes its effector and GTP is cleaved into GDP, so that resulting inactive Ras is recycled, transforming Ras do not hydrolyze GTP, so that Ras remains permanently in the active form (Redrawn, with permission, from Gupta, P.K. 2009. *Genetics*. 4th edition. Meerut: Rastogi Publications)

which bind GTP and thus resemble α subunit. While bound to GTP, Ras becomes active and acts upon its target molecule. Following this interaction, GTPase activity of Ras protein hydrolyzes GTP molecule to GDP, thus returning itself to inactive condition. (Figure 4.36).

REFERENCES

Agris, C.H., K.R. Blaje, P.S. Miller, M.P. Reddy, and P.O.P. Tso'o, 1986. Inhiition of vesicular stomatitis virus protein synthesis and infection by sequence-specific oligodeoxyribonucleoside methylphosphonates. Biochemistry 25: 6268-75.

Aravin, A., D. Gaidatzis, S. Pfeffer, et al. 2006. A novel class of small RNAs bind to MILI protein in mouse testes. Nature 442: 203-7.

Bartkova, J., N. Rezaei, M. Liontos, et al. 2006. Oncogene-induced senescence is part of the tumorigenesis barrier imposed by DNA damage checkpoints. Nature 444: 633-7.

Beachy, P.A., S.S. Karhadkar, and D.M. Berman, 2004. Tissue repair and stem cell renewal in carcinogenesis. Nature 432: 324-31.

Berns, A. 2008. A tRNA with oncogenic specificity. Cell 133: 29-30.

Boon, T. 1993. Teaching the immune system to fight cancer. Scient. Am. 266(3): 32-9.

Christofori, G. 2007. Cancer: Division of labour. Nature 446: 735-6

Chu, B.C.F., and E. Orgell. 1985. Nonenzymatic sequence-specific cleavage of single-stranded DNA. Proc. Natl. Acad. Sci. USA 82: 963-7.

Churkov, S., J.K. Kurash, J.R. Wilson, et al. 2004. Regulation of p53 activity through lysine methylation. Nature 432: 353-60.

Clarke, M.F., and M.W. Becker. 2006. Stem cells: The real culprits in cancer? Scient. Am. 295(1): 52-9.

Cloos, P.A.C., J. Christensen, K. Agger, et al. 2006. The putative oncogene GASC1 demethylates tri- and dimethylated lysine 9 on histone H3. Nature 442: 307-11.

Collins, F.S., and A.D. Barker. 2007. Mapping the cancer genome. Scient. Am. 296(3): 50-7.

Cooney, M., J. Czernussenic, E.M. Postel, S.O. Flint, and M.E. Hoogan. 1988. Site-specific oligonucleic acid binding represses transcription of the human *c-myc* gene in vitro. Science 241: 456-459.

Culotti, J., and L. H. Hartwell, 1971. Genetic control of the cell division cycle in yeast. Exp.Cell Res. 67: 389-401.

Di Micco, R., M. Fumagalli, A. Cicalese, et al. 2006. Oncogene-induced senescence is a DNA damage response triggered by DNA hyper-replication. Nature 444: 638-42.

Dominguez-Sola, D., C.Y. Ying, C. Grandori, et al. 2007. Non-transcriptional control of DNA replication by c-myc. Nature 448: 445-51.

Dreyer, C.H.B., and P.B. Dervan. 1985. Sequence-specific cleavage of single-stranded DNA: oligodeoxynucleotide-EDTA-Fe(II). Proc. Natl. Acad. Sci. USA 83: 968-72.

Duesberg, P. 2007. Chromosomal chaos and cancer. Scient. Am. 296(5): 52-9.

Evans, T., E.T. Rosenthal, J. Youngblom, D. Distel, and T. Hunt. 1983. Cyclin: a protein specified by maternal mRNA in sea urchin eggs that is destroyed at each cleavage division. Cell 33: 389-96.

Furchgott, R.F., and J.V. Zawadzki, 1980. The obligatory role of endothelial cells in the relaxation of arterial smooth muscle by acetylcholine. Nature 288: 373-6.

Gazin, C., N. Wajapeyee, S. Gobeil, C.-M. Virbasius, and M.R. Green. 2007. An elaborate pathway required for Ras-mediated epigenetic silencing. Nature 449: 1073-7.

Gilliland, D.G., Z. Steplewski, R.J. Collier, K.F. Mitchell, T.H. Chang, and H. Koprowski. 1980. Antibody directed cytototoxic agents: use of monoclonal antibody to direct the action of toxin chains to colorectical carcinon cells. Proc. Natl. Acad. Sci. USA 77: 4539-43.

Girard, A., R. Sachidanandam, G.J. Hannon, and M.A. Carmell. 2006. A germline-specific class of small RNAs bind mammalian Piwi proteins. Nature 442: 199-202.

Gorini, C., M. Squatrito, C. Luise, et al. 2007. Tip 60 is a haplo-insufficient tumor suppressor required for an oncogene-induced DNA damage response. Nature 448: 1063-1067.

Green, D.R., and G. Kroemer. 2009. Cytoplasmic functions of suppressor p53. Nature 458: 1127-30.

Green, M.R. 1989. When the products of oncogenes and antioncogenes meet. Cell 56: 1-3.

Greenman, C., P. Stephens, R. Smith, et al. 2007. Patterns of somatic mutation in human cancer genomes. Nature 446: 153-8.

Gupta, G.P., D.X. Nguyen, A.C. Chiang, et al. 2007. Mediators of vascular remodeling co-opted for sequential steps in lung metastasis. Nature 446: 765-70.

Haber, D.A., and J. Settleman. 2007. Cancer: Drivers and passengers. Nature 446: 145-6.

Halazonetis, T.D., V.G. Gorgoulis, and J. Bartek. 2008. An oncogene-induced DNA damage model for cancer development. Science 319: 1352-5.

Hartwell, L.H. 1971a. Genetic control of the cell division cycle in yeast. II. Genes controlling DNA replication and its inititation. J. Mol. Biol. 59: 183-94.

Hartwell, L.H. 1971b. Synchronization of haploid yeast cell cycles, a prelude to conjugation. Exp. Cell Res. 69: 265-76.

Hartwell, L.H. 1973. Three additional genes required for deoxyribonucleic acid synthesis in *Saccharomyces cerevisiae*. J. Bacteriol. 115: 966-74.

Hartwell, L.H., and T.A. Weinert. 1989. Checkpoints: controls that ensure the order of cell cycle events. Science 246: 629-34.

Hartwell, L.H., J. Culotti, and B. Reid. 1970. Genetic control of the cell-division cycle in yeast. Proc. Natl. Acad. Sci. USA 66: 352-9.

Hartwell, L.H., J. Culotti, J.R. Pringle, and B.J. Reid. 1974. Genetic control of the cell division cycle in yeast. Science 83: 46-51.

Hartwell, L.H., R.K. Mortimer, J. Culotti, and M. Culotti. 1973. Genetic control of the cell division cycle in yeast. Genetics 74: 267-86.

He, L., J.M. Thomson, M.T. Hemann, E. Hernando-Monge, et al. 2005. A microRNA polycistron as a potential human oncogene. Nature 435: 828-33.

Heikkila, R., G., Schwab, E. Wickstorm, et al. 1987. A *c-myc* antisense oligodeoxynucleotide inhibits entry into S phase but not from Go to G1. Nature 328: 445-9.

Holt, J.T., R.L. Redner, and A.W. Nienhuis. 1988. An oligomer complementary to *c-myc* mRNA inhibits proliferation of HL-60 promyelocyctic cells and induces differentiation. Mol. Cell Biol. 8: 963-73.

Huang, J., L. Perez-Burgos, B.J. Placek, et al. 2006. Repression of p53 activity by Smyd2-mediated methylation. Nature 444: 629-32

Hughes, S.H., F. Payvar, D. Spector, et al. 1979. Heterogeneity of genetic loci in chickens: analysis of endogenous viral and nonviral genes by cleavage of DNA with restriction endonuclease. Cell 18: 347-59.

Ignarro, L.J., K.S. Wood, R.E. Byrns, and G. Chanduri. 1987. Endothelial derived relaxing factor produced and released from atery and vein is nitric oxide. Proc. Natl. Acad. Sci. USA 84: 9265-9.

Johnson, D.L., and S.A.S. Johnson, 2008. RNA metabolism and oncogenesis. Science 320: 461-2.

Jones, P.A., and S.B. Baylin. 2007. The epigenomics of cancer. Cell 128: 683-92.

Kaelin, W.G. 2006. Divining cancer cell weaknesses. Nature 441: 32-4.

Kastan, M.B. 2007. Wild-type p53: Tumors can't stand it. Cell 128: 837-40.

Kastan, M.B., and J. Bertek. 2004. Cell cycle checkpoints and cancer. Nature 432: 316-23.

Kim, H., J. Chen, and X. Yu. 2007. Ubiquitin-binding protein RAP80 mediates BRCA1-dependent DNAage response. Science 316: 1202-5.

Koster, D.A., K. Palie, E.S.M. Bot, M.-A. Bjornsti, and N.H. Dekker. 2007. Antitumor drugs impede DNA uncoiling by topoisomerase I. Nature 448: 213-7.

Le Doan, T., L. Perrouault, D. Praseuth, et al. 1987. Sequence-specific recognition, photocrosslinking and cleavage of the DNA double helix by an oligo-[α]-thymidylate covalently linked to an azidoproflavin derivative. Nucl. Acids Res. 15: 7749-51.

Lowe, S.W., E. Cepero, and G. Evan. 2004. Intrinsic tumor suppression. Nature 432: 307-15.

Lu, J., G. Getz, E.A. Miska, et al. 2005. MicroRNA expression profiles classify human cancers. Nature 435: 834-8.

Luo, J., and S.J. Elledge. 2008. Deconstructing oncogenesis. Nature 453: 995-6.

Ma, L., J. Teruya-Feldstein, and R.A. Weinberg. 2007. Tumor invasion nd metastasis initiated by micro RNA-10b in breast cancer. Nature 449: 682-8.

Maeda, T., R.M. Hobbs, T. Merghoub, et al. 2005. Role of the proto oncogene *Pokemon* in cellular transformation and ARF repression. Nature 433: 278-85.

Mantovani, A. 2007. Cancer: An infernal triangle. Nature 448: 547-8.

Marshall, L., N.S. Kenneth, and R.J. White. 2008. Elevated $tRNA_i^{Met}$ synthesis can derive cell proliferation and oncogenic transformation. Cell 133: 78-89.

Massague, J. 2004. G1 cell cycle control and cancer. Nature 432: 298-306.

McMurray, H.R., E.R. Sampson, G. Compitello, et al. 2008. Synergistic response to oncogenic mutation defines gene class critical to cancer phenotype. Nature 453: 1112-6.

Meltzer, P.S. 2005. Cancer genomics: Small RNAs with big impacts. Nature 435: 745-6.

Meyerson, M. 2007. Cancer: Broken genes in solid tumors. Nature 448: 545-6.

Mizuno, T., M.Y. Chou, and M. Inouye. 1984. A unique mechanism regulating gene expression: translational inhibition by a complementary RNA transcript (micRNA). Proc. Natl. Acad. Sci. USA 81: 1966-70.

Moser, H.F., and P.B. Dervan. 1987. Sequence-specific cleavage of double helical DNA by triple helix formation. Science 238: 645-50.

Murad, F., C.K. Mittal, W.P. Arnold, S. Katsuki, and H. Kimura, 1978. Guanylate cyclase: Activation by azide, nitro compounds, nitric oxide, and hydroxyl radical and inhibition by hemoglobin and myoglobin. Adv. Cyclic Nucl. Res. 9: 145-58.

Murray, A.W., and T. Hunt. 1993. *The Cell Cycle*. New York: Freeman.

Nebreda, A. R., J.V. Gannon, and T. Hunt. 1995. Newly synthesized protein(s) must associate with p34cdc2 to activate MAP kinase and MPF, EMBO J. 14: 5597-607.

Ngo, V.N., R.E. Davis, L. Lamy, et al. 2006. A loss-of-function RNA interference screen for molecular targets in cancer. Nature 441: 106-10.

Nurse, P. 1975. Genetic control of cell size at cell division in yeast. Nature 256: 547-51.

Nurse, P. 1990. Universal control mechanism regulating onset of M-phase. Nature 344: 503-5.

Nurse, P. 2000. A long twentieth century of the cell cycle and beyond. Cell 100: 71-8.

Nurse, P., and P. Thuriaux, 1980. Regulatory genes controlling mitosis in the fission yeast *Schizosaccharomyces pombe*. Genetics 96: 627-37.

Nurse, P., and P. Thuriaux. 1977. Controls over the timing of DNA replication during the cell cycle of fission yeast. Exp. Cell Res.107: 365-75.

Nurse, P., Y. Masui, and L. Hartwell. 1998. Understanding the cell cycle. Nature Med. 4: 1103-6.

O'Donnell, K.A., E.A. Wentzel, K.I. Zeller, C.V. Dang, and J.T. Mendell. 2005. C-myc-regulated microRNAs modulate E2F1 expression. Nature 435: 839-43.

Pall, M.L. 1981. Gene amplification model of carcinogenesis. Proc. Natl. Acad. Sci. USA 78(4): 2465-8.

Rajagopalan, H., and C. Lengauer, 2004. Aneuploidy and cancer. Nature 432: 338-41.

Rous, P. 1911. A sarcoma of the fowl transmissible by an agent separable from the tumor cells. J. Exp. Med. 13: 397-411.

Shih, C., and R.A. Weinberg. 1982. Isolation of a transforming sequence from a human bladder carcinoma cell line. Cell 29: 161-9.

Sklar, M.D., E. Thpompson, M.J. Welsh, et al. 1991. Deletion of *c-myc* with specific antisense sequences reverse the transformed phenotype in *ras* oncogene-transformed NIH 3T3 cells. Mol. Cell. Biol.11: 3699-710.

Smith, C.C., L. Aurelian, M.P. Reddy, and P.S. Miller. 1986. Antiviral effect of an oligo(nucleoside methylphosphonate) complementary to the splice junction of herpes simplex virus type 1 immediate early pre-mRNAs 4 and 5. Proc. Natl. Acad. Sci. USA 83: 2787-91.

Smith, K.A. 1988. Interleukin-2: inception, impact and implications. Science 240: 1169-1176.

Sobhian, B., G. Shao, D.R. Lilli, et al. 2007. RP80 targets BRCA1 to specific ubiquitin structures at DNA damage site. Science 316: 1198-202.

Soda, M., Y.L. Choi, M. Enomoto, et al. 2007. Identification of the transforming EML4–ALK fusion gene in non-small-cell lung cancer. Nature 448: 561-6.

Spector, D.H., B. Baker, H.E. Varmus, and J.M. Bishop. 1978a. Characteristics of cellular RNA related to the transforming gene of avian sarcoma viruses. Cell 13: 381-6.

Spector, D.H., K. Smith, T. Padgett, et al. 1978b. Uninfected avian cells contain RNA related to the transforming gene of avian sarcoma viruses. Cell 13: 371-9.

Stanbridge, E.J. 1990. Human tumor suppressor genes. Annu. Rev. Genet. 24: 615-57.

Stites, E.C., P.C. Trampont, Z. Ma, and K.S. Ravichandran. 2007. Network analysis of oncogenic Ras activation in cancer. Science 318: 463-7.

Stratton, M.R., P.J. Campbell, and P.A. Futreal. 2009. The cancer genome. Nature 458: 719-24.

Suh, E.-K., A. Yang, A. Kettenbach, et al. 2006. P63 protects the female germline during meiotic arrest. Nature 444: 624-8.

Suzuki, H.I., K. Yamagata, K. Sugimoto, S. Kato, and K. Miyazono. 2009. Modulation of microRNA processing by p53. Nature 460: 529-33.

Toledo, F., and B. Bardot. 2009. Cancer: three birds with one stone. Nature 460: 466-7.

Tomizawa, J. 1986. Control of ColE1 plasmid replication: binding of RNA I to TNA II and inhibition of primer formation. Cell 47: 89-97.

Tomizawa, J., T. Itoh, T., G. Salzer, and T. Som. 1981. Inhibition of ColE1 RNA primer formation by a plasmid-specified small RNA. Proc. Natl. Acad. Sci. USA 78: 1421-5.

Vander Heiden, M.G., L.C. Cantley, and C.B. Thompson. 2009. Understanding the Warburg effect: The metabolic requirements of cell proliferation. Science 324: 1029-33.

Varambally, S., Q. Cao, R.-S. Mani, et al. 2008. Genomic loss of microRNA-101 leads to overexpression of histone methyltransferase EZH2 in cancer. Science 322: 1695-9.

Varmus, H.E., R.A. Weiss, R. Friis, W.E. Levinson, and J.M. Bishop, 1972. Detection of avian tumor virus-specific nucleotide sequences in avian cell DNAs. Proc. Natl. Acad. Sci. USA 69: 20-4.

Varmus, H.E., S. Heasley, and J.M. Bishop. 1974. Use of DNA-DNA annealing to detect new virus-specific dna sequences in chicken embryo fibroblasts after infection by avian sarcoma virus. J. Virol. 14: 895-903.

Varmus, H.E., W.E. Levinson, and J.M. Bishop. 1971. Extent of transcription by the RNA-dependent DNA polymerase of Rous sarcoma virus. Nat New Biol. 233(35): 19-21.

Ventura, A., D.G. Kirsch, M.E. McLaughlin, et al. 2007. Restoration of p53 function leads to tumor regression in vivo. Nature 445: 661-5.

Viale, A., F. De Franco, A. Orleth, et al. 2009. Cell cycle restriction limits DNA damage and maintains self-renewal of leukemia stem cells. Nature 457: 51-6.

Vujanovic, N.L ., R.B. Herberman, and J.C. Hiserodt. 1988a. Lymphokine activated killer cells in rats. 1. Analysis of tissue and strain distribution, ontogeny and target specificity. Cancer Res. 48: 878-83.

Vujanovic, N.L., R.B. Herberman, R.R. Salupet, et al. 1988b. Lymphokine activated killer cells in rats. 11. Analysis of progenitor and effector cell phenotype and relationship to natural killer cells. Cancer Res. 48: 884-90.

Wan, P.T.C., M.J. Garnett, S.M. Roe, et al. 2004. Mechanism of activation of the RAF-ERK signaling pathway by oncogenic mutations of B-RAF. Cell 116: 855-67.

Wang, Y., E.L. Botvinick, Y. Zhao, et al. 2005. Visualizing the mechanical activation of Src. Nature 434: 1040-5.

Weinberg, R.A. 1988. Finding the antioncogene. Scient. Am. 259(3): 34-41.

Weinert, T. A., and L.H. Hartwell. 1988. The RAD9 gene controls the cell cycle response to DNA damage in *Saccharomyces cerevisiae*. Science 241: 317-22.

Weiser, K.C., and M.J. Justice. 2005. Cancer biology: sleeping beauty awakens. Nature 436: 184-6.

Wickstorm, E.L., T.A. Bacon, A. Gonzalez, D.L. Freeman, G.H. Lyman, and E. Wickstorm. 1988. Human promyelocytic leukemia HL-60 cell proliferation and c-myc protein expression are inhibited by an antisense pentadecadeoxynucleotide targeted against *c-myc* mRNA. Proc. Natl. Acad. Sci. USA 85: 1028-32.

Xues, W., L. Zender, C. Miething, et al. 2007. Senescence and tumor clearance is triggered by p53 restoration in murine liver carcinomas. Nature. 445: 656-60.

Yunis, J. 1986. Chromosomal rearrangements, genes, and fragile sites in cancer: Clinical and biologic implications. In: *Important Advances in Oncology*. 1986. DeVita, V.T. Jr., Hellman, S. and Rosenberg, S.A. (ed.). pp. 93-128. Philadelphia: Lippincott.

Zamercnik, P.C., and M.L. Stephenson. 1978. Inhibition of Rous sarcoma virus replication and cell transformation by a specific oligonucleotide. Proc. Natl. Acad. Sci. USA 75: 280-284.

zur Hausen, H. 1976 . Condylomata acuminata and human genital cancer. Cancer Res. 36(2 pt 2): 794.

zur Hausen, H., H. Schulte-Holthausen, H. Wolf, K. Dörries, and H. Egger. 1974b. Attempts to detect virus-specific DNA in human tumors. II. Nucleic acid hybridizations with complementary RNA of human herpes group viruses. Int. J. Cancer 13: 657-64.

zur Hausen, H., H. Schulte-Holthausen, G. Klein, et al. 1970. EBV DNA in biopsies of Burkitt tumors and anaplastic carcinomas of the nasopharynx. Nature 228: 1056-8.

zur Hausen, H., L. Gissmann, W. Steiner, W Dippold, and I. Dreger. 1975. Human papilloma viruses and cancer. Bibl. Haematol. 43: 569-71.

zur Hausen, H., W. Meinhof, W. Scheiber, and G.W. Bornkamm. 1974a. Attempts to detect virus-secific DNA in human tumors. I. Nucleic acid hybridizations with complementary RNA of human wart virus. Int. J. Cancer 13: 650-6.

5

Gene Function

What function does gene perform? Answer to this question came by understanding relationship between gene and its product. First crucial step was to understand the genetic control of biochemical reactions. Relationship between genotype and phenotype was initially understood in *Drosophila*. Better understanding about function of gene came from work on nutritional mutants of *Neurospora*, *Salmonella*, and *Escherichia coli*. Final picture about gene function became clear as the fine structure of gene was known.

RELATIONSHIP BETWEEN GENE AND ENZYME

First clue to the nature of the primary gene function came from the studies on humans. Garrod (1909) noted that several hereditary human defects were produced by recessive mutations. Various human disorders of tyrosine metabolism, studied by Garrod are given in Figure 5.1. In case of phenylketonuria, large amounts of phenylalanine is accumulated and the affected person has lighter pigment but is not a complete albino since tyrosine is available in diet. In case of alkaptonuria, large amounts of homogentisic acid are released in urine. Urine turns black in presence of light. Various other metabolic disorders have been attributed to an absence or a defect in an enzyme. Thus, a relationship between genes and enzymes, which are responsible for conducting various biochemical reactions, was suggested.

GENETIC CONTROL OF BIOCHEMICAL REACTIONS

Lawrence and Sturgess (1957) began the biochemical analysis of flower pigments and eventually showed a precise correlation between genetic and biochemical changes. For example, in the anthocyanin pigments of the cape primrose (*Streptocarpus*), three individual biochemical effects were traced to separate genes. The three genes, called *R*, *O* and *D*, appeared to produce their effects by adding or subtracting hydroxyl (OH) units, methoxyl (O-CH_3) units, or sugar molecules. The *rroodd* triple recessive is salmon-colored, containing mainly the pigment whose structure is shown in Figure 5.2(A). Presence of the dominant allele *R* (*R-oodd*) results in hydroxylation or methoxylation at the 3' position and causes a change to primrose color (Figure 5.2(B). The addition of O causes hydroxylation or methoxylation at both the 3' and 5' positions, and the color changes to mauve (Figure 5.2C). In plants dominant for gene *D*, hexose sugars are found at the 3' and 5' positions instead of the hexose-pentose combination at the 3' position, and the anthocyanin appears bluer.

Investigations of numerous other plants have shown similar findings; the anthocyanin pigments are usually bluer with the addition of sugar and hydroxyl groups, and somewhat redder when the hydroxyl groups are methylated. Acidity of the cell, another hereditary factor in some plants, is also of

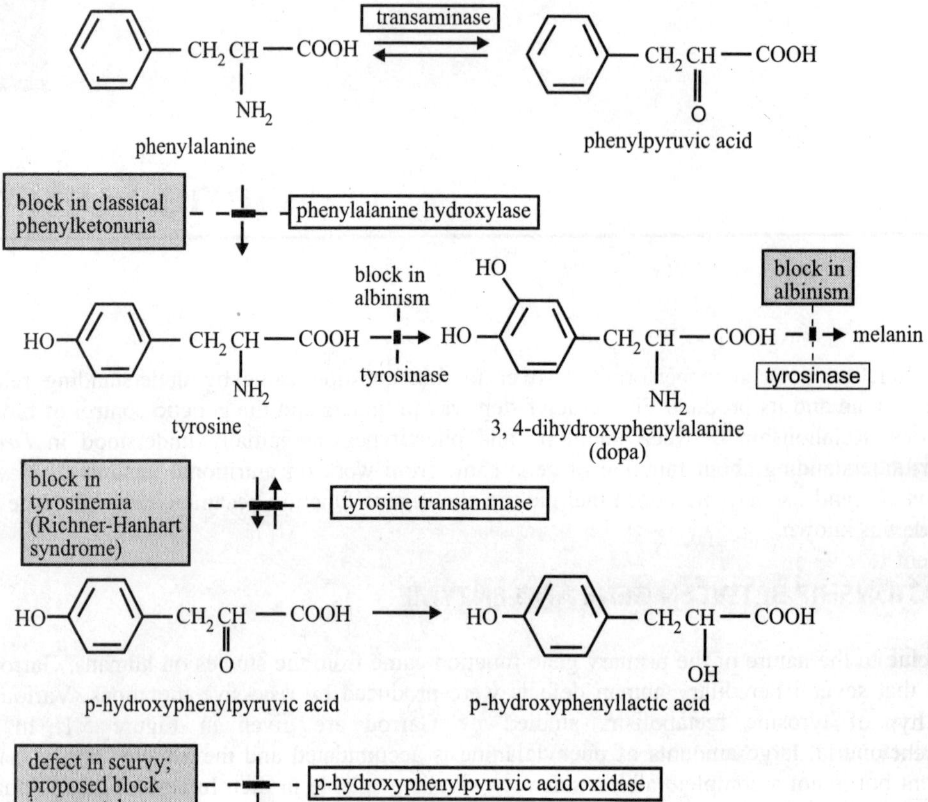

transaminase

phenylalanine

phenylpyruvic acid

block in classical phenylketonuria

phenylalanine hydroxylase

block in albinism

tyrosinase

tyrosine

3, 4-dihydroxyphenylalanine (dopa)

block in albinism

tyrosinase

melanin

block in tyrosinemia (Richner-Hanhart syndrome)

tyrosine transaminase

p-hydroxyphenylpyruvic acid

p-hydroxyphenyllactic acid

defect in scurvy; proposed block in neonatal tyrosinemia

p-hydroxyphenylpyruvic acid oxidase

block in albinism

tyrosinase

homogentisic acid

homogentisic acid oxidase

maleylacetoacetic acid

isomerase

block in tyrosinosis

hydrolase

acetoacetic acid

fumaric acid

fumarylacetoacetic acid

Figure 5.1 Some biochemical reactions in humans beings with metabolism of phenylalanine. Various metabolic blocks are shown in filled boxes. Defective enzymes responsible for a metallic disorder are mentioned in empty boxes

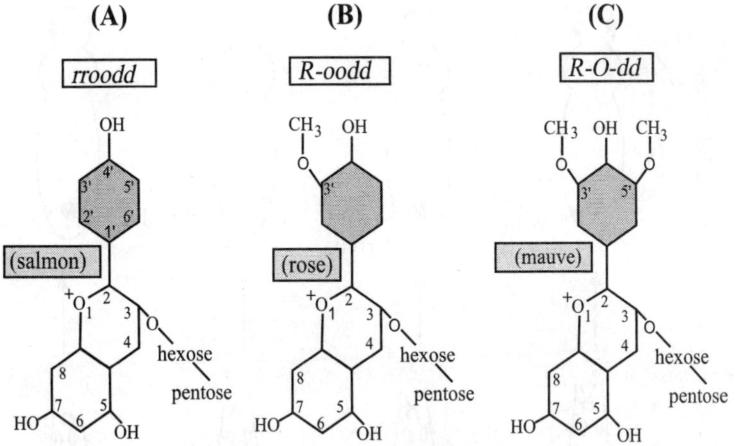

Figure 5.2 Anthocyanin pigments produced by some genotypes among garden forms of Streptocarpus

considerable importance, since dyes, such as cyanin, change from blue to red with increased acidity. Pigment inheritance, therefore, helped to show the effect of genes in controlling specific biochemical steps.

ONE GENE-ONE REACTION HYPOTHESIS

This hypothesis states that every specific biochemical reaction is under the ultimate control of a different gene. It was later refined as onc-gene one-enzyme hypothesis.

RELATIONSHIP BETWEEN GENOTYPE AND PHENOTYPE

Beadle and Ephrussi (1936) and Beadle and Ephrussi (1937) conducted eye disc transplantation experiments in *Drosophila melanogaster*. The experimental approach used by them is illustrated in Figure 5.3. The results of the experiment are explained in Table 5.1. Experiments 1-5 show non-autonomous development. In this case, *vermillion* (*v*), or *cinnabar* (*cn*) transplanted discs developed into normal eyes in wild-type host. Wild-type host in this case provided a diffusible substance that the discs used to bypass the genetic block and to make brown pigment. Experiment 6 is a case of autonomous development. Here wild type host was unable to provide some substance to the disc to enable the brown pigment to be produced. Experiments 7 and 8 showed that production of brown pigment involved a biochemical sequence with at least two precursors, v^+ and cn^+; wild-type having both v^+ and cn^+ substances, *cinnabar* having one (v^+), and vermillion having neither. The *vermillion* host is deficient in v^+ substance so no brown pigment can be produced. *Cinnabar* host makes up for the deficiency of *vermillion* disc by supplying it with diffusible substance.

Sequence of metabolic events suggested by the experiments of Beadle and Ephrussi (1937) are given in Figure 5.4. Two main types of pigments are known to be present in *Drosophila* eyes, the pterins (red pigments) and ommochromes (brown pigments), each effected by different set of genes. Both these pigments function by becoming attached to protein granules in *Drosophila*, the form in which they are deposited in ommatidia. A fly deprived of both types of pigments will have white eyes. A fly not capable of forming ommochromes will show only bright pterin pigments, e.g., *vermillion*, *scarlet*, or *cinnabar* mutants. A fly in which pterins are absent will have dull eye color, such as *brown*,

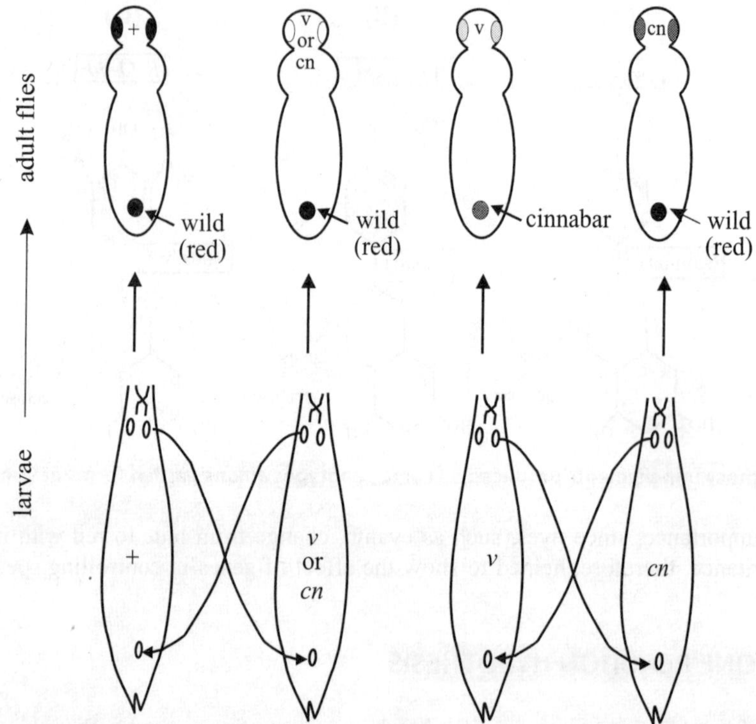

Figure 5.3 Eye transplantation experiment in *Drosophila* by Beadle and Ephrussi (1937) showing autonomous and nonautonomous development of implanted larval imaginal discs into various phenotypes (+, wild type; *v*, vermillion; *cn*, cinnabar)

Table 5.1 Results of eye disc transplantation experiment (Reprinted, with permission, from Beadle, G.W. and B. Ephrussi. 1937. Genetics 21: 225-47)

Experiment No.	*Source of eye disc*	*Host fly*	*Color of transplanted eye in adult*
1	+	*v*	+
2	*v*	+	+
3	+	*cn*	+
4	*cn*	+	+
5	+	*st*	+
6	*st*	+	*st*
7	*cn*	*v*	*cn*
8	*v*	*cn*	+

(*v*, vermillion; *cn*, cinnabar; *st*, scarlet; +, wild-type)

raspberry, or *garnet*. Flies deprived of both the ommochromes and the pterins, such as double mutants *cinnabar brown* and *scarlet brown* will have no eye color and hence will appear white. This work indicated a strong link between phenotype (eye color) and genotype (*v* and *cn*). Beadle found *Drosophila* too complex an organism to address this question of gene function further. He chose for his further work a simple unicellular eukaryote, *Neurospora*.

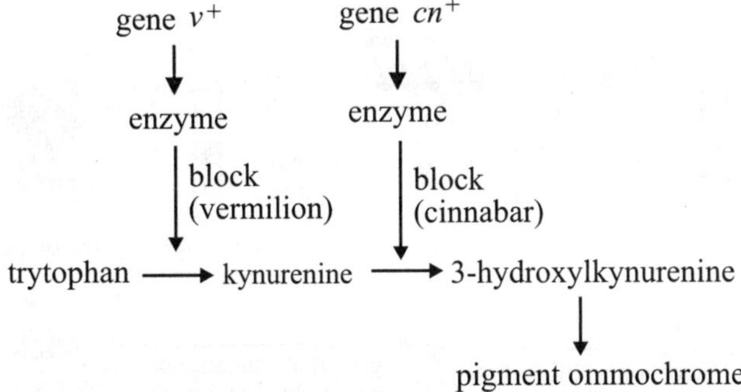

Figure 5.4 Some biochemical reactions for biosynthesis of eye

ONE GENE-ONE ENZYME HYPOTHESIS

Nutritional Mutants of *Neurospora*

The nature of gene function was substantially specified by Beadle and Tatum (1941), Srb and Horowitz (1944), and Beadle and Tatum (1945) when they showed, using *Neurospora crassa* (Ascomycetes) as their experimental organism, that genes control the synthesis of enzymes, and in particular that each individual gene is responsible for the synthesis of a single enzyme. G.W. Beadle and E.L. Tatum were awarded a Nobel Prize in Medicine or Physiology in 1958 for discovering that one gene regulates one definite chemical process. They induced nutritional mutants with X-rays in bread mould *N. crassa*. Procedure used in their experiments is shown in Figure 5.5. A nutritional mutant would grow only when minimal medium was supplemented with a substance that the organism itself could not synthesize. The inability of the mutant to synthesize a substance was found to be due to absence or defect in one enzyme. Genetic analysis showed that the defect in the enzyme was due to a change in single nuclear gene. This led to the proposal of one gene-one enzyme hypothesis. This hypothesis states that each gene controls the reproduction, function and specificity of a particular enzyme. In this way, Beadle and Tatum demonstrated relationship between gene and enzyme. Main ideas of this hypothesis were elaborated in the following form by Tatum: "All biochemical processes in all organisms are under genic control. The overall biochemical processes are resolvable into a series of stepwise reactions. Each single reaction is controlled in a primary fashion by a single gene, or in other terms, in every case a 1:1 correspondence of gene and biochemical reaction exists, such that mutation of a single gene results only in alteration in the ability of the cell to carry primary chemical reaction". Ample support to one gene-one enzyme hypothesis came from following studies on microorganisms.

Histidine Metabolism in *Salmonella*

More than 500 mutations for histidine metabolism in Salmonella were mapped. These were linked in nine closely linked loci, *A* to *H*, each affecting a different enzyme in the synthesis of histidine. The gene sequence was found to follow largely the metabolic sequence shown in Figure 5.6.

Arginine Synthesis in *Neurospora*

In *Neurospora*, following three kinds of mutants for arginine synthesis were isolated: (a) mutants which grow only when arginine is supplied and do not grow with the help of either ornithine or

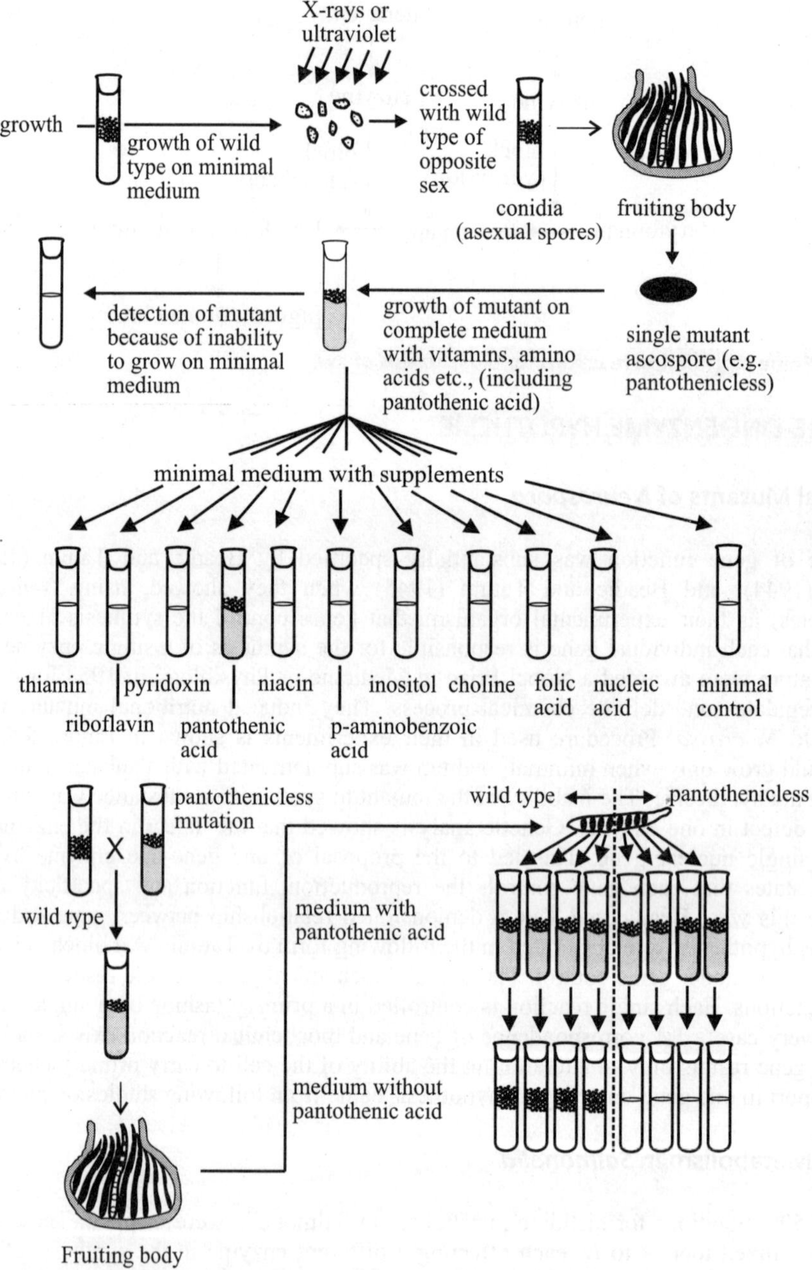

Figure 5.5 Procedure for induction and detection of nutritional mutations in *Neurospora crassa*

citrulline alone, (b) mutants which grow only when either citrulline or arginine is supplied but do not grow when only ornithine is supplied, and (c) mutants which can grow only when either ornithine or citrulline or arginine is supplied. These mutations indicate that in the first case, mutants are incapable of utilizing ornithine or citrulline; in the second case, mutants are incapable of utilizing ornithine; but

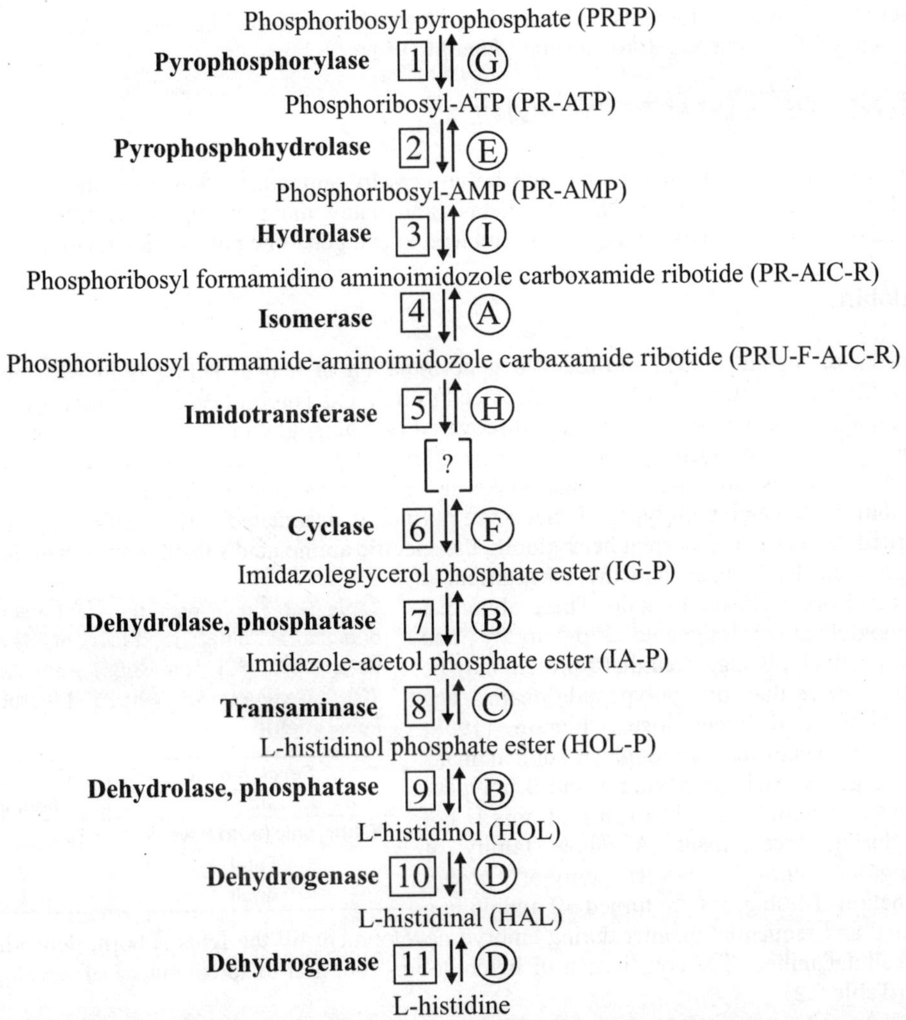

Phosphoribosyl pyrophosphate (PRPP)

Pyrophosphorylase [1] ⇅ (G)

Phosphoribosyl-ATP (PR-ATP)

Pyrophosphohydrolase [2] ⇅ (E)

Phosphoribosyl-AMP (PR-AMP)

Hydrolase [3] ⇅ (I)

Phosphoribosyl formamidino aminoimidozole carboxamide ribotide (PR-AIC-R)

Isomerase [4] ⇅ (A)

Phosphoribulosyl formamide-aminoimidozole carbaxamide ribotide (PRU-F-AIC-R)

Imidotransferase [5] ⇅ (H)

[?]

Cyclase [6] ⇅ (F)

Imidazoleglycerol phosphate ester (IG-P)

Dehydrolase, phosphatase [7] ⇅ (B)

Imidazole-acetol phosphate ester (IA-P)

Transaminase [8] ⇅ (C)

L-histidinol phosphate ester (HOL-P)

Dehydrolase, phosphatase [9] ⇅ (B)

L-histidinol (HOL)

Dehydrogenase [10] ⇅ (D)

L-histidinal (HAL)

Dehydrogenase [11] ⇅ (D)

L-histidine

Figure 5.6 Pathway of histidine biosynthesis. Encircled letters indicate genes and numbers in rectangles symbolize enzymes. There are 10 enzymes and 11 reactions in this pathway (Redrawn, with permission, from Miglani, G.S. 2006. *Developmental Genetics*. IKI International Publishing House, Pvt. Ltd., New Delhi)

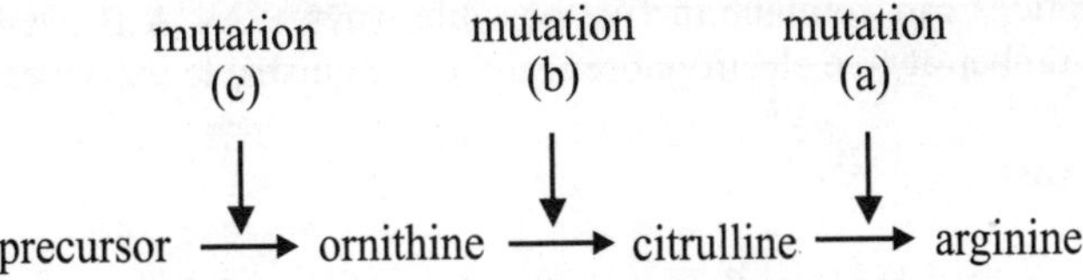

Figure 5.7 Different steps in synthesis of arginine and three different mutations blocking three steps

in the last case, only the precursor cannot be utilized (Newmeyer 1957). The biosynthetic pathway suggested on the basis of studies on these mutations is shown in Figure 5.7.

Later studies directed towards understanding of gene function had bearing on the validity of one gene-one enzyme hypothesis (Horowitz and Leupold 1951).

ONE GENE-ONE POLYPEPTIDE HYPOTHESIS

One gene-one enzyme hypothesis could not be universally applied. Studies on mutant hemoglobins, lactate dehydrogenase and tryptophan synthetase and many more examples available in literature challenged one gene-one enzyme hypothesis and lead to one gene-one polypeptide relationship.

Hemoglobin

Experiments of Ingram (1957) showed that gene mutation in human hemoglobin was the basis of chemical difference between normal and sickle-cell. Smith and Torbert (1958) and Itano and Robinson (1960) studied abnormal hemoglobins and discovered that hemoglobin was made up of two different polypeptides, each controlled by a separate gene. Identical pairs of chemically different polypeptide chains, the α-chains and the β-chains, make up protein portion of human adult hemoglobin. Hemoglobin A, in which both types of chains are normal, is designated $\alpha_2^A\beta_2^A$. Either type may carry the inherited defect of an abnormal hemoglobin; the specific amino acid substitution of hemoglobin S Or C occurs in the β-chain, and the substitution of hemoglobin I occurs in the α-chain. These abnormal adult hemoglobins are designated $\alpha_2^A\beta_2^S$, $\alpha_2^A\beta_2^C$, and $\alpha_2^I\beta_2^A$, respectively. It suggested that a protein may be made up of more than one polypeptide chain, each controlled by a different locus. Ingram (1958) explained how genes act. Although in adult humans only single genes, each specifying α and β polypeptide chains of hemoglobins are expressed, this is not enough during development. A whole family of related *α globin* genes and another family of *β globin* genes function. These genes are turned off and on in a coordinated and sequential manner during embryo development till the fetus is born, depending upon the intracellular milieu. The constitution of hemoglobins found at different stages of development is shown in Table 5.2.

Table 5.2 The constitution of hemoglobins in humans at different stages of development (Reprinted, with permission, from Gupta, P.K. 2009. Genetics. 4th edition. Meerut: Rastogi Publications)

Development stage	*Constitution of hemoglobin*
Embryonic (upto 8 weeks)	$\xi_2\varepsilon_2$, $\xi_2\gamma_2$ and $\alpha_2\varepsilon_2$
Fetal	$\alpha_2\gamma_2$
Adult	$\alpha_2\delta_2$, $\alpha_2\beta_2$

Lactate Dehydrogenase

Lactate dehydrogenase (LDH) in horse is a tetrameric enzyme which comprises of two polypeptides A and B. These polypeptides can combine in five possible ways – A_4, A_3B_1, A_2B_2, A_1B_3, and B_4, thus revealing maximum number of five electrophoretically distinguishable isozymes.

Tryptophan Synthetase

Tryptophan synthetase in, Neurospora, *Salmonella typhimurium* and *E. coli* (Table 5.3 and Figure 5.8) is made up of two polypeptides A and B, each synthesized by a gene (Brenner 1955).

Thus one gene-one enzyme relationship does not hold good for all the structural (protein-encoding) genes. This led to the concept of one gene-one polypeptide hypothesis. One gene-one polypeptide hypothesis states that one gene controls the synthesis of one polypeptide. Wagner and Mitchell (1964) gave a general account of genetics and metabolism.

Table 5.3 Effect of different media on the growth response of tryptophan mutations in *Salmonella typhimurium* (Reprinted, with permission, from Brenner, S. 1955. Proc. Natl. Acad. Sci. USA 41: 862-3)

Tryptophan gene mutation	*Minimal medium*	*Medium supplemented with*				*Accumulated substance*
		Ant	*IGP*	*I*	*Tryptophan*	
trp-8	–	+	+	+	+	–
trp-2, -4	–	–	+	+	+	Ant
trp, -3	–	–	–	+	+	IGP
trp, -1, -6, -7, -9, -10, -11	–	–	–	–	+	IGP & I

+, growth; –, no growth; Ant, Anthranilic acid; IGP, Indole glycerol phosphate; I, Indole

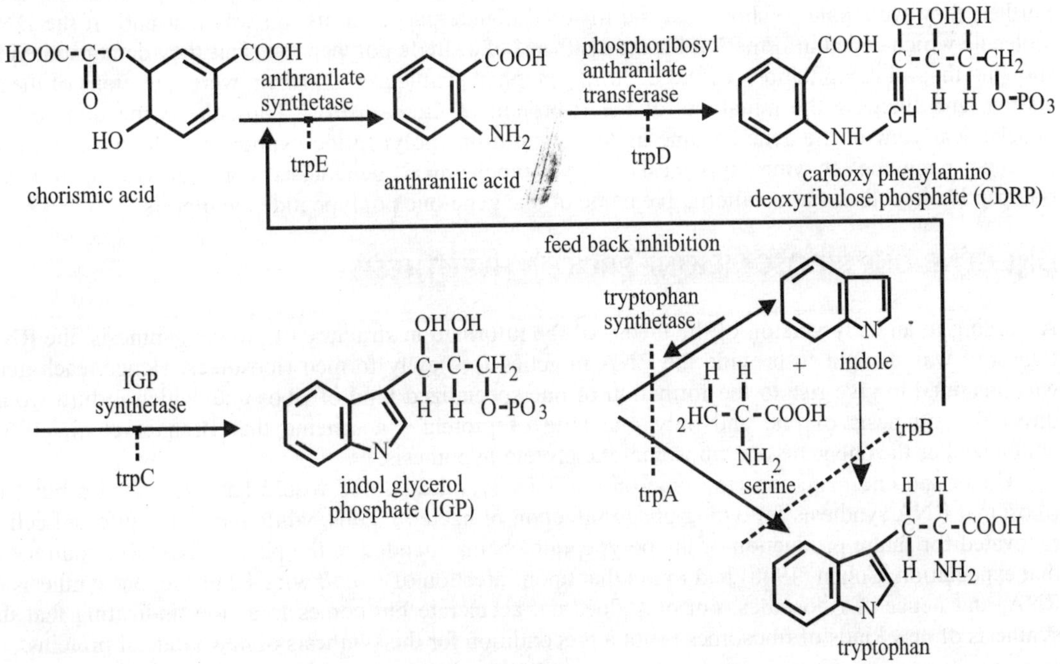

Figure 5.8 Biosynthetic pathway of tryptophan from chrismic acid showing the enzymes and genes at each step

ONE GENE-ONE CHROMOMERE HYPOTHESIS

In a very detailed study, Judd et al. (1972) identified and characterized a large number of mutations in one region of X chromosome. They identified 16 genes (complementation groups) in a segment of the chromosome containing 15 distinct bands. Similarly, the small chromosome 4 of *Drosophila* has 43 genes (units of function) and has between 33 and 50 bands indicating a good fit to the one gene-one band hypothesis. The conclusions drawn from the results obtained from cytogenetic analysis of the 3A-3C region of the X chromosome are: (a) There is one essential function expressed by genetic information contained in each chromomere. (b) The one chromomere-one function relationship holds regardless of the amount of DNA the chromomere contains. They thus proposed that the chromomere is a relatively short length of structural DNA together with a variable length of regulatory DNA.

ONE GENE-ONE ANTIGEN HYPOTHESIS

This hypothesis states that each of the genes in a tissue transplant governing the host's reaction to it is responsible for the manufacture of a particular transplant antigen alone (Cammack et al. 2006).

ONE CISTRON-ONE POLYPEPTIDE HYPOTHESIS

Yanofsky et al. (1964) compared the genetic map of the tryptophan synthetase gene of *E. coli* with the corresponding primary structure of the polypeptide. In other words, Yanofsky's group demonstrated the co-linearity of mutant nucleotide sites and the corresponding mutated aminoacyl polypeptide sites. Further, they were able to show that the material counterpart of a cistron was that part of the DNA molecule which coded information for the synthesis of a single polypeptide. This they demonstrated by showing that all the mutations of the *E. coli* tryptophan synthetase A-protein were mutations of the *A*-cistron, and likewise the mutations of the B-protein of the *B*-cistron. Thus, the cornerstone of the neoclassical view of the gene became the one cistron-one polypeptide hypothesis, which replaced the old one gene-one enzyme hypothesis. Some biochemical geneticists consider one cistron-one polypeptide hypothesis as an alternative name of one gene-one poplypeptide hypothesis.

ONE GENE-ONE RIBOSOME-ONE PROTEIN HYPOTHESIS

According to an early version of the theory of the information structure of protein synthesis, the RNA transcript was thought to provide the RNA moieties for newly formed ribosomes. Hence, each gene was imagined to give rise to the formation of one specialized kind of ribosome, which in turn would direct the synthesis of one and only one kind of protein - a scheme that Brenner et al. (1961) epitomized as the one gene-one ribosome-one protein hypothesis.

Under the one gene-one ribosome-one protein hypothesis, one would have expected a burst of ribosomal RNA synthesis following phage infection of bacterial cells, while the phage-infected cell is renovated for future production of the polypeptide chains encoded in the phage DNA. But contrary to that expectation, Cohen (1948) had found that upon infection of *E. coli* with T2 phage, net synthesis of RNA, and hence of ribosomes, not only does not accelerate but comes to a stop, indicating that the synthesis of new kinds of ribosomes is not a precondition for the synthesis of new kinds of proteins.

ONE GENE-ONE mRNA-ONE PROTEIN HYPOTHESIS

One gene-one ribosome-one protein hypothesis was rejected when it was reported that RNA is translated into protein by ribosome and this led to one gene-one mRNA-one protein hypothesis. Volkin et al (1958) found that the purine-pyrimidine base composition of the RNA of bacterial cells after T7 phage infection is significantly different than that of *E. coli* ribosomal RNA and instead resembles more closely that of the T7 phage DNA, indicating that the RNA was virus-specific. Subsequently, Volkin (1960) measured a rapid turnover of RNA formed after infection, it having a short half-life of the order of a few minutes, and found that phage production would not occur in the absence of this fraction. Thus, the hypothesis developed that protein synthesis was concluded by the unstable virus-specific RNA and not by the ribosomal RNA.

Brenner et al. (1961) then showed with the aid of heavy carbon isotopes and ultra-centrifugation that in phage infection it was indeed the RNA of the phage and not the bacterium which was responsible for the synthesis of the phage coat protein. Jacob and Monod (1961) had proposed the name messenger RNA (mRNA) for this protein-synthesis-conducting RNA. Thus, the neoclassical view of the gene culminated in a theory according to which one gene or cistron controls the synthesis of one mRNA molecule, which in turn controls the synthesis of one polypeptide.

ONE GENE-ONE PRIMARY CELLULAR FUNCTION HYPOTHESIS

Above mentioned hypotheses put forward to explain gene function were based on the studies on protein-encoding genes. In addition to the structural genes, which code for polypeptides, there are other genes (tRNA genes and rRNA genes) which produce non-genetic RNAs required in protein biosynthesis. The major three types of non-genetic RNAs (mRNAs, tRNAs and rRNAs) are considered primary cellular products of genes. Thus one gene was noted to perform one primary cellular function. This gave birth to one gene-one primary cellular function hypothesis. Woods (1973) reviewed biochemical aspect of gene.

GENE DISCOVERIES

Discoveries of new types of genes in the 1960s and later made the neoclassical concept of gene obsolete. The following observations led to a situation where none of the classical or the neo-classical criteria of the definition of the gene hold strictly true.

Repeated Genes

Waring and Britten (1966) and Britten and Kohne (1968) were the first to observe repeated DNA sequences. The genes of ribosomal RNA are repeated in several tandem copies. Each one consists of one transcription unit, but the gene cluster is usually transmitted from one generation to the next as a single unit. Thus, the units of transmission and transcription are not always the same. Likewise, the histone genes have been observed to be repeated in such tandem repeats in many higher eukaryotic organisms by Lewin (1980).

Multiple Polyadenylation Sites

During the maturation of messenger RNA, i.e., during the process in which the primary transcript ripens to form messenger RNA, about 200 adenosine nucleotides are added in a polyadenylation reaction at the 3′-end. These are not coded by the corresponding gene. In certain cases, there are multiple alternative polyadenylation sites in the primary transcript. This was first observed in adenoviruses. Alternative polyadenylation sites usually involve the untranslated trailer sequence in the messenger RNA, but they can also involve translated sequences, and in this case they can affect the structure of the encoded protein. Thus, multiple polyadenylation sites are one mechanism whereby a single gene can control the synthesis of more than one polypeptide.

Moveable Genes

Movable genes are DNA elements that can move from one location to another in the genome of an organism. Already in the 1940s, Barbara McClintock explained certain variegated phenotypes of maize by means of movable genes, which she called "control elements". At the time, however, these elements appeared so odd that nobody really knew what to think of them. Nowadays movable genes have been found in virtually all organisms and their molecular nature is quite well known. Consequently, mobile genetic elements became one of the most important discoveries of genetics, and Barbara McClintock was awarded the Nobel Prize for Physiology or Medicine in 1983 at the age of 81 years. The existence of movable genes shows that the hypothesis of a fixed location of the gene in the chromosome, adopted by both the classical and neoclassical view, does not necessarily hold true.

Pseudogenes

Pseudogenes are DNA sequences significantly homologous to a functional gene which, however, have been altered so as to prevent any normal function (Rieger et al. 1991). The first pseudogene was found in the 5S DNA gene family of *Xenopus laevis* by Jacq et al. (1977). Pseudogenes are a common feature of many multigene families in higher eukaryotes. In fact, pseudogenes are ancient genes which have lost their function, and pseudogenes which have in the past been protein coding genes usually contain many stop codons, i.e., their reading frame has been closed during the course of evolution.

Protein *Trans*-Splicing

An intein is a segment of a protein that is able to excise itself and rejoin the remaining portions (the exteins) with a peptide bond. Inteins have also been called protein introns. Intein-mediated protein splicing is a self-catalytic process in which the intervening intein sequence is removed from a precursor protein and the flanking extein segments are ligated with a native peptide bond. Splice junction proximal residues and internal residues within the intein direct these reactions. The protein *trans*-splicing poses problem for gene concept that start and end sites are not determined by genes.

Retrogene

In certain genes there is reverse transcription of an mRNA and insertion of the product into genome. This is a case of RNA to DNA flow of information. Some examples of retrogenes are *RPL36AL*, *HSPA2*, *SPIN2B*, and *FAM50B*.

TWO GENES-ONE POLYPEPTIDE HYPOTHESIS

The enormous versatility of antibodies was for a long time a difficult problem in genetics. How was it possible that in the genome there was room for the codes of millions of different antibodies? The matter was solved when it became clear that the functional genes of immunoglobulins mature by means of somatic recombination from a few units in the germline during the maturation of immune cells. An early theory regarding functioning of immunoglobin genes was the selective theory of antibody formation by Jerne (1955). According to Jerne, all the antibodies are preformed (constitutive) in the immune system, and will be combined with the antigens, as a result of which immune complexes will be formed. These complexes are then introduced into the macrophage cells where the antigen is cleaved from the complex, and the information of the antibody is presented to the lymphocytes, which then begin to form the antibodies in huge quantities. Burnet (1959) assumed that natural antibodies do not react with the antigens, but rather that specific clones of immunocompetent cells are selected. But this theory was discarded in favor of somatic recombination.

To determine the chemical basis for rabbit heavy chain allotypes, amino acid analyses were carried out on IgG and IgM antibodies isolated from rabbits which were homozygous a1 or a3 for the γ chain locus and homozygous b4 for the light chain locus. The compositional differences between a1 and a3 IgM antibodies were found to be identical to those between their IgG counterparts (Koshland et al. 1969). The identity of the allotypic amino acid replacements showed that the same genetic markers were present in the variable sequences of IgG and IgM heavy chains. Since the constant sequences of these heavy chains are controlled by different loci, these data demonstrated that rabbit heavy chains are coded by two separate germline genes, one producing the variable region and a second one of various genes producing the constant regions. The use of two antibodies, antiazophenylarsonate and antiazophenyl-β-lactoside, permits the compositions to be compared also on the basis of immunological specificity. The specificity amino acid replacements were found to be identical in the IgM and IgG heavy chains whether the two antibodies were isolated from and individual animal of a1 or a3 allotype. These results further supported the conclusions of the allotype measurements that at least two genes coded for the rabbit heavy chain.

The lambda chain seems to be encoded by two separate germline genes (a specificity region gene and a common region gene) which are expressed as a single, continuous polypeptide chain (Hood and Ein 1968). Antibody light chains appear to be an exception to the rule of one gene-one polypeptide chain hypothesis. It has been suggested to call these regions as “gene segments” or “genelets”.

Each antibody molecule is a tetramer, consisting of two identical light chains and two identical heavy chains. Each chain consists of a constant and a variable region. In the genome of the germline, there are many gene segments for the variable region and a few gene segments for the constant region. In somatic recombination, these can be combined during the maturation of the functional antibody gene into several thousands of different combinations whereby millions of different antibodies are formed. This phenomenon was first demonstrated by Hozumi and Tonegawa (1976). The immuoglobulin genes, which can be called assembled genes, do not fit any classical or neoclassical definition of the gene, since the genetic unit in the germline and in the mature immune cell is completely different. The somatic recombination in immunoglobin genes led to many genes-one polypeptide hypothesis. Hood (1972) discusses whether two genes-one polypeptide chain is fact or fiction.

STUDYING GENE FUNCTION BY KNOCKING OFF GENES

O. Smithies was awarded Nobel Prize in 2007 for development of ‘knockout’ technique which allows for shutting off genes in animals to study their function (Smithies et al. 1985). Knockout mice are genetically modified mice that have one or more genes silenced. Knockout mice are used to recreate human diseases in mice and to study the effect of individual genes on an organism's development. Gene targeting is a technique allowing scientists to block or alter a gene's function. It works by infusing strands of lab-made DNA into stem cells, where they latch on to their target genes. These cells are injected into a mouse embryo, spreading the new gene throughout. Embryonic stem cells are collected from fertilized eggs a few days after they have started to divide. The cells are the basic blocks that develop into all of the tissues in an animal. Scientists hope they may lead to treatments of serious medical conditions.

ONE GENE-MANY PROTEINS HYPOTHESIS

Alternative RNA splicing is a mechanism by which more than one protein is made from a single gene. This observation gave rise to one gene-many proteins hypothesis. It means that the old paradigm that one gene makes one protein is clearly in need of revision. Through mechanisms that include "alternative RNA splicing," one gene can direct the synthesis of many proteins. Victor A. McKusick, of the Johns Hopkins University School of Medicine, says, "It seems to be a matter of five or six proteins, on average, from one gene". McKusick suggests that people who now claim that the number of human genes is much higher, may be looking at and counting separate messenger RNAs—the molecules that take information from genes and direct the production of proteins. A new assay uses fiber-optic microarrays to measure alternative RNA splicing (Grabowski 2002). According to Barbara J. Culliton, "Ultimately it will be necessary to measure mRNA in specific cell types to demonstrate the presence of a gene" (Culliton 2001).

In viruses, very many genes encode for one single large polypeptide which, however, after translation, is cleaved enzymatically into smaller subunits. Such polyprotein genes are also known in multicellular eukaryotes. The neuropeptide genes of mammals and the proline-rich proteins of salivary glands are such examples. Polyprotein genes thus contradict the hypothesis adopted by the neoclassical view of the gene that each gene encodes for a single polypeptide. The discovery of polyproteins led to one gene-many proteins hypothesis.

ONE ENZYME-TWO FUNCTIONS CONCEPT

Another recent discovery is that of enzyme pon2 (human paraoxonase 2) which has two functions, enzymatic lactonase activity and reduction of intracellular oxidative stress. This led to one enzyme-two functions concept (Altenhofer 2010).

RECENT THOUGHTS ON GENE FUNCTION

Is Gene Concept Dead?

The gene is acknowledged as a material entity, its membership criteria are unclear and its boundaries are fuzzy indeed (Marks and Lyles 2005). In actual fact, more than one gene can occupy the same space at the same time, and consequently, even though the term "gene" is widely used and is central to the discipline discourse, the concept of the gene eludes rigorous definition. Some authors even claim that the whole concept of the gene is dead, comparable to the concept of 'phlogiston' in the early history of chemistry, while others have the opinion that the gene concept still survives and should survive (Falk 2004; Knight 2007).

Continuums of Genetic Transcription

In eukaryotic organisms, notably mammals, the structural boundaries of the gene as the unit of transcription are far from clear; in reality, the mammalian 'transcriptome' as well as that of other eukaryotic organisms is very complex. The human genome is pervasively transcribed from both DNA strands, such that the majority of its bases can be found in primary transcripts, including non-protein-coding transcripts, and those that extensively overlap one another. The complexity of the transcription of protein-coding and non-coding RNA sequences is evident: transcripts may be derived from either of both DNA strands, and they may be overlapping and interlaced, and the transcripts can even use the same coding sequences (Mattlick 2005). Moreover, many examples of transcripts are known in which protein-coding exons from one part of the genome combine with exons from another part that can be hundreds of thousands of nucleotides away, and separated by several other transcription units (Kapranov et al. 2007). This continuum of transcription might even spill over the boundaries of chromosomes, as is the case for certain human immune system genes that are controlled by regulatory regions from another chromosome (Spilianakis et al. 2005). This piece of evidence shows that whole chromosomes, if not the whole genome, seem, in reality to be continuums of genetic transcription, and genes can no longer be defined as units of transcription (Gingeras 2007).

Concordant Regulation of Sense/Antisense Pairs of Transcripts

Gene expression profiling in mammals reveals frequent concordant regulation of sense/antisense pairs of transcripts, and experimental evidence has been presented that perturbation of antisense RNA can alter the expression of sense messenger RNAs, suggesting that antisense transcription contributes to the control of transcriptional outputs of the genome in mammals.

Long Non-Coding (ln) RNA

Another discovery was that mouse transcriptome had 180,000 transcripts but only 20,000 were protein coding. Many of the RNAs synthesized were longer than 200 nucleotides. These are called long non-coding (ln) RNA. Some scientists consider these lnRNAs to be just noise but certain regulatory functions like controlling genome dynamics, cell biology and developmental programming have been associated with them (Amaral et al. 2008).

Gene Fusions

Another recent observation, many examples are known of gene fusions in multicellular eukaryotes, i.e., cases where certain exons can be members of at least two, and possibly even more, transcripts. For example in man these chimeric transcripts, formed by transcription of two consecutive genes into one RNA, were found in a systematic survey in over 200 cases involving 421 genes, and, furthermore, the authors also demonstrated that this number of genes was only a subset of the actual number of fused genes (Akiva et al. 2006).

Encrypted Genes

In the organelles of microbial eukaryotes and in the prokaryotes many examples of so-called encrypted genes are known; genes can be found as separate segments around the genome, so that, for example, all building blocks of a given mRNA molecule can be located, as modules, on separate chromosomes (Landweber 2007).

Epigenetic Factors Control Functional Status of a Gene

In addition to the structure of the gene, its functional status, regulated by epigenetic factors, can also be inherited from one cell to its daughter cells, or even from one individual of a given generation to the next generations. This significant finding is of paradigmatic importance because it may force us to rethink the central dogma of genetics, namely the theory that acquired characters are not heritable.

Genetic Restoration

The most revolutionary discovery in present day genetics which involves the concept of the gene is the discovery of epigenetic non-Mendelian inheritance of extra-genomic information, first found in *Arabidopsis thaliana*, and called genetic restoration (Lolle et al. 2005). Later, the same phenomenon was also found in mammals, suggesting common occurrence (Rassoulzadegan et al. 2006). Ultimately, the phenomenon of genetic restoration not only seriously challenges the conventional concept of the gene, but it also really forces us to change our comprehension of inheritance. What was observed is that organisms can sometimes rewrite their DNA on the basis of RNA messages inherited from generations past. Plants and animals can inherit allele-specific DNA sequence information that was not present in the genome of their parents but was present in previous generations. In *Arabidopsis*, this process was shown to occur in all DNA sequence polymorphisms examined and therefore it seems to be a general mechanism for extra-genomic inheritance of DNA sequence information. Moreover, it was postulated that these genetic restoration events are the result of a template-directed process that makes use of an ancestral RNA sequence cache.

A significant number of observations from recent years involving several aspects of the gene, its structure, function and regulation, and the phenomenon of inheritance itself, force us to renew our way of thinking about what really is a gene. These observations have seriously shaken every aspect of the conventional concept of the gene, and consequently attempts to formulate a new definition have been difficult, leading in most cases to a failure to account for every aspect of the gene. For this very reason, Resende et al. (2011) have discussed the implications of changing concept of gene. They have suggested a marker could or will not be able to identify the DNA region which, under given conditions, is responsible for a better phenotypic expression. Further quoting Lee they say phenotypic era of plant breeding is endless and irreplaceable and the real challenge is how to enable phenotypic selection and to make it more effective.

The current focus is on studying the gene function at genomic level. The definition of gene from function point of view needs to be re-written in light of the present observations. Another question is raised about defining the gene in terms of its function: Is it possible to understand gene function only from its nucleotide sequence? Transcriptomics (sum total of transcription profile of a cell) and proteomics (sum total of protein products of a cell) can provide better understanding about gene function. There is another dimension to gene function – epigenetic modifications involving DNA methylation and protein modification. Gene function cannot be understood satisfactorily without analyzing influence of these epigenetic modifications on action of the gene.

REFERENCES

Akiva, P., A. Toporik, S. Edelheit, Y. Peretz, A. Dider, R. Shemesh, A. Novik, and R. Sorek 2006. Transcription-mediated gene fusion in the human genome. Genome Res. 16: 30-6.

Altenhöfer, S., I. Witte, J.F. Teiber, et al. 2010. One enzyme, two functions: PON2 prevents mitochondrial superoxide formation and apoptosis independent from its lactonase activity. J. Biol. Chem. 285: 24398-403.

Amaral P.P., M.E. Dinger, T.R. Mercer, and J.S. Mattick. 2008. The eukaryotic genome as an RNA machine. Science 319: 1787-9.

Beadle, G.W., and B. Ephrussi. 1936. The differentiation of eye pigments in *Drosophila* as studied by transplantation. Genetics 21: 225-47.

Beadle, G.W., and B. Ephrussi. 1937. Development of eye colors in *Drosophila*: diffusible substances and their interrelations. Genetics 22: 76-86.

Beadle, G.W., and E.L. Tatum. 1941. Genetic control of biochemical reactions in *Neurospora*. Proc. Natl. Acad. Sci. USA 27: 499-506.

Beadle, G.W., and E.L. Tatum. 1945. Neurospora. II. Methods of producing and detecting mutations concerned with nutritional requirements. Amer. J. Bot. 32: 678-86.

Brenner, S. 1955. Tryptophan biosynthesis in *Salmonella typhimurium*. Proc. Natl. Acad. Sci. USA 41: 862-3.

Brenner, S., F. Jacob, and M. Meselsohn. 1961. An unstable intermediate carrying information from genes to ribosomes for protein synthesis. Nature 190: 576-80.

Britten R J, and D.E. Kohne. 1968. Repeated sequences in DNA. Science 161: 529-40.

Burnet, F.M. 1959. *The Clonal Selection Theory of Acquired Immunity*. Cambridge: Cambridge University Press.

Cammack, R., T.K. Attwood, P.N. Campbell, et al. 2006. *Oxford Dictionary of Biochemistry and Molecular Biology*. UK: Oxford University Press.

Chen, C., T. Malone, S.K. Beckendorf, R.L. Davis. 1987. At least two genes reside within a large intron of the *dunce* gene of *Drosophila.* Nature 329: 721-4.

Cohen, S.S. 1948. Synthesis of bacterial viruses: Synthesis of nucleic acid and protein in *Esherichia coli* B infected, with T2 bacteriophage. J. Biol. Chem. 174: 281-93.

Culliton, B.J. 2001. One Gene, Many Proteins. Genome News Network. Online publication of the J. Craig Venter Institute. http://www.genomenewsnetwork.org/articles/02_01/One_gene. shtml

Falk, R. 2004. Long live the genome! So should the gene. Hist. Philos. Life Sci. 26: 105-21.

Garrod, A.E. 1909. *Inborn Errors of Metabolism.* Oxford: Oxford Univ. Press.

Gingeras, T.R. 2007. Origin of phenotypes: genes and transcripts. Genome Res. 17: 682-90.

Grabowski, P. 2002. Alternative splicing in parallel. Nat. Biotechnol. 20: 346-7.

Hood, L. 1972. Two genes, one polypeptide chain – fact or fiction? Fed. Proc. 31: 177-87.

Hood, L., and D. Ein. 1968. Immunoglobulin Lambda Chain Structure: Two Genes, One Polypeptide Chain. Nature 220: 764-7.

Horowitz, N.H., and U. Leupold. 1951. Some recent studies bearing on the one gene-one enzyme hypothesis. Cold Sp. Harbor Symp. Quant. Biol. 16: 65-74.

Hozumi, N, and S. Tonegawa. 1976. Evidence for somatic rearrangement of immunoglobulin genes coding for variable and constant regions. Proc. Natl. Acad. Sci. USA 73: 3628-32.

Ingram, V.M. 1957. Gene mutation in human haemoglobin: The chemical difference between normal and sickle-cell haemoglobin. Nature 180: 326-8.

Ingram, V.M. 1958. *How Do Genes Act?* Scientific American. Available as Offprint 104. San Francisco: W.H. Freeman and Co.

Itano, H.A., and E. Robinson. 1960. Specific recombination of the subunits of hemoglobin. Ann. New York Acad. Sci. 88: 642-54.

Itano, H.A., and E.A. Robinson. 1960. Genetic control of the α- and β-chains of hemoglobin. Proc. Natl. Acad. Sci. USA 46: 1492-501.

Jacob, F., and J. Monod. 1961. Molecular and biological characterization of messenger RNA. J. Mol. Biol. 3: 318-56.

Jacq, C., J.R. Miller, and G.G. Brownlee. 1977. A pseudogene structure in 5S DNA of *Xenopus laevis.* Cell 12: 109-20.

Jerne, N. K. 1955. The natural selection theory of antibody formation. Proc. Acad. Sci. Paris 41: 849-57.

Judd, B.H., M.W. Shen, and T.C. Kaufman. 1972. The anatomy and function of a segment of the X chromosome of *Drosophila melanogaster.* Genetics 71: 139-56.

Kapranov, P., A.T. Willingham, and T.R. Gingeras. 2007. Genome wide transcription and the implications for genomic organization. Nat. Rev. Genet. 8: 413-23.

Knight, R. 2007. Reports of the death of the gene are greatly exaggerated. Biol. Philos. 22: 293-306.

Koshland, M.E., J.J. Davis, and N.J. Fujita. 1969. Evidence for multiple gene control of a single polypeptide chain: the heavy chain of rabbit immunoglobulin. Proc. Natl. Acad. Sci. USA 63:1274-81.

Landweber, L.F. 2007. Why genomes in pieces? Science 318: 406-7.

Lawrence, W.J.C., and V.C. Sturgess. 1957. Studies on *Streptocarpus.* III. Genetics and chemistry of flower color in the garden forms, species, and hybrids. Heredity 11: 303-36.

Lewin, B. 1980. *Gene expression.* Volume 2. Eukaryotic chromosomes. 2nd ed. New York: Wiley.

Lolle, S.J., J.L. Victor, J.M. Young, and R.E. Pruitt. 2005. Genome-wide non-mendelian inheritance of extra-genomic information in *Arabidopsis.* Nature 434: 505-9.

Marks, J., and R.B. Lyles. 2005. Rethinking genes. Evol Anthropol: Issues, News and Reviews 3: 139-46.

Mattlick, J.S. 2005. The functional genomics of noncoding RNA. Science 309: 1527-8.

Newmeyer, D. 1957. Arginine synthesis in *Neurospora crassa*: genetic studies. J. Gen. Microbiol. 16: 449-62.

Rassoulzadegan, M., V. Grandjean, P. Gounon, S. Vincent, I. Gillot, and F. Cuzin. 2006. RNA mediated non-mendelian inheritance of an epigenetic change in the mouse. Nature 441: 469-74.

Resende, K.F.M., F.M.C Santos, M.A.D. Dias, and M.A.P. Ramalho. 2011. Implication of the changing concept of genes on plant breeder's work. Crop Br. App. Biotech. 11: 345-51.

Smith, E.W., and J.V. Torbert, 1958. Study of two abnormal hemoglobins with evidence for a new genetic locus for hemoglobin formation. Bull. Johns Hopkins Hosp. 101: 38-52.

Smithies, O., R.G. Gregg, S.S. Boggs, M.A. Doralewski, and R.S. Kucherlapati. 1985. Insertion of DNA sequences into the human chromosomal beta-globin locus by homologous recombination. Nature 317: 230-4.

Spilianakis, C.G., M.D. Lalioli, T. Town, G.R. Lee, and R.A. Flavell. 2005. Inter-chromosomal associations between alternative expressed loci. Nature 435: 637-45.

Srb, A.M., and N.H. Horowitz. 1944. The ornithine cycle in Neurospora and its genetic control. J Biol. Chem. 154: 129-39.

Volkin, E. 1960. The function of RNA in T2-infected bacteria Proc. Natl. Acad. Sci. USA 46: 1336-48.

Volkin, E., L. Astrachan, and J.L. Countryman. 1958. Metabolism of RNA phophorus in *Escherichia coli* infected with bacteriophage T7. Virology 6: 545-55.

Wagner, R.P., and Mitchell, H.K. 1964. *Genetics and Metabolism*. New York: Wiley.

Waring M, and R.J. Britten. 1966. Nucleotide sequence repetition: A rapidly reassociating fraction of mouse DNA. Science 154: 791-4.

Woods, R.A. 1973. *Biochemical Genetics*. London: Chapman & Hall.

Yanofsky, C., B.C. Carlton, J.R. Guest, D.R. Helinski, and U. Henning. 1964. On the colinearity of gene structure and protein structure. Proc. Natl. Acad. Sci. USA 51: 266-72.

6

Transcription in Bacteria and Viruses

Transfer of information from DNA to protein begins with the synthesis of RNA molecules during a process called transcription. The first step of gene expression is transcription. Transcription more or less defines those segments of DNA that contain information since at present the best operational definition of a gene is a DNA segment that is transcribed and specifies a single product. On the basis of gene expression, genes may be divided into three classes: tRNA genes synthesize pre-tRNA, rRNA genes synthesize pre-rRNA and structural genes synthesize mRNA (in prokaryotes) or pre-mRNA (in eukaryotes); these transcripts after processing become functional molecules tRNA, rRNA and mRNA, respectively. Whereas tRNA and rRNA do not encode for polypeptides, mRNA contains codons and acts as a template for transcription. The transfer of information from DNA to protein requires the participation of three different classes of RNA molecules. The role of the first type of RNA is as a messenger (mRNA) to carry the sequence information of DNA to particles in the cytoplasm known as ribosomes, where the messenger will be translated. The role of the second type of RNA is that of a transfer molecule (tRNA) to bring the amino acids to the ribosomes, where protein synthesis actually takes place. The role of the third type of RNA is as a structural part of the ribosome. This last type of RNA is called ribosomal RNA (rRNA).

Transcription is a vital process in biological life forms. It is through this process that the biological road map encoded in a strand of DNA is used to produce a complementary mRNA copy. During transcription, only one of the two DNA strands is transcribed into mRNA. The DNA template strand for a given mRNA is termed the antisense strand. mRNA sequence is complementary to antisense strand while it is similar to sense strand. Synthesis of RNA proceeds into 5′→3′ direction which is catalyzed by RNA polymerase (RNAP). Several aspects of gene expression are described by Jasny and Roberts (2004).

CENTRAL DOGMA AND ITS MODIFICATION

Genetic information from DNA is transcribed to RNA and that information from RNA is translated to polypeptide chain. This concept about flow of biological information from DNA to RNA and then to protein was given by Crick (1958) and is known as central dogma. This concept of central dogma presented in Figure 6.1 holds true for those organisms where DNA is the genetic material. Coding and non-coding strands of DNA were first described by Alberts et al. (1989) (Figure 6.2). Crick's concept of central dogma was modified with increasing knowledge in the subject. In some plant viruses, for example, tobacco mosaic virus (TMV), where RNA is the genetic material, genetic information flows from RNA to DNA with the help of an enzyme reverse transcriptase. This phenomenon was termed as reverse transcription by Temin and Mizutani (1970) in mice leukemia virus and Baltimore (1970) in Rous sarcoma virus. The enzyme responsible for this type of transcription was named as reverse trans-

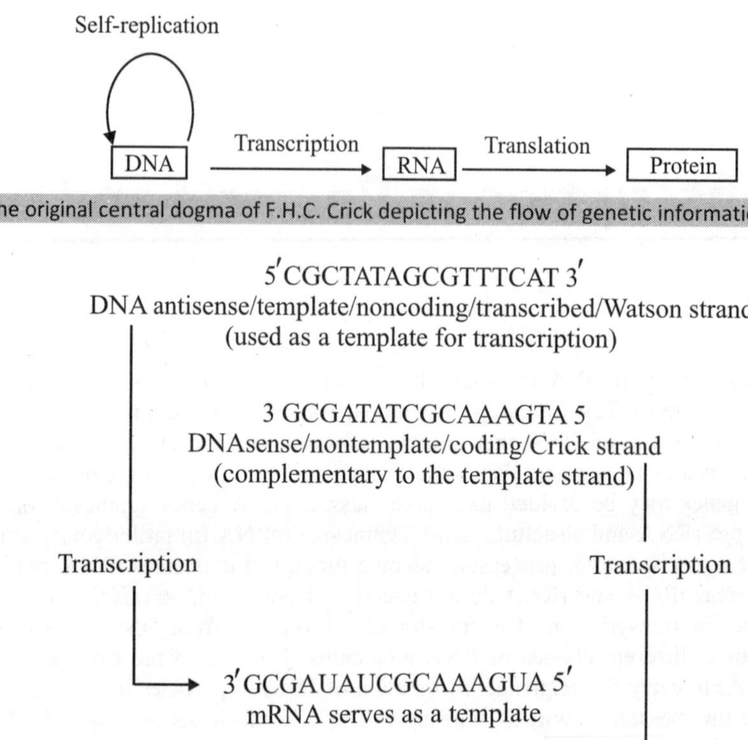

Figure 6.1 The original central dogma of F.H.C. Crick depicting the flow of genetic information

Figure 6.2 Nomenclature of the two strands of double-stranded DNA

criptase. It has been amply demonstrated that the DNA product of RNA-directed DNA polymerase system has base sequence complementary to the viral RNA.

RNA can undergo self-replication. The modified central dogma is shown in Figure 6.3. The second modification to the original central dogma is that RNA can act as a template for its own replication, a process observed in a small class of phages. These RNA phages, such as R17, f2, MS2 and QB, are the simplest phages known. MS2 contains about 3,500 nucleotides which code for only three proteins; a coat protein, an attachment protein (responsible for attachment to and subsequent penetration of the host), and a subunit of the enzyme RNA replicase. The RNA replicase subunit combines with three of the cell's proteins to form the enzyme RNA replicase that allows the single-stranded RNA of the phage to replicate itself. Since the new protein needed to construct the RNA replicase enzyme must be synthesized before the phage can replicate its own RNA, the phage RNA must first act as a messenger when it infects the cell. Thus we have the situation of protein synthesis without the process of transcription ever taking place. The viral genetic material, RNA, is first used as a messenger in the process of translation and then used as a template for RNA replication.

Under laboratory conditions proteins can be synthesized directly using DNA as a template. B.J. McCarthy and J.J. Holland showed that under certain experimental conditions, denatured (single-stranded) DNA could bind to ribosomes and be translated into proteins (McCarthy and Holland 1965; McCarthy and Holland 1966). The experimental conditions usually involved the addition of antibiotics that interacted with the DNA or the ribosome. Direct translation of DNA is not known to occur naturally. Even in updated dogma (Figure 6.3), there are no arrows originating at protein. In other words, protein cannot self-replicate, nor can it use amino acid sequence information to reconstruct RNA or DNA. Crick (1970) has called these arrows "forbidden transfers". We know of no cellular machinery to produce these forbidden processes. Many small non-coding RNAs are used in the process of maturation of messenger RNA from precursor of messenger RNA.

Non-coding RNA central dogma of molecular biology is presented in Figure 6.4.

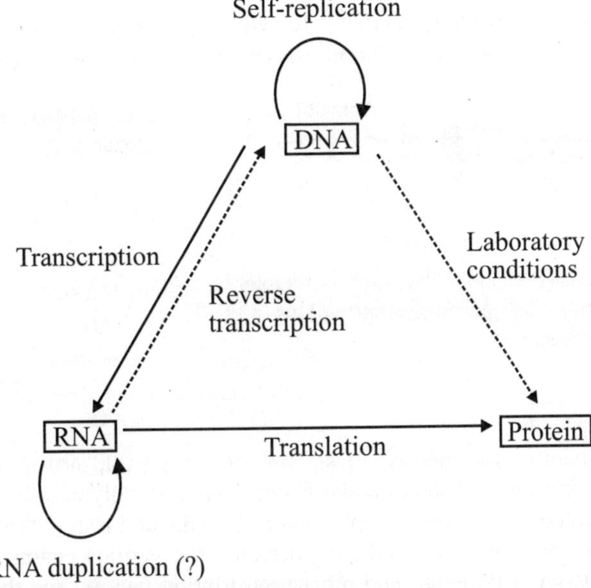

Figure 6.3 Updated central dogma of molecular biology, showing all known pathways of genetic information transfer. Path added to Fig. 16.1 are shown in dotted lines

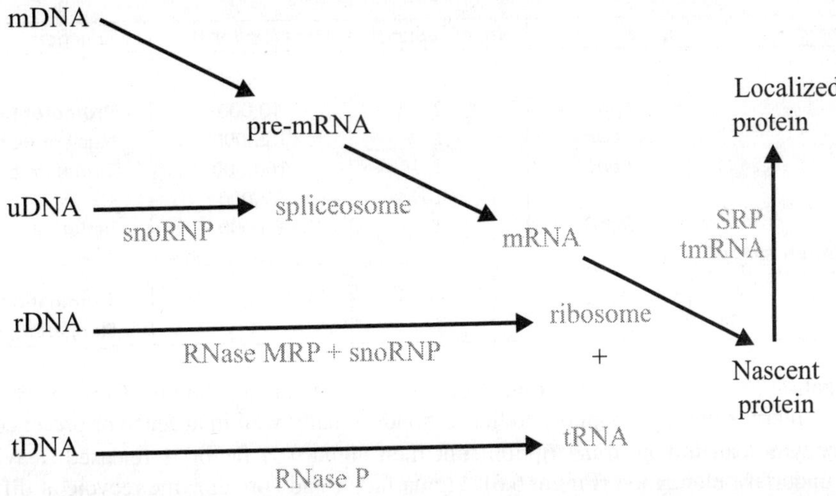

Figure 6.4 Non-coding RNA central dogma (Redrawn from http://en.wikipedia.org/wiki/Non-coding_RNA)

TRANSCRIPTION IN BACTERIA

Process of transcription requires: (a) four types of ribonucleotides of A, U, C and G, (b) template DNA strand — two complementary strands of double helical DNA separate and only one of them (3'←5'), in relation to a particular gene, is used as a template for RNA synthesis. This strand of DNA is called sense strand. RNA chain grows in 5'→3' direction, as shown in Figure 6.5. A unit of DNA that acts as a template for RNA transcription is called an operon in prokaryotes. An operon contains a promoter, an operator, and structural gene(s). (c) DNA-dependent RNA polymerase (symbolized as RNAP or Pol) — in bacteria, there is only one species of this enzyme (Table 6.1).

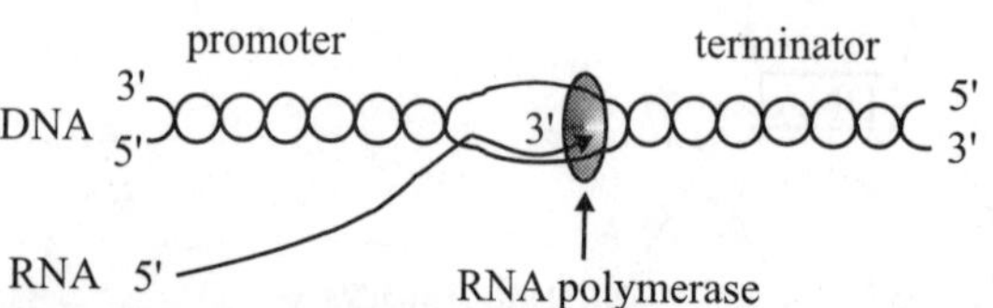

Figure 6.5 A transcribed piece of RNA and its template DNA showing position of promoter, terminator and direction of transcription

Table 6.1 Sizes of different transcripts in prokaryotes

Transcript	*Size (nucleotides)*
16S rRNA	1,542
23S rRNA	2,904
5S rRNA	120
tRNAs	73-93
mRNAs	300-4,000

RNA Polymerase

Bacterial RNA polymerase is composed of six polypeptides ($\alpha_2\beta\beta'\omega\sigma$) of five different kinds. This means five different genes are needed to make *Escherichia coli* polymerase. Active form of enzyme is called holoenzyme. Holoenzyme is made up of a core enzyme and a sigma (σ) factor. Core enzyme has five subunits ($\alpha_2\beta\beta'\omega$). No covalent bond runs between the various chains of RNAP. Genes coding different subunits of RNA polymerase and other transcription factors, and their functions are given in Table 6.2.

Table 6.2 Bacterial RNA polymerase

RNA polymerase	*Gene*	*No. of subunits*	*Mass (Daltons)*	*Function*
Subunits				
α	*RpoA*	2	40,000	Promoter binding
β	*RpoB*	1	155,000	Nucleotide binding
β'	*RpoC*	1	160,000	Template binding
ω	-	1	12,000	-
σ	*RpoD*	1	85,000	Initiation
Transcription factors				
ρ	*Rho*	6	-	Termination
NusA	*NusA*	1	-	Elongation; termination

Aggregation of RNAP results from formation of secondary bonds. Core enzyme catalyzes formation of internucleotide 3'-5' phosphodiester bonds equally well in absence or presence of σ factor. Only holoenzyme can initiate transcription; but then sigma (σ) factor is released, leaving the core enzyme to undertake elongation (Figure 6.6). Sigma factor and core enzyme recycle at different points in transcription. Rho (ρ) factor plays role in termination of transcription (Figure 6.7). Role of NusA subunit is also shown. RNAP has two unwinding activities – the leading unwindase opens the DNA

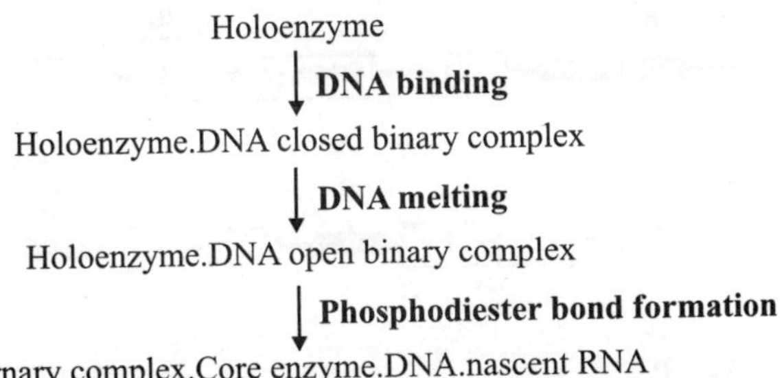

Figure 6.6 Initial reactions involved in conversion of RNA polymerase holoenzyme to core enzyme or transcription start

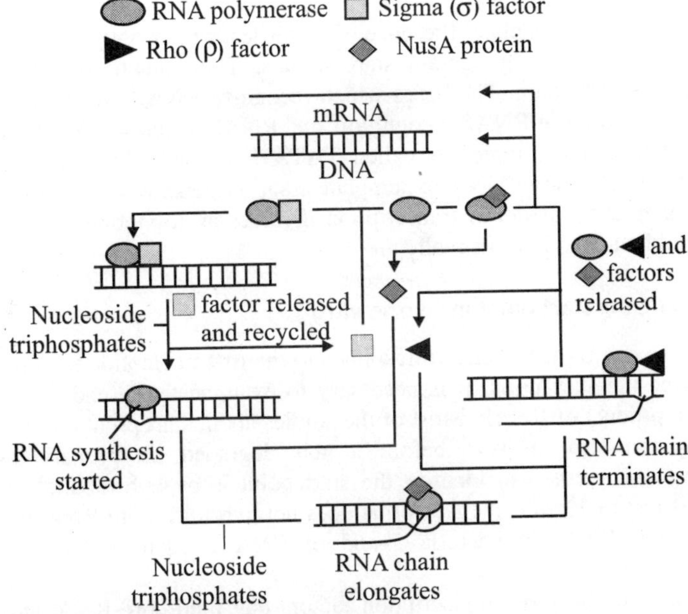

Figure 6.7 Start, elongation and termination of transcription

helix, which is then subsequently reformed by lagging rewindase activity. Three operons, namely σ operon, β operon and α operon, contain genes for σ, β, β′ and α subunits of RNA polymerase. These operons are shown in Figure 6.8. Positions of promoters in the operons are also indicated by arrows. RNAP binds to specific sites on the DNA molecule. It partially unwinds the template DNA and synthesizes an RNA primer for elongation. It catalyzes processive chain elongation by unwinding and rewinding the DNA. It also terminates transcription after copying the gene.

Transcription Initiation

Initiation of transcription requires startise, a point at which first nucleotide is incorporated during transcription. Sigma (σ) factor recognizes start signals along the DNA molecules and also the correct

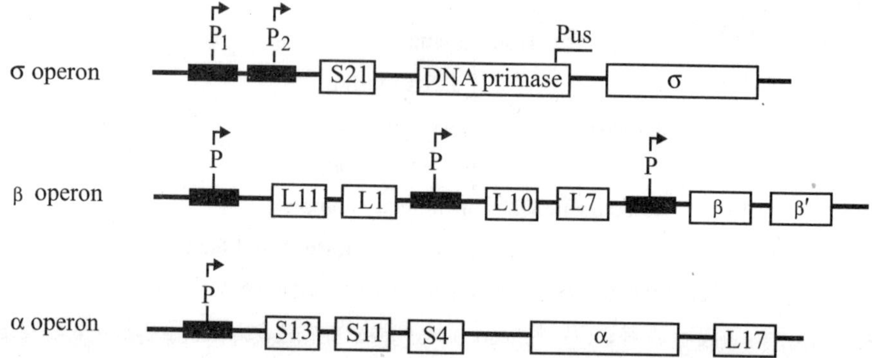

Figure 6.8 Three operons containing genes for σ, β and β′, and α subunits of RNA polymerase

strand of DNA to be used as template. Presence of factor leads to very tight binding of holoenzyme to promoter regions of DNA which contain start signals. Then much localized unwinding occurs, allowing synthesis of RNA chain. The bases which recognize RNAP are not themselves transcribed. Soon after this initial binding of RNAP recognition site, RNAP diffuses to RNAP binding site which is A=T rich region. This binding region is called TATA box. Twin-domain model of active RNAP suggests that RNAP tracks along DNA as template rotation generates twin domains of supercoiling (Cook 1990). An alternative model of transcription involves an immobile polymerase. It needs to be determined how mobile active RNAPs really are.

Determining/defining the start point in vivo/in vitro

Start point is the base pair on DNA that corresponds to the first nucleotide incorporated into transcript by RNAP. To define the start point, it is necessary to examine the 5′-end of the primary transcript which is immediate product of RNAP. Most of the studies about start point are in vitro as it is difficult to isolate primary transcript in vivo before it gets degraded (in prokaryotes) or modified (in eukaryotes). A common method to identify the start point is by hybridizing the transcript with its template DNA. It is just to degrade the DNA that does not hybridize with RNA and then determine the sequence of surviving DNA. In bacteria 5′-end of RNA is identified very easily as it carries triphosphate terminus.

Sigma (σ) factors are bacterial transcription factors that bind core RNA and direct transcription initiation at cognate promoter sites. However, most of their functions have been investigated in the context of RNAP. This has made the exact function of σ and the importance of core RNAP in modulating σ function ambiguous. Hsu et al. (2006) identify a *Bacillus subtilis* mutant σ^A that is independently capable of specific binding and melting of the promoter DNA. Interestingly, specific and independent promoter binding of σ is sufficient for temperature- and Mg^{2+}-independent melting of promoter DNA around the transcription site, in contrast to the temperature- and Mg^{2+}-dependent melting by RNAP around the promoter –10 element. Thus core RNAP is able to negatively modulate the σ-initiated melting of the transcription start site and, by sensing the changes in temperature and Mg^{2+} concentration, to regulate the efficiency of promoter –10 melting.

Determining DNA binding sites for RNA polymerase

The binding sites at which RNAP forms a stable initiation complex with DNA lie within promoter. The threev different methods are used for the purification of the RNA polymerase binding sites (Figure 6.9)

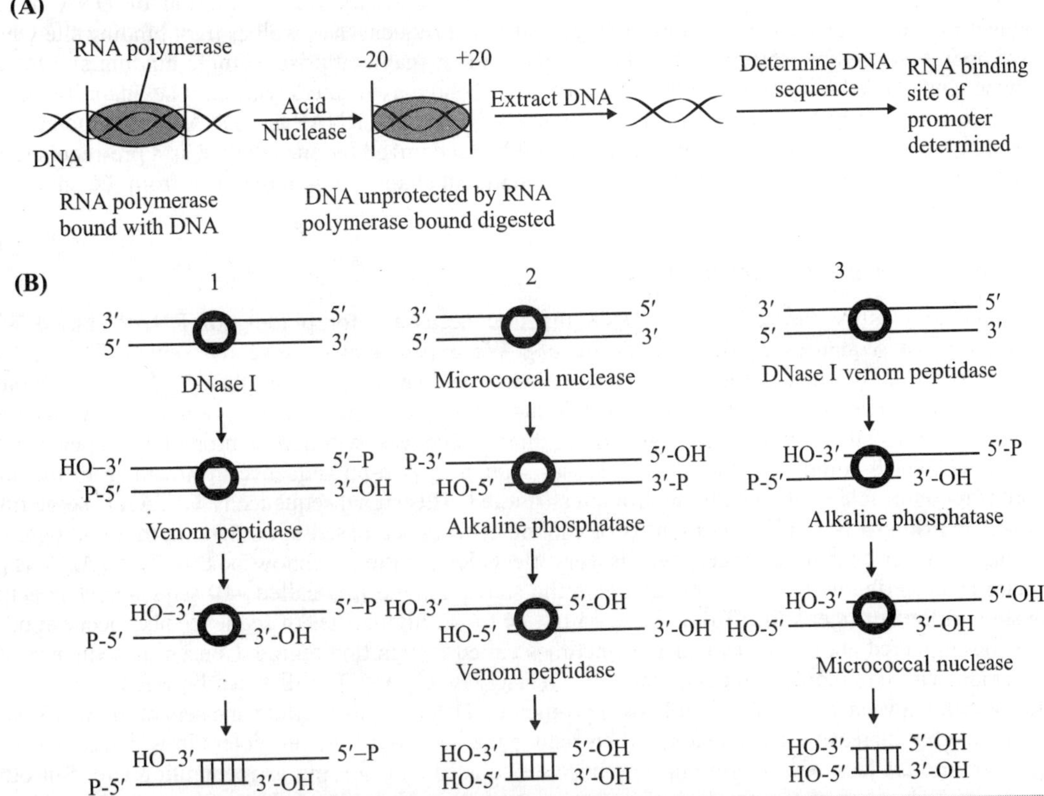

Figure 6.9 Determination of RNA binding sites for RNA polymerase (Redrawn, with permission, from Giacomoni, P.U. et al. 1974. Proc. Nat. Acad. Sci. USA 71: 3091-5)

(Giacomoni et al. 1974). In Method 1, the pancreatic DNase leaves 3'-OH and 5'-P termini. The exonucleolytic attack is carried out by viper's venom phosphodiesterase alone, which digests from the 3'-OH termini; it is, therefore, expected that single-stranded segments terminated by 5'-P are left undegraded. Method 2 has been developed to eliminate the single-stranded parts, micrococcal nuclease acts as an endonuclease, leaving 3'-P and 5'-OH, and as an exonuclease digesting from 5'-OH termini. The alkaline phosphatase is added to yield free 3'-OH termini, thus allowing the viper's venom phosphodiesterase to act. Method 3 has been developed to demonstrate the possibility of cutting out the protruding segments left in Method 1, thus establishing the relation between the segments prepared by Methods 1 and 2.

These sequences can be recovered by degrading all the regions of DNA that are not protected by the RNAP. RNAP is bound in vitro to a particular transcription unit and then recovering the promoter after digestion. Sequence of the protected DNA is then determined. Using this approach, the protected fragments recovered are 41-44 base pairs in length. Polymerase left attached to this region can synthesize short RNA of 17-20 bases and would terminate when it runs off the fragment end which shows that the protected fragment carries the start point sequence which locate it in the centre of the binding site (promoter). The upstream sequences are prior to this start point (written as minus) and downstream sequences are after the start point (written as plus). Degradation protected fragment extends from –20 to +20 roughly. Interestingly, the recovered fragment cannot bind to polymerase which shows that other sequences beyond it must be necessary for recognition of DNA by the polymerase. Hence the promoter includes these additional sequences as well as tight binding site which leave open two possible structures for the promoter. (a) It may comprise a single binding site for the enzyme and the additional sequences may be less tightly associated with it. The inability of the protected fragment reflects the geometry of the reaction between RNAP and DNA. (b) Promoters may actually contain two types of sequences: first could be recognized by one enzyme as a pre-requisite for binding to the second site. There may be a conformational change, e.g., movement from recognition to binding site.

Sequence homologies in *E. coli* promoters

Attempts to identify the features of DNA that are necessary for binding of RNAP started with comparison of sequences of different promoters. We expect that an essential sequence should be present in all the promoters, which is said to be conserved sequence. More than 50 *E. coli* promoters have been sequenced but essential feature is that they lack any extensive conservation of sequence over the 60 bp associated with RNAP but some short sequences within the promoters appear to be conserved, which could be referred to as signal sequence. A 6-bp sequence upstream from the start point is recognizable in almost all the promoters studied. This signal sequence is TATAAT, sometimes known as Pribnow box. It is a consensus or canonical sequence based upon maximum homology. On the bases of per cent occurrence, there is very clear demarcating Pribnow box $T_{89}T_{89}T_{50}A_{65}A_{65}T_{100}$. Centre of the Pribnow box is 10 bp upstream the start point, so it is called –10 sequence. The actual location of the hexamer varies from –11 to –5 to –14 to –8. Similarities of sequence also occur at other locations centered about –35 sequence, sometimes called recognition region. Consensus sequence of –35 region with commonly occurring bases is $T_{85}T_{83}G_{81}A_{61}C_{69}A_{52}$. The distance between –35 and –10 sites varies between 16 and 19 bp in known promoters. Deletions that reduce the separation to 15 bp or insertions that increase the distance to 20 bp reduce the activity of the promoters in which they occur. Some promoters lack –35 region but they have an adjacent site serving as recognition site. But other promoters subject to the same assistance retain the –35 region, which indicates that –35 region is involved in the efficiency of the polymerase recognition. Few promoters lack –10 region which shows neither of the conserved sequences is absolutely necessary for promoter functioning. But a typical promoter can use –35 and –10 sequence to be recognized by RNAP.

Up and down promoter mutations

Some mutations in promoters affect the level of transcription of the genes they control without altering the gene products. Bacterial strains that have lost or reduced transcription of the adjacent gene are known as down mutations and, less frequently, mutations with increased transcription level are known as up mutations. Strikingly, both these types of mutations can be produced involving only a single base pair, which shows that RNAP reaction with these sites must be very much specific. Down mutations, however, also could be due to deletion of an extragenic part of the promoter.

RNA polymerase breaks its initial bounds with DNA

RNAP stores energy to break its initial bounds with DNA by scrunching the single strands of DNA that were unwound in the region where the polymerase started RNA synthesis (Roberts 2006).

Abortive transcription initiation

During transcription initiation in vitro, prokaryotic and eukaryotic RNAP can engage in abortive initiation – the synthesis and release of short (2-15 nucleotides) RNA transcripts – before productive initiation. It has not been known whether initiation occurs in vivo. Employing single-molecule DNA nanomanipulation, Revyakin et al. (2006) show that abortive initiation involves DNA "scrunching" – in which RNAP remains stationary and unwinds and pulls downstream DNA into itself – and that scrunching requires RNA synthesis and depends on RNA length. They show that promoter escape involves scrunching, and that scrunching occurs in most or all instances of promoter escape. These results support the existence of an obligatory stressed intermediate, with approximately one turn of additional DNA unwinding, in escape and are consistent with the proposal that stress in this intermediate provides the driving force to break RNAP-promoter and RNAP-initiation-factor interaction in escape. Model for RNAP-active-center translocation in abortive initiation, as proposed by Revyakin et al. (2006), is shown in Figure 6.10. The "scrunching" model invokes a flexible element in DNA. In each cycle of abortive initiation, RNAP unwinds downstream DNA and pulls it into itself, accommodating the accumulated DNA as single-stranded bulges in the unwound region; upon release of the abortive RNA, RNAP extrudes the internalized DNA. The "inchworming" model invokes a flexible element in RNAP. In each cycle of abortive initiation, a module of RNAP containing the active center detaches from the remainder of RNAP and translocated downstream; upon release of the abortive RNA, this module of RNAP reverse translocates. The "transient excursions" model invokes abortive cycles that are transient – too short in lifetime and too infrequent in occurrence to be detected in a time-averaged, population-averaged approach, such as DNA footprinting. In each cycle of abortive initiation, RNAP translocates downstream as a unit; upon release of the abortive RNA, RNAP reverse translocates as a unit.

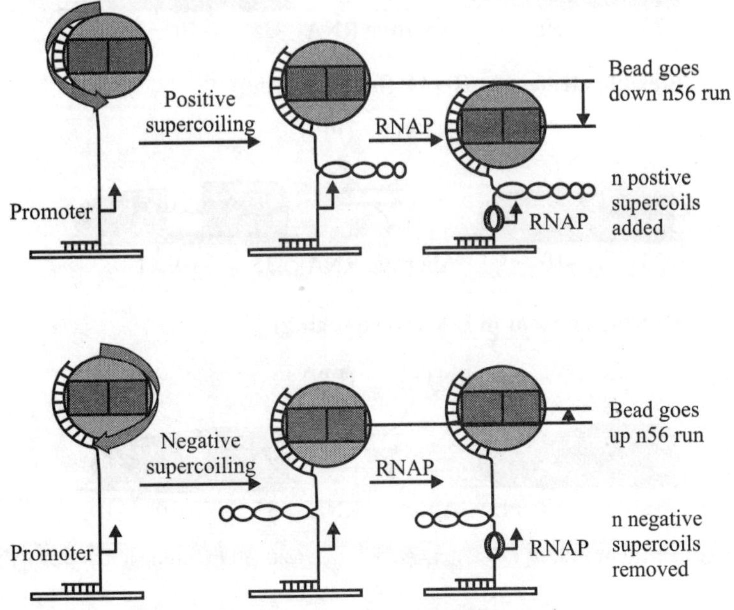

Figure 6.10 Model for RNAP-active-center translocation in abortive initiation

Using hybridization with locked nucleic acid probes, Goldmen et al. (2009) directly detected abortive transcripts in bacteria. In addition, they show that in vivo abortive initiation shows characteristics of in vitro abortive initiation: Abortive initiation increases upon stabilizing interactions between RNAP and either promoter DNA or sigma factor, and also upon deleting elongation factor GreA. Abortive transcripts may have functional roles in regulating gene expression in vivo.

Initial transcription involves scrunching

Kapanidis et al. (2006) show that initial transcription proceeds through a "scrunching" mechanism, in which RNAP remains fixed on promoter DNA and pulls downstream DNA into itself and past its active center. They show further that putative alternative mechanism for RNAP active-center translocation in initial transcription, involving "transient excursions" of RNAP relative to DNA or "inchworming" of RNAP relative to DNA, do not occur. The results support a model in which a stressed intermediate, with DNA-unwinding stress is used to derive breakage of interactions between RNAP and promoter DNA and between RNAP and interaction factors during promoter escape. Three models have been proposed for RNA active-center translocation during initial transcription: transient excursions, inchworming, and scrunching (Kapanidis et al. 2006) (Figure 6.11). Initial transcription does not involve transient excursions Initial transcription also does not involve inchworming. Initial transcription involves scrunching.

Forward-reverse translocations(transient excursions)

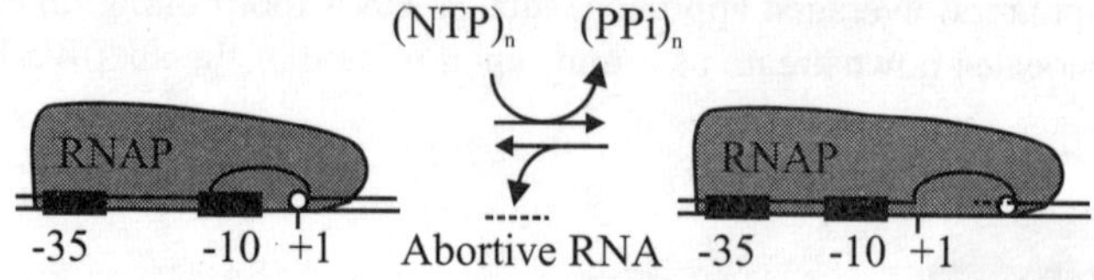

Flexible element in RNAP (inchworming)

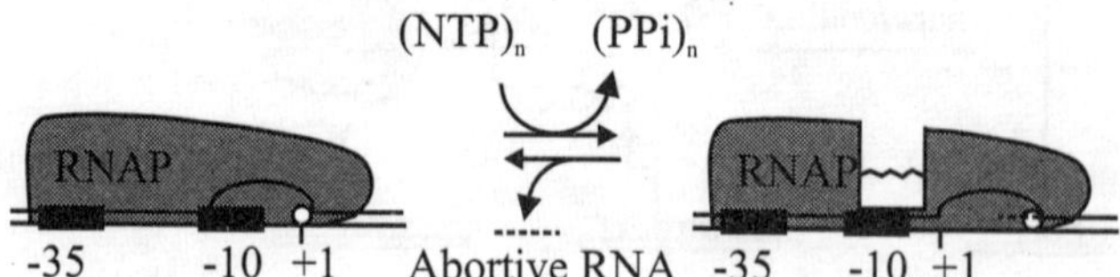

Flexible element in DNA (scrunching)

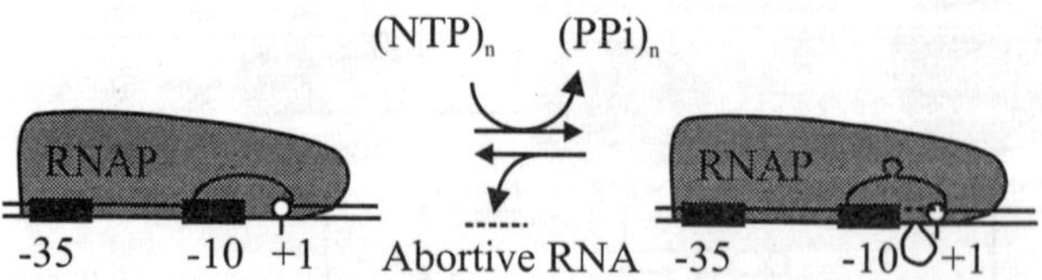

Figure 6.11 Three models proposed for RNAP active center translocation during initial transcription

Transcription initiation – promoter melting

Young et al. (2004) determined the minimal portion of *E. coli* RNAP holoenzyme able to accomplish promoter melting, the crucial step in transcription initiation that provides RNAP access to the template strand. Upon duplex DNA binding, the N terminus of the β' subunit (amino acids 1-314) and amino acids (94-507) of the σ subunit, together less than one-fifth of RNAP holoenzyme, were able to melt an extended −10 promoter in a reaction remarkably similar to that of authentic holoenzyme. These results support the model that capture of non-template bases extruded from the DNA helix underlies the melting process.

Transcription Elongation

Elongation of transcription proceeds in alternating laps of monotonous and inchworm-like movement with the flexible DNA polymerase configuration being subject to direct sequence control (Nudler et al. 1994). During this step, RNAP catalyzes nucleophilic attack by the 3′-OH group of the growing RNA chain on α phosphorous of the ribonucleoside triphosphate, resulting in formation of a new phosphodiester linkage and release of pyrophosphate. 12 nucleotides of the template strand base-pair with newly synthesized RNA to form a hybrid RNA:DNA helix. Then a new nucleotide joins the 3′-end of the growing RNA chain and lengthens the RNA: DNA hybrid by following steps:

- The length of the RNA:DNA hybrid must be reduced by melting one bp at the upstream end of the transcription bubble.
- The DNA bp must be melted at the downstream end of bubble.
- The DNA bp must be reformed at the upstream end of bubble.
- RNAP must advance along the DNA by one nucleotide, positioning its active site over the next unpaired nucleotide on the template strand.

Elongation continues till a termination complex is formed. Elongation may be impeded by pause sites which induces temporary reversible block to nucleotide addition or by arrest or dead ends which stops the transcription which could only be resumed by factors like Gre A and Gre B. Gre A and Gre B are elongation factors involved in transcription elongation including suppression of transcription arrest, enhancement of transcription fidelity and facilitating transcription from abortive initiation to productive elongation.

During transcription, RNAP moves processively along a DNA template, creating a complementary RNA. Abbondanzieri et al. (2005) present the development of an ultra-stable optical trapping system with Angstrom-level resolution, which they used to monitor transcriptional elongation by single molecules of *E. coli* RNAP. It is shown that RNAP advances along DNA by a single base pair per nucleotide addition to the nascent RNA. They also determined the force-velocity relationship for transcription at saturating and subsaturation nucleotide concentrations.

During transcription, RNAP catalyzes the addition of nucleotides of growing RNA chain. High-resolution structural snapshots indicate that the polymerase first identifies its substrate, and then incorporates it (Cramer 2007). Two-step mechanism of RNA chain elongation during transcription is illustrated in Figure 6.12. During transcription from a DNA template, elongation of a RNA sequence involves the binding of a nucleoside triphosphate (NTP) to the RNAP enzyme in an active 'preiinsertion' state; this is characterized by partial folding of the RNAP trigger loop. Complete folding of the trigger loop brings the two metal ions, I and II, of RNAP close together, delivering the NTP to the insertion site where catalysis leads to nucleotide incorporation and the release of pyrophosphate.

Bacterial GreA and GreB promote transcription elongation by stimulating an endogenous, endonucleolytic transcription cleavage activity of the RNAP. The structure of *E. coli* core RNAP bound to GreB was determined to a nominal resolution of 15Å, allowing fitting of high-resolution RNAP and GreB structures (Opalka et al. 2003). In the resulting model, the GreB N-terminal coiled-coil domain extends 45Å through a channel directly to the RNAP active site. The model leads to

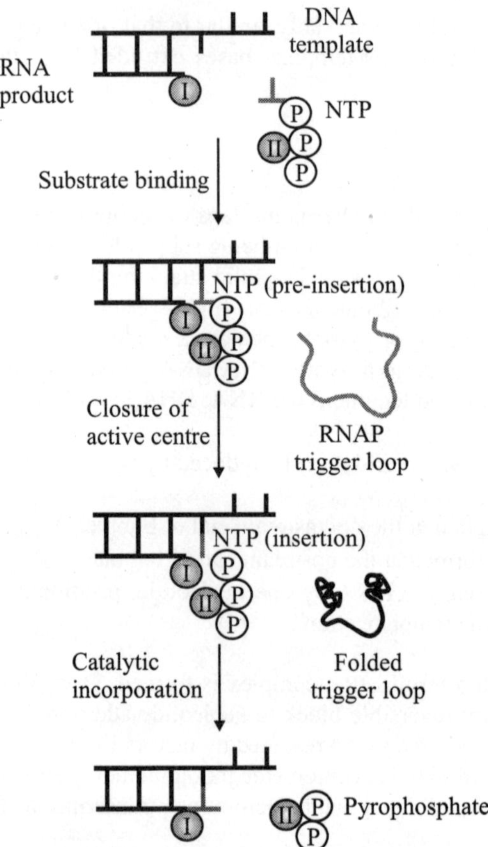

Figure 6.12 Two-step mechanism of RNA chain elongation during transcription

detailed insight into the mechanism of Gre factor activity that explains a wide range of experimental observations and points to a key role for conserved acidic residues at the tip of the Gre factor coiled coil in modifying the RNAP active site to catalyze the cleavage reaction. Mutational studies confirm that these positions are critical for Gre factor function.

Tuning of transcription elongation

RNA chain elongation is a highly processive and accurate process that is finely regulated by numerous intrinsic and extrinsic signals. Bar-Nahum et al. (2005) describe a general mechanism that governs RNAP movement and response to regulatory inputs such as pause, terminators, and elongation factors. *E. coli* RNAP moves by a complex Brownian ratchet mechanism, which acts prior to phosphodiester bond formation. The incoming substrate and flexible F bridge domain of the catalytic center serve as two separate ratchet devices that function in concert to drive forward translocation. The adjacent G loop domain controls F bridge motion, thus keeping the proper balance between productive and inactive states of the elongation complex. This balance is critical for cell viability since it determines the rate, processivity, and fidelity of transcription. A two Pawl ratchet model of transcription elongation, as proposed by Bar-Nahum et al. (2005), is shown in Figure 6.13.

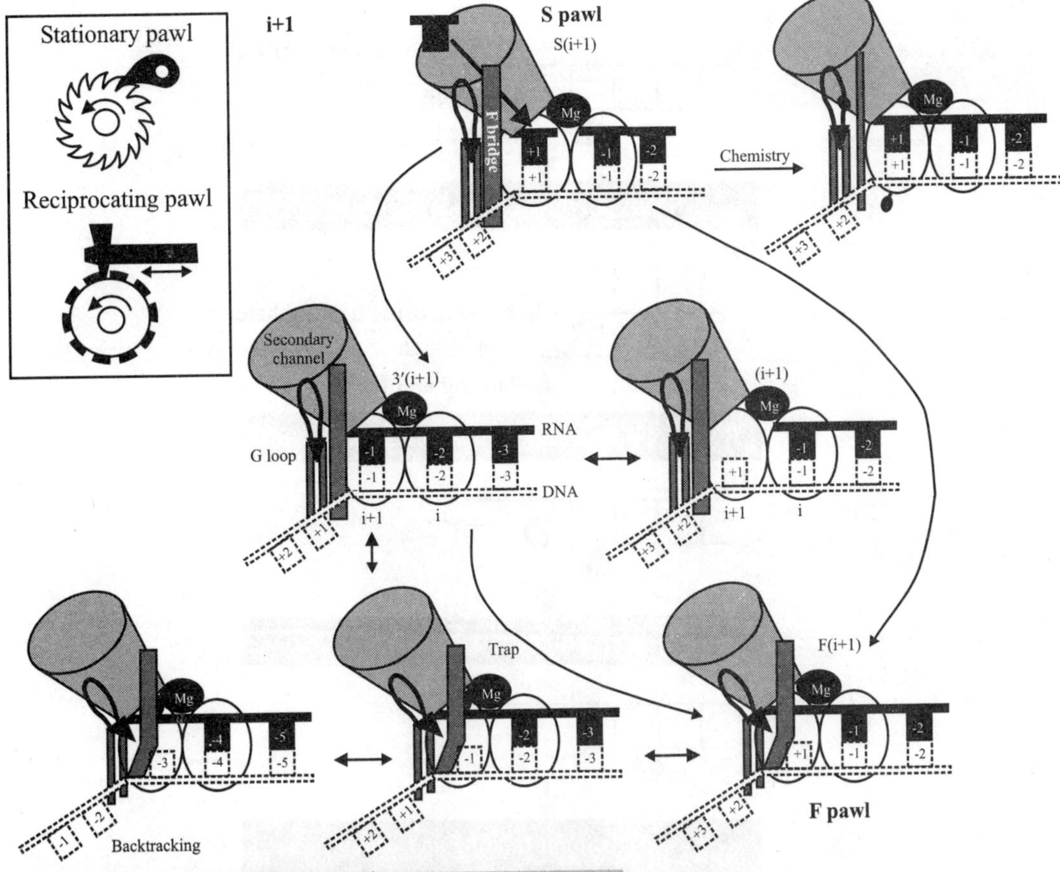

Figure 6.13 A two Pawl ratchet model of transcription elongation

Erroneous RNA itself triggers proofreading mechanism

Mistakes can occur as RNAP copies DNA into transcripts. A proofreading mechanism that removes incorrect RNA is triggered by the erroneous RNA itself (Cramer 2006). RNA-assisted transcriptional proofreading is shown in Figure 6.14A. Correction of misincorporated errors at the growing end of the transcribed RNA is stimulated by the misincorporated nucleotide. Mg^{2+} ions are bound to the catalytic region of RNAP.

Sydow and Cramer (2009) have put forth error removal model of the RNAP proofreading cycle (Figure 6.14B). Crystal structures of RNAP II ECs in different functional states suggest a model for transcriptional proofreading. The vertical dashed line indicates register +1, the nucleotide addition site. Different functional states shown are: (a) post-translocation state (PDB 1Y1W), (b) pre-translocation state (PDB 1I6H, downstream DNA was modeled from 1Y1W, (c) paused state with a frayed 3′-RNA guanine (PDB 3HOW), (d) backtracked state (PDB 3GTJ), (e) post-translocation state. In this structure, dinucleotide cleavage occurred after the crystallization setup (PDB 3HOY).

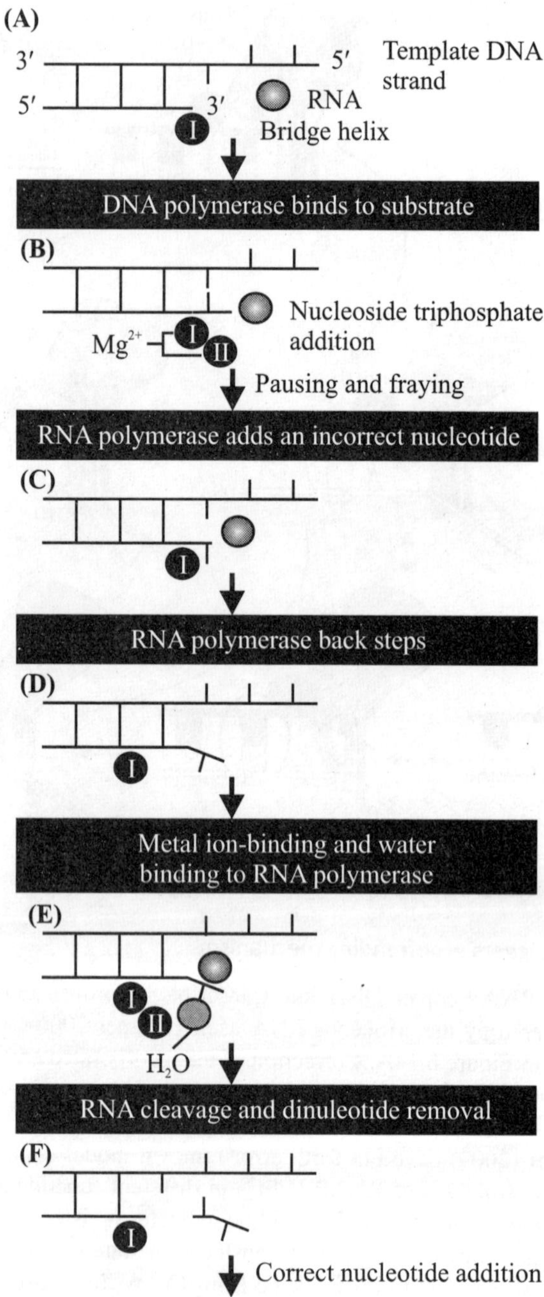

Figure 6.14 RNA-assisted transcriptional proofreading

Fidelity of transcription

Fidelity of template-dependent nucleic acid synthesis is main determinant of stable heredity and error-free gene expression. The mechanism (or mechanisms) ensuring fidelity of transcription by DNA-dependent RNAPs is not fully understood. Zenkin et al. (2006) show that the 3' end-proximal nucleotide of the nascent transcript stimulates hydrolysis of the penultimate phosphodiester bond by providing active groups and coordination bonds to the RNAP active center. This stimulation is much higher in the case of misincorporated nucleotides. They show that during transcription elongation, the hydrolytic reaction stimulated by misincorporated nucleotides proofreads most of the misinccorporation events and thus serves as an intrinsic mechanism of transcriptional fidelity. Cleavages of misincorporated elongation complexes (MECs) and correct elongation complexes (CECs) are shown in Figure 6.15 (Zenkin et al. 2006). Catalytic reactions characteristic of transcription elongation complexes in different stages are shown.

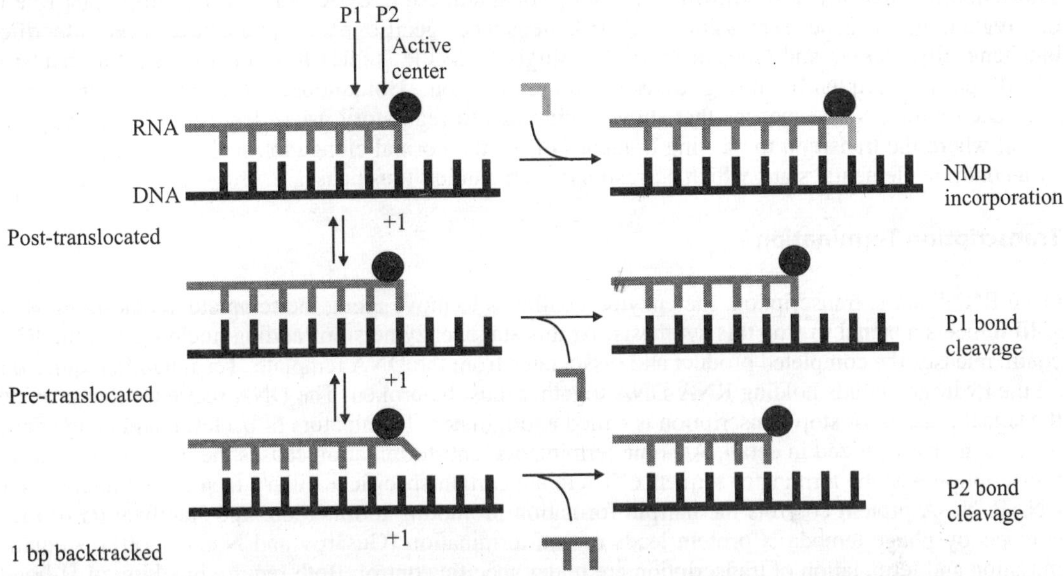

Figure 6.15 Catalytic reactions characteristic of transcription elongation complexes in different stages

Transcription elongation complex

The mechanism of substrate loading in multisubunit RNAP is crucial for understanding the general principles of transcription. Vassylyev et al. (2007a) report the 3.0Å resolution structures of *Thermus thermophilus* elongation complex (EC) with a non-hydrolysable substrate analog, adenosine-5'-[(α, β)-methyleno]-triphosphate (AMPcPP), and with AMPcPP plus the inhibitor streptolydigin. In the EC/AMPcPP structure, the substrate binds to the active ('insertion') site closed through refolding of the trigger loop (TL) in to two α-helices. In contrast, the EC/AMPccPP/streptolydigin structure reveals an inactive ('preinsertion') substrate configuration stabilized by streptolydigin-induced displacement of the TL. Refolding of the TL is vital for catalysis and has three main implications. First, despite differences in the details, the two-step preinsertion/insertion mechanism of substrate loading may be universal for all RNAPs. Second, freezing of the preinsertion state is an attractive target for the design of novel antibiotics. Last, the TL emerges as a prominent target whose refolding can be modulated by regulatory effects. The RNAP elongation complex (EC) is both highly stable and processive, rapidly extending RNA chains for thousands of nucleotides. Understanding the mechanisms of elongation and its regulation requires detailed information about the structural organization of EC. Vessylyev et al. (2007a,b) report the 2.5Å resolution structure of *Thermus thermophilus* EC; the structure reveals the post-translocated intermediate with the DNA template in the active site available for pairing with the substrate. DNA strand separation occurs one position downstream the active site, implying that only one substrate at a time can specifically bind to the EC. The upstream edge of the RNA/DNA hybrid stacks on the β′-subunit 'lid' loop, whereas the first displaced RNA base is trapped within a protein pocket, suggesting a mechanism for RNA displacement. The RNA is threaded through the RNA exit channel, where it adopts a conformation mimicking that of a single-strand within a double helix, providing insight into a mechanism for hairpin-dependent pausing and termination.

Transcriptional pausing

Transcriptional pausing, also known as transcriptional stuttering, by RNAP plays an important role in the regulation of gene expression. Defined, sequence-specific pause sites have been identified biochemically (Adhya and Gottesman 1978). Single-molecule studies have also shown that bacterial RNAP pauses frequently during transcription elongation. Ubiquitous pauses were found to be associated with DNA sequences that show similarities to regulatory pause sequences. Data support a model where the transition to pausing branches off of the normal elongation pathway and is mediated by a common elemental state, which corresponds to the ubiquitous pause.

Transcription Termination

Once RNAP starts transcription, the enzyme continues to move along the template synthesizing RNA until it meets a signal to stop this synthesis. At this stage, enzyme stops adding nucleotide to the RNA chain, releases the completed product and dissociates from the DNA template. Terminators require that all the hydrogen bonds holding RNA-DNA together must be broken. The DNA sequence that provides the signal to cease or stop transcription is called a terminator. Terminators of bacterial and phage genes have been characterized in detail. At some terminators, anti-termination causes the enzyme to continue transcription past the terminator sequence. There are certain specific ancillary factors that interact with RNAP. NusA protein controls the hairpin formation promoting termination, and stabilization of these contacts by phage lambda N protein leads to anti-termination (Gusarov and Nudler 2001). Actually, initiation and termination of transcription are under specific control. Both require breaking of H-bonds and both require additional proteins to interact with the enzyme.

In vivo identification of a terminator particular to an RNA molecule that has been synthesized in the living cell is difficult. In both prokaryotes and eukaryotes, a 3′-end of the RNAP is generated by the cleavage of the primary transcript and therefore does not represent the actual site at which RNAP terminated transcription. There is no marker at the 3′-end comparable to 5′-end that had triphosphate. The best identification of termination site comes from in vitro studies because binding of the enzyme to DNA is strongly influenced by ionic strength. The termination site on DNA can be identified by comparing the template with 3′-terminal sequence of its RNA product. Many prokaryotic terminators have been examined in this way. There are no similarities of sequence beyond the point at which last base is added to RNA. Termination of transcription occurs because the elongation complex is less stable when transcribing certain specific DNA sequences. Certain termination signals on DNA are involved which are called terminators (Figure 6.16). The termination of transcription occurs at specific terminator sequen-

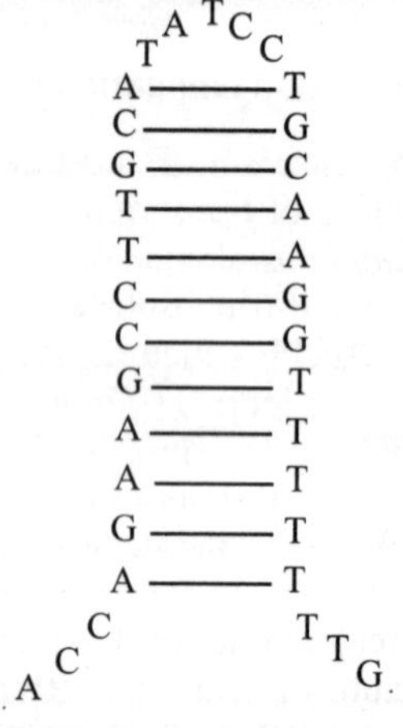

Figure 6.16 A typical terminator in a tRNA operon of *B. subtilis*

ces in DNA. Termination process in prokaryotes involves formation of 'hairpin structure'. There are of two kinds of terminators – rho-dependent terminators and rho-independent terminators. Rho-independent or simple terminator core enzyme can terminate in vitro at certain sites in absence of any other factor. These terminators have two structural features — a hairpin in secondary structure and a run of 6 uracils (U's) at the very end of the unit. Both features are needed for termination. The hairpins contain a G-C rich region near base of the stem. Probably all the hairpins that form in the RNA product cause the polymerase to slow or pause in RNA synthesis. Pausing creates an opportunity for termination to occur. Rho-independent terminators include palindromic regions that form hairpins varying in length from 7-20 bp. The stem-loop structure is followed by a run of U residues (Figure 6.17(A). The string of U residues probably provides the signal that allows RNA polymerase to dissociate from the template when it pauses at the hairpin. In case of rho-dependent terminators, there is a need for addition of rho (ρ) factor for termination in vitro. Rho factor is required for termination in vivo also. Most of the known rho-dependent terminators are found in phage genomes. A rho-dependent terminator is shown in Figure 6.17(B).

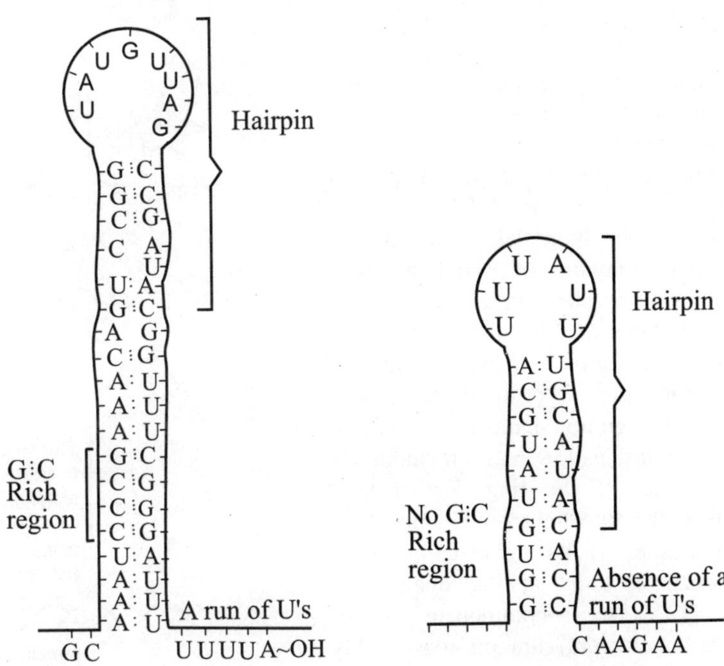

Figure 6.17 (A) A rho-independent terminator. (B) A rho-dependent terminator

An example of palindromic sequences present near the rho-dependent terminator site is given in Figure 6.18. There is similarity in the secondary structure generated by palindromic sequences in both ρ-dependent and ρ-independent termination. There is complementation between bases on same strand leading to hairpin structure.

Figure 6.18 An example of palindromic sequences present near the rho-dependent terminator site (Redrawn from http://www.biology.ewu.edu/aHerr/Genetics/Bio310/Pages/ch11pges/ch11fs 12. html)

Working of *E. coli* rho factor is shown in Figure 6.19. Rho simply moves along transcript faster than RNAP moves along DNA. When rho catches RNAP at a terminator, the enzyme pauses and termination reaction occurs. Alternative possibility is that RNA polymerase follows pauses during transcription as an intrinsic part of its action. Rho factor causes termination when it catches polymerase at a pause site. In the absence of RNAP, rho can use an RNA-DNA hybrid to unwind. ATP is used to provide energy for this reaction. Unwinding reaction proceeds in 5′→3′ direction. Termination is completed by release of rho and RNAP from DNA template. Attenuation is a mechanism that links the supply of an aminoacyl~tRNA to the ability of RNAP to read through termination site. Terminator is located at the beginning of cluster of structural genes coding for enzyme that synthesizes amino acid carried by tRNA. Attenuators are a very important mechanism of termination of transcription.

In bacteria, one of the major transcriptional termination mechanisms requires a RNA/DNA helicase known as the Rho factor. Skordalakes and Berger (2003) determined two structures of Rho complexed with nucleic acid recognition site mimics in both free and nucleotide bound states to 3.0Å resolution. Both structures show that Rho forms a hexameric ring in which two RNA binding sites – a primary one responsible for target mRNA recognition and a secondary one required for mRNA translocation and unwinding – point towards the center of the ring. Rather than forming a closed ring, the Rho hexamer is split open, resembling a "lock washer" in its global architecture. The distance between subunits at the opening is sufficiently wide (12Å) to accommodate single-stranded RNA. This open configuration most likely resembles a state poised to load onto mRNA and suggests how related ring-shaped enzymes may be breached to bind nucleic acids.

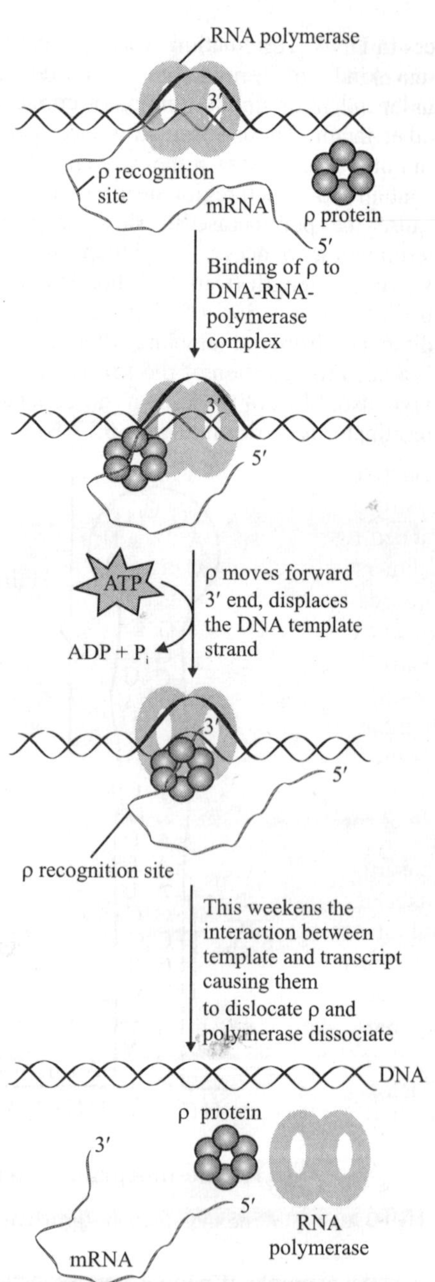

Figure 6.19 Rho factor pursues RNA polymerase along the RNA and can cause termination when it catches up the enzyme pausing at a rho-dependent terminator (Redrawn from http://www2.hawaii.edu/~johnb/micro/micr230/micr230_lectures/lectu re8.htm)

RNA translocation and unwinding cycles of a helicase

Helicases are a ubiquitous class of enzymes involved in nearly all aspects of DNA and RNA metabolism. Despite recent progress in understanding their mechanism of action, limited resolution has left inaccessible the detailed mechanisms by which these enzymes couple the rearrangement of nucleic acid structures to the binding and hydrolysis of ATP. Observing individual mechanistic cycles of these motor proteins is central to understanding their cellular functions. Dumont et al. (2006) follow in real time, at a resolution of two base pairs and 20 ms, the RNA translocation and unwinding cycles of a hepatitis C virus helicase (NS3) monomer. NS3 is a representative superfamily-2 helicase essential for viral replication, and therefore a potentially important drug target. They show that the cyclic movement of NS3 is coordinated by ATP in discrete steps of 11±3 bp, and that actual unwinding occurs in rapid smaller substeps of 3.6±1.3 bp, also triggered by ATP binding, indicating that NS3 might move like an inchworm. This ATP-coupling mechanism is likely to be applicable to other non-hexameric helicases involved in many essential cellular functions.

Rho transcription termination factor

Hexameric helicases and translocases are required for numerous essential nucleic-acid transactions. To better understand the mechanisms by which these enzymes recognize target substrates and use nucleotide hydrolysis to power molecular movement, Skordalakes and Berger (2006) have determined the structure of Rho transcription terminator factor, a hexameric RNA/DNA helicase, with single-stranded RNA bound to the motor domains of the protein. The structure reveals a closed-ring "trimer of dimers" conformation for the hexamer that contains an unanticipated arrangement of conserved loops required for nucleic-acid translocation. RNA extends across a shallow intersubunit channel formed by conserved amino acids. Required for RNA-stimulated ATP hydrolysis and translocation and directly contacts a conserved lysine, just upstream of the catalytic GKT triad, in the phosphate-binding (P loop) motif of the ATP binding pocket. The structure explains the molecular effects of numerous mutations and provides new insights into the links between substrate recognition, ATP turnover, and coordinated strand movement.

Role of rho factor on a genome-wide scale

Transcription of the bacterial genome by the RNAP must terminate at specific points. Transcription can be terminated by Rho factor, an essential protein in eubacteria. Cardinale et al. (2008) used antibiotic bicyclomycin, which inhibit Rho, to access its role on a genome-wide scale. Rho is revealed as a global regulator of gene expression that matches *E. coli* transcription to translational needs. They also find that genes in *E. coli* that are most repressed by Rho are prophages and other horizontally acquired portions of the genome. Elimination of these foreign DNA elements increases resistance to bicyclomycin. Although rho remains essential, such reduced-genome bacteria no longer require Rho-cofactors NusA and NusG. Deletion of the cryptic rac prophage in wild-type *E. coli* increases bicyclomycin resistance and permits deletion of *musG*. Thus, rho termination, supported by NusA and NusG, is required to suppress the toxic activity of foreign genes.

DNA Topoisomerases

The enzymes that introduce or remove turns from the double helix by transient breakage of one or both polynucleotides are known as DNA topoisomerases. These enzymes are important during transcription. RNAP generates positive supercoiling ahead and leaves negative supercoiling behind. Type I DNA topoisomerase rectifies the situation behind by making a transient break in one strand of DNA. Type II DNA topoisomerase relaxes negative supercoiling during transcription by introducing a transient double-stranded break in DNA (Wang 1985).

Cross-talk between rho family GTPases

Many feature of cell behavior are regulated by Rho family GTPases, but the most profound effects of these proteins are on the actin cytoskeleton and it was these that first drew the attention to this family of signaling proteins. Focusing on Rho and Rac, Burridge and Wennerberg (2004) explain how their effectors regulate the actin cytoskeleton. The describe how the activity of Rho proteins is regulated downstream from growth factor receptors and cell adhesion molecules by guanine nucleotide exchange factors and GTPase activating proteins. There is cross-talk between family members. Various bacterial pathogens have developed strategy to manipulate Rho protein so as to enhance their own survival.

Coordinated regulation of gene expression

The coordinated regulation of gene expression is required for homeostasis, growth and development in all organisms. Such coordination may be partly achieved at the level of messenger RNA stability, in which the targeted destruction of subset of transcripts generates the potential for cross-regulating metabolic pathways. In *E. coli*, the balance and composition of of the transcript population is affected by RNase E, an essential endonuclease that not only turns over RNA but also processes certain key RNA precursors. RNase E cleaves RNA internally, but its catalytic power is determined by the 5′ terminus of the substrate, even if this lies at a distance from the cutting site. Callaghan et al. (2005) report the crystal structure of the catalytic domain of RNase E as trapped allosteric intermediates with RNA substrates. Four subunits of RNase E catalytic domain associate into an interwoven quaternary structure, explaining why the subunit organization is required for catalytic activity. The subdomain encompassing the active site is structurally congruent to a deoxyribonuclease, making an expected link in the evolutionary history of RNA and DNA nucleases. The structure explains how the recognition of the 5′ terminus of the substrate may trigger catalysis and also sheds light on the question of how RNase might selectively process, rather than destroy, specific RNA precursors. The structure of RNase E catalytic domains, as worked out by Callaghan et al. (2005), is given in Figure 6.20.

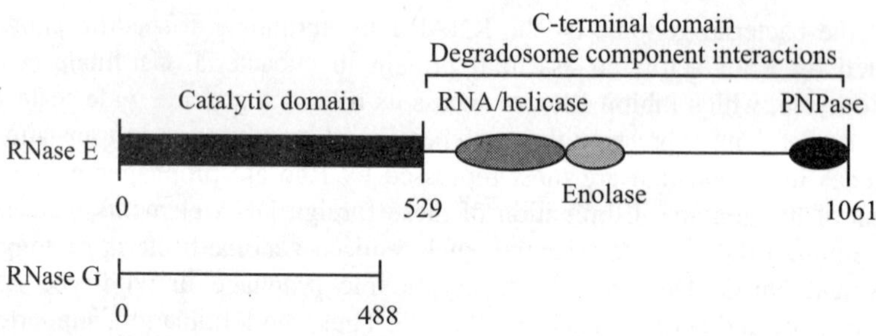

Figure 6.20 The structure of RNase E catalytic domains

Antibiotic myxopyronin inhibits bacterial RNA polymerase

Myxopyronin inhibits bacterial RNAP. The antibiotic binds to a pocket deep inside the RNAP clamp head domain, which interacts with the DNA template in the transcription bubble. Binding of dMyx (a desmethyl derivative of myxopyronin) stabilizes refolding of the β′-subunit switch-2 segment, resulting in a configuration that might indirectly compromise binding to, or directly clash with, the melted template DNA strand. Antibiotic binding does not prevent nucleation of the promoter DNA melting but instead blocks its propagation towards the active site (Balogurov et al. 2009). Switch-2 might act as a molecular checkpoint for DNA loading in response to regulatory signals or antibiotics. The universally conserved switch-2 may have the same role in all mutisubunit RNAPs. A schematic drawing of the dMyx action is given in Figure 6.21.

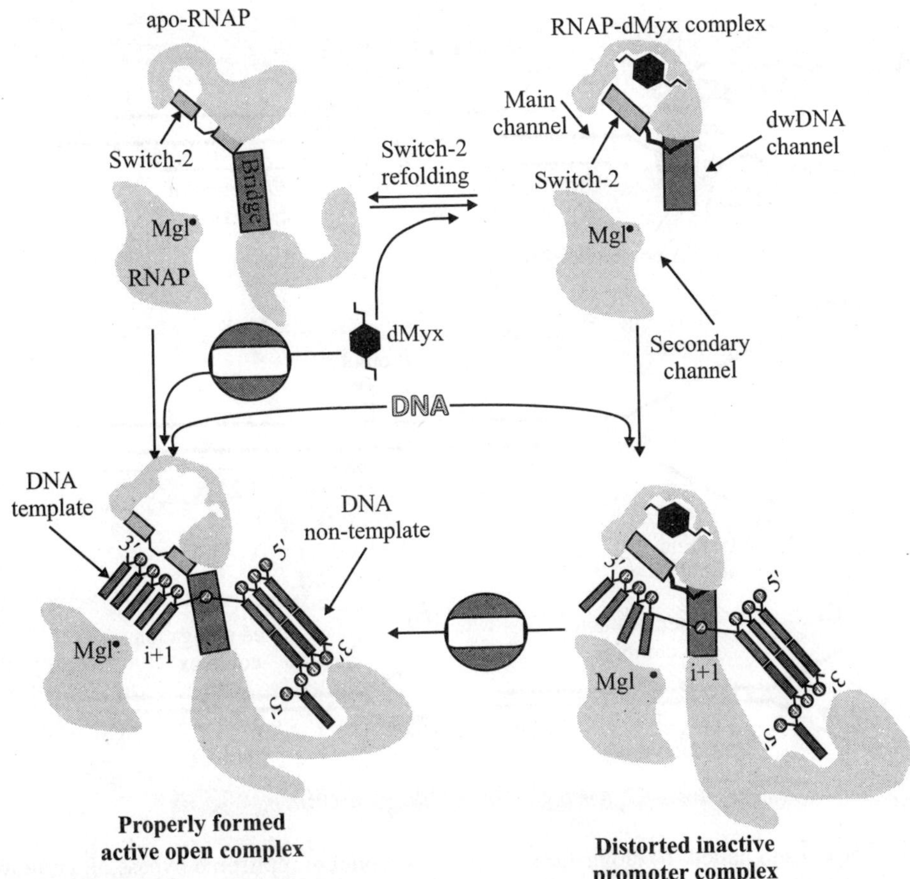

Figure 6.21 A mechanism of the dMyx action

Collision of the replisome with RNA polymerase

Replication forks are impeded by DNA damage and protein-nucleic acid complexes such as transcribing RNAP. For example, head-on collision of the replisome with RNAP results in replication fork arrest. However, co-directional collision of the replisome with RNAP has little or no effect on fork progression. Pomerantz and O'Donnell (2008) examine co-directional collisions between a replisome and RNAP in vitro. They show that the *E. coli* replisome uses the RNA transcript as a primer to continue leading-strand synthesis after the collision with RNAP that is displaced from the DNA. This action results in a discontinuity in the leading strand, yet the replisome remains intact and bound to DNA during the entire process. These findings underscore the notable plasticity by which the replisome operates to circumvent obstacles in its path and may explain why the leading strand is synthesized discontinuously in vivo. Leading strand synthesis is interrupted by a co-directional RNAP Figure 6.22 (Pomerantz and O'Donnell 2008). (a) Replisome proteins include RNAP III core, β-clamp, DnaB and the clamp loader. Primase was omitted from reaction and lagging-strand is not shown. (b) A 2.2-kb template was constructed that supports the leading-strand synthesis and co-directional transcription. Leading strand synthesis was performed in the presence of increasing concentrations of

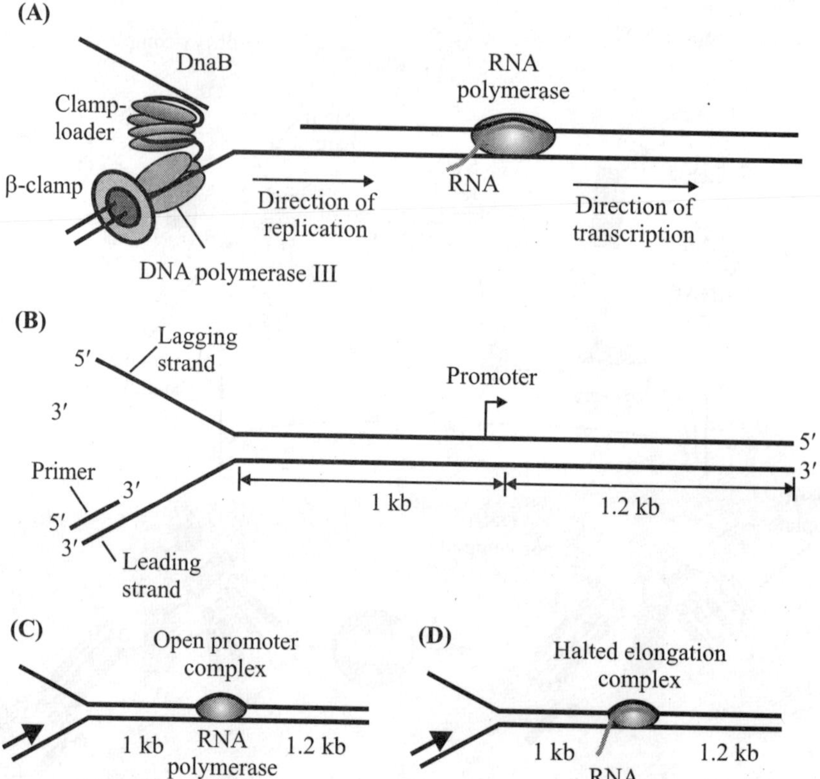

Figure 6.22 Leading strand synthesis is interrupted by a co-directional RNA polymerase

an RNAP open (c) and halted (d) elongation complex. A model of replisome bypass of a co-directional RNAP is presented in Figure 6.23 (Pomerantz and O'Donnell 2008). (a) The replisome encounters a co-directional RNAP. (b) RNAP is displaced from DNA. The lagging-strand polymerase dissociates from the β-clamp and DNA but remains bound to the lagging strand. (c) The clamp-loader assembles a new β-clamp at the 3′ terminus of the RNA-DNA hybrid. (d) The leading-strand polymerase binds to the newly assembled β-clamp. (e) The leading-strand polymerase extends the mRNA leaving behind a nick or gap in the leading strand.

Interchromosomal contacts yield transcriptional hub

Why would two distinct genes on separate chromosomes and from different nuclear locations unite in response to signals for gene expression? They might be seeds for formation of transcriptional hubs (Lonard and O'Malley 2008). Nunez et al. (2008) find that, in the absence of estrogen, two genes (TEF1 and GREB1) that are activated by this hormone and found on different chromosomes, reside in different locations within the nucleus (Figure 6.24) (Lonard and O'Malley 2008). On exposure to estrogen, these two genes make interchromatin contacts, facilitated by the nuclear receptor ERα, transcriptional coactivators (CoAs), actin and molecular motor proteins. Subsequently, the gene pair interacts with a nuclear speckle in an LSD1-dependent manner, creating a multiprotein complex that might act as a transcriptional hub for DNA transcription into mRNA. A model of actin/myosin1/

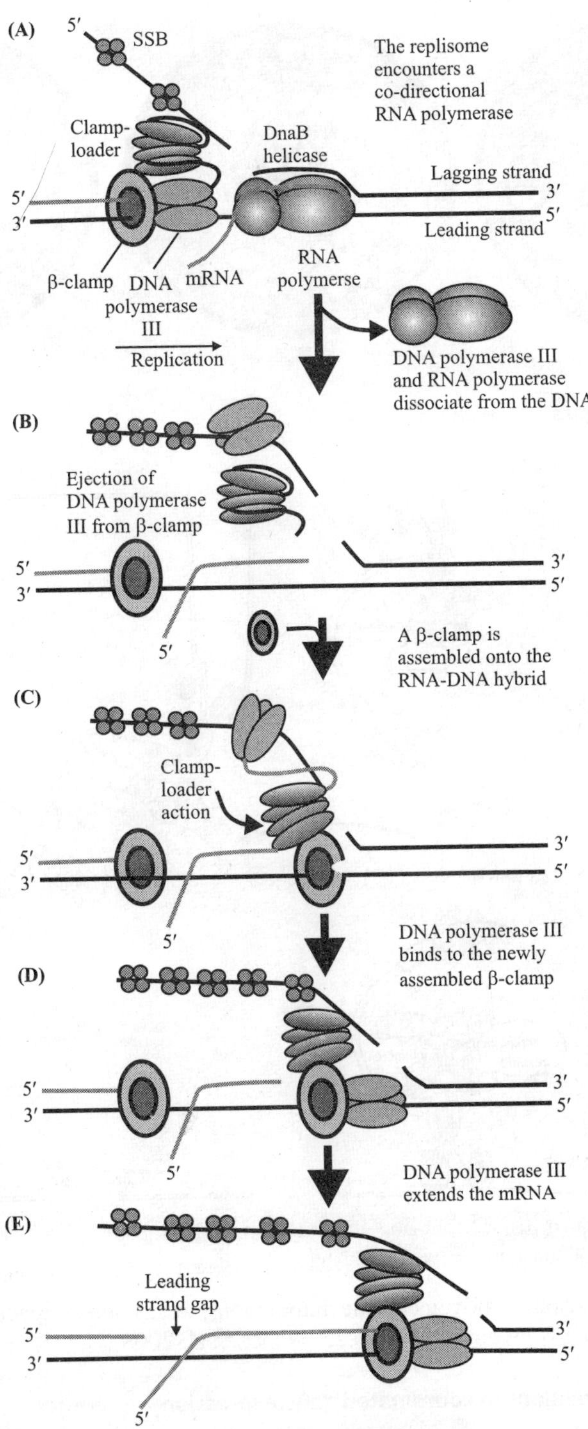

Figure 6.23 Model of replisome bypass of co-directional RNA polymerase

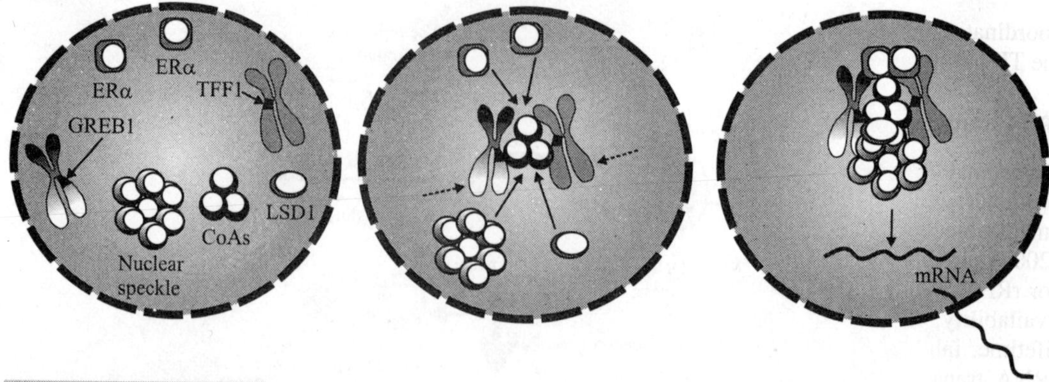

Figure 6.24 Rambling genes

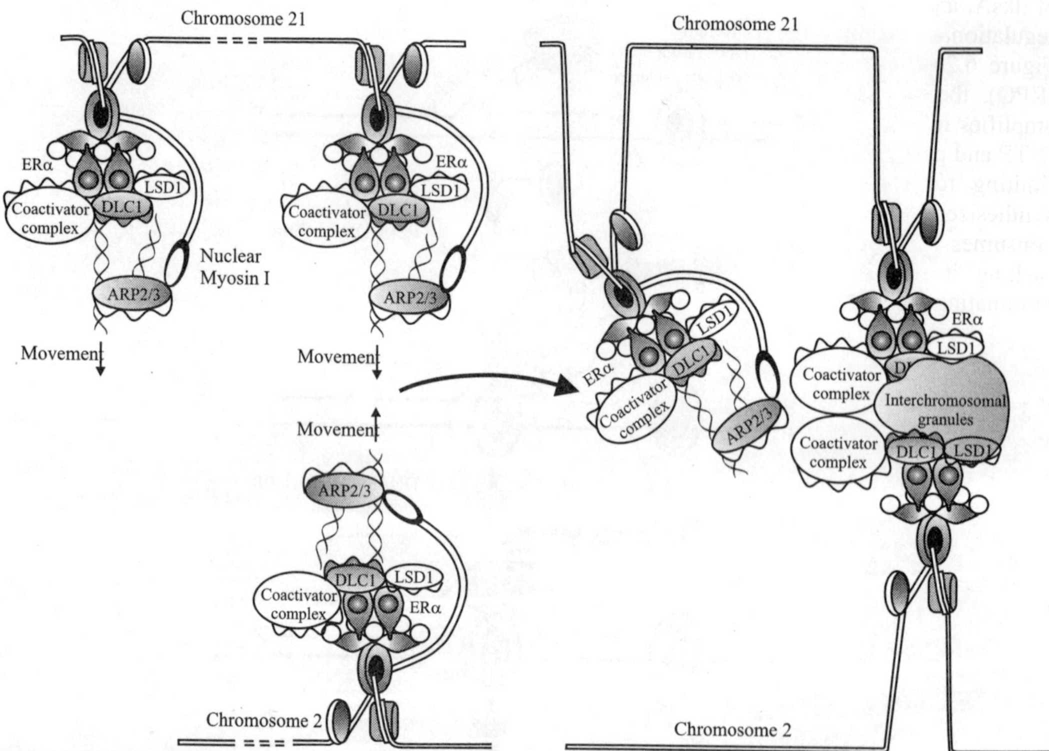

Figure 6.25 Model of actin/myosin1/DLC1-dependent chromosomal movement and interactions with nuclear speckles in response to the nuclear hormone signaling

DLC1-dependent chromosomal movement and interactions with nuclear speckles in response to the nuclear hormone signaling is given in Figure 6.25 (Nunez et al. 2008).

Interchromosomal interactions in coordinated gene expression

Spilinakis et al. (2005) provide an example of eukaryotic genes (cytokine genes) located in separate chromosomes associating physically in the nucleus via interactions that may have a function in coordinating gene expression. IFN-γ gene is located in chromosome 10 and the regulatory regions of the TH2 cytokine locus in chromosome 11; these two loci show interchromosomal interactions.

rRNA transcription regulation

Ribosomal RNA (rRNA) transcription is regulated primarily at the level of initiation from rRNA promoters. The unusual kinetic properties of these promoters result in their specific regulation by two small molecule signals, ppGpp and the initiating NTP, that bind to RNAP at all promoters. Paul et al. (2004) show that DskA, a protein previously unsuspected as transcription factor, is absolutely required for rRNA regulation. In ΔdksA mutants, rRNA promoters are unresponsive to changes in amino acid availability, growth rate, or growth phase. In vitro, DksA binds to RNAP, reduces open complex lifetime, inhibits rRNA promoter activity and amplifies effects of ppGpp and the initiating NTP on rRNA transcription, explaining the dskA requirement in vivo. These results expand our molecular understanding of rRNA transcription regulation, may explain previously described pleiotropic effects of dksA, and illustrate how transcription factors that bind DNA can nevertheless potentiate RNAP for regulation. A model depicting contribution of DksA to regulation of rRNA expression is given in Figure 6.26 (Paul et al. 2004). DksA binds to RNAP, decreases the half-life of the open complex (RPO), thereby increasing the concentration of the iNTP required for transcription initiation, and amplifies inhibition by ppGpp. rRNA promoters are sensitive to changes in the concentrations of the iNTP and ppGpp because DksA brings the lifetime of the open complex into the range where it is rate limiting for transcription in vivo. Two feedback loops control rRNA transcription: ppGpp is synthesized by RelA in response to uncharged tRNAs in the ribosomal A site, and the translation consumes ATP and GTP. In a ΔdksA mutant, the rRNA promoter open complex is longer-lived, making it insensitive to physiologically relevant changes in iNTP and ppGpp concentrations, eliminating regulation of rRNA transcription.

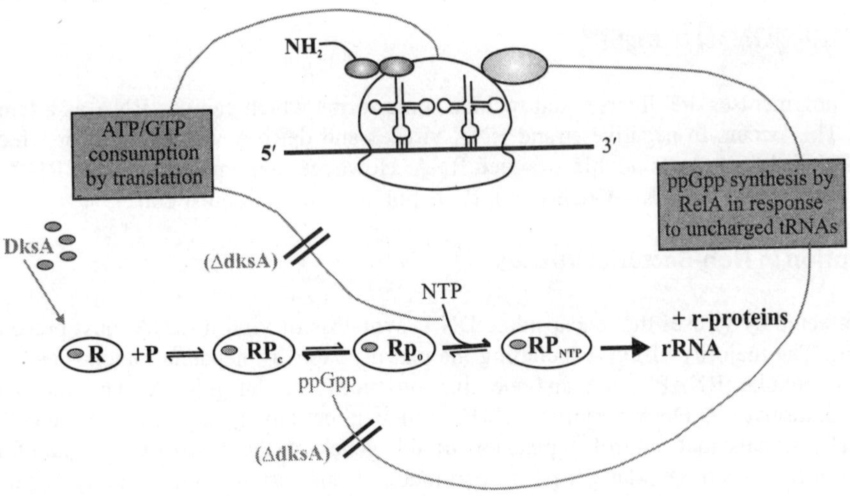

Figure 6.26 Model depicting contribution of DksA to regulation of rRNA expression

RNA-DIRECTED RNA SYNTHESIS

Non-coding small RNAs regulate gene expression in all organisms, in some cases through direct association with RNAP. Wassarman and Saeker (2006) report that the mechanism of 6S RNA inhibition of transcription through specific, stable interactions with the active site of *E. coli* RNAP that exclude promoter DNA binding. In fact, the DNA-dependent RNAP uses bound 6S RNA as a template for RNA synthesis, producing 14- to 20-nucleotide RNA product (pRNA). These results demonstrate that 6S RNA is functionally engaged in the active site of RNAP. Synthesis of pRNA destabilizes 6S RNA-RNAP complexes leading to the release of the pRNA-6S RNA hybrid. In vivo, 6S RNA-directed RNA synthesis occurs during outgrowth from the stationary phase of and is likely responsible for liberating RNAP from 6S RNA in response to nutrient availability.

TRANSCRIPTIONAL APPARATUS IN ARCHAEA

The transcriptional apparatus in Archaea can be described as a simplified version of its eukaryotic RNA polymerase II (RNAPII) counterpart, comprising of an RNAPII-like enzyme as well as two general transcription factors, the TATA-binding protein (TBP) and the eukaryotic TFIIB ortholog TFB. It has been widely understood that that precise comparison of cellular RNAP crystal structure could reveal elements common to all enzymes that that these insights would be useful in analyzing components of each enzyme that enable it to perform domain-specific gene expression. However, the structure of archeal RNAP has been limited to individual subunits. Hirata et al. (2008) report the first crystal structure of archaeal RNAP from *Sulfolobus solfataricus* at 3.4Å resolution, completing the suite of multi-subunit RNAP structures from all three domains of life. They also report the high-resolution (at 1.76Å) crystal structure of the D/L subcomplex of archaeal RNAP and provide the first experimental evidence of any RNAP possessing an iron-sulfur (Fe-S) cluster, which may play a structural role in a key subunit of RNAP assembly. The striking structural similarity between archaeal RNAP and eukaryotic RNAPII highlights the simpler archaeal RNAP as an ideal model system for dissecting the molecular basis of eukaryotic transcription.

TRANSCRIPTION IN VIRUSES

The viral polymerases are diverse, and include some forms which can use RNA as a template instead of DNA. This occurs in negative strand RNA viruses and dsRNA viruses, both of which exist for a portion of their life cycle as double-stranded RNA. However, some positive strand RNA viruses, such as polio, also contain these RNA-dependent RNA polymerases (Ahlquist 2002).

Transcription in Non-Bacterial Viruses

In cells infected by viruses that encapsidate DNA, synthesis of viral m-RNA must precede production of proteins. The majority of DNA including the parvo-, papova-, hepadna-, adeno- and herpesviruses, depend on cellular RNAPII, the enzyme that produces cellular mRNA. The retroviruses, which establish themselves as DNA proviruses early in their infections cycles, also use the cellular RNAPII system. The signals that control expression of the genes of these viruses are therefore similar in organization to those of cellular genes. In fact, much of our current understanding of the mechanisms of cellular transcription stems from study of the transcription of viral DNA templates. Some of these viruses, notably simple retroviruses, depend exclusively on the cellular transcriptional machinery for expression of their genetic information. The genomes of poxviruses are transcribed by a viral DNA-dependent RNAP and accessory viral proteins, which control the recognition of viral promoters. In general, viral enzymes and regulatory proteins are made during the initial period of infection, whereas structural proteins from which virus particles are built are synthesized only after viral DNA synthesis begins. Such orderly expression of viral genes is primarily the result of transcriptional regulation by viral proteins. Thus, viral genes are transcribed in a fixed sequence, and transcription of specific genes is activated during specific periods in the infection cycle. This pattern of gene expression is quite different from that characteristic of the majority of RNA viruses.

In DNA viruses, RNAPII is responsible for transcription in majority of the cases but several viral proteins stimulate viral gene transcription (Zhou and Knipe 2002). Viral genes in DNA viruses are transcribed in fixed sequences during specific period. Regulation in viruses is under control of viral proteins along with local and distant regulatory sequences.

Transcription cycle

All cellular RNAPs synthesize RNA from DNA templates by a common set of reactions. The transcriptional machinery binds to the promoter and includes local unwinding of the double-stranded DNA template so that copying of the DNA into RNA can begin. These reactions comprise initiation of transcription. The transcribing complex progressively unwinds the template as it reads the DNA sequence and adds nucleotides to the 3′-end of the nascent RNA chain. These elongation reactions continue until a termination signal is encountered in the template. Transcription then ceases, and both the RNA transcript and the transcriptional machinery are released from the DNA template during termination.

Regulation of transcriptional and viral proteins

Cellular transcriptional components acting alone transcribe the viral gene for viral protein A. Once synthesized and returned to the nucleus viral protein A can stimulate transcription either of the same transcription unit (A) or of a different viral transcription unit (B). In either case, viral protein A acts in concert with components of the cellular transcriptional machinery. The autoregulatory strategy (A) is exemplified by complex retroviruses such as human immunodeficiency virus type 1. The transcriptional programs of viruses with DNA genomes are variations of the sequential transcriptional cascade (B).

Transcription in Bacteriophages

When a temperate bacteriophage such as lambda (λ) infects a bacterium, it can follow either of two pathways of development. A temperate phage can either (1) enter the lytic cycle, during which it reproduces and lyses the host cell, just like a virulent phage or (2) enter the lysogenic pathway, during which its chromosome is inserted into the chromosome of the host and thereafter is replicated like any other segment of that chromosome. When integrated in the chromosome of the host cell, the temperate phage chromosome is called a prophage. In a lysogenic bacterium, the genes of the prophage that encode products involved in the lytic pathway must not be expressed. The prophage genes specifying enzymes involved in the replication of phage DNA, structural proteins required for phage morphogenesis, and the lysozyme that catalyzes cell lysis must be kept turned off to maintain a stable lysogenic state. The lytic pathway genes of the prophage are kept turned off in lysogenic cells by a simple repressor-operator-promoter circuit, much like the negative regulatory circuits of bacterial operons. However, the really interesting aspect of phage lambda is how the virus makes the initial choice between the lytic pathway and the lysogenic pathway. Whether a lambda phage will enter lysogenic state or undergo lytic development is determined by which of the two regulatory proteins, λ repressor or cro-protein, occupies key operator site. Lysogenic state is maintained by λ repressor. Lytic development is controlled by regulatory cascade, e.g., *Stx* genes are expressed only during lytic growth of bacteriophage 933W (Koudelka 2004).

Temporal sequences of gene expression during phage infection

Regulation of gene expression during the life cycles of virulent bacteriophages is quite different from the reversible on-off switches characteristic of bacterial operons. In phage-infected bacteria, the viral genes are expressed in genetically preprogrammed sequences or cascades. Temporal sequence of gene expression is discussed here briefly in bacteriophages SP01, T4, and T7.

Phage SPO1: *Bacillus subtilis* phage SP01 exhibits a slightly more complex pathway of sequential gene expression, including three sets of genes – the early, middle, and late genes, in reference to their time of expression during the phage reproductive cycle, e.g. products of SPO1 genes *44*, *50* and *51* are required for normal transition from early to middle expression during infection of *Bacillus subtilis* by bacteriophage SPO1 (Sampath and Stewart 2004). The phage SP01 early genes are transcribed by the *B. subtilis* RNAP. One of the early gene products is a polypeptide that binds to the host cell's RNAP, changing its specificity so that the modified RNAP transcribes the middle genes of SP01. Two of the products of the middle genes are, in turn, polypeptides that associate with the *B. subtilis* RNAP, further changing its specificity so that it then transcribes the SP01 late genes.

Bacteriophage T4: Phage T4 exhibits an even more complex pattern of sequential gene expression, involving several different modifications of the host cell's RNAP. Thus, in the case of these bacterial viruses, the control of the observed sequential gene expression occurs primarily at the level of transcription and is mediated by modifications of the specificity of RNAP for different promoter sequences.

Bacteriophage T7: Many viruses also encode for RNA polymerases. Perhaps the most widely studied viral RNAP is found in bacteriophage T7. Structural studies of the T7 bacteriophage DNA-dependent RNA polymerase (T7 RNAP) have shown that the conformation of the amino-terminal domain changes substantially between the initiation and elongation phases of transcription. Durniak et al. (2008) report crystal structure of T7 RNAP bound to promoter DNA containing either 7- or 8-nucleotide (nt) RNA transcript that illuminate intermediate states along the transition domain (PBD), which consists of two segments separated by subdomain H. The structures of the intermediate complex reveal that the PBD and the bound promoter rotate by ~45° upon synthesis of an 8-nt RNA transcript. This allows the promoter contacts to be maintained while the active site expanded to accommodate a growing heteroduplex. The C-helix subdomain moves modestly toward its elongation conformation, whereas subdomain H remains in its initiation rather than its elongation-phase location, more than 70Å away. Crystal structures of transcription complexes formed by bacteriophage T7 RNAP reveal a nucleotide-addition cycle driven by active-site conformational changes similar to those observed in DNA polymerases, and suggest provocative hypotheses for the more complex multisubunit RNAPs of free-living organisms. This single-subunit RNAP is related to that found in mitochondria and chloroplasts, and shares considerable homology to DNA polymerase (Hedtke et al. (1997). It is believed that most viral polymerases therefore evolved from DNA polymerase and are not directly related to the multi-subunit polymerases.

DNA Replication and Transcription Collision

An in vitro system reconstituted from purified proteins has been used to examine what happens when the DNA replication apparatus of bacteriophage T4 collides with an *Escherichia coli* RNA polymerase transcription complex that is poised to move in the direction opposite to that of the moving replication fork. In the absence of a DNA helicase, the replication fork stalls for many minutes after its encounter with the RNA polymerase. However, when T4 gene 41 DNA helicase is present, the replication fork passes the RNA polymerase after a pause of a few seconds (Liu and Alberts 1995). This brief pause is longer than the pause observed for a co-directional collision between the same two polymerases suggesting that there is an inherent disadvantage to having replication and transcription direction oriented head to head. As for a co-directional collision, the RNA polymerase remains competent to resume faithful RNA chain elongation after DNA replication passes; most strikingly, the RNA polymerase has switched from its original template strand to use the newly synthesized daughter DNA strand as the template.

REVERSE TRANSCRIPTION IN DNA VIRUSES

We do not expect reverse transcription in DNA viruses but Varmus (1982a,b; 1983) proposed reverse transcription step in replication of cauliflower mosaic virus (CaMV), a double-stranded DNA virus, which infects plants, on the basis of structural studies of encapsidated CaMV DNA and of intracellular DNA and RNA species judged to be intermediates in the replication of CaMV. These studies suggested that (a) CaMV has a capacity similar to retroviruses to produce a major RNA species that contains all viral information and have short repeat (R) sequences at both the ends. (b) CaMV uses primer for synthesis of DNA like retroviruses. Analogous species of CaMV has a 14-nucleotide priming binding site for host $tRNA^{Met}$ and polypurine tract (PPT) for (+) priming site. The priming binding sites are positioned similarly to retroviral priming binding sites. But the genome of this virus is like that of HBV, i.e., DNA.

Reverse transcriptase activity was confirmed in CaMV by direct expression of gene coding for this enzyme in yeast and *E. coli*. The sequence of CaMV genome contains a number of open reading frames (ORFs) and one of these, *ORF-V* encodes a protein which is homologous to retroviral transcriptase. The ORF-V was inserted immediately downstream from acid phosphatase promoter of yeast expression vector pAM82 to form a plasmid pAMOFR-V; the integrity of *ORF-V* gene in this plasmid was confirmed by fine restriction mapping of a rescued pAMORF-V from AH22/pAMORF-V cells. This plasmid, simultaneously with pAM82 vector, was transferred to yeast AH22 cells. These AH22/pAMORF-V and AH22/pAM82 cells were grown in liquid medium and the induced in phosphate deprived medium. The extract from A22/pAMORF-V cells revealed an inducible RNA-dependent DNA polymerase activity but A22/pAM82 cell extracts did not show any such activity. So, RNA-dependent DNA polymerase activity induced specifically in AH22/pAMORF-V cells was, therefore, identified as the ORF-V gene which encoded reverse transcriptase. Molecular weight of this enzyme is 70 kDa.

The most important implication of reverse transcription is to understand cancer induction in human beings (Varmus 1987; Varmus 1989). The DNA transcripts of viral RNA are attached to the genome of the cell and can, when genetically active, cause spontaneous cancer. The hepatitis B virus (HBV), cauliflower mosaic virus (CaMV) and retroviruses are thought to have the same origin. CaMV is an important vector for gene transfer in plants. All reactions involving reverse transcription of RNA are segregated from cytosol within a subviral particle or capsid composed of major capsid protein, the polymerase and the RNA template. A key step in the formation of these particles is selective encapsidation of RNA template. Encapsidation has been carefully studied only in retroviruses. Hirsch et al. (1990) have examined encapsidation reaction in enveloped DNA virus HBV that replicates through reverse transcription. HBV gene product is required for RNA packaging and encapsidation function of the enzyme can be separated from DNA polymerase activity.

Nature has devised remarkable solutions to the problems inherent in the conversion of single-stranded viral RNA into viral DNA. Figure 6.27 suggests that DNA synthesis requires molecular

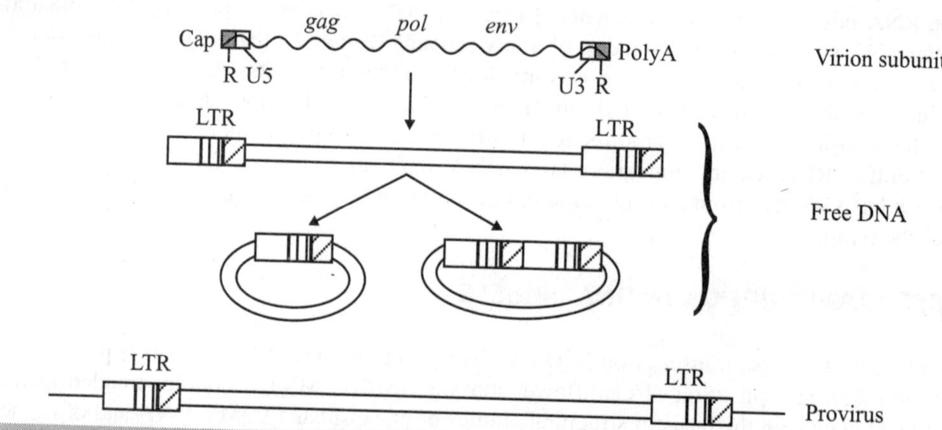

Figure 6.27 Replication and expression of retroviral genomes (See text for amplification)

aerobatics, the transfer of nascent DNA strand twice between templates during reverse transcription (Coffin et al. 1997). One of the two identical subunits of viral RNA genome with its major structural and genetic features is shown: the 5′-cap nucleotide added to the first nucleotide of the transcript, the short sequence repeated at both termini (R-filled boxes), the sequence unique to the 5′-terminus and repeated in the viral DNA (U5 shaded box; the host tRNA hydrogen bonded to the genome at the U5 boundary, the coding domain for virion structural proteins (*gag*, *pol* and *env*); the sequence unique to the 3-terminus and repeated in vial DNA (U3 open box) and the tract of polyadenylic acid [Poly(A)]. The major primary product of reverse transcription linear duplex DNA with its long terminal repeats (LTRs) composed of U3, R and U5 is also shown. The closed circular DNA with one or two copies of the LTR is illustrated. Proviral DNA is also shown.

Synthesis of first (–) strand initiated with a host to RNA primer, where the 3′-end is hydrogen bonded to an 18-base sequence more than the 5′-end of viral RNA subunit (Varmus 1982) (Figure 6.28(A). The point at which synthesis begins defines the 3'-boundary of U5. When the template is exhausted 100 to 180 bases from the initiation site, the first jump occurs, with the use of DNA complementing to R to form a bridge at the 3′-end of one of the RNA subunit (Figure 6.28(B). Once safely repositioned, the nascent (–) strand is extended, apparently continuously, over the major expense of the genome. Meanwhile, synthesis the second, or (+) strand begins at a position that defines the 5′ boundary of U3 (Figure 6.28(C). Although the primer of this event has not been identified, a purine-rich tract ending in AATG always spans the priming site. Extension of (+) strand is also limited by the amount of available template, after the U3, R and U5 regions of (–) strand DNA are copied, the linked tRNA is probably transcribed upto the position of the first modified base, setting for the next major event. To execute the second jump, duplex can be formed between the 3′-end of the growing (–) strand, one it had copied from the tRNA binding site, and the 3′-end of the arrested (+) strand (Figure 6.28(D) and Figure 6.28(E). Extension of both strands will then produce a blunt-ended, linear duplex, the major product of retroviral DNA synthesis in the cytoplasm of infected cells (Figure 6.28(F). How do genes perform their phenotypic functions, i.e., how do genes exert their effects on the phenotype of a virus, a cell, or an organism? Different genes exert their effects in different ways; genes may express themselves in several ways. As we have seen earlier, gene is a linear sequence of nucleotides. Polypeptide is a linear sequence of amino acids. What is the relationship between the two? Sequence of amino acids in a polypeptide chain is determined by sequence of nucleotides in a gene in a linear fashion. This is termed as colinearity of gene and its polypeptide product.

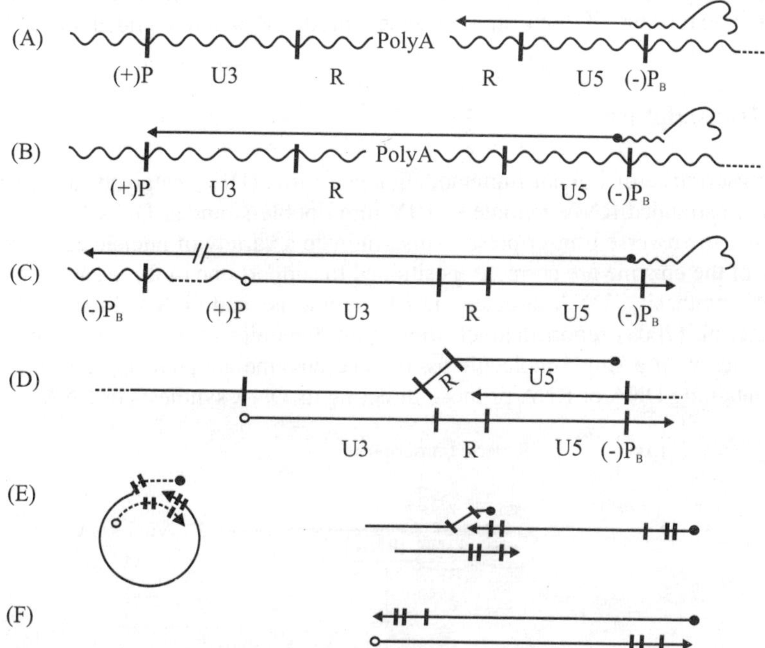

Figure 6.28 Critical steps in the synthesis of retroviral DNA (See text for amplification)

Some DNA viruses also undergo reverse transcription. Summers and Mason (1982) and Mason et al. (1982) proved that Hepatitis-B virus (HBV) replicates with the aid of reverse transcription. These viruses have double-stranded circular DNA. This reverse transcription in the virus occurs due to asymmetry of DNA strands. Synthesis of negative strand begins with DR1 sequence of progenomic RNA, and this initiation can occur either from 5′- or 3′-end of the RNA template. If it occurs from 3′-end then negative strand elongates without any interruption. However, if synthesis initiates from 5′-end, it needs transfer of nascent to the 3′-end of the same RNA molecule. Initiation of plus strand synthesis takes place from 3′-end of DR2 and the extension of plus strand takes place upto the terminal redundancy of minus strand to form the complete DNA molecule.

M. Temin and D. Baltimore were awarded Nobel Prize in 1975 for discovering reverse transcriptase in tumor viruses.

REVERSE TRANSCRIPTION IN BACTERIA

Reverse transcriptase encoded by cellular genes has also been described in bacteria – myxobacteria and *E. coli* (Temin 1989). This enzyme is thus not unique to eukaryotes. This suggests that reverse transcriptase existed before retroviruses. Retrons with reverse transcriptase became retroposons, which with their ability to transpose became reterotransposons, which with long terminal repeats became retroviruses, which with the ability to form virus became pararetroviruses, which lost the ability to integrate and transpose, became DNA viruses. During reverse transcription, the synthesis of DNA requires the transfer of nascent DNA strand twice between templates. Synthesis of negative (minus) strand is initiated at 3′-end of U5 sequence of viral RNA with the help of host template RNA primer. The first jump occurs and it forms a bridge like structure with 3′-end of second RNA subunit extended over the major expanse of the genome. The synthesis of positive (plus) strand starts at 5′-boundary of U3 at purine-rich tract. (AATG) and its extension are also limited by the amount of available template. Now for the second jump to occur, the duplex can be fórmed with the 3′-end of the negative strand. This results into extension of both strands which produces a blunt ended linear duplex of DNA molecule.

HIV Reverse Transcriptase

The reverse transcriptase of human immunodeficiency virus (HIV) catalyzes a series of reactions to convert the single-stranded RNA genome of HIV into double-stranded DNA for host-cell integration. This task requires the reverse transcriptase to discriminate a variety of nucleic acid substrates such that the active sites of the enzyme are correctly positioned to support one of three catalytic functions: RNA-directed DNA synthesis, DNA-directed DNA synthesis and DNA-directed RNA hydrolysis. Abbondanzieri et al. (2008) report distinct orientation dynamics of reverse transcriptase observed on different substrates with a single-molecule assay. The enzyme adopted opposite binding orientations on duplexes containing DNA or RNA primers, directing its DNA synthesis or RNA hydrolysis activity,

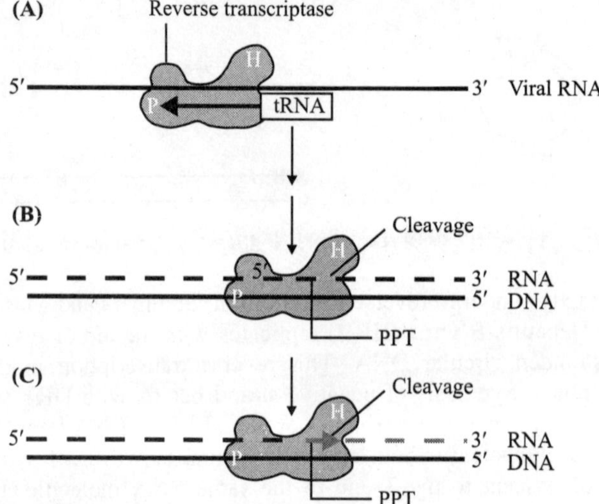

Figure 6.29 Reverse transcription in HIV

respectively. On duplexes containing the unique polypurine RNA primers for plus-strand DNA synthesis, the enzyme can rapidly switch between the two orientations. The switching kinetics was regulated by cognate nucleotide. With two catalytic activities and many substrates, how does HIV's reverse transcriptase enzyme know what to do to which substrate? Zooming in on the enzyme's molecular interactions provides a tantalizing clue (Arnold and Sarafianos 2008). Like other reverse transcriptase enzymes, that of HIV mediates transcription of the virus's RNA genome into double-stranded DNA (Arnold and Sarafianos 2008) (Figure 6.29). To form the first DNA strand (the minus strand) the enzyme uses a transfer RNA as a primer and interacts with the tRNA 3' end in a polymerase (P) binding mode. As the complementary minus-DNA sequence is being synthesized the enzyme cleaves the RNA template (but leaves the PPT sequence of the RNA intact) by binding to it in an RNase H (H) mode. Finally, to initiate the synthesis of the second (plus) DNA strand, the reverse transcriptase uses the PPT sequence as a primer, once again binding in the polymerase mode.

REFERENCES

Abbondanzieri, E.A., G. Bokinsky, J.W., Rausch, J.X. Zhang, and S.F.J. Le Grice. 2008. Dynamic binding orientations direct activity of HIV reverse transcriptase. Nature 453: 184-9.

Abbondanzieri, E.A., W.J. Greenleaf, J.W. Shaevitz, R. Landick, and S.M. Block. 2005. Direct observation of base-pair stepping by RNA polymerase. Nature 438: 460-5.

Adhya, S., and M. Gottesman. 1978. Control of transcription termination. Annu. Rev. Biochem. 47: 967-96.

Ahlquist, P. 2002. RNA-dependent RNA polymerases, viruses, and RNA silencing. Science 296: 1270-3.

Alberts, B., D. Bray, J. Lewis, M. Raff, K. Roberts, and J.D. Watson. 1989. *Molecular Biology of the Cell.* New York: Garland Publications.

Allison, L.A. 1988. *Fundamental Molecular Biology*. Oxford: Blackwell Publishing.

Arnold, E., and S.G. Sarafianos. 2008. A HIV secret uncovered. Nature 453: 169-70.

Balogurov, G.A., M.N. Vassylyeva, A. Sevostyanova, et al. 2009. Transcription inactivation through local refolding of the RNA polymerase structure. Nature 457: 332-5.

Baltimore, D. 1970. RNA dependent DNA polymerase in virions of RNA tumour viruses. Nature 226: 1209-11.

Bar-Nahum, G., V. Epshtein, A.E. Ruckenstein, R. Rafikov, A. Mustaev, and E. Nudler. 2005. A retchet mechanism of transcription elongation and its control. Cell 120: 183-93.

Burridge, K., and K. Wennerberg. 2004. Rho and Rac take center stage. Cell 116: 167-79.

Callaghan, A.J., M.J. Marcaida, J.A. Stead, K.J. McDowell, W.G. Scott, and B.F. Luisi. 2005. Structure of *Escherichia coli* RNase E catalytic domain and implications for RNA turnover. Nature 437: 1187-91.

Cardinale, C.J., R.S. Washburn, V.R. Tadigotia, L.M. Brown, M.E. Gottesman, and E. Nudler. 2008. Termination factor Rho and its cofactors NusA and NusG silence foreign DNA in *E. coli*. Science 320: 935-8.

Cook, P.R. 1990. How mobile are active RNA polymerases? J. Cell Sci. 96: 189-92.

Cramer, P. 2006. Self-correcting messages. Science 313: 447-8.

Cramer, P. 2007. Gene transcription: extending the message. Nature 448: 142-3.

Crick, F.H.C. 1958. On protein synthesis. Soc. Exptl. Biol. 12: 138-63

Crick, F.H.C. 1970. Central dogma of molecular biology. Nature 227: 561-3.

Dumont, S., W. Cheng, V. Serebrov, et al. 2006. RNA translocation and unwinding mechanism of HCV NS3 helicase and its coordination by ATP. Nature 439: 105-8.

Durniak, K.J., S. Bailey, and T.A. Steitz. 2008. The structure of a transcribing T7 RNA polymerase in transition from initiation to elongation. Science 322: 553-5.

Giacomoni, P.U., J.Y. Talaer, and J.B. L. Pecq. 1974. *Escherichia coli* RNA-polymerase binding sites on DNA are only 14 base pairs long and are located between sequences that are very rich in A+T. Proc. Nat. Acad. Sci. USA 71: 3091-5.

Goldmen, S.R., R.H. Ebright. and B.E. Nickets. 2009. Direct detection of abortive RNA transcripts in vivo. Science 324: 927-8.

Gusarov, I., and E. Nudler. 2001. Control of intrinsic transcription termination by N and NusA. Cell 107: 437-49.

Hedtke, B., T. Börner, and A. Weihe. 1997. Mitochondrial and chloroplast phage-type RNA polymerases in Arabidopsis. Science 227: 809-11.

Herr, C. 2006. Charlie's Genetics – Genetics 310. Topic 11. Transcription and RNA processing. http://www.biology.ewu.edu/aHerr/Genetics/Bio310/Pages/ch11pges/ch11fs12.html.

Hirata, A., B.J. Klein, and K.S. Murakami. 2008. The X-ray crystal structure of RNA polymerase from Archaea. Nature 451: 851-4.

Hirsch, R.C., J.E. Lavine, L.-J. Chang, H.E. Varmus, and D. Ganem. 1990. Polymerase gene products of hepatitis B viruses are required for genomic RNA packaging as well as for reverse transcription. Nature 344: 552-5.

Hsu, H.-H., K.-M. Chung, T.-C. Chen, and B.-Y. Chang. 2006. Role of the sigma factor in transcription initiation in the absence of core RNA polymerase. Cell 127: 317-27.

Jasny, B.R., and L. Roberts. 2004. Solving Gene Expression. Science 306: 629. DOI: 10.1126/science.306.5696.629.

Kapanidis, A.N., E. Margeat, S.O. Ho, E. Kortkhonjia, S. Weiss, and R.H. Ebright. 2006. Initial transcription by RNA polymerase proceeds through a DNA-scrunching mechanism. Science 314: 1144-7.

Koudelka, A.P., L.A. Hufnagel, and G.B. Koudelka. 2004. Purification and characterization of the repressor of the Shiga toxin-encoding bacteriophage 933W: DNA binding gene regulation and autocleavage. J. Bacteriol. 186: 7659-69.

Lonard, D.M., and B.W. O'Malley. 2008. Gene transcription: Two worlds emerged. Nature 452: 946-7.

Mason, W.S., G. Seal, and J. Summers. 1982. Replication of the genome of a hepatitis B-like virus by reverse transcription of an RNA intermediate. Cell 29: 403-15.

McCarthy, B.J., and J.J. Holland. 1965. Denatured DNA as a direct template for in vitro protein synthesis. Proc. Natl. Acad. Sci. USA 54: 880-6.

McCarthy, B.J., J.J. Holland, and C.A. Buck. 1966. Single-stranded DNA as a template for in vitro protein synthesis. Cold Sp. Harb. Symp. Quant. Biol. 31: 683-92.

Nudler, E., A. Goldfarb, and M. Keshleu. 1994. Discontinuous mechanism of transcription elongation. Science 265: 793-6.

Nunez, E., Y.-S. Kwon, K.R. Hutt, et al. 2008. Nuclear receptor-enhanced transcription requires motor- and LSD1-dependent gene networking in interchromatin granules. Cell 132: 996-1010.

Opalka, N., M. Chlenov, P. Chacon, W.J. Rice, W. Wriggers, and S.A. Darst. 2003. Structure and function of the transcription elongation factor GreB bound to bacterial RNA polymerase. Cell 114: 335-45.

Passarge, E. 2001. Color *Atlas of Genetics*. New York: Georg Thieme Verlag.

Paul, B.J., M.M. Barker, W. Ross et al. 2004. DksA: A critical component of the transcription initiation machinery that potentiates the regulation of rRNA promoters by ppGpp and the initiating NTP. Cell 118: 311-22.

Pomerantz, R.T., and M. O'Donnell. 2008. The replisome uses mRNA as a primer after colliding with RNA polymerase. Nature 456: 762-6.

Revyakin, A., C. Liu, R.H. Ebright, and T.R. Strick. 2006. Abortive initiation and productive initiation by RNA polymerase involving DNA-scrunching. Science 314: 1139-43.

Roberts, J.W. 2006. RNA polymerase, a scrunching machine. Science 314: 1097-8

Sampath, A., and C.R. Stewart. 2004. Roles of Genes 44, 50, 51 in regulating gene expression and host take over during infection of Bacillus subtilis by bacteriophage SPO1. J. Bacteriol. 186: 1785-92.

Skordalakes, E., and J.M. Berger, 2003. Structure of the rho transcription terminator: Mechanism of mRNA recognition and helicase loading. Cell 114: 135-46.

Skordalakes, E., and J.M. Berger. 2006. Structural insights into RNA-dependent ring closure and ATPase activation by the rho termination factor. Cell 127: 553-64.

Spilianakis, C.G., M.D. Lalioti, T. Town, G.R. Lee, and R.A. Flavell, 2005. Interchromosomal associations between alternatively expressed loci. Nature 435: 637-45.

Summers, J. and Mason, W.S. 1982 Replication of the genome of a hepatitis B-like virus by reverse transcription of an RNA intermediate. Proc. Natl. Acad. Sci. USA 79: 3997-4001.

Sydow, J.F. and P. Cramer. 2009. RNA polymerase fidelity and transcriptional proofreading. Curr. Op. Str. Biol. 19(6) 732-9.

Temin, H.M. 1989. Reverse transcriptase – retrons in bacteria. Nature 339: 254-5.

Temin, H.M., and S. Mizutani, 1970. RNA-dependent DNA polymerase in virions of Rous sarcoma virus. Nature 226: 1211-3.

Varmus, H. E. 1983. Retroviruses. In: *Mobile Genetic Elements*. Pp. 41 1-503. Shapiro, J. ed. New York: Academic Press.

Varmus, H.E. 1982a. A growing role for reverse transcription. Nature 299: 204-5.

Varmus, H.E. 1982b. Form and function of retroviral provirus. Science 216: 812-20.

Varmus, H.E. 1987. Reverse transcription. Scient. Am. 257(3): 56-66.

Varmus, H.E. 1989. Reverse transcription in bacteria. Cell 56: 721-4.

Vessylyev, D.G., M.N. Vassylyeva, A. Perederina, T.H. Tahirov, and I. Artsimovitch. 2007a. Structural basis for transcription elongation by bacterial RNA polymerase. Nature 448: 157-62.

Vessylyev, D.G., M.N. Vassylyeva, J. Zhang, M. Palangat, and I. Artsimovitch. 2007b. Structural basis for substrate loading in bacterial RNA polymerase. Nature 448: 163-8.

Wang, J.C. 1985. DNA topoisomerase. Annu. Rev. Bio. Chem. 54: 665-97.

Wassarman, K.M., and R.M. Saecker. 2006. Synthesis-mediated release of a small RNA inhibitor of RNA polymerase. Science 314: 1601-3.

Young, B.A., T.M. Gruber, and C.A. Gross. 2004. Minimal machinery of RNA polymerase holoenzyme sufficient for promoter melting. Science 303: 1382-4.

Zenkin, N., Y. Yuzenkova, and K. Severinov. 2006. Transcript-assisted transcriptional proofreading. Science 313: 518-20.

Zhou, C., and D.M. Knipe. 2002. Association in herpes simplex virus type 1 ICP8 and ICP 27 proteins with cellular RNA polymerase II holoenzyme. J. Virology 76: 5893-904.

7

Transcription in Eukaryotes

Transcription apparatus of eukaryotic cells is more complex than that of prokaryotes. In prokaryotes, there is only one species of RNAP. In eukaryotes, three major types of RNA polymerases (RNAPs) transcribe nuclear genes. Transcription and translation are integrated processes in prokaryotes. Translation on mRNA begins before transcription is complete. After translation, mRNA is released and degraded. In eukaryotes, there are certain post-transcriptional processes that occur in most of the transcripts before translation begins. These steps are known as processing of RNA.

PROMOTER SEQUENCES FOR EUKARYOTIC RNA POLYMERASES

Promoters of Class I Genes

A eukaryotic rDNA unit, as given by Dover (1988) and presented in Figure 7.1, is usually a tandemly arrayed multigene family. Promoter sequences for RNAPI are located upstream (before start point of transcription) and showed three regions at start point. These three regions are TATA or 'Hogness box' (7-bp long) which is located at position -20 bp, CAAT box, which is located between –70 bp to –80 bp

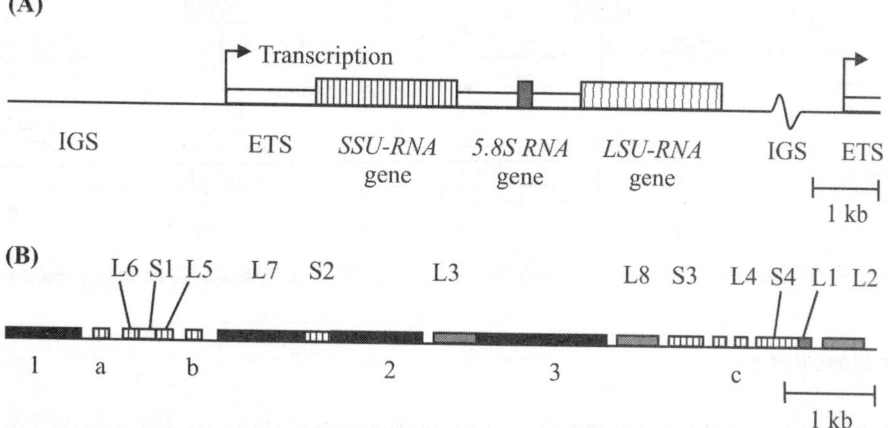

Figure 7.1 (A) Eukaryotic rDNA unit. ETS, external transcribed spacer; LSU and SSU, large- and small-subunit RNA genes; IGS, intergenic spacer. In prokaryotes, there is no spacer between the 5.8 RNA gene and LSU RNA gene. Conserved core segments are shaded and the variable expansion segments are open boxes within each of the main genes. (B) Modular organization of the SSU (S1-4) and of the LSU (L1-8) RNA genes together with three RNA genes (a, b, c) and three protein-encoding genes (1-3) in the mitochondrial genome of *C. reinhardtii*

and GC box, which is located between -60 bp to –100bp. Mammalian gene promoters for transcription by RNAPII are typically organized in the following order: upstream sequence motif(s)/TATA box/initiation site. Xu et al. (1992) conducted studies in which the order, orientation and DNA sequence of these components were varied to determine how these affected polarity of transcription. They determined that the polarity of transcription is primarily determined by the linear order of an upstream sequence relative to a TATA box, rather than by the individual orientations of either of these two elements. Promoter sequences for RNAPIII they are located downstream with exception of 7S genes of signal recognition particles where they are present both upstream and downstream.

Promoters of Class II Genes

Many class II genes contain a region located between 19 and 27 bp upstream from the transcription start site whose consensus sequence TATANA is similar (except in location) to the prokaryotic TATA box, also known as Hogness box, where N means any nucleotide. In mammal, TATA box is bound by a general transcription factor TFIID. This binding helps in selection of proper start site. In some cases TFIIA is required to facilitate the binding of TFIID to the TATA box. Another common promoter element is called the CAAT box. It is found between –80 and –60 nucleotides upstream of the transcription start site for RNAPII. The consensus sequence for this region is GGNCAATCT. Although this sequence is not palindromic, it occurs naturally in both orientations. This site is bound by transcription factor CTF (CAAT box transcription factor). Class II promoters contain yet another upstream element with consensus sequence GGGCGG. This sequence functions in either direction and is often present in multiple copies and binds the transcription factor SP1. Class II genes contain still other binding sites for transcription factors in a variety of locations. Certain sequences are present only in some classes of genes, e.g., those for heat shock proteins). These sequences bind to regulatory factors. Other factors required to initiate transcription in mammalian cells are TFIIB, TFIIE and TFIIF. General pattern that has emerged for class II gene promoter is shown in Figure 7.2 (Rawn 1989).

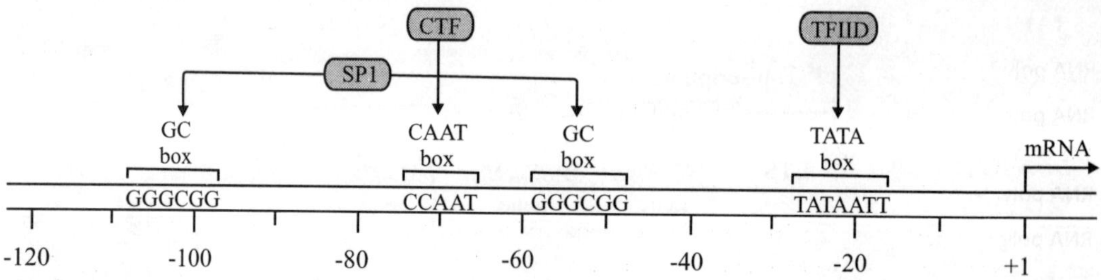

Figure 7.2 Hypothetical class II gene promoter. Three factors – SP1, CTF and TFIID – bind to the promoters of class II genes

Promoters of Class III Genes

The promoters of most class III genes include discontinuous intragenic structures, termed internal control regions (ICRs) that are composed of essential sequence blocks separated by nonessential nucleotides. The ICRs of 5S rRNA genes are sometimes referred to as type I. These comprise two functional domains: an A block, B block (a domain consisting of an intermediate element), and a C-block. Most class III genes, including tRNA, VA, Alu, EBER, 7SL, 4.5S, B1, and B2 genes, have type II ICRs: these again have two domains, an A-block and a B-block. The A blocks of types I and II are homologous and can substitute for one another in *Xenopus*, although not in *Neurospora*. The A-block is located much further from the start site in type I than it is in type II promoters. As well as the ICR, extragenic sequences can also affect the strength of type I and II promoters. However, substitutions in the extragenic regions are generally well tolerated, unlike mutations in the ICR. In contrast, with type III promoters, such as those of the vertebrate U6 and 7SK genes, transcription is independent of intragenic elements and is dictated solely by 5′ flanking regions.

Repeats-Containing Promoters

A search for tandem repeats in the *S. cerevisiae* genome revealed that the nucleosome-free region directly upstream of genes (the promoter region) is enriched in repeats. One-fourth of all gene promoters contain tandem repeat sequences. Genes driven by these repeat-containing promoters showed significantly higher rate of transcriptional divergence. Variation in repeat length results in changes in expression and local nucleosome positioning. Tandem repeats are variable elements in promoters that may facilitate evolutionary tuning of gene expression by affecting local chromatin structure.

EUKARYOTIC RNA POLYMERASES

Eukaryotic RNA polymerases characterized by type of RNA they synthesize. Five types of RNA polymerases are well known. Table 7.1 mentions type of RNA synthesized by these eukaryotic RNA polymerases. The five types of eukaryotic RNAPs are described here briefly.

Table 7.1 Eukaryotic RNA polymerases characterized by type of RNA they synthesize

RNA polymerase	*RNA synthesized*	*Reference*
RNA polymerase I	pre-rRNA 45S (35S in yeast), which matures into 28S, 18S and 5.8S rRNAs which form the major RNA sections of the ribosome	Grummt (1999)
RNA polymerase II	precursors of mRNAs and most snRNA and microRNAs	Lee et al. (2004)
RNA polymerase III	tRNAs, rRNA 5S and other small RNAs found in the nucleus and cytosol	Willis (1993)
RNA polymerase IV	siRNA in plants microRNAs	Herr et al. (2005)
RNA polymerase V	RNAs involved in siRNA-directed heterochromatin formation in plants	Wierzbicki et al. (2005)

RNA Polymerase I

In eukaryotic cells, the genes for different rRNA molecules are transcribed as a single precursor. In most eukaryotes, the rRNA genes are found in clusters composed of 100 to 5,000 repeats of these genes. RNAPI, found in the nucleolus, transcribes them in a region of the nucleus called nucleolar organizing region (NOR), the site where nucleolus is attached. A eukaryotic cell typically contains 40,000 RNAPI molecules which may be enough for transcribing 100 genes. rRNA precursor is long (6,000 to 10,000 nucleotides) and as many as 50 polymerase molecules may transcribe each rRNA gene simultaneously. This intense transcriptional activity reflects the need of the cell for rRNA; 10 million copies of rRNA are required per cell division.

Structure of yeast RNA polymerase I

Kuhn et al. (2007) report the complete structure of the 14-subunit yeast RNAPI enzyme at 12Å resolution using cryo-electron microscopy (cryo-EM). This study reveals that three subunits of RNAP I perform functions in transcription elongation that are outsourced to the transcription factors TFIIF and TFIIS in the analogous RNAP II transcription system. RNAPI subunits are given in Table 7.2 (Kuhn et al. 2007). Hybrid structure subcomplex A14/43 is a clamp and the dock domain which contributes a unique surface interacting with promoter-specific initiation factors. Subunits A49 and A34.5 form a heterodimer near the enzyme tunnel that acts as a built-in elongation factor and is related to the RNAP II-associated factor TFIIF. In contrast, to RNAP II, RNAP I has a strong intrinsic 3′-RNA cleavage activity, which requires the C-terminal domain of subunit A12.2 and, apparently enables ribosomal RNA proofreading and 3′-end trimming.) This piece of work is considered a breakthrough paper (Haag and Pikaard 2007).

Table 7.2 RNA polymerase I subunits

Polymerase part	*Pol I subunit*	*MW (kDa)*	*Corresponding Pol II subunit*	*Subunit type*	*Seqμence identity (%)*	*Conserved Pol II fold (%)*
Core	A100	186.4	Rpb1	Homolog	22.3	47.8
	A135	135.7	Rpb2	Homolog	26.0	62.1
	AC40	37.7	Rpb3	Homolog	21.2	53.5
	AC19	16.2	Rpb11	Homolog	17.6	77.5
	A12.2	13.7	Rpb9	Homolog	19.2	35.2
	Rpb5 (ABC27)	25.1	Rpb5	Common	100	100
	Rpb6 (ABC23)	17.9	Rpb6	Common	100	100
	Rpb8 (ABC14.5)	16.5	Rpb8	Common	100	100
	Rpb10 (ABC10β)	8.3	Rpb10	Common	100	100
	Rpb12 (ABC10α)	7.7	Rpb12	Common	100	100
Subcomplex A14/43	A14	14.6	Rpb4	Counterpart	4.5	25.0
	A43	36.2	Rpb7	Counterpart	8.0	78.4
Subcomplex A40/34.5	A40	46.7	RAP74	Specific	7.6	57.2
	A34.5	26.9	RAP30	Specific	8.3	80.5

Amplification of ribosomal RNA gene repeats

Organisms maintain ribosomal RNA gene repeats (rDNA) at stable copy number by recombination; the loss of repeats results in gene amplification. Kobayashi and Goranley (2005) have shown that amplification is dependent on transcription from a non-coding bidirectional promoter (E-pro) within the rDNA spacer. E-pro transcription stimulates the dissociation of cohesion, a DNA binding protein complex that suppresses sister-chromatid-based changes in rDNA copy number. Transcription-induced cohesion dissociation may be a general mechanism of recombination regulation. Transcription-induced cohesion dissociation model of rDNA amplification is given in Figure 7.3. (A) In normal situation as well as wild-type rDNA copy number, SIR2 represses E-pro activity, allowing cohesion to associate throughout the IGS. DSBs, formed by replication forks pausing at the RFB site, are repaired by equal sister-chromatid recombination, with no change in the rDNA copy number. (B) When SIR2 repression is removed, such as with sir2 mutation or low copy number, E-pro becomes active and transcription displaces cohesion. Unequal sister chromatids can be used as templates for DSB repair, resulting in the changes in the rDNA copy number. Lines represent single chromatids (double-stranded DNA). The IGS in which the replication fork is paused is expended in the bracket.

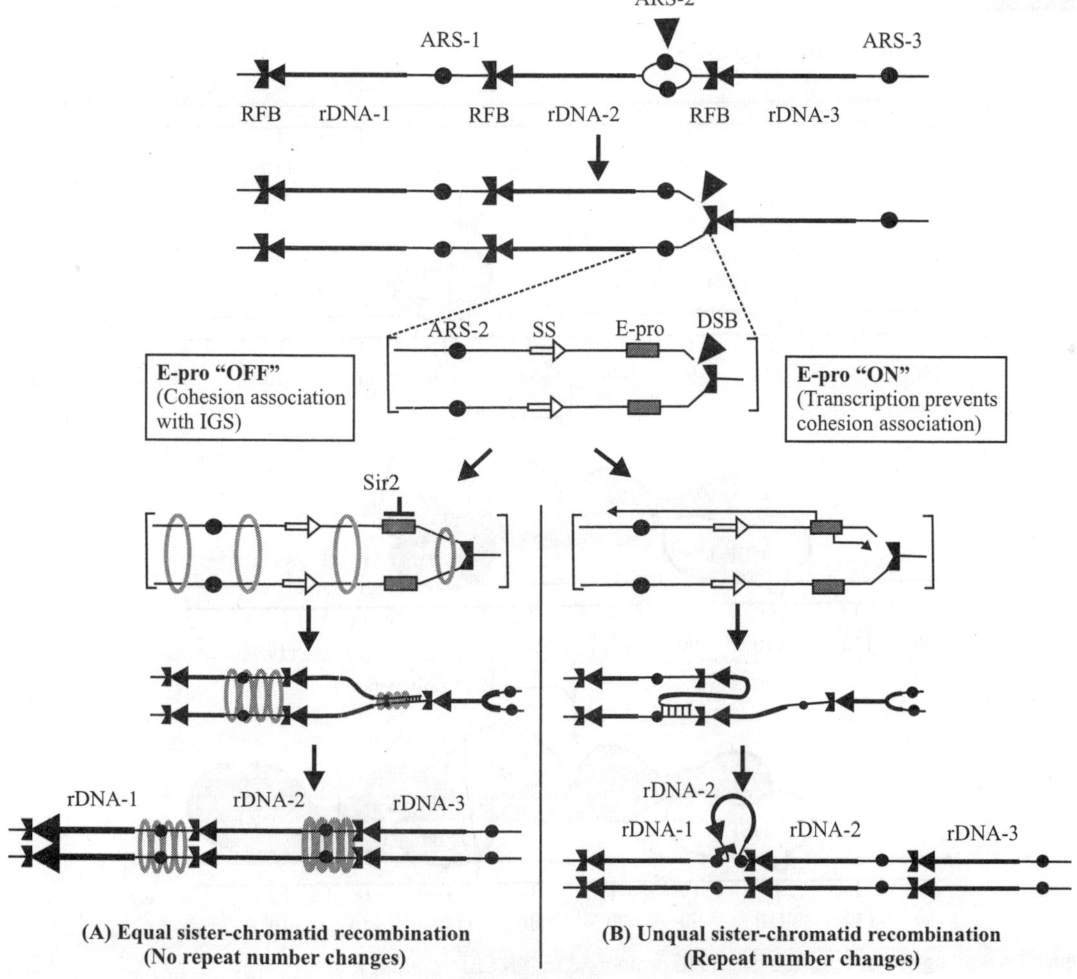

Figure 7.3 Transcription-induced cohesion dissociation model of rDNA amplification

The promoter sequences for class I genes are located just upstream of the transcription start site, in a region of the repeating unit called non-transcribed spacer (NTS). The class I promoter sequences of distantly related organisms share little homology and the proteins that bind the more distal promoter elements work in a species-specific manner. All the class I promoters analyzed consist of several binding sites that can be subdivided into three important regions. The first region usually between –40 and +20 spans the transcription initiation site by RNAPI. The second region, between –40 and –15, is necessary for binding of the transcription factor TFID, which is analogous to TFIID required for transcription of class II genes, assists the polymerase in accurately selecting a start site. The third region lies further upstream, typically between –50 and –100, and is involved in binding transcription factors. Binding of transcription factors to promoters for RNAPI, as illustrated by Rawn (1989), is shown in Figure 7.4. Transcription factors involved are TFID, UBF-1, TBIC and SL1.

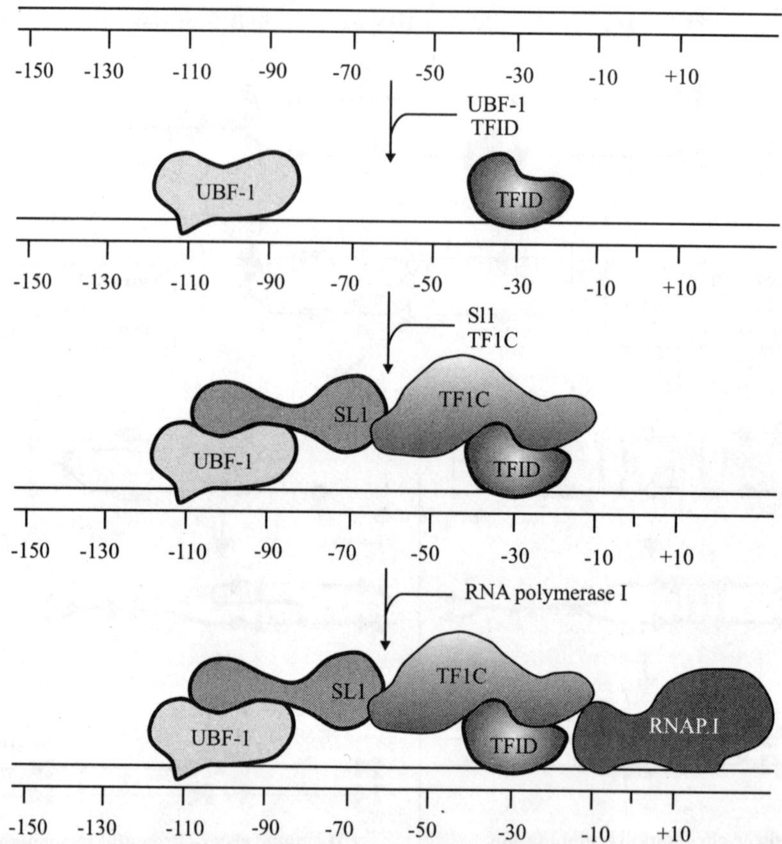

Figure 7.4 Binding of transcription factors to promoters for RNA polymerase I

RNA Polymerase II

Carboxy-terminal domain of the RNA polymerase II

A tandem repeat of seven amino acids, Thr-Ser-Pro-Thr-Ser-Pro-Ser (YSPTSPS, using single-letter amino acid symbols), is found at C-terminus, known as C-terminus domain (CTD) in the large subunit of RNAPII (Suzuki 1990). 11-23 units are necessary for normal functioning of the enzyme. This C-terminus tandem repeat (heptad)has 26 units in yeast and 52 units in mammals) which plays part in the normal functioning of the polymerase These sequences of RNAPII are proposed to bind with transcription factors. *Saccharomyces cerevisae* cells are inviable if RNAPII large subunit genes encode less than 10 complete heptapeptide repeats, if they encode 10 to 12 complete repeats, cell are temperature sensitive and cold sensitive, but 13 or more complete repeats will allow wild-type growth at all temperatures (Scafe et al. 1990). Different CTD phosphorylation patterns act as recognition sites for the binding of various messenger RNA processing factors, thereby coupling transcription and mRNA processing. Polyadenylation factors are co-transcriptionally recruited by phosphorylation of CTD serine 2 and these factors are also required for transcription termination. RNAPII transcribes past the poly(A) site., the RNA is cleaved by the polyadenylation machinery, and the RNA downstream of the cleavage site is degraded. Kim et al. (2004) suggest a model in which poly(A) site cleavage and subsequent degradation of the 3′-downstream RNA by Rat1 trigger transcription termination.

Crystal structure of RNAPII

The crystal structure of RNAPII in the act of transcription was determined at 3.3Å resolution by Gnatt et al. (2001). Protein-nucleic acid contacts help explain DNA and RNA strand separation, the specificity of RNA synthesis, "abortive cycling" during transcription initiation and RNA and DNA translocation during transcription elongation. No nuclear RNAP appears to bind specifically and directly to DNA. Instead, interaction between polymerase and DNA requires the presence of sequence-specific accessory proteins called transcription factors (TFs). General TFs are required for transcription of all the genes of a class whereas other TFs are required for certain genes of gene families. RNAPII transcribes all the genes that are destined to be translated into proteins as well as those genes that transcribe for small RNA molecules involved in RNA processing. About 40,000 molecules of RNAPII are found in every eukaryotic cell. The activity of this enzyme accounts for about 20-40 per cent of all cellular RNA synthesis.

Five sets of intermediate complexes containing distinct transcription factors have been identified (Buratowski et al. 1989). Binding of transcriptional factors to promoter region by RNAPII requires TFIID, TFIIA, TFIIB, TFIIF and TFIIE in a sequential order.

Salient properties of transcriptional factors

In eukaryotes, transcription of the diverse array of tens of thousands of protein-encoding genes is carried out by RNAPII. The control of this process is predominantly mediated by a network of thousands of sequence-specific DNA binding transcription factors that interpret the genetic regulatory information, such as in transcriptional enhancers and promoters, and transmit the appropriate response to the RNAPII transcriptional machinery (Kadonaga 2004). Some early advances in the discovery and characterization of the sequence-specific DNA binding transcription factors as well as some of the properties of these regulatory proteins are described here: (1) Sequence-specific factors are composed of functional modules. Sequence-specific factors regulate transcription via recruitment of coactivators and corepressors. (3) Sequence-specific factors can be regulated by post-translational modifications. (4) Sequence-specific factors are often members of mutiprotein family. (5) Chromatin is an integral component in the function of sequence-specific factors. (3) Recognition sites for sequence-specific factors tend to be located in clusters.

Chromatin regulation during transcription initiation

Chromatin structure imposes significant obstacles in all aspects of transcription that are mediated by RNAPII. The dynamics of chromatin structure are tightly regulated through multiple mechanisms including histone modification, chromatin remodeling, histone variant incorporation, and histone eviction. Chromatin regulation affects the binding of transcription factors as well as the initiation and elongation steps of transcription (Li et al. 2007). Model of chromatin regulation during transcription initiation is presented in Figure 7.5. At the silent promoter Htz1-containing nucleosomes flank a 200 bp NFR on both sides. Upon targeting to the upstream-activation sequence (UAS), activators recruit various coactivators (such as Swi/Snf or SAGA). SAGA stands for Spt-Ada-Gcn5-Acetyl transferase complex. This recruitment further increases the binding of activators, particularly for those bound within nucleosomal regions. More importantly, histones are acetylated at promoter-proximal regions, and these nucleosomes become much more mobile. In one model (left), a combination of acetylation and chromatin remodeling directly results in the loss of Htz-containing nucleosomes, thereby express-

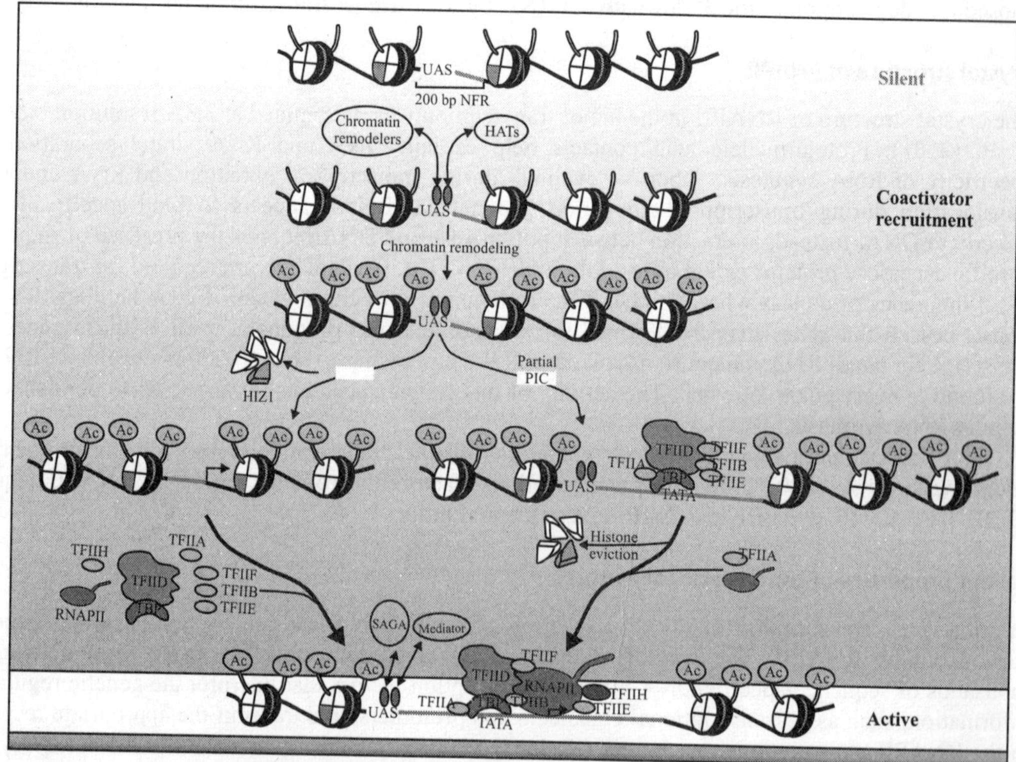

Figure 7.5 Model of chromatin regulation during transcription initiation

ing the entire core promoter to the general transcription factors (GTFs) and RNAPII. SAGA and mediator then facilitate preinitiation complex (PIC) formation through direct interactions. In the other model (right), which represents the remodeled state, partial PICs could be assembled at the core promoter without loss of Htz-1. It is the binding of RNAPII and TFIIH that leads to the displacement of Htz1-containing nucleosomes and the full assembly of PIC.

Regulation of nucleosome dynamics during transcription elongation is shown in Figure 7.6 (Li et al. 2007). (A) The chromatin landscape is determined by the factors associated with different forms of RNAPII. PAF facilitates the binding of FACT, COMPASS, and Rad6/Bre1 to the Ser5-phosphorylated CTD, which results in H2B ubiquitination and accumulation of trimethylation of H3K4 at the 5′-end of ORF. Set2 directly interacts with Sef2-phosphorylated CTD, thus methylating H3K36 at the 3′-end. (B) Maintenance of nucleosomal stability during transcription is shown. When RNAPII migrates into promoter-distal regions where the influence of activator-dependent HATs is diminishing, RNAPII requires other HATs (elongators or those associated with RNAPII) to acetylase the nucleosome in front of elongation machinery. The passage of RNAPII causes histone displacement. Subsequently, these histones are redeposited onto the DNA behind RNAPII via concerted actions of histone chaperones. Alternatively, the free forms of histones in the nucleus are also available for reassembly. These newly deposited nucleosomes are somehow hyperacetylated and are immediately methylated by Set2. Methylation of H3K36 is then recognized by the chromodomain of Eaf3, which in turn recruits the Rpd3S deacetylase complex. Rpd3S removes the acetyl marks and leaves the nucleosome in a stable state. Methylation of H3K36 is eventually eliminated by a histone demethylase when the gene turns off.

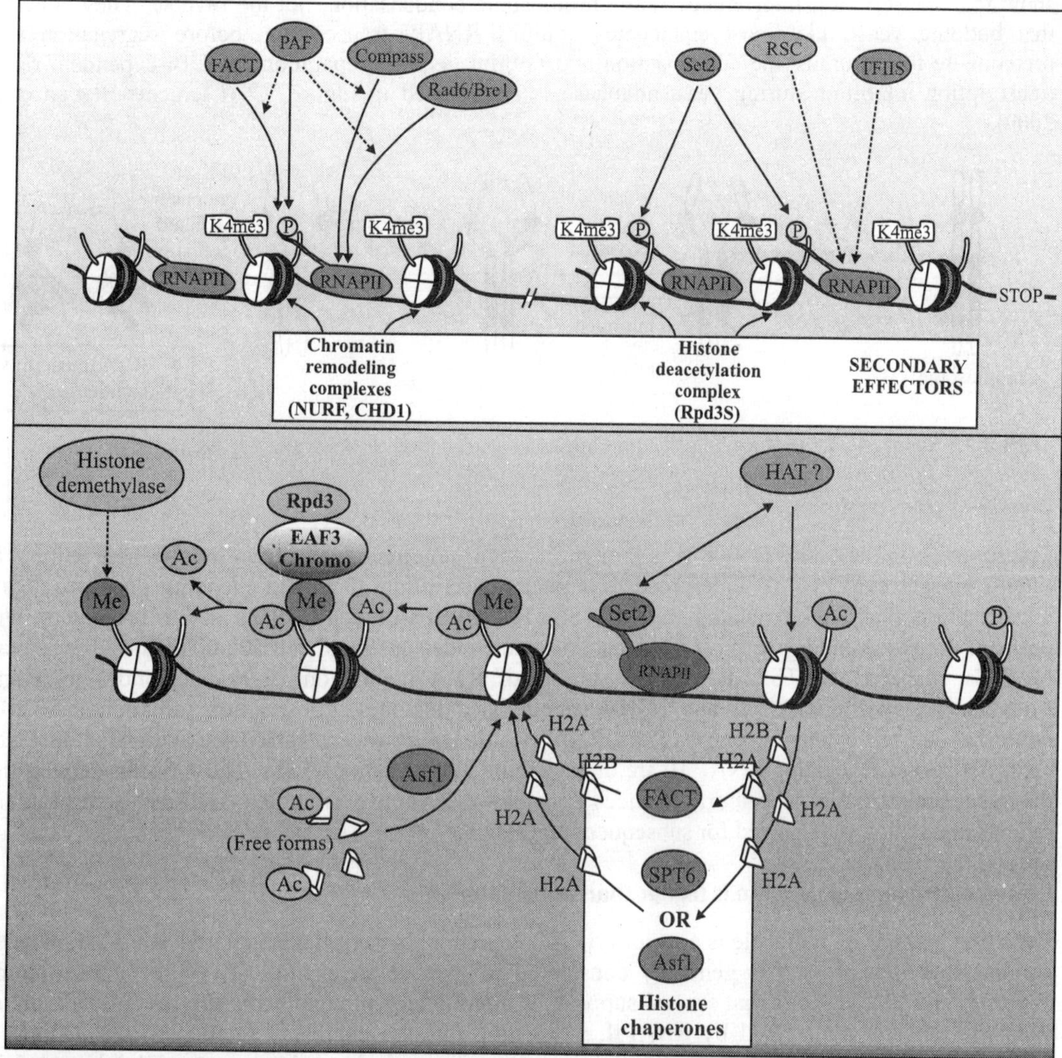

Figure 7.6 Regulation of nucleosome dynamics during transcription elongation

Chromosome condensation and repression of transcription

Chromosome condensation and the global repression of gene transcription are features of mitosis in most eukaryotes. The logic behind this phenomenon is that chromosome condensation prevents the activity of RNAPs. In budding yeast, however, transcription was proposed to be continuous during mitosis. Clemente-Blanco et al. (2009) have shown that Cdc14, a protein phosphatase required for nucleolar segregation and mitosis exit, inhibits transcription of yeast ribosomal genes (rDNA) during anaphase. The phosphatase activity of Cdc14 is required for RNAPI inhibition in vitro and in vivo. Moreover, Cdc14-dependent inhibition involves nucleolar exclusion of RNAPI subunits. They demonstrate that transcription inhibition is necessary for complete chromosome disjunction, because ribosomal RNA (rRNA) transcripts block condensin binding to rDNA, and show that bypassing the role of Cdc14 in nucleolar segregation requires in vivo degradation of nascent transcripts. These results show that transcription interference with chromosome condensation, not the reverse. They conclude that budding yeast, like most eukaryotes, inhibits RNAPI transcription before segregation as a prerequisite for chromosome condensation and faithful genome separation. CDC14-dependent rRNA transcription inhibition during yeast anaphase is represented in Figure 7.7 (Clemente-Blanco et al. 2009).

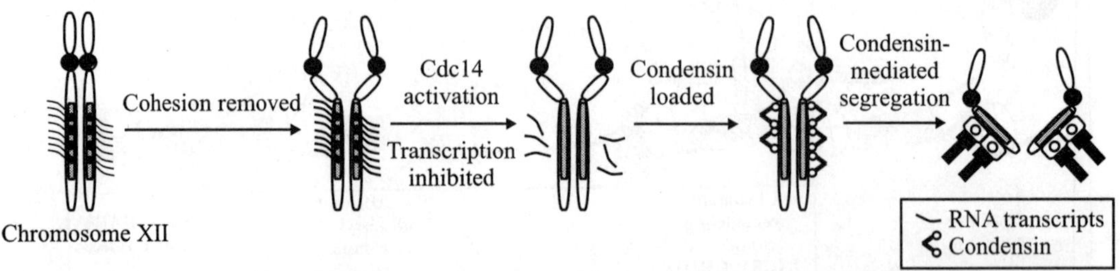

Figure 7.7 Condensin localization to rDNA and nucleolar segregation removes rRNA transcripts

Divergent transcription

Transcription initiation by RNAP II is thought to occur unidirectionally from most genes. Seila et al. (2008) present evidence of widespread divergent transcription at protein-encoding gene promoters. Transcription start site-associated RNAs (TSSa-RNAs) non-randomly flank active promoters, with peaks of antisense and sense short RNAs at 250 nucleotides upstream and 50 nucleotides downstream of TSSs, respectively. TSSa-RNAs are subsets of RNA population 20-90 nucleotides in length. Promoter-associated RNAP II and H3K4-trimethylated histones, transcription initiation hallmarks, colocalize at sense and antisense TSSa-RNA positions; however, H3K79-demethylated histones, characteristic of elongating RNAP II, are only present downstream of TSSs. These results suggest that divergent transcription over short distances is common for active promoters and may help promoter regions maintain a state poised for subsequent regulation.

Expressed fraction of genome is higher than anticipated

Detection of an RNA molecule is a major criterion to define transcriptional activity. The fraction of the genome that is expressed is generally considered to parallel the complexity of the transcriptome. Wyers et al. (2005) reports that several supposedly silent intergenic regions in the genome of budding yeast are actually transcribed by RNAPII, suggesting that the expressed fraction of the genome is higher than anticipated. However, RNAs originating from these regions are rapidly degraded by combined action of the exosome and a new poly(A) polymerase activity that is defined by the Trf4 protein and one of two RNA binding proteins Air1p or Air2p. They show that such a polyadenylation-assisted degradation mechanism is also responsible for the degradation of several RNAPI and RNAPIII transcripts. These data strongly support the existence of post-transcriptional quality control mechanism limiting inappropriate expression of genetic information.

RNA quality-control systems

Eukaryotic cells contain numerous RNA quality-control systems that are important for shaping the transcriptome of eukaryotic cells (Doma and Parker 2007). These systems not only prevent accumulation of non-functional RNAs, but also regulate normal mRNAs, repress viral and parasitic RNAs,

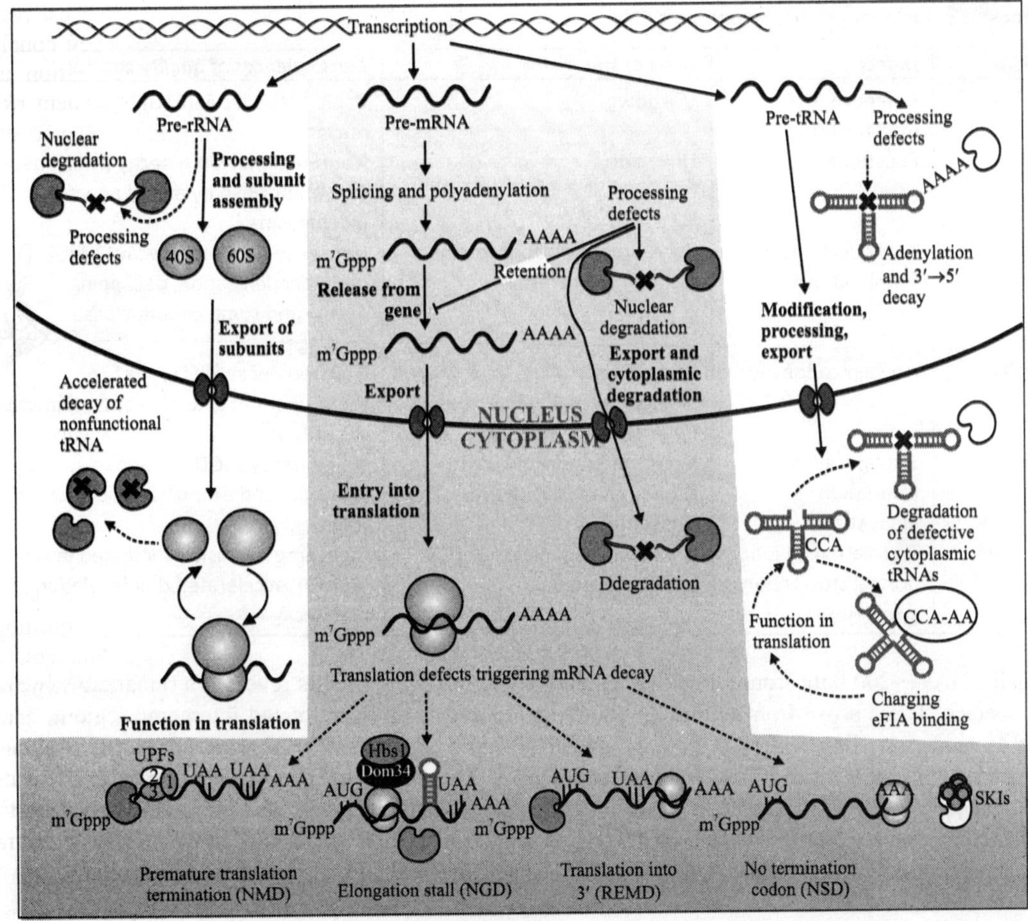

Figure 7.8 Diversity of RNA-control systems in eukaryotes

and potentially contribute to the evolution of new RNAs and hence proteins. These quality control circuits can be viewed as a series of kinetic competitions between steps in normal RNA biogenesis or function and RNA degradation pathways. These RNA quality-control circuits depend on specific adaptor proteins that target aberrant RNAs for degradation as well as coupling of individual steps in mRNA biogenesis and function. Diversity of RNA-control systems in eukaryotes is given in Figure 7.8.Some of the known RNA quality control systems for aberrant rRNA, mRNA and tRNA are depicted in eukaryotic cells. Quality control of cytoplasmic and nuclear RNAs is given in Table 7.3 and Table 7.4.

Transcriptomes of eukaryotic cells

Transcriptome is the set of all transcripts (coding as well as non-coding RNAs) produced in one or a population of cells. Transcriptome is highly dynamic as it is influenced by cell type, developmental stage, and environment. The transcriptomes of eukaryotic cells are incredibly complex. Individual non-coding RNAs dwarf the number of protein-coding genes, and include classes that are well understood as well as classes for which the nature, extent and functional roles are obscure. Deep sequencing of

Table 7.3 Quality control of cytoplasmic RNAs

RNA	*Defect*	*Basis of specificity*	*Consequences of quality control*
tRNA	Defective RNA modification	Unknown	Rapid tRNA decay (RTD) by an unknown nuclease
rRNA	Functional defect in rRNA	Unknown	Nonfunctional rRNA decay (NRD), decay of mutant rRNA by an unknown mechanism
mRNA	Aberrant translation termination	Recruitment of Upf proteins to termination complex	Nonsense-mediated decay (NMD), rapid deadenylation, decapping, 3'-5' decay, and some endonuclease cleavages
mRNA	No stop codon	Recruitment of Sk17p to elongation stall with no mRNA in A site of the ribosome	Nonstop decay (NSD), , Sk17p recruitment of exosome and rapid 3'-5' degradation
mRNA	Strong stall in translation elongation	Recruitment of Hbs1p and Dom34p to A site of the ribosome	No-go decay (NGD), endonucleolytic cleavage, and exonucleolytic decay of fragments
mRNA	Translation beyond normal stop codon into 3'-UTR	Unknown, removal of 3'-UTR binding proteins?	Ribosome extension-mediated decay (REMD), accelerated deadenylation, and decay

small RNAs (<200 both from human Hela and HepG2 nucleotides cells revealed a remarkable breadth of species. These arose from within annotated genes and from unannotated intergenic regions. Small RNAs tended to align with CAGE (cap-analysis of gene expression) tags, which mark the 5' ends of capped, long-range transcripts. Many small RNAs, including the previously described promoter-associated small RNAs, appear to possess cap structures. Affymetrix/Cold Spring Harbor Laboratory ENCODE Transcriptome Project (2009) show that processing of mature mRNAs may generate complex populations of both long and short RNAs whose apparently capped 5' ends coincide.

Transcriptome analysis

Recent transcriptome analysis using high-density tiling arrays and data from large-scale analysis of full-length complementary DNA libraries by the FANTOM3 consortium demonstrate tht many transcripts are non-coding RNAs (ncRNAs). These transcriptome analyses indicate that many of the non-coding regions, previously thought to be functionally inert, are actually transcriptionally active regions with various features. Furthermore, most relatively large (~several kilobase) polyadenylated messenger RNA transcripts are transcribed from regions harboring little coding potential. However, the function of such ncRNAs is mostly unknown and has been a matter of debate. Hirota et al. (2008) have shown that RNAPII transcription of ncRNAs is required for chromatin remodeling at the fission yeast *fbp1*$^+$ locus during transcriptional activation. The chromatin at *fbp1*$^+$ is progressively converted to an open configuration, as several species of ncRNAs are transcribed through *fbp1*$^+$. This is coupled with the translocation of RNAPII through the region upstream of the eventual *fbp1*$^+$ transcriptional start site. Insertion of a transcription terminator into this upstream region abolishes both the cascade of transcription of ncRNA and the progressive chromatin alteration. These results demonstrate that transcription through the promoter region is required to make DNA sequences accessible to transcriptional activators and to RNAPII. A model of chromatin remodeling by a cascade of ncRNAs is shown in Figure 7.9 (Hirota et al. 2008).

Table 7.4 Quality control of nuclear RNAs

RNA	*Defect*	*Consequences of quality control*
tRNA (yeast)	Missing modification, processing defects	TRAMP-dependent adenylation and 3'-5' decay by the exosome
rRNA (yeast and plants)	Stochastic errors or defects in rRNA processing and/or assembly	TRAMP-dependent adenylation and 3' to 5' decay by the exosome; retention of immature ribosomes in the nucleus; adenylation by poly(A) polymerase and Rrp6-dependent dacay; nuclear 3' to 5' decay by Rat1p
rRNA (yeast)	Generation of aberrant rRNA by incorporation of 5-fluorouracil	Adenylation and Rrp6p-dependent degradation
5S rRNA (yeast)	Mutations or defective processing	Ro protein-dependent decay by unknown mechanism; adenylation and 3' to 5'degradation
snRNAs and snoRNAs (yeast and plants)	Stochastic errors or mutant forms	TRAMP-dependent adenylation and 3' to 5' decay by the exosome; adenylation and Rrp6p-dependent degradation
mRNA (yeast)	Hyperadenylation/Hypoadenylation, defects in THO/Sub2 complex, defects in 3'-end processing	Rrp6p and/or core exosome-dependent nuclear retention and degradation of RNA
mRNA (mammals)	Failure of polyadenylation or splicing defects	Retention of the mRNA near or at the transcription site
mRNA (mammals)	Absence of introns in a gene that normally contains introns	Accelerated nuclear degradationdependent on 3' poly(A) tail
mRNA (yeast)	Splicing defect; not recognized by spliceosome	Export and cytoplasmic decapping and 5' to 3' decay; retention in nucleus by MLP proteins
mRNA (yeast)	Splicing defect; trapped lariat intermediate	Nuclear degradation by exosome; debranching, export and cytoplasmic 5' to 3' decay by Xrn1p
mRNA (yeast)	Defect in capping	Export and 5' to 3' decay by cytoplasmic Xm1p
dsRNA (mammals)	Double-stranded RNA	RNA editing and nuclear retention
Intergenic transcripts (yeast and plants)	No known function after transcription	TRAMP-dependent adenylation and 3'-5' decay by the exosome; export and 5' to 3' degradation by decapping and Xrn1p

Reciprocal binding of poly(ADP-ribose) polymerase-1 and histone H1

Nucleosome binding proteins act to modulate the promoter chromatin architecture and transcription of target genes. Krishnakumar et al. (2008) used genomic and gene-specific approaches to show that two such factors, histone H1 and poly(ADP-ribose) polymerase-1 (PARP-1), exhibit a reciprocal pattern of chromatin binding at many RNAPII-transcribed promoters. PARP-1 was enriched and H1 was depleted at these promoters. This pattern of binding was associated with actively transcribed genes. Furthermore, they showed that PARP-1 acts to exclude H1 from a subset of PARP-1-stimulated promoters, suggesting a functional interplay between PARP-1 and H1 at the level of nucleosome binding. Thus, although H1 and PARP-1 have similar nucleosome-binding properties and effects on chromatin structure in vitro, they have roles in determining gene expression outcomes in vivo.

Multiple steps in transcription

Transcription by RNAPII is thought to be predominantly regulated by recruitment of RNAPII to promoters. Recent genome-wide analyses demonstrate that many genes are in fact regulated after

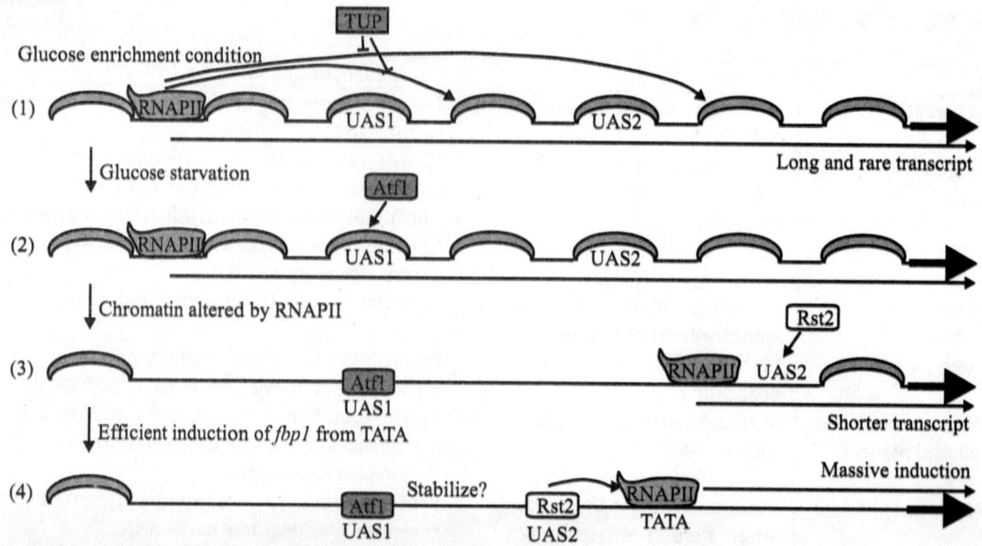

Figure 7.9 Model of chromatin remodeling by a cascade of ncRNAs

recruitment of RNAPIII, by mechanisms such as pausing of RNAPII proximal to promoters (Margaritis and Hoistege 2008). Multiple steps that occur during transcription by RNA polymerase II are described from top to bottom in Figure 7.10. The other transcription factors are indicated. The arrow depicts the start site of transcription. The numbers refer to the positions in the DNA template reached by the active site of RNAPII. Different steps can be rate limiting and regulated for different genes. Recruitment of the preinitiation complex is the most well-established mechanism by which genes are known to be recruited. A large number of genes are also regulated through promoter-proximal pausing, shown in this figure as the fifth step. Most of these steps can be further subdivided. For example, initiation requires a promoter opening step and the abortive initiation step includes at least one additional transition.

Purified RNA polymerase II transcription system

Guermah et al. (2006) reconstituted a highly purified RNAPII transcription system containing chromatin templates assembled with purified histones and assembly factors, the histone acetyltransferase, p300, and components of the general transcription machinery that by themselves suffice for activated transcription (initiation and elongation) on DNA templates. They show that this system mediates activator-dependent initiation, but not productive elongation, on chromatin templates. They further report the purification of a chromatin transcription-enabling activity (CTEA) that, in a manner dependent upon p300 and acetyl-CoA, strongly potentiates transcription elongation through several contiguous nucleosomes as must occur in vivo. The transcription elongation factor SII is a major component of CTEA and strongly synergizes with p300 (histone acetylation) at a step subsequent to preinitiation complex formation. The purification of CTEA also identifies HMGB2 as a coactivator that, while inactive on its own, enhances SII and p300 functions.

RNA polymerase II in "backtracked" state

Transcribing RNAPs oscillate between three stable states, two of which, pre- and post-translocated, were previously subjected to X-ray crystal structure determination. Wang et al. (2009) report the

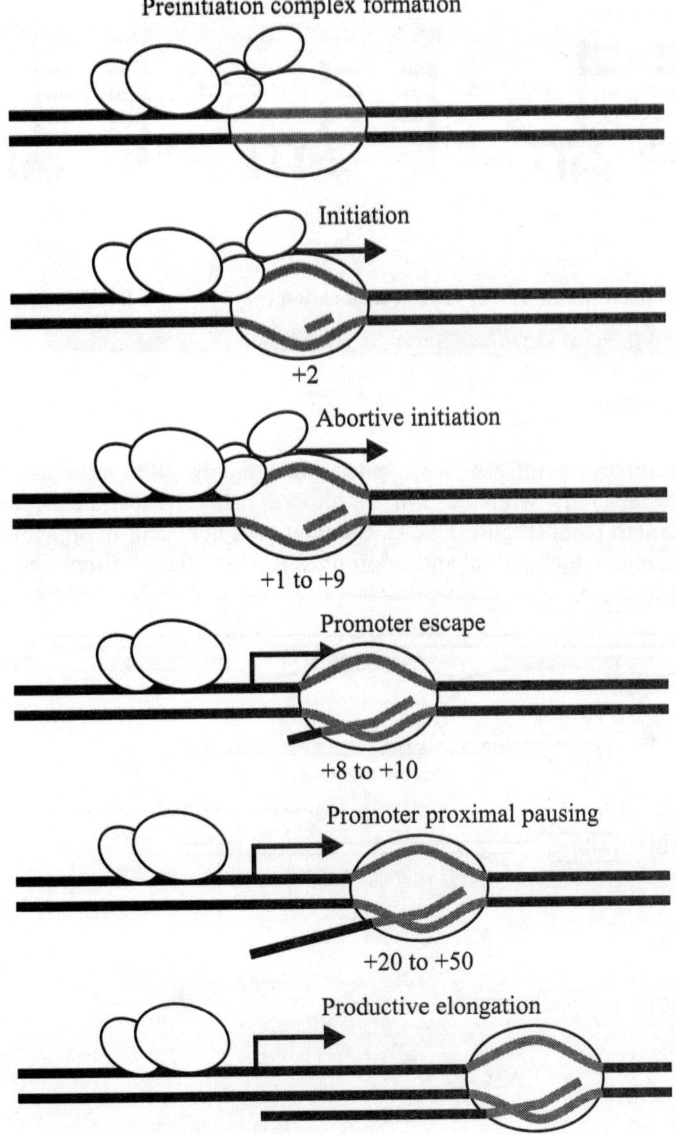

Figure 7.10 Multiple steps in transcription are shown

crystal structure of RNAPII in the third state, the reverse translocated or "backtracked" state. The defining feature of the backtracked structure is a binding site for the first backtracked nucleotide. This binding site is occupied in case of nucleotide misincorporation in the RNA or damage to the DNA, and is termed the "P" site because it supports proofreading. The predominant mechanism of proofreading is the excision of a dinucleotide in the presence of the elongation factor SII (TFIIS). Structure determination of a cocrystal with TFIIS reveals a rearrangement whereby cleavage of the RNA may take place. The three states of a RNAPII transcription elongation complex are shown in Figure 7.11 (Wang et al. 2009).

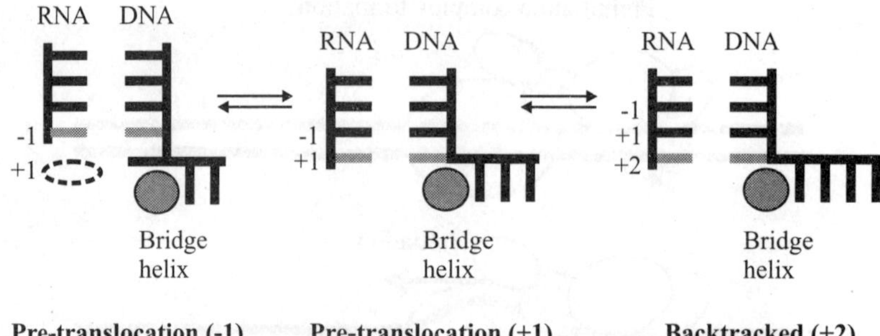

Figure 7.11 The three states of a RNA polymerase II transcription elongation complex

Transcription Start Site

In eukaryotes, precursors to mRNA are capped immediately after initiation of transcription. As capping reaction occurs only with the 5′-tri or diphosphate as substrate hence capping site must correspond with the start point (Figure 7.12A). Capping does not occur in organelles. Hence there is no problem in identifying mitochondrial and chloroplast RNA products. Simplified protocol for trans-

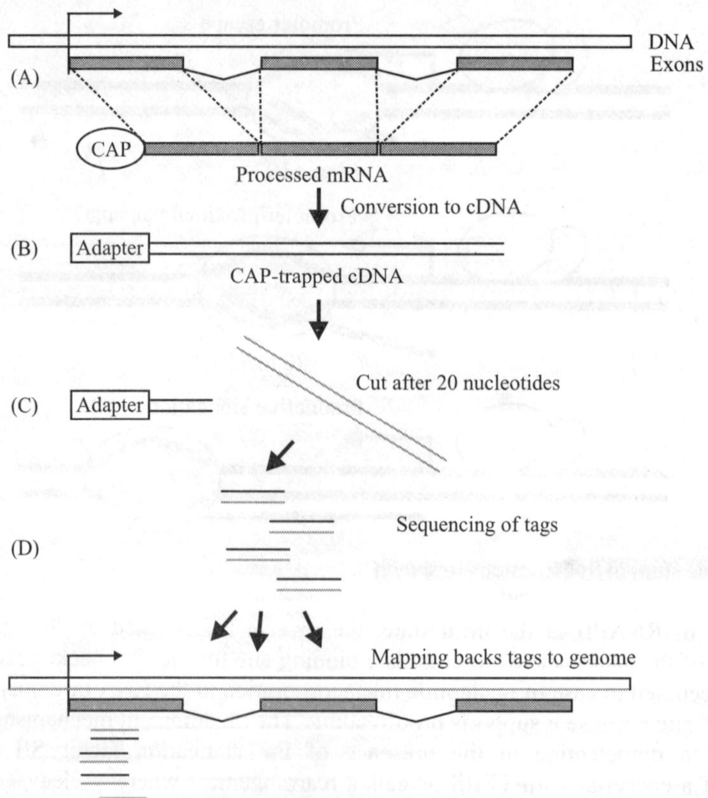

Figure 7.12 Method to identify start point of transcription. Simplified protocol for transcription start site tagging (Developed, with permission, from http://www.biokemi.org/biozoom/issues/526/articles/2401)

cription start site tagging involves the following steps (Figure 7.12B): (A) A gene on the genome is transcribed, spliced and processed into mature mRNA. A cap structure is added at the start of the RNA by the cell (B) The RNA is converted into double-stranded, complementary DNA by the use of reverse transcriptase. Only those cases where the transcriptase reaches the cap are retained. An adapter replaces the cap. (C) The complementary DNA is cut 20 nucleotides downstream of the adapter - the remaining DNA is degraded. (D) Each captured RNA results in a tag. These tags are sequenced, and mapped back to the genome, where they will indicate the start of the RNA that was transcribed in (A).

RNA polymerase II transcription preinitiation complex

Biochemical probes positioned on the surface of the general transcription factor TFIIB were used to probe the architecture of the RNAPII transcription preinitiation complex (PIC) (Chen and Hahn 2004). In PICs, the TFIIB linker and core domains are positioned over the central cleft and wall of the RNAPII. This positioning is not observed in the smaller RNAPII-TFIIB complex. These results lead to a new model for the structure of PIC, which agrees with most previously documented protein-DNA interactions within RNAPII and archaea PICs. Specific interaction of the TFIIB core domain with RNAPII positions and orients the promoter DNA over the RNAPII central cleft, and TBP-DNA bending leads to bending of the promoter around the surface of RNAPII. The TFIIF subunit Tfg1 was found in close proximity to the TFIIB finger, linker, and core domains, suggesting that these two factors closely cooperate during initiation (Chen and Hahn, (2004). TFIIB domains in budding yeast are shown in Figure 7.13 (Chen and Hahn 2004).

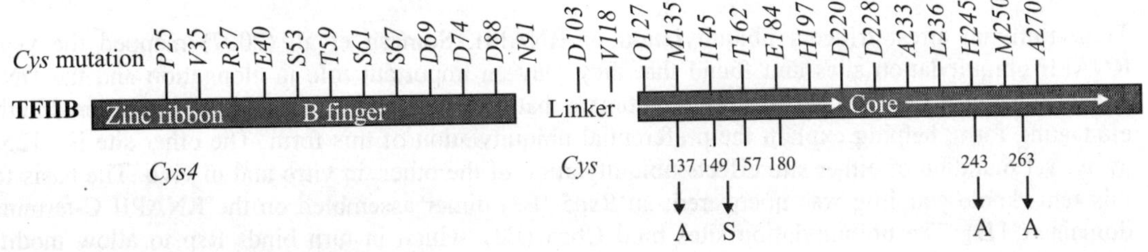

Figure 7.13 TFIIB domains in budding yeast

RNA polymerase II ubiquitylation and degradation

In order to study mechanisms and regulation of RNAPII ubiquitylation and degradation, highly purified factors were used to reconstitute RNAPII ubiquitylation in vitro. Somesh et al. (2005) show that arrested RNAPII elongation complexes are preferred substrates for ubiquitylation. Accordingly, not only DNA damage-dependent but also DNA damage-independent transcriptional arrest results in RNAPII ubiquitylation in vivo. Defl, known to be required for damage-induced degradation of RNAPII, stimulates ubiquitylation of RNAP II only in an elongation complex. Ubiquitylation of RNAPII is dependent on its C-terminal repeat domain (CTD). Moreover, CTD phosphorylation at serine 5, a hallmark of the initiating polymerase, but not of serine 2, a hallmark of the elongating polymerase, completely inhibits ubiquitylation. In agreement with this, ubiquitylated RNAPII is hypophosphorylated at serine 5 in vivo, and mutation of the serine 5 phosphatase SSU72 inhibits RNAPII degradation. These results identify several mechanisms that confine ubiquitylation of RNAPII to the forms of the enzyme that arrest during elongation. Elongating RNAPII is the preferred target for

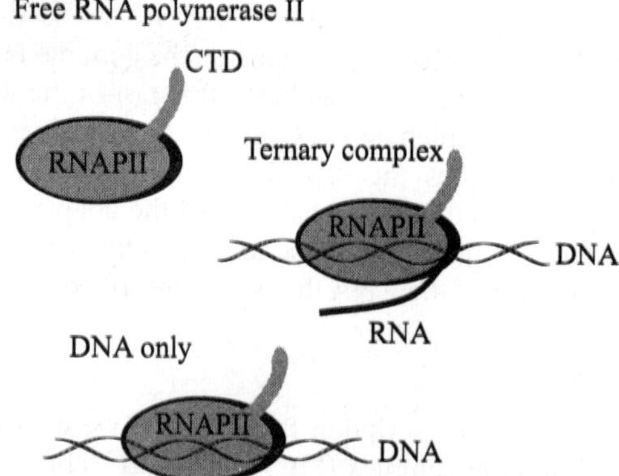

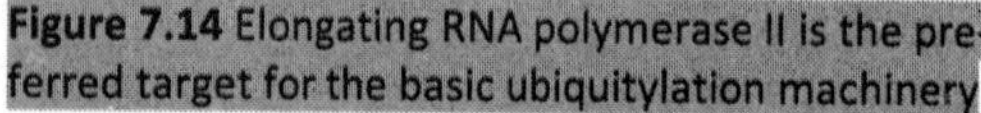

Figure 7.14 Elongating RNA polymerase II is the preferred target for the basic ubiquitylation machinery

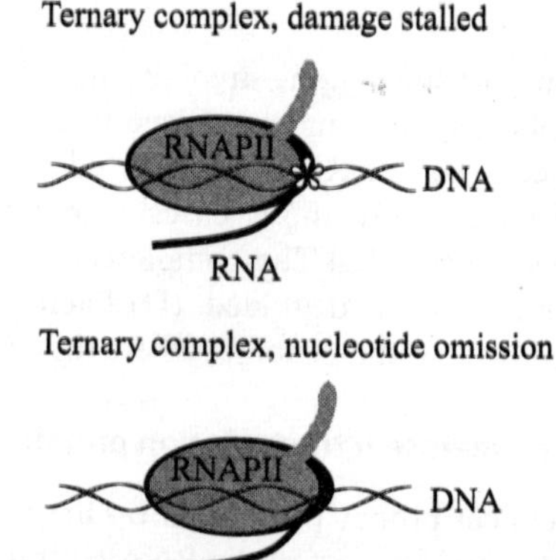

Figure 7.15 The effect on DNA damage and Def1 on the efficiency of RNA polymerase II ubiquitylation

the basic ubiquitylation machinery (Figure 7.14) (Somesh et al. 2005). Different forms of RNAPII tested are shown. The effect on DNA damage and Defl on the efficiency of RNAPII ubiquitylation is shown in Figure 7.15 (Somesh et al. 2005). Different forms of RNAPII tested are shown. Asterisks represent an AAF-mediated transcription-blocking lesion in the RNAPII active sites.

RNA polymerase II ubiquitylation sites

Transcriptional arrest triggers ubiquitylation of RNAPII. Somesh et al. (2007) mapped the yeast RNAPII ubiquitylation sites and found that they play an important role in elongation and the DNA damage response. One site lies in a protein domain that is unordered in free RNAPII, but ordered in the elongating form, helping explain the preferential ubiquitylation of this form. The other site is >125Å away, yet mutation of either site affects ubiquitylation of the other, in vitro and in vivo. The basis for this remarkable coupling was uncovered: an Rsp5 (E3) dimer assembled on the RNAPII C-terminal domain (CTD). The ubiquitylation sites bind Ubc5 (E2), which in turn binds Rsp to allow modification. Evidence for folding of the CTD compatible with this mechanism of communication between distant sites is provided. These data reveal the specificity and mechanism of RNAPII ubiquitylation and demonstrate that E2 can play a crucial role in substrate recognition. A model for the mechanism of RNAPII ubiquitylation is given in Figure 7.16 (Somesh et al. 2007). As shown in the left panel, with wt RNAPII, Ubc5 can bind body of the polymerase around the ubiquitylation sites, and two molecules of Rsp5 bind the CTD. These are brought to the sites of ubiquitylation (thereby ordering the CTD across the surface of the RNAPII) by the Rsp5-Ubc5 interaction. As shown in the right panel, in the absence of K330, Ubc5 binding to that site is reduced, and the correct placement of (both) CTD bound Rsp5 is hindered. No ubiquitylation takes place. As shown in the lower panel, in the absence of K695, Ubc5 binding to that site is reduced, although some association at K330 still takes place, a fully functional Ubc5-Rsp5-substrate complex is not formed. Only little ubiquitylation therefore takes place. Gray circles indicate mutated ubiquitylation sites.

RNA polymerase II-transcribing complex

The structure of RNAPII-transcribing complex has been determined in the post-translocation state, with a vacancy at the growing end of the RNA-DNA hybrid helix (Westover et al. 2004a). At the

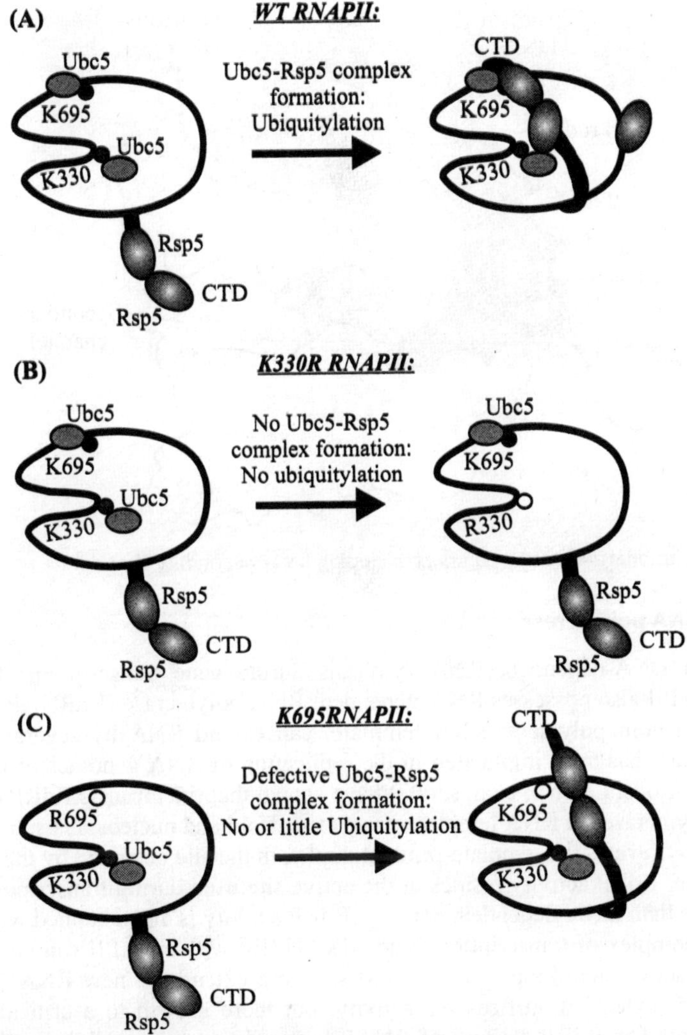

Figure 7.16 Model for the mechanism of RNA polymerase II ubiquitylation

opposite end of the hybrid helix, the RNA separates from the DNA template. This separation of the nucleic acid strands is brought about by interaction with a set of protein loops in a strand/loop network. Formation of the network must occur in the transition from abortive initiation to promoter escape.

RNA polymerase secondary channel

High-resolution crystal structures have highlighted functionally important regions in multisubunit RNAPs, including the secondary channel, or pore, which is postulated to allow the diffusion of small molecules both into and out of the active center of the enzyme. Regulatory factors and small molecules can exploit the secondary channel to gain access to the active site and modify the transcription properties of RNAP (Nickels and Hochschild 2004). Transcriptional elongation complex, including RNAP secondary channel, is shown in Figure 7.17.

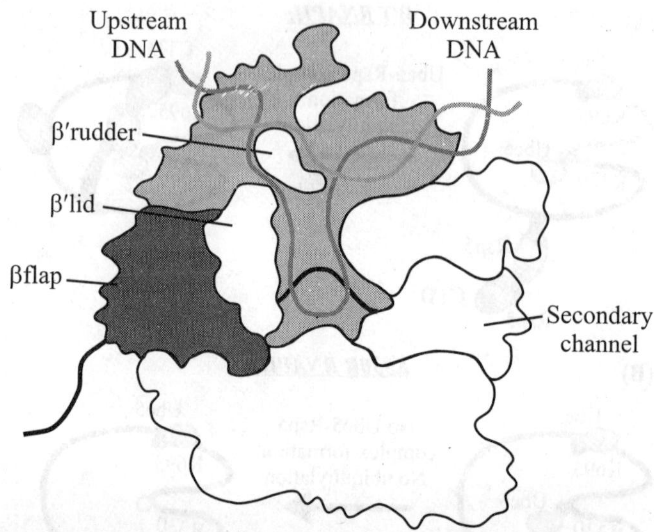

Figure 7.17 Transcriptional elongation complex, including RNAP secondary channel

RNA-dependent RNA polymerase

RNAPII catalyzes DNA-dependent RNA synthesis during gene transcription. There is, however, evidence that RNAPII also possesses RNA-dependent RNA polymerase (RdRNAP, or RdRP) activity. RNAPII can use a homopolymeric RNA template, can extend RNA by several nucleotides in the absence of DNA, and has been implicated in the replication of RNA genomes of hepatitis delta virus (HDV) and plant viriods. Lehman et al. (2007) have shown that the intrinsic RdRP activity of RNAPII with only pure polymerase, an RNA template-product scaffold and nucleoside triphosphates (NTPs).

Crystallography reveals the template-product duplex in the site occupied by the DNA-RNA hybrid during transcription. RdRP activity resides at the active site used during transcription, but it is slower and less processive than DNA-dependent activity. RdRP activity is also obtained with part of the HDV antigenome. The complex of transcription factor IIS (TFIIS) with RNAPII can cleave only one HDV strand, create a reactive stem-loop in the hybrid site, and extend the new RNA 3′ end. Short RNA stem-loop with a 5′ extension suffices for activity, but there growth to a critical length apparently impairs processivity. The RdRP activity of RNAPII provides a missing link in molecular evolution, because it suggests that RNAPII evolved from an ancient replicase that duplicated RNA genome. RNA-dependent RNAPII activity is shown in Figure 7.18 (Lehman et al. 2007). Structure-based design of the RdRP scaffold is shown.

Transcription dynamics of single RNA polymerase II molecule

RNAPII is responsible for transcribing all messenger RNAs in eukaryotic cells during a highly regulated process that is conserved from yeast to human, and this serves as a central control point for cellular function. Galburt et al. (2007) investigate the transcription dynamics of single RNAPII molecules from budding yeast. Against force and in the presence and absence of TFIIS, a transcription elongation factor is known to increase transcription through nucleosomal barriers. Using a single-molecule dual-trap optical-tweezers assay combined with a novel method to enrich for active complexes, they found that the response of RNAPII to a hindering force is entirely determined by enzyme backtracking. Surprisingly, RNAPII molecules ceased to transcribe and were able to recover

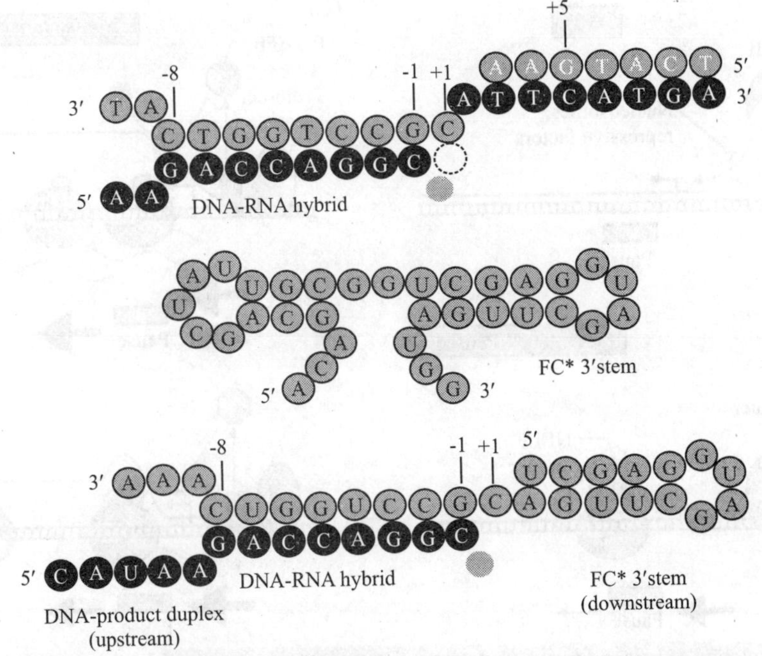

Figure 7.18 RNA-dependent RNA polymerase II activity

from backtracks at a force of 7.5±2 pN, only one-third of the force determined for *Escherichia coli* RNAP. They show that backtrack pause duration follows a t–3/2 power law, implying that during backtracking RNAPII diffuses in discrete base-pair steps, and indicating that backtracks may account for most of RNAPII pauses. Significantly, addition of TFIIS rescued backtracked enzymes and allowed transcription to proceed upto a force of 16.9±3.4 pN. Taken together, these involve a regulatory mechanism of transcription elongation in eukaryotes by which transcription factors modify the mechanical performance of RNAPII, allowing it to operate against higher loads.

Transcription elongation factor TFIIS

The transcription elongation factor TFIIS induces mRNA cleavage by enhancing the intrinsic nuclease activity of RNAPII. Kettenberger et al. (2003) diffused TFIIS into RNAPII clusters and derived a model of the RNAPII-TFIIS complex from X-ray diffraction data to 3.8Å resolution. TFIIS extends from the polymerase surface via a pore to the internal active site, spanning a distance of 100Å. Two essential and invariant acidic residues in a TFIIS loop complement the RNAPII active site and position a metal ion and a water molecule for hydrolytic RNA cleavage. TFIIS also induces extensive structural changes in RNAPII that realign nucleic acids in the active center. These results support the idea that RNAPII contains a single tunable active site for RNA polymerization and cleavage, in contrast to DNA polymerases with two separate active sites for DNA polymerization and cleavage.

Transcriptional pausing by RNA polymerase II

Recent work has shown that the RNAPII enzyme pauses at a promoter-proximal site of many genes in *Drosophila* and mammals (Core and Lis 2008). This rate-limiting step occurs after recruitment and initiation of RNAPII at a gene promoter. This stage in early elongation appears to be an important

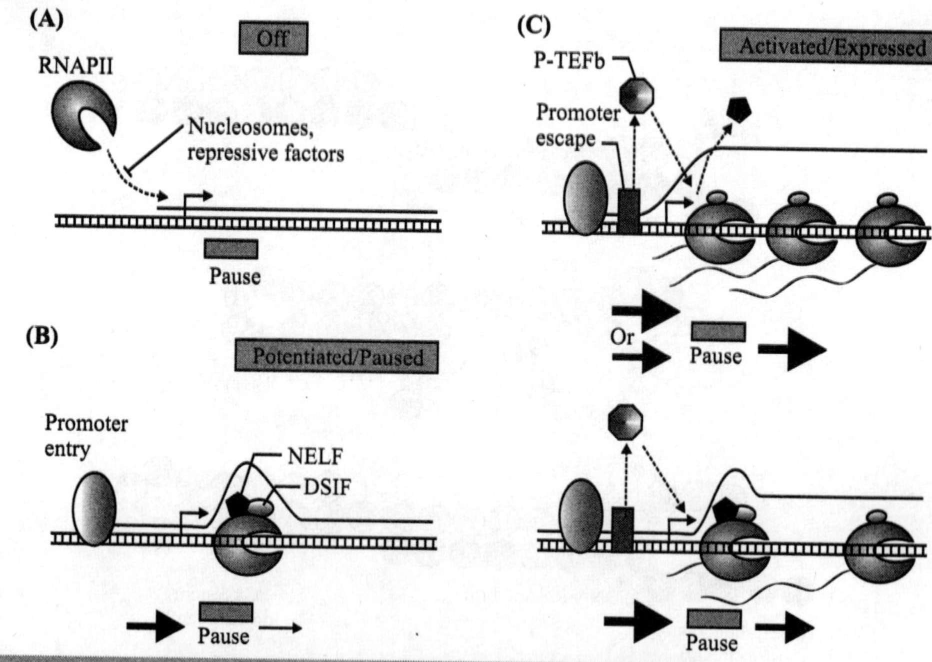

Figure 7.19 Regulation of early and escape of RNA polymerase II at pause sites

broadly used target of gene regulation. Regulation of early and escape of RNAPII at pause sites is shown in Figure 7.19. The rate of pause site entry is defined at a rate at which RNAPII enters a pause site when it is freely accessible. (A) RNAPII cannot access the promoter and transcription is "off". (B) A potential state through the set up of a promoter-proximal paused RNAPII by factors that promote entry. NELF and DSIF stabilize the paused RNAPII. (C) Fully activated transcription requires factors that promote escape. Also, single factors can have one or both types of activation domains that in turn can be regulated by reversible modifications and associations.

RNA Polymerase III

It carries out about 10 per cent of cellular transcription in eukaryotes. The genes transcribed by this enzyme are those for small non-coding RNA molecules such as 5S rRNA, tRNAs, the U6 snRNA found in spliceosmes and the 7S RNA component of signal recognition particle. (a) Transcription of 5S rRNA gene by this enzyme requires 3 TFs – TFIIIA, TFIIIB and TFIIIC. The promoter sequences for 5S rRNA and tRNA gene lie within the transcribed region of the genes. Such promoter sequences are called internal control regions (ICRs). Comparison of tRNA and 5S rRNA genes from a variety of organisms have revealed consensus sequences, designated as blocks, that correspond to the ICRs. The 5S rRNA gene also contains a conserved intermediate element (IE), which plays a role in formation of transcription complex Figure 7.20 (Rawn 1989). In 5S rRNA gene transcription, TFIIIA is the first protein to bind to DNA. It recognizes and binds tightly to specific nucleotides in the C block and intermediate element and aligns itself on the gene by containing nucleotides in the C block. Once TFIIIA is bound, TFIIIC enters the complex by making both protein-protein contacts with TFIIIA and protein-DNA contact with 5S rRNA gene. TFIIIB then binds to the TFIIIA-TFIIIC-5S rRNA gene complex, again via protein-protein contacts. RNAPIII then joins the stable transcription complex and begins transcription at the +1 site.

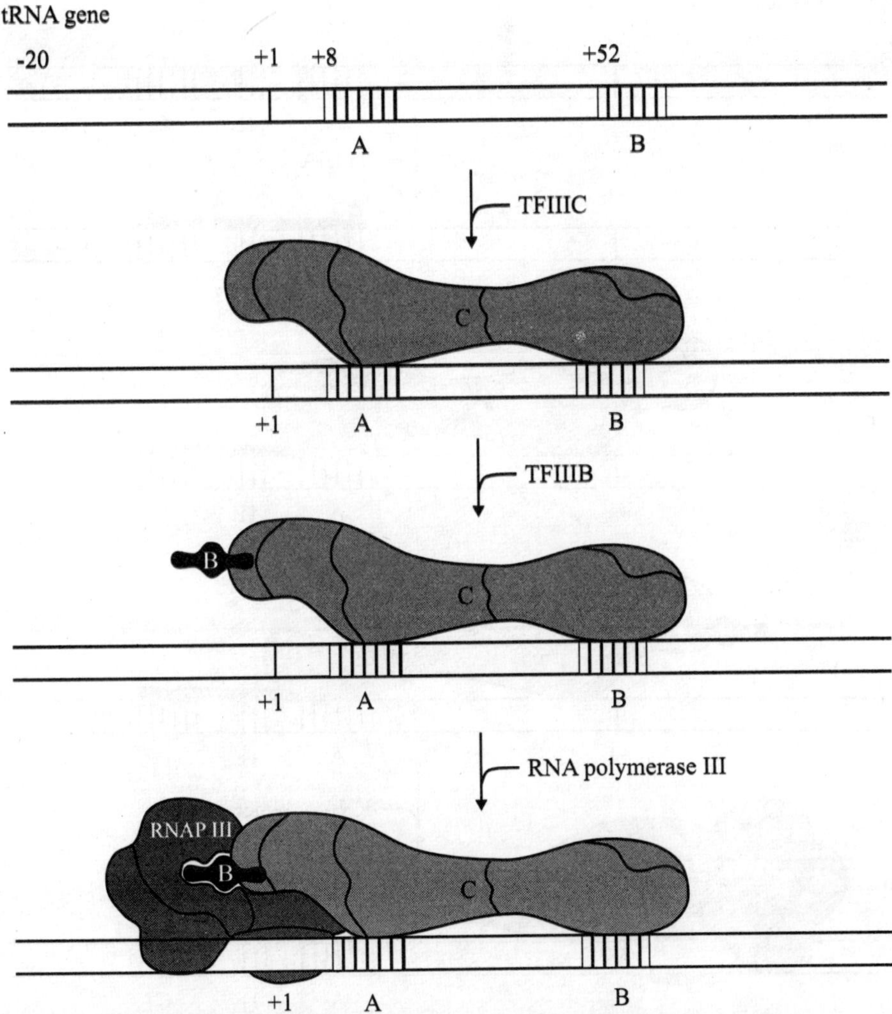

Figure 7.20 Binding of transcription factors to the internal control region (ICR) of a tRNA gene

Two transcription factors, TFIIIB and TFIIIC must form a stable transcription complex with the ICR before transcription can be initiated from a tRNA gene. The large, multimeric protein TFIIIC first binds tightly to the B block and then aligns itself on the gene by containing nucleotides in A block (Figure 7.21) (Rawn 1989). Once TFIIIC is found, TFIIIB joins the TFIIIC-tRNA gene complex. TFIIIB does not actually contact DNA but makes protein-protein contact with TFIIIC. The presence of this protein complex on the tRNA gene allows RNAPIII to initiate transcription at the start site. Once formed, transcription complex remains stably associated with the gene for multiple rounds of initiation.

Thus it is not necessary to assemble initiation complex each time these short genes are transcribed. The U6 and 7S RNA have slightly more complicated promoters, which seem to rely more on uncharacterized upstream sequences than on the consensus A and B blocks within the genes. However, these genes appear to use some of the same transcription factors as do the better understood genes for tRNA and 5S rRNA. How does the large transcription complex remain on the 5S rRNA and

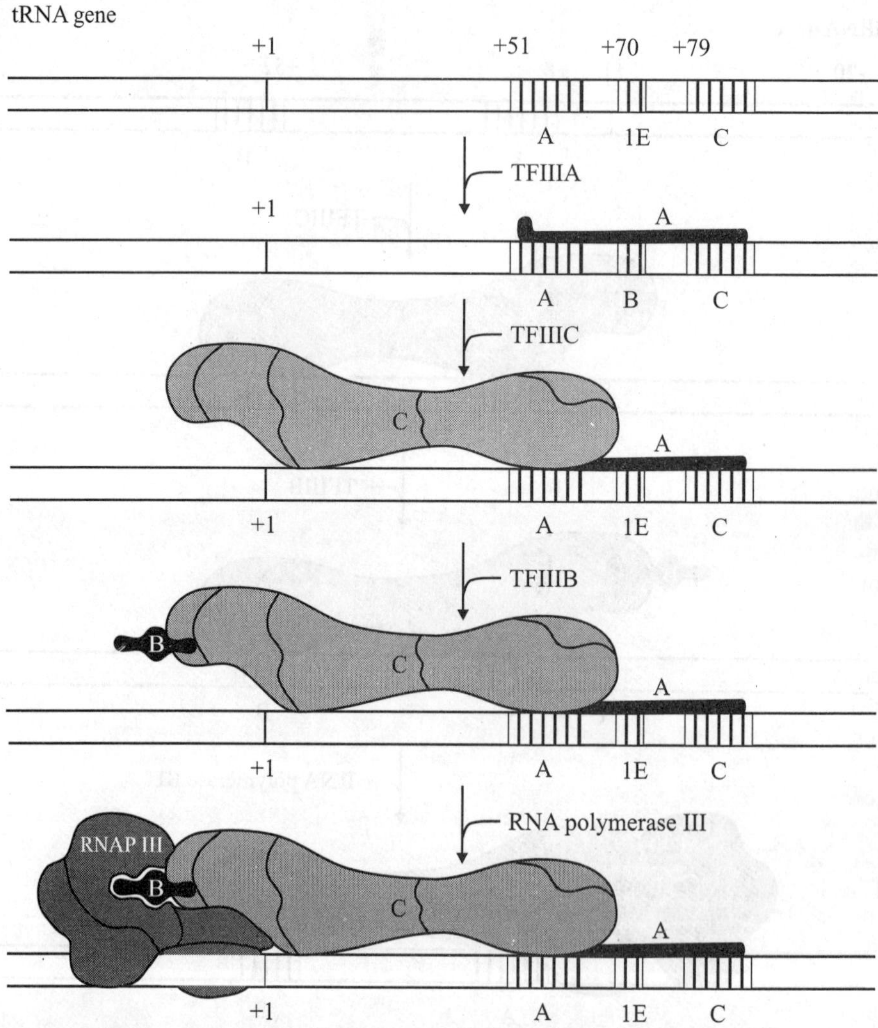

Figure 7.21 Binding of transcription factors to the internal control region of a 5S rRNA gene

tRNA genes while RNA passes through their binding sites? A possible answer to this is: when TFIIIA binds to the 5S rRNA gene it recognizes determinants that are primarily located on the non-transcribed strand of the 5S rRNA gene. As RNAPIII reads through the genes, the transcribed and non-transcribed strands separate. While polymerase interacts with the transcribed strand, the huge transcription complex of over 1,000 kDa can remain bound to the non-transcribed DNA. Thus, when the strands anneal after the polymerase has passed, the transcription complex is still bound to the same site.

RNA Polymerases IV and V (in plants)

Plants have distinct RNA polymerase complexes with largely unknown roles in maintaining small RNA-associated gene silencing. Plants have catalytic subunits for a nuclear RNAPIV, which mediates siRNA and DNA methylation-dependent heterochromatin formation. Curiously, the eudicot

Arabidopsis thaliana is not affected when either function is lost. By use of mutation selection and positional cloning, Erhard et al. (2009) showed that the largest subunit of the presumed maize RNAPIV is involved in paramutation, an inherited epigenetic change facilitated by an interaction between two alleles, as well as normal maize development. Nuclear run-on transcription assays indicate that RNAPIV does not engage in the efficient RNA synthesis typical of the three major eukaryotic DNA-dependent RNA polymerases (DdRPs, usually written as RNAPs). These studies indicate that RNA polymerase IV (RNAPIV or PolIV) employs abnormal RNAP activities to achieve genome-wide silencing and that its absence affects both maize development and heritable epigenetic changes.

RNAPIV does not functionally overlap with RNAPI, RNAPII, or RNAPIII and is non-essential for viability. However, disruption of RNAPIV catalytic subunit genes *NRPD1* or *NRPD2* inhibits heterochromatin association into chromocenter, coincident with losses in cytosine methylation at pericentromeric 5S gene clusters and AtSN1 retroelements. Loss of CG, CNG, and CNN methylation in RNAPIV mutants implicates a partnership between RNAPIV and the methyltransferse responsible for RNA-directed de novo methylation. Consistent with this hypothesis, 5S gene and AtSN1 siRNA are essentially eliminated in RNAPIV mutants. The data suggest that RNAPIV helps to produce siRNAs that target de novo cytosine methylation events required for facultative heterochromatin formation and high-order heterochromatin associations.

By mutation of the two largest subunits (NRPD1a and NRDP2), Herr et al. (2005) have shown that Pol IV silences certain transposons and repetitive DNA in a short interfering RNA pathway involving RdRNAPII and Dicer-like 3. The existence of this distinct silencing polymerase may explain the paradoxical involvement of an RNA silencing pathway in maintenance of transcriptional silencing.

Kravchenko et al. (2005) show that transcription of some mRNAs in humans and rodents is mediated by previously unknown single-polypeptide nuclear RNA polymerase (spRNAPIV). spRNAP-IV is expressed from an alternative transcript of the mitochondrial RNA polymerase gene (*POLRMT*). The spRNAP-IV lacks 262 amino-terminal amino acids of mitochondrial RNA polymerase, including the mitochondrial-targeting signal, and localizes into the nucleus. Transcription by spRNAPIV is resistant to RNAPII inhibitor α-amanitin but is sensitive to short interfering RNA specific for the POLRMT gene. The promoters for spRNA-IV differ substantially from those used by RNAPII, do not respond to transcriptional enhancers and contain a common functional sequence motif.

Nuclear RNA polymerase V (RNAPV or PolV) is an RNA silencing enzyme recently shown to generate non-coding transcripts at loci silenced by 24-nt siRNAs. Nuclear RNA polymerase V (PolV) collaborates with DRD1 (deficient in RNA-dependent DNA methylation 1) to generate transcripts at heterochromatic loci that are hypothesized to bind to siRNA-AGO4 complexes and subsequently recruit the de-novo DNA methylation and/or histone modifying machinery.

GENE CLASSIFICATION BASED ON RNA POLYMERASE USED

The genes transcribed by three major RNA polymerases, RNAPI, RNAPII, and RNAPIII, are classified as class I, class II and class III genes, respectively.

Class I genes

Binding of transcriptional factors to promoters for RNAPI requires protein-proteins as well as protein-DNA interaction. Transcription factors bind to several sites upstream from the start site of transcription. TFID binds between –40 and –15, and UBF-I binds between –120 and –105. The TFID:rRNA gene complex is then bound by TFIC. The UBF-1:rRNA gene complex is bound by SLI. RNAPI then joins these complexes, completing formation of the transcription machine.

Class II genes

Van Dyke et al. (1988) reconstituted transcription system composed of HeLa cell-derived from transcription factors TFIIB, TFIID, TFIIE and USF and RNAPII. They investigated the transcription factor requirement for the assembly of complete preinitiation complexes. Assembly of the transcription machine during initiation in eukaryotic is a complex process (Figure 7.22). So, no eukaryotic RNAP appears to bind specifically and directly to DNA. Instead, interactions between polymerase and DNA require the presence of sequence specific accessory proteins called transcription factors (TF). TFIID or its part is common for all three RNAPs. It involves formation of pre-initiation complex (PIC) of TATA-binding protein (TBP) alongwith TBP associated factors (TAFIIs). PIC includes a set of general transcription factors (TFIIA B, D, E, F and H) along with RNAPII (Walker et al. 2004).

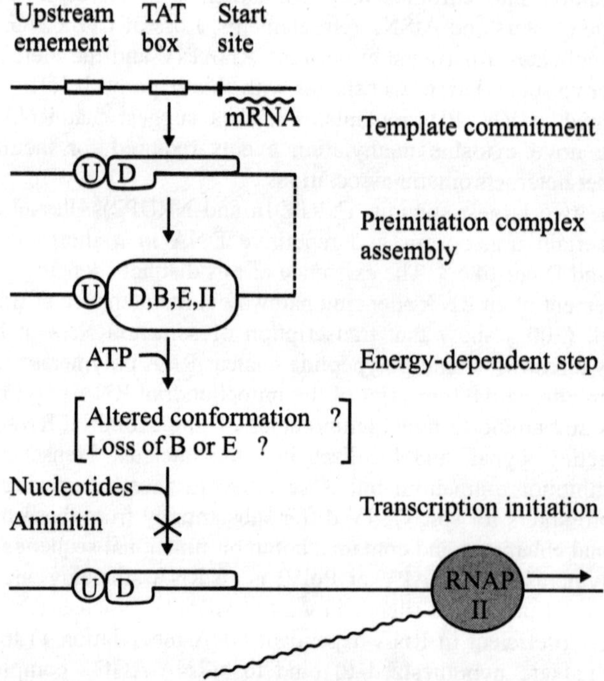

Figure 7.22 Model for the mechanism of specific transcription by RNA polymerase II

Transcription of protein-encoding genes by human RNAPII requires multiple ancillary proteins (transcription factors). Interactions between these proteins and the promoter DNA of a viral class II gene (the major late transcription unit of adenovirus) were investigated by enzymatic and chemical footprinting (Van Dyke et al. 1988). The experiments indicated that the assembly of functionally active RNAPII-containing transcription preinitiation complexes requires a complete set of transcription factors, and that both specific protein DNA and protein-protein interactions are involved. This allows individual steps along the transcription reaction pathway to be tested directly, thus allowing a basis for understanding basic transcription initiation mechanisms as well as the regulatory processes that act on them. Information stored in DNA must be expressed so as to give a particular phenotype and this process requires the production of mRNA. The best understood mechanism exerts control of transcription, in which the production of mRNA is regulated according to need. As we move from lower to higher form of life transcription complexity increases.

The potent transcriptional domain of herpes simplex virus protein VP16 was found to bind strongly and highly selectively to the human and yeast TATA box-binding factors (Stringer et al. 1990). This implies that the principal target for acidic activation domains is the TATA box factor TFIID. Transcription factor IID (TFIID) binds to the TATA box promoter element and regulates the expression of most eukaryotic genes transcribed by RNAPII. cDNA encoding a human TFIID protein has been cloned. TFIID polypeptide has 339 amino acids (Kao et al. 1990). Carboxyl terminal 181 amino acids of the human TFIID protein share 80 per cent identity with TFIID protein from yeast. The TATA-binding protein, TFIID, plays a central role in initiation of eukaryotic mRNA synthesis (Peterson et al. 1990). Human (Davison et al. 1983; Sawadogo and Roeder 1985), *Drosophila* (Parker and Topol 1984) and yeast (Buratowski et al. 1988; Horikoshi et al. 1989; Cavallini et al. 1988) TFIID reveals a highly conserved carboxyl-terminal 180 amino acids. Human TFIID gene when expressed in *E. coli* and HeLa cells produces a protein that binds specifically to a TATA box and promotes basal transcription.

Class III genes

Transcription factors for class III genes bind to internal control regions (ICRs). RNAPIII carries out about 10 per cent of cellular transcription in eukaryotes. Transcription of class III genes requires 3 TF: TFIIIA, TFIIIB and TFIIIC. A high degree of homology occurs in RNAPIII, TFIIIB, TFIIIA and TFIIIC (Huang and Maraia 2001). The two TFs TFIIIB and TFIIIC must form a stable transcription complex with the ICR before transcription can be initiated from a tRNA gene. Transcription of 5S rRNA gene also begins with formation of a stable transcriptional complex which includes TFIIIB, TFIIIC and 3rd protein TFIIIA. In tRNA genes, TFIIIC binds with DNA sequence followed by TFIIIB and then RNAP binds and transcription occurs, while in 5SrRNA an additional transcriptional factor TFIIIA binds first with DNA sequence and then other transcriptional factors, i.e., TFIIIC and TFIIIB joins followed by RNAP binding and transcription takes place.

PROCESS OF EUKARYOTIC TRANSCRIPTION

Transcription Initiation

Transcription of protein-encoding genes by human RNAPII requires the coordinated action of multiple protein factors. These can be classified into two categories: general transcription factors and upstream element-binding proteins. Following four steps have been defined in the transcription mechanism: (1) A commitment of the template after binding of a subset of transcription factors. (2) Formation of an activated state through the action of other transcription factors and RNAPII. (3) Fulfillment of an energy requirement. (4) Initiation of transcription (formation of first phosphodiester bond).

Eukaryotic Transcriptional Factors

Leucine Zipper Family: of eukaryotic transcriptional factors. It includes CCAAT-enhancer-binding proteins (C/EBPs), Fos, Jun and GCN4 (Sauer 1990). The C/EBP family uses a bipartite structural motif to bind DNA. The C/EBP family consists of six transcription factors (α, β, γ, δ, ε and ζ). Two protein chains dimerize through a set of amphipathic α-helices termed the leucine zippers. "Scissors grip" model has been given to explain how C/EBP binds DNA in the major groove (Shuman et al. 1990).

Zinc Fingers: Zinc fingers constitute important eukaryotic DNA binding domains, being present in many transcription factors (Nardelli et al. 1991). These factors have now been analyzed in detail.

Yeast Protein RAP1: The yeast protein RAP1, initially described as a transcriptional regulator, binds in vitro to sequences found in a number of seemingly unrelated genomic loci (Lusting et al. 1990). These include the silencers at the transcriptionally repressed mating-type genes, the promoters of many genes important for cell growth and the poly[(cytosine)1-3adenine] repeats of telomeres. RAP1 may be involved in telomere formation in vivo.

Yeast Transcription Activator GCN4: "bZIP", a DNA-binding protein, consists of a basic region that contacts DNA and an adjacent "leucine zipper" that mediates protein dimerization (McKnight and Kim 1990). A peptide model for basic regions of yeast transcription activator GCN4 has been developed in which leucine zipper has been replaced by disulphide bond. The 34-residue peptide dimer binds DNA with nanomolar affinity at 4 °C.

Crucial features of transcription initiation

The structure of the general transcription factor IIB (TFIIB) in a complex with RNAPII reveals three features crucial for transcription initiation; an N-terminal zinc ribbon domain of the TFIIB that contacts the "dock" domain of the polymerase, near the path of RNA exit from a transcribing enzyme; a "finger" domain of TFIIB that is inserted into the polymerase active center; and a C-terminal domain whose interaction with both the polymerase and with a TATA box-binding protein (TBP)-promoter DNA complex orients the DNA for unwinding and transcription. TFIIB stabilizes an early initiation complex, containing an incomplete RNA-DNA hybrid region. It may interact with the template strand, which sets the location of the transcription start site, and may interfere with RNA exit, which leads to abortive initiation or promoter escape. The trajectory of promoter DNA determined by the C-terminal domain of TFIIB traverses sites of interaction with TFIIE, TFIIF, and TFIIH, serving to define their roles in the transcription initiation process (Bushnell et al. 2004). Components of RNAP II machine in yeast are given in Table 7.5. Different domains of TFIIBN are shown in Figure 7.23.

Table 7.5 RNA polymerase II (also written as pol II or RNAPII) transcription machines. Mass data are for yeast proteins

Component	*Subunit*	*Mass (kDa)*	*Function*
Pol II	12	520	RNA synthesis
TFIIB	1	38	Start site determination
TFIID (TBP)	1	27	Bending TATA box DNA around TFIIB and Pol II
(TAFs)	14	749	Promoter recognition
TFIIE	2	92	Coupling Pol II-promoter interaction to recruitment of TFIIH
TFIIF	3	156	Interaction with non-template DNA strand
TFIIH	9	525	Promoter opening Pol II phosphorylation
Mediator	20	1003	Regulatory signal transduction

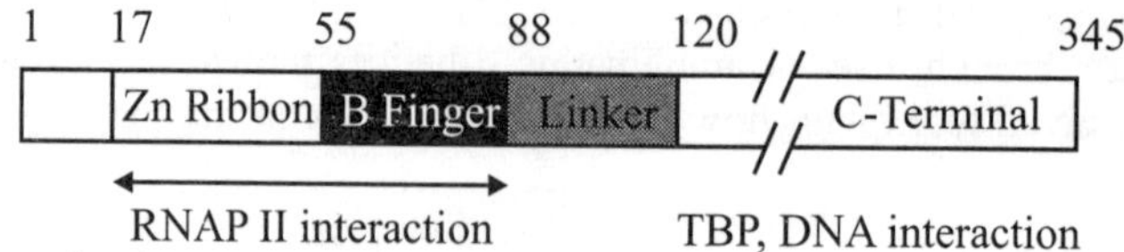

Figure 7.23 Domain structure of TFIIB

The structure of the initiator binding protein from the primitive eukaryote *Trichomonas viganilis* reveals how a single initiator binding protein (IBP) can orchestrate promoter recognition, RNAPII recruitment, and start site selection, and provide an evolutionary link with transcription initiation in higher eukaryotes (Marmorstein 2003). IBP has two domains – Inr and C. Inr domain is used for DNA recognition while C-domain mimics the CTD of RNAPII.

Transcription Elongation

Transcription elongation with some transcription components remaining committed to the template. Elongation factors enhance overall activity of RNAPII leading to increase in elongation rate. Two such factors are TFIIF and TFIIS. TFIIF accelerates RNA chain growth uniformly and TFIIS helps in elongation by relieving obstacles. TFIIS causes hydrolytic cleavage at 3′-end of RNA chains which are blocked. Its action is similar to cleavage by GreA and GreB in *E. coli*. Rpb9 is small subunit of yeast RNAPII participating in elongation (Van Mullem et al. 2002). TFIIS plays an important role in initiation of transcription at GAL1 gene of *S. cerevisiae* in addition to its well-characterized role in transcription elongation (Prather et al. 2004).

Transcription Termination

Termination of transcription is an important regulatory step closely related to transcriptional interference, and even transcriptional initiation. Telxeira et al. (2004) report a new phenomenon, co-transcriptional cleavage (CoTC). This primary cleavage event within β-globin pre-messenger RNA, downstream of the poly(A) site, is critical for efficient transcriptional termination by RNAPII. They show that CoTC process involves a self-cleaving activity. They show that autocatalytic core of the CoTC ribozyme has functional role in efficient termination in vivo. This CoTC may be a general phenomenon and functionally it may provide an entry point for exonuclease involved in mRNA maturation, turnover and, in particular, transcriptional termination.

Much less is known about termination of transcription in eukaryotes than in prokaryotes because (a) eukaryotes are more difficult to analyze experimentally for comparison of consensus sequences that are required for transcription termination, (b) 3′-ends of mature mRNA molecules result not from termination of transcription but from post-transcriptional processing. (c) site at which transcription complex dissociates from most eukaryotic genes is not precisely known. All three eukaryotic RNAPs terminate at the sites that include a stretch of Ts on the non-template strand of the gene. These T-runs may be the only sequence required for termination of transcription in mammalian cells as there is no RNA hairpin structure formed like in prokaryotes.

'Torpedo' model of transcriptional termination

Eukaryotic protein-encoding genes possess poly(A) signals that define the end of the messenger RNA and metabolic downstream transcriptional termination by RNAPII. Termination could occur through an 'anti-termination' mechanism whereby elongation factors dissociate when the poly(A) signal is encountered, producing termination-competent RNAPII. An alternative 'torpedo' model postulated that poly(A) site cleavage provides an unprotected RNA 5'-end that is degraded by 5'→3' exonuclease activities (torpedoes) and so induces dissociation of RNAPII from the DNA template (West et al. 2004). This model has been questioned because unprocessed transcripts read all the way to the site of transcriptional termination before upstream polyadenylation. However, nascent transcripts located 1 kilobase downstream of the human *β-globin* gene poly(A) signal are associated with a co-transcriptional cleavage (CoTC) activity that acts with the poly(A) signal to elicit efficient transcriptional termination. The CoTC sequence is an autocatalytic RNA structure that undergoes rapid self-cleavage. They show that CoTC acts as a precursor to termination by presenting a free 5'-end that is recognized by the human 5'→3' exonuclease Xrn2. Degradation of the downstream cleavage product by Xrn2 results in transcriptional termination, as evidenced in the torpedo model, which is presented in Figure 7.24 (West et al. 2004). Autocatalytic CoTC acts as a precursor to termination by presenting a free 5′-end that is recognized by the 5'→3' exonuclease Xrn2. Subsequent degradation of this downstream cleavage product by Xrn2 leads to RNAPII transcriptional termination analogous to the original torpedo model. The 3'-end of the cleavage RNA is likely to be degraded by the exosome (Exo).

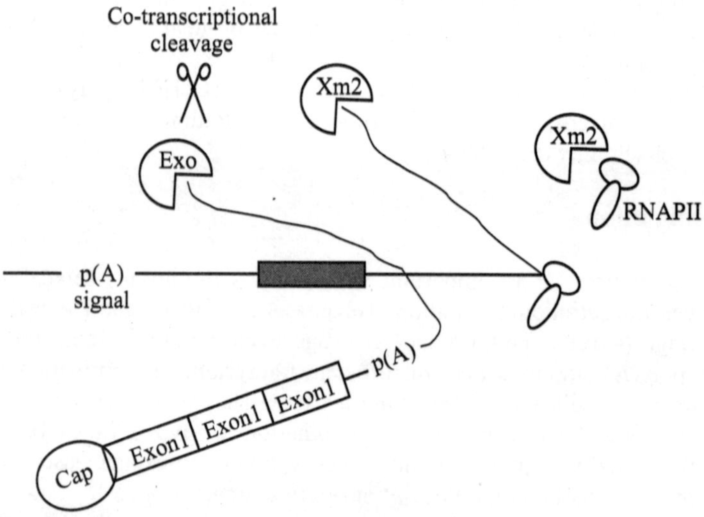

Figure 7.24 Model for RNA polymerase II transcriptional termination

Transcription termination in yeast

The information encoded in our genes must be copied into messenger RNA, which will program the protein-synthesis machinery. New results support intriguing mechanisms for ending the copying process (Tollervey 2004). Transcription termination in yeast has been shown in Figure 7.25. (A) Transcription of a gene by RNA polymerase produces a nascent messenger RNA (mRNA). mRNAs eventually have two distinctive features: a 7-methyl-guanosine cap (m7G) at one end and a polyadenosine (poly(A)) tail at the other. The carboxy-terminal domain (CTD) of the polymerase comprises repeats of a seven-amino-acid motif and undergoes reversible modification by phosphorylation (not shown). In particular, phosphorylation of a serine amino acid at position 2 in each repeat increases as the polymerase moves along the mRNA. (B) the Rtt103 protein binds a CTD peptide carrying the serine 2 phosphorylation, and also associates with the exonuclease Rat1 (Xrn2 in humans) and its cofactor Rai1. Thus, Rat1 may be recruited to the nascent mRNA via Rtt103 bound to the polymerase CTD, and by an independent system that involves Rai1. Following cleavage of the poly(A) site, Rat1 binds and eats away at the free end of the RNA, halting transcription when it catches the polymerase (C).

Transcription termination at the β-globin gene in human cells is shown in Figure 7.26 (Tollervey 2004). (A) The newly synthesized human -globin mRNA will be cut at two positions: at the normal site

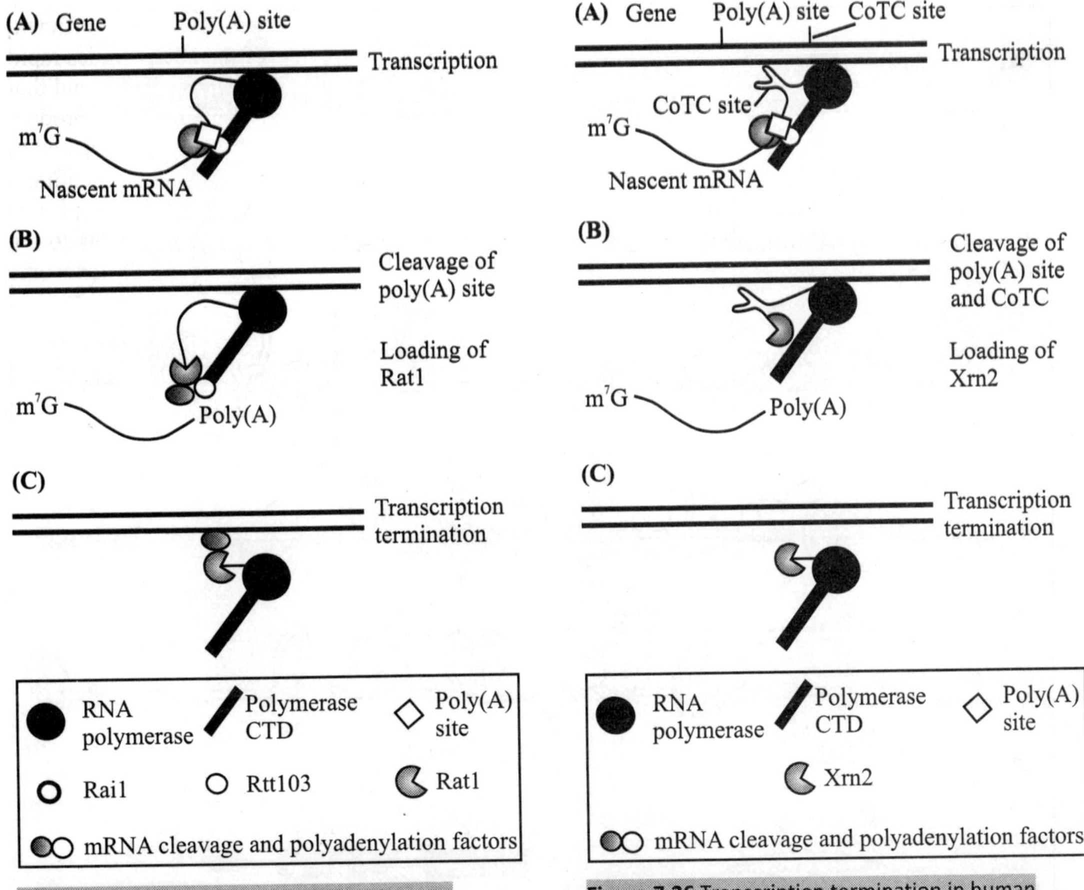

Figure 7.25 Transcription termination in yeast

Figure 7.26 Transcription termination in human

of poly(A) addition and within the co-transcriptional cleavage (CoTC) site, which Teixeira et al. (2004) show to be self-cleaving. (B) West et al. (2004) propose that this allows the entry of Xrn2 (the human Rat1 counterpart), which again eates away at the mRNA until it catches up with the RNA polymerase (C). Replacing the CoTC region with a different self-cleaving RNA, a 'hammerhead ribozyme', inhibits transcription termination, possibly reflecting specific recruitment of Xrn2 to the CoTC. However, the hammerhead cleavage leaves a product with a 5'-hydroxyl group, which is a poorer substrate for Rat1 than the 5'-phosphate produced by cleavage of the CoTC, so it might be a poorer substrate for Xrn2 as well.

Cohesion distribution

During cell division the cohesion complex mediates the pairing of sister chromatids. Emerging evidence shows that cohesion also has roles in interphase cells. New studies, including that of Gullerova and Proudfoot (2008), reveal how cohesion is targeted to specific sites on chromosomes and implicate cohesion in the regulation of gene expression (Peric-Hupkes and van Steensel 2008). Cohesin distribution and gene regulation is shown in Figure 7.27(A).

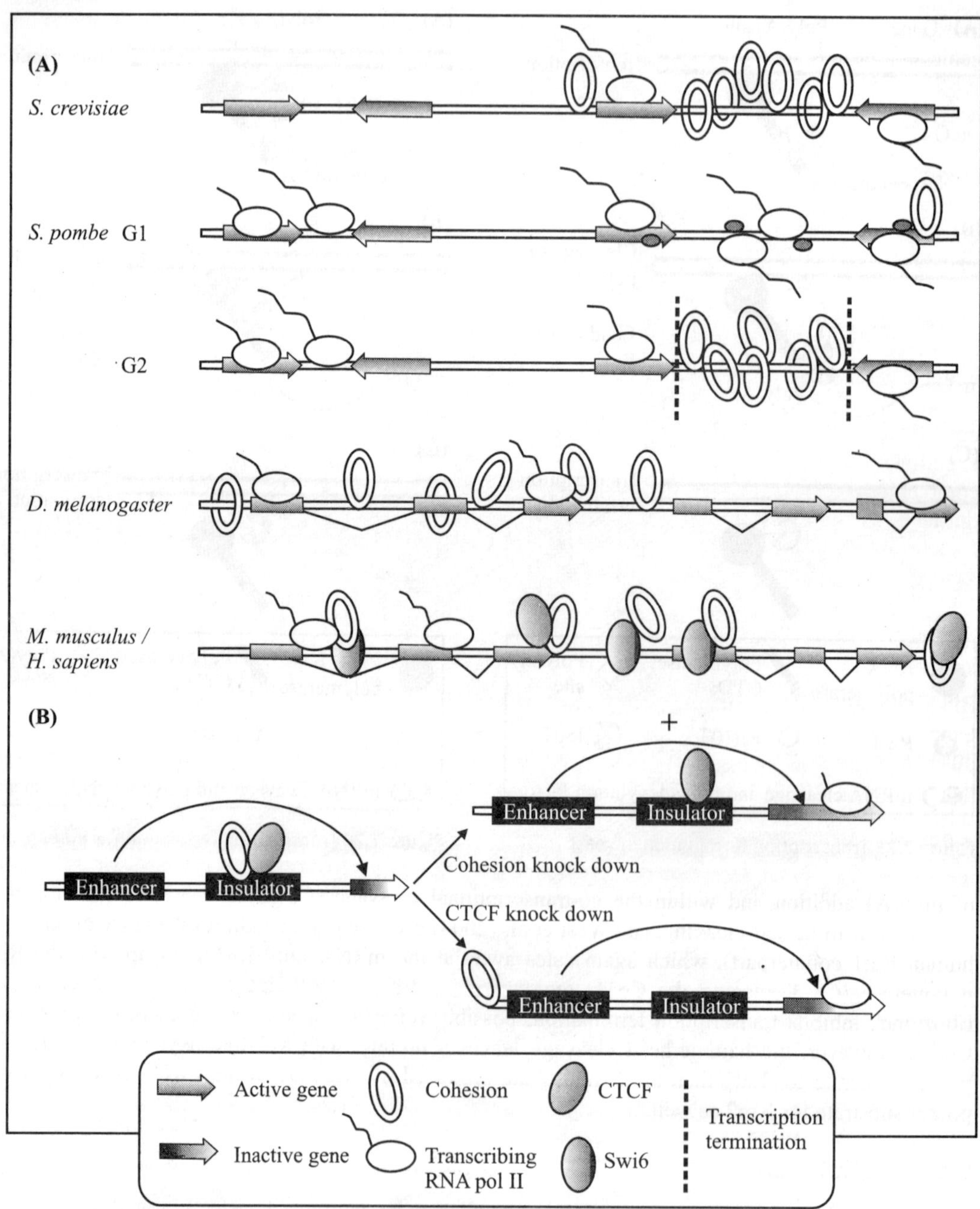

Figure 7.27 Cohesin distribution and gene regulation

The genomic distributions of cohesin in various eukaryotes are depicted. In *S. cerevisiae* cohesin associates with regions located between convergently transcribed genes. During G1 in *S. pombe* readthrough transcription at convergent genes leads to H3K9 methylation and recruitment of Swi6. In G2 these heterochromatin marks are mostly lost; cohesin accumulates at sites of convergent transcription in G2 and mediates transcription termination. In *D. melanogaster* cohesin is found both in genic and intergenic regions. The population that is bound to genic regions is preferentially located at a subset of active genes and depleted from inactive genes. In mammalian cells cohesin shows a high degree of colocalization with the insulator protein CTCF. Figure 7.27(B) shows that removal of either cohesin or CTCF causes a loss of CTCF-mediated insulator activity, resulting in increased enhancer-promoter interaction. Cohesin is shown as turquoise rings; it is not known whether in each case the cohesin ring embraces chromatin or associates through a different mechanism.

TRANSCRIPTIONAL COMPLEXITY

Transcribed portions of several organisms are larger and more complex than expected, and many functional properties of transcripts are based not on coding sequences but on regulatory sequences in untranslated regions or non-coding RNAs. Alternative start and polyadenylation sites and regulation of intron splicing add additional dimensions to the rich transcriptional output. This transcriptional complexity has been sampled mainly using hybridization-based methods under one or a few experimental conditions. Wilhelm et al. (2008) applied direct high-throughput sequencing of complementary DNAs (RNA-Seq) in fission yeast. Hundreds of introns showed regulated splicing during cellular proliferation or differentiation.

SOME FEATURES OF EUKARYOTIC TRANSCRIPTION

Binding of Ribonucleoside Triphosphate RNA Polymerase II Transcribing Complex

Binding of a ribonucleoside triphosphate to an RNAPII transcribing complex, with base pairing to the template DNA, was revealed by X-ray crystallography (Westover et al. 2004b). Binding of a mismatched nucleoside triphosphate was also detected, but in an adjacent site, inverted with respect to the correctly paired nucleotide. These results are consistent with a two-step mechanism of nucleotide selection, with initial binding to an entry (E) site beneath the active center in an inverted orientation, followed by rotation into the nucleotide addition (A) site for pairing with the template DNA. This mechanism is related to that of single subunit RNAPs and so defines a new paradigm for the large, multisubunit enzymes. Additional findings from these studies include a third nucleotide binding site that may define the length of backtracked RNA; DNA double helix unwinding in advance of the polymerase active center; and extension of the diffraction limit of RNAPII crystals to 2.3Å.

Cohesion Association along Budding Yeast Chromosomes for Convergent Transcription

Sister chromatids, the products of eukaryotic DNA replication, are held together by the chromosomal cohesion complex after their synthesis. This allows the spindle in mitosis to recognize pairs of replication products for segregation into opposite directions. Cohesin forms large protein rings that may bind DNA strands by encircling them, but characterization of cohesion binding to chromosomes in vivo has remained vague. Lengrome et al. (2004) have performed high-resolution analysis of cohesion association along budding yeast chromosomes III-IV. Cohesin localizes almost exclusively between genes that are transcribed in converging directions. They find that active transcription positions cohesins at these sites, not the underlying DNA sequence. Cohesin is initially loaded onto chromosomes at separate places, marked by the Scc2/Scc4 cohesin loading complex, from where it appears to slide to its more permanent locations. But even after sister chromatid cohesion is established, changes in transcription lead to repositioning of cohesion. Thus the sites of cohesion binding and therefore probably sister chromatid cohesion, a key architectural feature of mitotic chromosomes, display surprising flexibility. Cohesin localization to places of convergent transcription is conserved in fission yeast, suggesting that it is a common feature of eukaryotic chromosomes.

Cohesion Binding Adapts to Changes in Transcription

Cohesin complexes have a central role in cell division, mediating the association between sister chromosomes. It now seems that cohesion binding is dynamic, adapting to changes in gene transcription (Ross and Cohen-Fix 2008). Cohesion association with zones of convergent transcription is shown in Figure 7.28: (A) Genes occur in several orientations along the chromosomes. Arrows indicate direction of transcription. Genes *A* and *B* are arranged tail to tail; genes *B* and *C* are head to tail; and genes *C* and *D* are head to head. Cohesins are generally found in the spaces between genes that are arranged tail to tail. (B) Transcription leads to cohesion rearrangement. When transcription of a gene switches from off to on, cohesins are lost from the coding region and instead associate near the end of the gene. RNAP is shown in beige and the newly synthesized RNAs are shown as thick black lines.

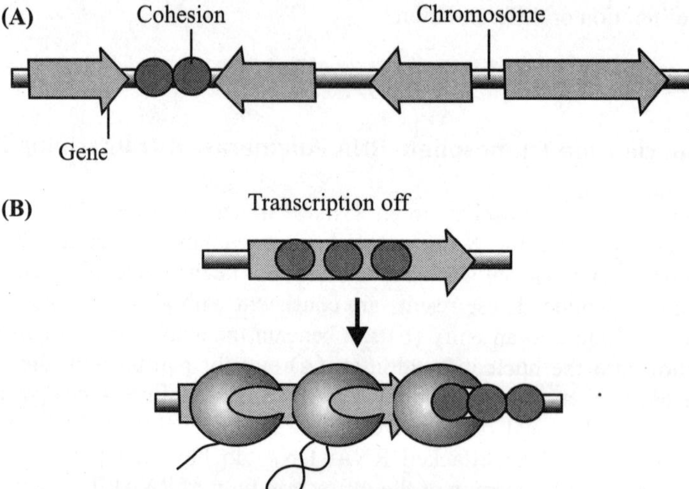

Figure 7.28 Cohesion association with zones of convergent transcription

Extrinsic Variability in Gene Expression

Variable gene expression within a clonal population of cells has been implicated in a number of important processes including mutation and evolution, determination of cell fates and development of genetic disease. A significant component of expression variability arises from extrinsic factors thought to influence multiple genes simultaneously. Volfson et al. (2006) have identified two major sources of extrinsic variability in budding yeast. One unavoidable source arising from the coupling of gene expression with population dynamics leads to a ubiquitous lower limit for expression variability. A second source, which is modeled as originating from a common upstream transcription factor, exemplifies how regulatory networks can convert noise in upstream regulator expression into extrinsic noise at the output of a target gene. These results thus highlighted the importance of interplay of gene regulatory with population heterogeneity for understanding the origin of cellular diversity.

Cryptic Unstable Transcripts in Eukaryotes

Pervasive and hidden transcription is widespread in eukaryotes. Cryptic unstable transcripts (CUTs) were recently described as a principal class of RNAPII transcripts in budding yeast. These transcripts are targeted immediately for degradation immediately after synthesis by the action of the Nrd1-exosome-TRAMP complexes. CUT degradation mechanisms have been analyzed in detail. Neil et al. (2009) report the high-resolution genomic map of CUTs in yeast, revealing a class of potentially functional CUTs and intrinsic bidirectional nature of eukaryotic promoters. CUT clusters have been classified according to the type of associated genomic intergenic regions as tandem sense (TS), tandem antisense (TA), divergent (D) and convergent (C), as shown in Figure 7.29 (Neil et al. 2009). Arrows indicate genomic organization of CUTs relative to surrounding features. The numbers of CUTs in each of four categories defined is indicated on the right.

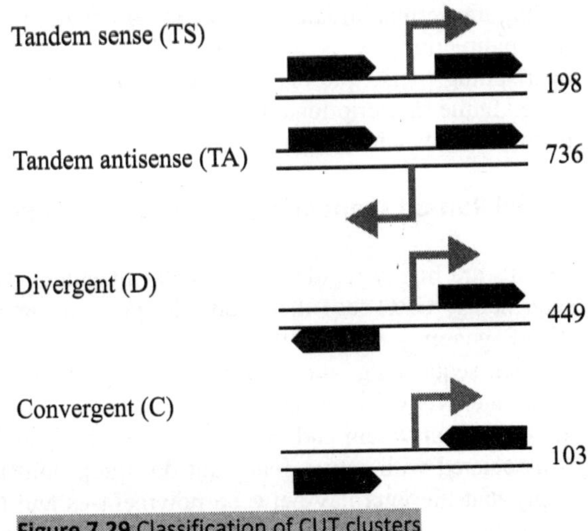

Figure 7.29 Classification of CUT clusters

Proteins Interact with DNA to Interpret Genome

Spontaneous preferences of DNA binding proteins are a primary mechanism by which cells interpret the genome. Despite the central importance of these proteins in physiology, development, and evolution, comprehensive DNA binding specificities have been determined experimentally for only a few proteins. Badis et al. (2009) used microarrays containing all 10-bp sequences to examine the binding specificities of 104 distinct mouse DNA binding proteins representing 22 structural classes. The results reveal a complex landscape of binding, with virtually every protein analyzed possessing unique preferences. Roughly half of the proteins each recognized multiple distinctly different sequence motifs, challenging our molecular understanding of how proteins interact with their DNA binding sites. This complexity in DNA recognition may be important in gene regulation and in the evolution of transcriptional regulatory networks

Genome-Wide Pervasive Transcription

Genome-wide pervasive transcription has been reported in many eukaryotic organisms, revealing a highly interleaved transcriptome organization that involves hundreds of previously unknown non-coding RNAs. These recently identified transcripts either existed stably in cells (stable unannotated transcripts, SUTs) or are rapidly degraded the RNA surveillance pathway (cryptic unstable transcripts, CUTs). One characteristic of pervasive transcription is the extensive overlap of SUTs and CUTs with previously annotated features, which prompts question regarding how these transcripts are generated, and whether they exert function. Single-gene studies have shown that transcription of SUTs and CUTs can be functional, through mechanisms involving the generated RNAs or their generation itself. Xu et al. (2009) provide a comprehensive analysis of these transcripts. They show that both SUTs and CUTs display distinct patterns of distribution at specific locations. Most of the newly identified transcripts initiate from nucleosome-free regions (NFRs) associated with the promoters of other transcripts mostly protein-coding genes), or from NFRs at the 3′-ends of protein-coding genes. Likewise, about half of all coding transcripts initiate from NFRs associated with promoters of other transcripts. The data suggest that bidirectionality is an inherent feature of promoters. Such an arrangement of divergent and overlapping transcripts may provide a mechanism for local spreading of regulatory signals – that is, coupling the transcriptional regulation of neighboring genes by means of transcriptional interference or histone modification.

Global Run-on sequencing of engaged RNA Polymerase Genome-Wide

RNAPs are highly regulated molecular machines. Core et al. (2008) present a method (global run-on sequencing, GRO-seq) that maps the position, amount and orientation of transcriptionally engaged RNAP genome-wide. In this method, nuclear run-on RNA molecules are subjected to large-scale parallel sequencing and mapped to the genome. They show that transcription extends beyond messenger RNA 3' cleavage, and antisense transcription is prevalent. Most promoters have an engaged polymerase upstream and in an orientation opposite to the annotated gene. This divergent polymerase is associated with active genes but does not elongate effectively beyond the promoter. These results imply that the interplay between polymerases and regulators over road promoter regions dictates the orientation and efficiency of productive transcription.

Global Transcription Machinery Engineering

Global transcription machinery engine-eering (gTME) is an approach for reprogramming gene transcription to elicit cellular phenotypes important technological applications. Alper et al. (2006) show the application of gTME to *S. crevisiae* for improved glucose/ ethanol tolerance, a key trait for many biofuel programs. Mutagenesis of the transcripttion factor Spt15p and selec-tion led to dominant mutations that conferred increased tolerance and more efficient glucose conversion to ethanol. The desired phenotype results from the combined effect of three separate mutations in the SPT15 gene (serine substiuted for phenylalanine (Phe177Ser) and, similarly, Tyr195-His, and Lys218Arg can provide a route to complex phenol-types that are not readily accessible by traditional methods. These three mutations are mapped on the global transcriptional machinery) (Figure 7.30). Collectively, these three mutations lead to a mechanism involving Spt3p.C.

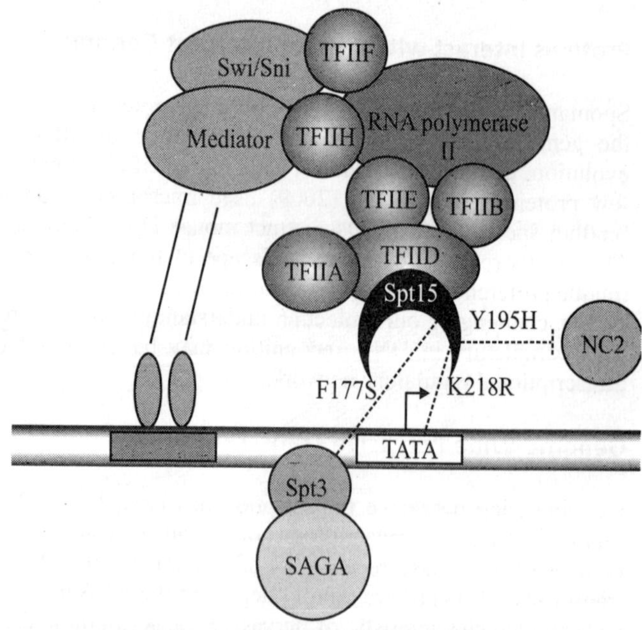

Figure 7.30 Elucidation and validation of a mechanism partially mediated by the Spt3-SAGA complex

Transcription Associated Mutagenesis

Highly activated transcription is associated with eukaryotic genome instability, resulting in increased rate of mitotic recombination and mutagenesis. The association between high transcription and genome instability is probably due to a variety of factors including an enhanced accumulation of DNA damage, transcription-associated supercoiling, collision between replication forks and the transcription machinery, and the persistence of RNA-DNA hybrids. In the case of transcription-associated mutagenesis, it has been previously shown that there is a direct relationship between the level of transcription and the mutation rate in budding yeast, and that the molecular nature of mutation is affected by highly activated transcription. Kim and Jinks-Robertson (2009) show that accumulation of apurinic/apyrimidinic sites is greatly enhanced in highly transcribed yeast DNA. They further demonstrate that most apurinic/pyrimidinic sites in highly transcribed DNA are derived from the removal of uracil, the presence of which is linked to direct incorporation of dUTP in place of dTTP. These results show an unexpected relationship between transcription and the fidelity of DNA synthesis, and raise intriguing cell biological issues with regard to nucleotide pool compartmentalization. Figure 7.31 presents a model for complex interaction at the 6A hotspot (Kim and Jinks-Robertson 2009). Thick lines and large letters correspond to newly synthesized DNA. **0** denotes AP site.

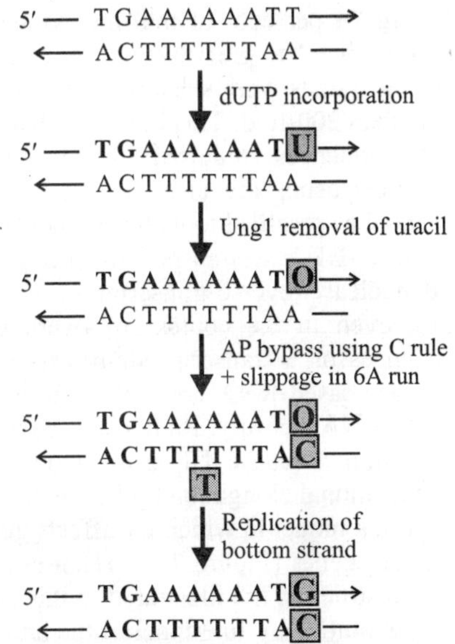

Figure 7.31 Model for complex interaction at the 6A hotspot

Highly Unstable RNAs

The bulk of eukaryotic genomes is transcribed. Transcriptome maps are frequently updated, but low-abundant transcripts probably go unnoticed. To eliminate RNA degradation, Preker et al. (2008) depleted the exonucleolytic RNA exosome from human cells and then subjected the RNA to tiling microarray analysis. This revealed a class of short, polyadenylated and highly unstable RNAs. These promoter upstream transcripts (PROMPTs) are produced ~0.5 to 2.5 kb upstream of active transcription start sites. PROMPT transcription occurs in both sense and antisense directions with respect to the downstream gene. In addition, it requires the presence of the gene promoter and is positively correlated with gene activity. PROMPT transcription is a common characteristic of RNAPII transcribed genes with a possible regulatory potential.

TRANSCRIPTION OF LINE-1 ELEMENTS

LINE-1 (L1) elements are retrotransposons that comprise large fractions of mammalian genomes. L1 elements account for about 17 per cent of the human DNA. Transcription through L1 open reading frames is inefficient owing to an elongation defect, inhibiting the robust expression of L1 RNA and proteins, the substrate and enzyme(s) for retrotransposition. This elongation defect probably controls

L1 transposition frequency in mammalian cells. Han and Boeke (2004) report bypassing this transcriptional defect by synthesizing the open reading frames of L1 from synthetic oligonucleotides, altering 24 per cent of the nucleic acid sequence without changing the amino acid sequence. Such resynthesis led to greatly enhanced steady-state L1 RNA and protein levels. When these synthetic open reading frames were substituted for the wild-type open reading frames, transposition levels increased more than 200-fold. This indicates that there are probably no large, rigidly conserved *cis*-acting nucleic acid sequences required for retrotransposition within L1 coding regions. These synthetic retrotransposons are also the most highly active L1 elements known so far. These regions have potential as practical tools for manipulating mammalian genomes.

The L1 retrotransposon encodes two proteins, open reading frame (ORF)1 and ORF2 endonuclease/reverse transcriptase. L1 RNA and ORF2 proteins are difficult to detect in mammalian cells, even in the context of overexpression systems. Han et al. (2004) show that inserting L1 sequences on a transcript significantly decreases RNA expression and therefore protein expression. This decreased RNA concentration does not result from major effects on the transcription initiation rate or RNA stability. Rather, the poor RNA expression is primarily due to inadequate transcriptional elongation. Because L1 is an abundant and broadly distributed mobile element, the inhibition of transcriptional elongation by L1 might profoundly affect expression of endogenous human genes. They propose a model in which L1 affects gene expression genome-wide by acting as a 'molecular rheostat' of target genes (Figure 7.32) (Han et al. 2004). Effects on transcription and on messenger RNA and protein structure are illustrated in this model. Data are consistent with the hypothesis that L1 can serve as an evolutionary fine-tuner of the human transcription.

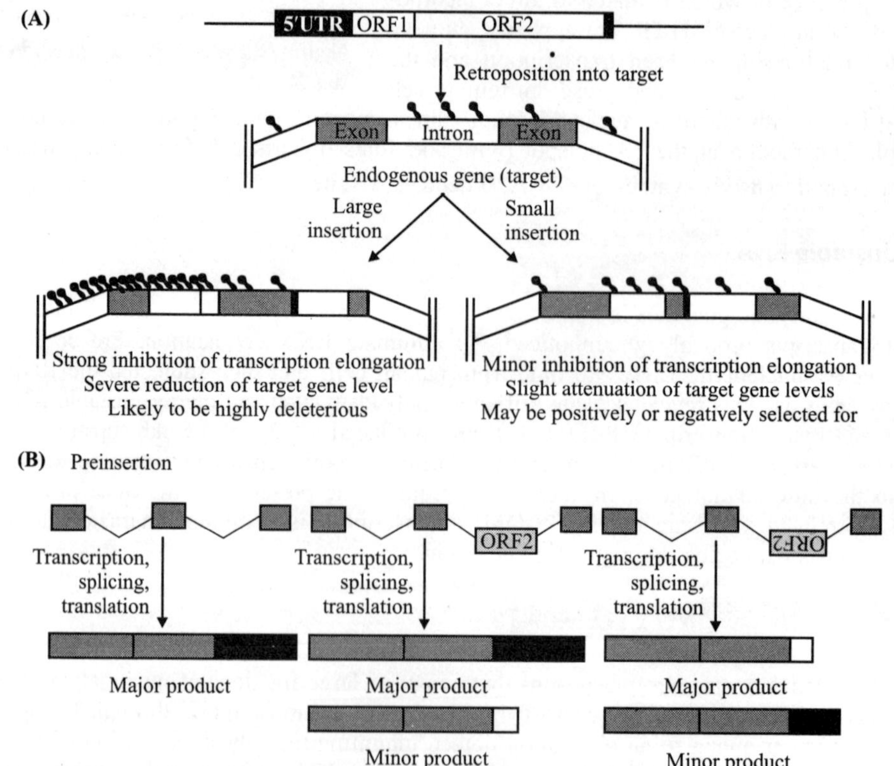

Figure 7.32 Model for L1-mediated modulation of gene expression/structure

TRANSCRIPTION OF CENTROMERIC REPEATS

Heterochromatin in eukaryotic genomes regulates diverse chromosomal processes including transcriptional silencing. However, in fission yeast RNAPII transcription of centromeric repeats is essential for RNA interference-mediated heterochromatin assembly. Chen et al. (2008) studied heterochromatin dynamics during the cycle and its effect on RNAPII transcription. They describe brief period during the S phase of the cell cycle in which RNAPII preferentially transcribes centromeric repeats. This period is enforced by heterochromatin, which restricts RNAPII accessibility at centromeric repeats for most of the cell cycle. RNAPII transcription during S phase is linked to the loading of RNA interference and heterochromatin factors such as the Ago1 subunit of the RITS complex and the Clr4 methyltransferse complex subunit Rik1. Moreover, Set2, an RNAPII-associated methyltransferase that methylates histone H3 lysine 36 at repeat loci during S phase acts in a pathway parallel to promote heterochromatin assembly. Phosphorylation of histone H3 serine 10 alters heterochromatin during mitosis, correlating with the recruitment of condensin that affects silencing of centromeric repeats. They suggest at least two distinct modes of heterochromatin targeting to centromeric repeats, whereby RNAPII transcription of repeats and chromodomain proteins bound to methylated histone H3 lysine 9 mediate recruitment of silencing factors. Together, these processes probably facilitate heterochromatin maintenance through successive cell divisions. A model of heterochromatin dynamics during the cell cycle has been shown in Figure 7.33 (Chen et al. 2008).

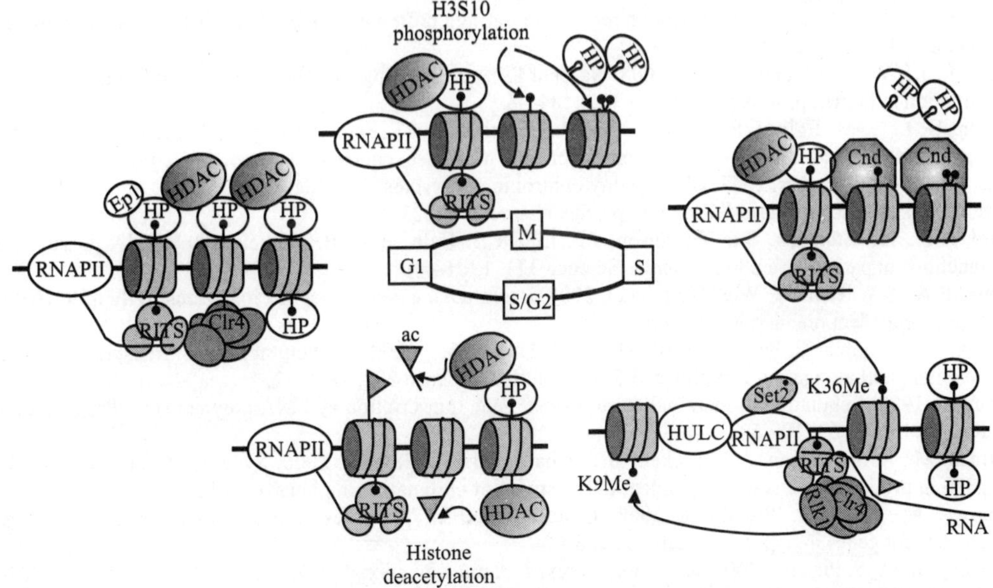

Figure 7.33 Heterochromatin assembly during S phase requires RNA polymerase II-associated activities

During G2, heterochromatin proteins (HPs) Swi6 and Chp2 bound to H3K9me not only recruit silencing factors (HDACs) but also anti-silencing factors (Epe1). H3S10ph coincides with the decrease in Swi6 levels and recruitment of condensing (Cnd). HULC might facilitate RNAPII transcription during S phase. RNAPII targets H3K36me by Set2 but might also recruit Clr4, presumably through Rik1, to methylate H3K9, and RIRS via Ago1 for siRNA production. H3K36me and H3K9me directly or indirectly (via HPs) target HDACs to restore G2 heterochromatin.

REFERENCES

Affymetrix/Cold Spring Harbor Laboratory ENCODE Transcriptome Project. 2009. Post-transcriptional processing generates a diversity of 5'-modified long and short RNAs. Nature 457: 1028-32.

Alper, H., J. Moxley, E. Nevoigt, G.R. Fink, and G. Stephanopoulos. 2006. Engineering yeast transcriptional machinery for improved ethanol tolerance and production. Science 314: 1565-68.

Badis, G., M.F. Berger, A.A. Philippakis, et al. 2009. Diversity and complexity in DNA recognition by transcription factors. Science 324: 1720-3.

Buratowski, S., S. Hahn, L. Guarente, and P.A. Sharp. 1989. Five intermediate complexes in transcription initiation by RNA polymerase II. Cell 56: 549-61.

Buratowski, S., S. Hahn, P.A. Sharp, and L. Guarente. 1988. Function of a yeast TATA element-binding protein in a mammalian transcription system. Nature 334: 37-42.

Bushnell, D.A., K.D. Westover, R.E. Davis, and R.D. Korenberg. 2004. Structural basis of transcription: an RNA polymerase II-TFIIB cocrystal at 4.5 Angstoms. Science 303: 983-8.

Cavallini, B., J. Huet, J.L. Plassat, A. Sentenac, J.-M. Egly, and P. Chambon. 1988. A yeast activity can substitute for the HeLa TATA box factor. Nature 334: 77-80.

Chen, E.S., K. Zhang, E. Nicolas, H.P. Cam, M. Zofall, and S.I.S. Grewal. 2008. Cell cycle control of centromeric repeat transcription and heterochromatin assembly. Nature 451: 734-7.

Chen, H.-T., and S. Hahn, 2004. Mapping the location of TFIIB within the RNA polymerase II transcription preinitiation complex: A model for the structure of PIC. Cell 119: 169-80.

Clemente-Blanco, A., M. Mayán-Santos, D.A. Schneider, et al. 2009. Cdc14 inhibits transcription by RNA polymerase I during anaphase. Nature 458: 219-22.

Core, L.J., and J.T. Lis. 2008. Transcription regulation through promoter-proximal pausing of RNA polymerase II. Science 319: 1791-2.

Core, L.J., J.J. Waterfall, and J.T. Lis. 2008. Nascent RNA sequencing reveals widespread pausing and divergent initiation at human promoters. Science 322: 1845-8.

Davison, B. L., J.-M. Egly, E.R. Mulvhill, and P. Chambon. 1983. Formation of stable preinitiation complexes between eukaryotic class B transcription factors and promoter sequences. Nature 301: 680-6.

Doma, M.K., and R. Parker, 2007. RNA quality control in eukaryotes. Cell 131: 660-8.

Dover, G.A. 1988. rRNA world falling into pieces. Nature 336: 623-4.

Erhard, K.F., J.L. Stonaker, S.E. Parkinson, J.P. Lim, C.J. Hale, and J.B. Hollick. 2009. RNA polymerase IV functions in paramutation in *Zea mays*. Science 323: 1201-4.

Galburt, E.A., S.W. Grill, A. Wiedmann, et al. 2007. Backtracking determines the force sensitivity of RNAPII in a factor-dependent manner. Nature 446: 820-4.

Gnatt, A. L., P. Cramer, J. Fu, D.A. Bushnell, and R.D. Kornberg. 2001. Structural basis of transcription: an RNA polymerase II elongation complex at 3.3 Å resolution. Science 292: 1876-82.

Grummt, I. 1999. Regulation of mammalian ribosomal gene transcription by RNA polymerase I. Prog. Nucl. Acid Res. Mol. Biol. 62: 109-54.

Guermah, M., V.B. Palahan, A.J. Tackett, B.T. Chait, and R.G. Roeder. 2006. Synergistic functions of SII and p300 in productive activator-dependent transcription of chromatin templates. Cell 125: 275-86.

Gullerova, M., and N.J. Proudfoot. 2008. Cohesin complex promotes transcriptional termination between convergent genes in *S. pombe*. Cell 132: 983-95.

Haag, J.R., and C.S. Pikaard. 2007. RNA polymerase I: A multifunctional molecular machine. Cell 131: 1224-5.

Han, J.S., and J.D. Boeke. 2004. A highly active synthetic mammalian retrotransposon. Nature 429: 314-318.

Han, J.S., S.T. Szak, and J.D. Boeke. 2004. Transcriptional disruption by the L1 retrotransposon and implications for mammalian transcriptomes. Nature 429: 268-74.

Herr, A.J., M.B. Jensen, T. Dalmay, and D.C. Baulcombe. 2005. RNA polymerase IV directs silencing of endogenous DNA. Science 308: 118-20.

Hirota, K., T. Miyoshi, K. Kugou, C.S. Hoffman, T. Shibata, and K. Ohta. 2008. Stepwise chromatin remodeling by a cascade of transcription initiation on non-coding RNAs. Nature 456: 130-4.

Horikoshi, M., C.K. Wang, H. Fuji, J.A. Cromlish, P.A. Weil, and R.G. Roeder. 1989. Purification of a yeast TATA box-binding protein that exhibits human transcription factor IID activity. Proc. Natl. Acad. Sci. USA 86: 4843-7.

Huang, Y., and R.J. Maraia. 2001. Comparison of the RNA polymerase III transcription machinery in *Schizosaccharomyces pombe*, *Saccharomyces cerevisiae* and human. Nucl. Acids Res. 29: 2675-90.

Kadonaga, J.T. 2004. Regulation of RNA polymerase II transcription by sequence-specific DNA binding factors. Cell 116: 247-257.

Kao, C.C., P.M. Lieberman, M.C. Schmidt, Q. Zho, R. Pei, and A.J. Berk. 1990. Cloning of a transcriptionally active human TATA binding factor. Nature 248: 1646-50.

Kettenberger, H., K.-J. Armache, and P. Cramer. 2003. Architecture of the RNA polymerase II-TFIIS complex and implications for mRNA cleavage. Cell 114: 347-57.

Kim, M., N.J. Krogan, L. Vasiljeva, et al. 2004. The yeast Rat1 exonuclease promotes transcription termination by RNA polymerase II. Nature 432: 517-22.

Kim, N., and S. Jinks-Robertson. 2009. dUTP incorporation into genomic DNA is linked to transcription in yeast. Nature 459: 1150-3.

Kobayashi, T., and A.R.D. Ganley. 2005. Recombination regulation by transcription-induced cohesion dissociation in rDNA repeats. Science 309: 1581-4.

Kravchenko, J.E., I.B. Rogozin, E.V. Koonin, and P.M. Chumakov. 2005. Transcription of mammalian messenger RNAs by a nuclear RNA polymerase of mitochondrial origin. Nature 436: 735-9.

Krishnakumar, R., M.J. Gamble, K.M. Frizzel, J.G. Berrocal, M. Kininis, and W.L. Kraus. 2008. Reciprocal binding of PARP-1 and histone H1 at promoters transcribing complexes specifies transcriptional outcomes. Science 319: 819-21.

Kuhn, C., S.R. Geiger, S. Baumil, et al. 2007. Functional architecture of RNA polymerase I. Cell 131: 1260-72.

Lee, Y., M. Kim, J. Han, K.H. Yeom, S. Lee, S.H. Baek, and V.N. Kim. 2004. MicroRNA genes are transcribed by RNA polymerase II. EMBO J. 23: 4051-60.

Lehmann, E., F. Brueckner, and P. Cramer. 2007. Molecular basis of RNA-dependent RNA polymerase II activity. Nature 450: 445-9.

Lengrome, A., Y. Katon, S. Mori, et al. 2004. Cohesion relocation from sites of chromosomal loading to places of convergent transcription. Nature 430: 573-8.

Li, B., M. Carey, and J.L. Workman. 2007. The role of chromatin during transcription. Cell 128: 707-19.

Lusting, A.J., S. Kurtz, and D. Shore. 1990. Involvement of the silencers and UAS binding protein RAP1 in regulation of telomere length. Science 250: 549-53.

Margaritis, T., and F.C. Holstege. 2008. Poised RNA polymerase II gives pause for thought. Cell 133: 581-4.

Marmorstein, R. 2003. Transcription initiation at its most basic level. Cell 115: 370-2.

McKnight, C.J., and P.S. Kim. 1990. Sequence-specific DNA binding by a short peptide dimer. Science 249: 769-71.

Nardelli, J., T.J. Gibson, C. Vesque, and P. Charnay. 1991. Base sequence discrimination by zinc-finger DNA-binding domains. Nature 349: 175-8.

Neil, H., C. Malabat, Y. d'Aubenton-Carafa, Z. Xu, L.M. Steinmetz, and A. Jacquier. 2009. Widespread bidirectional promoters are the major source of cryptic transcripts in yeast. Nature 457: 1038-42.

Nickels, B.E., and A. Hochschild. 2004. Regulation of RNA polymerase (RNAP) through secondary channel. Cell 118: 281-4.

Parker, C. S., and J. Topol. 1984. A *Drosophila* RNA polymerase II transcription factor contains a promoter-specific DNA-binding activity. Cell 36: 357-69.

Peric-Hupkes, D., and B. van Steensel, 2008. Linking cohesion to gene regulation. Cell 132: 925-8.

Peterson, M.G., N. Tanese, B.F. Pugh, and R. Tjian. 1990. Functional domains and upstream activation properties of cloned human TATA binding protein. Science 248: 1625-30.

Prather, D.M., E. Larschan, and F. Winston. 2004. Evidence that the elongation factor TFIIS plays a role in transcription initiation a GAL1 in *Saccharomyces cereviseae*. Molecul. Cellu. Biol. 25: 2650-9.

Preker, P., J. Nielsen, S. Kammler, et al. 2008. RNA exosome depletion reveals transcription upstream of active human promoters. Science 322: 1851-4.

Rawn, J.D. 1989. *Biochemistry*. Burlington, NC: Neil Patterson Publishers.

Ross, K.E., and O. Cohen-Fix. 2004. Cohesins slip slides away. Nature 430: 520-1.

Sauer, R.T. 1990. Transcriptional control: scissors and helical forks. Nature 347: 5114-5.

Sawadogo, M., and R.G. Roeder. 1985. Interaction of a gene-specific transcription factor with the adenovirus major late promoter upstream of the TATA box region. Cell 43: 165-75.

Scafe, C., D. Chao, J. Lopes, J.P. Hirsch, S. Henry, and R.A. Young. 1990. RNA polymerase II C-terminal repeat influences response to transcriptional enhancer signals. Nature 347: 491-4.

Seila, A.C., J.M. Calabrese, S.S. Levine, et al. 2008. Divergent transcription from active promoters. Science 322: 1849-51.

Shuman, J.D., C.R. Vinson, and S.L. McKnight, 1990. Evidence of changes in protease sensitivity and subunit exchange rate on DNA binding by C/EBP. Science 249: 771-4.

Somesh, B.P., J. Reid, W.-F. Liu, et al. 2005. Multiple mechanisms confining RNA polymerase II ubiquitylation to polymerase undergoing transcriptional arrest. Cell 121: 913-23.

Somesh, B.P., S. Sigurdsson, H. Saeki, H. Erdjument-Bromage, P. Pempst, and J.Q. Svejstrup. 2007. Communication between distant sites in RNA polymerase II through ubiquitylation factors and the polymerase CTD. Cell 129: 57-68.

Stringer, K.F., C.J. Ingles, and J. Greenblatt. 1990. Direct and selective binding of an acidic transcriptional activation domain to the TATA-box factor TFIID. Nature 345: 783-6.

Suzuki, M. 1990. The heptad repeat in the largest subunit of RNA polymerase II binds by intercalating into DNA. Nature 344: 562-5.

Telxeira, A., A. Tahiri-Alaoui, S. West, et al. 2004. Autocatalytic RNA cleavage in the human b-globin pre-mRNA promotes transcription termination. Nature 432: 526-30.

Tollervey, D. 2004. Termination by torpedo. Nature 432: 456-7

Van Dyke, M.W., R.G. Roeder, and M. Sawadogo. 1988. Physical analysis of transcription preinitiation complex assembly on a class II gene promoter. Science 241: 1335-8.

Van Mullem, V.V., M. Wery, M. Werner, J. Vandenhaute, and P. Thuriaux. 2002. The Rpb9 subunit of RNA polymerase II binds transcription factor TFIIE and interferes with the SAGA and elongator histone acetyl transferases. J. Biol. Chem. 227: 10220-5.

Volfson, D., J. Marciniak, W.J. Blake, L.S. Tsimring, and J. Hasty. 2006. Origins of extrinsic variability in eukaryotic gene expression. Nature 439: 861-4.

Walker, A.K., Y. Shi, and T.K. Blackwell. 2004. An extensive requirement for transcription factor IID-specific TAF-1 in *Caenorhabditis elegans* embryonic transcription. J. Biol. Chem. 279: 15339-47.

Wang, D., D.A. Bushnell, X. Huang, K.D. Westover, M. Levitt, and R.D. Korenberg. 2009. Structural basis of transcription: backtracked RNA polymerase II at 3.4 Angstrom resolution. Science 324: 1203-6.

West, S., N. Gromak, and N.J. Proudtoot. 2004. Human 5'→3' exonuclease Xm2 promotes transcription termination at co-transcriptional cleavage sites. Nature 432: 522-5.

Westover, K.D., D.A. Bushnell, and R.D. Korenberg. 2004a. Structural basis of transcription: nucleotide selection by rotation in the RNA polymerase II active center. Cell 119: 481-9.

Westover, K.D., D.A. Bushnell, and R.D. Korenberg. 2004b. Structural basis of transcription: separation of RNA from DNA by RNA polymerase II. Science 303: 1014-6.

Wierzbicki, A.T., T.S. Ream, J.R. Haag, and C.S. Pikaard. (2009). RNA Polymerase V transcription guides ARGONAUTE4 to chromatin. Nat. Genet. 41: 630-4.

Wilhelm, B.T., S. Marguerat, S. Watt, et al. 2008. Dynamic repertoire of a eukaryotic transcriptome surveyed at single-nucleotide resolution. Nature 453: 1239-43.

Willis, I.M. 1993. RNA polymerase III. Genes, factors and transcriptional specificity. Eur. J. Biochem. 212: 1-11.

Wyers, F., M. Rougenmaille, G. Badis, et al. 2005. Cryptic pol II transcripts are degraded by a nuclear quality control pathway involving new poly(A) polymerases. Cell 121: 725-37.

Xu, L., M. Thali, and W. Schaffner. 1992. Upstream box/TATA box order is the major determinant of the direction of transcription. Nucl. Acids Res. 19: 6699-704.

Xu, Z., W. Wei, J. Gagneur, et al. 2009. Bidirectional promoters generate pervasive transcription in yeast. Nature 457: 1033-7.

8

Intricacies of Eukaryotic Transcription

The 2006 Nobel Prize in Chemistry was awarded to R. Kornberg for elucidating the molecular basis of eukaryotic transcription ((Bushnell et al. 2004; Cramer et al. 2001; Gnatt et al. 2001; Kelleher et al. 1990). New structures of RNA polymerase II (RNAPII) revealed a likely key to transcription. The trigger loop swings beneath a correct nucleoside triphosphate (NTP) in the nucleotide addition site, closing off the active center and forming an extensive network of interactions with the NTP base, sugar, phosphates, and additional RNAPII residues. A histidine side chain in the trigger loop, precisely positioned by these interactions, may literally "trigger" phosphodiester bond formation. Recognition and catalysis are coupled, ensuring the fidelity of transcription.

MAPPING TRANSCRIBED REGIONS

The identification of untranslated regions, introns, and coding regions within an organism is quite challenging. Nagalakshmi et al. (2008) developed a quantitative sequencing-based method called RNA-Seq for mapping transcribed regions, in which complementary DNA fragments are subjected to high throughput sequencing and mapped to the genome. They applied RNA-Seq to generate a high-resolution transcriptome map of the yeast genome and demonstrated that most (74.5%) of the non-repetitive sequence of the yeast genome is transcribed. They confirmed many known and predicted introns and demonstrated that others are not actively used. Alternative initiation codons and upstream open reading frames also were identified for many yeast genes. They also found unexpected 3'-end heterogeneity and the presence of many overlapping genes. These results indicate that the yeast transcriptome is more complex than previously appreciated.

TRANSCRIPTION COEXPRESSION OF INTERACTING GENES

Transcription coexpression of interacting gene products is required for complex molecular processes; however, the function and evolution of *cis*-regulatory elements that orchestrate coexpression remain largely unexplored. Brown et al. (2007) mutated 19 regulatory elements that derive coexpression of *Ciona* muscle genes and obtained quantitative estimates of *cis*-regulatory activity of the 77 motifs that comprise these elements. They found that individual motif activity ranges broadly within and among elements, and different instantiations of the same motif type. The activity of the orthologous motifs is strongly constrained, although motif arrangement, type, and activity vary greatly among the elements of different co-regulated genes. Thus, syntactical rules governing this regulatory function are flexible but become highly constrained evolutionarily once they are established in a particular element.

TRANSCRIPTIONAL FORESTS

This study describes comprehensive polling of transcription start and termination sites and analysis of previously unidentified full-length complementary DNAs derived from the mouse genome. The FANTOM Consortium and RIKEN Genome Exploration Research Group and Genome Science Group (Genome Network Project Core Group) (2005) identify 5' and 3' boundaries of 181,047 transcripts with extensive variation in transcripts arising from alternative promoter usage, splicing and polyadenylation. There are 16,247 new mouse protein-coding transcripts, including 5,154 encoding previously unidentified proteins. Genomic mapping of the transcriptome reveals transcriptional forests, with overlapping transcription on both strands, separated by deserts in which few transcripts are observed. The data provide a comprehensive platform for the comparative analysis of mammalian transcriptional regulation in differentiation and development.

TRANSCRIPTIONAL PROCESSIVITY

The affinity of RNA polymerase for template DNA is referred to as transcriptional processivity. The interactions that occur during elongation process come under this phenomenon (Tjian 1995). Anatomy of transcription apparatus is shown in Figure 8.1. Repressors bind to selected sets of genes at sites called silencers. They interfere with functioning of activators and slow down transcription. Three types of factors play role in transcriptional processivity. Basal factors respond to activators and position RNA polymerase at the start of the protein-encoding region of a gene and send the enzyme on its way. Coactivators are the adaptor molecules that integrate signals from activators and repressors and relay the result to basal factors. Activators bind to genes at sites called enhancers. They help to determine which genes will be switched on, and they speed up rate of transcription.

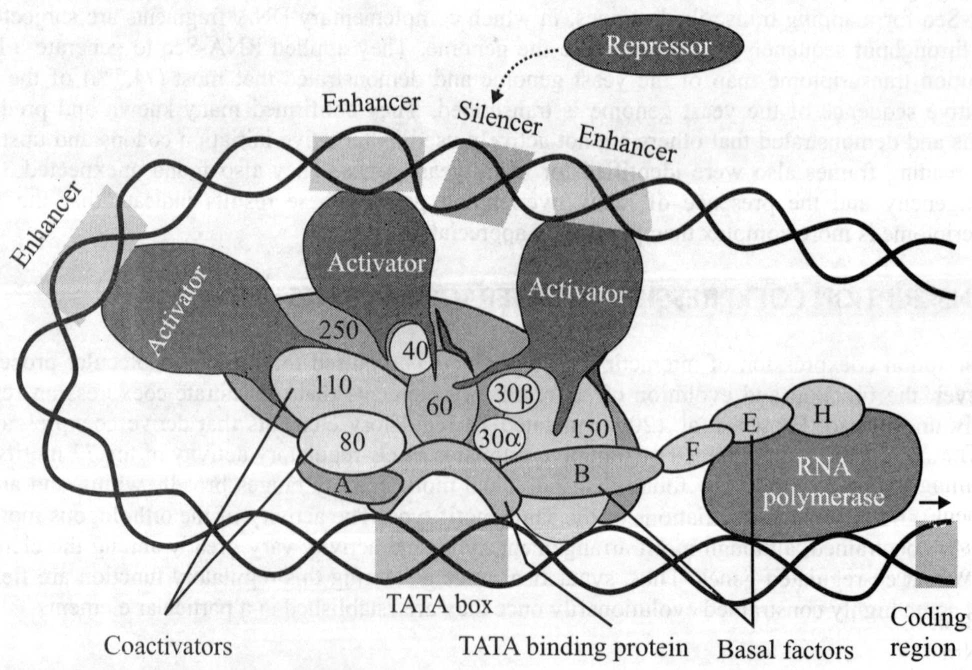

Figure 8.1 Anatomy of transcription apparatus

Once the primer reaches 10-nucleotide length, the RNA holoenzyme undergoes a transition from the initiation to the elongation mode and the transcription complex moves away from the promoter along the DNA template. Transition to elongation stage of transcription is associated with a conformational change in holoenzyme that releases the σ factor. An accessory protein NusA binds to the core enzyme and prevents σ factor from reassociating during elongation. NusA also interacts with other accessory proteins and plays a role in transcription. NusA-core enzyme elongates the RNA chain. In advancing elongation complex, RNA polymerase combines two contradictory biochemical features: (a) exceptional stability for dissociation and (b) the ability to easily translocate along DNA. Thus, elongating RNA polymerase simultaneously behaves as a strong DNA binding protein and as a protein with no affinity for particular DNA sites. The combination of these features ensures processivity of RNA polymerase.

TRANSCRIPTIONAL ELEMENTS

Transcriptional elements play crucial role in gene expression. Transcriptional elements of heat shock protein *hsp70* gene have been studied in detail by Cohen and Meselson (1988). Maximal transcription of heat shock protein *hsp70* gene requires two copies of a 14-bp heat shock element (HSE) with consensus sequence CnnGAAnnTTCnnG. In response to heat stock, HSEs bind heat shock transcription factor (HSTF). Protein first binds to strong site which is proximal to *hsp70* transcription initiation site (–62 to –49) then protein binds successively and cooperatively to adjacent weak site (–85 to –72). The strong site matches the consensus HSE at all 8 positions while weak site matches the consensus HSE only at 6. Cooperative binding of HSTF to two HSEs led these authors to investigate the possibility that there is a requirement for specific rotational alignment of the two elements. Map of *hsp70-cat* transfection vector is shown in Figure 8.2.

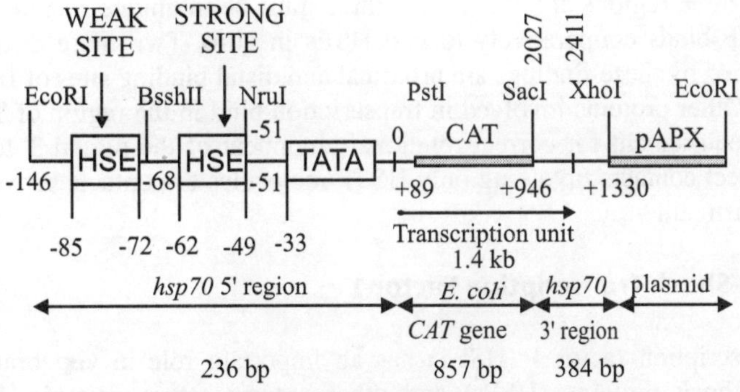

Figure 8.2 Map of the parental *hsp70-CAT* transvection vecor (not to the scale)

The *hsp70-Cat* gene is cloned into the XhoI-EcoRI site of plasmid pAX. Transcription vector containing *hsp70-λ* gene is identical to *hsp70-Cat* vector except that the former contains 1051 PstI-SacI fragment of phage λ in place of PstI-SacI CAT fragment. BsshII cleaves *hsp70-Cat* gene immediately downstream of –68 and Nru-I cleaves this gene immediately downstream of -51. Sequences of varying sizes were inserted at BsshII site and Nru-I site. Sizes (in bp) of sequences inserted are as follows:

Bssh II - 4, 10, 14, 20, 24, 26, 63, 77, 295, 296, 300, 302, 304
Nru-I - 6, 10, 16, 20, 26, 30, 67, 73, 291, 292, 298, 300, 302

All insertions less than 63 bp were verified by sequencing the final construct transcription of a cloned *hsp70* gene in transfected heat shocked *Drosophila* tissue culture cells and were determined as (a) a function of spacing between the two HSE's and (b) a function of spacing between proximal HSE and sequences downstream. The purpose of the insertion of varying sizes of DNA fragments was to increase the intervals at (a) and (b). Insertions contributed either a near integral or non-integral multiple of 10.5 bp, i.e., one turn of B form of the DNA helix. All short insertions that differed from an integral multiple of 10.5 by 1 bp or less (10. 20, 63, 73) reduced transcription by no more than a factor of 2. All insertions that differed from an integral multiple of 10.5 by at least 3 bp (4, 6, 14, 16, 24, 26, 67, 77) reduced transcription 4- to 15-fold. Each of the ten large insertions (291-304 bp) reduced transcription to about 50 per cent of the wild-type level, with no indication of periodicity. The conclusions from these studies were: The striking periodicity of 10.5 bp that is seen for short insertions indicates that maximal *hsp70* transcription requires specific, rotational alignment of DNA-protein interaction sites flanking the Bssh II and Nru I sites. The lack of periodicity for long insertions suggests that spacings of 300 bp are sufficient to achieve correct rotational alignment by DNA twisting. The periodic effect of insertions at the Bssh II and NruI sites could reflect a requirement for unique rotational alignment of 3 binding regions — first, upstream of the BsshII site: second, downstream of the NruI site; and third, between the two restriction sites.

Another possibility also was thought to exist. Only the two outside binding regions required such alignment. This, if true, an *hsp70* gene with half integral insertion at the BsshII site and another half integral insertion at the NruI site should be transcribed with approximately wild-type efficiency. To test this possibility, northern blot analysis of transcripts of *hsp70-Cat* genes was done. The blot was hybridized with RNA probes for transcripts of *hsp70-Cat* and hsp70-λ.

The numbers of base pairs inserted at the BsshI and NruI sites of *hsp70-Cat* are shown in Figure 8.3 (Cohen and Meselson 1988). Constructs 4/6 are transcribed only about 1/20th as efficiently as wild-type. Conclusion from this experiment is that for maximal transcription, protein must interact with DNA in each of the three regions and that all the three must be in unique rotational alignment. It was concluded that HSTF binds cooperatively to two HSEs in vitro. Two more distal regions of DNA-protein binding implied by these findings are proximal and distal binding sites of HSTF. HSTF binding may extend to –40. Other proteins involved in transcription bind in the region of TATA motif and the initiation site. The requirement for correct rotational alignment of the region 3' to the proximal HSE could, therefore, reflect contacts involving only HSTF molecules or contacts between HSTF and other proteins bound downstream.

Activation of Heat-Shock Transcription Factor 1

The heat-shock transcription factor 1 (HSF1) has an important role in vertebrates by inducing the expression of heat-shock proteins (HSPs) and other cytoprotective proteins. HSF1 is present in unstressed cells in an inactive monomeric form and becomes activated by heat and other stress stimuli. HSF1 activation involves trimerization and acquisition of a site-specific DNA-binding activity, which is negatively regulated by interaction with certain HSPs. Shamovsky et al. (2006) show that HSF1 activated by heat shock is an active process that is mediated by a ribonucleoprotein complex containing translation elongation factor eEF1A and a previously unknown non-coding RNA that has been termed HSR1 (heat shock RNA-1). HSR1 is constitutively expressed in human and rodent cells and its homologs are functionally interchangeable. Both HSR1 and eEF1A are required for HSF1 activation in vitro; antisense oligonucleotides or short interfering RNA (siRNA) against HSR1 impair the heat-shock response in vivo, rendering cells thermosensitive. The central role of HSR1 during heat shock implies that targeting this RNA could serve as a new therapeutic model for cancer, inflammation and other conditions associated with HSF1 deregulation.

BsshII insertions (bp)	NruI insertions (bp)
4 CGCG	6 GGTACC
10 CGCGGGTACC	10 GGTACGTACC
14 CGCGGGTACGTACC	10 CGGAATTCCG
14 CGCGCGGAATTCCG	10 CGGGATCCCG
20 CGCGGGTACGGTACCGTACC	16 GGTACGGTACCGTACC
24 CGCGGGTACCGGAATTCCGGTACC	20 GGTACCGGAATTCCGGTACC
26 CGCGGGTACCCGGAATTCCGGGTACC	20 GGTACCGGGATCCCGGTACC
63 CGCGGGTAC[a]GTACC	26 GGTACGGTACGTACGGTACCCGTACC
77 CGCGGGTAC[b]GTACC	30 GGTACCGGAATTGGTACCAATTCCGGTACC
295 CGCGGGTAC[*]GTACC	67 GGTACC[c]GTACC
296 CGCGGGTAC[*1]GTACC	73 GGTACG[b]GTACC
300 CGCGGGTACGTAC[*1]GTACC	291 GGTAC[*]GTACC
302 CGCGGGTAC[*2]GTACC	292 GGTAC[*1]GTACC
304 CGCGGGTAC[*3]GTACC	298 GGTAC[*2]GTACC
	300 GGTAC[*3]GTACC
	302 GGTAC[*4]GTACC

a, b, c, * represent the 49, 63, 57 and pBR322 *Alu* fragments, respectively; *1, *2, *3 and *4 represent the 281 bp pBR322 *Alu* fragment containing the insertions G, GGGTACC, GGCTCGAGC, and GCGGAATTCCG, respectively.

Figure 8.3 DNA sequences inserted between the distal and proximal HSEs (BssHI insertions) and at the 3' end of the proximal HSE (NruI insertions)

SYNERGISTIC TRANSCRIPTION

Gene expression is said to be synergistic when combined quantitative effect of two transcriptional activation elements is more than the sum of their individual expression. Thus phenomenon is explained by the following example. GAL4 is a sequence-specific DNA-binding protein that activates transcription in yeast. It has 881 amino acids. GAL4(1-147) protein bears first 147 amino acids which bind to four 17-mer site known as upstream activating sequence of galactose (UAS_G). This part does not activate transcription. Glucocorticoid receptor (GR) is also a sequence-specific DNA-binding protein. It activates transcription of mammalian genes, e.g. in the presence of dexamethasone, GR binds to several glucocorticoid responsive elements (GREs) upstream of the transcriptional start site in the mouse mammary tumor virus (MMTV) promoter and activates transcription. Kakidani and Ptashne (1988) showed that GAL4 and two deletion derivatives bearing one or both activating regions attached to GAL4 (-147) activates the MMTV promoter in mammalian cells, but GAL4 (1-147) did not. A synergistic effect of GR and GAL4 was observed. For this, mammalian cells were co-transfected with two plasmids: effector plasmid and reporter plasmid.

Effector and Reporter Plasmids

Effector plasmids carry a gene that expresses a regulatory protein which in turn regulates the expression of another gene carried in a reporter plasmid. Effector plasmid expresses the mouse dihydrofolate reductase (DHFR) gene driven by adenovirus major late promoter (AdMLP) (Kakidani and Ptashne 1988) (Figure 8.4). This plasmid also contains tripartite leader sequences from AdMLP (dotted box) and the intron sequences from the immunoglobulin heavy chain gene (filled box). The bars shown below denote the GAL4 derived sequences present in each plasmid.

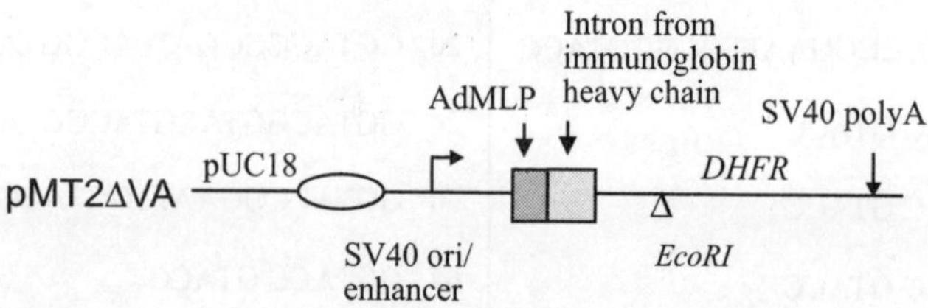

Figure 8.4 Structure of effector plasmid. GAL4-derived sequences present in different effector plasmids are given in Table 8.1

Table 8.1 GAL sequences present in different plasmids

Plasmid	*GAL sequences*	*Regions included*
pAGH	1-881	D + I + II
pAG147	1-147	D
pAG236	1-47, 768-881	D + II
pAG242	1-238, 768-881	D + I + II

D, DNA-binding region; I and II, the two activation regions

DNA fragments encoding GAL4 and its derivatives were inserted into the EcoRI site of the plasmid pMT2 VA (Table 8.1). Reporter plasmid bears a chloramphenicol acetyltransferase (*CAT*) gene fused to the *MMTV* promoter (Figure 8.5A). *CAT* gene is fused downstream of the MMTV-LTR, which includes two GREs and a NF1 binding site. Two additional GR-binding sites lie between the SacI site and TATA, one of which overlaps the NF1 site, the sites alone showed barely detectable activity. Out of the plasmid pGMCS, four reporter plasmids were generated (Figure 8.5B). There are several advantages of using these plasmids. CAT gene produced is easily assayed. For all transcription experiments, Chinese hamster ovary (CHO) cell line was used. These cells synthesized no GR, and so in experiments exploring the effect of this receptor, CHO cells stably transfected with a plasmid that direct constitutive synthesis of receptor. In yeast, transcriptional activation by GAL4 (although not DNA-binding) depends on growth in galactose because galactose dissociates the inhibitor GAL80, which otherwise covers the activating surface(s). Thus in experiments reported here, no GAL80 was present and so galactose was not required for transcription activation by GAL4. Important results of these experiments, depicted in Figure 8.5, are described here.

(A)

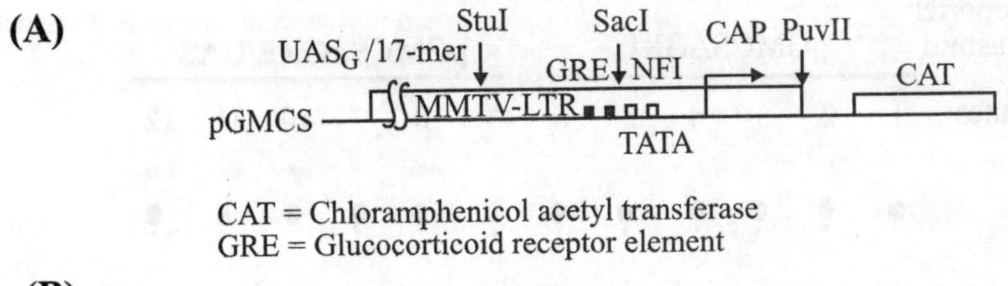

CAT = Chloramphenicol acetyl transferase
GRE = Glucocorticoid receptor element

(B)

Plasmid	Description
pGMCS/17	a single 17-mer contained in 36-bp fragment was insrted into SacI site 104 bp upstream of the transcription site
pGMCS/UAS	an intact UAS$_G$ carrying 365-bp fragment into the same site UAS$_G$
pGMCSΔGRE	GREs have been deleted from pGMCS
pGMCSΔGRE/UAS	GREs have been deleted from pGMCS/UAS

Figure 8.5 (A) Structure of reporter plasmid. (B) Four reporter plasmids containing GAL4 regulatory sequences

Two deletion derivatives, GAL4 (1-147, 768-881) and GAL4 (1-238, 768-881) that activated transcription in yeast were also active in mammalian cells. Cells transfected with GAL4 or one of its active derivatives synthesized CAT at least 20-fold above background (lanes 8, 10, 11). No detectable CAT activity was elicited by GAL4 (1-147) (lane 9) nor by reporter plasmid alone (lane 1) (Figure 8.6). GAL4 derivatives bearing both activating regions directed more CAT activity than did the other two active GAL4 species (compare lanes 8, 10, 11). In other experiments, usually no differences were detected between the two GAL4 deletion derivatives each of which gave rise to slightly more activity than did the wild-type GAL4. In some experiments, a small amount of CAT activity was stimulated by an effector plasmid encoding no GAL4 (lane 12). RNA analysis showed that unlike the GAL4-stimulated activity, this activity did not arise from transcripts initiating near the MMTV transcriptional start site. In the experiment on stimulation of MMTV promoter lacking GRE, RNase protection mapping of MMTV/CAT mRNA was done. The reporter plasmids lacking UASG produced only barely detectable CAT activity in response to GAL4 or any of its derivatives (lanes 2 through 5). GAL4 and its active derivatives elicited transcripts that initiated at the same position as that stimulated by active GR in wild-type MMTV promoter (lanes 6, 8, 9). A reference plasmid bearing the SV40 enhancer synthesizes α-globin mRNA from its own promoter. In Figure 8.7, 135-mark indicates MMTV/CAT transcript protected and 95-mark indicates α-globin mRNA protected. In this figure, M = labeled size marker, an Msp I digest of pBR322. No transcripts initiating at the correct position in the MMTV promoter were elicited by effector plasmid encoding only the DNA-binding portion of GAL4 or no GAL4 (lanes 7, 10). Lanes 1 through 5 showed that reporter plasmids lacking UAS$_G$ synthesized no detectable MMTV/CAT mRNAs, whether or not GAL4 or its derivatives were present. GAL4 (1-147) is stable and binds to the 17-mer.

Galactose 4-Glucocorticoid Receptor Synergism

For demonstrating synergism, Kakidani and Ptashne (1988) inserted reporter plasmid bearing a 17-mer into the intact MMTV promoter where GREs were present. CHO/GR1 cells used here expressed rat GR constitutively. Reporter plasmids used were pGMCS (GRE) and pGMCS/17 (GRE and 17mer).

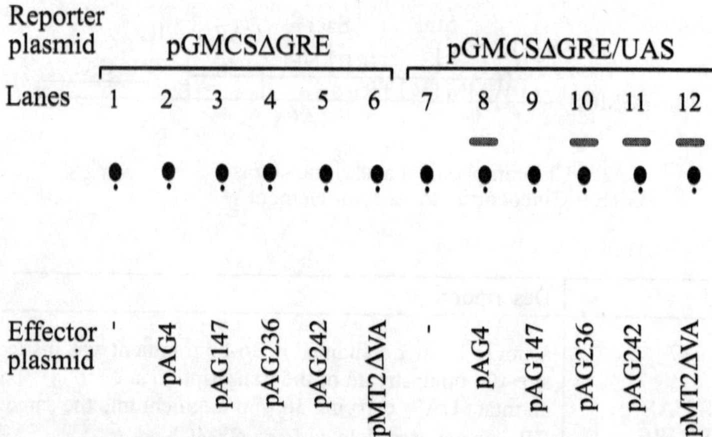

Figure 8.6 Mammalian gene activation by GAL4. The MMTV promoter lacks GRE and contains UAS which is activated by GAL4. CAT activity is measured

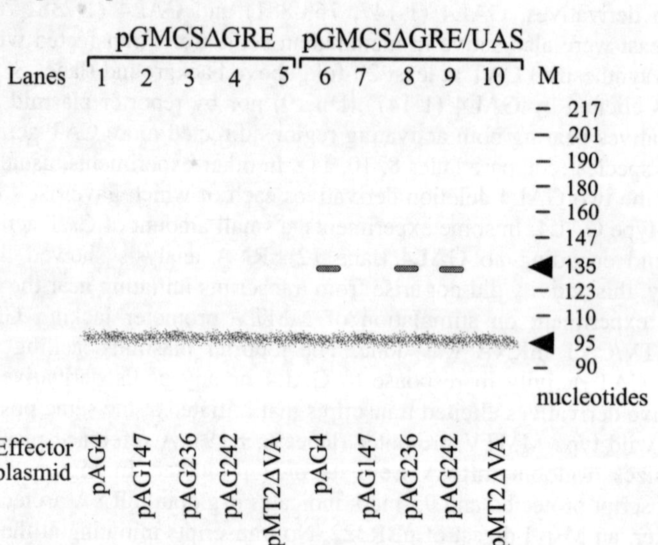

Figure 8.7 Stimulation of an MMTV promoter lacking GREs measured by RNase protection mapping of MMTV/CAT mRNA

Effector plasmids were as indicated in Figure 8.8. In lanes marked "+", cells were treated with dexamethasone. CAT activity in the extracts of transfected cells was analyzed. Insertion of the 17-mer diminished the ability of GREs to activate transcription in response to dexamethasone in the absence of GAL4 (compare lanes 1 and 7). In absence of dexamethasone, GAL4 did not stimulate transcription for this 17-mer (lane 11). In yeast, the 17-mer mediates GAL4-dependent transcriptional activity significantly less efficiently than does the intact UAS_G. So this result is not surprising. The presence of both dexamethasone (and hence active GR) and GAL4 markedly stimulated transcription (lane 8). As a control, it was found that addition of dexamethasone had no effect on the activity of GAL4 if the

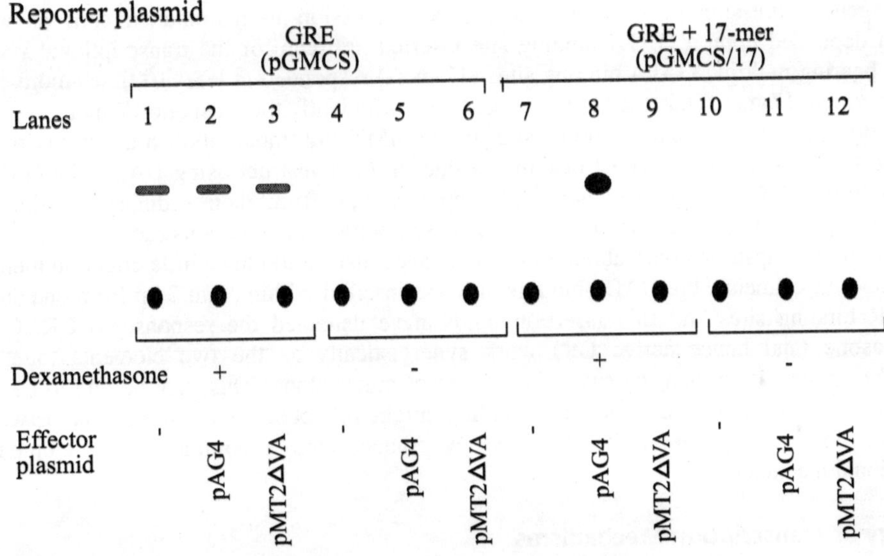

Figure 8.8 Stimulation of an MMTV promoter bearing both GREs and 17-mer sequences

reporter plasmid bore a 17-mer but was deleted for the adjacent GREs. Stimulation of an MMTV promoter bearing both GREs and UAS_G was observed using CHO (GR1) cells (Figure 8.9). Synergy between GAL4 and GR was also observed with a reporter plasmid bearing UAS_G plus GREs. Similar to but more extreme than the case with the 17-mer, insertion of UAS_G abolished the ability of the GREs to mediate stimulation by dexamethasone in the absence of GAL4 (lane 1). GAL4 stimulated expression from this plasmid in the absence of dexamethasone (lane 5); a reset similar to that seen earlier in experiment presented in Figure 8.6. Addition of dexamethasone to cells synthesizing GAL4 increased the CAT activity to a level about 4-fold higher than that seen with GAL4 alone (lane 2). Thus, GAL4 and two deletion derivatives, GAL (1-147, 768-881) and GAL (1-238, 768-881), which

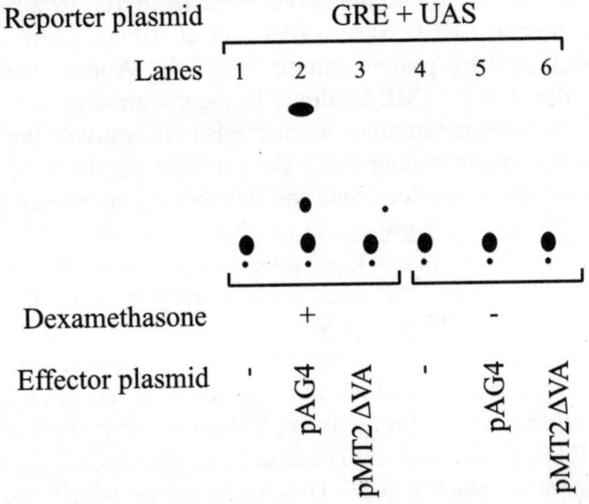

Figure 8.9 Stimulation of an MMTV promoter bearing both GREs and UAS_G. Chinese hamster ovary cells

activated gene expression in yeast activated gene expression in mammalian cells as well. This activation depended upon a GAL4 binding site inserted upstream of the transcriptional start site. A promoter bearing multiple GAL4 binding sites (17-mers) responded at least 10 times more efficiently than did a simple 17-mer. Thus the UAS_G worked more efficiently than a single 17-mer.

Why does insertion of GAL4 binding sites in the MMTV promoter abolish the ability of the latter to respond to GR in the presence of dexamethasone. In the construct using UAS_G, the GREs shown earlier in Figure 8.6 were moved about 360 bp upstream from their ordinary position and this displacement per se might account for their decreased activity. In the construct using 17-mer, these GREs were moved upstream only about 30 bp, a distance that should have little effect on their activity. In both these experiments, the GAL4-binding site was inserted within , and 2 bp from one end of, one of the GR binding sites and this insertion might have damaged the response to GR. GAL4 and dexamethasone (and hence active GR) work synergistically as the two elements together were considerably more active than the sum of activities of either alone. This may be due to co-operative binding of GR and GAL4. Co-operative binding might reflect a direct interaction between these proteins or, more likely, might be mediated by contact with a common, third element of the transcriptional machinery.

Homology of Transcription Mechanisms

TATA-binding protein (TBP) is a classless factor (Sharp 1991) essential for all transcription in eukaryotic cell nuclei and Archaebacteria showing that transcription mechanisms of Archaebacteria and eukaryotes are fundamentally homologous (Rowlands et al. 1994).

SMART TRANSCRIPTION FACTORS

Transcription factors are those proteins that are required for initiation of transcription. But smart transcription factor cAMP response element binding (CREB) protein first gets phosphorylated and then it binds to co-activator CREB binding protein (CBP). After that this complex binds to DNA element known as cAMP-regulated enhancer and this leads to initiation of transcription (D'Arcangelo and Curran 1995). Chromosomal rearrangements and point mutations in chromosome 16p caused Rubinstein-Tyabi syndrome (RTS). These mutations were restricted to a region in chromosome 16p which contained gene for human CREB protein (Petrij et al. 1995). Chirivia et al. (1993) found that CREB was phosphorylated only by protein kinase A (PKA). Amino acids 101-160 within CREB activation domain were sufficient for CBP binding. Tetracarcinoma cells were infected with reporter plasmid pGAL-CAT and various combinations of plasmids to investigate functional properties of CBP. It was concluded that transactivation potential of CBP could be regulated by PKA (Kwok et al. 1994). Petrij et al. (1995) found that RTS translocations and microdeletions were found to range between 130 and 650 kb and RTS break points were present in an area of 150 kb. With the help of fluorescence in situ hybridization, they observed that human CBP gene was present in chromosome 16p-13.3.

Haploinsufficiency, i.e., reduced CBP gene dosage, may be one of the causes of RTS. Greg syndrome also results from haploinsufficiency. Vortkamp and Grzeschik (1991) reported that mutation disturbing normal expression of GL13 gene might have a causative role in Greg syndrome. In RTS patients, point mutations and chromosomal rearrangements removed the CREB binding site of CBP. There will be no translocation which leads to dysregulation of gene expression. Although CBP is the most promising gene for RTS, its identity still needs to be confirmed. RTS patients can be identified at birth by morphological characteristics of RTS. Drugs that act on cAMP signaling pathway might be beneficial to RTS patients. All transcription factors do not recognize DNA sequences; some may recognize other transcription factors or RNA polymerase. Some transcription factors are smart enough to make their contacts with proteins through their phosphorylation in turn to initiate transcription (Arcangelo and Curran 1995). CREB is one such transcription factor. It first gets phosphorylated then to phosphorylated CREB, CREB binding protein (CBP), which is a co-activator, binds and leads to initiation of transcription.

Petrij et al. (1995) reported that RTS was associated not only with a rearrangement of 16p chromosome but also with a point mutation in this chromosome. The break points involved were restricted to a region that contained gene for human CBP. This region, shown in Figure 8.10 (Petrij et al. 1995), consisted of CREB binding domain, two zinc fingers, PKA site, and carboxy-terminal glutamine-rich region. CBP binds to phosphorylated form of CREB transcription factor and increases the expression of genes containing cyclic AMP response enhancer. A palindromic consensus DNA sequence TGACGTCA functioned as a cAMP responsive transcriptions enhancer (CRE). An increase in the level of intracellular cAMP activated protein kinase A which in turn phosphorylated CREB on serine 133 and increased its association with CBP and stimulated CRE-dependent gene expression. In addition to this cAMP-signal transduction pathway, CBP may also mediate transcriptional activity in response to growth factors. CBP is closely related to E1A-associated protein p300 and both induced gene expression and cell immortalization as well as CREB-dependent gene activation. CBP may regulate transcription through interactions with as yet unidentified proteins. CBP may also bind to TFIIB, a component of basic transcription machinery. A. Carlsson, P. Greengard and E. Kandel were awarded Nobel Prize in 2000 for their discoveries concerning signal transduction (Carlsson (1995; Carlsson and Fornstedt 1991; Greengard 1978; Greengard et al. 1993; Kandel and Abel 1995; Kandel 1997; Kandel and Pittenger 2000).

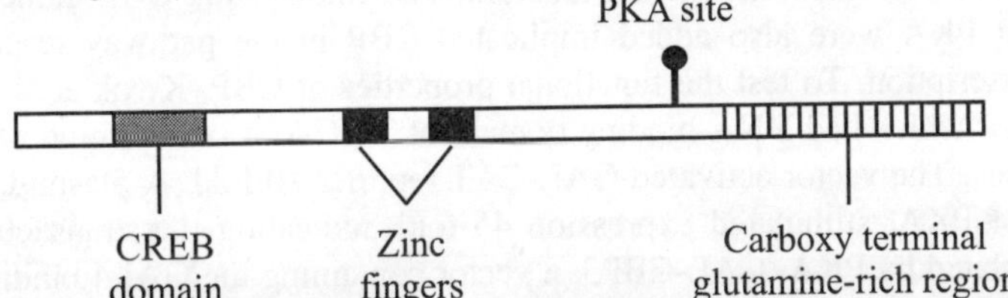

Figure 8.10 Region containing gene for human CREB binding protein (CBP), which consists of cAMP response element binding (CREB) binding domain, two zinc fingers, a PKA site and carboxy-terminal glutamine-rich region

Chirivia et al. (1993) screened human thyroid cDNA library to find proteins that interacted with phosphorylated CREB. Screening was done with recombinant CREB labeled at PKA site. One positive clone identified was used to screen a mouse brain cDNA library. The screened sequence of the predicted mouse CBP contained several kinase II phosphorylation sites, two regions resembling zinc fingers and a carboxy-terminal glutamine-rich region. To determine how phosphorylation influences CREB-CBP interactions, PKA phosphorylated or non-phosphorylated CREB was incubated with ^{32}P-labeled CRE oligonucleotide and was used to probe a 58-kDa fragment of CBP blotted onto nitrocellulose filters. When non-phosphorylated CREB was used, no CBP was detected. In contrast, phosphorylated CREB clearly detected the CBP fragment. This indicated that binding occurs only if CREB is phosphorylated. Co-immunoprecipitation was used to determine the specificity of CBP-CREB interactions. Caesin kinase II phosphorylated CREB on three serine residues but this phosphorylation did not result in transcriptional activation. Chirivia et al. (1993) phosphorylated CREB with either PKA or caesin kinase II and incubated it with a radiolabeled 78-kDa CBP fragment and immunoprecipitated the complexes with anti-CBP antibodies. Only PKA-phosphorylated CREB co-precipitated with CBP indicating that binding to CBP requires the specific phosphorylation of CREB at serine 133. CBP did not bind to other proteins phosphorylated by PKA.

To identify the regions of CREB involved in CBP binding, Chirivia et al. (1993) expressed CREB residues 1-160 in bacteria as a glutathione-S-transferase fusion protein. The CREB 1-160 peptide was purified and radiolabeled. PKA was used to probe intracellular electroblotted CBP. CREB 1-160 detected immobilized CBP, indicating that CREB activation domain was sufficient for binding. To define this domain more precisely, they examined a series of fusion proteins containing various activation-domain deletions. Transcription properties of these deletion mutants were determined. CREBΔQ1, which lacked N-terminal 87 amino acids of CREB, still showed binding to CBP. This deletion did not block PKA activation of CREB in vivo. CREBΔQ2, which lacked amino acids 160-283, also retained ability to bind to CBP. The fusion protein GST-CREB101-255 recognized CBP. These results revealed that amino acids 101-160 within CREB activation domain were sufficient for CBP binding. Deletions between amino acids residues 136-160 bp abolished transactivation properties of CREB. It was further investigated whether these mutations also lack CBP binding. CRBP peptide lacking amino acids 140-160 can be phosphorylated but did not bind to CBP. A phosphopeptide representing only residues 128-141 of CREB did not bind to CBP indicating that structural motifs outside PKA phosphorylation site were required for CBP binding.

In the presence of CREB and PKA, components of transcription machinery should become limiting. In the absence of CBP, both CREB and PKA increased transcriptional activity by 20-fold. In the presence of CREB and PKA, CBP increased the transcriptional activity by 90-fold. If CREB was phosphorylated by PKA but PKA was mutated then CBP would not bind to it. In the absence of PKA, CREB did not permit CBP-mediated gene induction. The finding that CBP induced gene expression only if CREB and PKA were also added implicated CBP in the pathway responsible for cAMP-activated gene transcription. To test the functional properties of CBP, Kwok et al. (1994) constructed an expression vector which had DNA-binding domain of TF GAL4 (with amino acids 1-147) fused to CBP amino terminus. The vector activated GAL-CAT reporter 10-fold. A plasmid, which encodes the catalytic subunit of PKA, stimulated expression 45-fold, indicating that transactivation potential of CBP could be regulated by PKA. GAL-CBP3, a vector containing the GAL4 binding domain fused to CREB binding portion of CBP was only minimally active when transfected either alone or with PKA. GAL-CBP3 was slightly more active in presence of CREB and PKA.

To determine what portion of CBP allowed transactivation, a fusion gene containing residues 1,678-2,441 of CBP fused to the DNA binding domain of transcription factor GAL4 was constructed. This region of CBP contained zinc-finger motif, PKA phopsphorylation site and glutamine-rich region. Plasmid pGAL-CBP (1,678-2,441) was more than 100-fold more active than pGAL1-147 in inducing expression of a GAL-CAT reporter and was thus considerably more than the full length GAL-CBP fused gene. PKA increased activity of the full length GAL-CBP by 40- to 50-fold and a similar level of induction by PKA was seen using smaller GAL-CBP gene. However, mutating the consensus PKA site in CBP did not affect either basal or PKA-stimulated expression of the reporter. Thus, although the interaction of phospho-CREB with CBP appeared to be the key regulatory step in PKA-activated transcription, indirect effects of PKA might influence CBP function as well. Having determined that the C-terminal portion of CBP allowed transcriptional activation, the next question arises whether this region interacts with basal transcription factors. Using GST-TFIIB fusion proteins linked to glutathione agarose beads tested the binding of CBP to TFIIB. Specific binding was observed with the 1,680-2,441 fragment of CBP, although a small amount of non-specific binding to GST alone was also detected. When the C-terminal region of CBP was separated into fragments extending from 1,680-1,891 and 1,892-2,441, specific binding was detected only with the former portion. A small fragment encompasses only residues 1,680-1,812 also bound specifically. This 133 amino-acid region of CBP contained the most 3′-mCBP fragment hybridizes to cosmid RT1. Thus human *CBP* gene spans at least 150-kb genome area containing RTS breakpoints. In 8 RTS patients (A-H) it was observed that the smallest deletions were of size 130 kb in patients A and B and the largest deletions were of size exceeding 650 bp. It was also noticed that in patients A-F some part of CBP gene area was present but microdeletions in patients G and H extended more than 200 kb beyond the 5′-end of *CBP* probably removing the entire gene. Petrij et al. (1995) found cases with truncated mRNA. To search for mutations in RTS patients without rearrangements they used protein truncation test (PTT). PTT is a mutation detection system, which is based on in vitro transcription and translation. It was performed on a 1.1-kb CBP product which corresponded to 42 kDa proteins. This test was performed on 11 (1-11) patients. Two patients (2 and 9) showed truncated protein 15kDa and 39.5 kDa in size. To study point mutations, sequence analysis was done. In both the patients who had RTS phenotype, a codon for glutamine was changed to a stop codon by C→T substitution. In patient 9, the mutations destroyed a PvuII site.

The occurrence of point mutations in a single gene implied that a contiguous gene syndrome, in which the deletion of several adjacent genes caused a composite set of abnormalities, did not cause RTS. The microdeletions, translocations and point mutations, all implicated CBP dearrangement as cause of RTS. Petrij et al. (1995) found that RTS might be caused by haplo-insufficiency, i.e. reduced amount of normal *CBP* gene due to hemizygous deletion in chromosomal region. Greg syndrome is one of the examples of haplo-insufficiency Vortkamp and Grezeschik (1991) reported that *GL13* zinc-finger gene was interrupted by translocations in Greg syndrome families. The Greg Cephalo Polysyndactyly Syndrome (GCPS) is an autosomal dominant disorder affecting limbs and craniofacial development in humans. The genetic locus has been pinpointed to 7p13 chromosome. Two out of three translocations disrupted the *GL13* gene. In the third translocation, chromosome 7 was broken at about 10 kb downstream of 3′-end of *GL13*. So their results indicated that mutations disturbing normal *GL13* expression might have a causative role in GCPS. Petrij et al. (1995) argued that the reduced amount of normal CBP protein in the hemizygous condition was responsible for RTS. Due to microdeletions and point mutations the CREB binding site or entire CBP gene gets removed. Then CREB transcription factor will not show its smartness in initiation of transcription and causes dysregulation of gene expression.

CHLOROPLAST DNA TRANSCRIPTION

A template-binding polypeptide has been identified in pea chloroplast transcriptional complex (Khanna et al. 1992). It has a molecular weight of 150 kDa and binds to both chloroplast ribosomal (16S rRNA) and messenger (*psbA*) promoters and it exhibits some degree of correlation to total transcriptional element under various salt concentrations. The results demonstrate the 150 kDa polypeptide is a functional template binding polypeptide of the pea.

MITOCHONDRIAL DNA TRANSCRIPTION

Regulation of mammalian mtDNA gene expression is critical for altering oxidative phosphorylation capacity in response to physiological demands and disease processes. The basal machinery for initiation of mtDNA transcription has been molecularly defined. Park et al. (2007) have shown that MTERF3 is a negative regulator of mtDNA transcription initiation. The *MTERF3* gene is essential because homozygous knockout mouse embryos die in midgestation. Tissue-specific inactivation of MTERF3 in the heart causes aberrant mtDNA transcription and severe respiratory chain deficiency.

MTERF3 binds the mtDNA promoter region and depletion of MTERF3 increases transcription initiation on both mtDNA strands. This increased transcription initiation leads to decreased expression of critical promoter-distal tRNA genes, which is possibly explained by transcriptional collision on the circular mtDNA molecule. MTERF3 seems to be the first example of a mitochondrial protein that acts as a specific repressor of mammalian mtDNA transcription in vivo.

TRANSCRIPTION IN RELATION TO OTHER BIOLICAL PROCESSES

Transcription and Recombination

Effect of RNA polymerase II-dependent transcription on recombination between directly related sequences of *GAL10* in *S. cerevisiae* have been reported by Thomas and Rothstein (1989). Direct repeat recombination leading either to plasmid loss or conversion was examined in isogenic strains containing null mutations in the positive activation *GAL4*, or the repressor, GAL80. A15-fold increase in the rate of plasmid loss is observed in cells constitutively expressing the construct compared with cells that are not. Stimulation in recombination is mediated by events initiating within the integrated plasmid sequences.

DNA Supercoiling and Transcription

Many genes are responsive to DNA supercoiling (Preiss and Drlica 1989). Does the cell use supercoiling to regulate gene expression? How does supercoiling enhance the expression of some genes while inhibit the expression of some others. Transcription itself can be major contribution to the level of supercoiling.

Heterologous DNA Binding Domains and Transcription

Amyloid-β precursor protein (APP), a widely expressed cell surface protein, is cleaved in transmembrane region by γ-secretase. The cleavage of APP produces the extracellular ameyloid β-peptide of Alzheimer's disease and released an intracellular tail fragment of unknown function. Cao and Siidhaf (2001) have demonstrated tail from complex with Fe65 and Tip60. This complex stimulates transcription via heterologous GAL4 or LexA DNA binding domains. The released tail of APP may function in gene expression.

RNA-DNA Hybrid in Biological Functions

Secondary structures of RNA, including double-stranded duplex hairpins, internal loops, bulged bases and pseudoknotted structures were intimately connected with biological functions including splicing reactions and ribozyme activity (Bhattacharyya et al. 1990). Formation of RNA-DNA hybrids is important in transcription of DNA, reverse transcription of viral RNA and DNA replication. Bulged bases in RNA helices are potentially significant in RNA folding and in providing sites for specific protein-RNA interactions and in coat proteins. Bulges introduce pronounced kicks into RNA and RNA-DNA helices, depending upon the number and types of bases in the bulge and its position in the fragment. By varying the spacing between two bulge-induced kinks, Bhattacharyya et al. (1990) have measured the periodicity of RNA and RNA-DNA helices in solution.

HORMONES AND TRANSCRIPTION

Study of sex differentiation and role of hormones shows that realization of genotypic capability depends upon and can be modified by hormones, while the production of hormones is itself guided by the genetic constitution of the individual. A further proof that hormones may influence gene expression is obtained by influence of sex hormones on sex-influenced and sex-limited traits. Puffing in salivary chromosomes of dipteran flies are regarded as sites of gene activity. In *Chironomus*, two puffs A (on chromosome 1) and B (on chromosome 4) appear during molting of third and fourth instars. Ecdysone injected into larvae that are between molts produces puffs similar to normal molting. These puffs are not restricted to salivary chromosomes but are also formed in cells of other tissues. Temperature, oxygen, pressure, nutrition, etc. act as inducers of gene activity. It has been demonstrated that sex steroid hormones, namely estrogen and progesterone, can regulate gene activity. Glucocorticoids, another set of steroids, found endogenously within the cells and used for treatment of leukemia, asthma and arthritis also regulate gene expression. These glucocorticoids have a multitude of effects in various tissues. Classical example of a sex steroid inducible protein is chick ovalbumin and glucocorticoid-inducible proteins are the enzymes tryosine aminotransferase (TAT) and tryptophan oxydase. The glucocorticoids induce synthesis of these enzymes through receptor molecules. The steroid-receptor complex interacts each with a specific DNA sequence, called steroid receptor element (SRE) or glucocorticoid receptor element (GRE) or hormone receptor element (HRE). Through recombinant DNA technology, gene constructs were prepared by fusing different lengths of DNA sequences flanking the 5'-end of *TAT* gene with the marker gene for *CAT* (chloramphenicol acetyltransferase). These cloned gene constructs were introduced into animal cells and through glucocorticoid induced expression of *CAT*, steroid receptor elements (SRE) were identified (Figure 8.11).

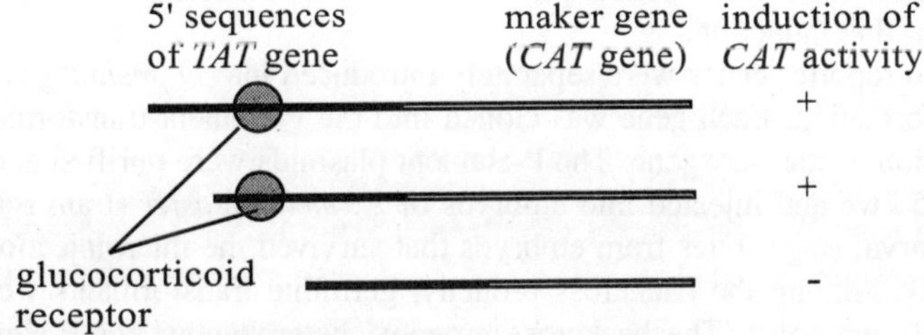

Figure 8.11 Use of deletions through recombinant DNA technology for identification of response elements

E.W. Sutherland Jr. was awarded Nobel Prize in 1971 for discovery of mechanism of hormone action. Role of cyclic adenosine 3′,5′-monophosphate (cAMP) in this process was understood from their work (Rall and Sunderland 1958; Sutherland and Rall 1958; Sutherland 1992).

TISSUE-SPECIFIC GENE EXPRESSION

Effector and reporter genes have been used to study tissue-specific gene expression. An effector gene is a gene that produces a regulatory molecule that drives the expression of another gene. A reporter gene is a gene that researchers attach to a regulatory sequence of another gene of interest in bacteria, cell culture, animals, or cells.

Fischer et al. (1988) show that galactose 4 (GAL4), a yeast regulator, when expressed in particular tissue *Drosophila* larvae, stimulates tissue-specific transcription of a *Drosophila* promoter linked to GAL4 binding sites. For this purpose, they made use of effector and reporter genes. The effector gene used is diagrammatically represented in Figure 8.12. This is a *Adh/GAL4* hybrid gene, which is a 4.8-kb XbaI fragment. This effector gene derives transcription of the yeast GAL4 coding sequences with the help of *Adh* promoter. GAL4 regulatory protein binds to galactose upstream activating sequences (UAS_G) and activates transcription of linked genes. GAL4 has two functions: (1) DNA-binding domain, located near amino terminus, binds GAL4 protein to DNA and (2) one or more activating domains, acidic in nature, located at the carboxyl end, activate transcription. This domain can be replaced by a short peptide designed to form a negatively charged amphipathic α-helix.

Effector gene *Adh/GAL4*

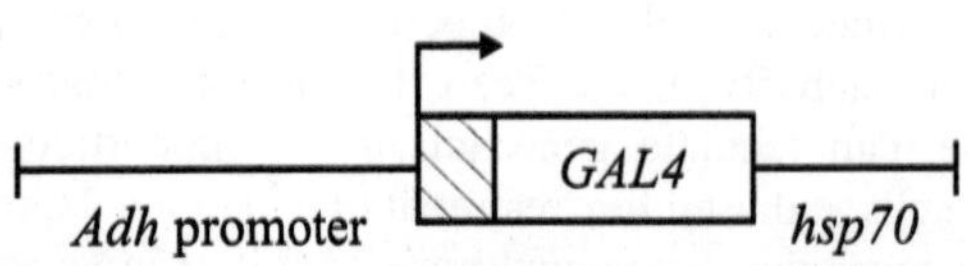

Figure 8.12 Effector gene *Adh/GAL4* introduced into the *D. melanogaster* genome by P-element transformation

Reporter gene 17-*hsp70/lac z*

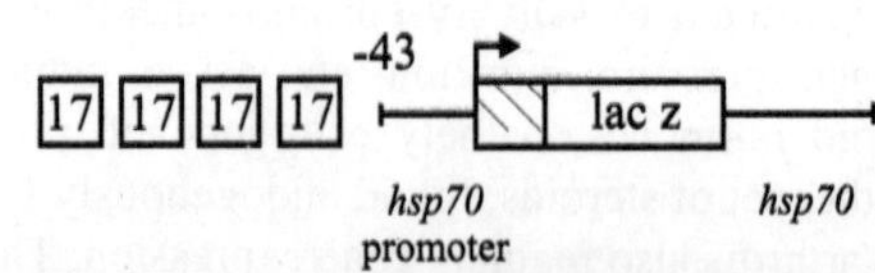

Figure 8.13 Reporter gene *17 hsp70/lacZ GAL4* introduced into the *D. melanogaster* genome by P-element transformation

The reporter gene used in this study is *17-hsp70/lacZ*, a 3.6 kb XbaI fragment (Figure 8.13). In this gene, 17 represents four copies of GAL4 17-mers (5′-CGGAGTACTGTCCTCCG-3′), a high affinity binding site for GAL4 protein. These 17-mers are closely related to the GAL4 binding sites in UAS_G. The reporter gene derives transcription of *E. coli* gene encoding β-galactosidase (*lacZ*) gene. This gene is transcriptionally inactive; the *hsp70* promoter is normally heat inducible but sequences upstream of –43 from the *hsp70* transcription site were deleted, thus eliminating the regulatory elements required for that induction.

The effector and reporter genes were separately introduced into *D. melanogaster* genome element by P-element transformation. Each gene was cloned into the P-element transformation vector C70TX in the same orientation as the *rosy* gene. The P-element plasmids were purified and co-joined with the helper plasmid pπ25.7we and injected into embryos of *D. melanogaster* strain rosy. Very few larvae survived past first larval stage. Flies from embryos that survived the microinjection were individually back crossed to rosy. Among the backcross progeny, germline transformants were distinguished by their wild-type (ry^+) eye color. The backcross progeny, heterozygous for P-element, were used to perform the effector × reporter crosses. Three independent effector transformant lines (each heterozygous for P-element) were separately crossed with nine independent reporter transformant lines (each heterozygous for P-element). Thus 27 crosses were made. Twelve larvae from the progeny of each cross were tested for expression of reporter gene by histochemical staining assay for β-galactosidase activity. β-galactosidase activity was detected in approximately one-fourth of the larval progeny. Enzyme activity levels observed in 13 different tissues (Figure 8.14) are presented in Table 8.2.

No β-galactosidase activity was detected with a reporter gene lacking 17-mers, Also, no activity was observed in larvae transformed with the reporter gene alone. Enzyme activity in progeny larvae of all the crosses was observed at high levels in fat body and anterior midgut where *Adh* promoter was expected to be active. A third instar larva carrying effector and reporter genes expressed β-galactosidase in fat body and anterior midgut at high levels. Also, as expected, variable levels of enzyme activity were detected in hindgut and tracheae. In middle midgut and Malpighian tubules, promoter was expected to be active. Enzyme activity was detected in small number of larvae and in a few cells. Perhaps, *Adh/GAL4* transcript or GAL4 protein itself is unstable in these two tissues. No β-

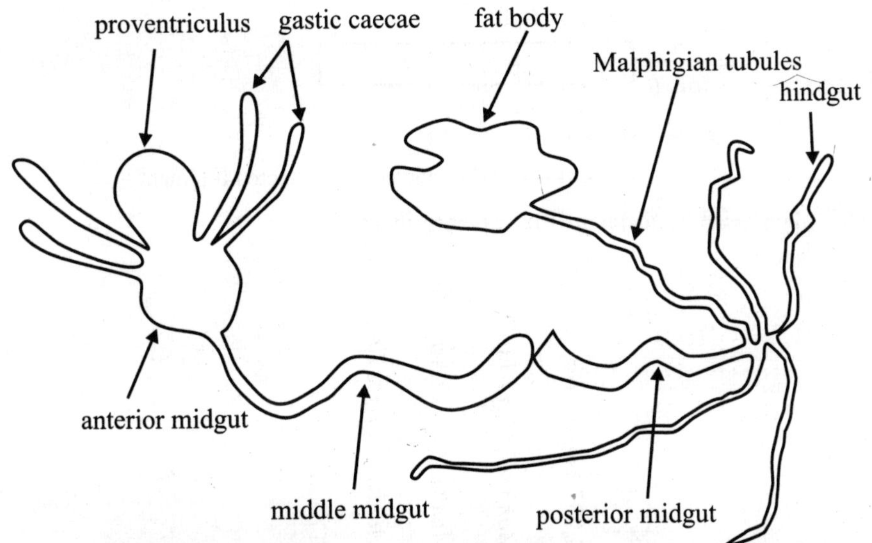

Figure 8.14 Diagrammatic representation of the *D. melanogaster* tissues studied. Results of the tissue-specificity experiment are given in Table 8.2

Table 8.2 β-galactosidase activity levels observed in 13 different tissue of *Drosophila melanogaster*

Tissue	*Fat body*	*Malpighian tubules*	*Anterior midgut*	*Middle midgut*	*Hindgut*	*Tracheae*	*Gastric caecae*
**Adh/lacZ* (6)	+	+	+	+	Var	Var	–
Adh/GAL4 (3): *17-hsp70/lacZ* (9)	+	+/–	+	+/–	Var	Var	–
Tissue	*Proven-triculus*	*Posterior midgut*	*Imaginal discs*	*Brain*	*Salivary glands*	*Ovaries/ testis*	
**Adh/lacZ* (6)	–	–	–	–	–	–	
Adh/GAL4 (3): *17-hsp70/lacZ* (9)	–	–	–	–	–	–	

*Hybrid *Adh/lacZ* gene contains same Adh promoter as in the effector gene but here fused in-frame to the *E. coli lacZ* gene rather than to *GAL4* gene; Numbers in parantheses indicate the number of different transformant lines assayed; Histochemical staining intensity of β -galactosidase enzyme is represented as: +, high intensities in all lines; +/–, staining observed in some larvae, and only at a low level in a few cells; –, staining never seen; Var, staining intensities varied from high to none within or between different lines or crosses

galactosidase activity was detected in tissues other than those in which *Adh* promoter of the effector gene was expected to be active. Reporter gene transcripts in transformed larvae were assayed for quantitative RNase protection using two probes – one, a uniformly ^{32}P-labelled RNA probe complementary to the 5′-end of the reporter transcripts and, two, an RNA probe complementary to *α-1 tubulin* gene as an internal control. SP6 RNA polymerase was used for transcription from plasmids *SP6-αtub* and *SP6-hsp-lac*. SP6 probe was ~450 nucleotides and protected a 408-nucleotide RNA fragment (Figure 8.15). SP6-αtub probe protected 90-nucleotide fragment of the endogenous *α-tubulin* gene.

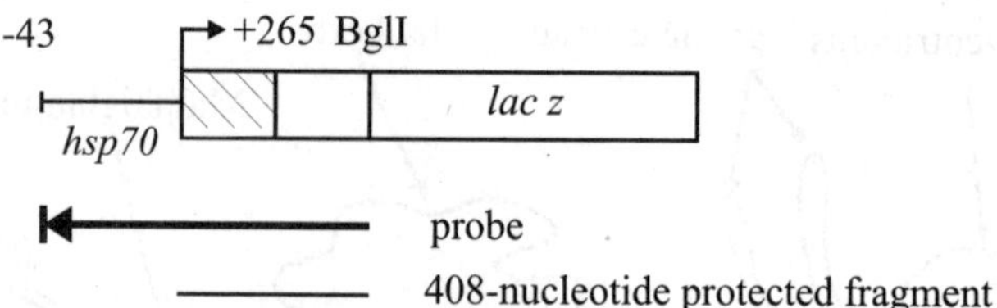

Figure 8.15 RNA probe used for quantitative RNase protection assay

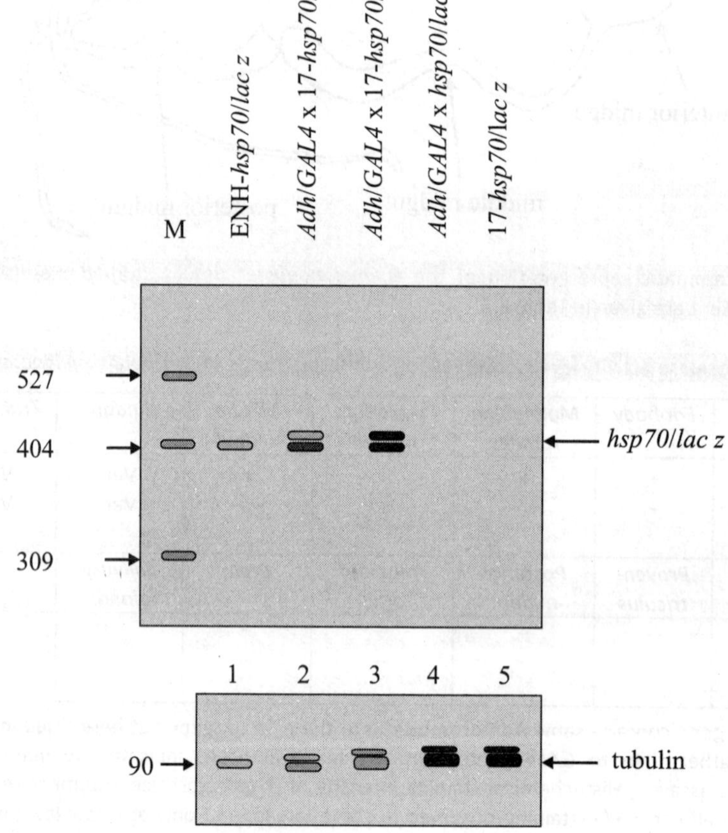

Figure 8.16 Reporter gene transcripts in transformed *Drosophila* larvae

The results of above experiment are shown in Figure 8.16. Transcripts from following five lines were assayed: *ENH-hsp70/lacZ* is the gene like the reporter except enhancer of the *D. melanogaster AdhI* gene distil promoter (-660 to –128) from the distil transcript site was installed upstream of truncated *hsp70* promoter instead of the GAL4 17-mers (lane 1). Two independent effector *Adh/GAL4* transformant lines were crossed with the same reporter *17-hsp70/lacZ* lines (lanes 2 and 3). Cross of effector *Adh/GAL4* and reporter *hsp70/lacZ* with no 17-mers (lane 4). From lanes 1-3, identical fragments of the reporter transcripts were protected. Appearance of 2 bands is probably an artifact of the RNA digestion. From lines 4 and 5, no protected transcripts were detected.

Densitometric studies representing hsp70/lacZ : tubulin ratio showed that GAL4 protein bound to 17-mers is comparable with *Drosophila Adh* enhancer in its ability to activate transcription of the reporter gene, i.e., both are equally efficient. GAL4 activated transcription of the reporter gene was initiated at the same start point as was transcript of the reporter gene activated by a *Drosophila Adh* enhancer. Thus a single yeast protein GAL4, when expressed in a tissue-specific manner, generates a tissue-specific expression in *Drosophila.* Apparent complexity of tissue-specific enhancers is not necessary for their function as transcription activators. Perhaps this complexity is exploited to obtain intricate patterns of gene control from a relatively small number of regulatory proteins. GAL4 protein has ability to activate mammalian and *Drosophila* promoters. This suggests that protein with which DNA-bound GAL4 interacts to stimulate transcription in yeast has a close homolog in higher eukaryotes. It seems likely that the mechanism of action of at least one class of gene activators is conserved between yeast, mammalian cells and *Drosophila.*

EXPLOITING GENE EXPRESSION

Here are given some terms related to gene expression analysis.

Transcriptome – The set of all RNA molecules, including mRNA,rRNA,tRNA, and non-coding RNA produced in one or a population of cells.

Microarray – A 2D array on a solid substrate (usually a glass slide or silicon thin-film cell) that assays large amounts of biological material using high-throughput screening methods.

SAGE (Serial Analysis of Gene Expression) – A molecular biology technique to produce a snapshot of the mRNA population in the form of small tags that correspond to fragments of those transcripts.

ESTs (Expressed sequence tag) – A short sub-sequence of a transcribed cDNA sequence produced by one-shot sequencing of a cloned mRNA mostly used to identify gene transcripts.

Virtual Southern blot – An early method of gene expression profiling, where ESTs derived from particular tissues are counted to provide the insight in which EST libraries (i.e., tissues) corresponding gene is expressed.

Exon array – Arrays designed to interrogate exon-level expression. They differ significantly from expression arrays in the number and placement of the oligonucleotide probes and in the design of control probes for background correction.

Mass spectrometry – An analytical technique for the determination of the elemental composition or chemical structure of a molecule such as peptides and other chemical compounds on the basis of charge and mass ratio.

Microarray quality controls – A method of low-level inspection of quality of results obtained in microarrays. Usually proper background/signal ratio must be achieved as well as 3'- to 5'-end probes signals cannot differ above some sert level. Other parameters include a percentage of absent calls or so called scaling factor.

Microarray data normalization – A basic algorithm of data transformation which should enable intra- and intrersample comparisons. The most frequently used algorithms of microarray normalization are: MAS5.0, RMA, GCRMA, LVS-GCRMA, and GSEA.

MAS5.0 (Affymetrix MAS5) – Algorithm used for making expression summaries uses both perfect match and mismatch probes.

RMA (Robust Multi-Array) normalization – Retains probe level information, it consists of three steps: a background adjustment, quantile normalization and finally summarization.

GCRMA – Uses GC content of probes in normalization with RMA, gives one value for each probe set instead of keeping probe level information.

LVS-GCRMA – More conservative GCMRA protocol, designed for the purpose of analysis of gene sets with minor changes in expression.

GSEA (Gene Set Enrichment Analysis) – A simple method for annotation of microarray data where a number of selected genes (e.g., differentially expressed) is compared with a total number of elements (genes of given function). Further, using binomial probability it is assessed against expected values (i.e., all differentially expressed genes vs total number of genes). This approach is used for identification of potentially changed pathway or more than expected altered functional groups of genes. Notable GSEA is a knowledge-based approach for interpreting genome-wide expression profiles, focuses on gene sets, chromosomal location. GSEA yields insights into several cancer related data sets like revealing biological pathways.

The techniques of gene expression analysis have been exploiting in several ways: (a) global gene expression profiling (b) identification of novel minimal residual disease markers in curing neuroblastoma, (c) explicit use of the temporal order in the data by fitting polynomial functions to the temporal profile of each gene, (d) quantitative transcript expression profiling, (e) microarray gene expression data analysis by using support vector machines.

Global Gene Expression Profiling

The term gene expression relates to any phenotypic expression of gene products. It means: from transcription, through splicing/edition to translation and post-translational modifications. In general concept - the most complete description of cell status should give us the best source of data on activity of particular elements of cell. This description should cover majority of elements (genes). The idea of high-scale profiling of gene expression is exploited on level of DNA, RNA and protein. The level of DNA is the least informative. In some situations we deal with gene copy number variants (CNV) where some genes (segments of genome) are multiplied and as such are more actively transcribed (the effect of gene number). It finds a special application in cancer biology, where CNV of some oncogenes are gained or lost from genome. The most reliable methods for gene-scale mapping of CNV are copy-number arrays (genomic microarrays of high density of probed regions).

The most exploited way to determine activity of genes is probing on level of RNA. This fact is a result of few aspects: (a) technological – synthesis of probes or sequencing technologies can facilitate full-genome scale analyses, (b) replicability – application of highly sensitive quantitative methods can identify even a slight in level of expression, (c) accuracy and relatively good dynamic range of measurements – allows to test the least abundant transcripts in the same time as the most active genes.

Technologies here were evolving from a repetitive single gene measurements, through library sequencing methods to all-in-one readouts performed with DNA microarrays which are based on hybridization. Less quantitative technique is a sequence-based sampling – serial analysis of gene expression (SAGE). The problem which we deal here is suboptimal relation between measured gene product (mRNA) and activity. Therefore we must consider that for some genes the amount of mRNA in cell is related to the amount of corresponding active protein (true functional gene product), but this relation is not clear in full. We also don't even know the list of genes which for which correlation between mRNA and active protein is little.

The third level of gene expression measurements is cellular proteins. This basic material of cell is in the fact in the closest association with true function of particular gene products. Proteins (enzymatic or structural) execute most of the functions. The basic technology is a mass spectrometry coupled identification of cell proteins. The major obstacle for this technology is still low coverage of sampled proteins. It causes some additional problems for good analysis of results: (a) low dynamic range which do not allow to identify reliable absolute quantitative measurements, (b) low reproducibility due to the effect of sampling (random finding of a narrow repertoire of proteins), (c) lack of ability to identify the active form of proteins (mostly: phosphorylation status).

Identifying Novel Minimal Residual Disease Markers

Minimal residual disease (MRD) presents a significant hurdle to curing metastatic neuroblastoma (NB). Biologic therapies directed against MRD can improve outcome. Evaluating treatment efficacy requires MRD measurement, which serves as surrogate endpoint. Because of tumor heterogeneity, no single marker will likely be adequate. Genome-wide expression profiling can uncover potential MRD markers differentially expressed in tumors over normal marrow/blood (Cheung et al. 2008). Marker discovery based on differential gene expression profiling, stringent sensitivity and specificity assays, and well-annotated patient samples can rapidly prioritize and identify potential MRD markers of neuroblastoma.

Exploiting Temporal Gene Expression Profiling

The identification of biologically interesting genes in a temporal expression profiling dataset is challenging and complicated by high levels of experimental noise. Most statistical methods used in the literature do not fully exploit the temporal ordering in the dataset and are not suited to the case where temporal profiles are measured for a number of different biological conditions. Vinciotti et al. (2006) present a statistical test that makes explicit use of the temporal order in the data by fitting polynomial functions to the temporal profile of each gene and for each biological condition. A Hotelling T2-statistic is derived to detect the genes for which the parameters of these polynomials are significantly different from each other. They validate the temporal Hotelling T2-test on muscular gene expression data from four mouse strains which were profiled at different ages: dystrophin-, beta-sarcoglycan and gamma-sarcoglycan deficient mice, and wild-type mice. The first three are animal models for different muscular dystrophies. Extensive biological validation shows that the method is capable of finding genes with temporal profiles significantly different across the four strains, as well as identifying potential biomarkers for each form of the disease. The added value of the temporal test compared to an identical test which does not make use of temporal ordering is demonstrated via a simulation study and through confirmation of the expression profiles from selected genes by quantitative PCR experiments. The proposed method maximizes the detection of the biologically interesting genes, whilst minimizing false detections. The temporal Hotelling T2-test is capable of finding relatively small and robust sets of genes that display different temporal profiles between the conditions of interest. The test is simple, it can be used on gene expression data generated from any experimental design and for any number of conditions, and it allows fast interpretation of the temporal behavior of genes.

Quantitative Transcript Expression Profiling

Measurement precision determines the power of any analysis to reliably identify significant signals, such as in screens for differential expression, independent of whether the experimental design incorporates replicates or not. With the compilation of large-scale RNA-Seq datasets with technical replicate samples, however, we can now, for the first time, perform a systematic analysis of the precision of expression level estimates from massively parallel sequencing technology. This then allows considerations for its improvement by computational or experimental means. Łabaj et al. (2011) report on a comprehensive study of target identification and measurement precision, including their dependence on transcript expression levels, read depth and other parameters. They introduced a new approach for mapping and analyzing sequencing reads that yields substantially improved performance in gene expression profiling, increasing the number of transcripts that can reliably be quantified to over 40%. Extrapolations to higher sequencing depths highlight the need for efficient complementary steps.

Microarray Gene Expression Data Analysis by using Support Vector Machines

Brown et al. (2000) introduce a method of functionally classifying genes by using gene expression data from DNA microarray hybridization experiments. The method is based on the theory of support vector machines (SVMs). SVMs are considered a supervised computer learning method because they exploit prior knowledge of gene function to identify unknown genes of similar function from expression data. SVMs avoid several problems associated with unsupervised clustering methods, such as hierarchical clustering and self-organizing maps. SVMs have many mathematical features that make them attractive for gene expression analysis, including their flexibility in choosing a similarity function, sparseness of solution when dealing with large data sets, the ability to handle large feature spaces, and the ability to identify outliers. They test several SVMs that use different similarity metrics, as well as some other supervised learning methods, and find that the SVMs best identify sets of genes with a common function using expression data.

REFERENCES

Abelson, J., C.R. Trotta, and H. Li. 1998. RNA splicing. J. Biol. Chem. 273: 12685-8.

Bhattachararyya, A., A.I.H. Murchie, and D.M.J. Lilley. 1990. RNA bulges and the helical periodicity of double-stranded RNA. Nature 343: 484-7.

Brown, C.D., D.S. Johnson, and A. Sidow. 2007. Functional architecture and evolution of transcriptional elements that derive gene expression. Science 317: 1557-60.

Brown, M.P., W.N. Grundy, D. Lin, N. Cristianini, C.W. Sugnet, T.S. Furey, M. Ares Jr., and D. Haussler. 2000. Knowledge-based analysis of microarray gene expression data by using support vector machines. Proc. Natl. Acad. Sci. USA 97: 262.

Bushnell, D.A., K.D. Westover, R.E. Davis, and R.D. Kornberg. 2004. Structural basis of transcription: An RNA polymerase II – TFIIB cocrystal at 4.5 angstroms. Science 303: 983-8.

Cao, X., and T.C. Südhof. 2001. A Transcriptively active complex of APP with Fe65 and histone acetyltransferase Tip60. Science 293: 115-20.

Carlsson, A. 1995.Towards a new understanding of dopamine receptors. Clin. Neuropharmacol. 18 (suppl. 1): S6-I3

Carlsson, A., and B. Fornstedt. 1991. Possible mechanisms underlying the special vulnerability of dopaminergic neurons. Acta Neurol. Scand. 84: 16-18.

Cheng, H., K. Dufu, C.S. Lee, J.L. Hsu, A. Dias, and R. Reed. 2006. Human mRNA export machinery recruited to the 5'-end of mRNA. Cell 127: 1389-400.

Cheung, I.Y., Y. Feng, W. Gerald, and N.-K.V. Cheung. 2008. Exploiting gene expression profiling to identify novel minimal residual disease markers of neuroblastoma. Clin. Cancer Res. 14: 7020-7.

Chirivia, J.C., R.P.S. Kwok, N. Lamb, H. Hagiwara, M.R. Montmigg, and R.H. Goodman. 1993. Phosphorylated CREB binds specifically to the nuclear protein CBP. Nature 365: 855-9.

Cohen, R.S., and M. Meselson. 1988. Periodic interactions of heat shock transcriptions elements. Nature 332: 856-8.

Cramer, P., D.A. Bushnell, and R.D. Kornberg. 2001. Structural basis of transcription: RNA polymerase II at 2.8 Angstrom resolution. Science 292: 1863-76.

D'Arcangelo, G., and T. Curran. 1995. Smart transcription factors. Nature 376: 292-3.

Fischer, J.A., E. Giniger, T. Maniatis, and M. Ptashne. 1988. GAL4 and activates transcription in Drosophila. Nature 332: 853-6.

Global gene expression profiling. http://lib.bioinfo.pl/courses/view/1023.

Gnatt, A.L., P. Cramer, J. Fu, D.A. Bushnell, and R.D. Kornberg. 2001. Structural basis of transcription: An RNA polymerase II elongation complex at 3.3 Å resolution. Science 292: 1876-82.

Greengard, P. 1978. Phosphorylated proteins as physiological effectors. Science 199: 146-52.

Greengard, P., F. Valtorta, A.J. Czernik, and F. Benfenati. 1993. Synaptic vesicle phosphoproteins and regulation of synaptic function. Science 259: 780-5.

Kakidani, H., and M. Ptashne. 1988. GAL4 activates gene expression in mammalian cells. Cell 52: 161-7.

Kandel, E., and T. Abel. 1995. Neuropeptides, adenylyl cyclase, and memory storage. Science 268: 825-6.

Kandel, E.R. 1997. Genes, synapses, and long-term memory. J. Cell. Physiol. 173: 124-5.

Kandel, E.R., and C. Pittenger. 2000. The past, the future and the biology of memory storage. Phil. Trans. R. Soc. Lond. B 354: 2027-52.

Kelleher III, R.J., P.M. Flanagan, and R.D. Kornberg. 1990. A novel mediator between activator proteins and the RNA polymerase II transcription apparatus. Cell 61: 1209-15.

Khanna, N.C., S. Lakhani, and K.K. Tiwari. 1992. Identification of the template binding polypeptide in the pea chloroplast transcriptional complex. Nucl. Acids Res. 20: 69-74.

Kwok, R.P.S., J.R. Lundblad, and J.C. Chirivia. 1994. Nuclear protein CBP is a co-activator for the transcription factor CREB. Nature 370: 223-6.

Łabaj, P.P., G.G. Leparc, B.E. Linggi, L.M. Markillie, H.S. Wiley, and D.P. Kreil1. 2011. Characterization and improvement of RNA-Seq precision in quantitative transcript expression profiling. Bioinformatics 27(13): i383-i391.

Liu, B., and B.M. Alberts. 1995. Head on collision between a DNA replication apparatus and RNA polymerase transcription complex. Science 267: 1131-7.

Nagalakshmi, U., Z. Wang, K. Waern, et al. 2008. The transcriptional landscape of the yeast genome defined by RNA sequencing. Science 320: 1344-9.

Park, C.B., J. Asin-Cayuela, Y. Camara, et al. 2007. MTERF3 is a negative regulator of mammalian mDNA transcription. Cell 130: 273-85.

Petrij, F., R.H. Giles, and H.G. Dauwerse et al. 1995. Rubinstein-Tyabi syndrome caused by mutation in the transcriptions co-activator CBP. Nature 376: 348-51.

Preiss, G.J., and K. Drlica. 1989. DNA supercoiling and prokaryotic transcription. Cell 56: 521-3.

Rall, T W., and E.W. Sutherland. 1958. Formation of a cyclic adenine ribonucleotide by tissue particles. J. Biol. Chem. 232: 1065-76.

Rowlands, T., P. Baumann, and S.P. Jackson. 1994. The TATA-binding protein: A general transcription factor in eukaryotes and Archaebacteria. Science 264: 1326-9.

Shamovsky, I., M. Ivvanikov, E.S. Kandel, and D. Gershon. 2006. RNA-mediated response to heat shock in mammalian cells. Nature 440: 556-60.

Sharp, P.A. 1991. TATA-binding protein is a classless factor. Cell 68: 819-21.

Sheth, U., and R. Parker, 2006. Targetting of aberrant mRNAs to cytoplasmic processing bodies. Cell 125: 1095-109.

Sutherland, E.W. 1992. Studies on the mechanism of hormone action. *Nobel Lectures, Physiology or Medicine.* 1971–1980. Lindsten, J., ed. Singapore: World Scientific Publishing Co.

Sutherland, E.W., and T.W. Rall. 1958. Fractionation and characterization of a cyclic adenine ribonucleotide formed by tissue particles. J. Biol. Chem. 232: 1077-92.

The FANTOM Consortium and RIKEN Genome Exploration Research Group and Genome Science Group (Genome Network Project Core Group). 2005. The transcriptional landscape of the mammalian genome. Science 309: 1559-63.

Thomas, B.J., and R. Rothstein. 1989. Elevated recombination rates in transcriptionally active DNA. Cell 56: 619-30.

Tjian, R. 1995. Molecular machines that control genes. Scient. Am. 272(2): 38-44.

Vinciotti, V., L. Xiaohui, R. Turk, E.J. de Meijer, and P.C. t'Hoen. 2006. Exploiting the full power of temporal gene expression profiling. BMC Bioinformatics, 7:183 doi:http://dx.doi.org/10.1186/1471-2105-7-183.

Vortkamp, A.G.M., and K.H. Grzeschik, 1991. GL13 zinc-finger gene interuppted by translocations in Greg syndrome families. Nature 352: 539-40.

9

RNA Processing in General

All the primary transcripts produced in the nucleus must undergo processing steps to produce functional RNA molecules for export to the cytosol. RNA processing may be defined as modification, mainly through cleavage and/or splicing, of RNA transcripts so as to release functional transfer RNA (tRNA), ribosomal RNA (rRNA) and messenger RNA (mRNA) and other species of RNA molecules from them.

Intron removal is one important step in processing of precursors of mRNAs, tRNAs and ribosomal RNAs. There are various types of introns with respect to their splicing strategies: (1) Introns in nuclear protein-coding genes that are removed by spliceosomes. (2) Introns in nuclear and archaeal genes that are removed by tRNA splicing enzymes. (3) Self-splicing introns – (a) self-splicing group I introns that are removed by RNA catalysis. Group I introns are large self-splicing ribozymes. They catalyze their own excision from mRNA, tRNA and rRNA precursors in a wide range of organisms. Twin ribozyme introns are formed by two ribozymes belonging to the group I family and occur in some ribosomal RNA transcripts and (b) self-splicing group II introns that are removed by RNA catalysis. Group II introns are large class of self-catalytic ribozymes as well as mobile genetic elements found within the genes of all three domains of life. Group II introns are type of RNA enzyme (ribozyme) that catalyzes their own excision from RNA transcripts and insertion into new genetic locations.

The group I-like ribozyme, GIR1, liberates the 5'-end of a homing endonuclease messenger RNA in the slime mold *Didymium iridis*. Nielsen et al. (2005) report that this cleavage occur by a transesterification reaction with the joining of the first and the third nucleotide of the messenger by 2',5'-phosphodiester linkage. Thus, a group I-like ribozyme catalyzes an RNA branching reaction similar to the first step of splicing in group II introns and spliceosomal introns. The resulting short lariat, by forming a protective 5' cap, might have been useful in a primitive RNA world.

The crystal structure of a group II intron from *Oceanobacillus theyensis* at 3.1-Å resolution shows a complex architecture with metal ions at the catalytic center (Piccirilli 2008; Toor et al. 2008). Structural and functional analogies support the hypothesis that group II introns and the spliceosome share a common ancestor.

Here we discuss processing of precursors of mRNAs, rRNAs and tRNAs.

PRE-MESSENGER RNA PROCESSING

Messenger RNA is not made directly in a eukaryotic cell. RNA transcribed from eukaryotic genes is called heterogeneous nuclear RNA (hnRNA) one of them is known as pre-mRNA which is a large primary transcript RNA and is processed into mRNA. It is 5-10 per cent of total RNA in the cell. It codes for amino acid sequence. Therefore, it is also known as coding RNA. Thus messenger RNA is a copy of the information carried by a gene on the DNA. The role of mRNA is to move the information contained in DNA to the translation machinery. mRNA is heterogeneous in size and sequence. About 90 per cent of length of pre-mRNA is removed to produce mRNA Most eukaryotic genes are split into segments – exons and introns. Exons are those sequences in DNA that are complementary to those present in finished mRNA. Exons may be translatable (e.g. sequence from AUG......UAG) or untranslatable (sequence upstream of initiation codon in mRNA) and sequence downstream of termination codon in RNA). Introns are the sequences that are present in DNA but are lacking in finished mRNA. Since, in eukaryotes, nuclear membrane separates the site where mRNA is synthesized and where protein is to be synthesized, mRNA is transported to cytoplasm in order for the translation process to start. Thus transcription and translation are separate processes in eukaryotes. In decoding the open reading frame of a gene for a known protein, one usually encounters periodic stretches of DNA calling for amino acids that do not occur in the actual protein product of that gene. Various steps of pre-mRNA processing are: removal of parts of leader and train, addition of "Cap" at 5'-end, addition of poly(A) tail at 3'-end, and splicing of non-coding sequences (introns) and retaining only some sequences (exons) and methylation of some adenines present.

Processing of Leaders

Addition of cap at 5′-end

Eukaryotic mRNA always has a 5' cap composed of a 5'→5' triphosphate linkage between two modified nucleotides: a 7-methylguanosine and a 2' O-methyl purine. This cap is a modified guanine (m^7GpppN_1) which serves to identify this RNA molecule as an mRNA to the translational machinery. The 5' cap contributes to the stability of the mRNA in the cell. Cap is attached to the 5′-end of the pre-mRNA as it emerges from RNA polymerase II. The cap protects the RNA from being degraded by enzymes that degrade RNA from the 5′-end. Addition of 5' cap takes place at early stages of transcription in eukaryotes by RNA polymerase II. Cap is suggested to be derived by endonucleolytic cleavage and subsequent addition. Different types of caps on mRNA, their frequencies and location are given in Table 9.1. In certain mRNAs (e.g., histone mRNA) caps may be absent and may not be required for translocation.

Table 9.1 Different types of caps of mRNA

Type of cap	*Frequency*	*Location*
Cap 0	100% cases	Cap with a single methyl group at 5′-end of mRNA (m^7G)
Cap 1	Most of the cases	Methyl group may be present on penultimate base at 2′ O position of the sugar moiety
Cap 2	10-50% of the cases	Methyl group may be present on the third base also at the 2′ O position of the sugar moiety

Processing of Trains

3'-end processing of histone pre-mRNAs

Most mRNA molecules contain a poly(Adenosine), symbolized as poly(A), tail at the 3'-end after transcription is completed. The 3' tail also contributes to the stability of the mRNA in the cell. During 3'-end processing, histone pre-mRNAs are cleaved 5 nucleotides after a conserved stem loop by an endonuclease dependent on the U7 small nuclear ribonucleoprotein (snRNP). The upstream cleavage product is degraded by a 5'-3' exonuclease, also dependent on the U7 snRNP. To identify the nuclease activities, Dominski et al. (2005) carried out UV-crosslinking studies using both the complete RNA substrate and the downstream cleavage product, each containing a single radioactive phosphate and a phosphorothioate modification at the cleavage site. They detected a 85-kDa protein that cross-linked to each substrate in a U7-dependent manner. This protein was identified as CPSF-73, a known component of cleavage/polyadenylation machinery. These studies suggest that CPSF-73 is both the endonuclease and 5'-3' exonuclease in histone-pre-mRNA processing and reveal an evolutionary link between 3'-end formation of histone mRNAs and polyadenylated mRNAs. The 3'-end processing of histone pre-mRNAs, as illustrated by Dominski et al. (2005), is shown in Figure 9.1.

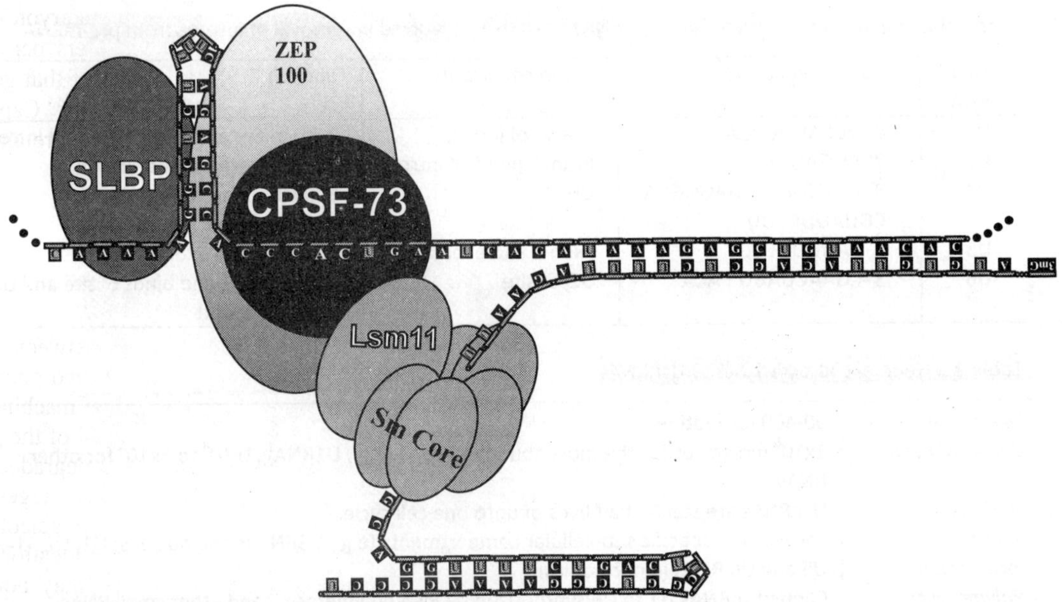

Figure 9.1 3′-end processing of histone pre-mRNAs

Mandel et al. (2006) report the crystal structure of human CPSF-73 at 2.1-Å resolution, complexed with zinc ions and a sulfate that might mimic the phosphate group of the substrate, and the related yeast protein CPSF-100 (Ydh1) at 2.5-Å resolution. Both CPSF-73 and CPSF-100 contain two domains, a metallo-β-lactamase domain and a novel β-CASP domain. The active site of CPSF-73, with two zinc ions, is located at the interface of the two domains. Purified recombinant CPSF-73 possesses RNA endonuclease activity, and mutations that disrupt zinc binding in the active site abolish this activity. These studies provide first direct experimental evidence that CPSF-73 is the pre-mRNA 3′-end-processing endonuclease.

Addition of poly(A) tail

Polyadenylation signal in the train is AAUAAA in pre-mRNA. Polyadenylation involves two steps — cleavage and addition of poly(A) tail. Multiple copies of polyadenylation signal are of common occurrence. Tropomyosin of *Drosophila* has a cluster of 5′ poly(A) signals followed by pair of about 250 residues downstream. First of these was a dual structure AAUAAUAAA. Different signals are assisting in transport during processing and translation, and direct involvement in protein synthesis.

Removal of Introns

Small RNAs, known as U RNAs, are used in removal of introns. Partial sequence, complementarity, and function of five small RNAs, according to Madhani and Guthrie (1994), are given in Table 9.2. Some general characteristics of U snRNAs and other RNAs are enlisted in Table 9.3. The snRNAs have well defined role in splicing of eukaryotic mRNA (Lerner et al. 1980) and also in its polyadenylation (Berget 1984).

Table 9.2 The five small nuclear RNAs (snRNAs or U RNAs) involved in removal of introns from pre-mRNA

Small RNA	*Partial sequence*	*Complementarity*	*Function*
U1	3'-UCCAUUCAUA	5'-end of intron	Recognizes and binds 5'-end of intron
U2	3'-AUGAUGU	Branch point of intron	Binds branch point of intron
U4	3'-UUGGUGU...AAGGGCA CGUAUUCCUU	U6	Binds to (inactivates) U6
U5	3'-CAUUUUCCG	Exon 1 and exon 2	Binds to both exons;
U6	3'-CGAACUAGU...ACA	U2, 5' site	Displaces U1 and binds 5' site and U2 at branch point

Table 9.3 General characteristics of UsnRNAs

Size range	90-400 nucleotides.
Copies per cell	$1x10^6$ molecules for the most abundant RNAs (e.g., U1RNA); $1x10^4$ to $5x10^5$ for other RNAs.
Half-lives	UsnRNAs are stable; half lives of upto one cell cycle.
Specific localization	Localized to specific sub-cellular compartments (e.g., U3RNA in the nucleus; U1, U2, U4, U5 and U6 RNAs in nucleoplasm
Polymerases	Capped snRNAs U1 to U6 synthesized by RNA polymerase II and other small RNAs synthesized by RNA polymerase III.
Cap structure	U1 to U6 snRNAs are capped; U1 to U5 rRNAs contain a trimethyl-guanosine cap and U6 has a different cap.
RNP particles	All UsnRNAs exist in form of RNP particles
Associated with precursor RNAs	The UsnRNAs are associated with precursor hnRNAs and pre-rRNA
Genes coding for mRNAs	No IVS in U1, U2 and U3 genes; genes dispersed throughout the genome. Putative adenylation signal (AATAAA) and the termination signal (TTT at the 3′ end of the RNA) are not found for the U1 gene. U1 RNA is not polyadenylated. Suggested number of genes for each snRNA is 2,000.

Process of pre-mRNA splicing is shown in Figure 9.2. Introns in many eukaryotic genes begin with dinucleotide 5′-GT-3′ and with dinucleotide 5′-AG-3′. There is a limited structural similarity at and involved in addition of poly(A) tail in mRNA processing in different tissues. This is a stretch of adenine (A) nucleotides (200-250 nucleotides of adenine). When a special poly(A) attachment site in the pre-mRNA emerges from RNA polymerase II, the transcript is cut there, and the poly(A) tail is attached to the exposed 3′-end. Some times polyadenylation plays a role in generation of the translational stop codons. Other functions of polyadenylation are providing for stability of mRNAs, Various mechanisms of RNA splicing are known (Smith et al. 1989). The detailed process of intron removal is shown diagrammatically (Figure 9.3). Certain *cis* and *trans* requirements are there. All the

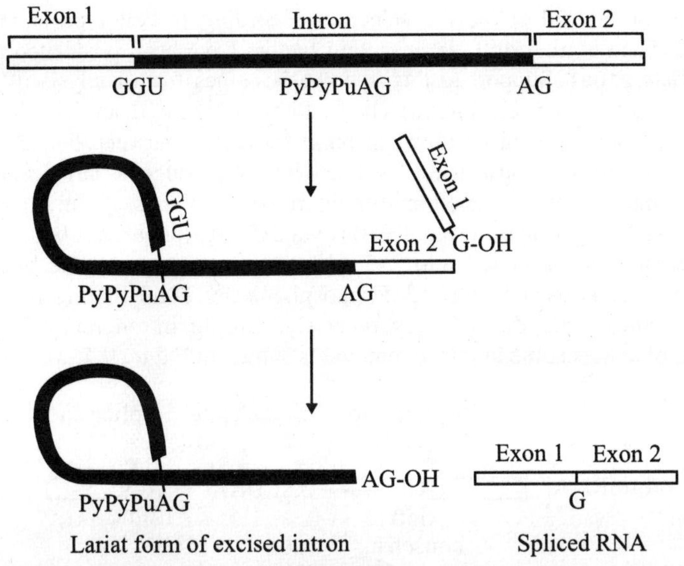

Figure 9.2 Two-step pathway of excision of introns from nuclear pre-mRNAs. This process occurs in spliceosomes (Redrawn from http://what-when-how.com/molecular-biology/rna-splicing-molecular-biology)

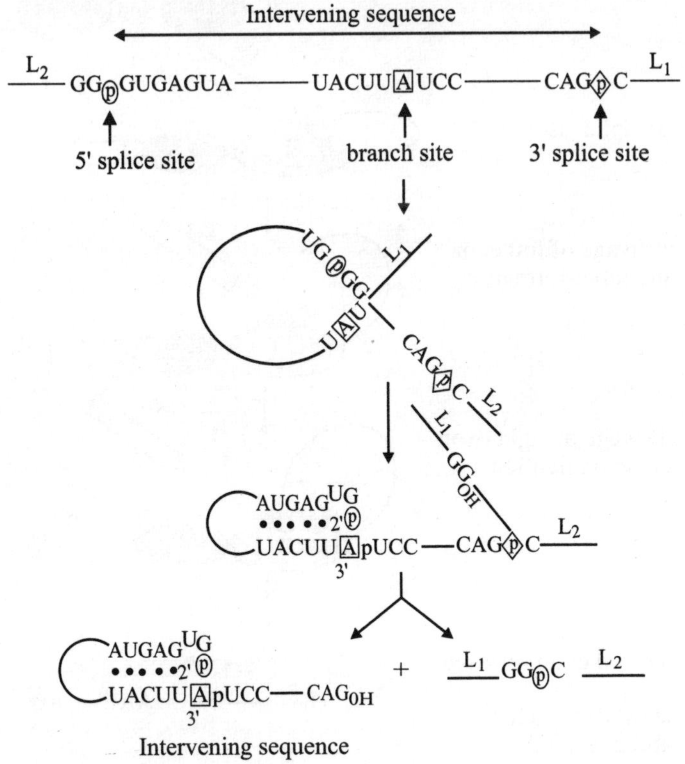

Figure 9.3 Detailed process of intron removal from pre-mRNA

information required for faithful splice-site selection according to cell type is retained within the resulting pre-mRNA (Saga et al. 1990). *Trans*-acting factors have been postulated to account for the selection of appropriate exons. Support to this hypothesis comes from analysis of hybrid T-B cells. Introns are removed by a two step reaction. (a) The first covalent modification of the substrate RNA is cleaved at a precise 5' site followed by covalent bond formation between 5'-end of the intervening sequence and adenosine residue near the 3' splice site to produce a lariat molecule containing intervening sequence and 3' exon. (b) The bipartite intermediate is subsequently resolved by cleavage at 3' splice site followed by ligation of 5' and 3' exons via 3',5'-phosphodiester bond. The phosphates at the 5' splice site becomes incorporated into 2'-5' phosphodiester bond in the branch structure and phosphate at 3' splice site is used to form 3'-5' phosphodiester bond linking the two exons in the produced RNA. The introns are deleted very precisely. During intron removal, a spliceosome is formed. Involvement of spliceosome in intron removal is shown in Figure 9.4.

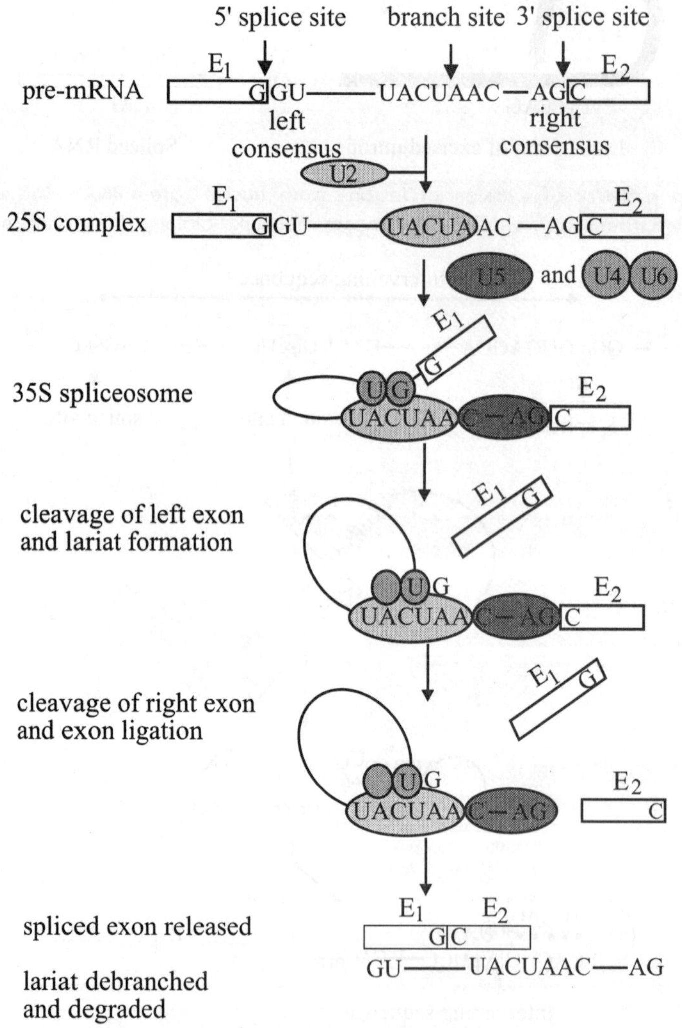

Figure 9.4 Spliceosome formation involves the interaction of small U RNAs with consensus sequences

Conserved sequences exist at exon-intron boundaries in pre-mRNA (Figure 9.5) (Watson et al. 1987). Small nuclear RNAs (snRNAs), called U1, U2, U3, U4, U5 and U6, which range in size from 100-215 nucleotides, are present in cells of all vertebrate species. They possess unusual trimethylated (m32.2.7G) cap structure at 5′-end, except for U6. SnRNAs complex with nucleoproteins to form small nuclear ribonucleoproteins (snRNPs). Involvement of U1 snRNP is known in pre-mRNA splicing. Exons sometimes correspond to protein folding domains. This is particularly true in case of *β-hemoglobin* gene. Nearly all RNA processing is nuclear. Histone gene expression is cell cycle-dependent. Turning on of heat-shock genes is controlled at both transcriptional and translational levels.

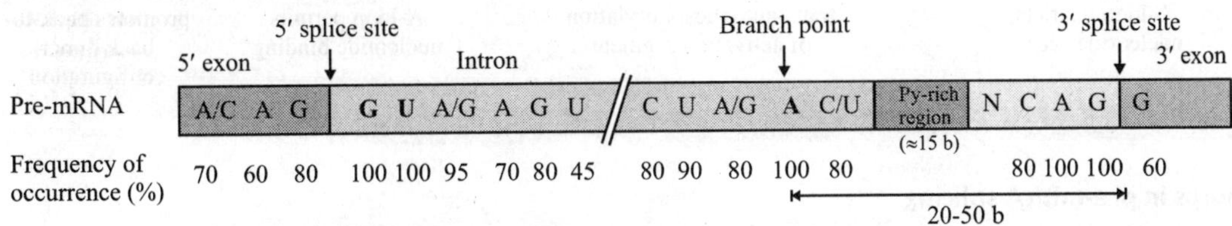

Figure 9.5 Exon-intron boundaries in pre-mRNA (Redrawn from http://oregonstate.edu/dept/biochem/hhmi/hhmiclasses/bb451/lectnotesgdp/EukaryoticmRNA.html)

These RNAs have been named as U1, U2, U3, U4, U5 and U6 RNA, and range in size from 110 to 220 bases. One, U3, is found in the nucleolus, the site of rRNA synthesis. All these U RNAs are very abundant and there may be as many as one million copies per nucleus. Their base sequences are highly conserved between organisms, and all seem to contain unusual trimethylguanosine structures at their 5′-end. These RNAs are most likely involved in processing of RNA. Precursor mRNA splicing may be catalyzed by RNA and that U6 may be one of the key RNA species involved (Guyer 1990). One of the processing ribonucleases in *Escherichia coli*, RNase P, contains a small RNA complexed to the protein. This RNA is about 350 nucleotides long and is essential for enzyme activity. It is not clear whether it plays a strictly structural role or contributes some specificity in helping the nuclease recognize the processing site.

Catalysis of mRNA via a spliceosome-independent manner

Ire1 is an ancient transmembrane sensor of ER stress with dual protein kinase and ribonuclease activities. In response to ER stress, Ire1 catalyzes the target mRNA via a spliceosome-independent manner. Lee et al. (2008) have determined the crystal structure of the dual catalytic region of Ire1 at 2.4-Å resolution, revealing the fusion of a domain, which they term the KEN domain, to the protein kinase domain. Dimerization of the kinase domain composes a large catalytic surface on the KEN domain which carries out ribonuclease function. They further show that signal induced trans-autophosphorylation of the kinase domain permits unfettered binding of nucleotide, which in turn promotes dimerization to compose ribonuclease active site.

Comparison of Ire1 to a topologically disparate nuclease reveals the convergent evolution of their catalytic mechanism. These findings provide a basis for understanding the mechanism of action of RNase L and other pseudokinases, which represent 10 per cent of the human kinome. A model of dimerization induced signaling by Ire1 across the ER membrane, as proposed by Lee et al. (2008), is given in Figure 9.6.

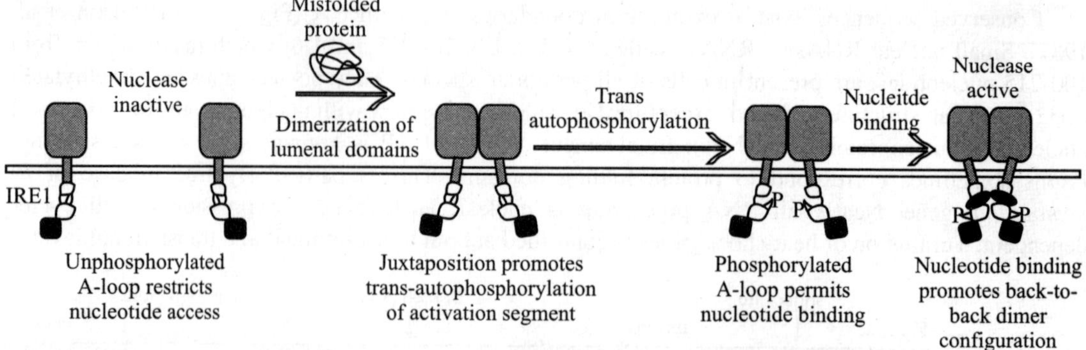

Figure 9.6 Model of dimerization induced signaling by Ire1 across the ER membrane

Snurps in pre-mRNA splicing

The introns of nuclear mRNA precursor transcripts are spliced out in a two-step reaction carried out by complex ribonucleoprotein particles called spliceosomes. They also attach new non-coding segments to the leading and trailing ends of the mRNA. This process is called splicing.

HnRNA and mRNA are never found free in the cell. Like DNA, they are bound by cations and proteins. These complexes are termed ribonucleoproteins or RNPs. Snurps stands for small nuclear ribonucleoprotein particles that help to remove meaningless introns from the messages issues by a cell's genes. Without snurps cellular activity would come to a halt. Steitz and her coworkers discovered snurps in 1978 (Lerner and Steitz 1979). These small complexes of RNA and a variety of proteins are critical components of a sophisticated molecular assemblage called spliceosome. Watson et al. (1987) proposed role of snurps in nuclear function. Both their protein components and the sequences of their RNA are highly conserved across all higher eukaryotic species. Since U3 snRNP has been localized to the nucleolus, it is expected to participate in ribosome biogenesis. The remaining snurps, namely U1, U2, U4, U5, U6, etc., reside in the nucleoplasm and are often found associated with hnRNA are therefore thought to contribute to various aspects of pre-mRNA processing. The major chemical constituent of snurps in uracil and they are thus named as U1 to U12. U1 to U4 are most abundantly found in all eukaryotic cells. One ribonucleoprotein (RNP) particle that contains a small RNA, U7, is necessary in processing the non-polyadenylated histone H3 mRNA. Still another RNP that contains small RNA, U1, is involved in the formation of the poly(A) tail.

Genes coding for snRNAs

No intervening sequences are found in U1, U2, and U3 genes. These genes dispersed throughout the genome. Putative adenylation signal (AATAAA) and the termination signal (TTT, at the 3'-end of the RNA) are not found for the U1 gene. U1 RNA is not polyadenylated. Suggested number of genes for each snRNA is 2,000. Termini and chain lengths of snRNAs are given in Table 9.4.

Mechanism of snRNP action

The exact mechanism of snRNP action is not known; however, Maniatis and Reed (1987) gave three possibilities of their mechanism of action: (a) they help to arrange the intron in a configuration that encourages self-splicing, (b) in some organism precursor mRNA forms lariat structure and splicing of intron occurs without the aid of protein or other RNA factors and (c) splicing may provide missing intron sequences necessary for spontaneous splicing. There are certain interesting questions about

Table 9.4 Termini and chain lengths of snRNAs

RNA	5′ terminus	Chain length	Alternative nomenclature
UsRNA			
U1	m3GpppAmUmA	165	D
U2	m3gPPPAmUmC	188-189	C
U3	m3GpppAmAG	210-214	A, D2
U4	m3GpppAmGmC	142-146	F
U5	m3GPPPAMUmA	116-118	G, 5SIII
U6	XpppGUG	107-108	H1, 4.5III
Other snRNAs			
tRNA	pN	80	
5S	pppN	121	G
5.8S	pN	158	E
La 4.5	pppG	90-94	
La 4.5I	pppG	98-99	H2
7S	pppG	294-925	L
VIA,II	pppG	157-160	
7-1	pppN	260	M
7-2	pppN	290	M
7-3	pppN	300	K
8S	pppN	400	
Y1-Y3	pppN	100	

splicing which need to be answered: (a) Splicing is subject to regulation. The same pre-mRNA can be spliced in different ways at different developmental stages or in different tissues. It still needs to be understood how snRNPs or other spliceosome factors could mediate such a regulation. (b) The pathway of spliceosome assembly and the total number of components involved are not clear. (c) The agent that holds the upstream exon until it is joined to its downstream exon has not yet been identified. The mechanism that ensures the fidelity of splicing is also not clear. Further, Inouye and Delihas (1988) discussed the diverse roles of small RNAs in the prokaryotes. Mowry and Steitz (1987) described the role of human U7 snRNP in the 3′-end maturation of histone pre-mRNAs. Snurps-like particles identified in cytoplasm and nucleolus are, respectively, named as "scRNPs" (or "scyrps") and "snorps" (Lerner and Steitz, 1981).

Spliceosomes

Spliceosome is a dynamic structure assembled on pre-mRNA templates in a stepwise fashion and then dissociates as the splicing reaction is completed. Even a small error in splicing will be intolerable resulting into non-functional polypeptide. Thus RNA processing involves high degree of specificity. Maniatis and Reed (1987) proposed a model for spliceosome formation. The multiple subunit structure of spliceosome has implications for regulation of a splicing event. The presence of multiple snRNPs in the spliceosome suggests that snRNP-snRNP interaction may dictate the backbone of this structure. The multiple snRNP nature of spliceosome also raises the possibility that different introns may be recognized by different combinations of snRNPs. This provides a mechanism by which the splicing of different introns could be individually regulated.

Spliceosome Assembly: A schematic working model of spliceosome assembly, splicing and disassembly is shown in Figure 9.7 (Ohi et al. 2007). The structure of substrate RNA is shown on the first line. Addition of this RNA to a nuclear extract results in the recognition of the 5′ splice site by U1 snRNP. The precursor RNA is probably also associated with hnRNP core proteins. It has been suggested that U5 snRNP binds to the 3′ splice site at this stage. These interactions are indicated in the second line. Incubation of substrate RNA in the presence of ATP results in the rapid formation of a U2 snRNP complex as shown on the third line. Further incubation yields assembly of the spliceosome containing at least U2, U4, U5 and U6 snRNAs. The U4 and U6 snRNAs can be associated in one snRNP particle. The two RNAs characteristics of the intermediate in splicing, the 5′ exon and the lariat intervening sequence terminating in the 3′ exon RNAs, are found exclusively in the spliceosome.

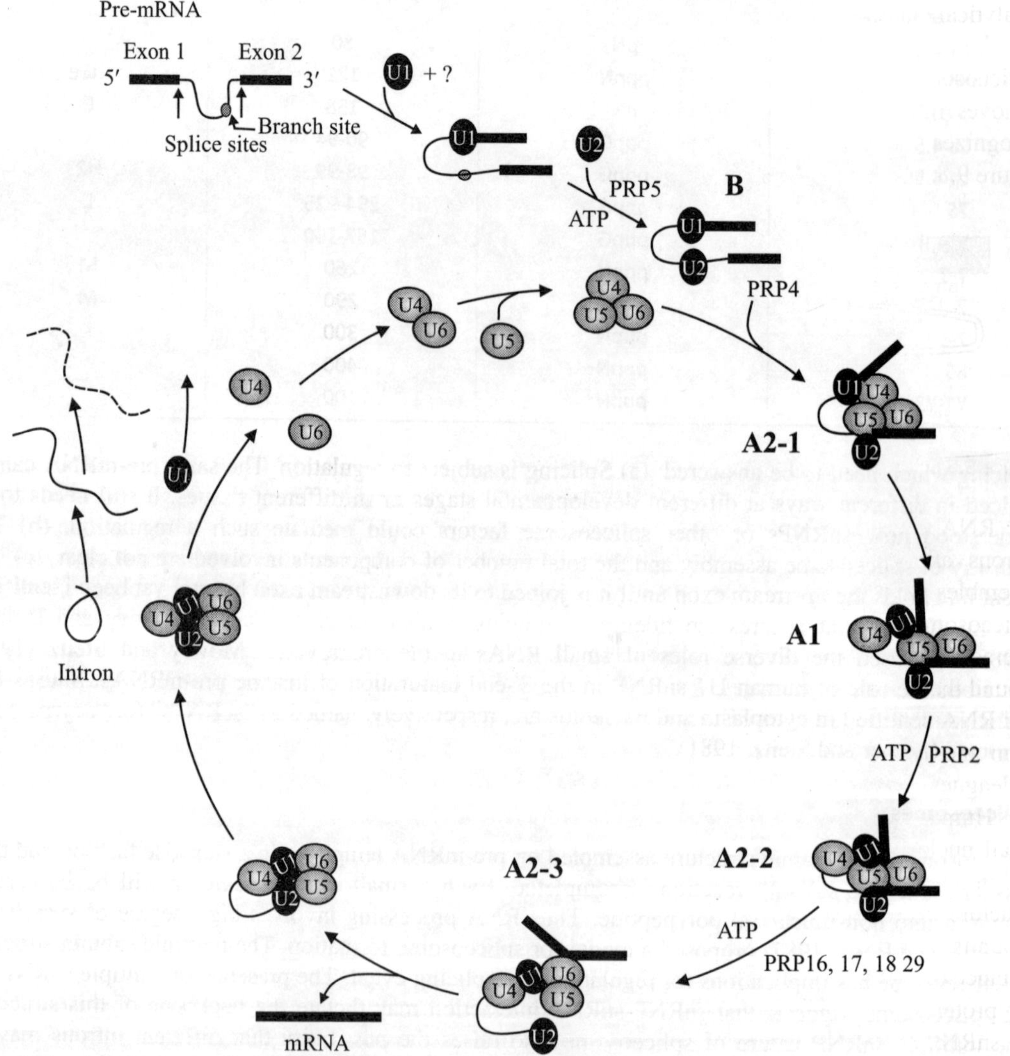

Figure 9.7 A schematic working model of spliceosome assembly, splicing and disassembly (Redrawn, with permission, from Ohi, M.D. et al. 2007. Proc. Natl. Acad. Sci. USA 104: 3195-200)

Active Site of Spliceosome: Formation of catalytically active RNA structures within the spliceosome requires the assistance of proteins. Number and nature of proteins needed to establish and maintain the spliceosome's active site is not fully known. Bessonov et al. (2008) purified human spliceosomal C complexes and showed that they catalyze exon ligation in the absence of added factors. Marked exchange of proteins is observed during the transition from the B to C complex, with apparent stabilization of Prp19-CDC5 complex proteins and destabilization of SF3a/b proteins. Disruption of purified C complexes led to the isolation of a salt-stable ribonucleoprotein (RNP) core that contained both splicing intermediates and U2, U5 and U6 small nuclear RNA plus predominantly U5 and human Prp19-CDC5 proteins and Prp19-related factors. These data provide insights into the spliceosome's catalytic RNP domain and indicate a central role for the aforementioned proteins in sustaining its catalytically active structure.

Spliceosome Crystal Structure: The spliceosome enzyme binds to RNA transcripts at splice sites and removes intron sequences. The crystal structure of a spliceosome subunit shows how the enzyme recognizes one end of the intron (Query 2009). Spliceosome assembly and RNA splicing is shown in Figure 9.8.

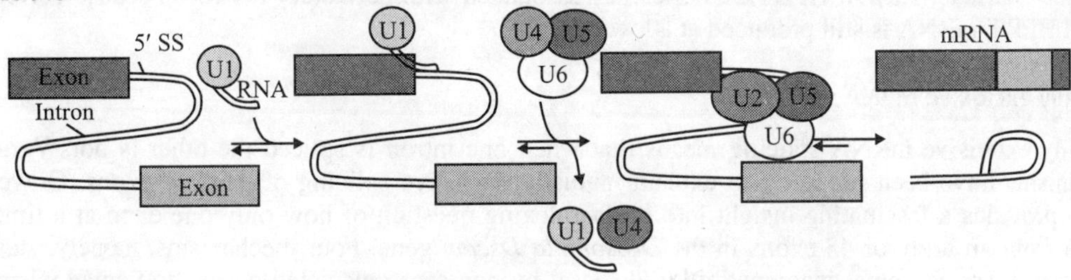

Figure 9.8 Spliceosome assembly and RNA splicing

RNA transcripts consist of regions that encode proteins (exons) and non-coding regions (introns). Introns are excised by a multi-component ribonucleoprotein known as the spliceosome, which assembles at both ends of the intron. (a) In the first step of assembly, the U1 component of the spliceosome binds to the 5' splice site (5'SS) through a combination of RNA–RNA and RNA–protein interactions. (b) The remaining subunits of the spliceosome (U2, and a complex of U4/5/6) assemble around the U1–intron complex, and then U1 and U4 are displaced, activating the catalytic spliceosome and RNA splicing. (c) The spliceosome removes the intron as a lariat, and joins the exons together to form messenger RNA ready for translation. The crystal structure of U1 reported by Nagai and colleagues1 offers clues to how U1 recognizes 5'SS and how U1 assembles from its own subunits.

Human spliceosomal U1 small nuclear ribonucleoprotein particles (snSNPs), which consist of U1 small nuclear RNA and ten proteins, recognizes the 5' splice site within precursor messenger RNAs and initiate the assembly of spliceosome for intron excision. Krummel et al. (2009) present the detailed structure of a spliceosomal snRNP, revealing a hierarchical network of intricate interactions between subunits. A striking feature is the amino (N)-terminal polypeptide of U1-70K, which extends over a distance of 180Å from its RNA binding domain, wraps around the core domain consisting of the seven Sm proteins, and finally contacts U1-C, which is crucial for 5'-splice-site recognition. The structure of U1 snRNP provides insights into U1 snRNP assembly and suggests a possible mechanism of 5'-splice-site recognition.

Spickles in RNA splicing

Pre-mRNA splicing is a predominantly co-transcriptional event, which involves a large number of essential splicing factors within the mammalian cell nucleus. Most splicing factors are concentrated in 30-40 distinct domains called specklets. The function of the specklets and the organization of cellular transcription and pre-mRNA splicing in vitro are not well understood. Misteli et al. (1997 show that speckles are highly dynamic structures that respond specifically to activation of nearby genes. These dynamic events are dependent on RNA polymerase II transcription and are sensitive to inhibitors of protein kinases and Ser/Thr phosphatases. When single genes are transcriptionally activated in living cells splicing factors leave speckles in peripheral extensions and accumulate at new sites of transcription. One function of speckles thus suggested is to supply splicing factors to active genes. Thus interphase nucleus is far more dynamic in nature than previously thought.

RNA splice mutations

Among patients suffering from the Lesch-Nyhan syndrome or gouty arthritis, the class of splice mutations amounts only to 7 per cent (Rossi et al. 1990). Carriers of splice mutations often do not show the characteristics of HPRTase deficiency associated with hereditary diseases because correctly spliced HPRT mRNA is still produced at a low rate.

Mutually exclusive mRNA splicing

Mutually exclusive mRNA splicing means that when one intron is spliced the other is not. Various mechanisms have been proposed to explain mutually exclusive splicing of pairs of exons. Graveley (2005) provides a fascinating insight into the perplexing question of how only one exon at a time is chosen from an array of 48 exons in the *Drosophila Dscam* gene. Four mechanisms, namely, stearic interference, spliceosomal incompatibility, disposal by non-sense-mediated decay, and antagonism of repressor by Docker: selector pairing, of mutually exclusive splicing are depicted in Figure 9.9 (Smith

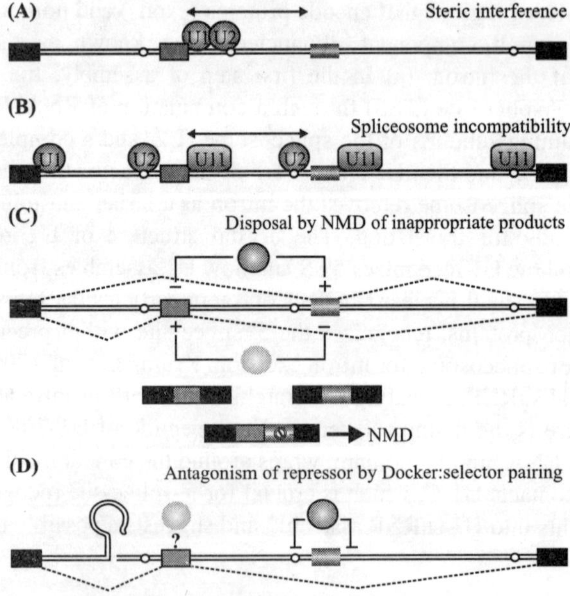

Figure 9.9 Mechanisms of mutually exclusive splicing

2005). Stearic interference occurs when the brach point (white circle) of the downstream mutually exclusive (ME) exon is too close to the upstream 5' splice site, as is the case in the coding of α-tropomyosin. ME exons are shown as filled boxes and constitutive exons as empty boxes. In case of spliceosomal incompatibility, the splice sites used by the "U1/U2" and "U11/U12" snRNP containing spliceosomes have distinct consensus sequences and are incompatible. For example, an intron U1 5' splice site and a U12 3' splice site cannot be spliced, as is the case of exons 6 and 7 of the human JNK1gene. Double headed arrows show impossible splice pathways. Non-sense-mediated decay (NMD) can dispose of mRNAs containing both ME exons if a premature termination codon is introduced as in the *FGFR2* gene. In case of antagonism of repressor by Docker:selector pairing, of mutually exclusive splicing, base-pairing between the docker element and one of the selector elements that precedes each ME exon encounters the action of a repressor that acts on all of the exon variants. Because only one selector can base pair with the docker, only the associated exon will be spliced. Structural organization of the *Drosophila Dscam* gene is presented in Figure 9.10 (Graveley 2005). This gene contains 115 exons, 95 of which are alternatively spliced. The exons 4, 6, and 9 clusters contain 12, 48, and 33 alternative exons, respectively. Exon 17 cluster contains two exons that encode alternative versions the transmembrane domain. The exons within each cluster are alternatively spliced in a mutually exclusive manner.

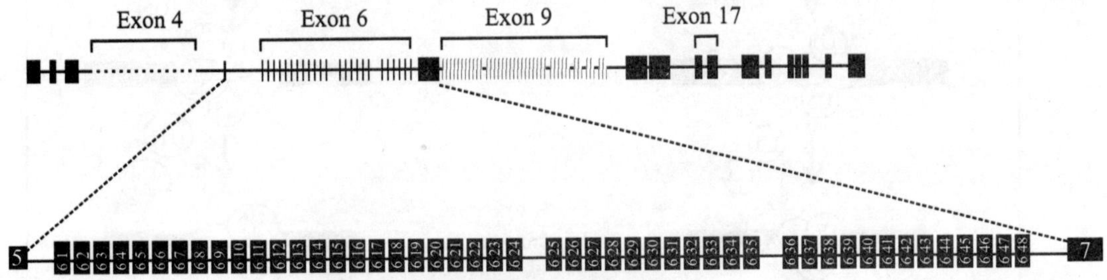

Figure 9.10 Structural organization of the *Drosophila Dscam* gene

Blocking of processing of primary mRNA transcripts

MicroRNAs (miRNAs) play critical role in development and dysregulation of miRNA expression has been observed in human malignancies. Recent evidence suggested that the processing of several primary miRNA transcripts (pri-miRNAs) is blocked post-transcriptionally in embryonic stem cells, embryonic carcinoma cells, and primary tumors. Viswanathan et al. (2008) show that Lin28, a developmentally-regulated RNA binding protein, selectively blocks the processing of pri-let-7 miRNAs in embryonic cells. Lin28 is necessary and sufficient for blocking microprocessor-mediated cleavage of pri-let-7 miRNAs. These results identify Lin28 as a negative regulator of miRNA biosynthesis and suggest that Lin28 may play a central role in blocking miRNA-mediated differentiation in stem cells and in certain cancers.

Major and minor splicing mechanisms

Two splicing mechanisms, known as major and minor splicing, have been identified (Will and Rührmann 2005; Caceres and Mistell 2007) (Figure 9.11). Accordingly, major and minor spliceosomes have been identified. This segregating splicing is shown in Figure 9.12. Koning et al. (2007) demonstrate that the two splicing pathways are spatially separated in the cell and may have distinct functions. Subcellular localization and function of major and minor pre-mRNA splicing pathways

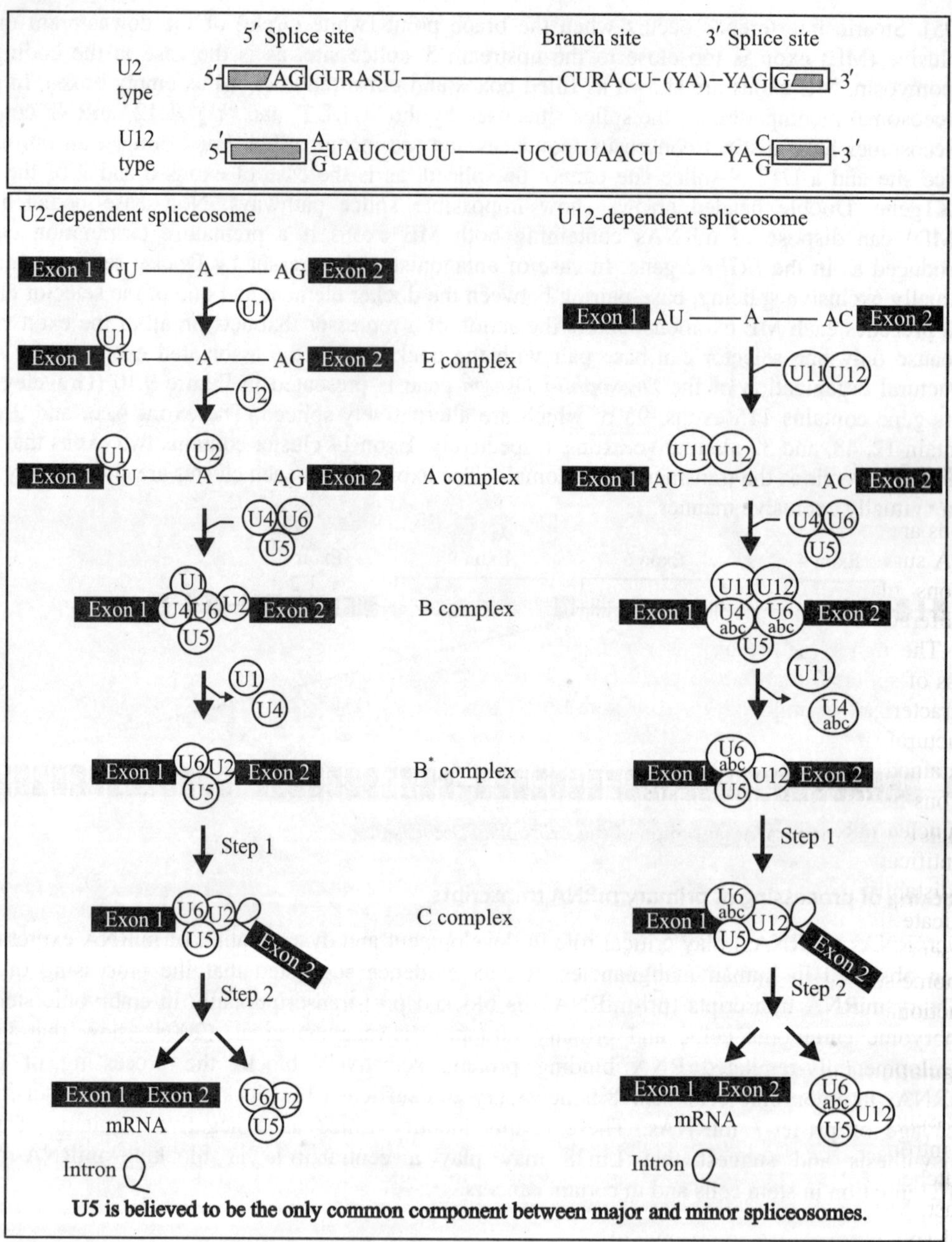

Figure 9.11 Major and minor splicing mechanisms (Redrawn from http://en.wikipedia.org/wiki/Minor_spliceosome)

are shown. The major splicing pathway removes U2-type introns allowing the adjacent exons to be spliced together in the nucleus before export of the fully spliced transcripts to the cytoplasm. The transcripts that also contain U12-type introns are exported as partially spliced transcripts and their

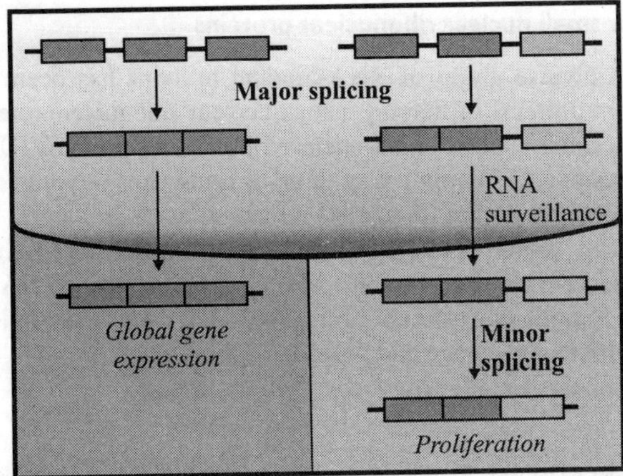

Figure 9.12 Segregating splicing

exons are spliced together by the U12 pathway in the cytoplasm. The mechanism by which they evade RNA surveillance in the nucleus is unknown. Although U2-type introns are found globally, U12-type introns may occur preferentially in particular sets of genes such as those involved in cellular proliferation.

The minor spliceosome is ribonucleoprotein complex that catalyzes the removal of an atypical class of spliceosomal introns (U12-type) from eukaryotic messenger RNAs. It was first identified and characterized in animals, where it was found to contain several unique RNA constituents that share structural similarity with and seem to be functionally analogous to the small nuclear RNAs (snRNAs) contained in the major spliceosome. Subsequently, minor spliceosomal components and U12-type introns have been found in plants but not in fungi. The evolutionary history of the minor spliceosome is unclear because there is evidence of it in so few organisms. Russell et al. (2006) report the identification of homologs of minor-spliceosome-specific proteins and snRNAs, and U12-type introns, in distantly related eukaryotic microbes (protists) and in a fungus (*Rhizopus oryzae*). These results indicate that the minor spliceosome had an early origin: several of its characteristic constituents are present in representative organisms from all eukaryotic subgroups for which there is any substantial genome sequence information. In addition, these results revealed marked evolutionary conservation of functionally important sequence elements contained within U12-type introns and snRNAs.

Regulation of mRNA splicing

Messenger RNAs don't usually correspond exactly to DNA – portions of the primary transcript, known as introns, are removed by splicing. A study reveals new ways in which splicing can be regulated (Futcher and Leatherwood 2008). Rem1 is a cyclin that is only expressed during meiosis in fission yeast. Molden et al. (2008) show that Rem1 expression is regulated at the level of both transcription and splicing, encoding two proteins with different functions depending on the intron retention. The regulation of rem1 splicing is not dependent on any transcribed region of the gene. Furthermore, when the rem1 promoter is fused to other intron-containing genes, the chimaeras show a meiosis specific regulation of splicing, exactly the same as endogenous rem1. This regulation is dependent on two transcription factors of the forkhead family, Mei4 and Fkh2. Whereas Mei4 induces both transcription and splicing of rem1, Fkh2 is responsible for the intron retention of the transcript during vegetative growth and the pre-meiotic S phase.

RNA-binding domain for small nuclear ribonuclear proteins

A domain common to a diverse group of RNA-binding proteins has been described (Mattaj 1989). Included in this group are proteins of heterogeneous nuclear ribonucleoproteins in hnRNPs, poly(A)-binding protein (PABP), some U class small nuclear ribonuclear proteins (UsnRNP), and *Drosophila* proteins that regulate sex-specific alternative splicing of transcripts required for sexual differentiation. These proteins have a common region of size 80-90 amino acids in length, known as "RNA-binding domain". A short stretch of 8-amino-acid in this domain is much more tightly conserved, known as "RNP consensus" sequence. HnRNP C proteins are required for pre-mRNA splicing. These proteins also specifically bind to regions downstream of the AAUAAA polyadenylation signal that is required for in vivo and in vitro efficient cleavage and polyadenylation.

Reverse RNA splcing

RNA molecules have been found to catalyze the breakage and reunion of phosphodiester bonds (Grivell 1990). Introns are self-spliced in a very simple way. Finding that reverse splicing occurs in vivo provides a piece of jigsaw that one day may allow us to reconstruct the paths that introns take when on the move. Nuclear pre-messenger RNA (pre-mRNA) splicing is an essential processing step for the production of mature mRNAs from most eukaryotic genes. Splicing is catalyzed by a large nucleoprotein complex, the spliceosome, which is composed of five small nuclear RNAs and more than 100 protein factors. Despite the complexity of the spliceosome, the chemistry of the splicing reaction is simple, consisting of two consecutive transesterification reactions. The presence of introns in spliceosomal RNAs of certain fungi has suggested that splicing may be reversible. Tseng and Cheng (2008) have shown that both catalytic steps of splicing can be efficiently reversed under appropriate conditions. These results provide considerable insight into the catalytic flexibility of the spliceosome.

Processing of Telomerase RNA

The spliceosome is best known for shepherding primary messenger RNA transcripts to maturity. This enzyme complex also contributes to the synthesis of an enzyme that maintains chromosome ends (Bonnal and Valcarrcel 2008). Processing of telomerase RNA by the spliceosome is shown in Figure 9.13. Chromosomal ends contain telomeric repeats, which are added to the 3'-end of DNA molecules by the telomerase enzyme complex. Telomerase consists of a reverse transcriptase enzyme (T), which uses another component of this complex, the telomerase RNA, as a template for telomere synthesis. In fission yeast, telomerase RNA is initially transcribed as a precursor containing an intron. Mature telomerase RNA is then generated by the spliceosome, which cleaves the 5'-end of the intron (Box et al. 2008). The second step of the 'standard' splicing reaction catalyzed by the spliceosome, which would cleave the intron at the 3'-end and mediate the splicing of the exons, is inhibited during maturation of telomerase RNA.

This completes the mRNA molecule, which is now ready for export to the cytosol. The remainder of the transcript is degraded, and the RNA polymerase leaves the DNA.

PRE-TRANSFER (SOLUBLE) RNA PROCESSING

Cells produce at least 40 different kinds of tRNAs. Each of which is a copy of specific tRNA gene. First pre-tRNA is synthesized which after certain modifications become tRNAs. They contain a large number of unusual bases. tRNAs interact with specific amino acids. tRNAs have various loops that

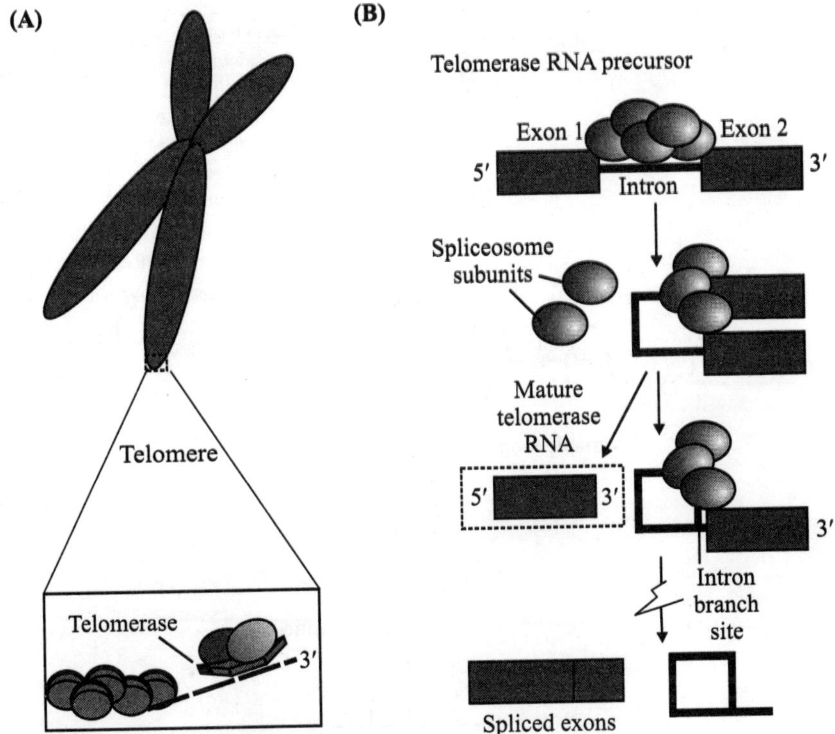

Figure 9.13 Processing of telomerase RNA by the spliceosome

serve useful functions. Some sequences in tRNA are complementary and they give stem-like structures. It is a small molecule of 75-85 ribonucleotides. The 3'-end of tRNA carries CCA sequence. It is with adenine of this sequence that an amino acid attaches itself with tRNA. tRNA contains an anticodon which permits temporary complementary pairing with codon on mRNA. tRNAs are of several different kinds and each type attaches with it a specific amino acid and brings it to the ribosome. Partial pairing of some complementary sequences gives tRNA a cloverleaf structure. tRNA for alanine ($tRNA^{Ala}$) has been studied in detail and its structure is given in Figure 9.14.

The presence or absence the 3'-terminal CCAUCA sequence in tRNA precursor substrate markedly affects the way in which these substrates interact with the catalytic RNA in the enzyme-substrate complex (Guerrier-Takada et al. 1989). There is a nucleotide, residue C92 in M1RNA that affects the site of cleavage by the enzyme. Actually, removal of intron is among the last steps in the processing of tRNAs. Different stages of tRNA maturation in fixed temporal order are:

- Trimming of the 5' and 3' ends in which RNase P acts on the 5' terminus endonucleolytically and generates mature 5'-terminus by characterizing mature tRNA sequence. The 3'-end maturation takes place by exonuclease activity by RNase D and 3'-CCA sequences of tRNA C (as in prokaryotes). When -3'-CCA is not encoded in DNA, this triplet is enzymatically added. But this triplet has to be added before removal of intron. Second situation is present in eukaryotes.
- Modification of the bases in mature coding region can occur on either the precursor molecular or cleavage products.

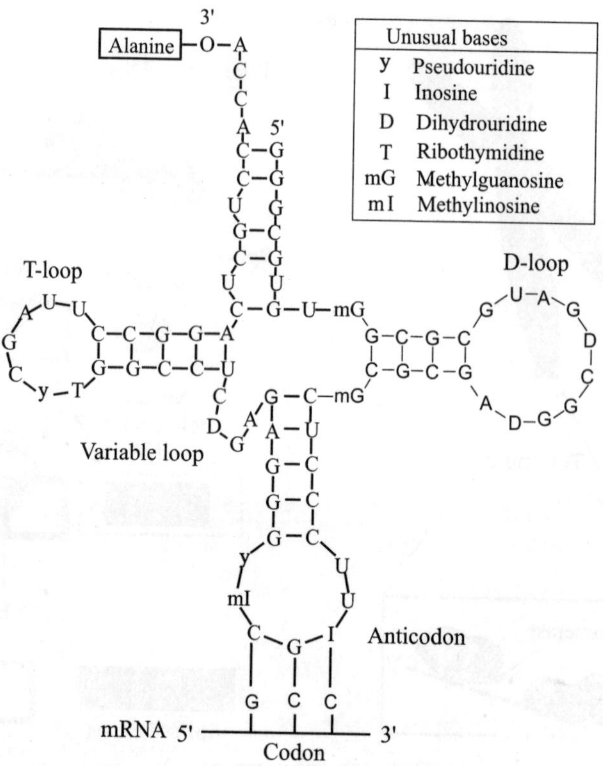

Figure 9.14 Structure and sequence of alanine transfer RNA in yeast. Unusual bases are also defined anticodon (on tRNA) and codon

- Removal of introns takes place when present. Introns are very short (13-60 nucleotides). The cleavage and splicing at intron junctions has been found to be highly dependent upon the presence of the -CCA and piece.

Ribonuclease P is the sole endonuclease responsible for processing the 5'-end of tRNA by cleaving the precursor and leading to RNA maturation. It was one of the first catalytic RNA molecules identified and consists of a single RNA component in all organisms and only one protein component in bacteria. It a true multi-turnover ribozyme and one of the only two ribozymes (the other being the ribosome) that are conserved in all kingdoms of life. Torres-Larios et al. (2005) show the crystal structure at 3.85-Å resolution of the RNA component of *Thermotonga maritime* ribonuclease P. The entire RNA catalytic component is revealed, as well as the arrangement of the two structural domains. The structure shows the general architecture of the RNA molecule, the inter- and intra-domain interactions, the location of the universally conserved regions, the regions involved in pre-tRNA recognition and the location of the active site. A model with bound tRNA is in agreement with all existing data and suggests the general basis for RNA-RNA recognition by this ribozyme.

tRNA CCA-adding Polymerase

CCA-adding polymerases mature the essential 3'-CCA terminus of transfer RNA without any nucleic-acid template. However, it remains unclear how the nucleotide triphosphate is selected in each reaction step and how the polymerization is driven by the protein and RNA dynamics. Tomita et al. (2006) present complete sequential snapshots of six complex structures of CCA-adding enzyme and four distinct RNA substrates with and without CTP (cytosine triphosphate) or ATP (adenosine triphosphate). The CCA-lacking RNA stem extends by one base pair to force the discriminator nucleoside into the active-site pocket, and then tracks back after incorporation of the first cytosine monophosphate (CMP). Accommodation of the second CTP clamps the catalytic cleft, including a reorientation of the β-turn, which flips C74 to allow CMP to be accepted. In contrast, after the second CMP is added, the polymerase and RNA primer are locked in the closed state, which directs the subsequent A addition. Between the CTP- and ATP-binding stages, the side-chain conformation of Arg224 changes markedly; this step is controlled by the global motion of the enzyme and position of the primer terminus, and is likely to achieve the CTP/ATP discrimination, depending on the polymerization stage. Throughout the CCA-binding reaction, the enzyme tail domain firmly anchors the TψC-loop of the tRNA, which ensures accurate polymerization and termination.

Transfer RNA nucleotidyltransferases (CCA-adding enzymes) are responsible for the maturation or repair of the functional 3'-end of tRNAs by means of the addition of the essential nucleotides CCA. However, it is unclear how nuceotidyltransferases polymerize CCA onto the 3' terminus of immature tRNAs without using a nucleic acid template. Xiong and Steitz (2004) describe the crystal structure of *Archaeoglobus fulgidus* tRNA nucleotidyltransferase in complex with tRNA. They also present ternary complexes of this enzyme with both RNA duplex mimics of the tRNA acceptor stem that terminate with the nucleotides C74 or C75, as well as appropriate incoming nucleoside 5'-triphosphates. A single nucleotide-binding pocket exists whose specificity for both CTP and ATP is determined by the protein side chain of Arg224 and backbone phosphates of the tRNA, which are non-complementary to and thus exclude UTP and GTP. Discrimination between CTP or ATP at a given addition step and at termination arises from changes in the size and shape of the nucleotide binding site that is progressively altered by the elongating 3'-end of the tRNA. The sequences of the AC74 and ACC75 RNA constructs are shown in Figure 9.15 (Xiong and Steitz 2004). The boxed regions indicate sequences identical to those of the acceptor stem of yeast tRNA^{Phe}.

AC74: GCGGAUAUCCGCAC
CACGCCUAUAGGCG

ACC75: CGGGAUCCGCACC
CCACGCCUAGGCG

Figure 9.15 The sequences of the AC74 and ACC75 RNA constructs

Intron Removal

RNase P for *E. coli* or its catalytic RNA subunit can evidently cleave small RNA substrates that lack conserved feature of natural substrates of RNase P if an additional small RNA is also present (Foster and Altman 1990). This small RNA must contain a sequence complementary to the substrate (external guide sequence, EGS) and a 3′-proximal CCA sequence to ensure cleavage. The universality of ribonuclease P (RNase P), the ribonucleoprotein essential for transfer RNA maturation, is challenged in the archaeon *Nanoarchaeum equitans*. Neither extensive computational analysis of the genome nor biochemical tests in cell extracts revealed the existence of this enzyme. Randau et al. (2008) showed that the conserved placement of its tRNA gene promoters allows the synthesis of leaderless tRNAs, whose presence was verified by the observation of 5′ triphosphorylated mature tRNA species. Initiation of tRNA gene transcription requires a purine, which coincides with the finding that tRNAs with a cytosine in position 1 display unusually extended 5' termini with an extra purine residue. These tRNAs were shown to be substrates for their cognate aminoacyl-tRNA synthetase. These findings demonstrate how nature can cope with the loss of the universal and supposedly ancient RNase P through genomic rearrangement at tRNA genes under the pressure of genome condensation.

Precursor tRNA structure is given in Figure 9.16 (Swerdlow and Guthrie 1984). Intervening sequences (IVS) are located in most of the tRNA precursors of eukaryotes and in advanced genera of prokaryotes, at a point, one site removed downstream from anticodon. However, in the gene for tRNA^{Leu}, the intron splits the anticodon sequence. For removal of intron from tRNA transcript, first step is cleavage of tRNA sections from the inserted sequence. These different tRNA precursors are separated by RNase E and III producing single tRNA precursors. The monomeric tRNA precursors interact with one or more RNases reducing the size of originally long tRNA precursor and permitting the mature tRNA precursor to assume its final secondary and tertiary structure. The introns of tRNA precursors are exercised by precise endonucleolytic cleavage and ligation reactions catalyzed by special splicing endonuclease and ligase activities. Splicing of tRNA precursors is shown in Figure 9.17 (Westaway and Abelson 1995).

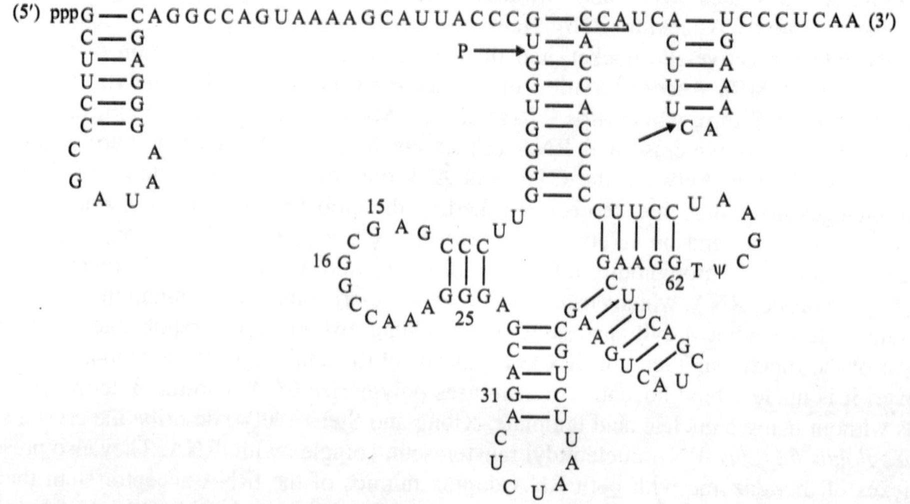

Figure 9.16 Generalized structure of tRNA precursor

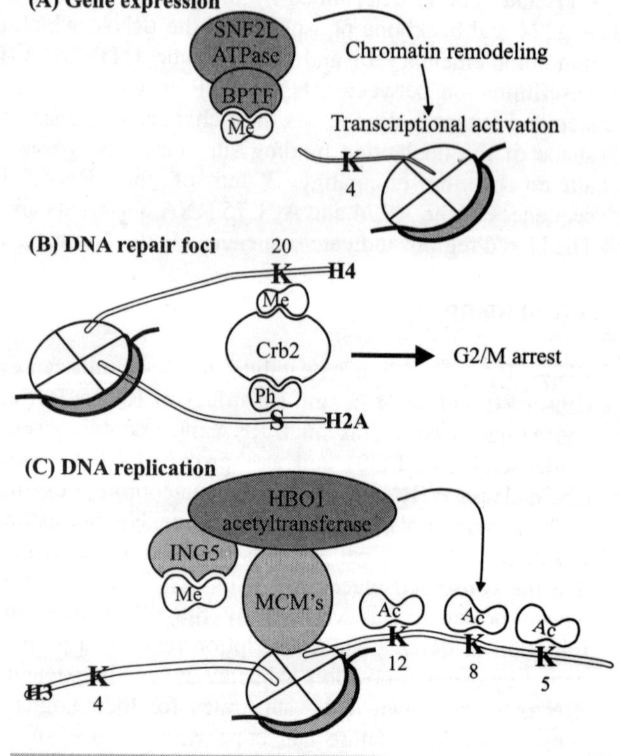

Figure 9.17 Splicing of tRNA precursors

Pathway of tRNA splicing is shown in Figure 9.18 (Abelson et al. 1998). The tRNA processing in eukaryotes has been shown in Figure 9.19. Pre-tRNAs introns

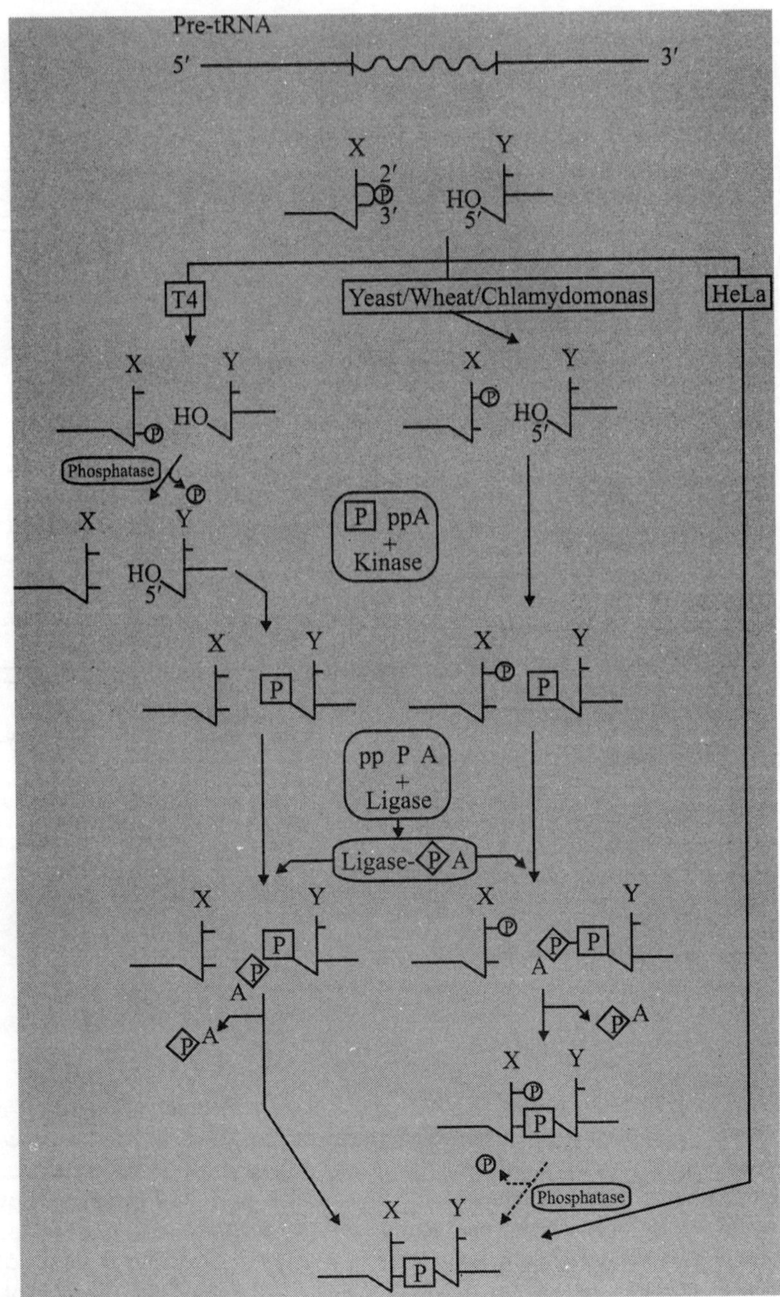

Figure 9.18 Pathway of tRNA splicing

are spliced by two enzymes: First, an endonuclease cleaves the phosphodiester bonds at the 3′ and 5′ splice sites, producing paired tRNA halves with 2′,3′-cyclic phosphate and 5′-OH ends as well as a linear intron. Secondly, the halves are joined by an ATP/GTP-dependent RNA ligase to generate

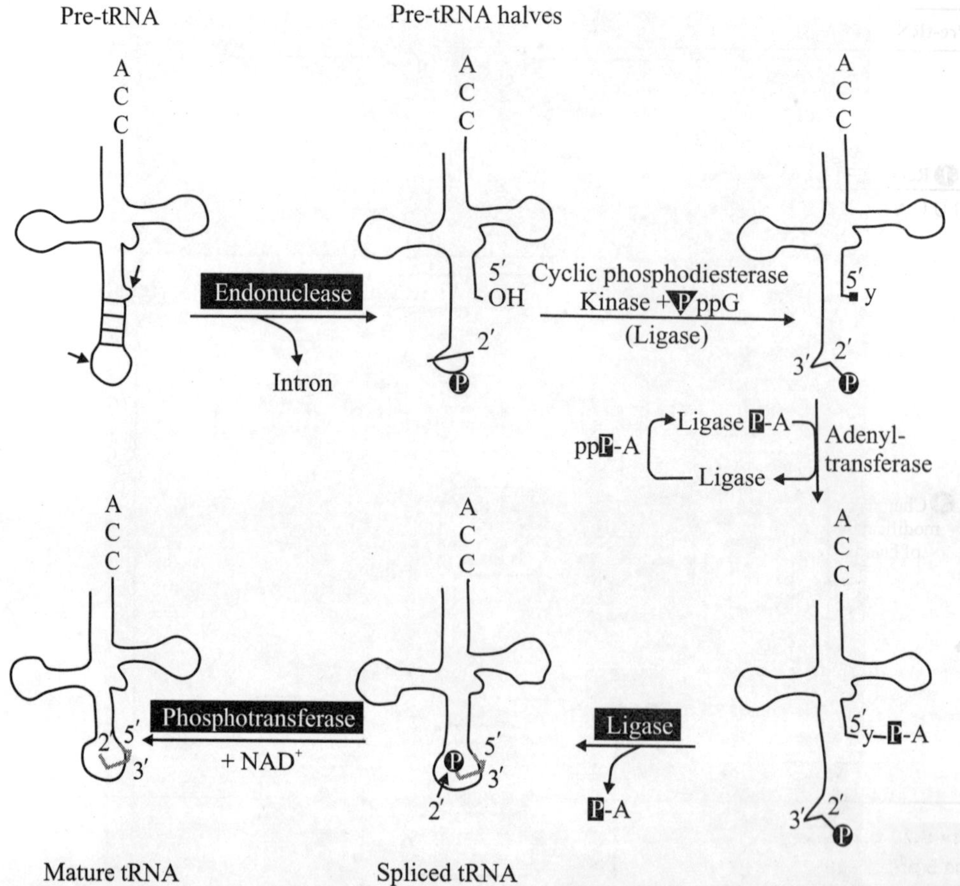

Figure 9.19 tRNA processing in eukaryotes. Pre-tRNAs introns are spliced by two enzymes: First, an endonuclease cleaves the phosphodiester bonds at the 3′ and 5′ splice sites, producing paired tRNA halves with 2′,3′-cyclic phosphate and 5′ OH ends as well as a linear intron. Secondly, the halves are joined by an ATP/GTP-dependent RNA ligase to generate mature tRNA (Redrawn, with permission, from http://www.biozentrum.uni-wuerzburg.de/fileadmin/REPORT/BIOCH/bioch007.htm)

mature tRNA. Processing for some yeast tRNAs (e.g., RNA^{Tyr}) involves (1) removal of the leader sequence at the 5′-end (2) replacement of two nucleotides at the 3′-end by the sequence CCA (with which all mature tRNA molecules terminate) (3) chemical modification of certain bases and (4) excision of an intron (Figure 9.20) (Wallace et al. 1980). tRNA in *E. coli* has been shown in Figure 9.21. The ends of tRNA are produced by the action of three nucleases that cleave the precursor to tRNA. A schematic of the pre-tRNA is shown at the top, with RNA extending from the 5′ and 3′ ends of the RNA that will become the mature tRNA (shown as a cloverleaf). The site of cleavage is indicated by the short vertical arrows above the lines denoting RNA, and they are labeled with the name of the enzyme cutting at that site. The enzymes catalyzing each reaction are listed above or adjacent to the reaction arrows.

The tRNA precursor splicing has been worked out in *Saccharomyces cerevisiae*. About 40 of the 400 nuclear tRNA genes in yeast are interrupted. Each tRNA precursor has a single intron, located just one nucleotide beyond the 3′-side of the anticodon. The introns vary in length from 14-46 bp. There is

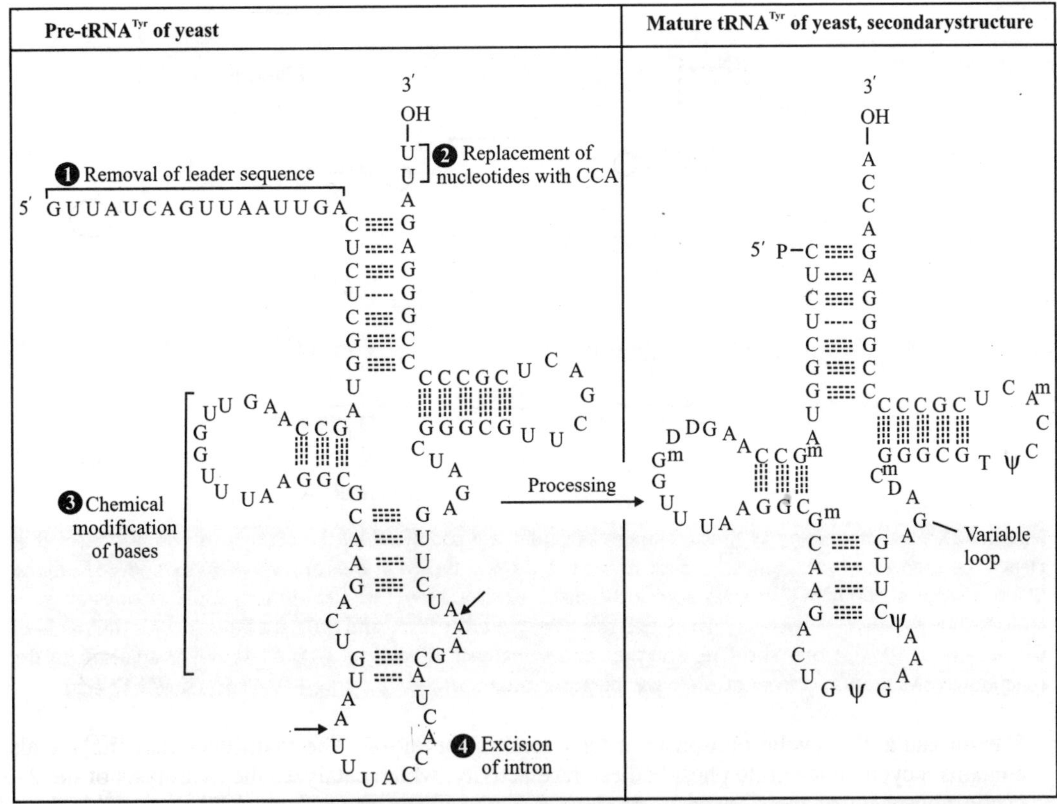

Figure 9.20 tRNA processing in yeast. Processing for some tRNA involves (1) removal of the leader sequence at the 5 prime end (2) replacement of two nucleotides at the 3 prime end by the sequence CCA (with which all mature tRNA molecules terminate) (3) chemical modification of certain bases and (4) excision of an intron (Redrawn from http://www.mun.ca/biology/desmid/brian/ BIOL2060/BIOL2060-21/2116. jpg)

no consensus sequence that could be recognized by splicing enzyme but all the introns include a sequence that is complementary to the anticodon of the tRNA. This creates an alternative conformation for the anticodon arm in which the anticodon is base paired to form an extension of the usual arm. This self-complementary region seems to form some kind of stem-loop configuration. That can be recognized by the enzyme or enzymes necessary for splicing. The excision of introns from yeast tRNA precursors occur in two stages. In splicing of yeast tRNA, a separate endonuclease and ligase act. First, a nuclear membrane bound splicing endonuclease makes two cuts precisely at the ends of the intron. Although after cleavage the two exons are no longer covalently attached to one another, they are held together by the territory structure of the tRNA precursors. In a second major step, a splicing ligase joins the two halves of the tRNA to produce the mature form of the tRNA molecule. The energy releases by hydrolysis of the phosphodiester bonds in the first step is lost as heat and the ligation reaction requires two molecules of ATP. The steps in the mechanism of pre-tRNA splicing of intron and ligation are as follows:

- Cleavage at the 5′ splice site leaves a 2′-3′ cyclic phosphate at the 3′-end of the 5′ exon and 5′-hydroxyl at the 5′-end of the intron. A similar cleavage leaves a hydroxyl group at the 5′-end of the

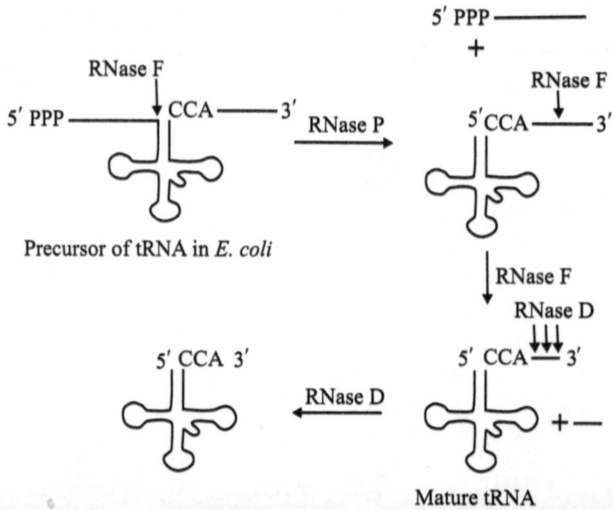

5′ PPP
+
RNase F
RNase F
5′ PPP CCA 3′
RNase P
5′CCA 3′
Precursor of tRNA in *E. coli*
RNase F
RNase D
5′ CCA 3′
5′ CCA 3′
RNase D
+—
Mature tRNA

Figure 9.21 tRNA processing in *E. coli*. The ends of tRNA are produced by the action of three nucleases that cleave the precursor to tRNA. A schematic of the pre-tRNA is shown at the top, with RNA extending from the 5′ and 3′ ends of the RNA that will become the mature tRNA (shown as a cloverleaf). The site of cleavage is indicated by the short vertical arrows above the lines denoting RNA, and they are labeled with the name of the enzyme cutting at that site. The enzymes catalyzing each reaction are listed above or adjacent to the reaction arrows (Redrawn from http://www.personal.psu.edu/rch8/ workmg/RNA ProcessingCh12.pdf)

3′ exon and a 2′-3′ cyclic phosphate at the 3′-end of the intron. The multifunctional enzyme also contains a cyclic nucleotide phosphodiesterase activity, which catalyzes the hydrolysis of the 2′-3′ cyclic phosphate of the 5′-exon to produce a 2′-phosphate group from ATP to the 5′-hydroxyl group left on the 3′ exon.

- The ATP-dependent RNA ligase catalyzes the transfer of AMP to the 5′-phosphate of the 3′-exon. The transferred AMP group is next displaced by the free 3′-hydroxyl group of the 5′ exon, thereby forming a 5′,3′ phosphodiester bond between the two exons. Phosphatase then catalyzes the hydrolysis of the 2′ phosphate ester to complete the splicing reaction. It may be noted that the phosphate group linking the two exons is not part of the original RNA but has provided by ATP during the reaction (Figure 9.22).
- A 2′,3′-cyclic phosphate is also generated during the tRNA-splicing reaction in plants and mammals. The human HeLa ligase directly joins an RNA end with a 2′,3′-cyclic phosphate group to an RNA end with a 5′-hydroxyl group. Thus the phosphate group that joins the two exons is the phosphate originally present at the end of the left exon.

The discovery of the RNA-self-splicing group I intron provided the first demonstration that not all the enzymes are proteins. Adams et al. (2004) report the X-ray crystal structure (3.1-Å resolution) of a complete group I bacterial intron in complex with both the 5'- and the 3'-exons. This complex corresponds to the splicing intermediate before the exon ligation step. It reveals how the intron used structurally unprecedented RNA motifs to select the 5'- and 3'-splice sites. The 5'-exon's 3'-OH is positioned inline nucleophilic attack on the conformationally constrained scissile phosphate at the intron-3'-exon junction. Six phosphates from the disparate RNA strands converge to coordinate two metal ions that are asymmetrically positioned on opposite sides of the reactive phosphate. This structure represents the first splicing complex to include a complete intron, both exons and an organized active site occupied with metal ions.

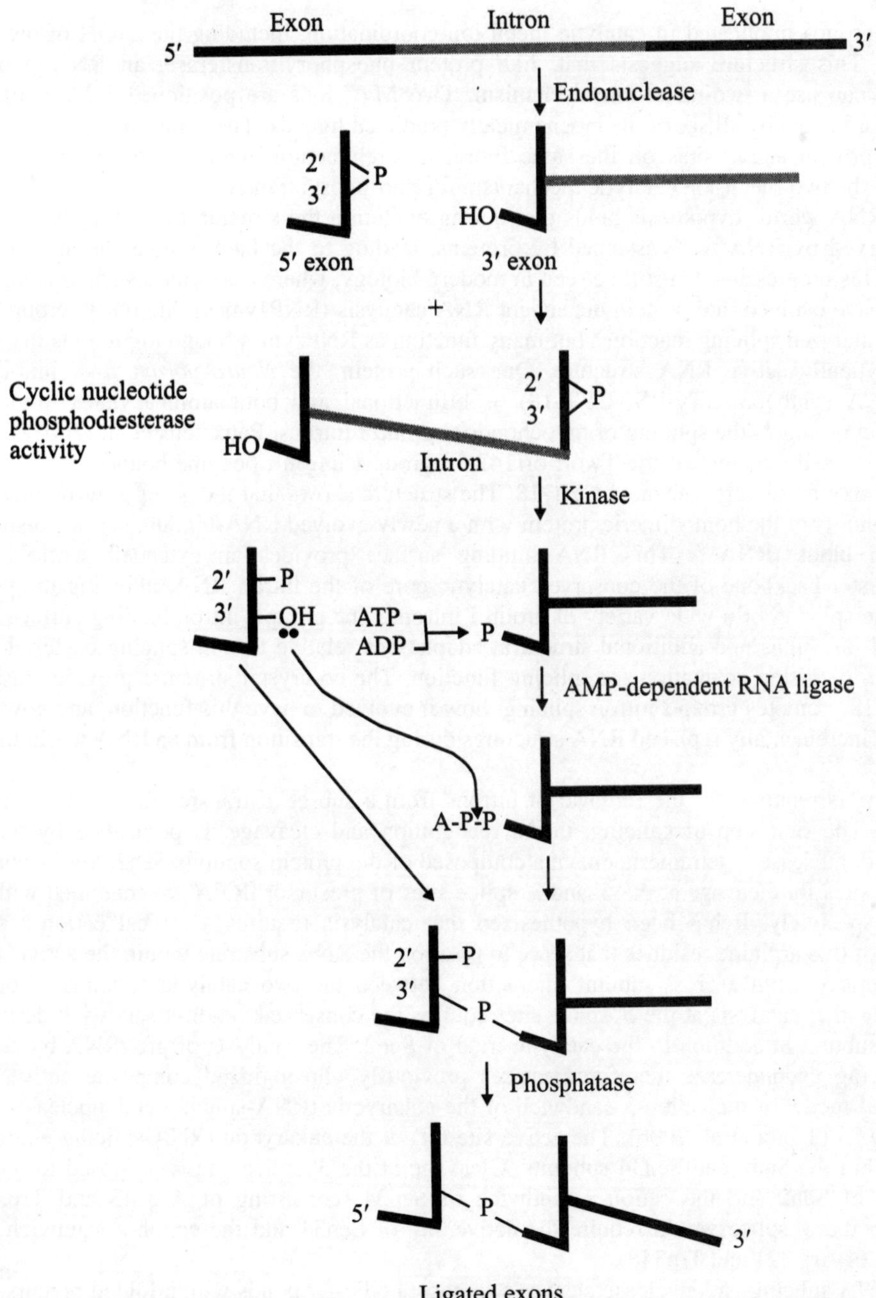

Figure 9.22 Mechanism of splicing of intron and ligation of two exons in precursor of tRNA (Redrawn from http:// www.bx.psu.edu/~ross/workmg/RNAProcessingCh12.htm)

Stahley and Strobel (2005) report 3.4Å crystal structure of a catalytically active group I intron splicing intermediate containing the complete intron, both exons, the scissile phosphate, and all of the functional groups implicated in catalytic metal ion coordination, including the 2'-OH of the terminal guanosine. This structure suggests that, like protein phosphoryltransferase, an RNA phosphoryltransferase can use a two-metal-ion mechanism. Two Mg^{2+} ions are positioned 3.9Å apart and are directly coordinated by all six of the biochemically predicted ligands. The evolutionary convergence of RNA and protein active sites on the same inorganic architecture highlights the intrinsic chemical capacity of the two-metal-ion catalytic mechanism for phosphoryl transfer.

The 'RNA world' hypothesis holds that during evolution the structural and enzymatic functions initially served by RNA were assumed by proteins, leading to the latter's domination of biological catalysis. This progression can still be seen in modern biology, where ribozymes, such as ribosome and RNase P, have evolved into protein-dependent RNA catalysis (RNPzymes). Similarly, group I introns use RNA-catalyzed splicing reactions, but many function as RNPzymes bound to proteins that stabilize their catalytically active RNA structure. One such protein, the *Neurospora crassa* mitochondrial tyrosyl-tRNA synthetase (TyrRS; CYT-18), is bifunctional and both aminoacylates mitochondrial $tRNA^{Tyr}$ and promotes the splicing of mitochondrial group I introns. Paukstelis et al. (2008) determine a 4.5-Å co-crystal structure of the Twort orf142-12 group I intron ribozyme bound to splicing active active, carboxy-terminally truncated CYT-18. The structure shows that the group I intron binds across the two subunits of the homodimeric protein with a newly evolved RNA-binding surface distinct from that which binds $tRNA^{Tyr}$. This RNA binding surface provides an extended scaffold for the phosphodiester backbone of the conserved catalytic core of the intron RNA, allowing the protein to promote the splicing of a wide variety of group I introns. The group I intron binding surface includes three small insertions and additional structural adaptations relative to non-splicing bacterial TyrRSs, indicating a multistep adaptation for splicing function. The co-crystal structure provides insight into how CYT-18 promotes group I intron splicing, how it evolved to have this function, and how proteins could have incrementally replaced RNA structures during the transition from an RNA world to an RNP world.

Splicing is required for the removal of introns from a subset of transfer RNAs in all eukaryotic organisms. The first step of splicing, intron recognition and cleavage, is performed by the tRNA-splicing endonuclease, a tetrameric enzyme composed of the protein subunits Sen2, Sen34 and Sen15. The active sites for cleavage at the 5' and 3' splice sites of precursor tRNA are contained within Sen2 and 34, respectively. It has been hypothesized that catalysis requires a critical cation-π sanswich composed of two arginine residues that serve to position the RNA substrate within the active site. This motif is derived from a cross-subunit interaction between the two catalytic subunits. Trotta et al. (2006) show that catalysis at the 5' splice site requires the conserved cation-π sandwich derived from the sen34 subunit in addition to the catalytic triad of Sen2. The catalysis of pre-tRNA by eukaryotic tRNA-splicing endonuclease therefore requires previously unrecognized composite active sites. A hypothetical model of the cation-π sandwich of the eukaryotic tRNA-splicing endonuclease is shown in Figure 9.23 (Trotta et al. 2006). The active site (H) of the eukaryotic tRNA-splicing endonuclease resides within the Sen2 and Sen34 subunits. Cleavage at the 3′ splice site is proposed to require the active site of Sen2 and the cation-π sandwich of Sen34 (consisting of Arg243 and Trp271, and cleavage at the 3' splice would require the active site of Sen34 and the cation-π sandwich of Sen2 (consisting of Arg 321 and Trp348).

The RNA splicing endonuclease cleaves two phosphodiester bonds within folded precursor RNAs during intron removal, producing the functional RNAs required for protein synthesis. Xu et al. (2006) describe at a resolution of 2.85Å the structure of a splicing enonuclease from *Archaeoglobus fulgidus* bound with a bulge-helix-bulge RNA containing non-cleaved and a cleaved splice site. The endonuclease dimer cooperatively recognized a flipped-out bulge base and stabilizes sharply bent bulge backbones that are poised for an in-line RNA cleavage reaction. Cooperativity arises because an arginine pair from one catalytic domain sandwiches a nucleobase within the bulge cleaved by the other

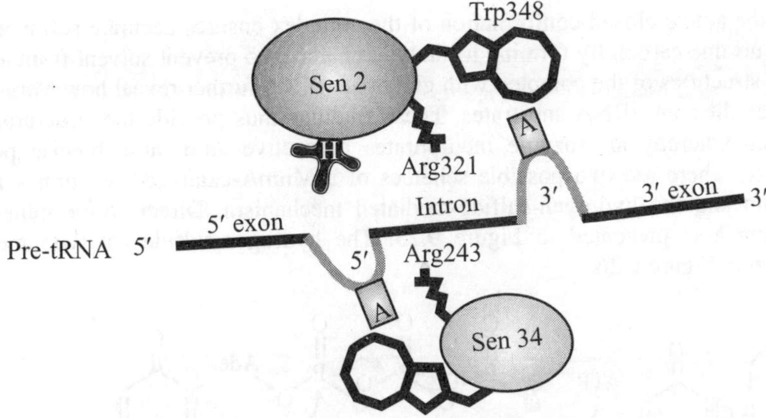

Figure 9.23 Hypothetical model of the cation-π sandwich of the eukaryotic tRNA-splicing endonuclease

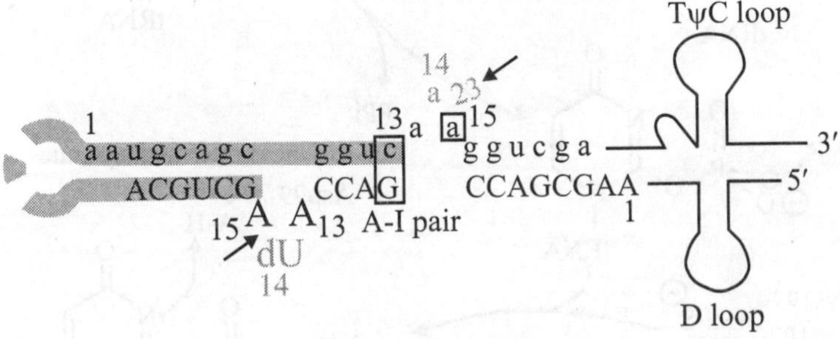

Figure 9.24 A precursor tRNA structure with the primary and secondary structure of the bulge-helix-bulge (BHB) RNA

catalytic domain. A precursor tRNA structure with the primary and secondary structure of the bulge-helix-bulge (BHB) RNA is presented in Figure 9.24 (Xu et al. 2006). Two arrows depict the two splice sites where the broken arrow indicates the modified 5' splice site. The anticidin-intron base pair (A-I pair) and the first 3′ exon nucleotide (a15) are highlighted by boxed letters. Small and capital letters of the same numbering are used for the strands containing the 3' splice site and the 5' splice sites, respectively.

Modification of Ribonucleotide Residues

Uridine at the first anticodon position (U34) of glutamate, lysine and glutamine transfer RNAs is universally modified by thiouridylase into 2-thiouridine (s2U34), which is crucial for precise translation by restricting codon-anticodon wobble during protein synthesis on the ribosomes. However, it remains unclear how the enzyme incorporates reactive sulfur into the correct position of the uridine base. Numata et al. (2006) present the crystal structure of the MnmA thiouridylase-tRNA complex in three discrete forms, which provide snapshots of the sequential chemical reactions during RNA sulfuration. On enzyme activation, α-helix overhanging the active site is restructured into an idiosyncratic β-hairpin-containing loop, which packs the flipped-out U34 deeply into the catalytic pocket and triggers the activation of the catalytic cysteine residues. The adenylated RNA intermediate is trapped. Thus, the active closed-conformation of the complex ensures accurate sulfur incorporation into the activated uridine carbon by forming a catalytic chamber to prevent solvent from accessing the catalytic site. The structures of the complex with glutamate tRNA further reveal how Mnm specifically recognizes its three different tRNA substrates. These findings thus provide the structural basis for a general mechanism whereby an enzyme incorporates a reactive atom at a precise position in a biological molecule. There are two possible schemes of a MnmA-catalyzed reaction – direct sulfur transfer mechanism and the hydrogen-sulfide-mediated mechanism. Direct sulfur transfer catalytic mechanism of MnmA is presented in Figure 9.25. The hydrogen-sulfide catalytic mechanism of MnmA is presented in Figure 9.26.

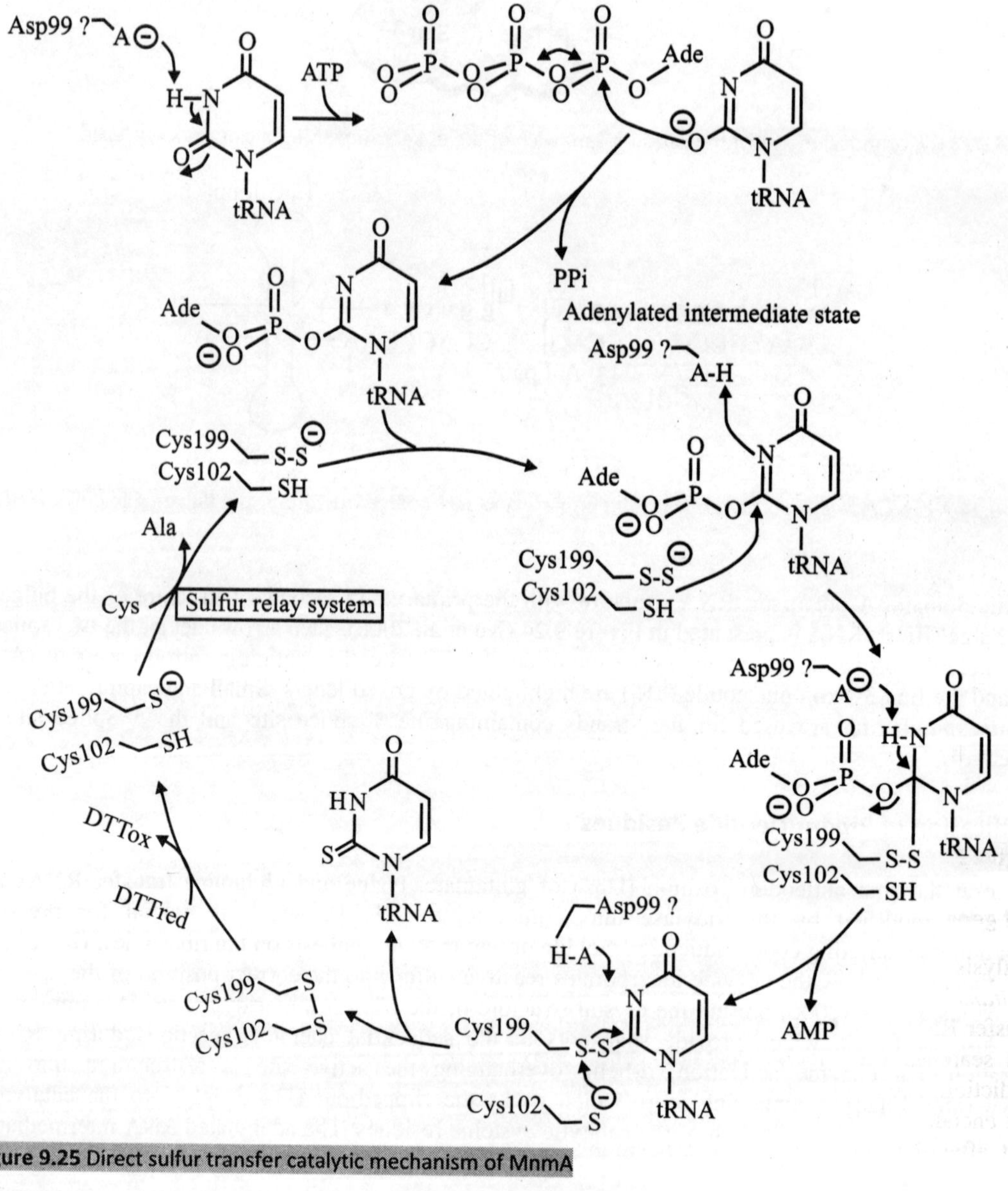

Figure 9.25 Direct sulfur transfer catalytic mechanism of MnmA

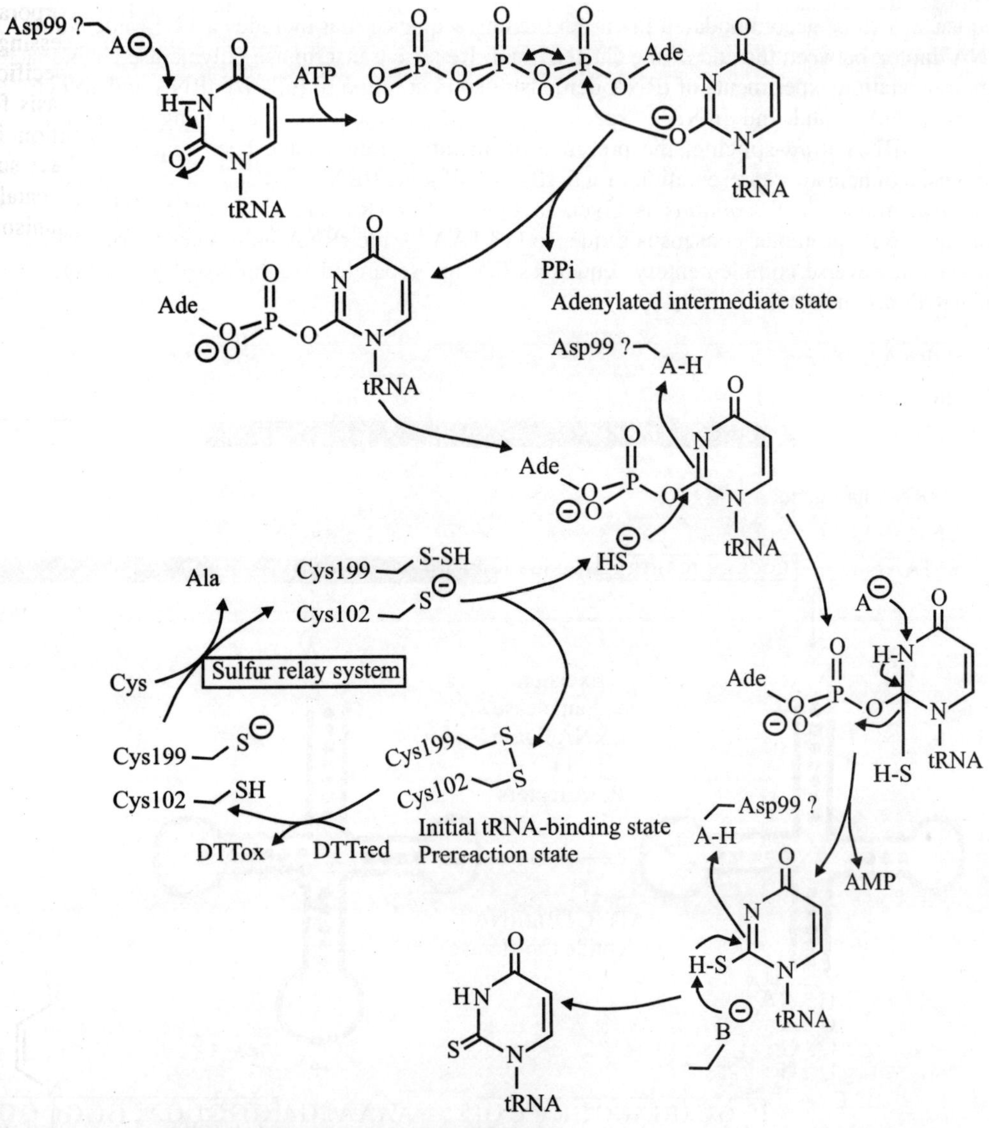

Figure 9.26 Hydrogen-sulfide-mediated mechanism of MnmA

Half genes

Analysis of genome sequence of the small hyperthermophilic archaeal parasite *Nanoarchaeum equitans* has not revealed genes encoding the glutamate, histidine, tryptophan and intiator methionine transfer RNA species. Randau et al. (2005) developed a computational approach to genome analysis that searched for widely separated genes encoding tRNA halves that, on the basis of structural prediction, could form intact tRNA molecules. A search of the *N. equitans* genome reveals nine genes that encode tRNA halves; together they account for missing tRNA genes. The tRNA sequences are split after the anticodon-adjacent position 37, the normal location of tRNA introns. The terminal sequences can be accommodated in an intervening sequence that includes a 12-14 nucleotide GC-rich RNA duplex between the end of the 5' tRNA half. Reverse transcriptase polymerase chain reaction and amino-acylation experiments of tRNA demonstrated maturation to full-size tRNA and acceptor activity of the $tRNA^{His}$ and and $tRNA^{Glu}$ species predicted in silico. So the joining mechanism possibly involves tRNA *trans*-splicing, the presence of an intron might have been required for early tRNA synthesis. Schematic representation of a 5' tRNA half gene ($tRNA^{Glu}$)) and the corresponding 3' tRNA half gene found in *N. equitans* is given in Figure 9.27 (Randau et al. 2005). The archaeal RNA polymerase III promoter consensus sequence (TTTAAA), the rRNA half genes (dark portion) and the intervening reverse complementary sequences that are separated are supposed to facilitate joining of the halves are indicated.

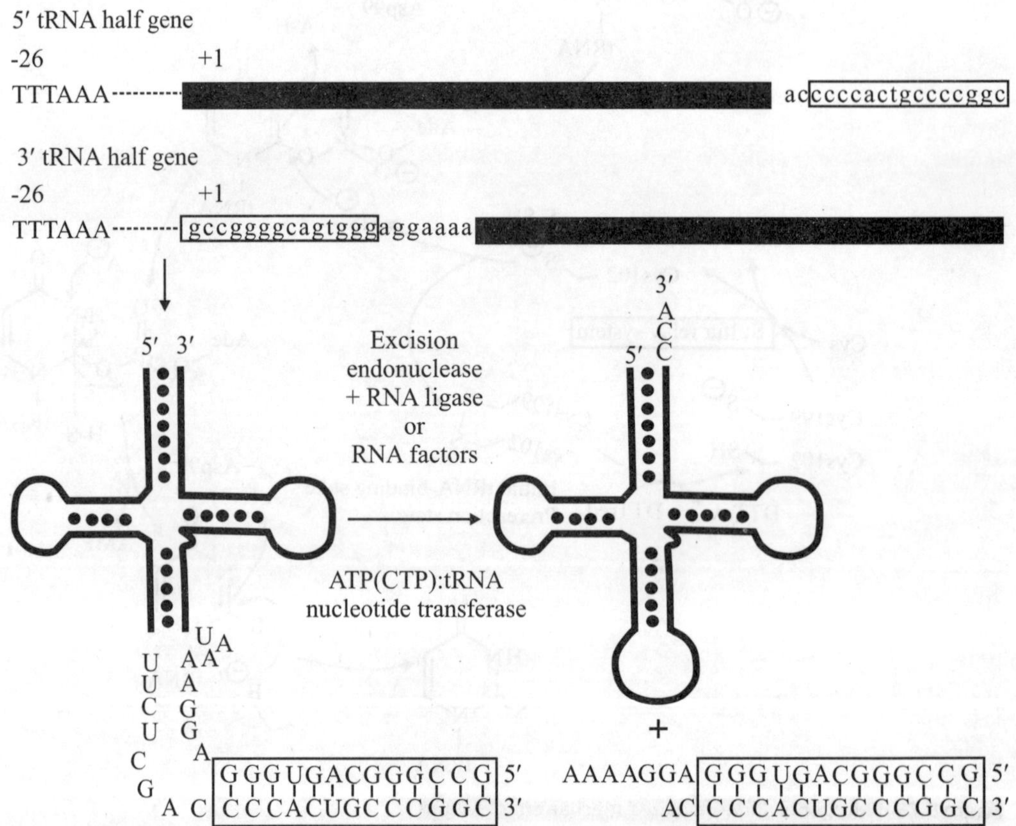

Figure 9.27 Schematic representation of a 5′ tRNA half gene ($tRNA^{Glu}$)) and the corresponding 3′ tRNA half gene found in *Nanoarchaeum equitans*

A computational analysis of the nuclear genome of a red alga, *Cyanidioschyzon merolae*, identified 11 transfer RNA genes in which the 3' half of the RNA lies upstream of the 5' half in the genome. Soma et al. (2007) verified that these genes are expressed and produce mature tRNAs that are aminoacylated. Analysis of tRNA-processing intermediates for these genes indicate an unusual processing pathway in which the termini of tRNA precursors are ligated, resulting in formation of a characteristic circular RNA intermediate that is then processed at the acceptor stem to generate the correct termini.

Editing site of aminoacyl-transfer RNA (tRNA) synthetases

Editing site of aminoacyl-transfer RNA (tRNA) synthetases, which catalyze the attachment of correct amino acid to its corresponding tRNA during translation of the genetic code, are proven antimicrobial drug targets. Rock et al. (2007) show that the broad-spectrum antifungal 5-flouro-1,3-dihydro-1-hydroxy-2,1-benzoxaborote (AN2690), in development for the treatment of onychomycosis, inhibits yeast cytoplasmic leucyl-tRNA synthetase by formation of a stable $tRNA^{Leu}$-AN2690 adduct in the editing site of the enzyme. Adduct formation is mediated through the boron atom of AN2690 and the 2'- and 3'-oxygen atoms of tRNA's 3'-terminal adenosine. The trapping of enzyme-bound $tRNA^{Leu}$ in the editing site prevents catalytic turnover, thus inhibiting synthesis of leucyl-$tRNA^{Leu}$ and consequently blocking protein synthesis. This result establishes the editing site as a bonafide target for aminoacyl-tRNA synthetase inhibition.

PRE-RIBOSOMAL RNA PROCESSING

Prokaryotic cells have three types of rRNA: 16S, 23S and 5S (S = Svedberg units). "S" value gives us an idea about size of a molecule. Genomic arrangement of genes in the sequence 5'-16S, tRNAs, 23S, 5S, tRNA(s)-3' is (in a single operon. The entire operon is often transcribed as a unit. Cleavage enzymes are required at a number of points even to separate several classes of pre-rRNAs. A typical rRNA operon is shown in Figure 9.28 (Rogers 1985).

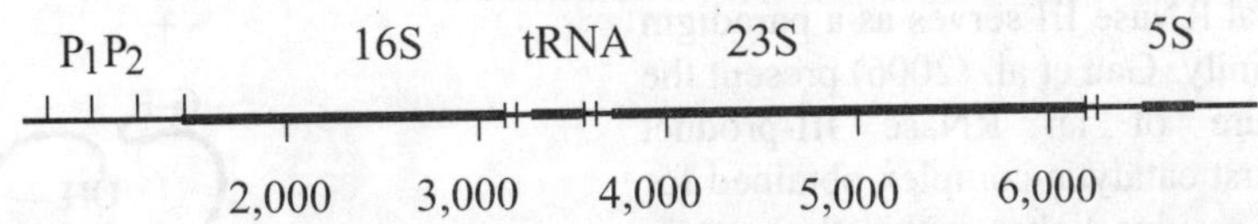

Figure 9.28 A typical RNA operon. Genomic arrangement is in the sequence 5'-16S, tRNA(s), 23S, tRNA(s)-3'. The entire operon is transcribed as a unit

Mechanism of splicing of rRNA precursors is shown in Figure 9.29. The secondary structure of the spacer sequences bordering 16S and 23S rRNA is shown in Figure 9.30 (King et al. 1986). Nucleolar organizer region is the actual site of rRNA synthesis. An intrinsic constituent of the ibosome particle provides structural framework on which protein synthesis proceeds. Ribosomal RNA transcripts are processed both in prokaryotes and eukaryotes to generate the mature rRNA molecules with unique 3′ and 5′ ends. The rRNA transcript undergoes the following two changes either during or, less likely, after its synthesis but before it's processing: (1) To each transcript, ~110 methyl groups are added to the ribose moieties; these methyl groups are retained in the mature molecules. Thus ribose methylation seems to act as a marker for the cleavage sites. (2) The RNA becomes bound to proteins so that it is processed as ribonulceoprotein complex rather than as free RNA.

The first processing reaction seems to be cleavage of the superfluous leader sequence at the 5′-end. The next cleavage occurs within the spacer sequence located between the 18S and 5.8 sequences; trimming may be needed to generate the 3′-end of 5.8S molecule. The 5.8 molecule is then cleaved from the 28S molecule and base-pairs with a segment of 28S RNA molecule. It is not clear if the cleavage occurs at the 5′- and 3′- ends of the three sequences or at some distance from them so that the 5′ and 3′ ends are generated through trimming. In prokaryotes, the primary rRNA transcript is 30S and has the sequences 16S-22S-5S in that order; these precursors may also contain sequences for tRNA. The tRNAs are released from the precursor RNA by cleavage; the remaining regions of the spacer are presumably degraded. The RNase III is responsible for the processing of 16S and 22S rRNAs. The rRNA molecules have several roles in protein synthesis. First, the 28S rRNA has a catalytic role, it forms part of the peptidyl transferase activity of the 60S subunit. Second, 18S rRNA has a recognition role, involved in correct positioning of the mRNA and the peptidyl tRNA. Finally, the rRNA molecules have a structural role. They fold into three-dimensional shapes that form the scaffold on which the ribosomal proteins assemble.

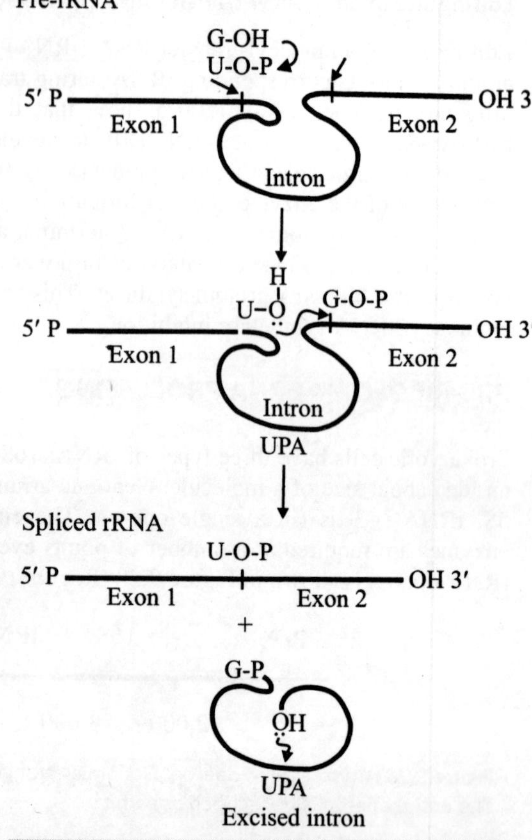

Figure 9.29 Autocatalytic splicing of rRNA precursors

Ribonuclease III

Members of the ribonuclease III (RNase III) family are double-stranded RNA (dsRNA)-specific endoribonucleases characterized by a signature motif in their active centers and a two-base 3' overhang in their products. While Dicer, which produces small interfering RNAs, has been the focus of intense interest, the structurally simpler bacterial RNase III serves as a paradigm of the entire family. Gan et al. (2006) present the crystal structure of an RNase III-product complex, the first catalytic complex obtained for the family. A7 residue linker within the protein facilitates induced fit in protein-RNA recognition. A pattern of RNA-protein interactions, defined by four RNA binding motifs in RNase III and three protein-interacting boxes in dsRNA, is responsible for substrate specificity, while conserved amino acid residues and divalent cations are responsible for scissile-bond cleavage. The structure reveals a wealth of information about the mechanism of RNA hydrolysis that can be extrapolated to other RNase III family members.

Early Steps in Processing 5S rRNA

After transcription, the primary transcript is cleaved into two parts, one bears the genes for the minor rRNA and one or two tRNAs present on the intergenic spacer and other contains the major and the supplemental rRNAs, depending upon whether or not a tRNA occurs on the 3'-end. Neither RNase III nor E is involved in initial enzymatic reaction.

Later Steps in Processing 5S rRNA

The 10S RNA contains 6 stem-and-loop structures generated by RNase III. Then by action of RNase P, fragment 7S RNA is generated which contains 5S rRNA and a termination stem. Removal of this stem is accomplished by RNase E. 5S rRNA from *B. subtilis* is shown in Figure 9.31 in a probable secondary structure. The two RNase M5 cleavage points are also indicated (Stahl et al. 1980).

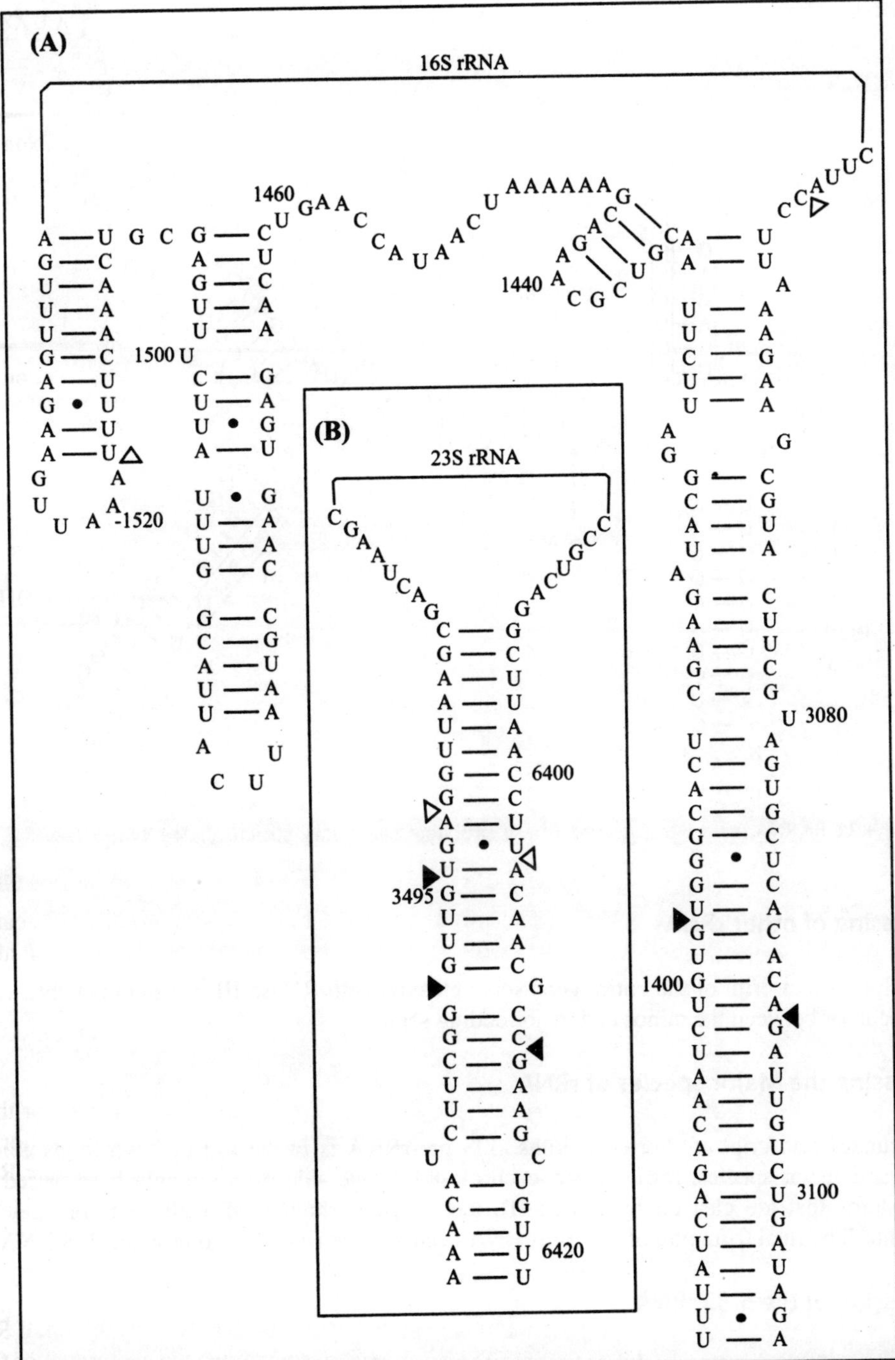

Figure 9.30 The secondary structure of the spacer sequences bordering (A) 16S and (B) 23S rRNA. Open triangles show the position of mature termini. Closed triangles show the positions of RNase III cleavage sites

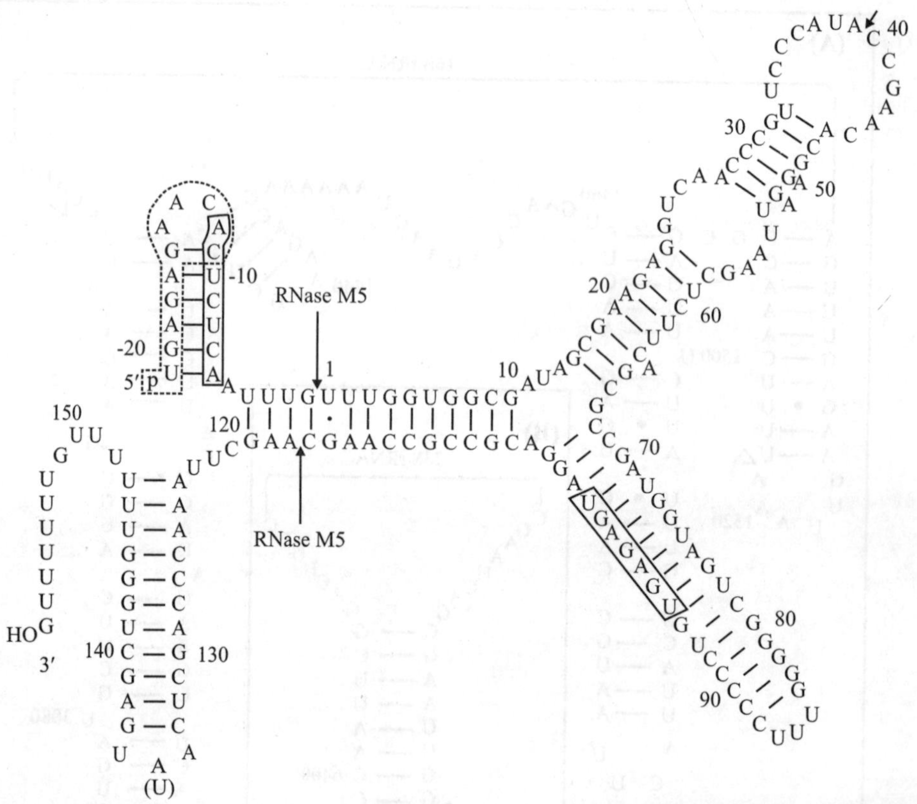

Figure 9.31 5S RNA from *Bacillus subtilis* in a probable secondary structure. The two RNase M5 points are shown

Processing of other rRNAs

The primary transcript of the entire gene set is cleaved with RNase III or its equivalent, which breaks the precursor between the minor and major coding sectors.

Processing the Major Species of rRNA

The primary transcript of shorter operons, 30S pre-rRNA is in the form of two loops containing the minor and major species, respectively, each closed at one side by a strongly base-paired stem. It is these stems that are cleaved by RNase III, the enzyme which specifically acts on double-stranded segments. The final trimming of 30S pre-rRNA occurs at the ribosome to produce 23S RNA.

Processing of the 5.8S rRNA

Processing of the supplemental species of rRNA is confined to the nucleolus and involved small RNA U3 and another type of RNA, 7-1. During much of its maturation, it is hydrogen bonded to the other pre-rRNAs. First product of this species is 12S RNA particle, followed by the one sedimenting at 8S. Finally, 5.8S rRNA is produced.

Synthesis of 28S and 18S rRNA

At the nucleolar organizer region, 45S RNA is synthesized which acts as a precursor molecule for the synthesis of 28S and 18S rRNAs. An endonulease cleaves 45S RNA into two short precursor molecules of size 41S and 20S, which are cleaved into 28S and 18S rRNAs, respectively. The 41S RNA undergoes three sequential cleavage steps by exonuclease to yield 28S rRNA whereas 20S RNA undergoes only one cleavage step by exonuclease to produce 18S rRNA. Schemes for processing of rRNA in eukaryotes are shown in Figure 9.32 (Lodish et al. 2000). Extrachromosomal rDNA of *Tetrahymena thermophila* is shown in Figure 9.33 (Engberg and Nielsen 1990). The wide bars represent the sequences coding for the primary transcript. Sequences present in mature rRNA are shown in black. Sequences removed from the precursor during processing are shown in white, except for intervening sequence which is stippled.

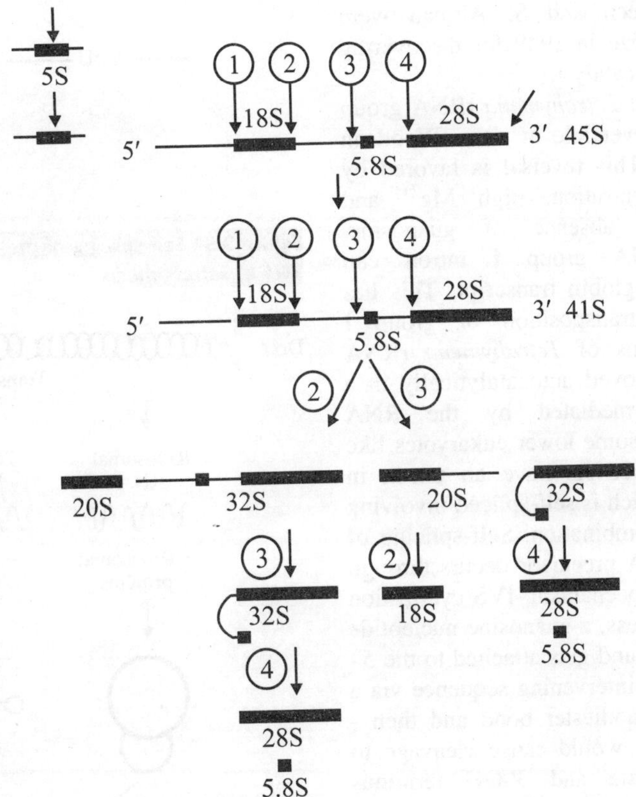

Figure 9.32 Schemes for processing of rRNA in eukaryotes

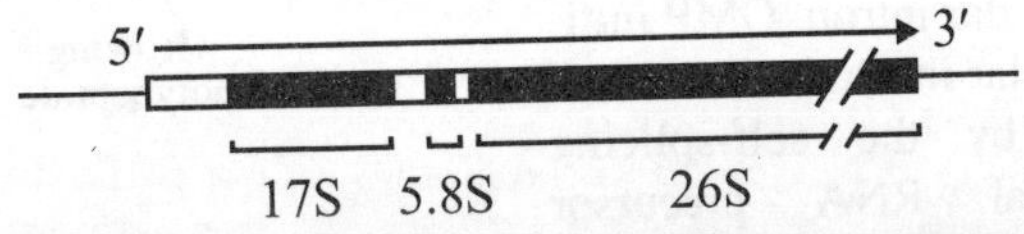

Figure 9.33 Extrachromosomal rDNA of *Tetrahymena thermophila*. The rRNAs are shown in black. Sequences removed from the precursor during processing are shown in white, except for IVS which is stippled

SELF-SPLICING OF rRNA IN TETRAHYMENA

It is known that during protein synthesis the information encoded in the DNA is first transcribed into mRNA which in turn directs the assembly noticed in a variety of single-celled parasites, including *Trypanosoma brucei* which causes African sleeping sickness, it appeared as if new information not encoded in the DNA, the cells master repository of information, was being added to certain RNA molecules. Only in the past few years molecular biologists began to crack the problem. T.R. Cech and S. Altman were awarded Nobel Prize in 1989 for discovering RNA as biological catalyst.

Splicing of the *Tetrahymena* rRNA group I intron is fully reversible in vitro (Woodson and Cech 1989). This reversal is favored by high RNA concentration, high Mg^{2+} and temperature, and absence of guanosine. *Tetrahymena* rRNA group I intron can integrate into a β-globin transcript. This has implications for transposition of group I introns. The introns of *Tetrahymena* rRNA precursors are removed autocatalytically in a unique reaction mediated by the RNA molecule itself. In some lower eukaryotes like *Tetrahymena*, pre-rRNAs have an intron in 28S component which is self-spliced involving intramolecular recombination. Self-splicing of intron in 28S rRNA precursor occurs through transesterification mechanism. IVS cyclization occurs. In this process, a guanosine nucleotide acts as nucleophile and gets attached to the 5'-end of the existing intervening sequence via a normal 3'-5' phosphodiester bond and then -OH of the 5' exon would cause cleavage to produce 5'-phosphate and 3'-OH terminus (Figure 9.34) (Rogers 1985). The ligation occurs without hydrolysis of ATP with the removal of covalent circular intron. GMP initiates the process. A circular intron is created. 5' splice site selection by the self-splicing *Tetrahymena* ribosomal RNA precursor requires the formation of a particular RNA helix by base pairing between an intron 'guide sequence' and bases surrounding the 5' splice

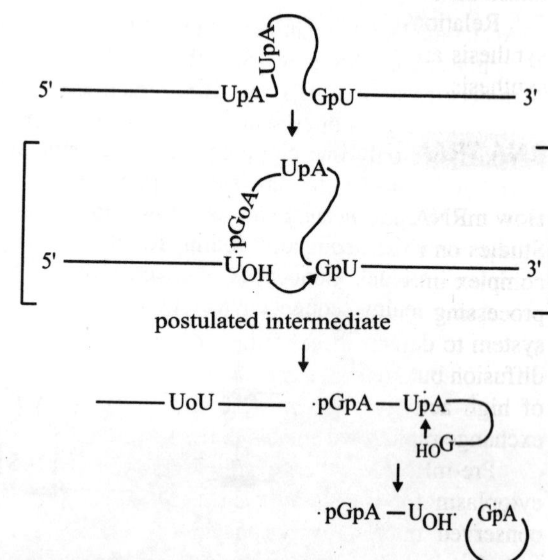

Figure 9.34 Self-splicing of pre-rRNA intron in 28S precursor RNA of *Tetrahymena*

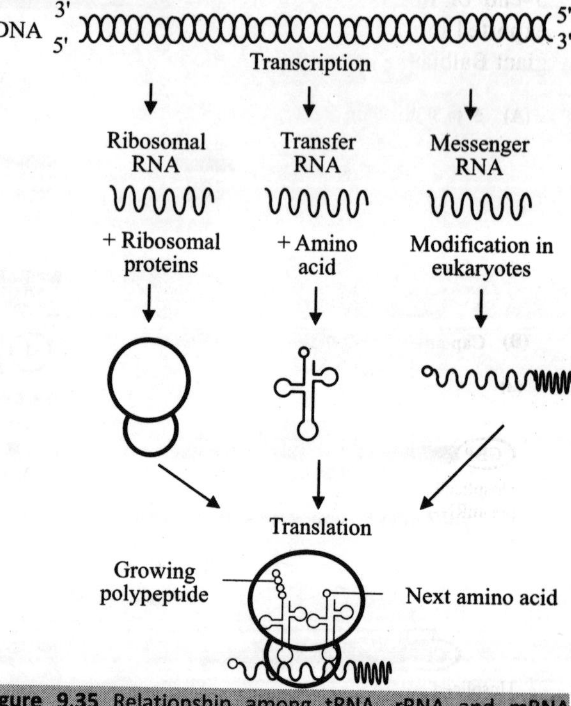

Figure 9.35 Relationship among tRNA, rRNA and mRNA during protein synthesis

junction. Part of the guide sequence subsequently holds the positions the 3′-end of the cut 5′ exon to attack the 3′ splice junction (Waring et al. 1986).

Relationship among the three types of RNA, ribosomal, transfer, and messenger, during protein synthesis are shown in Figure 9.35. All three types are found together at the ribosome during protein synthesis.

RNA TRANSPORT

How mRNA gets to the cytoplasm? The successive steps of mRNA maturation need to be emphasized. Studies on regulation of RNA transport by HIV-1 Rev protein have provided only a glimpse into the complex interplay between nuclear RNA splicing and RNA transport (Chang and Sharp 1990). RNA processing and assembly require many factors but the nucleus apparently lacks any active transport system to deliver these to the RNA (Lewis and Tollervey 2000). Instead, processing factors move by diffusion but are concentrated by transient association with functionally related components. At the site of high activity this can give rise to visible structures, the nuclear bodies with complements that exchange rapidly with pools in the surrounding nucleoplasm.

Pre-mRNAs undergo splicing to remove introns, and the spliced mRNA is exported to the cytoplasm for translation. Cheng et al. (2006) investigated the mechanism for recruitment of the conserved mRNA export machinery (TREX complex) to mRNA. They show that human TREX complex is recruited to a region near the 5'-end of mRNA, with the TREX component Aly bound closest to the 5' cap. Both TREX recruitment and mRNA export require the cap, and these roles for the cap are splicing dependent. CBP80, which is bound to the cap, associates efficiently with TREX to the 5'-end of mRNA, where it functions in mRNA export. As a consequence, the mRNA would be exported in a 5' to 3' direction through the nuclear pore, as observed in early electron micrographs of giant Balbiani ring mRNPs. A model for mRNA export is given in Figure 9.36 (Cheng et al. 2006). (A)

(A) 5′ to 3′ direction of mRNA export mediated by the hTREX complex bound to the 5′ end of mRNA

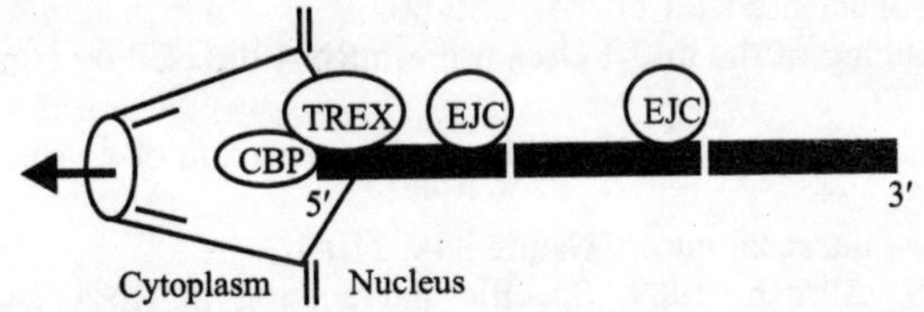

(B) Cap and splicing-depenent recruitment of the hTREX complex to mRNA

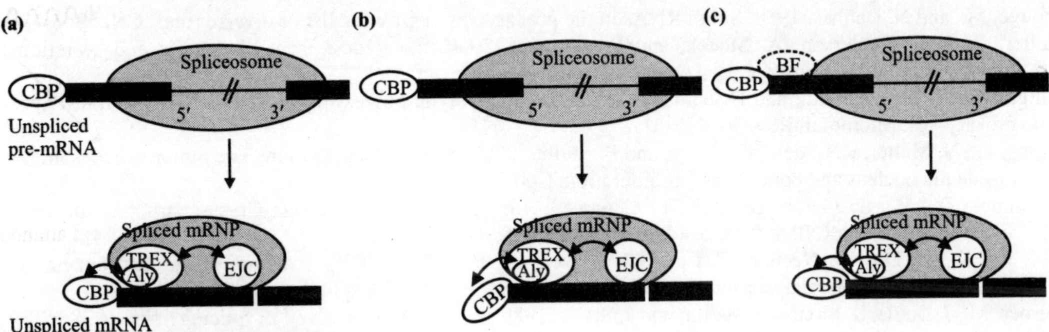

Figure 9.36 Model for mRNA export

A model for 5' to 3' translocation of mRNA is given. An mRNA containing three exons is shown with cap-binding complex (CBC) bound to the 5'-end. An exon junction complex (EJC) is bound at each exon-exon junction, and conserved mRNA export machinery, TREX, is bound at the 5'-end of the mRNA. The mRNA is exported in a 5' to 3' direction due to the presence of TREX at the 5'-end of the mRNA, and TREX may interact with the mRNA export receptor Tap/p15 located at the nuclear pore complex. (B) Model for cap- and splicing-dependent recruitment of the human TREX complex is shown.

REFERENCES

Abelson, J., C.R. Trotta, and H. Li. 1998. RNA splicing. J. Biol. Chem. 273: 12685-8.

Adams, P.L., M.R. Stahley, A.B. Kosek, J. Wang, and S.A. Strobel. 2004. Crystal structure of a self-splicing group I intron with both exons. Nature 430: 45-50.

Berget, S.M. 1984. Are U4 small nuclear ribonucleoproteins involved in polyadenylation? Nature 309: 179-82.

Bessonov, S., M. Anokhina, C.L. Will, H. Urlaub, and R. Luthrmann. 2008. Isolation of active step 1 spliceosome and composition of its RNP core. Nature 452: 846-50.

Bonnal, S., and J. Valcarcel. 2008. Spliceosome meets telomerase. Nature 456: 879-80.

Box, J.A., J.T. Bunch, W. Tang, and P. Baumann. 2008. Spliceosomal cleavage generates the 3' end of telomerase RNA. Nature 456: 910-4.

Caceres, J.F., and T. Mistell, 2007. Division of labor: Minor splicing in the cytoplasm. Cell 131: 645-7.

Chang, D.D., and P.A. Sharp. 1990. Messenger RNA transcript and HIV rev regulation. Science 249: 614-5.

Cheng, H., K. Dufu, C.-S. Lee, J.L. Hsu, A. Dias, and R. Reed. 2006. Human mRNA export machinery recruited to the 5′ end of mRNA. Cell 127: 1389-400.

Dominski, Z., X.-c. Yang, and W.F. Marzluff. 2005. The polyadenylation fact or CPSC-73 is involved in histone-pre-mRNA processing. Cell 123: 37-48.

Engberg, J., and H. Nielsen. 1990. Complete sequence of the extrachromosomal rDNA molecule from the ciliate *Tetrahymena thermophila* strain B1868VII. Nucl. Acids Res. 18: 6915-9.

Foster, A.C., and S. Altman. 1990. External guide sequences for an RNA enzyme. Science 249: 783-6.

Futcher, B., and J.K. Leatherwood. 2008. Bound to splice. Nature 455: 885-6.

Gan, J., J.E. Tropea, B.P. Austin, D.L. Court, D.S. Waugh, and X. Ji. 2006. Structural insights into mechanism of double-stranded RNA processing by ribonuclease III. Cell 124: 355-66.

Graveley, B.R. 2005. Mutually exclusive splicing of the insect Dscam pre-mRNA directed by competing intronic RNA secondary structure. Cell 123: 65-73.

Greer, C.L., C.L. Peebles, P, Gegenheimer, and J. Abelson. 1983 Mechanism of action of a yeast RNA ligase in tRNA splicing. Cell 32: 537-46.

Grivell, L.A. 1990. RNA catalysis: trailing the interanal intron. Nature 344: 110-1.

Guerrier-Takada, C., N. Lumelsky, and S. Altman. 1989. Specific interacdtion in RNA enzyme substrate complexes. Science 246:1578-84.

Guyer, R.L. 1990. Splicing RNA. Science 250: 551.

Inouye, M., and N. Delihas. 1988. Small RNAs in the prokaryotes: a growing list of diverse roles. Cell 53: 5-7.

Keller, S., M.P. Sanderson, A. Stoeck, and P. Altevogt. 2006. Exosomes: from biogenesis and secretion to biological function. Immunol. Lett. 107(2): 102-8.

King, T.C., R. Sirdeskmukh, and D. Schlessinger. 1986. Nucleolytic processing of ribonucleic acid transcripts in prokaryotes. Microbiol. Rev. 50: 428-51.

Konig, H., N. Matter, R. Bader, W. Thiele, and F. Muller. 2007. Splicing segregation: The minor spliceosome acts ouside the nucleus and controls cell proliferation. Cell 131: 718-29.

Krummel, D.A.P., C. Outbridge, A.K.W. Leung, J. Li, and K. Nagai. 2009. Crystal structure of human spliceosomal U1 snRNP at 5.5Å resolution. Nature 458: 475-80.

Lee, K.P.K., M. Dey, D. Neculai, C. Cao, T.E. Dever, and F. Sicheri. 2008. Structure of the dual enzyme tre1 reveals the basis for catalysis and regulation in nonconventional RNA splicing. Cell 132: 89-100.

Lerner, M., J. Boyle, S. Mount, S. Wolin, and J. Steitz. 1980. Are snRNPs involved in splicing? Nature 283: 22.

Lerner, M.P., and J.A. Steitz. 1979. Antibodies to small nuclear RNAs complexed with proteins are produced by patients with systemic lupus erythematosus. Proc. Natl. Acad. Sci. USA 76: 5495-9.

Lerner, M.R., and J.A. Steitz. 1981. Snurps and scyrps. Cell 25: 298-300.

Lewis, J.D., and D. Tollervey. 2000. Like attracts like: getting RNA processing together in the nucleus. Science 288: 1385-8.

Lodish, H., A. Berk, S.L. Zipursky, P. Matsudaira, D. Baltimore, and J. Darnell. 2000. *Molecular Cell Biology*. 4th edition. New York: W. H. Freeman.

Madhani, H.D., and C. Guthrie. 1994. Dynamic RNA-RNA interactions in the spliceosome. Annu. Rev. Genet. 28: 1-26.

Mandel, C.R., S. Kaneko, H. Zhang, et al. 2006. Polyadenylation factor CPSF-73 is the pre-mRNA 3′-end processing endonuclease. Nature 444: 953-6.

Maniatis, T. and Reed, R. 1987. The role of small nuclear ribonucleoprotein particles in pre-mRNA splicing. Nature 325: 673-8.

Mattaj, I.W. 1989. A bending consensus: RNA-protein interactions in splicing snRNP's and sex. Cell 57: 1-3.

Misteli, T., Cacers, J.F., and Spector, D.L. 1997. The dynamics of a pre-mRNA splicing factor in living cells. Nature 387: 523-7.

Molden, A., J. Malapeira, N. Gabrielli, et al. 2008. Promoter-driven splicing regulation in fission yeast. Nature 455: 997-1000.

Mowry, K.L., and J.A. Steitz. 1987. Identification of the human U7 snRNP as one of the several factors involved in the 3′ end maturaion of histone pre-mRNA. Science 238: 1682-7.

Nielsen, H., E. Westhof, and S. Johansen. 2005. An mRNA is capped by a 2',5' lariat catalyzed by a group I-like ribozyme. Science 309: 1584-6.

Numata, T., Y. Ikeuchi, S. Fuka, T. Suzuki, and O. Nureki. 2006. Snapshot of tRNA sulphuration via an adenylated intermediate. Nature 442: 419-24.

Ohi, M.D., L. Ren, J.S. Wall, K.L. Gould, and T. Walz.2007. Structural characterization of the fission yeast U5.U2/U6 spliceosome complex. Proc. Natl. Acad. Sci. USA 104: 3195-200.

Paukstelis, P., J.-H. Chen, E.Chase, A.M. Lambowitz, and B.L. Golden. 2008. Structure of a tyrosyl-tRNA synthetase splicing factor bound to a group I intron RNA. Nature 451: 94-7.

Piccirilli, J.A. 2008. Towards understanding self-splicing. Science 320: 56-7.

Query, C.C. 2009. Spliceosome subunit revealed. Nature 458: 418-9.

Randau, L., I. Schroder, and D. Soll. 2008. Life without RNase P. Nature 453: 120-2.

Randau, L., R. Munch, M.J. Holm, D. Jahn, and D. Soll. 2005. *Nanoarchaeum equitans* creates functional tRNAs from separate genes for their 5′- and 3′-halves. Nature 433: 537-40.

Rock, F.L., W. Mao, A. Yaremchuk, et al. 2007. An antifungal agent inhibits an aminoacyl-tRNA synthetase by mapping tRNA in the editing site. Science 316: 1759-61.

Rogers, H.J. 1985. Mechanism of RNA splicing. Int. Rev. Cytol. 99: 188-235.

Rossi, A.M., J.C.P. Thijssen, A.D. Tates, et al. 1990. Mutations affecting RNA-splicing in man are detected more frequently in somatic than in germ cells. Mutation Res 244(4) 353-7.

Russel, A.G., J.M. Charette, D.F. Spencer, and M.W. Gray. 2006. An early evolutionary origin for the minor spliceosome. Nature 443: 863-6.

Saga, Y, J.S. Lee, C. Saraiya, and E.A. Boyse. 1990. Regulation of alternative splicing in the generation of isoforms of the mouse Ly-5 (CD45) glycoprotein. Proc. Natl. Acad. Sci. USA 87: 3728-32.

Smith, C.W.J. 2005. Alternative splicing – when two s a crowd. Cell 123: 1-3.

Smith, C.W.J., J.G. Patton, and B. Nadal-Ginard. 1989. Alternative splicing in the control of gene expression. Annu. Rev. Genet. 23: 527-77.

Soma, A., A. Onodera, J. Sughahara, et al. 2007. Permuted tRNA genes expressed via a circular RNA intermediate in *Cyanidioschyzon merolae*. Science 318: 450-3.

Stahl, D.A., B. Meyhack, and N.R. Pace. 1980. Recognition of local nucleotide conformation in contrast to sequence by a rRNA processing endonuclease. Proc. Natl. Acad. Sci. USA77: 5644-80.

Stahley, M.R., and S.A. Strobel. 2005. Structural evidence for a two-metal-ion mechanism of group I intron splicing. Science 309: 1587-90.

Swerdlow H., and C. Guthrie. 1984. Structure of intron-containing tRNA precursors. Analysis of solution conformation using chemical and enzymatic probes. J. Biol. Chem. 259: 5197-207.

Tomita, K., R. Ishitani, S. Fukai, and O. Nureki. 2006. Complete crystallographic analysis of the dynamics of CCA sequence addition. Nature 443: 956-60.

Toor, N., K.S. Kealing, S.D. Taylor, and A.M. Pyle. 2008. Crystal structure of a self-spliced group II intron. Science 320: 77-82.

Torres-Larios, A., K.K. Swinger, A.S. Krasilnikov, T. Pan, and A. Mondragon. 2005. Crystal structure of the RNA component of bacterial ribonuclease. Nature 437: 584-7.

Trotta, C.R., S.V. Paushkin, M. Patel, H. Li, and S.W. Peltz, 2006. Cleavage of pre-tRNAs by the splicing endonuclease requires a composite active site. Nature 441: 375-7.

Tseng, C.-K., and Cheng, S.-C. 2008. Both catalytic steps of nuclear pre-mRNA splicing are reversible. Science 320: 1782-4.

Viswanathan, S.R., G.Q. Daley, and R.I. Gregory. 2008. Selective blockade of microRNA processing by Lin-28. Science 320: 97-100.

Waring, R.B., P. Towner, S.J. Minter, and R.W. Davies. 1986. Splice-site selection by a self-splicing RNA of Tetrahymena. Nature 321: 133-9.

Watson, J.D., H. Nancy, J. Hopkins, W. Roberts, J.A. Steitz, and A.M. Weiner. 1987. *Molecular Biology of the Gene*. California: The Benjamin/Cummings.

Westaway, S.K., and J. Abelson. 1995. Splicing of tRNA precursors Chapter 7. In: *tRNA: Structure, Biosynthesis, and Function*. Söll, D., and U. RajBhandary (Editors) Washington, D.C.: American Society for Microbiology.

Will, C.L., and R. Lührmann. 2005. Splicing of a rare class of introns by the U12-dependent spliceosome. Biol. Chem. 386: 713-24.

Woodson, S.A., and T.R. Cech. 1989. Reverse sef-splicing of the *Tetrahymena* group I intron: Implication for the directionality of splicing and for intron transposition. Cell 57: 335-45.

Xiong, Y., and T.A. Steitz. 2004. Mechanism of transfer RNA maturation by CGA-adding enzyme without using an oligonucleotide template. Nature 430: 640-5.

Xu, S., K. Calvin, and H. Li. 2006. RNA recognition and cleavage by a splicing endonuclease. Science 312: 906-10.

Alternative Splicing and RNA Editing

ALTERNATIVE RNA SPLICING

Alternative RNA splicing is the process through which more than one functional mRNA molecules are produced from one and the same primary transcript by differential removal of introns (Tamkun et al. 1984). Alternative RNA splicing was first observed in animal viruses (Berk and Sharp, 1978). In eukaryotes, alternative RNA splicing was first observed in murine immunoglobulin genes. The tissue specificity of alternative RNA splicing was first shown in the fibrinogen genes of rat and man (the gene which encodes for the fibrinogen protein, the precursor for fibrin) by Crabtree and Kant (1983). The first observations of developmental stage specificity concerned the alcohol dehydrogenase (*Adh*) gene of *Drosophila melanogaster*. Alternative RNA splicing makes it possible for a single gene to produce more than one messenger RNA molecule, which contradicts the basic conceptual framework of the neoclassical view of the gene structure and function.

In many eukaryotic genes, alternative paths of splicing can take place where different sites for splicing are chosen or avoided entirely. Thus, a single gene can produce several different proteins, depending upon splicing choice. For example, in yeast, the gene *RPL32* codes for a ribosomal protein. When there is an excess of this protein, it somehow causes the failure of intron removal. The result is non-functional mRNA and no further RPL32 protein is produced. For many genes, splicing is an invariant processing step. The same mature transcript is generated in all cells where gene is transcribed, and a single product is synthesized. This is constitutive splicing. For other genes, RNA splicing may be used as a mechanism of gene regulation, i.e., a gene may be expressed differently in two tissues even though it is transcribed in exactly the same manner and at the same rate. Such regulation may occur through processing or discard regulation or through differential or alternative RNA splicing. In the latter case, the primary transcript may be processed in different ways by alternative usage of exons which may be termed splice variants or splice isoforms.

In humans, it is estimated that 70 per cent of human protein-coding transcripts undergo alternative RNA splicing and that the 30,000 genes are therefore able to generate up to 150,000-200,000 mRNAs and proteins. Alternative RNA splicing events are regulated by varying levels of RNA interacting proteins such as RNA binding proteins and splicing factors. Intron/exon signals and trans-acting factors can act cooperatively or antagonistically to determine the fate of particular pre-mRNAs.

Exon Skipping

Most mutations in the *dystrophin* gene create a frameshift or a stop in the mRNA and are associated with severe Duchenne muscular dystrophy. Exon skipping that naturally occurs at low frequency sometimes eliminates the mutation and leads to the production of a rescued protein. Goyenvalle et al. (2004) have achieved persistent exon skipping that removes the mutated exon on the dystrophin messenger mRNA of the *mdx* mouse, by a single administration of an AAV vector expressing antisense sequences linked to a modified U7 small nuclear RNA. They report the sustained production of functional dystrophin at physiological levels in entire groups of muscles and the correction of the muscular dystrophy.

Types of Alternative RNA Splicing

Breitbart et al. (1987) have described six patterns of alternative RNA splicing: combinatorial exons, mutually exclusive patterns, retained introns, alternative donor and acceptor splice sites, alternative promoters and polyadenylation sites, and splice site compatibility. Alternative RNA splicing events in metazoan transcripts are depicted in Figure 10.1. Types of alternative RNA splicing for the generation of functionally distinct transcripts are depicted. Dark rectangles represent alternative exons.

Combinatorial exons

Inclusion or exclusion of a particular exon is independent of all other exons present in the transcript. For example, in Fast Skeletal Troponin T gene of rat, due to presence of 5 constitutive combinatorial exons, one primary RNA transcript has the potential of producing upto 32 different mRNAs.

Mutually exclusive patterns

In this type of alternative RNA splicing, selection of exons of a pair is such that if one member of a pair is spliced in, other is necessarily to be excised out. Two different protein isoforms are produced due to such behavior of exons 2 and 3 present in α-tropomyosin gene of rat.

Retained introns

Retained introns may also result in the production of different mRNAs with corresponding different translated products. For example, γ-fibrinogen gene of human and rat produce two different mRNA molecules due to retention or excision of an intron (Stuart et al. 1986).

Alternative donor and acceptor splice sites

Presence of alternative donor and acceptor splice sites for a given exon results in the excision of intron of different lengths with complementary variation in exon size. Use of such competing splice sites is found in adenovirus E1a transcript unit where mRNAs with different protein-coding regions are produced by use of alternate splice sites (Ziff 1980).

Alternative promoters and polyadenylation sites

Presence of alternative promoters and polyadenylation sites in the transcript may also yield different mRNAs molecules of different sizes by differential use of such sites. For example, in the gene for myosin light chain, presence of two promoters leads to differential use of 5′-end exons to yield different mRNAs (Breitbart et al. 1987). In immunoglobulin μ heavy chain gene and Calcitonin gene, different mRNAs are produced due to the presence of different polyadenylation sites which in turn produce different isomeric proteins (Amara et al. 1982).

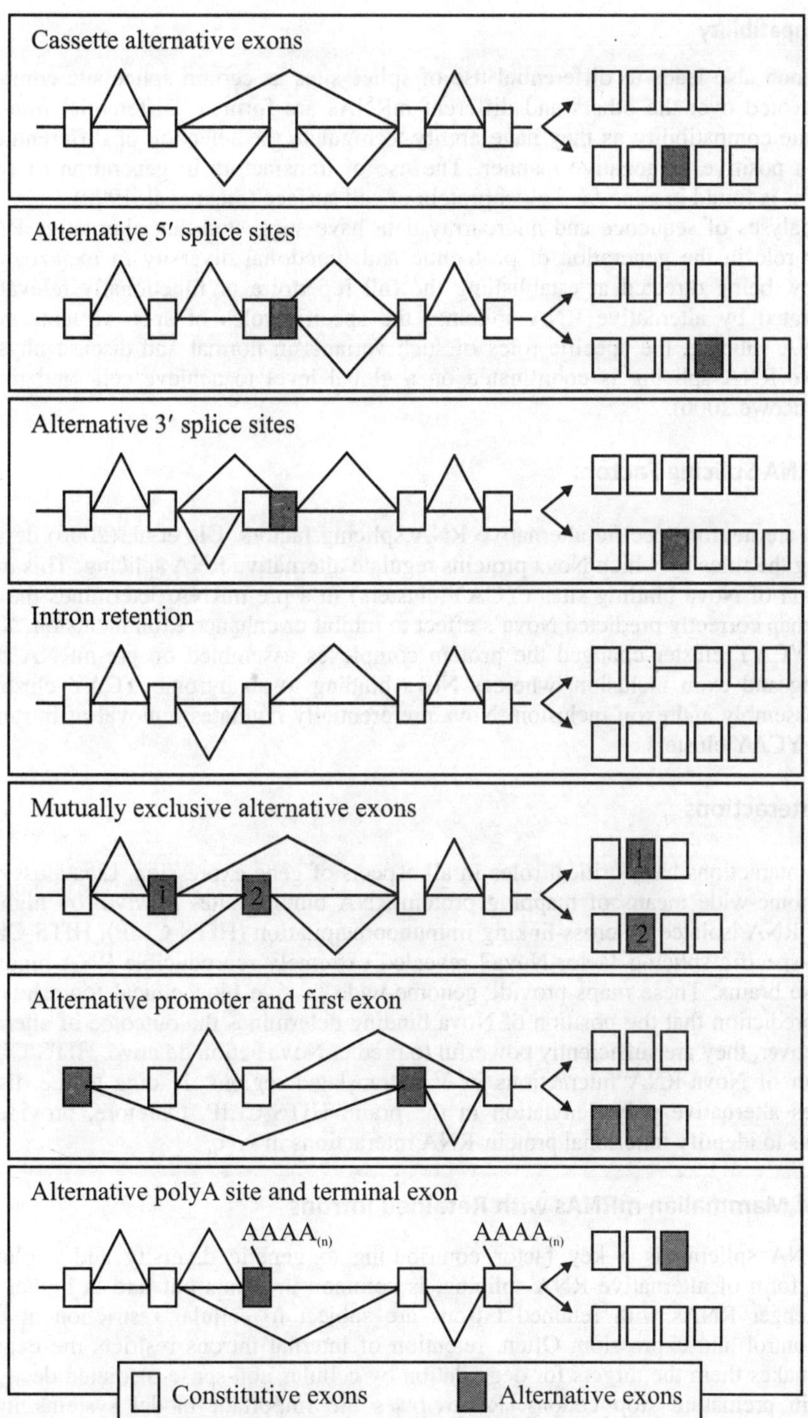

Figure 10.1 Alternative RNA splicing events in metazoan transcripts

Splice site compatibility

This phenomenon also leads to differential use of splice sites as certain splice site combinations are specifically favored over the others and different mRNAs are formed. Differential transfactors also affect splice site compatibility as they have ability to regulate the selection of differential splice site utilization in a positive or negative manner. The use of transfactors in generation of two different protein isoforms is found in gene *Ly-5* glycoprotein of cell surface (Saga et al. 1990).

Recent analyses of sequence and microarray data have suggested that alternative RNA splicing plays a major role in the generation of proteomic and functional diversity in metazoan organisms. Efforts are now being directed at establishing the full repertoire of functionally relevant transcript variants generated by alternative RNA splicing, the specific roles of such variants generated by alternative RNA splicing, the specific roles of such variants in normal and disease physiology, and how alternative RNA splicing is coordinated on a global level to achieve cell- and tissue-specific functions (Blencowe 2006).

Alternative RNA Splicing Factors

Nova proteins are neuron-specific alternative RNA splicing factors. Ule et al. (2006) derive an RNA map describing the rules by which Nova proteins regulate alternative RNA splicing. This map revealed that the position of Nova binding sites (YCAY clusters) in a pre-mRNA determines the outcome of splicing. The map correctly predicted Nova's effect to inhibit or enhance exon inclusion. Nova binding to an exonic YCAY cluster changed the protein complexes assembled on pre-mRNA blocking U1 snRNP binding and exon inclusion, whereas Nova binding to an intronic YCAY cluster enhanced spliceosome assembly and exon inclusion. Nova preferentially regulates removal of introns harboring (or closest to) YCAY clusters.

Nova-RNA Interactions

Protein-RNA interactions have critical roles in all aspects of gene expression. Licatalosi et al. (2008) develop a genome-wide means of mapping protein-RNA binding sites in vivo, by high-throughput sequencing of RNA isolated by cross-linking immunoprecipitation (HITS-CLIP). HITS-CLIP analysis of the neuron-specific splicing factor Nova2 revealed extremely reproducible RNA-binding maps in multiple mouse brains. These maps provide genome-wide in vivo biochemical footprints confirming the previous prediction that the position of Nova binding determines the outcome of alternative RNA splicing; moreover, they are sufficiently powerful to predict Nova action de novo. HITS-CLIP revealed a large number of Nova-RNA interactions in 3′ untranslated regions, leading to the discovery that Nova regulates alternative polyadenylation in the brain. HITS-CLIP, therefore, provides a robust, unbiased means to identify functional protein-RNA interactions in vivo.

Expression of Mammalian mRNAs with Retained Introns

Alternative RNA splicing is a key factor contributing to genetic diversity and evolution. Intron retention, one form of alternative RNA splicing, is common in plants but rare in higher eukaryotes, because messenger RNAs with retained introns are subject to cellular restriction at the level of cytoplasmic control and expression. Often, retention of internal introns restricts the export of these mRNAs and makes them the targets for degradation by cellular non-sense-mediated decay machinery if they contain premature stop codons. Retroviruses are important model systems in studies of regulation of RNAs with retained introns, because their genomic and their mRNAs contain one or more unspliced introns. Tap-CTE and MPMV-CTE structures are given in Figure 10.2. Mason-

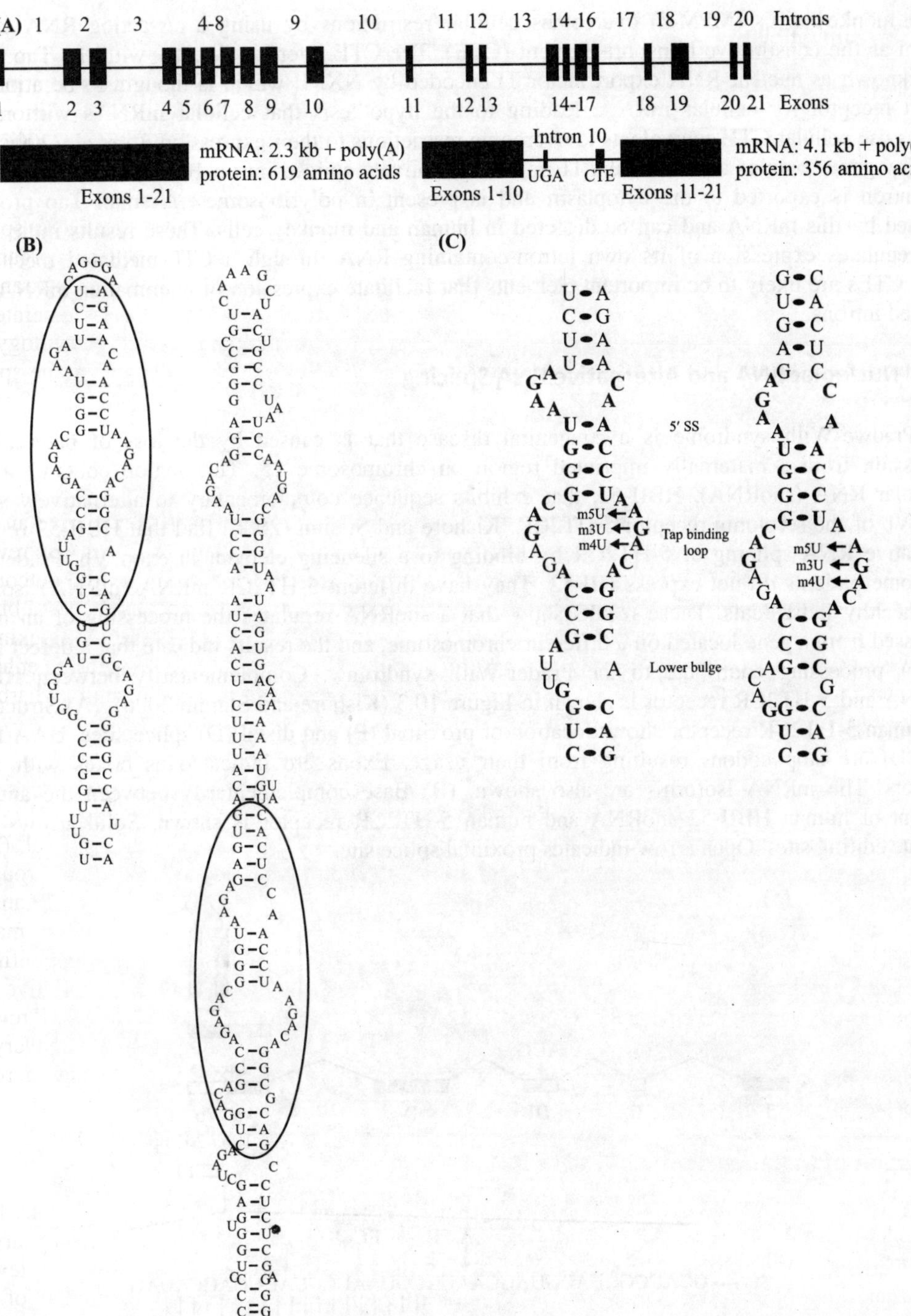

Figure 10.2 Genomic organization of the human Tap gene and comparison of the Tap-CTE and MPMV-CTE structures

Pfizer monkey virus (MPMV) overcomes cellular restrictions by using a *cis*-acting RNA element known as the constitutive transport element (CTE). The CTE interacts directly with the Tap protein (also known as nuclear RNA export factor 1) encoded by NXF1, which is thought to be a principal export receptor for cellular mRNA, leading to the hypothesis that cellular mRNAs with retained introns use cellular CTE equivalents to overcome restrictions to their expression. Li et al. (2006) show that *Tap* gene contains a functional CTE in its alternatively spliced intron 10. *Tap* mRNA containing this intron is exported to the cytoplasm and is present in polyribosomes. A small Tap protein is encoded by this mRNA and can be detected in human and monkey cells. These results indicate that Tap regulates expression of its own intron-containing RNA through a CTE-mediated mechanism. Thus, CTEs are likely to be important elements that facilitate expression of mammalian mRNAs with retained introns.

Small Nucleolar RNA and Alternative RNA Splicing

The Pradwe-Willi syndrome is a congenital disease that is caused by the loss of paternal gene expression from a maternally imprinted region on chromosome 15. This region contains a small nucleolar RNA (snoRNA), HBII-52, that exhibits sequence complementary to alternatively spliced exon Vb of the serotonin receptor 5-HT2CR. Kishore and Stamm (2006) find that HBII-52 regulates alternative RNA splicing of 5-HT2CR by binding to a silencing element in exon Vb. Prader-Willi syndrome patients do not express HII-52. They have different 5-HT2CR mRNA (mRNA) isoforms, than healthy individuals. These results show that a snoRNA regulates the processing of an mRNA expressed from a gene located on a different chromosome, and the results indicate that a defect in pre-mRNA processing contribute to the Prader-Willi syndromw. Complementarity between HBII-52 snoRNA and 5-HT2CR receptor is shown in Figure 10.3 (Kishore and Stamm 2006). (A) Structure of the human 5-HT2CR receptor shows location of proximal (P) and distal (D) splice sites. UAA(P) and UAA(D) are stop codons resulting from their usage. Exons are indicated as boxes with roman numbers. The mRNA isoforms are also shown. (B) Base complementarity between the antisense element of human HBII-52 snoRNA and human 5-HT2CR receptor is shown. Solid arrows (A-E) indicate editing sites. Open arrow indicates proximal splice site.

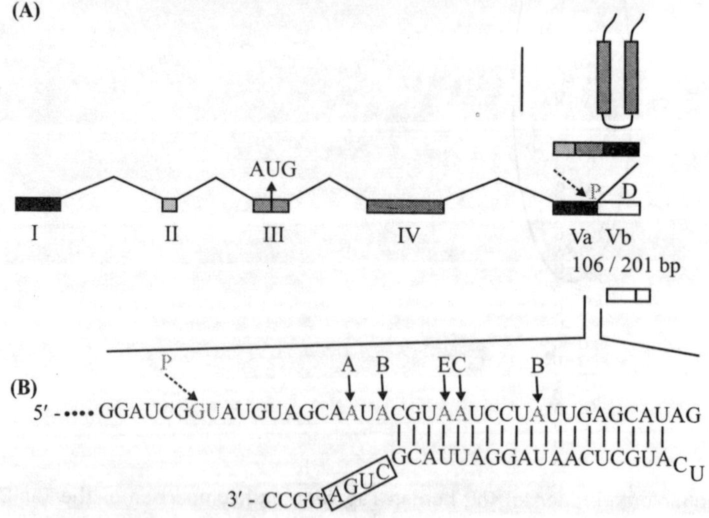

Figure 10.3 Complementarity between HBII-52 snoRNA and 5-HT2CR receptor

Coordinated Regulation of Both Splicing and Polyadenylation

Through alternative processing of pre-messenger RNAs, individual mammalian genes often produce multiple RNA and protein isoforms that may have related, distinct, or even opposing functions. Wang et al. (2008) report in-depth analysis of 15 diverse human tissues and human cell line transcriptomes on the basis of deep sequencing of complementary DNA fragments, yielding a digital inventory of gene and mRNA isoform expression. Analysis in which sequence reads are mapped to exon-exon junctions indicated that 92-94 per cent of human genes undergo alternative RNA splicing, ~86 per cent with a minor isoform frequency of 15 per cent or more. Differences in isoform-specific read densities indicated that most alternative RNA splicing and alternative cleavage and polyadenylation events vary between tissues whereas variation between individuals was approximately twofold to threefold less common. Extreme or 'switch-like' regulation of splicing between tissues was associated with increased sequence conservation in regulatory regions and with generation of full-length open reading frames. Pattern of alternative RNA splicing and alternative cleavage and polyadenylation were strongly correlated across tissues, suggesting coordinated regulation of both splicing and polyadenylation.

Gene Activity and Alternative RNA Splicing

The functional complexity of the human transcriptome is not yet fully elucidated. Sultan et al. (2008) report a high-throughput sequence of the human transcriptome from a human embryonic kidney (HEK) and a B cell line. They used shotgun sequences of transcripts to generate randomly distributed reads. Of these, 50 per cent mapped to the unique genomic locations, of which 80 per cent corresponded to known exons. They report that 66 per cent of the polyadenylated transcriptome mapped to known genes and 34 per cent to non-annotated genomic regions. On the basis of known transcripts, RNA-Seq can detect 25 per cent more genes than can microarrays. A global survey of messenger RNA splicing events identified 94,241 splice junctions (40% of which were previously unidentified) and showed that exon skipping is the most prevalent form of alternative RNA splicing. Comparison of number of expressed genes detected by RNA-Seq and microarrays is shown in Figure 10.4 (Sultan et al. 2008). Values for relaxed and stringent RNA-Seq parameters are in bold and parenthesis, respectively.

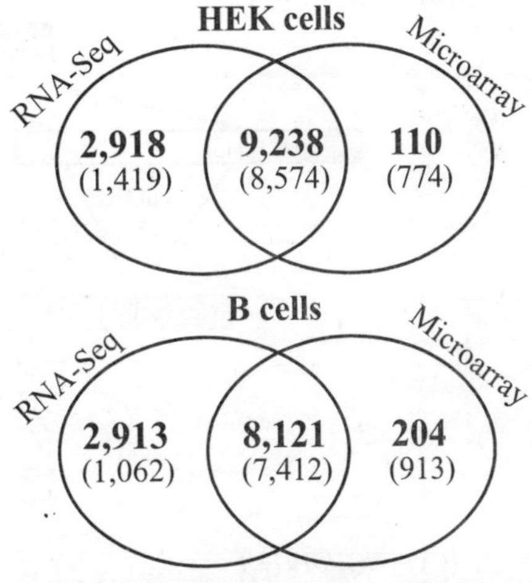

Figure 10.4 Comparison of number of expressed genes detected by RNA-Seq and microarrays

Riboswitch-Mediated Base-Pairing Changes and Alternative RNA Splicing

Bacteria make extensive use of riboswitches to sense metabolites and control gene expression, and typically do so by modulating premature transcription termination or translation initiation. The most widespread riboswitch class known in bacteria responds to the coenzyme thiamine pyrophosphate (TPP), which is a derivative of vitamin B_1. Representatives of this class have also been identified in fungi and plants, where they are predicted to control messenger RNA splicing or processing. Cheah et al. (2007) examined three TPP riboswitches in the filamentous fungus *Neurospora crassa*, and found that one activates to repress gene expression by controlling mRNA splicing. A detailed mechanism involving riboswitch-mediated base-pairing changes and alternative RNA splicing control was elucidated for precursor *NMT1* mRNAs, which code for a protein involved in TPP metabolism. These results demonstrate that eukaryotic cells employ metabolite-binding RNAs to regulate RNA splicing events that are important for the control of key biochemical processes. Three *N. crassa* genes, namely, *NMT1*, *TH14*, and *NCU01977.1*, carry TPP riboswitches in 5′ introns (Cheah et al. 2007) (Figure 10.5).

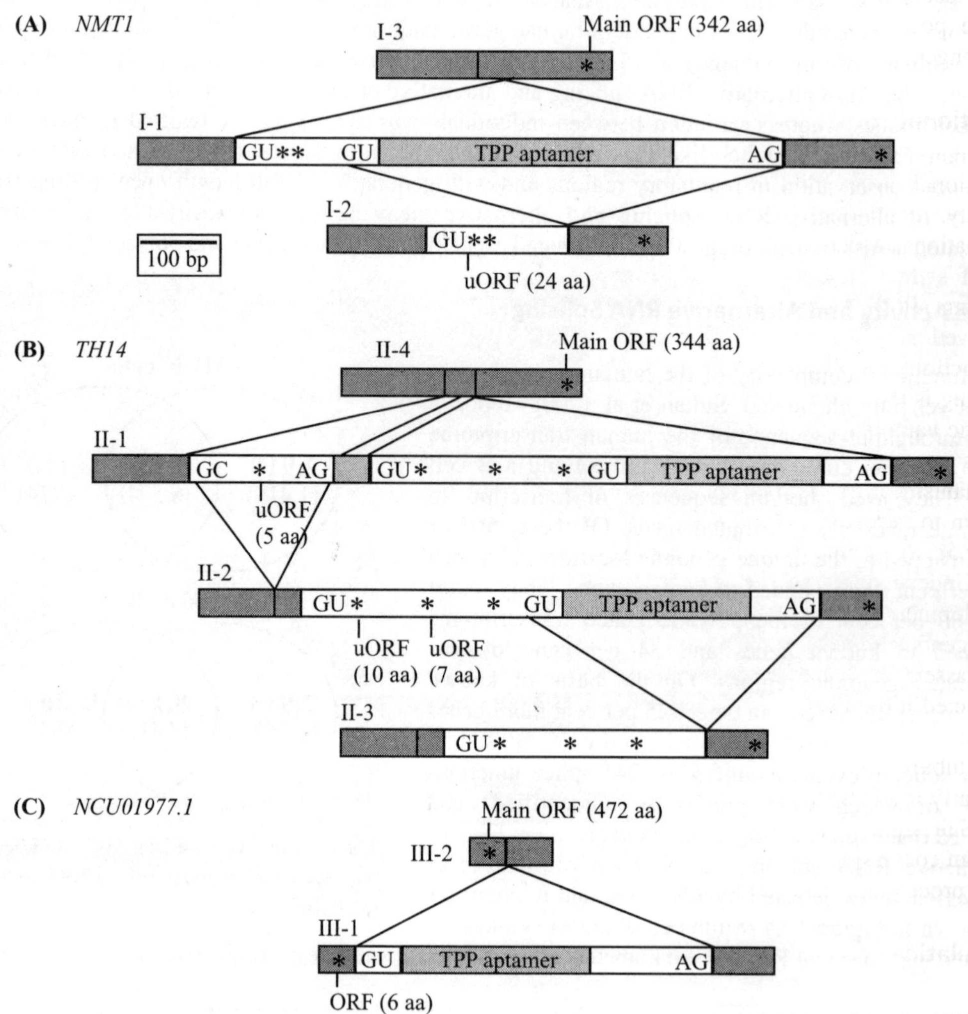

Figure 10.5 Three *Neurospora crassa* genes (*NMT1*, *TH14*, and *NCU01977.1*) that carry TPP riboswitches in 5′ introns

Splicing in mRNA Cytoplasmic Localization

Oskar messenger RNA localization at the posterior pole of the *Drosophila* oocyte is essential for germline and abdomen formation in the future embryo. The nuclear shuttling proteins Y14/Tsungi and

Mago nashi are required for *oskar* mRNA localization, and they co-localize with oskar mRNA at the posterior pole of the oocyte. Their human homologs, Y14/RBM8 and Magoh, are core components of the exon-exon junction complex (EJC). The EJC is deposited on mRNAs in a splicing-dependent manner, 20-24 nucleotides upstream of exon-exon junctions, independently of the RNA sequence. This indicates a possible role of splicing in *oskar* mRNA localization, challenging the established notion that the oskar 3′ untranslated region (3'UTR) is sufficient for this process. Hachet and Ephrussi (2004) show that splicing at the first exon-exon junction of *oskar* RNA is essential for *oskar* RNA localization at the posterior pole. The studies reveal an important new function for splicing: regulation of messenger ribonucleoprotein complex assembly and organization for mRNA cytoplasmic localization.

Functional Effects of Alternative RNA Splicing

Functional effects of alternative RNA splicing are: (a) protein localization, (b) deletion of protein activity, (c) modification of protein activity, (d) novel protein activities, and (e) RNA stability and translational efficiency. Alternative RNA splicing has role in developmental control by serving as on/off switch and through alternative RNA splicing of morphogenetic proteins. Alternative RNA splicing allows a high degree of protein diversity at low genetic cost. Alternative RNA splicing is involved in localization of different proteins in different cells, deletion of function of gene product and production of isomeric proteins having different functions (Smith et al. 1989). Alternative RNA splicing is the most efficient mechanism for generating protein diversity as it is reversible and allows genetic variation at a very low genetic cost (Breitbart et al. 1987). Through alternative RNA splicing, genes encode proteins with myriad functions. Alternative RNA splicing is an important regulatory mechanism in metazoans and their viruses (Breithart et al. 1987). More than 50 genes are currently known to generate protein diversity through use of alternative RNA splicing, e.g., *Drosophila* genes *Actin 5c*, *Adh*, *α-amylase* and human genes encoding growth hormones and γ-fibrinogen.

Understanding the mechanisms involved in the control of alternative RNA splicing will require the development of in vitro systems that accurately reflect the situation in vivo (Maniatis 1991). Autogenous regulation is found among proteins that are incorporated into macromolecular assemblies. The assembled particle may be unstable as a regulator, because it is large, too numerous, or too restricted in its locations. If assembly pathway is blocked for some reason, free sub-units accumulate and shut off unnecessary synthesis of further components. Eukaryotic cells have this type of regulation, e.g., tubulin synthesis. Free tubulins bind to nascent polypeptide or mRNA. mRNA is degraded. The T7 early region is transcribed into a single RNA that is cleaved into five individual mRNAs by RNase III. Hairpins in precursor RNA are recognized by the enzyme RNase III. When cleavage is prevented, certain coding regions are not translated. At least 12 ribonucleases exist in *E. coli*. Some are concerned with processing of tRNAs, some of rRNAs, and some of mRNAs.

Regulation of Alternative RNA Splicing

Splicing is regulated in a manner analogous to the transcriptional regulation, in that *trans*-acting proteins (repressors and activators) bind to *cis*-acting regulatory sites (silencers and enhancers) on the RNA. There are two major types of *cis*-acting RNA sequence elements present in pre-mRNAs, and specific *trans*-acting RNA binding proteins bind to each of these elements. Splicing silencers are sites to which splicing repressor proteins bind, reducing the probability that a nearby site will be used as a splice junction. These can be located in the intron itself (intronic splicing silencers, ISS) or in a neighboring exon (exonic splicing silencers, ESS). They vary in sequence, as well as in the types of proteins that bind to them. The majority of splicing repressors are heterogeneous nuclear ribonucleoproteins (hnRNPs) such as hnRNPA1 and polypyrimidine tract binding protein (PTB). Splicing enhancers are sites to which splicing activator proteins bind, increasing the probability that a nearby site will be used as a splice junction. These also may occur in the intron (intronic splicing enhancers, ISE) or exon (exonic splicing enhancers, ESE). Most of the activator proteins that bind to ISEs and ESEs are members of the SR protein family. Such proteins contain RNA recognition motifs and arginine and serine-rich (RS) domains. Most ESEs function by recruiting members of the SR protein family. These factors usually regulate splicing by binding ESEs through their N-terminal RRM domains and mediating protein–protein interactions that facilitate spliceosome assembly through C-terminal RS domains. ESSs are often bound by splicing repressors of the hnRNP class, a diverse group of proteins containing one or more RNA-binding domains and sometimes splicing inhibitory domains such as glycine-rich motifs. hnRNPs function by a wide variety of mechanisms. For example, PTB (hnRNP I) can block essential interactions between U1 and U2 snRNPs , whereas hnRNP A1 can inhibit splicing by binding on either side and looping out exons or by directly displacing snRNP binding.

A number of intronic SREs are also known. One well characterized ISE is the G triplet (GGG) or G run, which often occur in clusters and can enhance recognition of adjacent 5′ss or 3′ss. This ISE is common in GC-rich introns and is conserved between human and mouse. Intronic CA repeats in several cases can enhance splicing of upstream exons, probably through binding of hnRNP L. UGCAUG hexanucleotides or slight variations often occur downstream of neuron-specific exons and function as ISEs by binding to the brain- and muscle-specific splicing factors Fox-1/Fox-2. Characterized ISSs include binding sites for the splicing repressors PTB and hnRNP A1. Intronic elements (ISS and ISE) are likely of primary importance in regulating alternative RNA splicing events, as the intronic regions surrounding alternative exons are far more conserved in mammals than those surrounding constitutive exons, out to a distance of 150 bp or more (Burge and Wang 2008).

The first and the best characterized enhancers and silencers are those that control the earliest steps of spliceosome assembly, viz., the association of U1 snRNP ,U2AF, and the U2 snRNP with 5′ and 3′splice sites, respectively (House and Lynch 2008). The best-studied mechanisms of alternative RNA splicing regulation involve controlling splice site recognition by facilitating or interfering with the binding of the U1 or U2 snRNP to the splice sites.

Facilitating splice site recognition

SR proteins have important roles in facilitating splice site recognition. For example, they recruit the U1 snRNP to the 5′ splice site and the U2AF complex and U2 snRNP to the 3′ splice site by binding to an ESE and directly interacting with protein targets. These interactions are mediated by their RS (Arg–Ser repeat-containing) domains which have to be properly phosphorylated and dephosphorylated. SR proteins also cooperate with other positive regulatory factors to form larger splicing enhancing complexes by interacting with other RS domain-containing proteins such as transformer 2 (TRA2) and the SR-related nuclear matrix proteins SRm160 (also known as SRRM1) and SRm300 (also known as SRRM2).

Inhibiting splice site recognition

Inhibition of splice site recognition can be achieved in many ways. First, when splicing silencers are located close to splice sites or to splicing enhancers, inhibition can occur by stearically blocking the access of snRNPs or of positive regulatory factors. For example, polypyrimidine-tract binding protein (PTB; also known as PTB1 and hnRNP I), binds the polypyrimidine tract and blocks the binding of U2AF to regulated exons. In addition, hnRNP A1 binds ISSs that are located upstream of exon 3 in HIV Tat pre-mRNA and prevents binding of the U2 snRNP60. Finally, tissue-specific splicing factors

FOX1 and FOX2 inhibit the formation of the E′ complex by binding to an intronic sequence to prevent SF1 from binding to the branch site of *CAlCA* (calcitonin-relatedpolypeptide-α) pre-mRNA. Splicing inhibitors also stearically block the binding of activators to enhancers. Hu/ELAV family proteins inhibit U1 snRNP binding by competing with the binding of TIA1 to an AU-rich sequence downstream of 5′ splice site of exon 23a of neurofibromatosis type 1 pre-mRNA62. FOX1 and FOX2 also inhibit E complex formation by binding to an exonic sequence in the *CAlCA* pre-mRNA close to the ESE that TRA2 and upstream of the TRA2-dependent ESE in exon 7 in the SMN2 (survival of motor neuron 2) pre-mRNA, possibly inhibiting the formation or stabilization of the U2 snRNP complex. Some silencers can be over 100-200 bp away from enhancers, and a simple 'bind and block' model thus cannot explain their inhibitory effect. One explanation for the activity of such splicing inhibitors is that they function by masking splice site recognition through multimerization along the RNA.

Combinatorial effects of activators and inhibitors

Splicing of individual pre-mRNAs is frequently controlled by combinatorial or competitive effects of both activators and inhibitors. The final decision of whether an alternative exon is included is determined by the concentration or activity of each type of regulators often by SR proteins and hnRNPs (Chen and Manley 2009).

An excellent example of combinatorial effects of activators and inhibitors is provided by alternative RNA splicing of HIV-1 tat exon 2. The inclusion of tat exon 2 in the RNA is regulated by competition between the splicing repressor hnRNP A1 and the SR protein SC35. Within exon 2 an exonic splicing silencer sequence (ESS) and an exonic splicing enhancer sequence (ESE) overlap. If A1 repressor protein binds to the ESS, it initiates cooperative binding of multiple A1 molecules, extending into the 5′ donor site upstream of exon 2 and preventing the binding of the core splicing factor U2AF35 to the polypyrimidine tract. If SC35 binds to the ESE, it prevents A1 binding and maintains the 5′ donor site in an accessible state for assembly of the spliceosome. Competition between the activator and repressor ensures that both mRNA types (with and without exon 2) are produced.

Transcription-coupled alternative RNA splicing

Two models have been proposed to explain the role of RNAP II in the regulation of alternative RNA splicing. The first model is known as the recruitment model, in which RNAP II and transcription factors interact directly or indirectly with splicing factors, thereby increasing or decreasing the efficiency of splicing. A second, kinetic model proposes that the rate of transcription elongation influences the inclusion of alternative exons by affecting whether the splicing machinery is recruited sufficiently quickly for spliceosome assembly and splicing to occur (Chen and Manley 2009).

Regulation by histone modifications

Most human genes are alternatively spliced in a cell or tissue specific manner. Pre-mRNA splicing is influenced by RNA polymerase II elongation rate, chromatin remodelers and histone deacetylase inhibitors (Luco et al. 2010).

Applications of Alternative RNA Splicing

Alternative RNA splicing defines tissue specificity. It helps in understanding how mis-regulation in alternative RNA splicing leads to disease. Several abnormally spliced mRNAs are also found in a high proportion of cancerous cells. Until recently, it was unclear whether such aberrant patterns of splicing played a role in causing cancerous growth, or were merely a consequence of cellular abnormalities associated with cancer. It has been shown that there is actually a reduction of alternative RNA splicing in cancerous cells compared to normal ones, and the types of splicing differ; for instance, cancerous cells show higher levels of intron retention than normal cells, but lower levels of exon skipping. Some of the differences in splicing in cancerous cells may result from changes in phosphorylation of *trans*-acting splicing factors. Others may be produced by changes in the relative amounts of splicing factors produced; for instance, breast cancer cells have been shown to have increased levels of the splicing factor SF2/ASF.

One example of a specific splicing variant associated with cancers is in one of the human *DNMT* genes. Three DNMT genes encode enzymes that add methyl groups to DNA, a modification that often has regulatory effects. Several abnormally spliced *DNMT3B* mRNAs are found in tumors and cancer cell lines. In two separate studies, expression of two of these abnormally spliced mRNAs in mammalian cells caused changes in the DNA methylation patterns in those cells. Cells with one of the abnormal mRNAs also grew twice as fast as control cells, indicating a direct contribution to tumor development by this product.

Another example is the *Ron* proto-oncogene. An important property of cancerous cells is their ability to move and invade normal tissue. Production of an abnormally spliced transcript of *Ron* has been found to be associated with increased levels of the SF2/ASF in breast cancer cells. The abnormal isoform of the Ron protein encoded by this mRNA leads to cell motility.

Future Goals of Alternative RNA Splicing

Important goals of future studies of alternative RNA splicing regulation include: (a) understanding how regulators switch key splicing events during development and in response to environmental stimuli, (b) understanding how mis-regulation of alternative RNA splicing leads to disease, and (c) complete characterization of tissue-specific patterns of expression is of great importance to defining mechanisms of alternative RNA splicing regulation in different cell types (Chen and Manley 2009).

RNA EDITING

Central dogma has faced conceptual challenges since its formulation. Recent challenge has been phenomenon of RNA editing, proposed by Benne et al. (1986). Simpson and Shaw (1989) defined it as a process resulting in difference in nucleotide sequence of RNA from its template DNA as a result of any process other than splicing. The phenomenon of RNA editing was discovered in mid 1980's following some unexpected observations made by several groups of molecular biologists. These observations made particularly in the mitochondrial genes of *Trypanosoma* (which causes African sleeping sickness) and *Leishmania* (causing kala-azar), included the following: (a) Several genes having mutations, which were supposed to render these genes inactive (due to origin of premature stop signals, or due to loss of start signals), were still active. (b) The sequences of mRNA molecules in several cases were such which could not have been derived from the DNA sequence of the corresponding genes. These DNA sequences have sometimes been described as non-sensical 'cryptogenes'. These observations were later explained by RNA editing, which sometimes included addition of upto 550 and deletion of upto 41 uridine residues. It was shown that RNA editing mainly involved addition and deletion of uridines or insertions of cytidines, and that these changes took place, both in *Trypanosoma* and *Leishmania*, mtDNA sequences were found mainly in minicircle mtDNA, but rarely also in maxicircle mtDNA, that encode small RNA molecules, less than 40 nucleotides in length. This RNA apparently carried information for uridine insertions and deletions, and was, therefore, called guide RNA (gRNA). It was also shown that tails consisting of strings of uridine nucleotides added to the gRNA become the source of uridines inserted in mRNA.

In this process, uridine (U) residues are added or deleted at multiple precise sites. RNA editing also involves C to U change, multiple C to U or U to C changes, addition of G or C residues, A to G modification, and conversion of U to A, U to G and A to G. RNA editing takes place with the help of guide RNAs (gRNAs). The mRNAs are edited most frequently, whereas edits of tRNAs and rRNAs are extremely rare. In some cases, individual nucleotides can be added in the middle of the mRNA sequence by RNA editing. RNA editing challenges the gene as sole site of genetic information. It may represent a mechanism that evolved to repair mutant genes but objection is that it should not necessarily be restricted to C to U modifications.

RNA editing is thus a post-transcriptional process that results in radical changes in amino acid specified by a codon. Most of the editing events occur in the first two nucleotides of the coding sequences. RNA editing has been noticed in kinetoplastid, plant mitochondria, chloroplasts and mammalian nucleus. The nucleotides located immediately upstream of the edited cytosine is most often a pyrimidine (uridine) and very rarely guanidine. It has been shown that editing mainly occurs in the mRNA fraction rather than rRNA or tRNA.

Patterns of RNA Editing

Various patterns of RNA editing are: (a) Insertion or deletion of non-templated uridines, as in *Trypanosoma* mitochondrial transcripts (Blum et al. 1990). (b) Insertion of non-temlplated guanosines, as in paramyxoviral transcripts (Vidal et al. 1990a,b). (c) Creation of a stop codon by conversion of a specific cytosine to a uridine, as in apolipoprotein B transcripts in mammalian intestinal tissue (Chen et al. 1987). (d) Conversion of multiple cytosines to uridines or uridines to cytosines, as in higher plant mitochondrial transcripts (Gray 1989; Gualberto et al. 1989).

Characteristics of RNA Editing

General characteristics of RNA editing are: (a) Guide RNA editing is a post-transcriptional process. (b) Editing frequently results in radical changes in amino acid(s) specified by a codon. (c) Most editing events occur in the first two nucleotides of the codon. (d) RNA editing efficiently modifies the transcripts such that a relatively homogenous population of polypeptide is produced. (e) RNA editing generally occurs in the coding sequences.

Parallelism between RNA Editing and RNA Self-Splicing

Parallelism between RNA editing (done by gRNA) and self-splicing of introns in hnRNA (done by ribozymes) was demonstrated by Thomas Cech (Cech 1986, 1987). A parallelism between the function of IG-S in RNA splicing and the proposed role of guide RNAs in RNA editing in kinetoplastid mitochondria has been recognized (Mueller et al. 1993). They identified chimeric intermediate with gRNA covalently joined to 3' portion of the mRNA in vitro suggesting an evolutionary analogy to the catalytic mechanism involved in RNA splicing. Like self-splicing of introns, RNA-editing also involves transesterification, but with the help of guide RNA.

Certain species of mRNA have nucleotide sequences that differ greatly from the sequences of the genes from which they are presumably transcribed. The differences are usually due to addition of uridylate residues, but occasionally also to their deletion. The most spectacular case of RNA editing known occurs in mRNA of mitochondria mitochondrial gene for subunit III of cytochrome oxidase

(*COIII*) in *Tetrahymena brucei* (Volloch et al. 1990). The mRNA is twice the length of gene owing to the addition of several hundred uridylate residues and a few deletions spread over the entire length of the RNA molecule. The experiments showed that edited RNA was synthesized as a unit.

RNA editing involves addition and deletion of non-coded ribonucleotide residues to and from RNA transcripts, thus converting them into correctly translatable mRNA. RNA editing thus involves the modification the primary sequence of coding regions of mRNAs and provides yet another method of modulating genetic activity (Simpson and Shaw 1989). In kinetoplastid and perhaps the *Physarum*, RNA editing involves addition or deletion of nucleotides with coding regions of mRNA molecules. Another type of RNA editing is paramyxoviral type which involves a co-translational stuttering of the RNA polymerase on homopolymer stretches. In mammals, apolipoprotein type RNA editing is there which involves a chemical nucleotide modification. The question still remains what encodes the specificity of RNA editing.

In trypanosome mitochondrial genes, AUG (translation initiation codon) is absent, frameshifts and most conserved genes like ATP synthetase subunits 6, 8 and 9 and tRNA genes are absent. The genes whose transcripts get edited are called cryptogenes and portion of cryptogene that is edited is called pre-edited region. Depending on the position of editing in protein-encoding region, three types of crytogenes are found. Type I cryptogenes, e.g., *COII* in *Trypanosoma brucei*, *Crithidia fasiculata* and *Leishmania tarentolae* is edited in the internal portion. Type II cryptogene is edited at 5′-end, e.g., *Cyb* of *T. brucei*, *C. fasiculata* and *L. tarentolae*. Type III crytogene is extensively edited throughout the protein-encoding region, e.g., *COIII* in *Trypanosoma brucei*. Editing proceeds in a 3′→5′ direction. In 3′ poly(A) tails of transcripts of maxicircle genes in the trypanosomatid mitochondria also U's are added.

Mechanisms of RNA Editing

Various hypotheses proposed for mechanisms of RNA editing are: pseudogene hypothesis, variant of pseudogene hypothesis, edited template refractive to hybridization, cryptic signal sequences, self-replication edited RNA, alternative RNA/RNA heteroduplex model and multienzyme editing complex. Multienzyme editing complex hypothesis was proposed by Simpson (1990) as guide RNA (gRNA) model of RNA editing and is the most viable proposal. The unedited transcript and gRNA hybridize and the mis-matched nucleotides are edited using enzymes like ribonuclease, TUTase, and RNA ligase. The enzyme complex moves along the hybrid in 3′→5′ direction.

RNA editing adenosine-to-inosine

Primary transcripts of certain microRNA (miRNA) genes are subject to RNA editing that converts adenosine to inosine. However, the importance of miRNA editing remains largely undetermined. Kawahara et al. (2007) report that tissue-specific adenosine to inosine (A-to-I) editing of miR-376 cluster transcripts leads to predominant expression of edited miR-376 isoform RNAs. One highly edited site is positioned in the middle of the 5'-proximal half "seed" region critical for the hybridization of miRNAs to targets. They provide evidence that the edited miR-376 RNA silences specifically a different set of genes. Repression of phosphoribosyl pyrophosphate synthetase 1, a target of the edited miR-376 RNA and an enzyme involved in the uric-acid synthesis pathway, contributes to tight and tissue-specific regulation of uric-acid levels, revealing a previously unknown role for RNA editing in miRNA-mediated gene silencing.

A-to-I RNA editing leads to transcriptome diversity and is important for normal brain function. To date, only a handful of functional sites have been identified in mammals. Li et al. (2009) developed an unbiased assay to screen more than 36,000 computationally predicted non-repetitive A-to-I sites using massively parallel target capture and DNA sequencing. A comprehensive set of several hundred human RNA editing sites was detected by comparing genomic DNA with RNAs from seven tissues of a single individual. Specificity of their profiling was supported by observations of enrichment with known features of targets of adenosine deaminases acting on RNA (and validation by means of capillary sequencing. This efficient approach greatly expands the repertoire of RNA editing targets and can be applied to studies involving RNA editing-related human diseases. Schematic diagram of the padlock technology used for screening of RNA editing sites is given in Figure 10.6 (Li et al. 2009). The candidate RNA editing sites are specifically targeted by padlocks in both gDNA and cDNA samples from a single individual. Circles are formed when polymerase, deoxynucleotide triphosphate, and ligase are added, subsequently amplified, and sequenced with an Illumina genome analyzer.

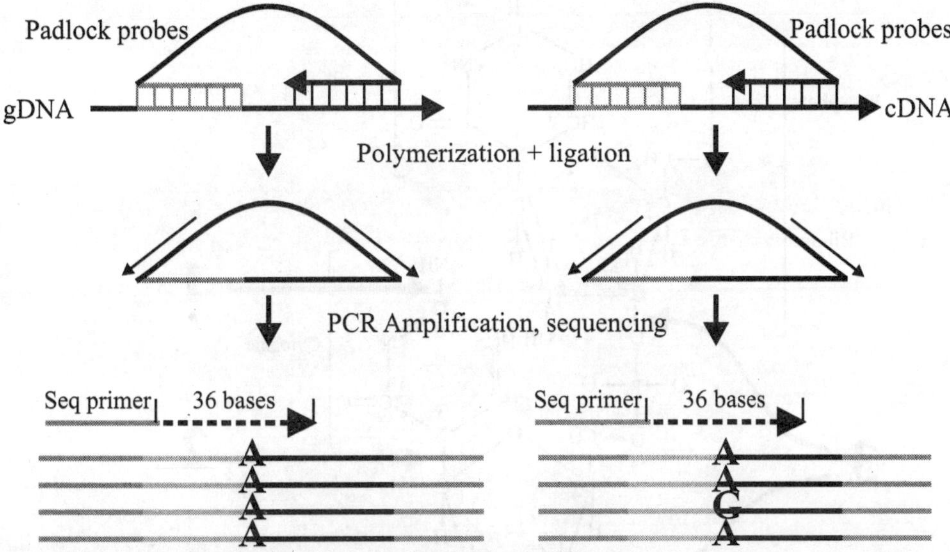

Figure 10.6 The padlock technology used for screening of RNA editing sites

RNA editing cytosine-to-uracil

All canonical transfer RNAs (tRNAs) have a uridine at position 8, involved in maintaining tRNA tertiary structure. However, the hyperthermophilic archeon *Methanopyrus kandleri* harbors 30 (out of 34) tRNA genes with cytidine at position 8. Randau et al. (2009) demonstrate cytosine-to-uracil (C-to-U) editing at this location in the tRNA's tertiary core, and present the crystal structure of a tRNA-specific deaminase, CDAT8, which has the cytidine deaminase domain linked to a tRNA-binding THUMP domain. CDAT8 is specific for C deamination at position 8, requires only the acceptor stem hairpin for activity, and belongs to a unique family within the "cytidine deaminase-like" superfamily. The presence of this C-to-U editing enzyme guarantees the proper folding and functionality of all *M. kandleri* tRNAs.

Three RNA editing mechanisms are potentially responsible for C-to-U conversion (Rajasekhar and Mulligan 1993) (Figure 10.7). The bonds cleaved by various RNA editing mechanisms are indicated by the numbered arrows. Mechanism 1 involves deamination (or transamination) of the C-6 amide of cytosine to convert the base to uracil. Mechanism 2 involves transglycosylation of the ribosyl moiety to replace the base. Mechanism 3 involves deletion and insertion of a new nucleoside monophosphate. These three mechanisms involving biochemical modifications are explained here only briefly.

Deamination/Amination: The simplest mechanism for C-to-U conversion is deamination at position U of cytosine leading to a uracil residue (Figure 10.8). This could be achieved by cytidine deaminase, an enzyme of the nucleotide metabolism (O'Donovan and Neuhard 1970). The reverse U-to-C modification found in plant mitochondria could be carried out by a CTP synthetase in an ATP-dependent reaction. However, existence of reverse editing from U-to-C implies the presence of two different enzymatic activities and not a reverse action of the same enzyme.

Base Exchange: Exchange of base without cleavage of the sugar-phosphate backbone is another possible way of biochemical modification (Figure 10.7). Such transglycosylate reaction has been desc-

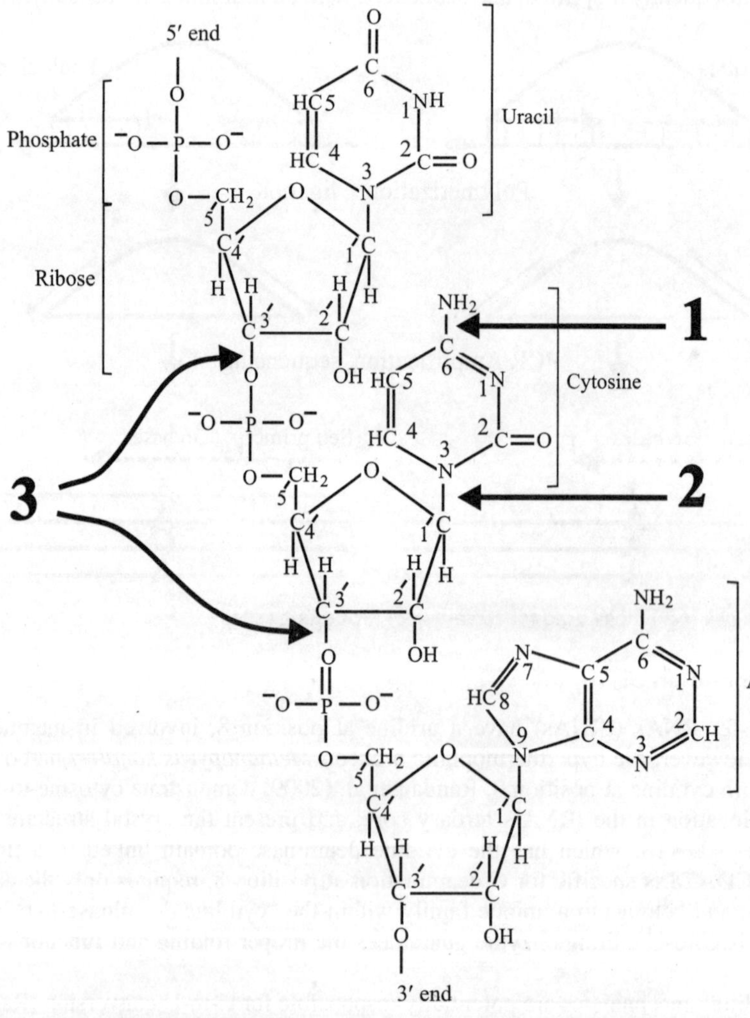

Figure 10.7 RNA editing mechanisms potentially responsible for C-to-U conversion. The bonds cleaved by various RNA editing mechanisms are indicated by the numbered arrows. *Mechanism 1* involves deamination (or transamination) of the C-6 amide of cytosine to convert the base to uracil. *Mechanism 2* involves transglycosylation of the ribosyl moiety to replace the base. *Mechanism 3* involves deletion and insertion of a new nucleoside monophosphate

ribed in transfer RNAs (Brooks et al. 1997). Pyrimidine replacement is also found in DNA repair metabolism (Olson et al. 1989). This process like deamination does not include any cut in the RNA chain.

Nucleotide Exchange: Nucleotide deletions at RNA editing sites have been reported in wheat mitochondrial cDNA clones (Gualberto et al., 1991). These cDNAs were interpreted to represent single-nucleotide deleted intermediates of the editing process, and a deletion and insertion mechanism was proposed by these investigators. Nucleotide exchange involves cleavage of the RNA chain, deletion of a cytosine, addition of uracil, followed by religation (Figure 10.8). The nucleotide exchange involves several steps and is similar to the one described for U addition and deletion in editing of mRNA of trypanosomes. This mechanism is most complex and would probably need an apparatus composed of multiple factors.

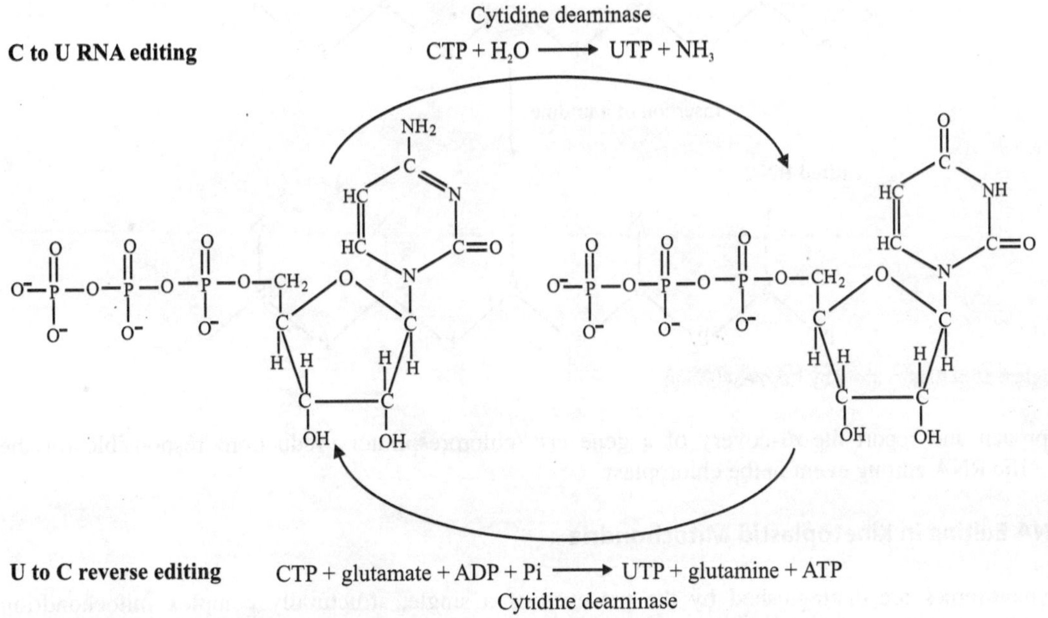

C to U RNA editing

Cytidine deaminase

$CTP + H_2O \longrightarrow UTP + NH_3$

U to C reverse editing

$CTP + glutamate + ADP + Pi \longrightarrow UTP + glutamine + ATP$

Cytidine deaminase

Figure 10.8 RNA editing by deamination/amination

RNA editing in the mitochondria and chloroplasts in protozoa has been reviewed by Altman (1989) and Simpson and Shaw (1989), Forster and Altman, (1990), and Altman et al. (1993).

RNA Editing in Chloroplasts

It has been reported that ACG codon appears at 5'-terminus of chloroplast genes where initiation codon ATG would be expected. Evidence of mRNA transcript from *rpl2* gene in maize plastome shows that mRNA editing also occurs in chloroplasts. In organellar transcripts of higher plants, specific cytidine residues are converted into uridine residues. In many cases, editing results in the restoration of conserved amino acid residues in a process that is essential for protein function in plastids. Despite the technical breakthrough in establishing systems in vivo and in vitro for analyzing RNA editing, its machinery still remains to be identified in higher plants. Kotera et al. (2005) introduce a genetic

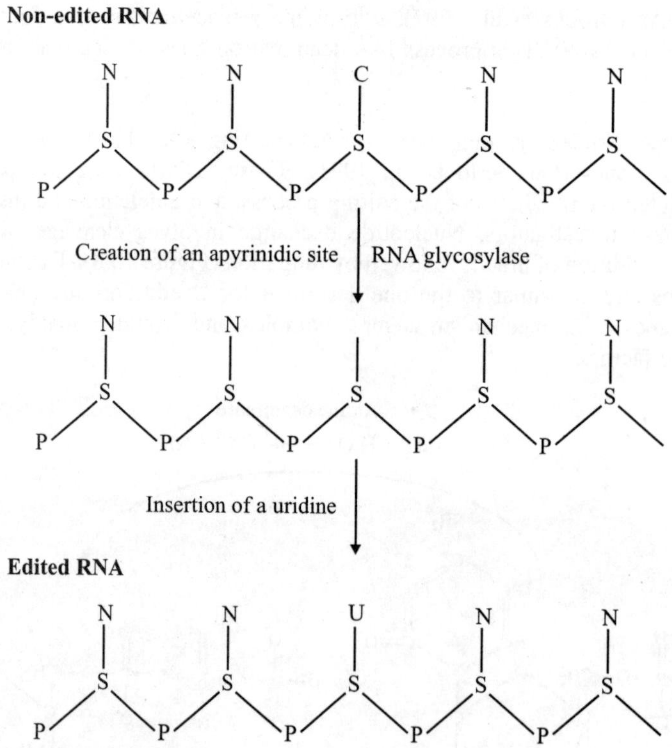

Figure 10.9 RNA editing by base exchange

approach and report the discovery of a gene *crr* (chlororespiratory reduction) responsible for the specific RNA editing event in the chloroplast.

RNA Editing in Kinetoplastid Mitochondria

Trypanosomes are distinguished by the presence of a single, structurally complex mitochondrion containing an unusual genome known as kinoplast DNA (kDNA). kDNA consists of two types of circular DNA molecules – one minicircle (465 to 2,500 bp) and maxicircles (23 to 36 kbp). No obvious protein-encoding genes are known in minicircles. Maxicircle molecules are homologs of informational mitochondrial DNA molecules in animals and fungal cells and possess various genes. Several of maxicircle genes represent cryptogenes, i.e., incomplete genes whose transcripts are edited to yield translatable sequences. Three such pan-edited cryptogenes are known in *Tryponosoma brucei* maxicircle DNA, which lack more than 50 per cent of the U residues present in mature mRNA. Blackburn (1990) has given a model of RNA editing of kinetoplast mRNAs.

Guide RNA in RNA editing

In the first step, the guide RNA aligns itself (by base pairing) with the unedited RNA, splits it into two, and makes a new bond between one of the broken ends and the uridine at the tip of the tail of gRNA at its 3′-end (Figure 10.11). This reaction is facilitated by mitochondrial uridyl transferase activity. In the next step, the broken end of other RNA segment which is not involved in addition of U, forms a bond with this RNA segment. The tail of gRNA is now released for another round of transesterification.

When a uridine needs to be deleted from mRNA, cleavage occurs at 5'-end of uridine (to be deleted) which is mis-matched during base pairing with gRNA. Subsequently, 3'-OH group in gRNA attacks 3' of the uridine to be deleted, leading to excision of uridine and production of gRNA that is one nucleotide longer. The above mechanism predicts an intermediate, in which gRNA is linked to mRNA. Such gRNA-mRNA chimeric molecules were actually found and their presence verified by polymerase chain reaction (PCR), in which 5'-primer specific to gRNA and the 3'-primer specific to mRNA were used. The PCR gave chimeric molecules as amplification products.

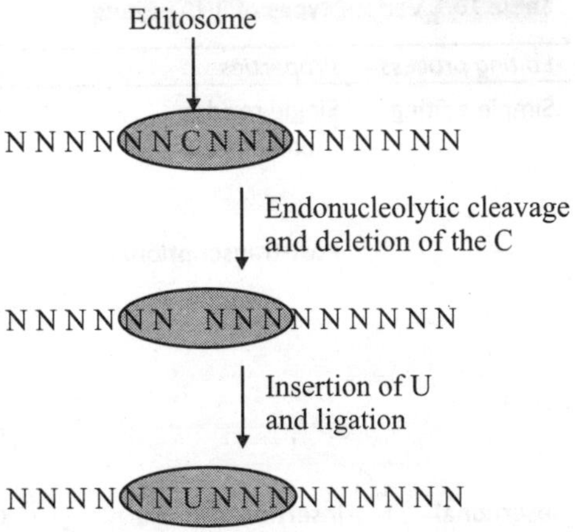

Blum et al. (1990) described small kinetoplastid molecules, known as guide RNAs (gRNAs). RNAs are required for RNA editing and for the transfer of genetic information to pre-mRNA. Guide RNA initially forms a duplex with the pre-mRNA that it edits (Seiwart and Stuart, 1994). Guide RNAs can form perfect hybrids with mRNA to be edited and they possess nucleotide sequences at their 5'-end that are complementary to the sequences of the mRNAs immediately downstream of the pre-edited regions (PER). gRNAs are of about 60 nucleotide length (Church et al. 1979; Blum et al. 1991). There are four categories of RNA editing which are described in Table 10.1. The guide RNAs do not represent classical templates for edited RNA due to the presence of abundant non-canonical G.U base pairs. The guide RNAs are found in the maxicircle genome and within variable regions of the minicircles, suggesting a function for these DNA molecules.

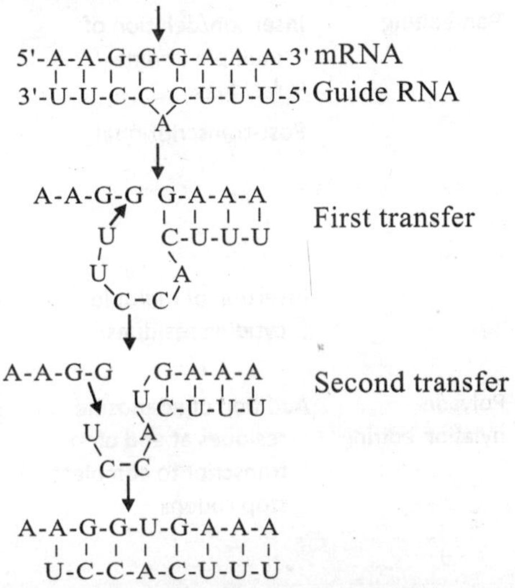

Editosome

N N N N N N C N N N N N N N N N

Endonucleolytic cleavage and deletion of the C

N N N N N N N N N N N N N N N

Insertion of U and ligation

N N N N N N U N N N N N N N N N

Figure 10.10 **RNA editing by nucleotide exchange**

5'-A-A-G-G-G-A-A-A-3' mRNA
3'-U-U-C-C C-U-U-U-5' Guide RNA
A

A-A-G-G G-A-A-A
U C-U-U-U
U A
C-C
First transfer

A-A-G-G G-A-A-A
U C-U-U-U
U A
C-C
Second transfer

A-A-G-G-U-G-A-A-A
U-C-C-A-C-U-U-U

Figure 10.11 RNA editing using guide RNA

The guide RNAs are specific for each edited mRNA and encode additional U residues as complementary A or G residues Blum et al. (1990). In the model proposed by Simpson (1990), a hybrid is formed between the 5′ ends of the gRNA and the region of the RNA that is adjacent to the pre-edited region (PER) (3′ anchor) (Figure 10.12). A stabilizing hybrid then forms between the 3′ oligo U tail of the guide RNAs and G.A-rich PER (5′ anchor). Editing occurs by specific endonuclease cleavage of the mRNA within the PER at a position 3′ to the first mis-matched nucleotide. Additional U residues to or, more rarely, deletion from the liberated 3′-hydroxyl terminus is followed by the formation of a base pair between the guide A or G and the added U residues, and religation of the cleaved mRNA molecule. The editing enzyme complex then migrates to the next mis-match and the cycle is repeated. For MURF2 genes in *Leishmania tarentolae*, there are two overlapping guide RNA sequences that together cover the entire edited regions. The 3′-most of the RNA sequences (with reference to the mRNA) is referred to as block I guide RNA (gRNA-I) and the 5′-most as block II gRNA (guide-II) in each case.

Table 10.1 Various types of RNA editing

Editing process	*Properties*	*Examples*	*Mechanism*
Simple editing	Single residue conversions	C→U transition in the mammalian apoliporprotein mRNA	Modification by specific cytidine deaminase
	Post-transcriptional	A→G transition in mRNA for mammalian glutamate receptor subunits and serotonin receptors. Editing also occurs in introns	Modification by dsRNA deaminase converts adenosine to inosine. Three genes have been identified
		C→U transitions in plant organelles	Unknown
Insertional editing	Insertion of single nucleotides or small runs of nucleotides Co-transcriptional	G insertions during transcription of the paramyxovirus *P* gene	Transcriptional strand slipping
Pan-editing	Insertion/deletion of multiple uridine residues Post-transcriptional	U insertions/deletions in trypanosome kinetoplast mRNA	Editing sequence provided by external antisense guide RNA which pairs with pre-edited mRNA in a ribonucleoprotein particle, the editosome, and identifies positions to be edited as mismatches.
	Insertion of multiple cytidine residues	C insertions in at least four *Physarum polycephalum* mitochondrial mRNAs	Unknown
Polyadenylation editing	Addition of adenosine residues at end of transcript to complete stop codons	Polyadenylation of several vertebrate mitochondrial mRNAs	Pre-mRNA lacking a stop codon is polyadenylated, with the first one or two adenylate residues providing the missing information

There are two possible ways these two types of gRNAs act. Guide RNA I acts first to complete editing and then guide RNA II acts. (ii) Both guide RNAs may act simultaneously according to their complementarity with different parts of mitochondrial RNA. There are circumstantial but convincing evidences for the second model. (a) The gRNAs for five cryptogenes in *L. terantolae* exist that can form perfect hybrids (assuming G.U base pairs) with edited mRNAs with the unique 5′ ends of the gRNAs located close to the beginning of the hybrid region. (b) Multple partially edited molecules have been detected. (c) Enzymes of several of the pre-edited activities exist in purified mitochondria of *L. terentolae* – a terminal uridylyl transferase (TUTase), an RNA ligase and a cryptic site-specific endonuclease. (d) Imprecise editing of the synthetic pre-edited mRNAs has been detected with crude mitochondrial lysates from *L.torentolae*.

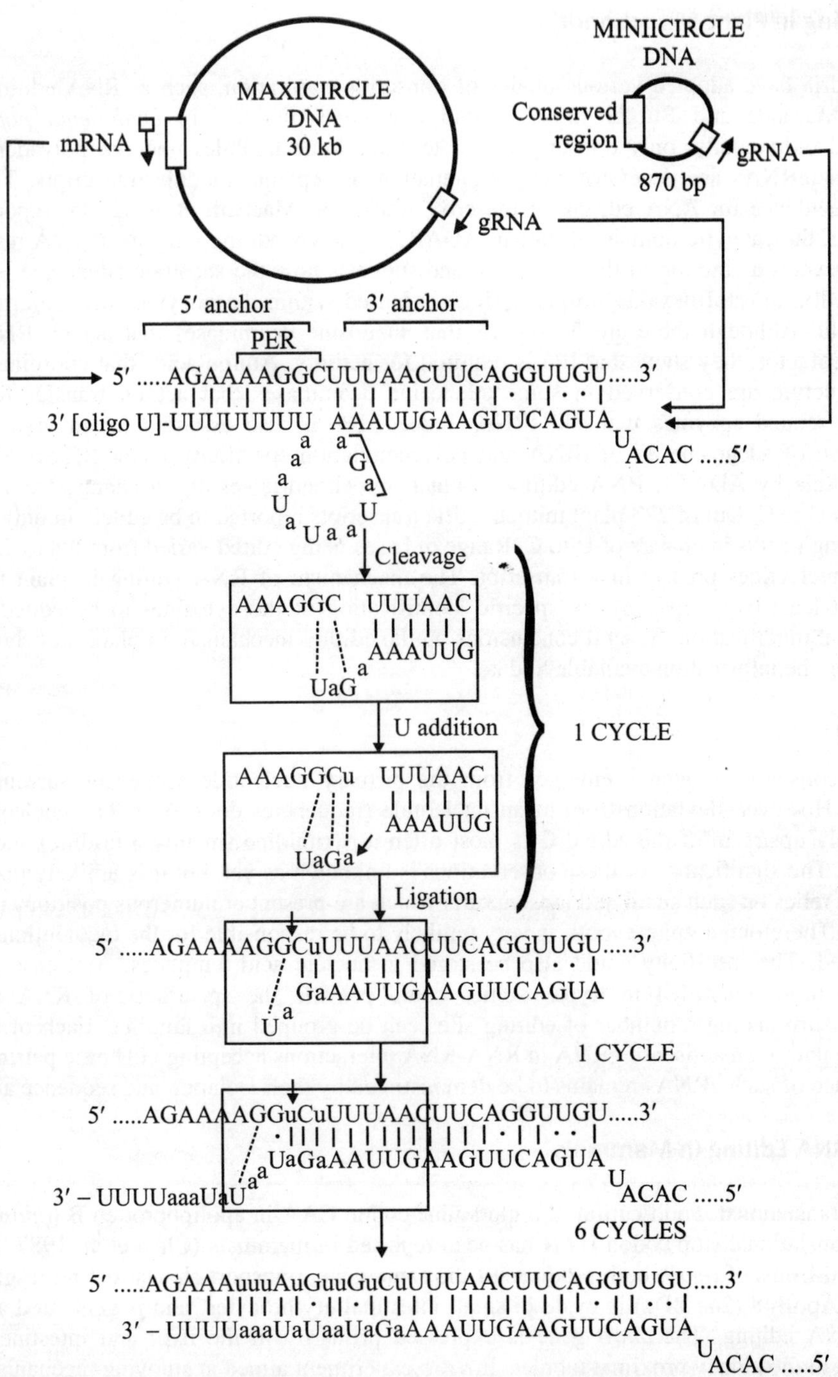

Figure 10.12 The gRNA model for RNA editing. One cycle is shown for the 5′ editing of the Cyb mRNA with guide RNA-I. The guide nucleotides in the gRNA are shown as a, and the added U residues as u. The cleavage sites are indicated by arrows. G.U base pairs are indicated by asterisk (*)

RNA Editing in Plant Mitochondria

Mitochondria have adopted various modes of transcript maturation, such as RNA editing and *trans*-splicing. Marande and Burger (2007) found that gene modules in *Diplonema pappilatum* are transcribed individually, only transcripts of C-terminal gene modules undergo polyadenylation, and contiguous mRNAs are generated via coccatenation of separate module transcripts. They detected potential evidence for RNA editing in the cox1 transcript. Macbeth et al. (2005) report the crystal structure of the catalytic domain of human ADAR2, an RNA editing enzyme at 1.7Å resolution. The structure reveals a zinc ion in the active site and suggests how the substrate adenosine is recognized. Unexpectedly, inositol hexakisphosphate (IP6) is buried within the enzyme core, contributing to the protein fold. Although there are no reports that adenosine deaminases that act on RNA (ADARs) require a cofactor, they show that IP6 is required for activity. Amino acids that coordinate IP6 in the crystal structure are conserved in some adenosine deaminases that act on transfer RNA (tRNA) (ADATs), related enzymes that edit tRNA. Indeed, IP6 is also essential for in vivo and in vitro deamination of adenosine 37 of tRNA and polyadenylation specificity factor (CPSF-73) subunit of cleavage Aala by ADAT1. RNA editing in plant mitochondria results in change from C to U and sometimes U to C. Out of 273 plant mitochondria transcripts reported to be edited, in only five of these cases editing results in change of U to C. Range of bases being edited varied from 0.8 to 5.8 per cent of the total nucleotides present in a transcript. The mechanism of RNA editing in plant mitochondria requires at least two steps: (a) the specific identification of the cytosines to be edited and (b) the biochemical modification. Several conclusions on the editing mechanism in plant mitochondria can be drawn from the information available to date.

Specificity

No clear consensus sequence emerges from the primary nucleotide sequences surrounding edited cytosines. However, deviation from mean nucleotide frequencies does exist. The nucleotides located immediately upstream of the edited C is most often a pyrimidine (mainly a uridine) and very rarely guanidine. The significance of these observations is unknown as yet, but it is unlikely that the editing specificity relies on such small and loose motifs which are present at numerous positions not amenable to edition. Therefore, a unique motif is very unlikely to be responsible for the recognition of cytosines to be edited. The specificity could also be stored in nucleic acid templates. Antisense guide RNAs similar to those identified in trypanosomes could provide the specificity of RNA editing. The sequences surrounding a number of editing sites can be grouped into families. Each of these groups could base pair to an antisense gRNA in RNA-RNA interactions accepting G.U base pairing. However, the existence of such gRNAs remains to be demonstrated by their isolation and sequence analysis.

Nuclear RNA Editing in Mammals

The post-translational modification of a glutamine codon CAA in apolipoprotien B (*apoB*) in RNA to create a translational stop codon UAA has been reported in mammals (Chen et al. 1987; Powell et al. 1987). Two froms of *apoB*, both products of the same gene, transport cholestrol and tryglyceroides in the blood ApoB48 (241 kDa) is made in small intestinal erythrocytes, and is generated as a result of apoB mRNA editing. The *apoB* gene is expressed primarily in the liver and intestine, with some expression in the kidney proximal tubules. In vitro experiment aimed at studying mechanism of editing showed that when ^{32}P dCTP is incorporated into an *apoB* mRNA substrate, it creates a ^{32}P UMP residue solely at the editing site (Kumar and Hedges, 1998). Several important conclusions can be drawn from this simple observation: (a) The production of RNA editing is simple uridine base. No complex hypermodified base is created. A hypermodified base is the one like lysidines, in which the attachment of a lysine amino acid to the 2 position of cytidine ring causes the base to be read as uridine. (b) There is no known polymerase that can remove a nucleotide and add a new nucleotide in its place without replacing the 5′ phosphate with the donor's phosphate. The editing reaction most probably involves site-specific cytidine deamination of RNA.

Site-directed mutagenesis of the editing site has been used to define the sequence requirements for editing in the nine nucleotides surrounding the edited C (nucleotides –3 to +5). Twenty out of 22 mutations allowed efficient editing in vitro, suggesting that the sequence specificity of apoB mRNA is lax (Chen et al. 1987). In most extensive mutagenesis, Shah et al. (1991) examined the 53-nucleotide sequence that undergoes efficient editing in vitro. They identified an 11-nucleotide sequence (5′-UGAUCAGUAUA-3′) downstream of the editing site (positions +5 to +15) where most mutations abolished or greatly reduced editing in vitro. Hypothesis is that the editing enzyme recognizes a downstream binding site and then searches for a cytidine at a fixed distance upstream. Some flexibility exists as other cytidines introduced in the vicinity of the edited C can be edited either independently or in concert with the natural site. This implied that once properly bound to a substrate RNA, the editing enzyme can catalyze multiple deaminations. A second RNA editing site has been discovered in human *apoB* mRNA (Navaratnam et al. 1995). The second site has a sequence similar to the primary editing site including 7 of 11 nucleotides of the proposed target sequence.

Comparison of Different Forms of RNA Editing

RNA editing has been divided into two classes – insertional editing and substitutional editing. Insertional editing involves editing of kinetoplastid mitochondrial mRNA. It involves base pairing to specific guide RNAs with poly(U) tails. Mechanism involves identification of editing sites in the mRNA by mis-matches with the guide RNA, cleavage of the mRNA and religation of mRNA halves with added or deleted U residues by TUTase. A similar mechanism underlies the insertion of C residues into slime mould mitochondrial pre-mRNA. The modification of apoB mRNA is different from insertional mRNA editing. No guide RNA is implicated. The phosphodiester backbone of the RNA is not cut and the new base is synthesized in situ It falls into second class of RNA editing – substitutional editing – that blurs the distinction between editing and RNA modification. ApoB editing appears to bear more mechanistic similarity to the C-to-U change reported in plant mitochondrial and chloroplast mRNA.

However, an interesting difference in the plant mitochondrial system is the back editing reaction, i.e., amination of U to form C. Multiple C residues are edited in most RNAs and introns can also be edited. The plant mitochondrial enzyme may also be cytidine deaminase perhaps being reversible or with a transaminase activity. However, plant mRNA target sequences are not edited by mammalian apoB editing extracts. The apoB editing system is nuclear controlled and seems to be a very specialized phenomenon since it is limited to one particular transcript and it works in a tissue-specific way.

Functional Significance of RNA Editing

RNA editing, which serves as a translational control mechanism, adds ribonucleotides for N-terminal amino acids in type II editing and created the whole protein-encoding region in type III editing. It also regulates maxicircle gene expression at developmental level. RNA editing seems to have originated long back and has selective advantage in evolution. It may be occurring in other organisms also. However, more comprehensive research is needed to understand this phenomenon. It has been seen that editing is ubiquitous in all gymnosperms and angiosperms except *Marchantia* (Covello and Gray 1989). The nucleotides located immediately upstream of the edited cytosine is most often a pyrimidine (uridine) and very rarely guanidine (Gualberto et al. 1989). On the basis of study of chimeric *C-atp6* gene of maize CMA-C cytoplasm, it has been found that editing is sequence-specific.

Consequences of RNA Editing

RNA editing challenges the genes as sole site of genetic information. RNA editing shows that genetic information resides not only in the genes per se but also in the apparatus that edits the transcript. RNA editing could be a relic of a RNA sequence correlation system that was utilized be for polymerase-dependent nucleic acid synthesis evolved (Simpson 1990). Alternatively, RNA editing may represent a mechanism that evolved to repair mutant genes and now has become entrenched in the gene expression system. However, if RNA editing is considered to correct mutation, it should not be restricted to C-to-U modifications. RNA editing could regulate gene expression, perhaps, in a tissue-specific manner. RNA editing in the 5′ ends of certain templates creates in frame translation initiation codon, AUG and thereby provides a translational control mechanism for the production of proteins. Regulation of translatable mRNA levels by RNA editing would have the advantage of keeping mRNA not required for translation accessible in an inactive pool. These mRNAs could be quickly activated by editing without massive de novo synthesis of entire mRNA.

RNA editing occurs in a relatively efficient manner and generally results in a homogenous population of transcripts that can direct the synthesis of functional polypeptides. However, many investigators have reported the presence of incompletely edited transcripts or cDNAs for various mitochondrial genes. In some *Oenothera* genes (*nad3*, *rps13*), incompletely edited transcripts are common. In some other cases, e.g., *atp9*, only completely edited transcripts are observed. Incompletely edited transcripts would direct the synthesis of polypeptides with radical amino acid substitutions. These aberrant polypeptides may have impaired or altered function. It is not known whether or not these incompletely edited transcripts are translated. The incompletely edited RNAs represent RNA editing intermediates. Further, RNA editing does not occur simultaneously with transcription – it is post-transcriptional process.

RNA editing of at least four mitochondrial cryptogenes in *T. brucei* are known to be developmental regulator (Feagin et al. 1987; Feagin et al. 1988) and is utilized as a translational control mechanism by regulating the abundance of translatable mRNAs. Thus RNA editing process originally evolved perhaps as developmental and translational regulatory mechanism and has subsequently further evolved to overcome the effect of accumulated mutations. RNA editing may be a key to the understanding of evolution of life. RNA editing has been nicknamed as "molecular fossil" by T. Cech. It has been suggested to be a living relic that provides a look backward to life's origin.

Species-Specific RNA Editing

Most RNA editing systems are mechanistically diverse, informally restorative, and scattershot in eukaryotic lineages. In contrast, genetic recoding by adenosine-to-inosine RNA editing seems common in animals; usually, altering highly conserved or invariant coding positions in proteins. Reenan (2005) reports striking variation between species in the recoding of *synaptotagmin I* (*sytI*). Fruitflies, mosquitoes and butterflies possess shared and species-specific sytI editing sites, all within a single exon. Honeybees, beetles and roaches do not edit *sytI*. The editing machinery is usually directed to modify particular adenosines by information stored in intron-mediated RNA structures. Combining comparative genomes of 34 species with mutational analysis reveals that complex, multidomain, pre-RNA structures solely determine species-appropriate RNA editing. One of these is a previously unreported long-range pseudoknot. He shows that small changes to intronic sequences, far removed from an editing site, can transfer the species specificity of editing between RNA substrates. These data support a phylogeny of *sytI* gene editing spanning more than 250 million years of hexapod evolution. The results also provide models for the genesis of RNA editing sites through the stepwise edition of structural domains, or by short walks through sequence space from ancestral structures.

RNA UNWINDING

RNA helicases of the DEAD-box family are involved in essentially all RNA-dependent cellular processes (Linder and Lasko 2006). Sangoku et al. (2006) solve the structure of the DEAD-box protein vasa in the presence of RNA and a non-hydrolysable ATP analog and provide important insights into how this family of helicases unwinds RNA. DEAD-box RNA helicases have two RecA-like domains as a catalytic core to alter higher-order RNA structures. They determined the 2.2-Å resolution struc-

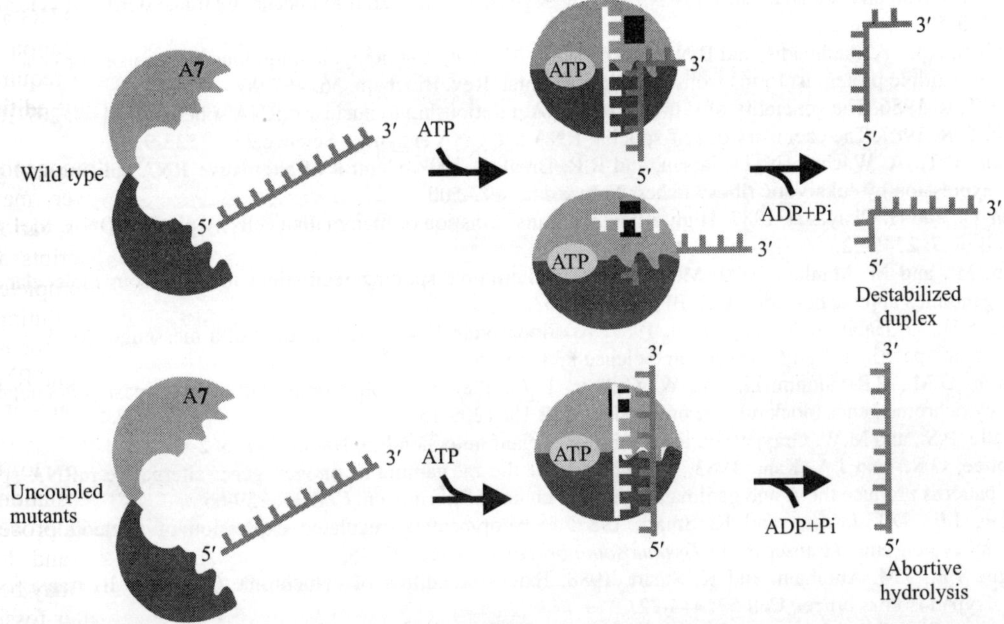

Figure 10.13 Proposed mechanisms of RNA unwinding by the DEAD-box protein and abortive ATP hydrolysis by the uncoupled mutants

tures of the core of the *Drosophila* DEAD-box protein vasa in complex with a single-stranded RNA and an ATP analog. The ATP analog intensively interacts with both of the domains, thereby bringing them into the closed form, with many interdomain interactions of conserved residues. The bound RNA is sharply bent, avoiding a clash with a conserved α helix in the N-terminal domain. This "wedge" helix should disrupt base pairs by bending one of the strands when a duplex is bound. Interdomain interactions couple ATP hydrolysis to RNA unwinding, probably through the fine positioning of the duplex relative to the wedge helix. This mechanism, which differs from those for canonical translocating helicases, may enable the targeted modulation of intricate RNA structures. Proposed mechanisms of RNA unwinding by the DEAD-box protein and abortive ATP hydrolysis by the uncoupled mutants are shown in Figure 10.13 (Sangoku et al. 2006).

REFERENCES

Altman, S. 1989. Ribonuclease P: an enzyme with a catalytic RNA subunit. Adv. Enzymol. 62: 1-36.

Altman, S., L. Kirsebom, and S. Talbot. 1993. Recent studies of ribonuclease P. FASEB J. 7: 7-14.

Amara, S.G., V. Johan, and M.G. Rosenfeld. 1982. Alternative RNA processing in calcitonin gene expression generates mRNAs encoding different polypeptide products. Nature 298: 240-4.

Benne, R., J.P.J. Van den Burg, P. Brakenhoff, P. Sloof, J.H. Van Boom, and M.C. Tromp. 1986. Major transcript of the frame shifted *coxII* gene from trypanosome mitochondria contains four nucleotides that are not encoded in the DNA. Cell 57: 355-66.

Berk, A.J., and P.A. Sharp. 1978 Splicing early mRNAs of SV40 Proc. Natl. Acad. Sci. USA 75: 1274-8.

Blackburn, E.H. 1990. RNA editing: copy editing – what is the U's? Nature 346: 609-10.

Blencowe, B.J. 2006. Alternative splicing: insights from global analysis. Cell 126: 37-47.

Blum, B., M. Bakalara, and L. Simpson. 1990. A model for RNA editing in kinetoplastic mitochondria: "Guide" molecules transcribed from maxicircle DNA provide the edited information. Cell 60: 189-196.

Blum, B., N.R. Strum, A.M. Simpson, and L. Simpson. 1991. Chimeric gRNA-mRNA molecules with oligo (U) tails covalently linked at sites of RNA editing suggested that U addition occurs by transesterification.Cell 65: 543-50.

Breitbart, R.E., A. Andreadis, and B.N. Ginard. 1987. Alternative splicing: a ubiquitous mechanism for generation of multiple protein isoforms from single gene. Annu. Rev. Biochem. 56: 467-96.

Cech, T. R. 1986. The generality of self-splicing RNA: relationship to nuclear mRNA splicing. Cell 44: 207-10.

Cech, T. R. 1987. The chemistry of self-splicing RNA and RNA enzymes. Science 236: 1532-9.

Cheah, M.T., A. Wachter, N. Dudarsan, and R.R. Breaker. 2007. Control of alternative RNA splicing and gene expression by eukaryotic riboswitches. Nature 447: 497-500.

Chen, C., and H. Okayama, 1987. High-efficiency transformation of mammalian cells by plasmid DNA. Mol. Cell Biol. 7: 2745-52.

Chen, M., and M. Manley. 2009. Mechanisms of alternative splicing regulation: insights from molecular and genomics approaches. Mol. Cell Biol. 10: 741-53.

Chen, S.H., G. Habib, C.Y.Yang, et al. 1987. Apolipoprotein B-48 is the product of a messenger RNA with an organ-specific in-frame stop codon. Science 238: 363-6.

Church, G.M., P.R. Slonimski, and W. Gilbert. 1979. Pleiotropic mutations within two yeast mitochondrial cytochrome genes block mRNA processing. Cell 18: 1209-15.

Covello, P.S., and M.W. Gray, 1989. RNA editing in plant mitochondria. Nature 341: 662-6.

Crabtree, G.R., and J.A. Kant. 1983. Organization of the rat gamma-fibrinogen gene: alternative mRNA splice patterns produce the A and gamma B (gamma) chains of fibrinogen. Cell 31: 159-66.

Feagin, J.E., D.J. Jasmer, and K. Stuart. 1987. Developmentally-regulated expression of nucleotides within apocytochrome b transcripts in *Trypanosoma brucei*. Cell 49: 337-45.

Feagin, J.E., J.M. Abraham, and K. Stuart. 1988. Extensive editing of cytochrome C oxidase III transcript in Trypanosomas brucei. Cell 53: 414-22.

Forster, A.C., and S. Altman. 1990. External guide sequences for an RNA enzyme. Science 249: 783-6.

Goyenvalle, A., A. Vulin, F. Fougerousse, et al. 2004. Rescue of dystrophic muscle through U7 snRNA-mediated exon skipping. Science 306: 1796-9.

Gray, M.W., P.J. Hanic-Joyce, and P.S. Covello. 1992. Transcription, processing and editing in plant mitochondria. Annu. Rev. Pl. Physiol. Pl. Molec. Biol. 43: 145-75

Gualberto, J.H., L. Lamattina, G. Bonnard, J.H. Weil, and J.H. Grienenberger. 1989. RNA editing in wheat mitochondria results in conservation of protein sequences. Nature 341: 660-2.

Gualberto, J.M., G. Bonnard, L. Lamattina, and J.M. Grienenberger. 1991. Expression of the wheat mitochondrial *nad3-rps72* transcription unit: Correlation between editing and mRNA maturation. Plant Cell 3 1109-20.

Hachet, O., and A. Ephrussi. 2004. Splicing of *oskar* RNA in the nucleus is coupled to its cytoplasmic localization. Nature 428: 959-63.

House, A.E., and Lynch, K.W. 2008. Regulation of alternative splicing: More than just the ABC's. Jour. Biol. Chem. 283: 1217-21.

Kawahara, Y., B. Zinshteyn, P. Sethupathy, H. Iizasa, A.G. Hatzigeorgious, and K. Nishikura. 2007. Redirection of silencing targets by adenosine-to-inosine editing of miRNAs. Science 315: 1137-40.

Kishore, S., and S. Stamm. 2006. The snoRNA HBII-52 Regulates Alternative Splicing of the Serotonin Receptor 2C. Science 311: 230-2.

Kotera, E., M. Tasaka, and T. Shikanai. 2005. A pentatricopeptide repeat protein is essential for RNA editing in chloroplasts. Nature 433: 326-30.

Kumar. S., and S.B. Hedges. 1998. A molecular time scale for vertebrate evolution. Nature 392: 917-20.

Li, J.B., E.Y. Levanon, J.-K. Yoon, et al. 2009. Genome-wide identification of human RNA editing sites by parallel DNA capturing and sequencing. Science 324: 1210-3.

Li, Y., Y.-c. Bor, Y. Misawa, Y. Xue, D. Rekosh, and M.-L. Hammarskjold. 2006. An intron with a constitutive transport element is retained in a Tap messenger RNA. Nature 443: 234-7.

Licatalosi, D.D., A. Mele, J.J. Fak, et al. 2008. HITS-CLIP yields genome wide insights into brain alternative RNA processing. Nature 456: 464-9.

Linder, P., and P. Lasko. 2006. Bent out of shape. RNA unwinding by the DED-box helicase vasa. Cell 125: 219-21.

Luco, R.F., Q. Pan, K. Tominaga , B.J. Blencowe, O.M. Perreira-Smith, and T. Misteli. 2010. Regulation of alternative splicing by histone modifications. Science 327: 996-9.

Macbeth, M.R., H.L. Schubert, A.P.V. VanDemark, A.T. Lingam, C. Hill, and B.L. Bass. 2005. Insitol hexakisphosphate is bound in the ADAR2 core and required for RNA editing. Science 309: 1534-9.

Maniatis, T. 1991. Mechanisms of alternative pre-mRNA splicing. Science 251: 33-4.

Marande, W. and Burger, G. 2007. Mitochondrial genome as a genomic jigsaw puzzle. Science 318: 415.

Mueller, M.W., M. Hetzer, and R.J. Schweyen. 1993. Group II intron RNA catalysis of progressive nucleotide insertion. A model for RNA editing. Science 261: 1035-8.

Navratnam, N., S. Bhattacharya, T. Fujino, D. Patel, A.L. Jarmuz, and J. Scott. 1995. Evolutionary origins of apoB mRNA editing: catalysis by a cytidine deaminase that has acquired novel RNA-binding motif at its active site. Cell 81: 187-95.

O'Donovan, G. A., and J. Neuhard. 1970. Pyrimidine metabolism in microorganisms. Bacteriol. Rev. 34: 278-343.

Olson, M., M.L. Hood, C. Cantor, and D. Botstein. 1989. A common language for physical mapping of the human genome. Science 245: 1434-5.

Powell, L.M., S.C. Wallis, R.J. Pease, Y.H. Edwards, T.J. Knott, and J. Scott. 1987. A novel form of tissue-specific RNA processing produces apolipoprotein-B48 in intestine. Cell 50: 831-40.

Rajasekhar, V.K., and R.M. Mulligan. 1993. RNA editing in plant mitochondria: phosphate is retained during C to U conversion in mRNA's. Plant Cell 5(12): 1843-52.

Randau, l., B.J. Stanley, A. Kohlway, S. Machta, Y. Xiong, and D. Soll. 2009. A cytidine deaminase edits C to U in transfer RNAs in Archaea. Science 324: 657-9.

Reenan R. 2005. Molecular determinants and guided evolution of species-specific RNA editing. Nature 434: 409-13.

Saga, Y., J.S. Lee, C. Sariya, and E.A. Boyse. 1990. Regulation of alternative splicing in the generation of isoforms of the mouse Ly-5 (CD45) glycoprotein. Proc. Natl. Acad. Sci. USA 87: 3728-32.

Sangoku, T., O. Nureki, A. Nakamura, S. Kobayashi, and S. Yokoyama. 2006. Structural basis for RNA winding by the DEAD-box protein *Drosophila* vasa. Cell 125: 287-300.

Seiwart, S.V., and K. Stuart, 1994. RNA editing – transfer of genetic information from gRNA to precursor mRNA in vitro. Science 266: 114-6.

Shah, R.R., T.J. Knott, J.E. Legross, N. Navratnam, J.C. Greeve, and J. Scott. 1991. Sequence requirements for the editing of apolipoprotein mRNA. J. Biol. Chem. 266: 16301-4.

Simpson, L. 1990. RNA editing – a novel geentic phenomenon. Science 250: 512-3.

Simpson, L., and J. Shaw. 1989. RNA editing and mitochondrial cryptogenes of kinetoplastid protozoa. Cell 57: 355-66.

Smith, C.W.J., Patton, J.G. and Nadal-Ginard, B. 1989. Alternative splicing in the control of gene expression. Annu. Rev. Genet. 23: 527-77.

Stuart, E.L., M.G. Rosenfeld, R.M. and Evans. 1986. Diversity in gene expression by alternative splicing. RNA processing. Annu. Rev. Biochem. 55: 1091-117.

Sultan, M., H.M. Schulz, H. Richard, et al. 2008. A global view of gene activity and alternative splicing by deep sequencing of human transcriptome. Science 321: 956-60.

Takenaka, M., J.A. van der Merwe, D. Verbitskiy, et al. 2008. The process of RNA editing in plant mitochondria. Mitochondrion 8: 35-46.

Tamkun, J.W., J.E. Schwarzbauer, and R.O. Hynes. 1984. A single rat fibronectin gene generates three different mRNAs alternating splicing of a complete exon. Proc. Natl. Acad. Sci. USA 81: 5140-4.

Ule, J., G. Stefani, A. Mele, et al. 2006. An RNA map predicting Nova-dependent splicing regulation. Nature 444: 580-6.

Vidal, S., J. Curran, and D. Kolakofsky. 1990a. Editing of the Sendai virus P/C mRNA by G insertion occurs during mRNA synthesis via a virus-encoded activity. J. Virol. 64: 239-46.

Vidal, S., J. Curran, and D. Kolakofsky. 1990b. A stuttering model for paramyxovirus P mRNA editing. EMBO. J. 9: 2017-22.

Volloch, V. B. Schweitzer, and S. Ris. 1990. Uncoupling of the synthesis of edited and unedited COIII RNA in *Tetrahymena brucei*. Nature 343: 482-4.

Wang, E.T., R. Sandberg, S. Luo, et al. 2008. Alternative isoform regulation in human tissue transcriptomes. Nature 456: 470-6.

Wang, Z., C.B. and Burge. 2008. Splicing regulation; From parts list of regulatory elements to integrated splicing code. RNA 14: 802-13.

Ziff, E.B. 1980. Transcription and RNA processing by the DNA tumour viruses. Nature 287: 491-9.

11

Updated Genetic Code

Genes carry information for determining the structure of proteins, which are responsible for directing cell metabolism through their activity as enzymes. Genetic information is specified by order of the four bases – adenine (A), cytosine (C), guanine (G) and thymine (T) that make up the DNA molecule. Proteins, in turn are the polymers of 20 amino acids, the sequence of which determines their structure and function. Genetic information from DNA is transcribed to RNA. After transcription, in the process of gene expression, genetic code serves as language of life, which is used for protein synthesis. Genetic language has four alphabets – A, T, G and C. Each word of this language is known as codon. Codon is defined as the number and sequence of bases specifying an amino acid. A set of all the codons that specify 20 amino acids is termed as genetic code. The genetic code is used for storing the information regarding proteins, which are synthesized under the control of nucleic acids.

We have erroneously believed that triplet genetic code is the only mechanism which gives meaning to ribonucleotide sequence in finished messenger RNA (mRNA). To this classical view of genetic code were added extended anticodon hypothesis, altered genetic code specificity, expansion of genetic alphabet, recoding, ribosome code, the second code. These relatively new aspects of genetic code are also discussed here.

Now code of life also includes histone code and nucleosome code. These epigenetic aspects of code, which are more relevant to regulation of gene expression, are not subject matter of this chapter and will, therefore, be discussed elsewhere.

GENETIC CODE

Gamow (1954) used mathematics to establish the number of nucleotides that should be necessary to code for one amino acid. According to him, if one-base were to be used on code word for one amino acid, 4 bases will specify only 4 different amino acids. If codons were two-base long, 16 different combinations of two bases are possible to code for 20 amino acids. This would require some ambiguity in genetic code. As ambiguity is undesirable; therefore 16 codons can code for only 16 different amino acids. On the other hand, if the codons are three-base long, 64 different codons are possible, which are enough to code for 20 different amino acids. It has now become clear that a sequence of 3 nucleotides on mRNA codes for an amino acid. Crick et al. (1961) designed an elegant experimental strategy to determine the nature of the genetic code. Remarkably, they reached the correct conclusion despite the absence of technology to analyze and compare DNA and protein sequences (Yanofsky 2007).

The genetic code is concerned with the processes involved in translating or decoding the information contained in the primary structure of DNA Genetic information flows from gene (DNA) to RNA to protein. RNA thus serves as an intermediate between DNA and proteins. The processes of genetic code are the basis of life or it may be considered a fundamental secret of nature.

GLOSSARY OF TERMS ON GENETIC CODE

Code letter	:	Nucleotides A, T, G, and C in DNA and A, U, G and C in mRNA.
Codon	:	Sequence of nucleotides specifying an amino acid.
Anticodon	:	Sequence of nucleotides on tRNA that complements the codon on mRNA.
Coding dictionary	:	A table of all code words that specify amino acids (See Table 11.1).
Codon length	:	When only as many amino acids are coded as there are code words in end-to-end sequence. (e.g., UUUCCC two code words; they code for only two amino acids). Code length is three code letters.
Overlapping code	:	When more amino acids are coded for than there are code words represented in end-to-end sequence. UUUCCC has two code words (UUU and CCC) but in case of overlapping code, codons will be four (UUU, UUC, UCC, and CCC). But actually code is not overlapping.
Non-overlapping code	:	When only as many amino acids are coded for as there are code words represented in end-to-end sequence (UUUCCC has two code and two codons actually present are UUU and CCC.
Non-degenerate code	:	When there is only one codon for each amino acid.
Degenerate code	:	When there is more than one codon for a particular amino acid.
Synonymous codons	:	Different codons that specify the same amino acid in a degenerate code.
Ambiguous code	:	When one codon can code for more than one amino acid.
Commaless code	:	When there are no spacer nucleotides between code words. There is no punctuation.
Reading frame	:	The particular nucleotide sequence that starts at a specific translation initiation point and is then partitioned into codons until the final word (in fact a termination codon) of that sequence is reached.
Sense word	:	A codon that specifies an amino acid normally present in that position in a protein.
Non-sense codon	:	A codon that does not code for an amino acid; also called a termination codon.
Universality	:	Utilization of the same genetic code in all organisms.
Initiation codon	:	First codon in translation (usually AUG) is the initiation codon.

PROPERTIES OF GENETIC CODE

Triplet Code

Since singlet and doublet codes are not enough to code for 20 amino acids, it was pointed out that triplet code is the minimum required. In triplet code, a series of 3 nucleotides specify one amino acid. In a triplet code of 64 codons, there is an excess of 44 codons and, therefore, more than one codon are present for the same amino acid. Direct experimental evidence for triplet code was given by Crick et al. (1961) by genetic analysis of T4 phage bearing mutations in an extensively studied gene called *rII*

Table 11.1 Genetic code dictionary. RNA codons (Reproduced, with permission, from Khorana, H.G. 1968. Nobel Lecture at: http://nobelprize.org/nobel prizes/medicine/laureates/1968/ khorana-lecture.html © The Nobel Foundation)

First Base ↓	← Second Base			→	Third Base ↓
	U	C	A	G	
U	Phe	Ser	Tyr	Cys	U
	Phe	Ser	Tyr	Cys	C
	Leu	Ser	Nonsense**	Nonsense*	A
	Leu	Ser	Nonsense**	Trp	G
C	Leu	Pro	His	Arg	U
	Leu	Pro	His	Arg	C
	Leu	Pro	Gln	Arg	A
	Leu	Pro	Gln	Arg	G
A	Ileu	Thr	Asn	Ser	U
	Ileu	Thr	Asn	Ser	C
	Ileu	Thr	Lys	Arg	A
	Met*	Thr	Lys	Arg	G
G	Val	Ala	Asp	Gly	U
	Val	Ala	Asp	Gly	C
	Val	Ala	Glu	Gly	A
	Val	Ala	Glu	Gly	G

*, Initiation codon; **, Termination codons

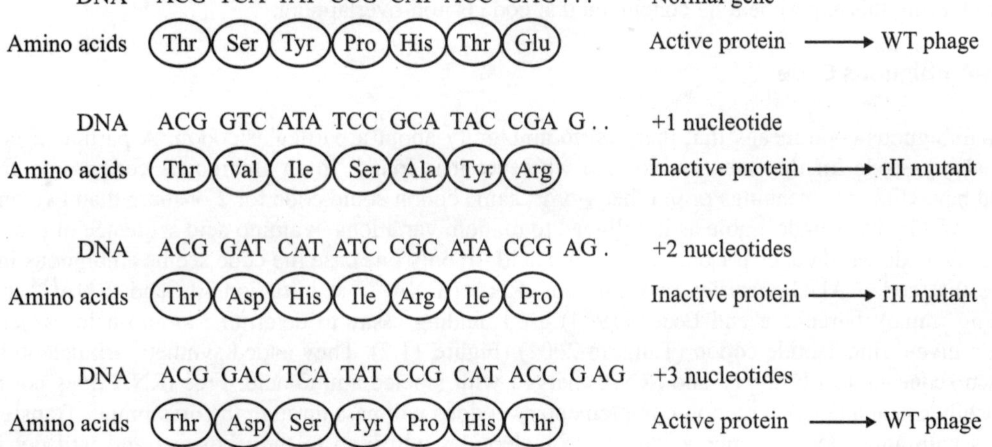

Figure 11.1 Genetic evidence for triplet code

(Figure 11.1) (Cooper 2000). These mutations were induced by acridine dyes. A series of mutations consisting of addition and deletion of one, two or three nucleotides were studied in *rII* locus of T4 bacteriophage. It was found that deletion or insertion of single nucleotide in a polynucleotide sequence led to change in reading frame. As a result, reading of subsequent codons was out of phase, i.e., the base sequence of each codon was changed. Amino acids incorporated were abnormal and an inactive protein was produced giving rise to mutant phage. Addition or deletion of 3 nucleotides altered only one amino acid, reading frame of remainder of gene was normal. An active protein giving rise to wild-type phage was produced. Since the addition or deletion of three nucleotides adds or deletes only one amino acid in the protein, codon is therefore triplet.

Commaless Code

Genetic code is commaless, which means that no punctuation is needed between any two words. In a punctuated code, some code letters are used as punctuation marks. But in a commaless code, after one amino acid is coded, the second amino acid will be automatically coded by the next three letters and no letters are wasted in telling that one amino acid has been coded and now second should be coded. RNA sequence UUUCCCGGGAAA has four code words and upon translation we will have a tetrapeptide chain of Phe-Pro-Gly-Lys. So all the letters are used to code for one or the other amino acid.

Non-Overlapping Code

Code is non-overlapping. There will be only two codons from six code letters if code is non-overlapping. It means that a base in mRNA is not used for two different codons. However, in overlapping genes it is shown that the same base can be used for different codons but only at different occasions of time, so that same base cannot be used for two different codons during the synthesis of same protein. Evidence of non-overlapping code was given by Wittmann (1962a,b). He treated tobacco mosaic virus (TMV) (having RNA genetic material) with nitrous acid which changed some of the nucleotides in its genetic material. Virus was then used to infect tobacco plants and so produce new viruses in leaves according to instructions of modified genetic material. He found that newly made protein showed change in only one amino acid. In no case two or three adjacent amino acid were altered. From this experiment he concluded that code is non-overlapping.

Non-Ambiguous Code

Non-ambiguous code means that there is no ambiguity about a particular codon. A particular codon will always code for the same amino acid wherever it is found. In an ambiguous code, same codon could have different meanings or in other words, same codon could code for 2 or more than two amino acids. Ambiguity is undesirable as it will lead to random variations in amino acid sequence of proteins. Thus one code word codes for only one amino acid. In only one case the code seems ambiguous in the sense that codon AUG specifies two kinds of tRNA ($tRNA_f^{Met}$ at initiation site and $tRNA^{Met}$ in the reading frame). Nirenberg and Leder (1964) used binding assay to determine amino acid associated with a given trinucleotide codon (Tamarin 2002) (Figure 11.2). They added synthetic trinucleotide of known sequence to ribosomes and tRNA charged with labeled amino acid. Free tRNA does not bind nitrocellulose and tRNA with non-complementary codons will pass through the membrane. Transfer RNAs with anticodons complementary to trinucleotides will bind to the ribosome and will not pass through nitrocellulose filter. When tRNA was charged with radioactive amino acid, the radioactivity is trapped in the filter. From this experiment, they concluded that CUC codes for leucine and not for Serine.

Degenerate Code

Genetic code is degenerate which means that different codons can specify one amino acid. Codons specifying the same amino acid are called synonyms, e.g., phenylalanine is specified by two codons,

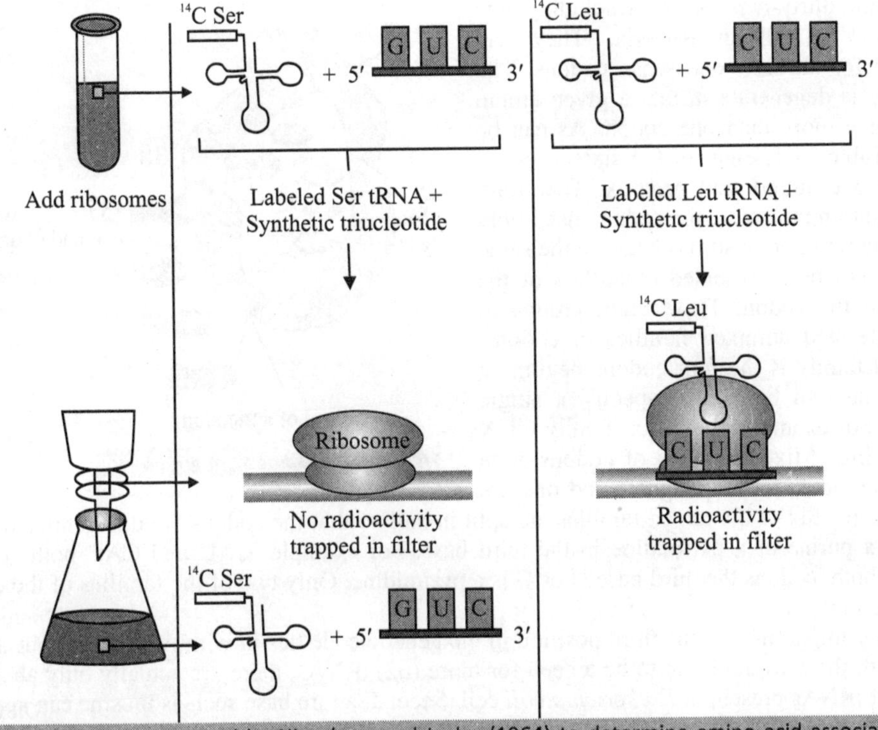

Figure 11.2 Binding assay used by Nirenberg and Leder (1964) to determine amino acid associated with a given trinucleotide codon

namely, UUU and UUC. Serine is specified by six codons – UCU, UCC, UCA, UCG, AGU, and AGC. Similarly, there are four codons (GCU, GCC, GCA, and GCG) for Alanine. This is explained like this: the first two positions of the triplet codon on mRNA pair precisely with the first two nucleotides of anticodon of tRNA but pairing at third position may not be precise. It shows that when first two nucleotides are identical, the third nucleotide can be either C or U and codon will code for same amino acid. A and G are similarly interchangeable.

Wobble Rules

Crick (1966) devised the wobble concept to explain degeneracy of genetic code. It states that the base at 5′-end of the anticodon is not as spatially confined as the other two, allowing it to form hydrogen bonds with any of several bases located at 3′-end of codon. For example, U at the wobble position can pair with either adenine or guanine, while I can pair with U, C or A. The pairing permitted by wobble rules are those that give ribose-ribose distances close to that of standard A:U or G:C pairings. Purine-purine or pyrimidine-pyrimidine pairs would give ribose-ribose distances that are too long or too short, respectively. Reason for wobble pairing is that, 5′-end of anticodon is conformationally flexible (Figure 11.3) (Dunham et al. 2007). In the three-dimensional structure of tRNA, the three anticodon bases as well as the two following (3′) bases in anticodon loop – all point roughly in the same direction. Stacking interactions between flat surfaces of bases are present. Since the first (5′) anticodon base is at the end of stack and is less restricted in its movement than other two anticodon bases and hence wobble in third position of codon.

There are thirty-two tRNAs (including one for initiation) in all in bacteria. They can complement for all sixty-one sense codons. The genetic code is degenerate in that a given amino acid may have more than one codon. As can be seen from Table 11.1, eight of the sixteen boxes contain just one amino acid per box. Therefore, for these eight amino acids, the codons need only be read in the first two positions because the same amino acid will be represented regardless of the third base of the codon. These eight groups of codons are termed unmixed families of codons. An unmixed family is the four codons beginning with the same two bases that specify a single amino acid. For example, the codon family GUX codes for valine. Mixed families of codons code for two amino acids for stop signals and one are two amino acids. Six of the mixed families are split in half so that the codons are differentiated by the presence of a purine or a pyrimidine in the third base. For example, CAU and CAC both code for histidine; in both, codons the third base U or C is a pyrimidine. Only two of the families of the codons are split differently.

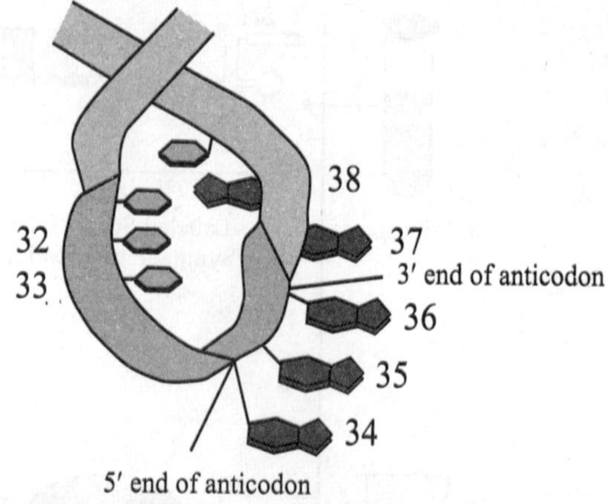

Figure 11.3 Enlarged anticodon loop

The lesser importance of the third position in the genetic code ties in with two facts about tRNAs. First, although there would seem to be a need for more (62) tRNAs, there are actually only about half (32) different tRNAs present in *Escherichia coli* cell. Second, a rare base such as inosine can appear in the anticodon, usually in the position that is complementary to the third position of the codon. These two facts lead to the concept that some kind of conservation of tRNAs is occurring and that rare bases may be involved. Since the first position of the anticodon (5') is not as constrained as the other two positions, a given base at that position may be able to pair with any of the several bases in the third position of the anticodon. Table 11.2 shows the possible pairings that would produce a tRNA system compatible with the known code. For example, if an isoleucine tRNA has the anticodon 3'-UAI-5', it is compatible with three codons for the amino acid: 5'-AUU-3', 5'-AUC-3' and 5'-AUA-3', i.e., inosine is the first (5') position of the anticodon can recognize U, C or A in the third (3') position of the anticodon, and thus one tRNA complements all three codons for isoleucine.

Table 11.2 Pairing combinations of the third codon position

Number-one base in tRNA (5'-end)	*Number-three base in mRNA (3'-end)*
G	U or C
C	G
A	U
U	A or G
I	A, U, or A

Colinearity of Code

The code is colinear, i.e., sequence of nucleotides in template strand of DNA determines sequence of amino acids in its protein in a linear fashion. This conclusion was drawn by Yanofsky et al. (1964) who studied tryptophan synthetase A polypeptide. Different mutants of Tryptophan A were crossed together and a genetic map was obtained showing the order of mutated sites along the DNA of gene.

Tryptophan synthetase A protein was purified from each mutant strain and they sequenced wild-type tryptophan synthetase A and also the mutants. The sequence of amino acids in mutants was compared to that of wild-type. The order of changes in amino acid sequence corresponds exactly to the order of genetic changes in DNA.

Universality of Code

Universal code means that a codon codes for same amino acid in all the organisms. Universality of code has had a huge impact on our understanding of evolution as it made it possible to directly compare protein coding sequences among all the organisms for which genome sequence is available. Universality of code also helped to create the field of genetic engineering by making it possible to express cloned copies of genes encoding useful protein products among host organisms. A synthetic ribonucleotide chain directs incorporation of the same amino acid into the polypeptides in cell free extracts derived from almost any organism. This shows that code is universal. Even non-sense codons appear to be universal in prokaryotes and eukaryotes.

In plant mitochondria, CGG codon was found to code for tryptophan. But universal genetic code for tryptophan is UGG. CGG is universal genetic code for arginine. So it was postulated that plant mitochondria do not follow universal genetic code. What actually happens is that CGG is edited to UGG after transcription by changing C to U. This editing occurs only at "tryptophan-conserved:" site. "Arginine-conserved" CGG codons are not edited. Therefore, higher plant mitochondria do use the universal genetic code. A wide variety of codon assignments, affecting both non-sense and sense codons, has been observed in mitochondrial genetic systems (Fox 1987). Mitochondrial genetic systems may tolerate relatively frequent changes in codon assignment since they produce few proteins. In some cases, termination codons are used to specify amino acids. There are some differences among standard and mitochondrial genetic code. UGA is not a stop signal but codes for tryptophan. Hence anticodon of mitochondrial $tRNA^{Trp}$ recognizes both UGG and UGA. In mitochondria, internal methionine is coded by both AUG and AUA. In mammalian mitochondria, AGA and AGG are not arginine codons but chain termination codons. Genetic code assignments in nuclear and yeast mitochondrial genes have been found to be different (Table 11.3). This questions universality of the genetic code. Phenomena of RNA editing and recoding to some extent resolve this dilemma.

Table 11.3 Genetic code in nuclear and yeast mitochondrial genes

Codon	*Nuclear gene*	*Yeast mitochondrial gene*
UGA	Termination	Tryptophan
CGG	Arginine	Tryptophan
AUA	Isoleucine	Methionine
CUN	Leucine	Threonine in yeast but not in fungi
AGA	Arginine	Stop codon in mammals and *Xenopus*
AGG	Arginine	Stop codon in mammals and *Xenopus*
AGA	Arginine	Serine in *Drosophila*
–	55 anticodons	22 anticodons
Termination codons	UAA, UAG, UGA	UAA, UAG, AGA, AGG
UAG codon	Aminoacylated by leucine	Accepts threonine

Initiation and Termination Codons

Specific codon is required for initiation of translation. Initiation codon is AUG or GUG. AUG functions as initiation codon both in vivo and in vitro. GUG acts as initiation codon (although it has been assigned for valine) in vitro but with lesser efficiency. UAA, UAG, and UGA do not code for any amino acid. They serve as termination codons, also known as non-sense codons. Two release factors, RF1 and RF2, have been recognized; RF1 is specific for UAA and UAG while RF2 is specific for UAA and UGA.

Invariant DNA Code

A mutation in the genetic code would place new amino acids in certain loci and entirely eliminate amino acids from the other loci of practically all proteins in an organism (Hinegardner and Engelberg 1963). It is reasonable to postulate that mutations of this kind cannot supplant the original code. The genetic code, once established therefore remains invariant.

Deep Division in DNA Code

The three-dimensional crystal structure of seryl-tRNA synthetase from *E. coli* refined at 2.5-Å resolution, has an N-terminal domain that forms an antiparallel a helical coiled-coil, stretching 60Å out into the solvent (Cusack et al. 1990). This structure bears no resemblance to aminoacyl-tRNA synthetase structures already described. This observation suggests a deep division in the genetic code.

DECIPHERING THE GENETIC CODE

To assign different codons to different amino acids, in vitro translation system was used. Cell extracts containing ribosomes, amino acids, tRNA and enzyme catalyze the incorporation of amino acids into proteins. Rate of protein synthesis was enhanced by addition of synthetic mRNA of known sequence in cell extracts. Since the added mRNA directed protein synthesis, DNA code was, therefore, deciphered by study of translation of mRNA of known base sequence. Formation of synthetic polynucleotides involves the enzyme polyribonucleotide phosphorylase. S. Ochoa was awarded Nobel Prize in 1959 for synthesis of polyribonucleotide in vitro (Ochoa 1963; Last et al. 1967). Nirenberg and Matthaei (1961) incubated synthetic polyribonucleotide polyuridylate with an *E. coli* extract, GTP and mixture of 20 amino acids in 20 different tubes (Figure 11.4) (Cooper 2000)). In each tube a different amino acid was radioactively labeled. Poly(U) can be regarded as an artificial mRNA, containing successive UUU triplets. A radioactive polypeptide was formed in only one of the 20 tubes that contained radioactive phenylalanine. Nirenberg and Matthaei concluded that triplet UUU codes for phenylalanine. Similarly, poly(C) codes for proline, Poly(G) codes for glycine.

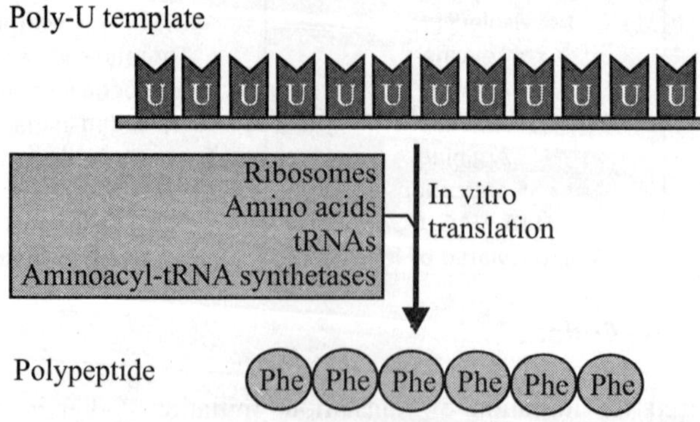

Figure 11.4 Nirenberg and Metthaei (1961) concluding that triplet UUU codes for phenylalanine

Mixed copolymers allowed additional codon assignments. To elucidate genetic code, an experiment was designed in which synthetic RNA containing only A and C residues in 5:1 ratio were used to direct polypeptide synthesis. Poly(AC) molecules contain 8 different codons: CCC, CCA, CAC, ACC, AAA, AAC, ACA, CAA. When AC copolymers attach to ribosomes, they cause the incorporation of asparagine, glutamine, histidine, threonine, proline and lysine. A complementary approach was provided by Har Gobind Khorana and his team (Khorana 1965; Khorana et al. 1967) who developed methods to synthesize polyribonucleotides with defined, repeating sequences of 2 to 4 bases. Ribosomes start protein synthesis at random points along these regular copolymers; yet they incorporate specific amino acids into polypeptides, e.g., repeating sequence CUCUCUCU is the messenger for regular polypeptide in which leucine and serine alternate. Polypeptides produced using these RNA as messengers had one or few amino acids in repeating patterns. When these patterns were combined with the information from random polymers used by Nirenberg, they permitted unambiguous codon assignments. For example, copolymer $(AC)_n$ has alternating CAC and ACA codons regardless of reading frame . Polypeptide synthesized in response to this polymer has equal amount of threonine and histidine. Similarly, RNA with three bases in a repeating pattern should yield three different types of polypeptides. Each polypeptide would be derived from a different reading frame and would contain a single kind of amino acid. Assignments of codons, having known sequences, with the help of copolymers having repetitive sequences of three bases, e.g., $(UUC)_n$, or four bases, e.g., $(GUAA)_n$, are given in Table 11.4.

Table 11.4 Assignments of codons, having known sequences, with the help of copolymers having repetitive sequences of three bases or four bases

Polynucleotide	*Polypeptide products*
$(UUC)_n$	$(Phe)_n$, $(Ser)_n$, $(Leu)_n$
$(AAG)_n$	$(Lys)_n$, $(Arg)_n$, $(Glu)_n$
$(UUG)_n$	$(Cys)_n$, $(Leu)_n$, $(Val)_n$
$(CAA)_n$	$(Gln)_n$, $(Thr)_n$, $(Asn)_n$
$(GUA_n$	$(Val)_n$, $(Ser)_n$, (chain terminator)
$(UAC)_n$	$(Tyr)_n$, $(Thr)_n$, $(Leu)_n$
$(AUC)_n$	$(Ileu)_n$, $(Ser)_n$, $(His)_n$
$(GAU)_n$	$(Met)_n$, $(Asp)_n$, (chain terminator)
$(UAUC)_n$	$(Tyr\text{-}Leu\text{-} Ileu\text{-}Ser\text{-})_n$
$(UUAC)_n$	$(Leu\text{-}Thr\text{-}Tyr)_n$
$(GUAA)_n$	Di- and tripeptides
$(AUAG_n$	Di- and tripeptides

When copolymer is made up of three bases, $(UUC)_n$, 3 types of polypeptides, having single amino acid repeat, i.e., $(Phe)_n$, $(Ser)_n$, $(Leu)_n$, are produced when termination codon is not formed but in certain cases, e.g., $(GUA)_n$, two types of polypeptides, having single amino acid repeat, i.e., $(Val)_n$, $(Ser)_n$, are produced, the third type not being synthesized when a termination codon is formed. When a coplymer is made up of four bases, e.g., $(UAUC)_n$, only one type of polypeptide, having four amino acids repeated, i.e., $(Tyr\text{-}Leu\text{-}Ser\text{-}Ile)_n$, may be formed but in certain cases, e.g., $(GUAA)_n$, a dipeptide and a tripeptide may be formed depending upon whether two or only one termination codon is formed. Results from all these experiments with polymers permitted the assignment of 61 of 64 possible codons. The basic question that we try to answer is: Which DNA code word specifies which specific amino acid? Answer to this comes from genetic code dictionary (Table 11.1).

M.W. Nirenberg and H.G. Khorana were awarded Nobel Prize in 1968 for deciphering the genetic code (Nirenberg 1963; Nirenberg et al. 1966; Khorana 1965; Khorana 1968).

GENETIC CODE AT WORK

If there is change in the nucleotide sequence of DNA, it would lead to a corresponding change in the mRNA. The change in nucleotide sequence of DNA may be brought about by a point mutation, insertion of nucleotide(s), deletion of nucleotide(s), etc. (Figure 11.5). Thus a change in nucleotide sequence of DNA is reflected in protein via a change in the nucleotide sequence of mRNA.

1 5 10 15	
3′ TAC UUU CCC **AAA** GGG 5′	DNA
5′ AUG AAA GGG **UUU** CCC 3′	mRNA
$_2$HN-Met - Lys - Gly - **Phe** - Pro-COOH	protein

(A) Assume nucleotide A at position 10 changes to C in DNA, new sequence will be:

1 5 10 15	
3′ TAC UUU CCC **CAA** GGG 5′	DNA
5′ AUG AAA GGG **GUU** CCC 3′	mRNA
$_2$HN-Met - Lys - Gly - **Val** - Pro-COOH	protein

(B) Assume nucleotide A at position 12 changes to C in DNA, new sequence will be:

1 5 10 15	
3′ TAC UUU CCC **AAC** GGG 5′	DNA
5′ AUG AAA GGG **UUG** CCC 3′	mRNA
$_2$HN-Met - Lys - Gly - **Leu** - Pro-COOH	protein

Figure 11.5 Genetic code at work. Effect of a mutation on amino acid sequence

EXTENDED ANTICODON HYPOTHESIS

The structure of anticodon loop and the proximal anticodon stem are related to the sequence of anitcodon. In other words, anticodon is extended into the nearby structure and consists of (a) two nucleosides at the 5′-end of the anticodon loop, (b) three nucleosides of anticodon, and (c) two nucleosides at the 3′-end of the anticodon loop (d) and five nucleoside pairs in the anticodon stem. Extended anticodons are involved in translation. The 3′ nucleotide of anticodon triplet is considered to be most important and is called cardinal nucleotide.

GENETIC CODE SPECIFICITY

A particular base modification in an *E. coli* isoleucine tRNA is important for reading the AUA isoleucine codon and for the specificity of its aminoacylation (Raj Bhandari 1988). Replacement of novel modified nucleoside Lysidine L(34) by unmodified Cytidine C(34) leads to substantial reduction in the rate of aminoacylation with isoleucine. Mutant isoleucine tRNA with CAU anticodon also reads AUA-A possibly. Change of L to C affects the coding specificity from AUA (isoleucine) to AUG (methionine).

One Codon can Code for Two Different Amino Acids

Strict one-to-one correspondence between codons and amino acids is thought to be an essential feature of the genetic code. However, Turanov et al. (2009) report that one codon can code for two different amino acids with the choice of the inserted amino acid determined by a specific 3′ untranslated region structure and location of the dual-function codon within the messenger RNA (mRNA). They found that the codon UGA specifies insertion of selenocysteine and cysteine in the ciliate *Euplotes crassus*; the dual use of this codon can occur within the same gene, and that the structural arrangements of *Euplotes* mRNA preserve location-dependent dual function of UGA when expressed in mammalian cells. Thus, the genetic code supports the use of one codon to code for multiple amino acids.

Synonymous Mutations Influence Gene Expression

Synonymous mutations do not alter the encoded proteins, but they can influence gene expression. To investigate how, Kudla et al (2009) engineered a synthetic library of 154 genes that varied randomly at synonymous sites, but all encoded the same green fluorescent protein (GFP). When expressed in *E. coli*, GFP protein levels varied 250-fold across the library. GFP messenger RNA (mRNA) levels, mRNA degradation patterns, and bacterial growth rates also varied, but codon bias did not correlate with gene expression. Rather, the stability of mRNA folding near the ribosomal binding site explained more than half the variation in protein levels. mRNA folding and associated rates of translation initiation play a predominant role in shaping expression levels of individual genes, whereas codon bias influences global translation efficiency and cellular fitness.

EXPANSION OF GENETIC ALPHABET

A new Watson-Crick base pair with a hydrogen pattern different from that in the A·T and G·C base pairs is incorporated into the duplex DNA and RNA by DNA and RNA polymerases, and expands the genetic alphabet from 4 to 6 letters (Piccirilli et al. 1990). The expansion could lead to RNAs with greater diversity in functional groups and greater catalytic potential. DNA and RNA polymerases seem to be sensitive only to the external geometry of a base pair and indifferent to the arrangement of hydrogen bonds that maintain that geometry. The genetic code can be extended artificially (Orgel 1990; Piccirilli et al. 1990).

RECODING

Studies conducted by Gesteland et al. (1992) have shown that certain instructions in mRNA may result in reprogramming of genetic information which mRNA receives from DNA. This reprogramming of mRNA is reflected in amino acid sequence of proteins. The phenomenon is called “recoding”. And the set of instructions in mRNA which brings about this recoding are called recoding signals. Two mechanisms for recoding have been proposed: first, change in linear read out of nucleotide sequence and, second, change in meaning of the code.

Alteration in the Linear Readout Frame

Studies of release factor 2 (RF2) of *E. coli*, certain retroviruses and T4 bacteriophage *hop60* gene have shown that recoding occurs due to alteration in linear read out of nucleotide sequence. In higher organisms, ornithine decarboxylase instability is overcome by +1 frameshift in mRNA. The second mechanism in which the meaning of the code is altered has been observed in case of selenocysteine amino acid. This amino acid is not naturally occurring; therefore, no code has been assigned to this amino acid. It is believed that UGA stop codon with altered meaning acts as a codon for this amino acid (Berry et al. 1991). This feature has also been observed in Moloney murine leukemia virus (Wills 1993).

There are several examples of this recoding mechanism: (a) In *E. coli*, mRNA of RF2 (protein meant for termination of translation = release factor) programs 30 per cent of ribosomes to change to the +1 reading frame after codon number 25 (codon 26 in UGS) (Figure 11.6) (Gesterland et al. 1992). The frameshift event is encouraged by an upstream sequence (termed stimulator) that pairs with 16S RNA in the ribosome. There are two components of recoding signals, a site of action (codons 25 and 26) and a stimulator. (b) In a class of retroviruses, many mRNAs exhibit a ribosomal frame shift to the –1 reading frame by tandem slippage of tRNA. The two components of recoding signal are a site of action consisting of a heptanucleotide in the mRNA and a stimulatory sequence in the form of stem-and-loop or pseudoknot structure (present downstream). A similar heptanucleotide for frameshift is found in *dna X* gene in *E. coli*, where 50 per cent of gene product is shortened due to frameshift. (c) In higher animals, a protein called antizyme renders the enzyme ornithine decarboxylase (ODC) unstable. Decoding of antizyme mRNA requires a +1 frameshift, which is regulated by the concentration of polyamines (product of ODC). (d) In the mRNA for T4 gene *60*, ribosomes hop a stretch of 50 nucleotides found between a pair of glycine codons. At the first glycine codon, 100 per cent ribosomes are released. These ribosomes hop and read the second glycine codon downstream and continue translation of the rest of mRNA. The stimulatory signals include the mRNA structure and the amino acid sequence of nascent peptide chain.

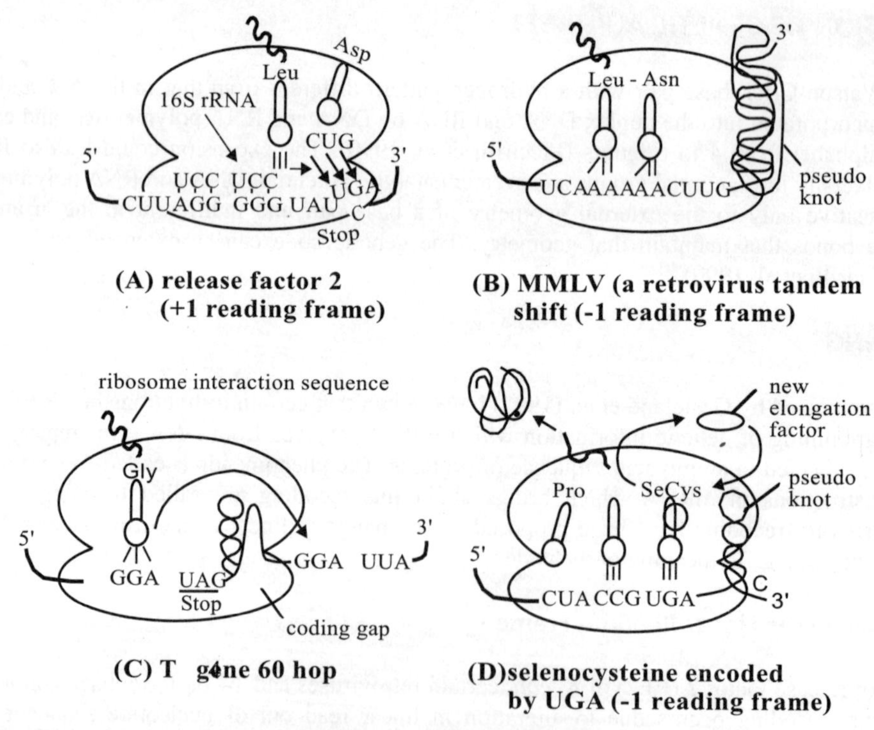

Figure 11.6 Different alternatives leading to recoding of the genetic code

Alteration in the Meaning of Code Words

In almost all organisms, there seem to be available "recoding signals" in mRNAs leading to altered meaning of codons. These are exceptions to the universal character of genetic code. In atleast two bacterial genes and three mammalian genes, an internal UGA codon codes for selenocysteine (SeCys) which has no unique codon in the genetic code dictionary. Identification of the enzyme that mediates insertion of a rare amino acid, pyrrolysine, into protein solves a puzzle and expands the rules of genetic code established nearly half a century ago (Schimmel and Beebe 2004). RNA structures signal incorporation of non-canonical amino acids. A hairpin structure called the 'selenocysteine insertion sequence' (SECIS) is present in the messenger RNA that corresponds to where selenocysteine (SeCys) is due to be inserted into the growing peptide chain. The SECIS element is an RNA element around 60 nucleotides in length that adopts a stem-loop structure (Walczak et al. 1996). In Moloney Murine Leukaemia Virus, MMLV), the SECIS interacts with the protein SBP2 and allows the recoding of UGA to selenocysteine. In bacteria, SelB binds to tRNA loaded with the selenocysteine and to SECIS, so that instead of UGA signaling a stop, selenocysteine is added into the peptide. In mammals, the gene for selenoprotein "P plasma protein" has more than 10 UGAs, 10 of them coding for SeCys, but presence of UGA codon is not enough (some UGA codons in the same mRNA code for termination). The requirements for UGA to code for SeCys include a specific minor tRNA a specific elongation factor (EF) and a particular downstream sequence. Sometimes this downstream sequence is 200 nucleotides away in the 3' non-coding region. Schematic secondary structures of SECIS elements in bacteria and eukaryotes and archea are shown in Figure 11.7A.

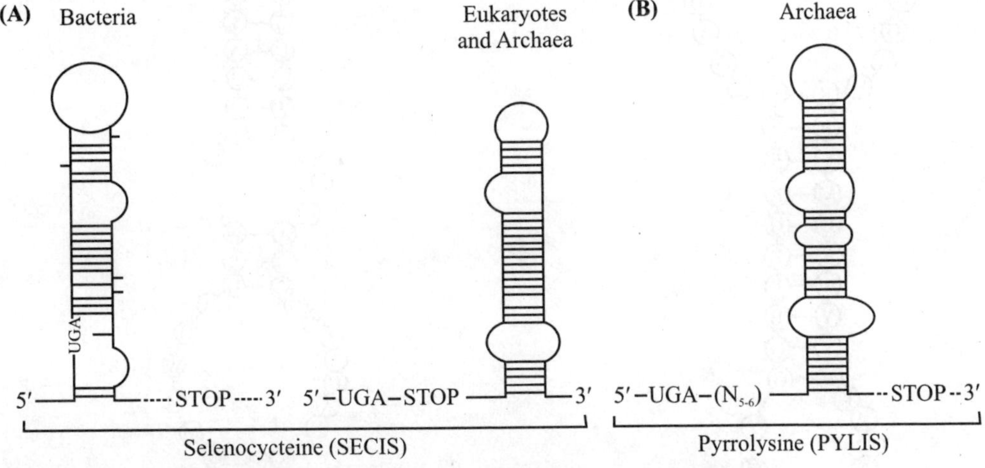

Figure 11.7 RNA structures signal incorporation of non-canonical amino acids

Pyrrolysine is the 22nd amino acid in proteins. UAG is normally the amber stop codon, but encodes pyrrolysine (Pyl) if a PYLIS element is present. The PYLIS downstream sequence (PYLIS: pyrrolysine insertion sequence) is a stem-loop structure which appears on some mRNA sequences. This structural motif causes the UAG (amber) stop codon to be translated to the amino acid pyrrolysine instead of ending the protein translation. In archaea the PYLIS downstream sequence is positioned straight after the UAG codon which is translated as pyrrolysine (Théobald-Dietrich et al. 2005; Zhang et al. 2005). Schematic secondary structure of PYLIS element in archea is shown in Figure 11.7B). Blight et al. (2004) establish that synthetic L-pyrrolysine is attached as free molecule to $tRNA_{CUA}$ by

PylS, an archaeal class II aminoacyl-tRNA synthetase. PylS activates pyrrolysine with ATP and ligates pyrrolysine to $tRNA_{CUA}$ in vitro in reactions specific for pyrrolysine. The addition of pyrrolysine to *E. coli* cells expressing pylT (encoding $tRNA_{CUA}$) and pulS results in the translation of non-sense UAG in vivo as sense codon. This is the first example from nature of direct aminoacylation of a tRNA with a non-canonical amino acid and shows that the genetic code of *E. coli* can be expanded to include UAG-directed pyrrolysine incorporation into proteins.

Sequence and secondary structures of SECIS and PYLIS elements are shown in Figure 11.8 and Figure 11.9, respectively.

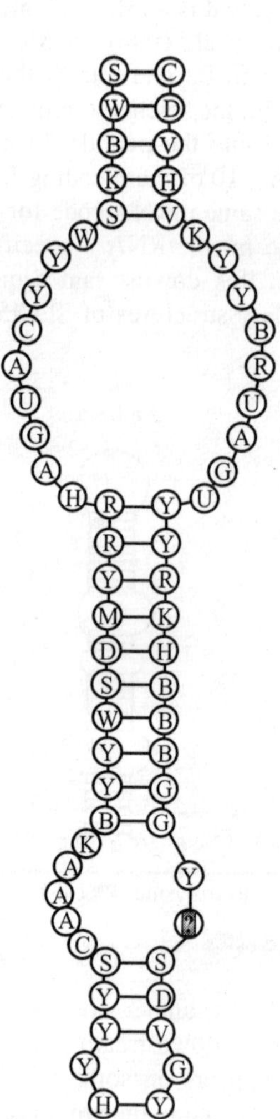

Figure 11.8 Sequence and secondary structure of SECIS element (Redrawn, from http://en.Wiki pedia. org/wiki/SECIS_element)

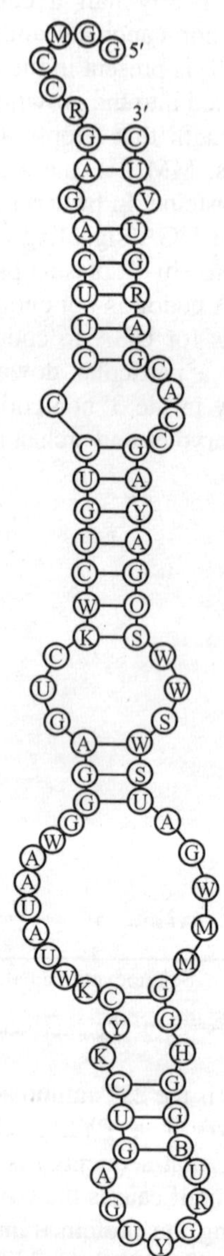

Figure 11.9 Sequence and secondary structure of PYLIS element) (Redrawn from http://en.wikipedia.org/wiki/ PYLIS_ downstream_sequence)

Mechanisms changing the meaning of codon

The encoding of two non-universal amino acids involves dynamic redefinition of 'stop' signals in the genetic code. Bacteria with multiple proteins containing these amino acids add to our appreciation of coding versatility (Atkins and Baranov 2007). Mechanism changing the meaning of codon translation is presented in Figure 11.10. (A) Stop codons (in reassigned form denoted as X) can be reassigned independently of the context wherever they occur in any messenger RNA in a cell. Three example mRNA sequences are shown. This mode of changing codon meaning occurs in certain specialized genomes, such as those of many mitochondria. (B) and (C), In context-dependent reassignment, either (B) site-specific codon redefinition can occur in response to a co-localized recoding signal (shown as a loop representing an RNA structure), or (C) mRNA-dependent redefinition of codons might happen in response to a recoding signal embedded outside the coding region.

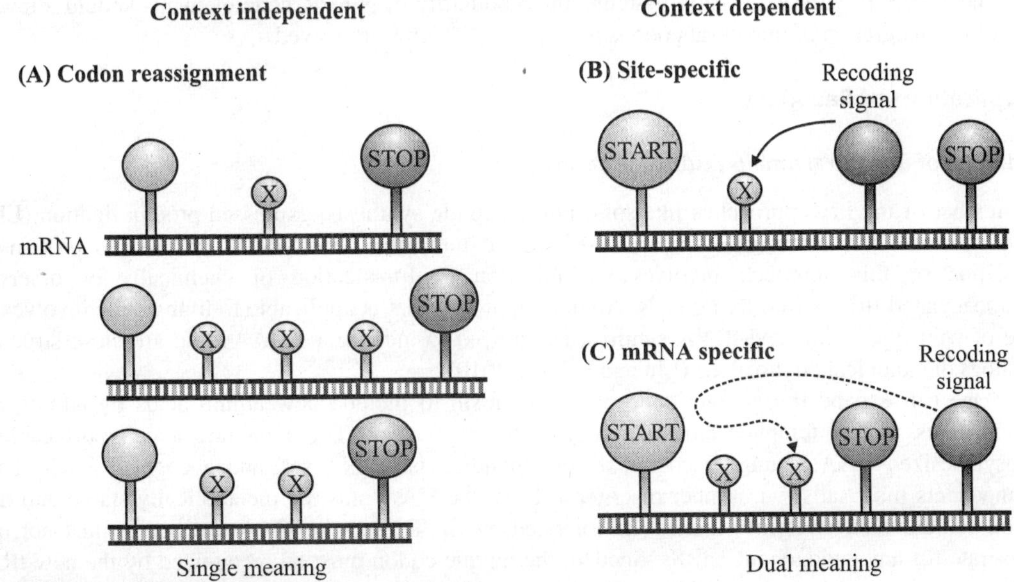

Figure 11.10 Changing the meaning of codon translation

Blank codons

Amber non-sense codons and, to a lesser extent, ochre, opal, or frameshift codons have been used to specify UAAs. In each case, suppression is in competition with other processes, such as RF binding to non-sense codons or recognition of frameshift codons by tRNAs with three-base anticodons, both of which lead to decreased co-translational incorporation efficiency. Although a number of efforts are employed to mitigate these undesired effects—examples include the overproduction of $tRNA_{SB}$ suppressors, synonymous mutation of codons surrounding codonBL, selection of efficient suppressors of frameshift codons that compete well with embedded rare three-base codons, and modulation of RF levels (Ryu and Schultz 2006; Bossi 1983; Taira et al. 2006; Magliery et al. 2001), and other efforts have tried to circumvent them, for example, a sense codon was used to encode an UAA in an auxotrophic strain by exploiting wobble position stability differences and limiting endogenous amino acid concentrations (Kwon et al. 2003), a general solution that involves the conversion of sense codons into blank codons would be ideal.

In theory, such a strategy is feasible with genome synthesis, whereby certain sense codons in an organism are replaced in a wholesale manner through synonymous mutations. Although effects on mRNA folding and translation initiation rates would have to be considered (Kudla et al. 2009; Coleman et al. 2008), the decreased degeneracy of the code would allow the reassignment of unused sense codons to UAAs. Such an approach is currently being carried out in *S. cerevisiae* and, if successful, would not only avoid the inefficiencies associated with non-sense and frameshift suppression, but would also allow several UAAs to be simultaneously encoded through mutually orthogonal UAARS/$tRNA_{SB}$ pairs. Such strains might allow completely unnatural polypeptides to be made ribosomally. An alternative approach that uses unnatural base pairs in addition to A-T and C-G could also provide new blank codons, in this case by extending the number of possible three-base codons beyond 64 (Bain et al. 1992; Hirao et al. 2002; Hirao et al. 2004). Although the inefficiency of unnatural base pairs in replication, transcription, and translation has kept this approach from moving beyond simple in vitro expression systems, the modularity of genetic code elements should allow for the direct integration of unnatural codons once these problems are solved.

Applications of Recoding

Addition of unnatural amino acids to proteins

A number of in vitro approaches like solid-phase peptide synthesis, expressed protein ligation (EPL), and cell free translation are being pursued to add unnatural amino acids (UAAs) to proteins. A variation of this approach involves the injection or transfection of chemically or otherwise aminoacylated tRNAs into living cells. Another approach that is applicable to living cells involves the use of wild-type aminoacyl-tRNA synthetases (aaRSs) to incorporate UAAs that are close structural analogs of canonical amino acids (Liu and Schultz 2010).

One can expand the genetic code of an organism to include new amino acids by adding new components to this template-directed biosynthetic machinery. These include a cell permeable or biosynthesized UAA, a unique codon, a corresponding iso-tRNA set, and a cognate aaRS. These components must satisfy a number of criteria: First, the UAA must be metabolically stable and have good cellular bioavailability; it must be tolerated by EF-Tu and the ribosome, but it must not be a substrate for any endogenous aaRSs. Second, the unique codon must be recognized by the new tRNA but not by any endogenous tRNAs. Third, the aaRS/tRNA pair must be specific for the UAA, functional in the host organism, and orthogonal in the context of all endogenous aaRS/tRNA pairs in the organism. In general, most UAAs added to the media are taken up by both prokaryotic and eukaryotic cells (Liu and Schultz 2010).

Clearly, the 20 common amino acids are sufficient for all known forms of life. However, if one assumes a substantial proteomic contribution to fitness during the solidification of the genetic code (Crick's "frozen accident" factor) (Sella and Ardell 2006; Vetsigian et al. 2006), then one must conclude that the identities of the 20 canonical amino acids are a consequence of factors present during early evolution (Lu and Freeland 2006; Weber and Miller 1981). Therefore, additional genetically encoded amino acids may offer evolutionary advantages to modern organisms and certainly extend our ability to manipulate the physicochemical and biological properties of proteins. Many researchers have genetically encoded UAAs representing an extensive range of structural and electronic properties not found in the common 20 amino acids. UAAs with chemically reactive groups can be used as bio-orthogonal handles for the site-specific in vitro, and in some cases intracellular, modification of proteins; they can also be used to introduce a new or enhanced catalytic function into proteins.

Importation of a heterologous aaRS/tRNA pair from a different domain of life

Although several strategies to generate orthogonal aaRS/tRNA pairs have been explored, ultimately, the most straightforward solution involves the importation of a heterologous aaRS/tRNA pair from a different domain of life. This is because tRNA recognition by aaRSs can be domain- or species-specific, a feature that can serve as the basis for orthogonality. The anticodon loop of the imported tRNA is then mutated to create a blank codon (codonBL) suppressor tRNA ($tRNA_{SB}$), and the orthogonality of the resulting aaRS/$tRNA_{SB}$ pair is assessed. If necessary, the orthogonality of this pair is improved by a two-step process, involving both positive and negative rounds of selection to identify functional optimized $tRNA_{SB}$s that exhibit no cross-reactivity with endogenous aaRSs (Wang and Schultz 2001).

Structure-based mutagenesis and a similar two-step selection strategy are used to alter the specificity of the heterologous aaRS so that it uniquely recognizes the UAA of interest (Santoro et al. 2002). This process has allowed for the systematic directed evolution of aaRS/tRNA pairs that are specific for a variety of UAAs and are orthogonal in bacteria, yeast, and mammalian cells. This approach should, in theory, make translation with UAAs accessible in any organism (Liu and Schultz 2010).

Probes of protein structure and function

Many biophysical and mechanistic studies require significant quantities of proteins with a probe incorporated at a unique site in a protein. UAA mutagenesis methodology is well suited to many such problems (Cellitti et al. 2008). UAAs with unique IR and X-ray diffraction signatures have also been efficiently incorporated into proteins to study structure and dynamics (Schultz et al. 2006). Because the genetic encoding of UAAs exploits the cells' own translational machinery, one can use genetically-encoded biophysical probes to study processes both in vitro and in living cells. For example, one can genetically encode small fluorescent UAAs at surface sites in a protein with minimal structural perturbation (this is in contrast to traditional fluorescent genetic tags such as GFP and its variants, which are limited by their large size). This labeling method can be used for the characterization of local protein conformation changes, protein folding, and biomolecular interactions (Guo et al. 2008; Wang et al. 2006; Lee et al. 2009a; Summerer et al. 2006). In addition to fluorescent probes, one can use photocaged and photoisomerizable UAAs to study protein function. In vitro these UAAs have been used to photoregulate ligand protein binding (Bose et al. 2006), site specifically cleave proteins (Peters et al. 2009), and probe protein-nucleic acid interactions (Lee et al. 2009b). Such genetically encoded UAAs can also allow the study of cellular processes. Another use of UAA incorporation is the characterization of protein-protein and protein-nucleic acid interactions in vitro or in living cells through photocross-linking. In many cases, biomolecular interactions are transient or unstable to in vitro isolation conditions, thus requiring covalent cross-linking of the interacting molecules to isolate relevant complexes.

Protein therapeutics

UAA mutagenesis is beginning to find many applications in the generation of therapeutic proteins, where the production of large quantities of homogenously modified protein is desired.

Protein evolution with an expanded genetic code

It is quite possible that the ability to encode additional amino acids with novel properties would be evolutionarily advantageous, especially since nature's choice of 20 could have been arbitrarily fixed at the point of transition between communal and Darwinian evolution paradigms and subsequently sustained by the code's inertia (Vetsigian et al. 2006).

THE SECOND CODE

Correct recognition of transfer RNAs by aminoacyl-tRNA synthetases is central to the maintenance of translational fidelity. The hypothesis that synthetases recognize anticodon nucleotides was proposed in 1964 and had considerable experimental support by mid-1970s. Nevertheless, the idea was not widely accepted. Until relatively recently, methodologies initially available for examining tRNA recognition proved hampering for adequately testing alternative hypotheses. Implementation of new technologies has led to a reasonably complete picture of how tRNAs are recognized (Sacks et al. 1994). The anticodon is indeed important for some of the 20 *E. coli* isoaccepting groups. For many of the isoaccepting groups, the acceptor stem or position 73 (or both) is important as well. For accurate translation of genetic messages the precision of two matchings is very important. Firstly that of amino acids with transfer RNA and secondly that of tRNA with mRNA. The later matching is a strait forward interaction of the codon of mRNA with anticodon in tRNA but the first matching is indirect and is mediated by specific enzymes, the aminoacyl synthetases (AAS). These enzymes are specific for each amino acid, there being 20 AASs for 20 essential amino acids.

Earlier it was thought that AASs simply recognized the anticodon on a tRNA and aminoacylated it accordingly. The discovery of suppressor tRNAs, however, challenged this concept as the amino acids carried by these tRNAs do not match the codon assignments made by the classical genetic code. The anticodon is thus obviously not the only parameter of a tRNA induced in recognition by its cognate AAS. The question then arose as to exactly what were the features of tRNA which were recognized by an AAS for specificity of aminoacylation. The AASs have been found to associate with the tRNA along and around inside diagonal of the three-dimensional structure of a tRNA. The regions of contact are the D stem, acceptor stem and the anticodon stem and loop (Schimell 1987). Although regions of contact have been found to be similar, the rules of recognition have been found to vary from case to case.

de Duve (1988) has proposed the existence of a second code which is imprinted in the structure of the AASs. It matches the amino acid with structural features of the tRNA, the phenomenon known as paracodons. The main features of this code are that it is non-degenerate, deterministic and older than the classical genetic code. Still lot of work is required to properly define paracodons. Schulman and Pelka (1983) studied the *E. coli* $tRNA^{Met}$-methionyl tRNA synthetase and found that the anticodon was important for recognition along with the stringent requirement being a cytidine at the wobble position. In case of $tRNA^{Gln}$, they found that the anticodon was important along with guanosine residue at the fourth position from the 3′-end. Large changes in the rates of aminoacylation of beef $tRNA^{Trp}$ yeast $tRNA^{Val}$, *E. coli* $tRNA^{Gly}$ and *E. coli* $tRNA^{Arg}$ have also been seen indicating the importance of the anticodon for recognition in these cases.

A number of cases where the anticodon is not instrumental in recognition have also been reported. Normanly et al. (1986) showed that the acceptor stem and the D stem and loop are important features of a $tRNA^{Ser}$ recognized by serine AAS. Hou and Schimmel (1988) have shown that the G3:U70 base pair in the acceptor helix is the most important determinant for alanine specificity. Other tRNAs where the anticodon has not been found to be vital for recognition are $tRNA^{Leu}$, $tRNA^{Cys}$ and $tRNA^{Phe}$.

Understanding the regulation of human gene expression requires knowledge of 'second genetic code', which consists of the binding specificities of transcription factors (TFs) and the combinatorial code by which TF binding sites are assembled to form tissue-specific enhancer elements. Using a novel high-throughput method, Hallicas et al. (2006) determined the DNA binding specificities of GLIs 1-3, Tcf4, and c-Ets1, which mediate transcriptional responses to the Hedgehog (Hh), Wnt, and Ras/MAPK signaling pathways. To identify mammalian enhancer elements regulated by these pathways, they developed a computational tool, enhancer element locator (EEL). They show that EEL can be used to identify *Hh* and *Wnt* target genes to to predict activated TFs based on changes in gene expression. Wei et al. (2006) describe a robust approach that couples chromatin immunoprecipitation (ChIP) with the paired-end ditag (PET) sequencing strategy to map p53 targets in human genome. Hallicas et al. (2006) and Wei et al. (2006) thus describe different ways of identifying direct targets of transcription factors and their corresponding regulatory sequences in the genome. Although still under development, these studies provide an efficient way to decipher regulatory networks (Holstege and Clevers 2006).

REFERENCES

Atkins, J.F., and P.V. Baranov. 2007. Translation: duality in the genetic code. Nature 448: 1004-5.

Bain J.D., C. Switzer, A.R. Chamberlin, and S.A. Benner. 1992. Ribosome-mediated incorporation of a non-standard amino acid into a peptide through expansion of the genetic code. Nature 356: 537-9.

Berry, M.J., L. Banu, Y. Chen, et al. 1991. Recognition of UGA as a selenocysteine codon in type I dexoxinase requires sequence in the 3′ untranslated region. Nature 353: 273-6.

Blight, S.K., R.C. Larue, A. Mahapatra, et al. 2004. Direct charging of tRNACUA with pyrrolysine in vitro and in vivo. Nature 431: 333-5.

Bose, M., D. Groff, J. Xie, E. Brustad, and P.G. Schultz. 2006. The incorporation of a photoisomerizable amino acid into proteins in *E. coli.* J. Am. Chem. Soc. 128: 388-9.

Bossi, L. 1983. Context effects: translation of UAG codon by suppressor tRNA is affected by the sequence following UAG in the message. J. Mol. Biol. 164: 73-87.

Cellitti, S.E., D.H. Jones, L. Lagpacan, et al. 2008. In vivo incorporation of unnatural amino acids to probe structure, dynamics, and ligand binding in a large protein by nuclear magnetic resonance spectroscopy. J. Am. Chem. Soc. 130: 9268-81.

Coleman J.R., D. Papamichail, S. Skiena, B. Futcher, E. Wimmer, and S. Mueller. 2008. Virus attenuation by genome-scale changes in codon pair bias. Science 320: 1784-7.

Crick, F.H.C. 1966. The genetic code: III. Scient. Am. 215: 55-62.

Crick, F.H.C., L. Bernett, S. Brenner, and R.J. Watts-Tobin. 1961. General nature of genetic code for proteins. Nature 192: 1227-32.

de Duve, C. 1988. The second genetic code. Nature 333: 117-8.

Fox, T.D. 1987. Natural variation in the genetic code. Annu. Rev. Genet. 21: 69-91.

Gamow, G. 1954. Possible relation between deoxyribonucleic acid and protein structure. Nature 173: 318-22.

Gesteland, R.F., R.B. Weiss, and J.F. Atkins. 1992. Recoding: reprogrammed genetic decoding. Science 254: 1640-1.

Guo, J., C.E. Melancon, H.S. Lee, D. Groff, and P.G. Schultz. 2009. Evolution of amber suppressor tRNAs for efficient bacterial production of proteins containing nonnatural amino acids. Angew. Chem. Int. Ed. Engl. 48: 9148-51.

Hallicas, O., K. Palin, N. Sinjushina, et al. 2006. Genome-wide prediction of mammalian enhancers based on analysis of transcription-factor binding affinity. Cell 124: 47-59.

Hinegardner, R.T., and J. Engelberg, 1963. Rationale for a universal genetic code. Science 142: 1083-5.

Hirao, I., T. Ohtsuki, T. Fujiwara, et al. 2002. An unnatural base pair for incorporating amino acid analogs into proteins. Nat. Biotechnol. 20: 177-82.

Hirao, I., Y. Harada, M. Kimoto, T. Mitsui, T. Fujiwara, and S. Yokoyama. 2004. A two-unnatural-base-pair system toward the expansion of the genetic code. J. Am. Chem. Soc. 126: 13298-305.

Holstege, F.C.P., and H. Clevers. 2006. Transcription factor target practice. Cell 124: 21-3.

Hou, Ya-M., and P. Schimell. 1988. A simple structural feature is a major determinant of a transfer RNA. Nature 335: 140-5.

Khorana, H.G. 1965. Polyumucleotide synthesis and the genetic code. Fed. Proc. 24: 1473-87.

Khorana, H.G. 1968. Nucleic acid synthesis in the study of the genetic code. In: *Nobel Lectures: Physiology or Medicine (1963–1970)*. Amsterdam: Elsevier, Science Ltd. 341-69.

Khorana, H.G., H. Buchi, H. Ghosh, et al. 1967. Polynucleotide synthesis and the genetic code. Cold Sp. Harb. Symp. Quant. Biol. 31: 39-49.

Kudla, G., A.W. Murray, D. Tollervey, and J.B. Plotkin. 2009. Coding-sequence determinants of gene expression in *Escherichia coli*. Science 324: 255-8.

Kwon, I., K. Kirshenbaum, and D.A. Tirrell. 2003. Breaking the degeneracy of the genetic code. J. Am. Chem. Soc. 125: 7512-3.

Last, J.A., W.M. Stanley, Jr., M. Salas, M.B. Hille, A.J. Wahba, and S. Ochoa. 1967. Translation of the genetic message. IV. UAA as a chain termination codon. Proc. Natl. Acad. Sci. USA 57: 1062-7.

Lee, H.S., J. Guo, A. Lemke , R.D. Dimla, and P.G. Schultz. 2009a. Genetic incorporation of a small, environmentally sensitive, fluorescent probe into proteins in *Saccharomyces cerevisiae*. J. Am. Chem. Soc. 131: 12921-3.

Lee, H.S., R.D. Dimla, and P.G. Schultz. 2009b. Protein-DNA photo-crosslinking with a genetically encoded benzophenone-containing amino acid. Bioorg. Med. Chem. Lett. 19: 5222-4.

Liu, C.C., and P.G. Schultz. 2010. Adding New Chemistries to the Genetic Code. Annu. Rev. Biochem. 79: 413-44.

Lu, Y., and S. Freeland. 2006. On the evolution of the standard amino-acid alphabet. Genome Biol. 7: 102.1-.6

Magliery T J, Anderson J C, Schultz P G (2001) Expanding the genetic code: selection of efficient suppressors of four-base codons and identification of "shifty" four-base codons with a library approach in *Escherichia coli*. J Mol Biol 307: 755-69.

Nirenberg, M.W. 1963. The genetic code: II. Scient. Am. 190: 80-94.

Nirenberg, M.W., and J.H. Matthaei. 1961. The dependence of cell-free protein synthsis in *E. coli* upon naturally occurring or synthetic polynucleotides. Proc. Natl. Acad. Sci. USA.47: 1588-602.

Nirenberg, M.W., and P. Leder. 1964. RNA code words and protein synthesis. Science 145: 1399-407.

Nirenberg, M.W., T. Caskey, R. Marshall, et al. 1966. The RNA codes and protein synthesis. Cold Sp. Harb. Symp. Quant. Biol. 31: 11-24.

Normanly, J., R.C. Ogden, S.J. Norvath, and J. Abelson. 1986. Changing the identity of a transfer RNA. Nature 321: 213-9.

Ochoa, S. 1963. Synthetic polynucleotides and the genetic code. Symp. Genet. Mechanics. Fed. Proc. 22: 62-74.

Orgel, L.E. 1990. Nucleic acids: adding to the genetic alphabet. Nature: 343: 18-20.

Peters, F.B., A. Brock, J. Wang, and P.G. Schultz. 2009. Photocleavage of the polypeptide backbone by 2-nitrophenylalanine. Chem. Biol. 16: 148-52.

Piccirilli, J.A., T. Krauch, S.E. Moroney, and S.A. Benner. 1990. Enzymatic incorporation of a new base pair into DNA and RNA extends the genetic alphabet. Nature 343: 33-7.

Ryu, Y., and P.G. Schultz. 2006. Efficient incorporation of unnatural amino acids into proteins in *Escherichia coli*. Nat. Methods 3: 263-5.

Sacks, M. E., J.R. Sampson, and J.N. Abelson. 1994. The transfer RNA identity problem: a search for rules. Science 263: 191-7.

Santoro, S.W., L. Wang, B. Herberich, D.S. King, and P.G. Schultz. 2002. An efficient system for the evolution of aminoacyl-tRNA synthetase specificity. Nat. Biotechnol. 20: 1044-48.

Schimell, P. 1987. Aminoacyl tRNA synthases. Annu. Rev. Biochem. 56: 125-58.

Schulman, L.H., and H. Pelka. 1983. Anticodon loop size and sequence requirement for recognition of formylmethionine tRNA by methionyl tRNA synthetase. Proc. Natl. Acad. Sci. USA 80: 6755-69.

Schultz, K.C., L. Supekova, Y. Ryu, J. Xie, R. Perera, and P.G. Schultz. 2006. A genetically encoded infrared probe. J. Am. Chem. Soc. 128: 13984-5.

Sella, G., and D.H. Ardell. 2006. The coevolution of genes and genetic codes: Crick's frozen accident revisited. J. Mol. Evol. 63: 297-313.

Summerer, D., S. Chen, N. Wu, A. Deiters, J.W. Ching, and P.G. Schultz. 2006. A genetically encoded fluorescent amino acid. Proc. Natl. Acad. Sci. USA 103: 9785-9.

Taira, H., T. Hohsaka, and M. Sisido. 2006. In vitro selection of tRNAs for efficient four-base decoding to incorporate non-natural amino acids into proteins in an *Escherichia coli* cell-free translation system. Nucl. Acids Res. 34: 1653-62.

Théobald-Dietrich, A., R. Giegé, and J. Rudinger-Thirion. 2005. Evidence for the existence in mRNAs of a hairpin element responsible for ribosome dependent pyrrolysine insertion into proteins. Biochimie 87: 813-7.

Turanov, A.A., A.V. Lobanov, D.E. Formenko, et al. 2009. Genetic code supports targeted insertion of two amino acids by one codon. Science 323: 259-61.

Vetsigian, K., C. Woese, and N. Goldenfeld. 2006. Collective evolution and the genetic code. Proc. Natl. Acad. Sci. USA 103: 10696-701.

Walczak, R., E. Westhof, P. Carbon, and A. Krol. 1996. A novel RNA structural motif in the selenocysteine insertion element of eukaryotic selenoprotein mRNAs. RNA 2: 367-79.

Wang, J., Xie, J., and P.G. Schultz. 2006. A genetically encoded fluorescent amino acid. J. Am. Chem. Soc. 128: 8738-9.

Wang, L., and P.G. Schultz. 2001. A general approach for the generation of orthogonal tRNAs. Chem. Biol. 8: 883-90.

Weber, A.L., and S.L. Miller. 1981. Reasons for the occurrence of the twenty coded protein amino acids. J. Mol. Evol. 17: 273-84.

Wei, C.-L., Q. Wu, V.B. Vega, et al. 2006. A global map of p53 transcription-factor binding sites in the human genome. Cell 124: 207-19.

Wills, P.R. 1993. Self-organization of genetic coding. J. Theor. Biol. 162: 267-87.

Wittmann, H.G. 1962a. Anätze zur Entschlussalung des genetischen Codes. Naturwissenchaften 481: 729-37.

Wittmann, H.G. 1962b. Proteinuntersuchungen an Mutatanten des Tabakmosaik virus als Beitrag zum Problem des genetischen Codes. Z. Vererbungsl. 93: 491-6.

Yanofsky, C. 2007. Establishing the triplet nature of the genetic code. Cell 128: 815-8.

Yanofsky, C., B.C. Carlton, J.R. Guest, D.R. Helinki, and U. Henning. 1964. On the colinearity of gene structure and protein structure. Genetics 51: 266-71.

Zhang, Y., P.V. Baranov, J.F. Atkins, and V.N. Gladyshev. 2005. Pyrrolysine and selenocysteine use dissimilar decoding strategies. J. Biol. Chem. 280: 20740-51.

12

Protein Biosynthesis

Although the process of protein synthesis is conceptually same in prokaryotes and eukaryotes yet some differences in the two machineries exist. For this reason we will first discuss protein synthesis in prokaryote *Escherichia coli* and then we will point out how translation process in eukaryotes differs from that in prokaryotes.

TRANSLATION IN PROKARYOTES

About 80 per cent of cellular RNA is ribosomal RNA (rRNA), 15 per cent is transfer RNA (tRNA) and 5 per cent is messenger (mRNA). Out of all different types of RNAs, only mRNA is destined to serve as a template for production of protein which in turn performs structural or functional role. It is the ribosome on which the mRNA and tRNA in conjunction with amino acids aggregate during protein synthesis and function as the sites of polypeptide synthesis. Structures of twenty-two standard amino acids that cell encodes during synthesis of proteins are given in Figure 12.1. These three types of RNAs are important component of the protein synthesizing machinery.

It is common, though it is probably a mistake, to think of RNA as a single-stranded molecule. Any single-stranded polynucleotide will form as many internal base pairs as it can. In vivo, we might think of two classes of RNA molecules. Unstable molecules such as mRNAs, whose function is to present a linear sequence of bases, are prevented from folding into compact internally base-paired structures by specific proteins whose function is to do just this. Stable molecules, such as rRNAs and tRNAs, fold into very characteristic secondary and tertiary structures that are an indispensable for their particular functions. These three types of RNA are produced from their respective precursor RNA molecules through RNA processing. Here we will discuss finished RNA molecules that participate in protein synthesis in *E. coli*.

Messenger RNA

Messenger RNA serves as a template on which a polypeptide is constructed and contains an initiation codon (AUG or GUG), at least one of the termination codons (UAA/UGA/UAG) and base sequence in form of triplet codons that dictate the order of amino acid in a polypeptide chain. mRNA also includes certain trailer and leader sequences that are not translated. Messenger RNA fraction is heterogeneous in size, ranging from 500 to 6,000 nucleotides in *E. coli*. mRNA acts as a template for translation. It provides the information that must be interpreted by translational machinery. Protein encoding regions of mRNA are composed of contiguous, non-overlapping string of codons called open reading frames (ORFs), each of which specifies single protein. Translation starts at 5′-end of open reading frame and proceeds one codon at a time to 3′-end. First and last codons of ORF are called start and stop codons, respectively.

(A) Amino acids containing one amino and one carboxyl group:

L-Glycine (Gly,G) L-Alanine (Ala,A) L-Valine (Val,V)

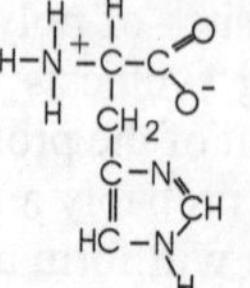

L-Isoleucine (Ile,I) L-Leucine (Leu,L) L-Serine (Ser,S) L-Threonine (Thr,T)

(B) Amino acids containing one amino and two carboxyls:

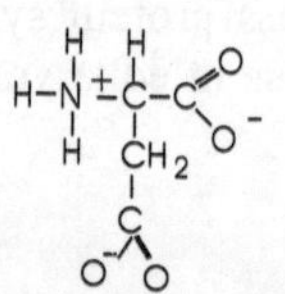

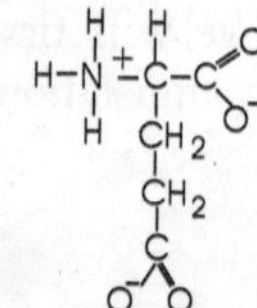

(C) Amides of dicarboxyl amino acids:

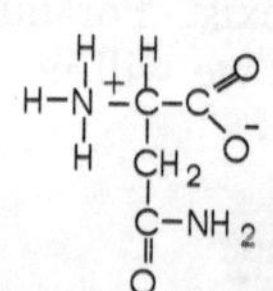

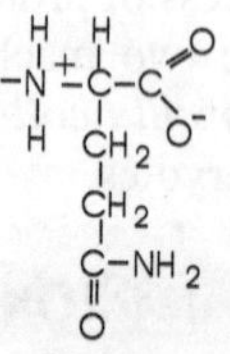

L-Aspartic acid (Asp,D) L-Glutamic acid (Glu,E) L-Asparagine (Asn,N) L-Glutamine (Gln, Q)

(D) Basic amino acids (additional amino group)

(E) Imino or cyclic amino acid

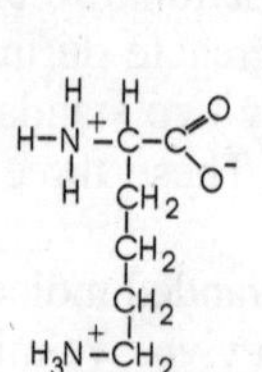

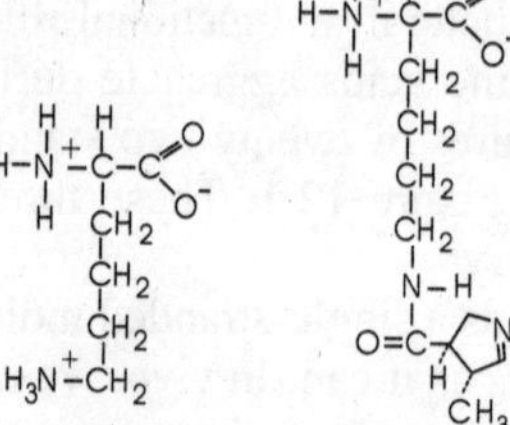

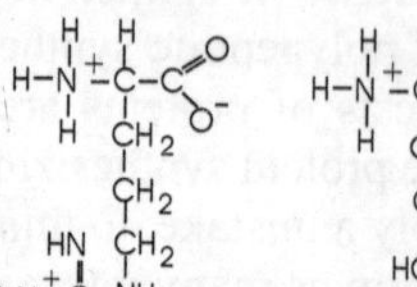

L-Lysine (Lys,K) L-Pyrrolysine (Pyl,O) L-Arginine (Arg,R) L-Histidine (His,H) L-Proline (Pro,P)

(F) Aromatic amino acids (containing ⬡ group)

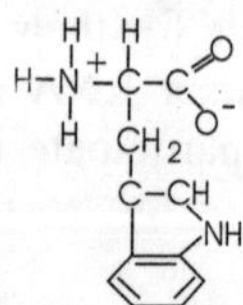

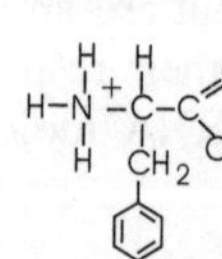

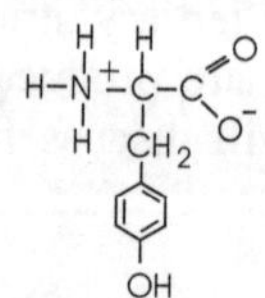

L-Tryptophan (Trp,W) L-Phenylalanine (Phe,F) L-Tyrosine (Tyr, Y)

(G) Sulphur-containing amino acids:

L-Methionine (Met,M) L-Cysteine (Cys,C) L-Selenocysteine (SeC,U)

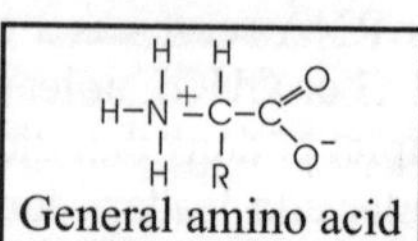

Figure 12.1 The twenty-two standard amino acids are found in proteins. At physiological pH, these amino acids usually exist as ions. Three and one letter abbreviations are also given. Asp and Glu are acidic, Lys, Pyl, Arg, and His are basic, Ala, Val, Pro, Leu, Phe, Trp, Met, Ile, Cys and Gly are nonpolar, and Ser, Thr, Asn, Gln and Tyr are polar (uncharged) amino acids

Prokaryotic mRNA frequently contains two or more ORFs and hence can encode multiple polypeptide chains. mRNA containing multiple ORFs are called polycistronic, which often encode proteins that perform related functions such as different steps in the biosynthesis of amino acids and nucleotides. To facilitate the binding by a ribosome, many prokaryotic ORFs contain a short sequence upstream on the 5′-end of start codon called ribosome binding site also called Shine-Dalgarno sequence (Shine and Dalgarno 1974, 1975). This sequence typically located 3-9 bp on the 5′ side of start codon is complementary to the sequence located near 3′-end of one of RNA components (16S rRNA) of ribosome. Ribosome binding site base-pairs with 16S rRNA component thereby aligning ribosome with beginning of ORF.

Transfer RNA

Crick (1966) proposed that prior to their incorporation into polypeptides amino acids must attach to special adapter molecule that is capable of directly interacting with and recognizing three-nucleotide long coding units of mRNA. Transfer RNA serves as adaptor molecule. The secondary structure of tRNA has a characteristic cloverleaf configuration. However, tRNA is further folded through alternative hydrogen bonding geometries (including Hoogsteen base pairs) into an L-shaped tertiary structure. Transfer RNA acts as adapter between codons on mRNA and amino acids that they specify. It seemed unlikely that direct interactions between mRNA template and amino acids could be responsible for specific and accurate ordering of amino acids in a polypeptide. It is because of the reason that unlike complementarity between DNA template and ribonucleotides of the mRNA, the side chains of amino acids have little or no specific affinity for purine and pyrimidine bases found in RNA. Hydrophobic side chains of amino acids cannot form hydrogen bonds with amino- or keto- groups of nucleotide bases. It is hard to imagine that several different combinations of three bases of RNA could form surfaces with unique affinities for aromatic amino acids.

The principal features of tRNA clover leaf are an accepter stem, three stem loops which are referred to as ψ loop, D loop and the anticodon loop, and fourth variable loop. Accepter stem is so named because it is the site of attachment of amino acid, is formed by pairing between 5′ and 3′ ends of tRNA molecule. The 5′-CCA-3′ at the extreme 3′ ends of molecule protrudes from this double-stranded stem. The ψU loop is no named because of the characteristic presence of unusual base ψU in loop. D loop takes its name from the characteristic presence of dihydrouridines in the loop. The anticodon loop contains the anticodon, a 3-nucleotide long decoding element that is responsible for recognizing the codon by base pairing with mRNA. How is a single amino acid recognized by its transfer RNA in protein synthesis? Each tRNA uses only small amount of information to specify its identity (Mlot 1989). Alanine with a single base in the acceptor site serves as a major determinant. The G3:U70 pair has apparently been constructed for alanine recognition. Direct chemical footprinting shows that translocation of tRNA occurs in the two discrete steps (Moazed and Noller 1989a,b). During the first step, which occurs spontaneously after the formation of peptide bond, the acceptor end of tRNA moves relative to the large ribosomal subunit, resulting in "hybrid state" of binding. During the second step, which is promoted by elongation factor EF-G, the anticodon end of tRNA, alongwith the mRNA moves relative to the small ribosomal subunit. Sequences important in recognition of tRNA have been mapped in vivo and in vitro (Moras 1990). A limited number of nucleotides are responsible for the specific recognition of a tRNA by its cognate synthetase.

Transfer RNA is the direct interface between amino-acid sequence of a protein and the information in DNA. It is a small 80-nucleotide long RNA. There are 50 to 60 different tRNA molecule classes in each cell. Each tRNA molecule is specific for one amino acid. There is specific joining of a tRNA to a specific amino acid. All tRNAs from all organisms have a similar structure, indeed a human tRNA can function in yeast cells. tRNA is synthesized in two parts. The body of the tRNA is transcribed from a tRNA gene. The acceptor stem is the same for all tRNA molecules and is added after the body is synthesized. It is replaced often during lifetime of a tRNA molecule. All the unusual bases present in the tRNA are produced by modifications of the four unusual bases (A, U, G, and C) after tRNA has been transcribed; none of the unusual bases, even if bound in the cell, is incorporated directly into the RNA. Structures of unusual bases found in tRNA are given in Figure 12.2. In ribothymidine, thymine is produced by methylation at position 5 of uracil.

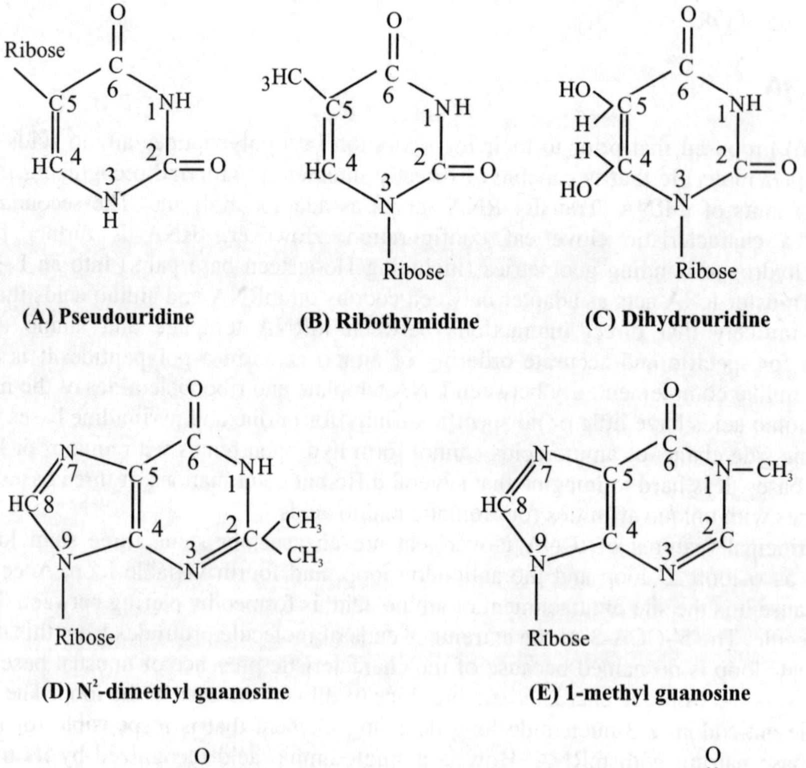

(A) Pseudouridine

(B) Ribothymidine

(C) Dihydrouridine

(D) N^2-dimethyl guanosine

(E) 1-methyl guanosine

(F) 1-methylinosine

(G) Inosine

Figure 12.2 Structures of unusual bases found in tRNA (Redrawn from http://rpi.edu/dept/bcbp/molbiochem/MBWeb/ mb2/part1/trna.htm)

Dihydrouridine (D) is generated by the saturation of a double bond in the ring of uracil thus having H-C-OH attached at C4 and C5. Pseudouridine is produced due to an interchange of the N and

C atoms in uracil. 4-Thiouridine represents a substitution of S for O in uridine. Inosine (I) is produced through a modification of adenine where at C2 $_2$HN- group is replaced by -H. N^2-dimethylguanosine (G-Me_2) is produced by attachment of $_3$HC-C-CH_3 at C2 of guanine. 1-methyl guanosine is formed by attachment of –CH_3 group at N1 of guanine. In 1-methyl inosine, $_2$HN- is replaced by –H at C2 and $_3$HC- is attached at N1 atom of guanine.

Some tRNA molecules have an extra or variable loop. Based on this, tRNAs have been classified into two groups. Class I tRNAs have only 3-5 bases in their extra loop; they represent ~75 per cent of all tRNAs. Class II RNAs have larger loop having 13-21 bases in their extra loop and 5 base pairs in the stem.

Diseases owing to tRNA defect

Wilson et al. (2004) demonstrated that a single change in a person's tRNA gene can contribute to a range of life-shortening risk factors, including high blood pressure, high cholesterol, and other metabolic disorders. Various combinations of these abnormalities affect up to a quarter of the U.S. population, and they are contributing to a public health epidemic of heart disease and stroke. The mutation affects the genes of the mitochondria – the energy-producing power plants of the cell that are passed from mother to offspring. They found one woman suffering from hypertension and low magnesium levels and when they spoke to her, she said that a number of other family members also had low magnesium. That suggested that she might have a new disease, because all the known genetic causes of low magnesium were autosomal recessives that would not occur so widely. Such a pattern immediately suggested a defect in the mitochondrial genome, because those genes are uniquely passed from mother to offspring, unlike the rest of the cell's genome, which is contained in the nucleus. Detailed sequencing of the mitochondrial genomes of family members revealed a specific mutation in all affected people. That defect was the substitution of a single base in the gene that coded for a specific tRNA in the mitochondria. The defective base the researchers pinpointed was in the gene for the tRNA that transports the amino acid isoleucine. That defect distorted the docking region of the tRNA, preventing it from recognizing and attaching to the messenger RNA to deposit its isoleucine cargo. Thus, the faulty tRNA could lead to defects in a vast array of proteins that normally contained isoleucine, thereby contributing to a broad range of cellular malfunctions.

Structure and Function of Ribosome

Ribosomal RNAs

Ribosomal RNAs are responsible for principal functions of the ribosome. Ribosome coordinates the correct recognition of mRNA by each tRNA and catalyzes the peptide bond formation between growing polypeptide chain and amino acid attached to selected tRNA. This is called second code. rRNA contributes directly to the catalytic properties of protein synthesis and proteins were added to the ribosome late in its evolution (Moore and Steitz 2002). Ribosome is a ribozyme and addresses the catalytic properties of its all RNA active sites (Nissen et al. 2000). The intact ribosome contains three tRNA binding sites that reach between two subunits; an A-site where charged tRNA enters the ribosome, a P-site that contains peptidyl-tRNA and E-site, where deacylated tRNA exits the ribosome. The ribosome consists of a large submit which contains peptidyl transferase center and small subunit which contains decoding centre. Each subunit is composed of one or more rRNAs and multiple proteins. It is the proteins in the ribosome that perform a largely structural function, not the RNA. Ribosomal RNA is of three sizes – 23S, 16S, and 5S. Ribosomal RNA is a component of the ribosomes, the protein synthetic factories in the cell. A large number of copies these rRNA molecules are present in each cell. They are always associated with a specific set of proteins called ribosomal proteins. Sometimes it functions as non-specific 'workbench" for the polypeptide synthesis. Genes coding for rRNA are present in nucleolar organizing regions (NOR) of the chromosome.

Ribosome – site of protein synthesis

The Royal Swedish Academy of Sciences awarded the Nobel Prize in Chemistry for 2009 jointly to Venkatraman Ramakrishnan, MRC Laboratory of Molecular Biology, Cambridge, United Kingdom; Thomas A. Steitz, Yale University, New Haven, CT, USA; and Ada E. Yonath, Weizmann Institute of Science, Rehovot, Israel, for studies of the structure and function of the ribosome (Yonath 2000; Yonath 2005; Yonath et al. 1987; Yonath et al. 1980). They showed what the ribosome looks like and how it functions at the atomic level. All three have used a method called X-ray crystallography to map the position for each and every one of the hundreds of thousands of atoms that make up the ribosome. The ribosome translates the DNA code into life. Ribosomes produce proteins, which in turn control the chemistry in all living organisms. As ribosomes are crucial to life, they are also a major target for new antibiotics.

Inside every cell in all organisms, there are DNA molecules. They contain the blueprints for how a human being, a plant or a bacterium, looks and functions. But the DNA molecule is passive. If there was nothing else, there would be no life. The blueprints become transformed into living matter through the work of ribosomes. Based upon the information in DNA, ribosomes make proteins: oxygen-transporting hemoglobin, antibodies of the immune system, hormones such as insulin, the collagen of the skin, or enzymes that break down sugar. There are tens of thousands of proteins in the body and they all have different forms and functions. They build and control life at the chemical level. An understanding of the ribosome's innermost workings is important for a scientific understanding of life. This knowledge can be put to a practical and immediate use; many of today's antibiotics cure various diseases by blocking the function of bacterial ribosomes. Without functional ribosomes bacteria cannot survive. This is why ribosomes are such an important target for new antibiotics. Above mentioned three Nobel Laureates have all generated 3D models that show how different antibiotics bind to the ribosome. These models are now used by scientists in order to develop new antibiotics, directly assisting the saving of lives and decreasing humanity's suffering.

Ribosomes have an asymmetrical organization. The size of ribosomes being expressed in terms of Svedberg units (S) based on the rate of sedimentation in an ultracentrifuge. Prokaryotic ribosome has sedimentation coefficient 70S. It contains 65 per cent rRNA and 35 per cent protein. Prokaryotic ribosomes have two subunits. Larger subunit has sedimentation coefficient 50S. It contains one molecule of 5S rRNA, one molecule of 23S rRNA and 34 different proteins. The 50S subunit has a fairly compact body (~150×200×200Å) from which a central protuberance and a stalk, stick out; this subunit is relatively more spherical than the smaller one. The partition between the head and body of the small subunit fix into the large subunit. There could be a space or tunnel between the two subunits. Smaller subunit (30S) has one molecule of 16S rRNA and 21 different proteins. The 30S subunit has an elongated and asymmetrical shape (about 55×220×220Å). It has a constricted region and cleft that separates the head from the base, while its platform projects out the base. The 30S and the 50S subunits may be held together by an association between discrete areas of these subunits. The ribosome is macromolecular machine that directs the synthesis of proteins. Large subunit contains peptidyl transferase center which is responsible for formation of peptide bonds. Small subunit contains decoding center in which charged tRNAs read or decode the codon units of mRNA.

The composition of 30S and 50S subunits is given in Table 12.1. rRNAs contributes to the binding of tRNA and mRNA to ribosome during protein synthesis. Ribosomes undergo a cycle of dissociation and reassociation during protein synthesis. Ratio of 70S:30S+50S subunit gives indication of the

Table 12.1 Subunit composition of bacterial ribosome

Ribosome size	*Subunit size*	*rRNAs*	*Proteins*
70S	Small - 30S	16S rRNA (1,541 bases)	21
	Large - 50S	23S rRNA + 5S rRNA (2,904 bases)	31

amount of protein synthesis. Protein synthesis is critical to recycling of subunits. Ribosomal RNAs are not simply structural components of ribosome rather they are directly responsible for key function of ribosome. Peptidyl transferase center is composed entirely of RNA. RNA also plays role in the function of small subunit of ribosome. Anticodon loops of charged tRNAs and the codons of mRNA contact 16S rRNA, not ribosomal proteins of small subunit. Small U-RNA molecules are involved in ribosome biosynthesis (Steitz and Tycowski 1995).

Self-assembling macromolecular machines derive fundamental cellular processes, including transcription, mRNA processing, translation, DNA replication and cellular transport. The ribosome, which carries out protein synthesis, is one such machine, and the 30S subunit of the bacterial ribosome is the preeminent model system for biophysical analysis of large RNA-protein complexes. Talkington et al. (2005) developed a method involving pulse-chase monitored by quantitative mass spectrometry (PC/QMS) to follow the assembly of 20 ribosomal proteins with 16S ribosomal RNA during formation of the functional particle. The assembly of the complex traverses a landscape dotted with various local conformational transitions.

Rapidly growing cells produce thousands of new ribosomes each minute, in a tightly regulated process that is essential to cell growth. How the *E. coli* 16S ribosomal RNA and the 20 proteins that make up the 30S ribosomal subunit can assemble correctly in a few minutes remains a challenging problem, partly because of the lack of real-time data on the earliest stages of assembly. By providing snapshots of individual RNA and protein interactions as they emerge in real time, Adilakshmi et al. (2008) show that 30S assembly nucleates concurrently from different points along the rRNA. Time-resolved hydroxyl radical footprinting was used to map changes in the structure of the rRNA within 20 milliseconds after the addition of total 30S proteins. Helical junctions in each domain fold within 100 milliseconds. In contrast, interactions surrounding the decoding site and between the 5′, the central and the 3′ domains require 2-200 seconds to form. Unexpectedly, nucleotides contacted by the same protein are protected at different rates, indicating that initial RNA-protein encounter complexes refold during assembly. Although early steps in the assembly are linked to intrinsically stable rRNA structure, later steps correspond to regions of induced fit between the proteins and the rRNA.

Moore and Steitz (2003) discuss three-dimensional architecture of the large ribosomal subunit and the mechanism by which it facilitates peptide bond formation. Ribosome is an RNA enzyme.

The crystal structure of the bacterial 70S ribosome refined to 2.8Å resolution reveals atomic details of its interaction with messenger RNA (mRNA) and transfer RNA (tRNA) (Selmer et al. 2006). A metal ion stabilizes a kink in the mRNA that demarcates the boundary between the A and P sites, which is potentially important to prevent slippage of mRNA. Metal ions also stabilize the intersubunit interface. The interactions of E-site tRNA with the 50S subunit have both similarities and differences compared to those in the archaeal ribosome. The structure also rationalizes much biochemical and genetic data on translation.

In addition to mRNA, tRNA, rRNA and ribosome, other factors are also necessary for protein synthesis. Peptidyl transferase forms peptide bonds between amino acids. Protein factors catalyze partial reactions in the initiation, elongation and termination of peptides. Miscellaneous factors, such as

ATP, GTP, Mg^{2+}, K^+, $NH4^+$ are important for various biochemical reactions. The enzymes required for protein synthesis are aminoacyl synthetases, phosphatase, transformylase, deformylase, methionine-specific peptidase, etc. No enzyme as complex as the modern ribosome could possibly have merged from primordial soup in a single step. The first ribosomes were very likely composed entirely of RNA. The evidence to this hypothesis is that the functional core of modern ribosome, its decoding site and its peptidyl transferase centre, consists primarily of RNA and the bulk of its proteins are found on its surface well removed from its functional centers (Moore and Steitz 2002). The peptidyl transferase centre, the most ancient of them all, has no proteins in it whatsoever. The decoding centre, the second to be added to the particle, is predominantly RNA but protein does make a modest contribution to its function. E site, which seems to be s a late addition during evolution, is protein rich.

A possible "ribosome code"

In the budding yeast *Saccharomyces cerevisiae*, 59 of the 79 cytoplasmic ribosomal proteins are encoded by two genes, stemming from an ancient genome duplication event (McIntosh and Warner 2007). Komili et al. (2007) report that these paralogous genes are not functionally equivalent, suggesting the possible existence of a "ribosome code". Duplicated genes escape gene loss by conferring a dosage benefit or evolving diverged functions. The budding yeast contains many duplicated genes encoding ribosomal proteins. Prior studies have suggested that these duplicated proteins are functionally redundant and affect cellular processes in proportion to their expression. In contrast, through the studies of ASH1 mRNA in yeast, they demonstrate paralog-specific requirements for the translation of localized mRNAs. Intriguingly, these paralog-specific effects are limited to a distinct subset of duplicated ribosomal proteins. Moreover, transcriptional and phenotypic profiling of cells lacking specific ribosomal proteins reveals differences between the functional roles of ribosomal protein paralogs that extend beyond effects on mRNA localization. Finally, they show that ribosomal protein analogs exhibit differential requirements for assembly and localization. These data indicate complex specialization of ribosomal proteins for specific cellular processes and support the existence of a ribosomal code.

Induced fit of ribosome structure by tRNA binding

The large ribosomal subunit catalyzes the reaction between the α-amino group of the aminoacyl-tRNA bound to the A site and the ester carbon of the peptidyl-tRNA bound to the P site, while preventing the nucleophilic attack of water on the ester, which would do unprogrammed deacylation of the peptidyl-tRNA. Schmeing et al. (2005a,b) report three new structures of the large ribosomal subunit of *Haloarcula marismortui* complexed with peptidyl transferase substrates analogs that reveal an induced-fit mechanism in which substrates and active site residues reposition to allow the peptidyl transferase reaction. Proper binding of an aminoacyl-tRNA analog to the A site induces specific movements of 23S rRNA nucleotides 2618-2620 (*E. coli* numbering 2583-2585) and 2541 (2506), thereby orienting the ester group of the peptidyl-tRNA and making it accessible for attack. In the absence of an appropriate A site substrate, the peptidyl transferase center positions the ester link of the peptidyl-tRNA in a conformation that precludes the catalyzed nucleophilic attack by water. Protein release factors may also function, in part, by inducing an active-site rearrangement similar to that produced by the A site aminoacyl-tRNA, allowing the carboxyl group and water to be positioned for hydrolysis.

Korostelev et al. (2006) present the crystal structure of the *Thermus thermophilus* 70S ribosome containing a model mRNA and two tRNAs at 3.7-Å resolution. Many structural details of the interaction between the ribosome, tRNA, and mRNA in the P and E sites and the ways in which tRNA structure is distorted by its interactions with the ribosome are seen. Differences between the conformations of the vacant and tRNA-bound 70S ribosomes suggest an induced fit of the ribosome structure in response to tRNA binding, including significant changes in the peptidyl transferase catalytic site.

Ribosome itself is an mRNA helicase

Most mRNAs contain secondary structure, yet their codons must be in single-stranded form to be translated. Until now no helicase activity has been identified which can account for ability of ribosomes to translate through downstream mRNA secondary structure. Takyar et al. (2005) show that ribosomes are able to disrupt downstream helices, including a perfect 27-bp helix of predicted Tm = 70°C. Using helices of different lengths and registers, the helices active site can be localized to the middle of the downstream tunnel, between the head and shoulder of the 30S subunit. Mutation of residues in proteins S3 and S4 that line the entry to the tunnel impairs helicase activity. They conclude that the ribosome itself is an mRNA helicase and that proteins S3 and S4 may play role in its processivity.

Ribosome profiling

Ingolia et al. (2009) present a ribosome profiling strategy that is based on the deep sequencing of ribosome-protected mRNA fragments and enables genome-wide investigation of translation with subcodon resolution. They used this technique to monitor translation in budding yeast under both rich and starvation conditions. These studies defined the protein sequence being translated and found extensive translational control in both determining absolute protein abundance and responding to environmental stress. They also observed distinct phases during translation that involve a large decrease in ribosome density going from early to late peptide elongation as well as widespread regulated initiation at non-adenine-uracil-guanine (AUG) codons. Ribosome profiling is readily adaptable to other organisms, making high precision investigation of protein translation experimentally accessible (Ingolia et al. 2009). Protocol for converting ribosome footprints or randomly fragmented mRNA into deep-sequencing library is given in Figure 12.3.

Process of Translation

The process by which genetic information contained within the order of nucleotides is used to generate linear sequence of amino acids in protein is called translation. Following four steps are involved in the process of translation – charging of tRNA, initiation , elongation and termination The machinery responsible for translating the language of mRNA to language of proteins is composed of four primary components - mRNA, ribosome, rRNA and aminoacyl-tRNA synthetase.

Translation preinitiation steps

Charging of transfer RNAs: Before translation commences, activation of amino acids takes place in the cytosol at the expense of ATP. During this charging reaction, amino acid is covalently attached to a specific tRNA by enzyme called aminoacyl-tRNA synthetases. Each tRNA can be charged only with the amino acid for which its anticodon is appropriate. During charging reaction, each of the 20 different amino acids gets attached to its corresponding tRNA by enzyme aminoacyl-tRNA synthetases. The amino acid is linked by an ester bond involving its carboxyl group to one of the last base of the tRNA (which is always adenine). The meaning of tRNA is determined by its anticodon and not by its amino acid. Charging reaction involves two steps, activation and transfer reactions, which

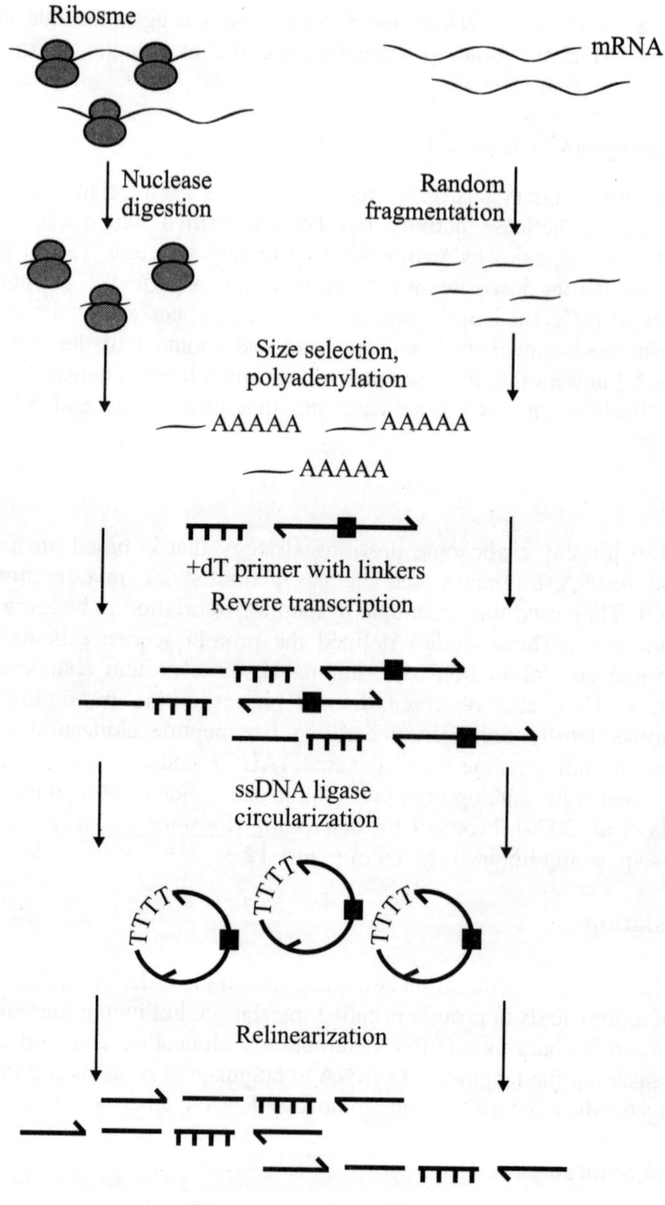

Figure 12.3 Quantifying mRNA abundance and ribosome footprints by means of deep sequencing

are crucial for functioning of the tRNAs. In the activation of amino acid, an amino acid reacts with ATP to become adenylated with concomitant release of pyrophosphate. As a result of adenylation, the amino acid is attached to adenylic acid via a high energy ester bond in which the carbonyl group of amino acid is joined to phosphoryl group of AMP. In transfer reaction, adenylated amino acid, which remains tightly bound to synthetase, reacts with tRNA. Each amino acid thus gets bound to accepter stem of tRNA by a specific aminoacyl-tRNA synthetase. These enzymes face two important challenges, they must recognize the correct set of tRNAs for a particular amino acid and they must charge all of these isoaccepting tRNAs with the correct amino acid. Some features of tRNA, called identity elements, help aminoacyl-tRNA synthetases to discriminate isoaccepting tRNA from rest of 19 amino acids. The accepter stem is an especially important determinant for the specificity of tRNA synthetase recognition.

The first important step in the initiation of translation is charging of $tRNA_f^{Met}$ with methionine The reactions involved in this step are shown in Figure 12.4. mRNA is the blue print of genetic information encoded in the gene. An initiation codon signals the start of a polypeptide chain. After the initiation triplet, the bases are read in group of three in sequence, until a chain terminating codon is reached at which time a polypeptide is released from the mRNA and ribosome.

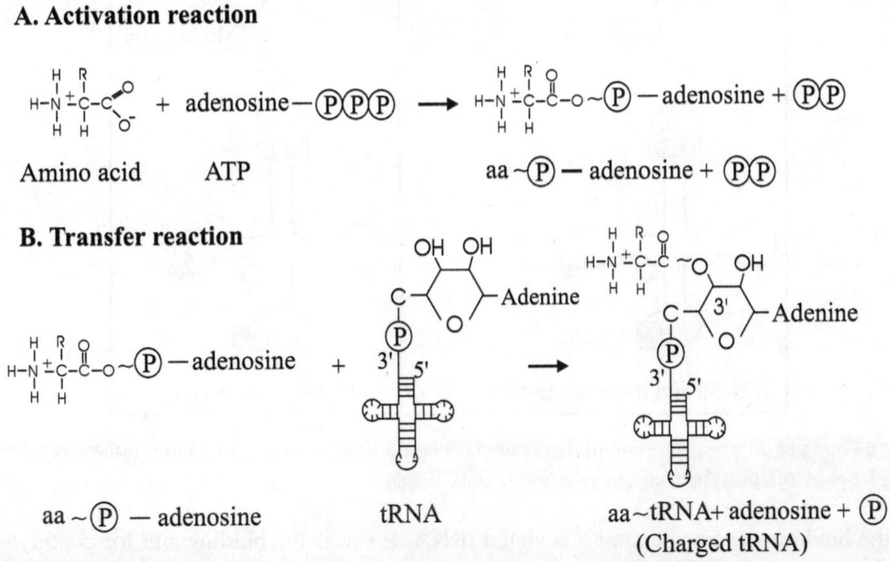

A. Activation reaction

Amino acid + adenosine—(P)(P)(P) (ATP) ⟶ aa ~(P) — adenosine + (P)(P)

B. Transfer reaction

aa ~(P) — adenosine + tRNA ⟶ aa~tRNA+ adenosine + (P) (Charged tRNA)

Figure 12.4 Charging of tRNA with an amino acid is a two step process: (A) Activation reaction in which an amino acid is attached to AMP with a high energy bond (~) and (B) Transfer reaction in which the activated amino acid is transferred to the RNA. Amino acyl-tRNA synthetase is used in these reactions

Polypeptide synthesis is initiated by amino acid N-formylmethionine in bacteria. N-formyl group allows amino acid to be bound at specific initiation site of ribosome and prevents it from entering interior positions. Structures of methionine and formylated methionine are compared in Figure 12.5. $tRNA^{Met}$ and $tRNA_f^{Met}$ differ from each other at three different points (Wu and Rajbhandary 1997) (Figure 12.6).

Binding Sites on the Ribosome for tRNA: Ribosome contains three tRNA binding sites, called acyl (A), peptidyl (P) and exit (E) sites. Both ribosomal subunits contribute to these sites; anticodon-mRNA interactions are small subunit functions, whereas peptidyl transfer which involves the aminoacyl-end of tRNA, is located on the large subunit. Movement from A to P site is achieved in two steps, not in one, and an intermediate state exists in which the anticodon end of tRNA is still in the A site on the small subunit while the aminoacyl end occupies the P site of the large subunit (Moore 1989). To carry out peptidyl transferase reaction, the ribosome must be able to bind at least two tRNAs simultaneously

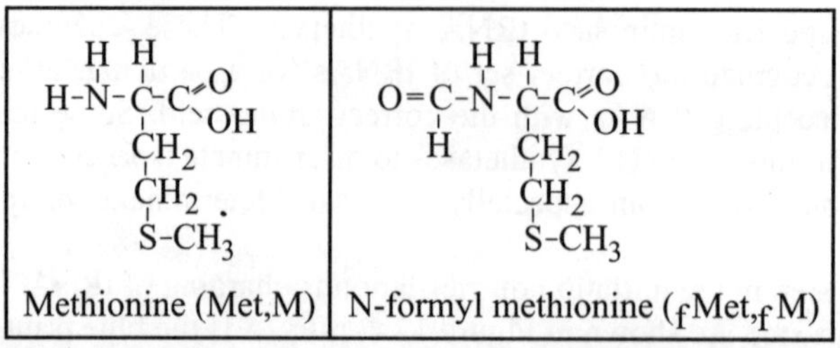

Methionine (Met,M)	N-formyl methionine ($_f$Met,$_f$M)
H H H-N-C-C=O, OH CH$_2$ CH$_2$ S-CH$_3$	H H O=C-N-C-C=O, OH H CH$_2$ CH$_2$ S-CH$_3$

Figure 12.5 Chemical structural differences between methionine and formylated methionine (Redrawn from http://sm.nwsuaf.edu.cn/mb/admin/upload/files/gzk/htm/chapter7_2.htm)

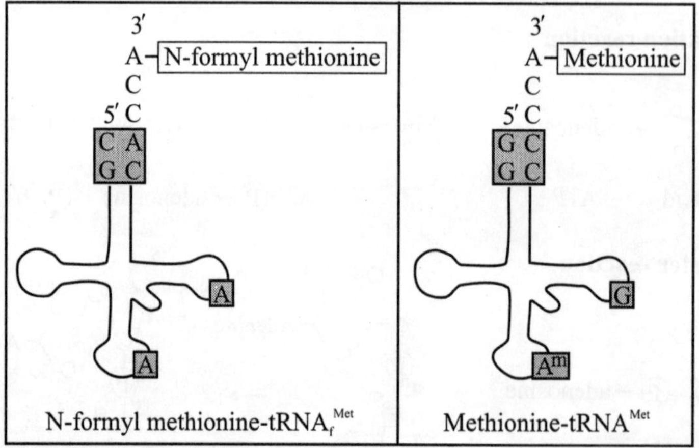

Figure 12.6 Points of difference between N-formylmethionine tRNA (tRNA$_f^{Met}$) and methionine tRNA (tRNA^{Met}) (Redrawn from http://proteopedia.org/wiki/index.php/TRNA)

A site is the binding site for the amino acylated tRNA. P site is the binding site for peptidyl-tRNA. E site is the binding site for tRNA that is released after growing polypeptide chain has been transferred to amino acyl tRNA. The E site is richer in proteins than the A and P sites (Moore and Steitz 2002). In addition to the several contacts that E-site-bound tRNAs make with both 16S and 23S rRNA, their anticodon stems interact extensively with S7 in the small subunit, and their T loops and T stems contact L1 in the large subunit.

Each tRNA binding site is formed at the interface between large and small subunits of ribosome. In this way bound tRNA can span the distance between peptidyl transferase center in large subunit and decoding center in small subunit. Channels through the ribosome allow the mRNA and growing polypeptide to enter and or exit the ribosome. The mRNA enters and exits the decoding center through two narrow channels in small subunit. The entry channel is only wide enough for unpaired RNA to pass through. This feature ensures that mRNA is in an extended form as it enters the decoding center by removing any intramolecular base pairing interactions. In between the two channels is a region that is accessible to tRNA and where adjacent codons can bind to aminoacyl-tRNA and peptidyl-tRNA at A and P sites, respectively. A second channel through the large subunit provides an exit path for newly synthesized polypeptide chain. Size of the channel limits the folding of growing polypeptide chain.

tRNA Signals its own Acceptance: During transfer RNA (tRNA) selection, a cognate codon-anticodon interaction triggers a series of events that ultimately results in the acceptance of that tRNA into the ribosome for peptide-bond formation. High-fidelity discrimination between the cognate tRNA and near- and cognate ones depends on their differential dissociation rates from the ribosome and on specific acceleration of forward rate constants by cognate species. Cochella and Green (2005) show that a mutant $tRNA^{Trp}$ carrying a single substitution in its D-arm achieves elevated levels of mis-coding by accelerating these forwarding rates constants independent of codon:anticodon pairing in the decoding center. These data provide evidence for a direct role for tRNA in signaling its own acceptance during decoding and supports its fundamental role during the evolution of protein synthesis. Kinetic scheme for tRNA selection on the ribosome identifying the two stages of initial selection and proofreading is presented in Figure 12.7 (Cochella and Green 2005).

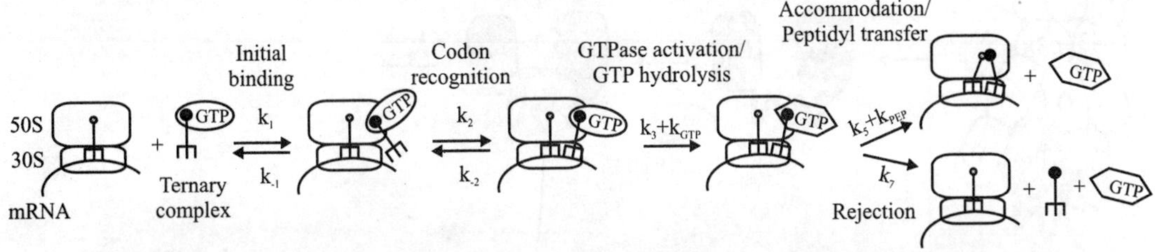

Figure 12.7 Kinetic scheme for tRNA selection on the ribosome identifying the two stages of initial selection and proofreading

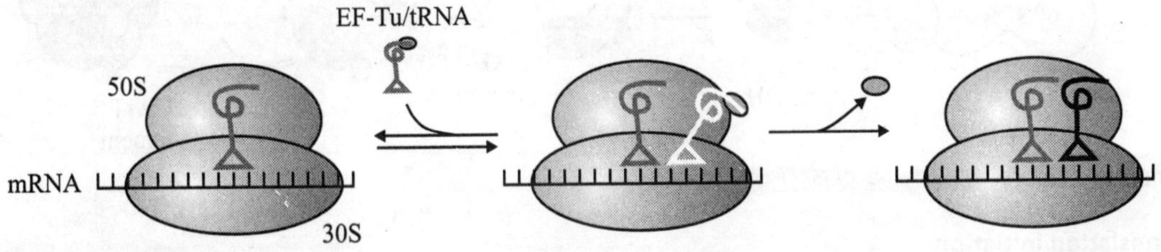

Figure 12.8 An active role for tRNA

An active role for tRNA has been depicted in Figure 12.8 (Daviter et al. 2005). A new aminoacyl-rRNA is brought into the A site of the ribosome in complex with elongation factor EF-Tu. Upon CTP hydrolysis, the aminoacyl-tRNA is released by EF-Tu, and swings into the peptidyl transferase center of the 50S subunit. The tRNA is distorted when bound with EF-Tu on the ribosome. The G24A mutation allows a tryptophenyl tRNA to read the UGA stop codon. The mutation allows GTP hydrolysis and the movement of tRNA into the peptidyl transferase center to proceed efficiently even on the stop codon.

RNA Multisynthetase Complex: Aminoacyl tRNA synthetases (ARSs) catalyze the ligation of amino acids to cognate tRNAs. Chordate ARSs have evolved distinct features absent from ancestral forms, including compartmentalization in a multisynthetase complex (MSC), non-catalytic peptide appendages, and ancillary functions unrelated to aminoacylation. Sampath et al. (2004) show that glutamyl-prolyl-tRNA synthetase (GluProRS), a bifunctional ARS of the MSC, has a regulated, non-canonical activity that blocks synthesis of a specific protein. GluProRS was identified as a component of the interferon (IFN)-γ-activated inhibitor of translation (GAIT) complex by RNA affinity chromatography using the ceruloplasmin (Cp) GAIT element as ligand. In response to IFN-γ, GluProRS is phosphorylated and released from the MSC, binds the Cp 3'-untranslated region in an mRNP containing three additional proteins, and silences Cp mRNA translation. Thus, GluProRS has divergent functions in protein synthesis: in the MSC, its aminoacylation activity supports global translation, but translocation of GluProRS to an inflammation-responsive mRNP causes gene-specific translational silencing. Two-stage model of GAIT complex formation is shown in Figure 12.9 (Sampath et al. 2004).

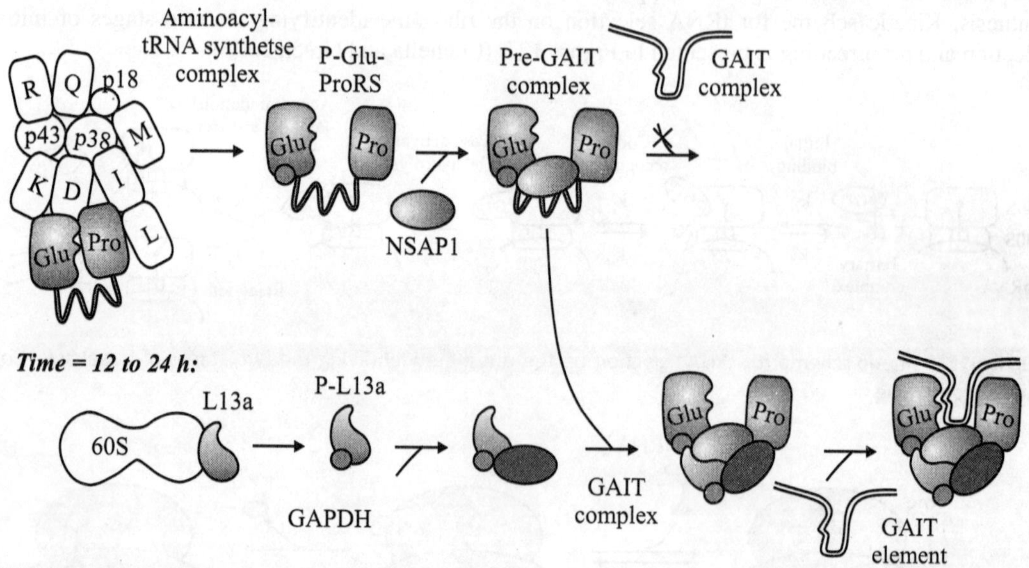

Figure 12.9 Two-stage model of GAIT complex formation

Translation initiation

Initiation involves reactions prior to forming the peptide bond between the first two amino acids of the protein which is a relatively slow step in protein synthesis. The initiation of polypeptide synthesis in prokaryotes requires (a) the 30S ribosomal subunit, which contains 16S rRNA (b) the mRNA coding for the polypeptide to be made, (c) the initiating fMet-$tRNA_f^{Met}$, (d) a set of three proteins called initiation factors (IF-1, IF-2 and IF-3), (e) GTP, (f) the 50S ribosomal subunit and (g) Mg^{2+}. Three steps in initiation of translation process are recognized. In the first step, the 30S initiation complex is formed. It requires mRNA, 30S subunit of ribosome, a special initiating species of aminoacyl~tRNA (that apparently starts all polypeptide chain, and three initiation protein factors — IF1, IF2, IF3. The initiation factor 3 (IF-3) prevents the 30S and 50S subunits from combining prematurely. Binding of the mRNA to the 30S subunit then takes place in such a way that the initiation codon (AUG) binds to a precise location on the 30S subunit. The initiating AUG is guided to the correct position on the 30S subunit by an initiating signal called the Shine-Dalgarno sequence in the mRNA. The first amino acid incorporated is formylated N-methionine (Figure 12.10). At this stage, formation of 30 initiation complex is complete. In the second step of the initiation process, the complex consisting of the 30S subunit, IF-3, and mRNA now forms a still larger complex by binding IF-2, which already is bound to GTP and the initiating fMet-$tRNA_f^{Met}$. The anticodon of this tRNA pairs correctly with the initiation codon in this step.

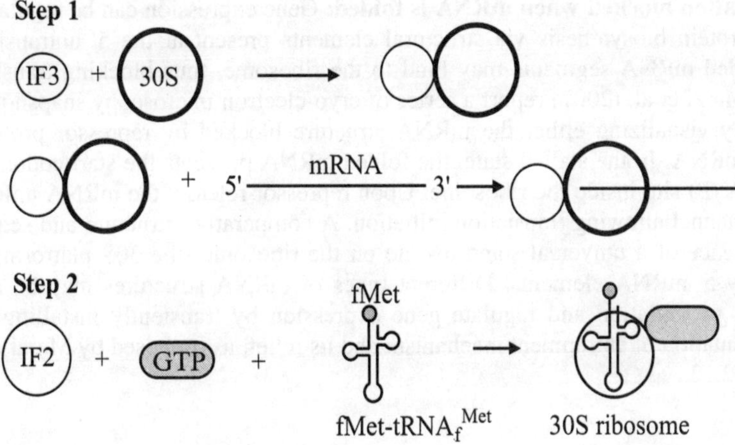

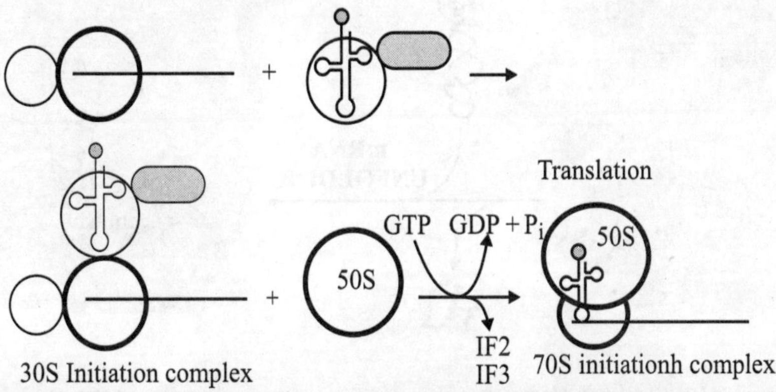

Figure 12.10 Formation of prokaryotic 70S initiation complex is a three-step process. In step 1, 30S ribosome and mRNA are combined. In step 2, initiator tRNA combines with IF2. In step 3, products of steps 1 and 2 are combined to form 30S initiation complex, followed by the formation of 70S complex

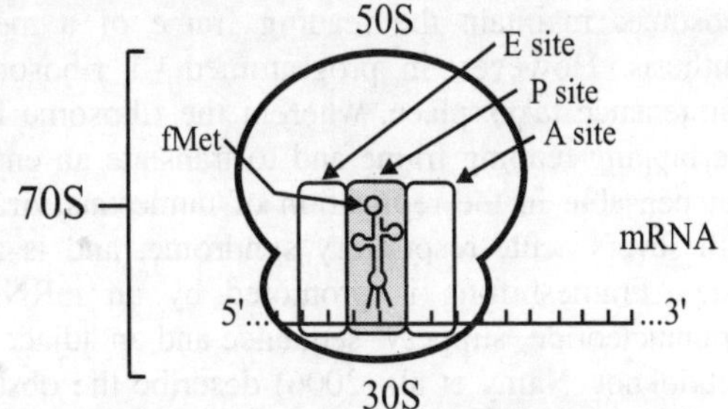

Figure 12.11 The 70S ribosome contains an aminoacyl site (A site) and peptidyl site (P site) and exit site (E site)

In the third step, this large complex combines with the 50S ribosomal subunit; simultaneously, the GTP molecule bound to IF-2 is hydrolyzed to GDP and Pi (which are released). IF-3 and IF-2 also depart from the ribosome. The 70S initiation complex is thus formed (Figure 12.11). During initiation, the mRNA bearing the code for polypeptide to be formed binds to smaller subunit of ribosome, this is followed by binding of initiating amino acyl-tRNA and large subunit to form initiation complex. The initiating aminoacyl-tRNA base-pairs with mRNA codon AUG at P-site that signals the beginning of polypeptide chain.

Translation Initiation blocked when mRNA is folded: Gene expression can be regulated at the level of initiation of protein biosynthesis via structural elements present at the 5' untranslated region of mRNAs. The folded mRNA segments may bind to the ribosome, thus blocking translation until the mRNA unfolds. Marzi et al. (2007) report a series of cryo-electron microscopy snapshots of ribosomal complexes directly visualizing either the mRNA structure blocked by repressor protein S15 or the unfolded, active mRNA. In the stalled state, the folded mRNA prevents the start codon from reaching the peptidyl-tRNA (P) site inside the ribosome. Upon repressor release, the mRNA unfolds the moves into the mRNA channel allowing translation initiation. A comparative structure and sequence analysis suggests the existence of a universal stand-by site on the ribosome (the 30S platform) dedicated for binding regulatory 5' mRNA elements. Different types of mRNA structures may be accommodated during translation preinitiation and regulate gene expression by transiently installing the ribosome. Schematic representation of entrapment mechanism and its relief, as proposed by Marzi et al. (2007), is shown in Figure 12.12.

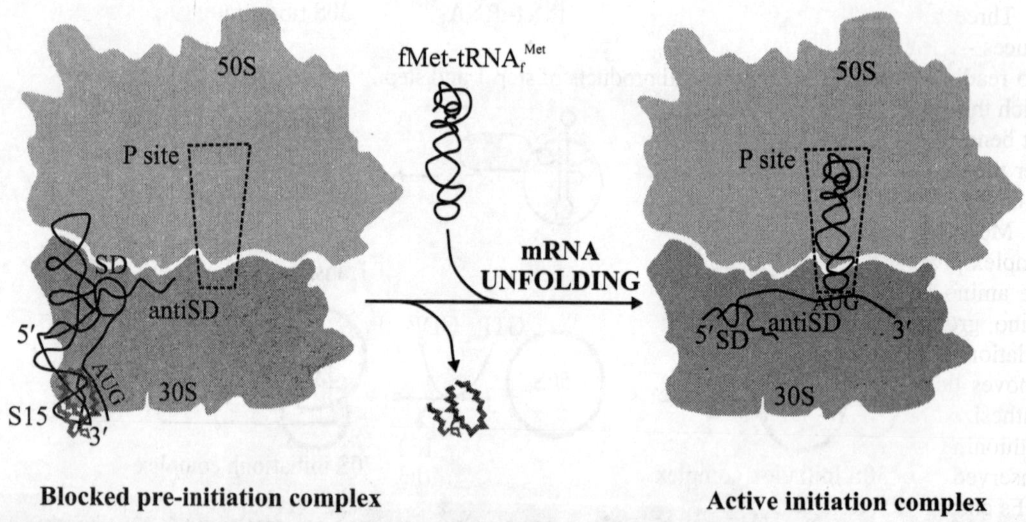

Figure 12.12 Schematic representation of the entrapment mechanism and its relief

mRNA Reading Frame Maintenance: The triplet-based genetic code requires that translating ribosomes maintain the reading frame of a messenger RNA faithfully to ensure correct protein synthesis. However, in programmed –1 ribosomal frameshifting, a specific subversion of frame maintenance takes place, wherein the ribosome is forced to shift one nucleotide backwards into an overlapping reading frame and to translate an entirely new sequence of amino acids. This process is indispensable in the replication of numerous viral pathogens, including HIV coronavirus associated with severs acute respiratory syndrome, and is also exploited in the expression of several cellular genes. Frameshifting is promoted by an mRNA signal composed of two essential elements: a heptanucleotide ‘slippery’ sequence and an adjacent mRNA secondary structure, most often an mRNA pseudoknot. Namy et al. (2006) describe the observation of a ribosome-mRNA pseudoknot complex that is stalled in the process of –1 frameshifting. A purified mammalian 80S ribosomes from rabbit reticulocytes paused at a cornovirus pseudoknot reveals an intermediate of the frameshifting process. The results provide a mechanical explanation of the pseudoknot manipulates the ribosome into a different reading frame Figure 12.13 (Namy et al. 2006).

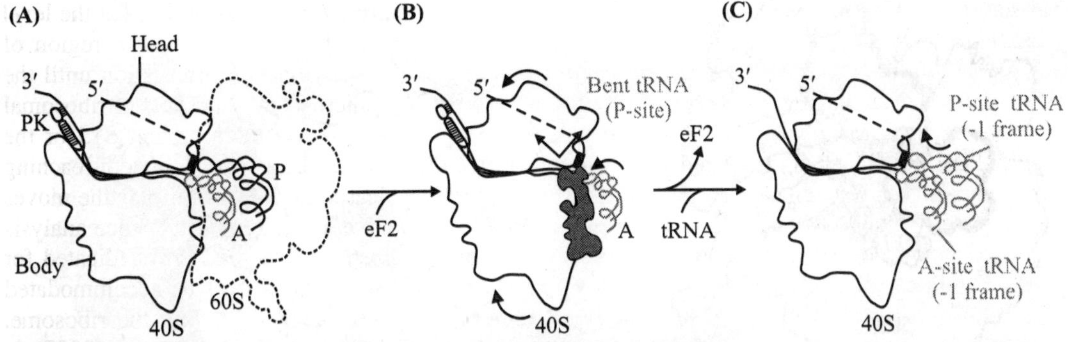

Figure 12.13 A mechanical model for pseudoknot-induced –1 frameshifting

Three different states of the small subunit translating an mRNA containing a pseudoknot that induces –1 frameshifting are shown. (a) The elongating ribosome approaching the pseudoknot in the zero reading frame. (b) Engagement with the pseudoknot, generating a frameshifting intermediate in which the small subunit is stalled during translocation with eEF2 bound, causing tension in the mRNA that bends the P-site tRNA in a (+) sense direction. As a result the anticodon-codon interaction breaks over the slippery sequence, allowing a spring-like relaxation of the tRNA in a (–) sense direction. (c) Re-engagement of the tRNA with the mRNA, leaving the ribosome translating in the –1 reading frame.

Messenger-RNA-directed protein synthesis is accomplished by the ribosome. In eubacteria, this complex process is initiated by a specialized transfer RNA charged with formylmethionine ($tRNA_f^{Met}$). The amino-terminal formylated methionine of all bacterial nascent polypeptides blocks the reactive amino group to prevent unfavorable side-reactions and to enhance the efficiency of translation initiation. The first enzymatic factor that processes nascent chains is peptide deformylase (PDF), it removes this formyl group as polypeptide emerges from the ribosomal tunnel and before the newly synthesized proteins can adopt their native fold, which may bury the N-terminus. Next, the N-terminal methionine is excised by methionine aminopeptidase. Bacterial PDFs are metalloproteases sharing a conserved N-terminal catalytic domain. All Gram-negative bacteria, including *E. coli*, possess class I PDFs characterized by a carboxy-terminal α-helical extension. Studies focusing on PDF as a target for antibacterial drugs have not revealed the mechanism of its co-translational mode of action despite indications in early work that it co-purifies with ribosomes. Bingel-Erlenmeyer et al. (2008) have provided biochemical evidence that *E. coli* PDF interacts directly with ribosome via its C-terminal extension. Crystallographic analysis of the complex between ribosome-interacting helix of PDF and the ribosome at 3.7-Å resolution reveals that the enzyme orients its active site towards the ribosomal tunnel exit for efficient co-translational processing of emerging nascent chains. The interaction of PDF with the ribosome enhances cell viability. The results provide the structural basis for understanding the coupling between protein synthesis and enzymatic processing of nascent chains, and offer insights into the interplay of PDF with the ribosome-associated chaperone trigger factor (TF). Footprints of PDF and trigger factor (TF) on the ribosome are shown in Figure 12.14 (Bingel-Erlenmeyer et al. 2008). A schematic representation of the trigger factor (TF) with its two arms and ribosome-binding domain forming a hydrophobic nascent chain folding and processing chamber and processing chamber and functioning as a router viewed from the ribosomal tunnel is shown in Figure 12.15 (Bingel-Erlenmeyer et al. 2008). Methionine-aminopeptidase (MAP) and PDF close the lateral openings of the trigger factor.

The ribosome of *Thermus thermophilus* was co-crystallized with initiator transfer RNA (tRNA) and a structural messenger RNA (mRNA) carrying a translational operator (Jenner et al. 2005). The

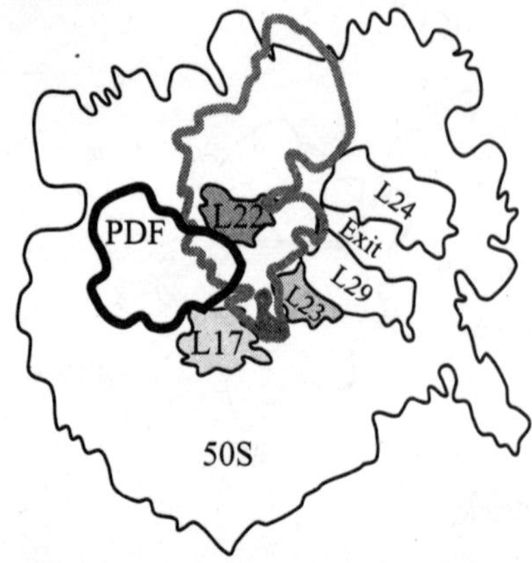

Figure 12.14 Model for the concerted mechanism of PDF and trigger factor – footprints of PDF

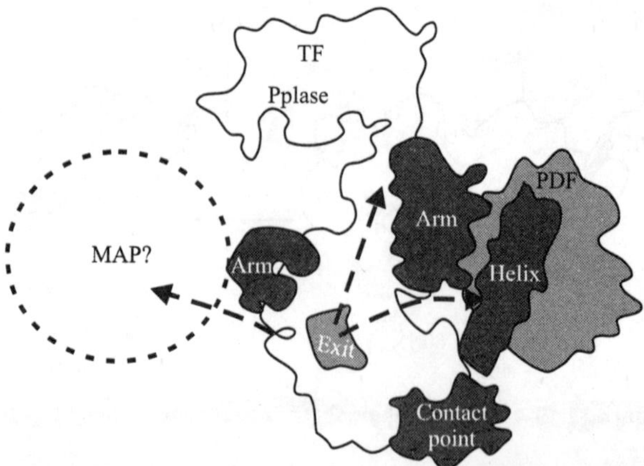

Figure 12.15 Model for the concerted mechanism of PDF and trigger factor – trigger factor with two arms and ribosome-binding domain

path of the mRNA was defined at 5.5Å resolution by comparing it with either the crystal structure of the same ribosomal complex lacking mRNA or with unstructured mRNA. A precise ribosomal environment positions the operator stem-loop structure perpendicular to the surface of the ribosome on the platform of the 30S subunit. The binding of the operator and of the initiator tRNA occurs on the ribosome with an unoccupied tRNA exit site, which is expected for an initiation complex. The positioning of the regulatory domain of the operator relative to the ribosome elucidates the molecular mechanism by which the bound repressor switches off the translation. These data suggest a general way in which mRNA control elements must be placed on the ribosome to perform their regulatory task.

Translation initiation is a major determinant of the overall expression level of a gene. The translation of functionally active protein requires the messenger RNA to be positioned on the ribosome such that the start/initiation codon will be read first in the correct frame. Recent crystal structures of the ribosomal subunits, the empty 70S ribosome containing functional ligands have provided information about the general organization of the ribosome and its functional centers. Yusupova et al. (2006) compared the X-ray structures of eight ribosome complexes modeling the translation initiation, post-initiation and elongation states. In the initiation and post-initiation complexes, the presence of the Shine-Dalgarno (SD) duplex causes strong anchoring of the 5′-end of mRNA onto the platform of the 30S subunit, with numerous interactions between mRNA and the ribosome. Conversely, the 5′-end of the 'elongator' mRNA lacking SD interaction is flexible, suggesting a different exit path for mRNA during elongation. After the initiation of translation, but while the SD interaction is still present, mRNA moves in the 3′→5′ direction with simultaneous clockwise rotation and lengthening of the SD duplex, bringing it into contact with ribosomal protein S2. mRNA motion on the ribosome is shown in Figure 12.16 (Yusupova et al. 2006).

Translation initiation, the rate limiting step of the universal process of protein synthesis, proceeds through sequential tightly regulated steps. In bacteria, the correct messenger RNA start site and the reading frame are selected when, with the help of initiation factors IF1, IF2 and IF3, the initiation codon is decoded in the peptidyl site of the 30S ribosomal subunit by the fMet-$tRNA_f^{Met}$ anticodon. This yields a 30S initiation complex (30SIC) that an intermediate in the formation of the 70S initiation complex (70SIC) that occurs on the joining of the 50S ribosomal subunit to the 30SIC and release of

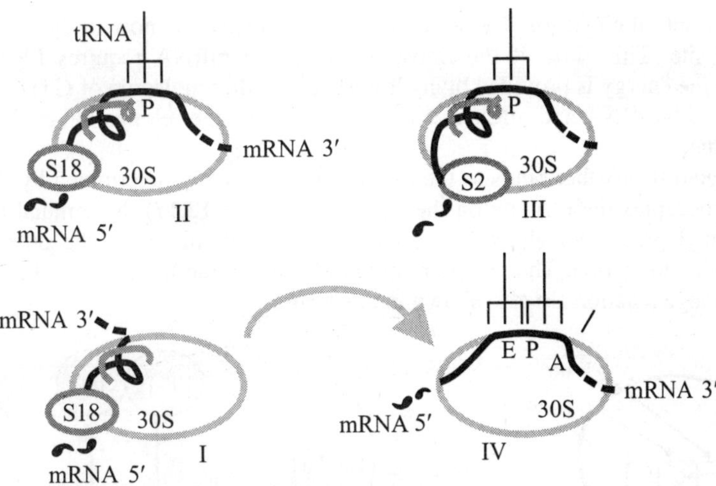

Figure 12.16 Different steps in mRNA motion on the ribosome

the initiation factors. The localization of IF2 in the 30SIC has proved to be difficult so far using biochemical approaches, but could now be addressed using cryo-elecrton microscopy. Yusupova et al. (2006) report direct visualization of a 30SIC containing mRNA, fMet-tRNA$_f^{Met}$ and initiation factors IF1 and GTP-bound IF2. The fMet-tRNA$_f^{Met}$ is held in a characteristic and precise position and conformation by two interactions that contribute to the formation of a stable complex: one involves the decoding stem which is buried in the 30S peptidyl site, and the other occurs between the carboxy-terminal domain of IF2 and the tRNA acceptor end. The structure provides insights into the mechanism of 70SIC assembly and rationalizes the rapid activation of GTP hydrolysis triggered on 30SIC-50S joining by showing that the GTP-binding domain of IF2 would directly face the GTPase activated centre of the 30S subunit.

Translation elongation

Polypeptide chain is lengthened by covalent attachment of successive amino acid units each carried to ribosomes by tRNAs, which base pairs with corresponding codon in mRNA at A-site. Elongation is promoted by elongation factors. Elongation requires (a) the initiation complex, (b) the next aminoacyl-tRNA, specified by the next codon in the mRNA, (c) a set of three soluble cytosolic proteins called elongation factors (EF-Tu, EF-Ts, and EF-G), and (d) GTP. Three steps are recognized in translation elongation. In the first step, the next aminoacyl-tRNA is first bound to a complex of a complex of EF-Tu containing a molecule of bound GTP. The resulting aminoacyl-tRNA-EF-Tu.GTP complex is then bound to the A site of the 70S initiation complex. The GTP is hydrolyzed, an EF-Tu.GDP complex is regenerated. In the second step, a new peptide bond is formed between the amino acids bound by their tRNAs to the A and P sites on the ribosome. This occurs by the transfer of the initiating N-formylmethionyl group from its tRNA to the amino group of the second amino acid now in the A site. The α-amino group of the amino acid in the A site acts as nucleophile, displacing the tRNA in the P site to form the peptide bond. This reaction produces a dipeptidyl-tRNA in the A site and the now "uncharged" (deacylated) tRNA$_f^{Met}$ remains bound to the P site. In the third step called translocation, the ribosome moves by the distance of one codon toward the 3′-end of the mRNA. Because the dipeptidyl-tRNA is still attached to the second codon of the mRNA, the movement of the ribosome shifts the dipeptidyl-tRNA from the A site to the P site, and the deacylated tRNA is released from the initial P site back into the cytosol. The third codon of the mRNA is now in the A site and the second codon in the P site. This shift of the ribosome along the mRNA requires EF-G (also called the translocase) and the energy is provided by hydrolysis of another molecule of GTP. The ribosome, with its attached dipeptidyl-tRNA and mRNA, is now ready for another elongation cycle to attach the third amino acid residue.

The amino acid-tRNA that binds to the (A) site is specifically determined by the triplet codon of the mRNA that occupies the (A) site on the ribosome (Figure 12.17). N-terminal of first amino acid (met) has a formyl group. So elongation can't take place at this end. The other end (-COOH) is available for chain elongation. The factors required for elongation are Tu, Ts, G and GTP. Tu and Ts involved in exchange reaction are also shown in this figure.

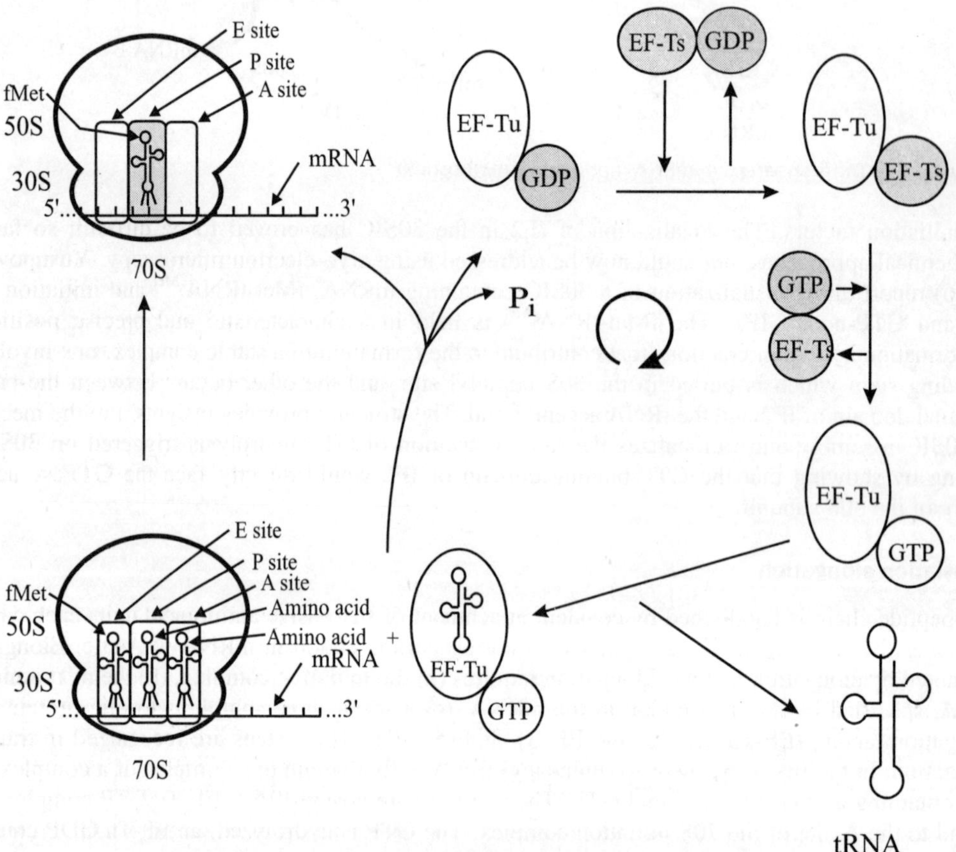

Figure 12.17 Elongation of polypeptide chain. The step used EF-Tu and EF-Ts GTP provides energy for this step

Peptide forming enzyme peptidyl transferase is required one copy for one 50S subunit (Figure 12.18). On completion of peptide bond formation, the (A) site is occupied by F-met-amino acid-tRNA A1 and the (P) site is occupied by $tRNA_f^{Met}$. The translocation step follows (Figure 12.19). The $tRNAf^{Met}$ is discharged from the ribosome. F-met-amino acid 1-tRNAA1 dipeptide is shifted from (A) to (P) site as ribosome moves by length of one codon along the mRNA in a 5'→3' thus bringing a new codon to (A) site. Thus polypeptide assembly begins at the amino-end and ends at carboxyl-end. mRNA is read in a 5'→3' direction.

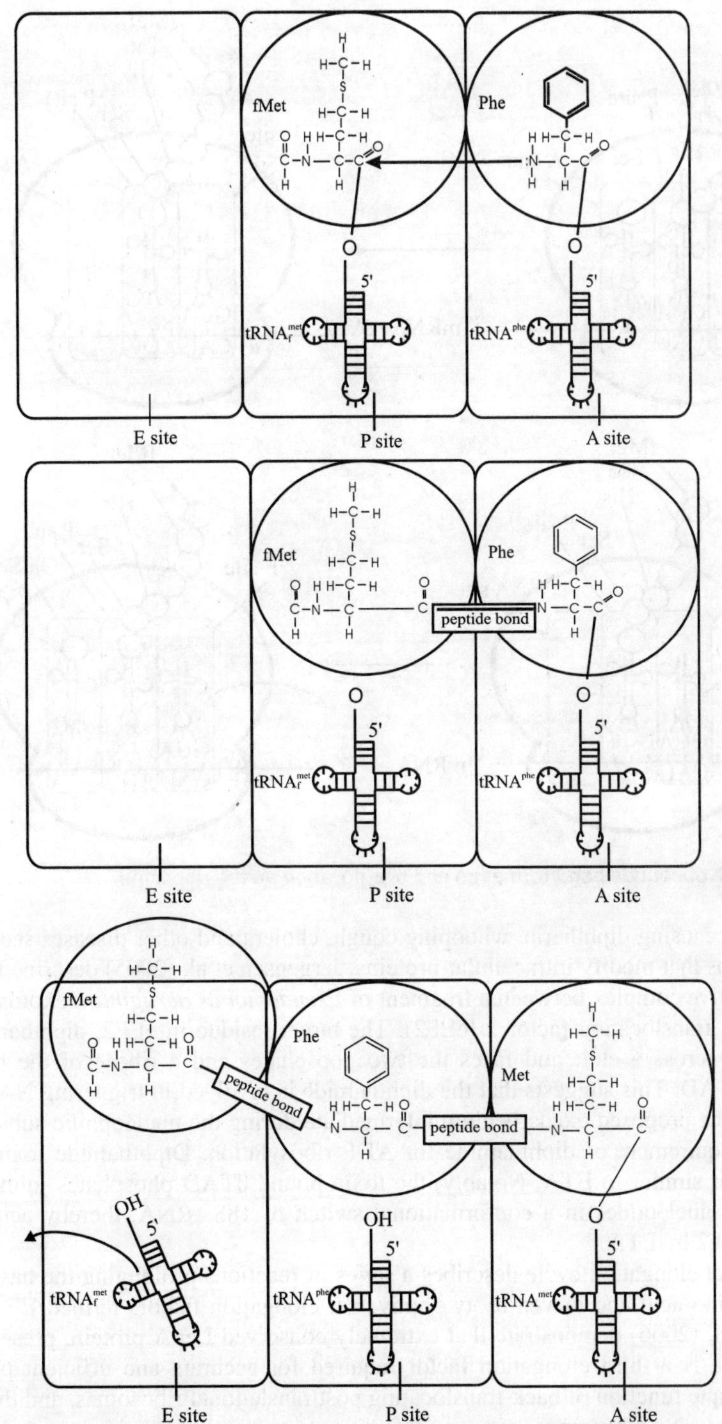

Figure 12.18 Peptide bond formation during protein synthesis requires peptidyl transferase

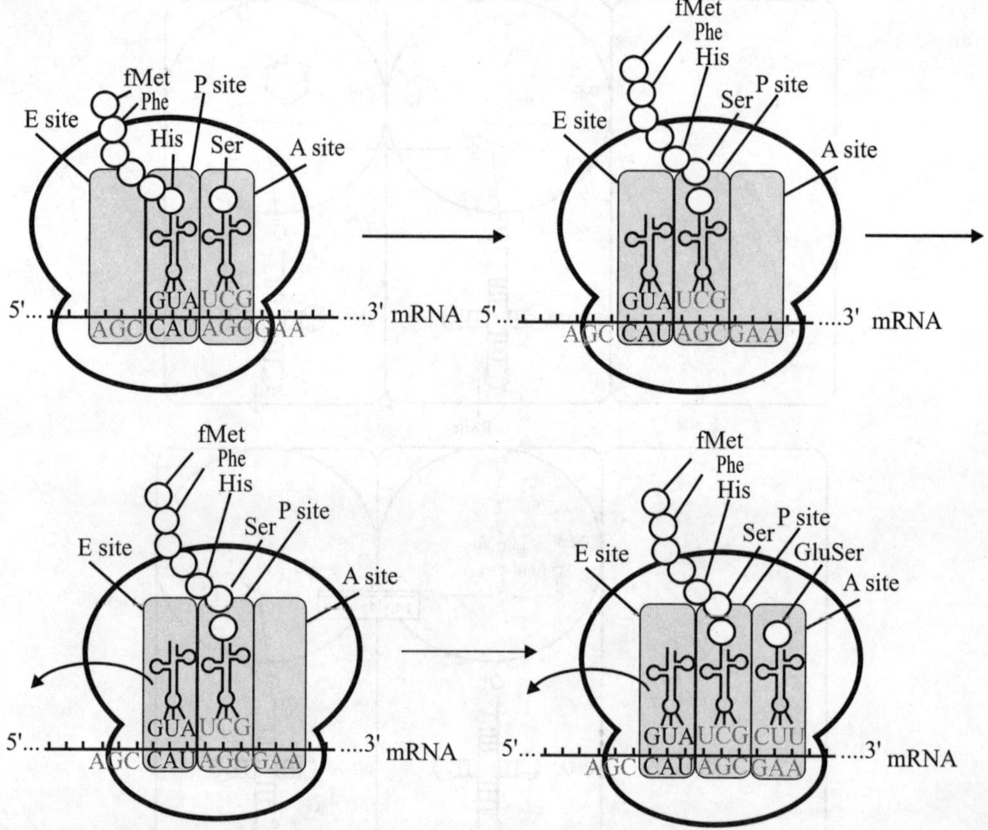

Figure 12.19 Cycle of peptide bond formation and translocation on the ribosome

The bacteria causing diphtheria, whooping cough, cholera and other diseases secrete mono-ADP-ribosylating toxins that modify intracellular proteins. Jergensen et al. (2005) describe four structures of a catalytically active complex between a fragment of *Pseudomonas aeruginosa* exotoxin (ETA) and its protein substrate, translocation factor 2 (eEF2). The target residue in eEF2, diphthamide (a modified histidine), spans across a cleft and faces the two phosphates and a ribose of the non-hydrolysable NAD^+ analog, βTAD. This suggests that the diphthamide is involved in triggering NAD^+ cleavage and interacting with the proposed oxacarbenium intermediate during the nucleophilic substitution reaction, explaining the requirement of diphthamide for ADP ribosylation. Diphthamide toxin may recognize eEF2 in a manner similar to ETA. Notably, the toxin-bound βTAD phosphates mimic the phosphate backbone of two nucleotides in a conformational switch of 18S rRNA, thereby achieving universal recognition of eEF2 by ETA.

The ribosomal elongation cycle describes a series of reactions prolonging the nascent polypeptide chain by one amino acid and driven by two universal elongation factors termed EF-Tu and EF-G in bacteria. Qin et al. (2006) demonstrate that extremely conserved LepA protein, present in all bacteria and mitochondria, is a third elongation factor required for accurate and efficient protein synthesis. LepA has the unique function of back-translocating posttranslational ribosomes, and the results suggest that it recognizes the ribosomes after a defective translocation reaction and induces a back-translocation, thus giving EF-G a second chance to translocate the tRNAs correctly. They suggest renaming LepA as elongation factor 4 (EF4). Model for LepA (EF-4) function is given in Figure 12.20. (A) Under optimal growth conditions the translocation has a very low rate of error, and therefore, EF4 is not so important under such conditions. Translocation involves the movement of tRNAs at the A and P sites (PRE state) to the P and E sites (POST state). This reaction is catalyzed by elongation factor G (EF-G) and GTP. After dissociation of EF-G, the A site is now free for binding of the next ternary complex aa-tRNA•EF-Tu•GTP, which leads to release of the E-tRNA. (B) In rare case if EF-G malfunctions or under high ionic strength, a defective translocation complex may result.

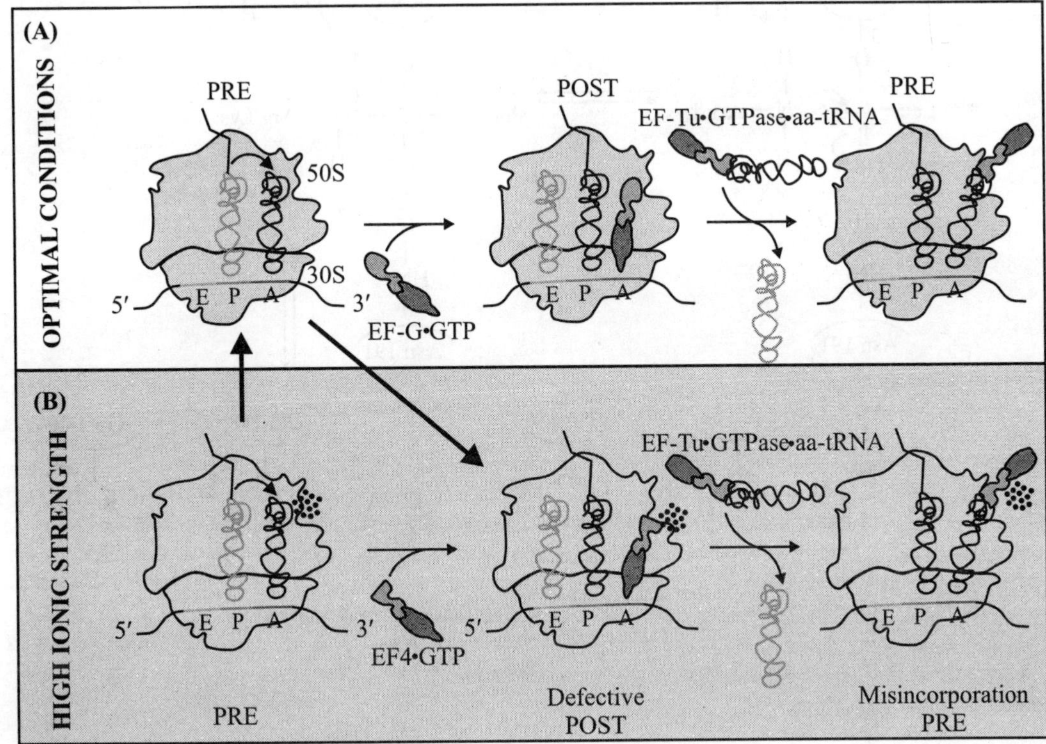

Figure 12.20 Model for LepA (EF-4) function

Peptide Bond Formation and Peptide Release: Peptide bond formation and peptide release are catalyzed in the active site of the large subunit of the ribosome where universally conserved nucleotides surround the CCA ends of the peptidyl- and aminoacyl-tRNA substrates. Youngman et al. (2004) reveal the catalytic center in which an inner shell of conserved nucleotides is pivotal for peptide release, while an outer shell is responsible for promoting peptide bond formation.

Eubacterial leucyl/phenylalanyl-tRNA protein transferase (LF-transferase) catalyzes peptide-bond formation by using Leu-tRNA^{Leu} (Phe-tRNA^{Phe}) and an amino-terminal Arg (or Lys) of a protein, as donor and acceptor substrates, respectively. Watanebe et al. (2007) determine the structure of complexes of LF-transferase and phenylalanyl adenosine, with and without a short peptide bearing an N-terminal Arg. Combining the two separate structures into one structure as well as mutation studies reveal the mechanism for peptide-bond formation by LF-transferase. The electron relay from Asp186 to Gln188 helps Gln188 to attract a proton from the α-amino group of N-terminal Arg of acceptor peptide. This generates the attacking nucleophile for the carbonyl carbon of the aminoacyl bond of the aminoacyl-tRNA, thus facilitating peptide-bond formation. The protein-based mechanism for peptide-bond formation by LF-transferase is similar to the reverse reaction of the acylation step observed in the peptide hydrolysis reaction by serine proteases. A model of the catalytic mechanism for peptide-bond formation by LF-transferase is given in Figure 12.21.

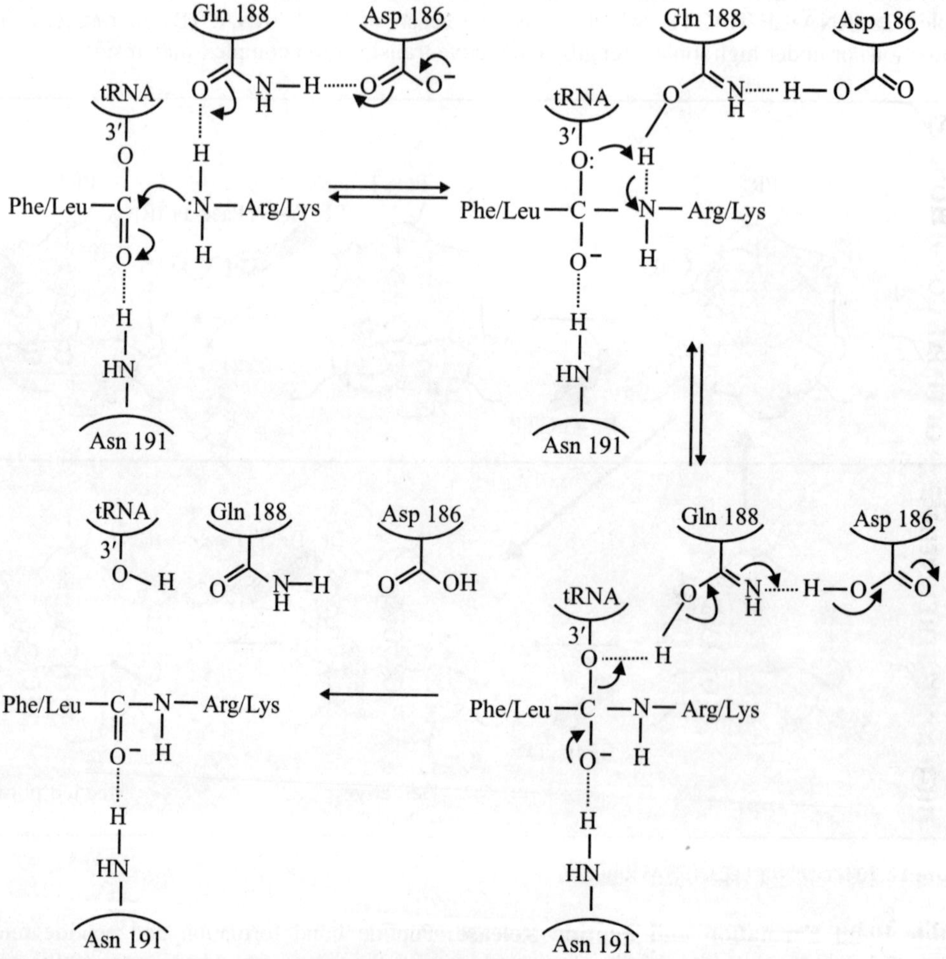

Figure 12.21 A model of the catalytic mechanism for peptide-bond formation by LF-transferase

Rapid protein synthesis in bacteria requires the G proteins IF2, EF-Tu, EF-G, and RF3. These factors catalyze all major steps in mRNA translation in a GTP-dependent manner. Zavialov and Ehrenberg (2003) show how the position of peptidyl-tRNA in the ribosome and presence of its peptide control the binding and GTPase activity of these transcription factors. These results explain how idling GTPase activity and negative interference between different translation factors are avoided and suggest that hybrid sites for tRNA on the ribosome play essential roles in translocation of tRNAs, recycling of class I release factors by RF3, and recycling of ribosomes back to a new round of initiation. They also propose a model for translocation of tRNAs in two separate steps, which clarifies the roles of EF-GTP and GTP hydrolysis in this process. A model explaining regulation of the activities of the G-proteins in bacterial synthesis is depicted in Figure 12.22.

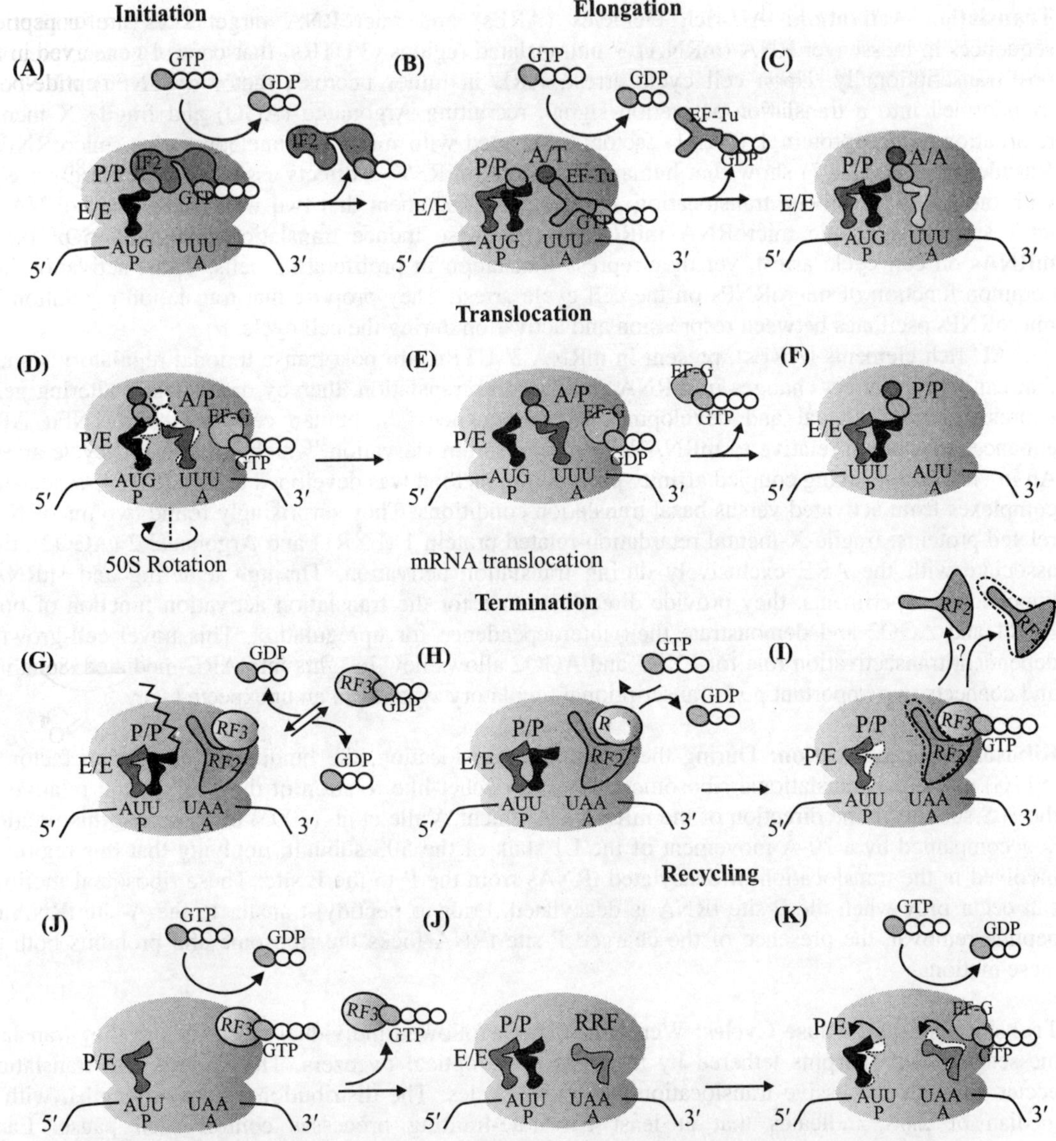

Figure 12.22 A model explaining regulation of the activities of the G-proteins in bacterial synthesis

Movement of mRNA and tRNAs during Translation: Schuwirth et al. (2005) describe two structures of the intact bacterial ribosome from *E. coli*. These structures provide a detailed view of the interface between the small and large ribosomal subunits and the conformation of the peptidyl transferase center in the context of the intact ribosome. Differences between the two ribosomes reveal a high degree of flexibility between the head and the rest of the small subunit. Swiveling of the head of the small subunit observed in the present structures, coupled to the rachet-like motion of the two subunits observed previously, suggests a mechanism for the final movement of messenger RNA (mRNA) and transfer RNAs (tRNAs) during translation.

Translation Activation: AU-rich elements (AREs) and microRNA target sites are conserved sequences in messenger RNA (mRNA) 3' untranslated regions (3'UTRs) that control gene expression post-transcriptionally. Upon cell cycle arrest, ARE in tumor necrosis factor α (TNFα) mRNA is transformed into a translation activation signal, recruiting Argonaute (AGO) and fragile X mental retardation-related protein 1 (FXR1) factors associated with micro-ribonucleoproteins (microRNPs). Vasudevan et al. (2007) show that human microRNA miR369-3 directs association of these proteins with the AREs to activate translocation. Further, they document that two well-studied microRNAs – let-7 and the synthetic microRNA miRcxcr4 – likewise induce translation upregulation of target mRNAs on cell cycle arrest, yet they repress translation in proliferating cells. Thus, activation is a common function of microRNPs on the cell cycle arrest. They propose that translation regulation by microRNPs oscillates between respression and activation during the cell cycle.

AU-rich elements (AREs), present in mRNA 3′-UTRs, are post-transcriptional regulatory signals that can rapidly effect changes in mRNA stability and translation, thereby dramatically altering gene expression with clinical and developmental consequences. In human cell lines, the TNFα ARE enhances translation relative to mRNA levels upon serum starvation, which induces cell cycle arrest. An in vitro cross-linking coupled affinity purification method was developed to isolate ARE-associated complexes from activated versus basal translation conditions. They surprisingly found two micoRNA-related proteins, fragile-X-mental retardation-related protein 1 (FXR1) and Argonaute 2 (AGO2), that associate with the ARE exclusively during translation activation. Through tethering and shRNA-knockdown experiments, they provide direct evidence for the translation activation function of both FXR1 and AGO2 and demonstrate their interdependence for upregulation. This novel cell-growth-dependent transactivation role for FXR1 and AGO2 allows new insights into ARE-mediated signaling and connects two important post-transcriptional regulatory systems in an unexpected way.

Ribosomal Translocation: During the ribosomal translocation, the binding of elongation factor G (EF-G) to the pre-translational ribosome leads to a ratchet-like rotation of the 30S subunit relative to the 50S subunit in the direction of the mRNA movement. Valle et al. (2003) observe that this rotation is accompanied by a 20-Å movement of the L1 stalk of the 50S subunit, implying that this region is involved in the translocation of deacylated tRNAs from the P to the E site. These ribosomal motions can occur only when the P-site tRNA is deacylated. Prior to peptidyl-transfer to the A-site tRNA or peptide removal, the presence of the charged P-site tRNA locks the ribosome and prohibits both of these motions.

Translocation-and-Pause Cycles: Wen et al. (2008) followed individual ribosomes as they translate messenger RNA hairpins tethered by the ends of to optical tweezers. They reveal that translation occurs through successive translocation-and-pause cycles. The distribution of pause lengths, with a median of 2.8S, indicates that at least two rate-limiting processes control each pause. Each translocation step measures three bases – one codon – and occurs in less than 0.1 s. Analysis of the times required for translocation reveals surprisingly that there are three substeps in each step. Pause lengths and thus the overall rate of translation depend on the secondary structure of the mRNA; the applied force destabilizes secondary structure and decreases pause duration, but does not affect translocation time. Translocation and RNA unwinding are strictly coupled ribosomal functions.

Proofreading in Translation: It is not clear what general level of accuracy is required in translating the genetic code. But the protective role of proofreading is evident from a case in which a small mistake has a catastrophic effect (Roy and Ibba 2006). Accuracy in gene expression is shown in Figure 12.23 (Roy and Ibba 2006). Transfer of information from the DNA sequence of a gene to the corresponding amino-acid sequence of a protein requires transcription of the sequence into messenger

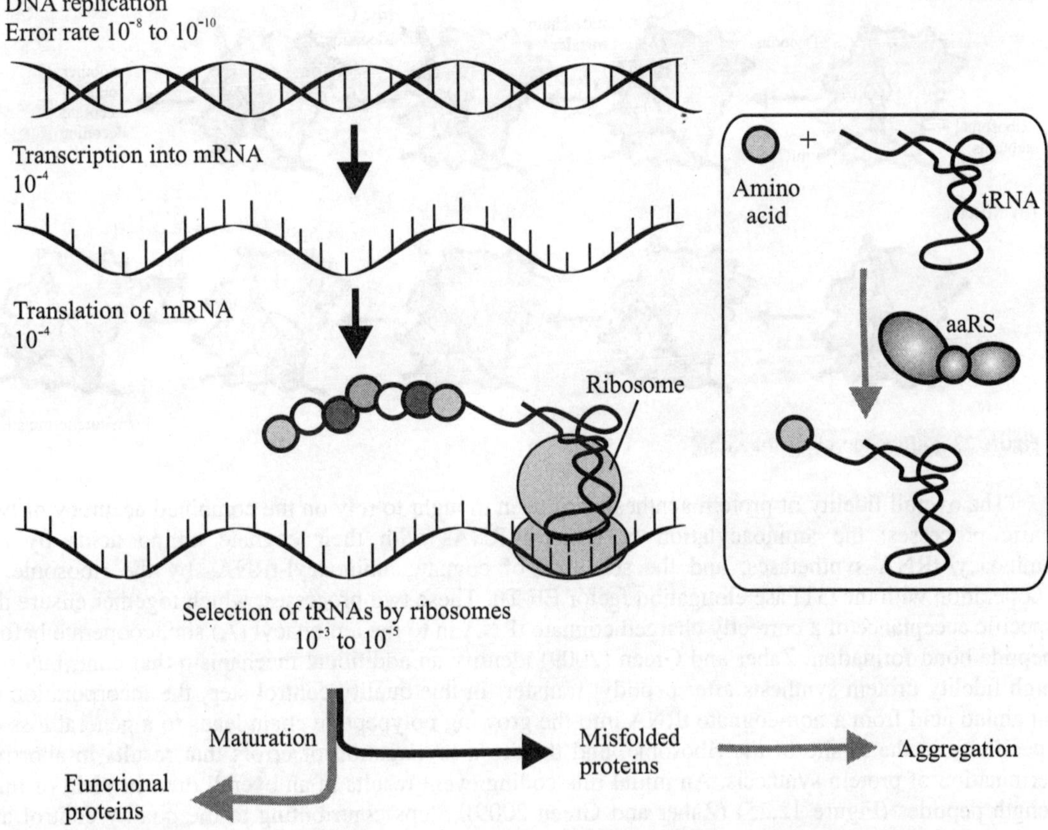

Figure 12.23 Accuracy in gene expression

RNA, then translation of that RNA into an amino-acid sequence at ribosomes, through the agency of amino-acid-specific transfer RNAs. These steps are all prone to error. The inset shows the aminoacylation of tRNA, in which an amino acid is paired with a tRNA by the enzyme aminoacyl-tRNA synthetase (aaRS). It is a failure of quality control at this stage that causes neurodegeneration in 'sticky mice'. Once synthesized, a protein usually becomes functional. In some cases, however, mis-folding occurs. Most mis-folded proteins are degraded by the cell — but some form insoluble aggregates that, as in the sticky mouse, can lead to disease.

Quality-Control Mechanism during Translation: During protein synthesis, mistakes in adding amino acids to growing chain are usually prevented. If they are not, a quality-control mechanism ensures premature termination of erroneous sequences (Fredrick and Ibba 2009) (Figure 12.24) (Fredrick and Ibba 2009). (A) Normally, the correct tRNA enters the A site of the ribosome and the appropriate amino acid is incorporated into the growing peptide chain, which transfers from tRNA in the P site to the tRNA at the A site. Both tRNAs as well as the mRNA, then shift towards the E site. (B) When mistakes are made and the mis-matched codon-anticodon helix (indicated by a cross) translocates to the P site, the ribosome complex becomes susceptible to premature termination by translation factors such as RF2, and the erroneous sequence is prematurely released.

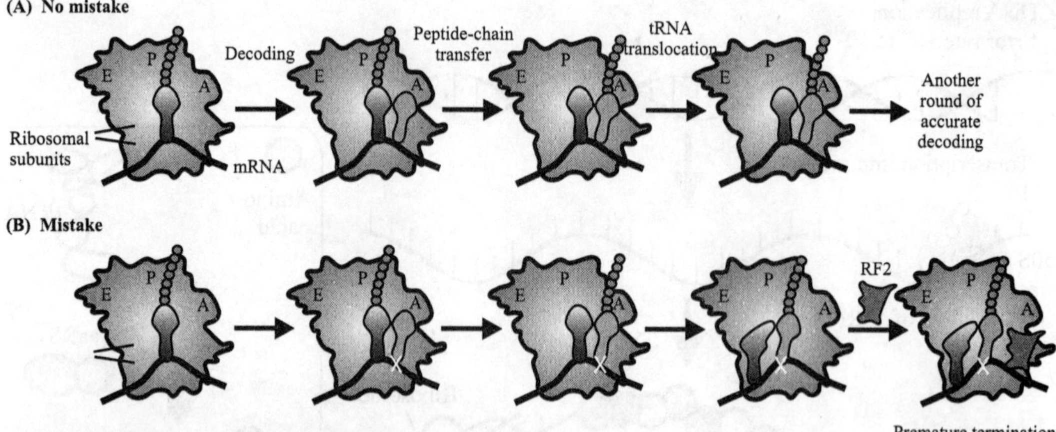

Figure 12.24 Ribosomal mismatching

The overall fidelity of protein synthesis has been thought to rely on the combined accuracy of two basic processes: the aminoacylation of transfer RNAs with their cognate amino acids by the aminoacyl-tRNA synthetases, and the selection of cognate aminoacyl-tRNAs by the ribosome in cooperation with the GTPase elongation factor EF-Tu. These two processes, which together ensure the specific acceptance of a correctly charged cognate tRNA in to the aminoacyl (A) site, cooperate before peptide bond formation. Zaher and Green (2009) identify an additional mechanism that contributes to high fidelity protein synthesis after peptidyl transfer. In this quality-control step, the incorporation of an amino acid from a non-cognate tRNA into the growing polypeptide chain leads to a general loss of specificity in the A site of the ribosome, and thus to a propagation of errors that results in abortive termination of protein synthesis. An initial mis-coding event results in an overall drop in yield of full-length peptides (Figure 12.25) (Zaher and Green 2009). Steps contributing to the quality control are indicated by broken arrows.

Editing Reactions during Translation: Synthesis of proteins containing errors (mis-translation) is prevented by aminoacyl tRNA synthetases through their accurate aminoacylation of cognate tRNAs and their ability to correct occasional errors of aminoacylation by editing reactions. A principle source of mis-translation comes from mistaking glycine or serine for alanine, which can lead to serious cell and animal pathologies, including neurodegeneration. A single-specific G.U base pair (G3.U70) marks a tRNA for aminoacylation by alanyl-tRNA synthetase. Mis-translation occurs when glycine or serine is joined to the G3.H70-containing tRNAs, and is prevented by the editing activity that clears the mis-charged amino acid. Previously it was assumed that specificity for recognition of $tRNA^{Ala}$ for editing was provided by the same structural determinants as used for aminoacylation. Beebe et al. (2008) show that the editing site of alanyl-tRNA synthetase, as an artificial recombinant fragment, targets mis-charged $tRNA^{Ala}$ using a structural motif unrelated to that aminoacylation so that, remarkably, two motifs, one for aminoacylation and one for editing) in the same enzyme independently can provide determinants for $tRNA^{Ala}$ recognition. The structural motif for editing is also found naturally in genome-encoded protein fragments that are widely distributed in evolution. These also recognize mis-charged $tRNA^{Ala}$. Thus, through evolution, three different complexes with the same tRNA can guard against mistaking glycine or serine for alanine. Multiple checkpoints of $tRNA^{Ala}$ recognition for prevention of mis-translation is presented in Figure 12.26 (Beebe et al. 2008). The first checkpoint occurs in the N-terminal domain in recognition of $tRNA^{Ala}$ and discrimination of alanine versus serine

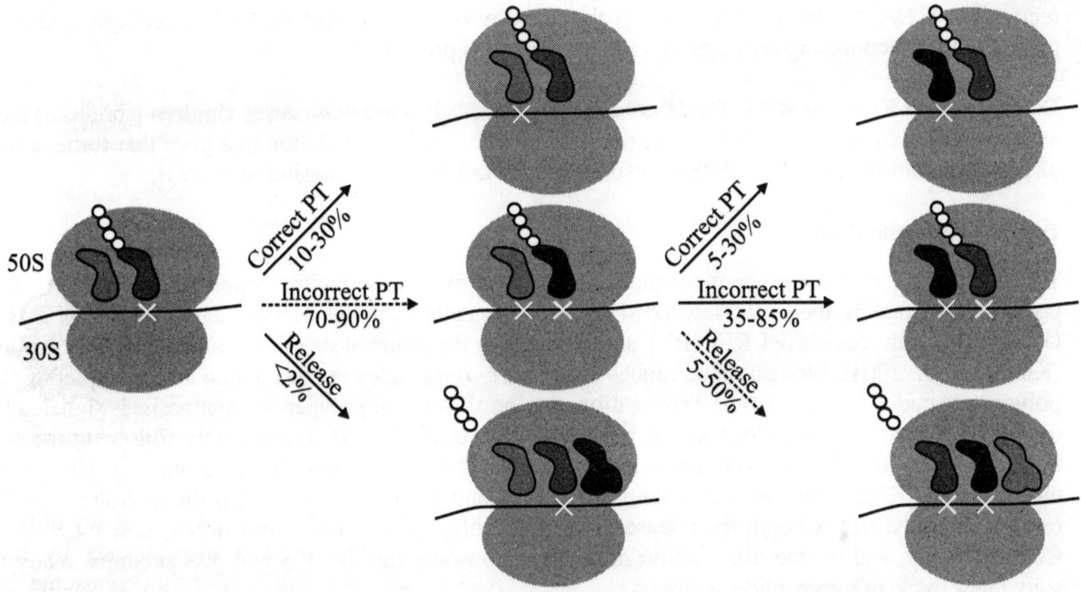

Figure 12.25 An initial miscoding event results in an overall drop in yield of full-length peptides

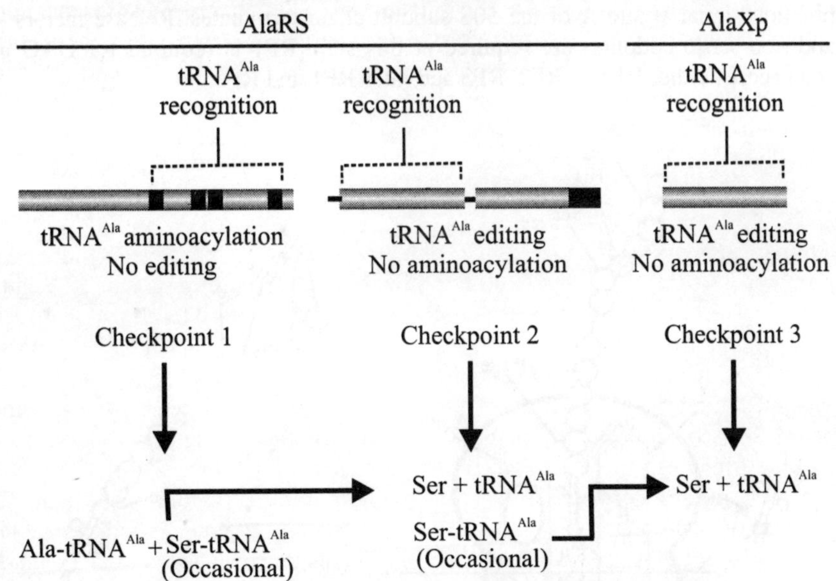

Figure 12.26 Multiple checkpoints of $tRNA^{Ala}$ recognition for prevention of mistranslation

(or glycine)14. The result is the occasional production of Ser–$tRNA^{Ala}$. This minor product is then recognized in a $tRNA^{Ala}$-dependent manner and cleared by a second checkpoint (the editing domain). Finally, any residual Ser–$tRNA^{Ala}$ that remains can be cleared by a third checkpoint (AlaXp) that also recognizes $tRNA^{Ala}$. The net effect is that the quantity of mis-acylated $tRNA^{Ala}$ is reduced, and mis-translation from confusing serine or glycine for alanine is prevented.

Disease due to Error in RNA Translation: A mysterious disease that causes children's brains to melt away is caused by errors in RNA translation (Ainsworth 2005). A mutation in a gene that forms a part of translation initiation factor eIF2B is responsible for failure in the translation process.

Translation termination

The completion of the polypeptide chain is signaled by a termination codon in the mRNA. The polypeptide chain is then released aided by proteins called release factors. The class II RF3 is a GTPase that removes class I RFs (RF1 and RF2) from ribosome after release of nascent polypeptide (Khaholz et al. 2004). Elongation continues until the ribosome adds the last amino acid, completing the polypeptide coded by the mRNA. Termination, the fourth stage of polypeptide synthesis, is signaled by one of three termination codons in the mRNA (UAA, UAG, UGA), immediately following the last amino acid codon. In bacteria, once a termination codon occupies the ribosomal A site, three termination or release factors, the proteins RF1, RF2 and RF3 contribute to (a) the hydrolysis of the terminal peptidyl-tRNA bond, (b) release of the free polypeptide and the last tRNA, now uncharged, from the P site, and (c) the dissociation of the 70S ribosome into its 30S and 50S subunits, ready to start a new cycle of polypeptide synthesis.

As polypeptide chain grows, it remains linked covalently to tRNA and bound to the ribosome. When complete, a polypeptide is released from both the components. Figure 12.27 shows the event when a termination signal at site A of the 50S subunit chain terminates. Release factors - RF1, RF2, RF3, GTP and non-sense codon(s) are required at this step. RF1 is required for UAG and RF2 for UGA. UAA can accept either RF1 or RF2. RF3 activates RF1 and RF2.

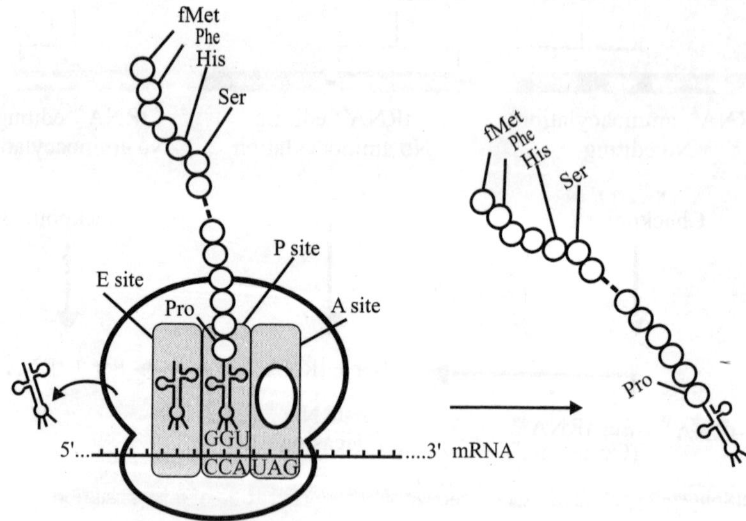

Figure 12.27 Chain termination at the ribosome

At termination of protein synthesis, type I release factors promote hydrolysis of the peptidyl-transfer RNA linkage in response to recognition of a stop codon. Lauberg et al. (2008) describe the crystal structure of the *Thermus thermophilus* 70S ribosome in complex with the release factor RF1, tRNA and a messenger RNA containing a UAA stop codon, at 3.2Å resolution. The stop codon is recognized in a pocket formed by conserved elements of RF1, including its P×T recognition motif, and 16S ribosomal RNA. The codon and the 30S subunit A site undergo an induced fit that results in stabilization of conformation of RF1 that promotes its interaction with the peptidyl transferase centre. Unexpectedly, the main chain amide group of Gln 230 in the universally conserved GGQ motif of the factor is positioned to contribute directly to the peptidyl-tRNA hydrolysis.

Petry et al. (2005) present the crystal structures of the ribosome from *Thermus thermophilus* with RF1 and RF2 bound to their cognate stop codons. The structures reveal details of interactions of the factors with the ribosome and mRNA, including elements previously implicated in decoding and peptide release. They also shed light on conformational changes both in the factors and in the ribosome during termination. Differences seen in the interaction of RF1 and RF2 with the L11 region of the ribosome allow us to rationalize previous biochemical data. This work demonstrates the feasibility of crystallizing ribosomes with bound factors at a defined state along the translational pathway.

During translation termination, class II release factor RF3 binds to ribosome to promote rapid dissociation of a class I release factor (RF) in a GTP-dependent manner. Gao et al. (2007) present the crystal structure of *E. coli* RF3.GDP, which has three-domain architecture strikingly similar to the structure of EF-Tu.GTP. Studies on RF3 mutants show that a surface region involving domain II and III is important for distinct steps in the action cycle of RF3. RF3.GTP binding induces large conformational changes in the ribosome, which break the interactions of the class I RF with both the decoding center and the GTPase-associated center of the ribosome, apparently leading to the release of the class I RF.

As GTP hydrolysis triggers release of RF3, Klaholz et al. (2004) trapped RF3 on *E. coli* ribosomes using a non-hydrolysable GTP analog. This complex can adopt two conformational states. In 'state 1', RF3 is pre-bound to the ribosome, whereas in 'state 2', RF3 contacts the ribosome GTPase center. The transfer RNA molecule translocates from the peptidyl site in state 1 to the exit site in state 2. This translocation is associated with a large conformational rearrangement of the ribosome. Because state 1 seems able to accommodate simultaneously both RF3 and RF2, whose position is known from previous studies, we can infer the mechanism of class I RFs.

The action of release factors on ribosomes has now been clarified by crystallography (Liljas 2008). The process of release from the ribosome is shown in Figure 12.28. The *Thermus thermophilus* 70S ribosome is shown with mRNA and two tRNA molecules bound to the E site and P site of the ribosome, respectively. Due to the presence of a stop codon in the decoding center, a release factor of class 1 (RF1 or RF2) binds to the A site. The release factor has an open conformation. Interaction between the stop codon in the mRNA and the release factor tripeptide at the decoding center of the ribosome is now clarified. At the opposite end of the release factor, the universally conserved Gly-Gly-Gln motif participates in hydrolyzing the peptide from the peptidyl-tRNA in the P site. This process occurs in the peptidyl transfer center of the ribosomal large subunit.

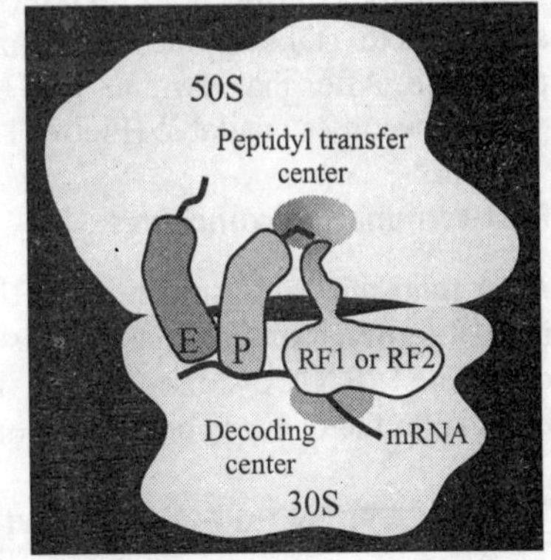

Figure 12.28 Process of release from the ribosome

The termination of protein synthesis occurs through the specific recognition of a stop codon in the A site of the ribosome by a release factor (RF), which then catalyzes the hydrolysis of the nascent protein chain from the P-site transfer RNA. Weixibaumer et al. (2008) present the crystal structure of RF2 in complex with its cognate UGA stop codon in the 70S ribosome. The structure provides insight into how RF2 specifically recognizes the stop codon; it also suggests a model for the role of a universally conserved GGQ motif in the catalysis of peptide release.

General shutdown of protein synthesis

In response to binding viral double-stranded RNA by-products within a cell, the RNA-dependent protein kinase, PKR, phosphorylates α subunit of the translation initiation factor eIF2 on a regulatory site, Ser51. This triggers the general shutdown of protein synthesis and inhibition of viral propagation. To understand the basis for substrate recognition by and the regulation of PKR, Dar et al. (2005) determine X-ray crystal structures of the catalytic domain of PKR in complex with eIF2α. The structures reveal that eIF2α binds to the C-terminal catalytic lobe while catalytic-domain dimerization is mediated by the N-terminal lobe. In addition to binding a local unfolding of the Ser51 acceptor site in eIFα, its mode of binding to PKR affords the Ser51 site full access to the catalytic cleft of PKR.

Polyribosome formation

If only one ribosome is attached to a single mRNA, protein synthesis will not be efficient. Instead, a single mRNA usually has several ribosomes attached to it, as many as 1 for every 90 nucleotides. This mRNA with its attached ribosomes is called a polyribosome or polysome (Figure 12.29). Ribosomes move in a 5'→3' direction, synthesizing polypeptide in an H_2N– to –COOH direction. In bacteria, RNA transcription and translation can occur simultaneously. At one mRNA, more than one initiation complexes can simultaneously engage themselves in synthesis of polypeptide. Since large number of ribosomes is simultaneously working, therefore, there is increase in efficiency of translation process

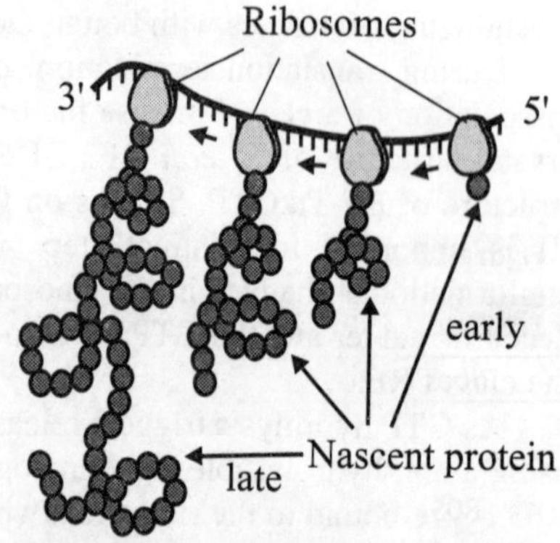

Figure 12.29 Protein synthesis at a polysome. Nascent proteins exist from a tunnel in the 50S subunit. The ribosome at the 5'-end of the messenger has the smallest polypeptide

Post-translational steps

From newly synthesized polypeptide, N-formyl group is removed by enzyme deformylase. Then the removal of first amino acid takes place by methionine-specific aminopeptidase. After that terminal tRNA is removed and complete protein becomes available (Figure 12.30).

Post-termination complexes

After translational termination, mRNA and P site deacylated tRNA remain associated with ribosomes in post-termination complexes (post-TCs), which must therefore be recycled by releasing mRNA and deacylated tRNA by dissociating ribosomes into subunits. Recycling of bacterial post-TCs requires elongation factor EF-G and a ribosome recycling factor RRF.

TRANSLATION IN EUKARYOTES

Although the process of translation is conceptually same in prokaryotes and eukaryotes yet some differences exist (Table 12.2). Ribosomes of eukaryotes and prokaryotes differ in size and other details. The cytoplasmic ribosomes of eukaryotes are of 80S, size, contain 60 per cent rRNA and 40 per cent protein and dissociate into a smaller 40S subunit and a larger 60S subunit. 60S subunit has 5S, 5.8S and 28S rRNA and 40S subunit has 18S rRNA.

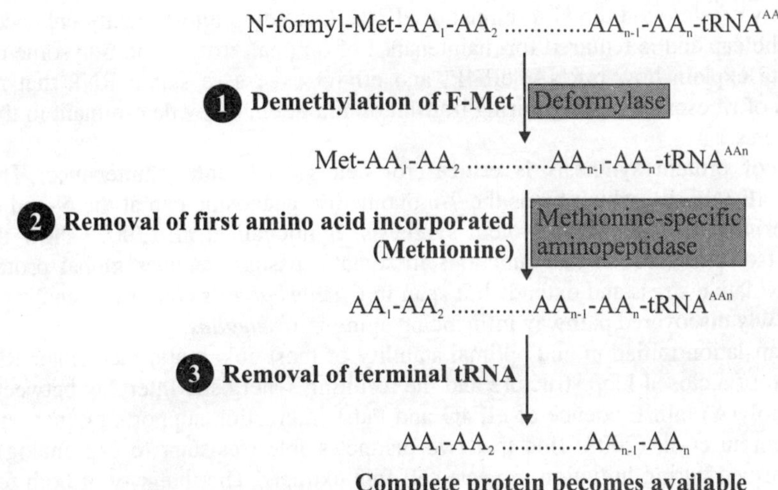

Figure 12.30 Post-translation steps in bacterial protein biosynthesis

Table 12.2 Subunit composition of eukaryotic ribosome

Ribosome	*Subunit*	*rRNAs*	*Proteins*
80S	Small - 40S	18S rRNA (1,900 bases)	33
	Large - 60S	28S rRNA + 5S rRNA + 5.8S rRNA (4,700 bases + 120 bases + 155 bases)	45

Translation in the Cell Cytoplasm

Eukaryotic translation initiation

A major difference in protein synthesis between prokaryotes and eukaryotes is the existence of at least 11 eukaryotic initiation factors (eIF1, eIF2, eIF3, eIF4A, eIF4B, eIF4C, eIF4D, eIF4E, eIF4F, eIF5, eIF6). There are two cap binding proteins (CBPI and CBP2)). CBP1 binds to the 5′ cap (m^7GpppX) of mRNA and facilitates formation of a complex between the mRNA and the 40S ribosomal subunit. CBP2 performs some yet unknown function. The 5′ cap is required for efficient translation (Lazaris-Karatzas et al. 1990). The eIF1 assists mRNA binding, eIF2 binds met-tRNA. It has 3 subunits - α (binds ATP), β (may be recycling factor) and γ (binds met-$tRNA_f^{Met}$). The eIF3 binds mRNA. The initiation factor eIF4A assists mRNA binding and also binds ATP, eIF4B assists mRNA binding and unwinding, eIF4C binds 60S subunit. Function of eIF4D and eIF4E is unknown. Initiation factor eIF4F mediates the function of cap. Initiation factor eIF4E binds the cap. This step is thought to be regulated by phosphorylation. Initiation factor eIF5 releases eIF2 and eIF3 while eIF6 prevents 40S-60S joining. The eukaryotic initiation factor 4G (eIF4G) is the core of a multicomponent switch controlling gene expression at the level of translation initiation. It interacts with the small ribosomal subunit interacting protein, eIF3, and the eIF4E/cap-mRNA complex in order to load the ribosome onto mRNA during cap-dependent translation. Gross et al. (2003) describe crystal structure of the complex between yeast eIF4E/cap and eIF4G (393-490). Binding triggers a coupled folding transition of eIF4G (393-490) and the eIF4E N terminus resulting in a molecular bracelet whereby eIF4G (393-490) forms a right-handed helical ring that wraps around the N terminus of eIF4E. Cofolding allosterically enhances association of eIF4E with the cap and is required for maintenance of optimal growth and polysome distribution in vivo. These data explain how mRNA, eIF4E, and eIF4G exist as a stable RNP that may facilitate multiple rounds of ribosomal loading during translation initiation, a key determinant in the overall rate of protein synthesis.

Regulation of protein synthesis is critical for cell growth and maintenance. The eukaryotic initiation factor 4E (eIF4E), which binds the 7-monomethyl guanosine cap at the 5′-end of all nuclear mRNAs, is a principal regulator of protein synthesis. Syntichali et al. (2007) show that loss of a specific eIF4E isoforms (IFE-2) that functions in somatic tissues, reduces global protein synthesis, protects from oxidative stress and extends life span in *Caenorhabditis elegans*. Signaling via eIF4E in the soma is a newly discovered pathway influencing aging in *C. elegans*.

Efficient translation initiation and optimal stability of most eukaryotic messenger RNAs depends on the formation of a closed-loop structure and the resulting synergistic interplay between the 5' m7G cap and the 3' poly(A) tail. Evidence of eIF4G and Pab1 interaction supports the notion of a closed-loop mRNP. Amrani et al. (2008) find that two distinct stable (resistant to cap analog) closed-loop structures are formed during initiation in yeast cell-free extracts. The integrity of both forms requires the mRNA cap and poly(A) tail, as well as eIF4E, eIF4G, Pab1 and eIF3, and is dependent on the length of both the mRNA and the poly(A) tail. Formation of the first structure requires the 48S ribosomal complex, whereas the second requires an 80S ribosome and the termination factors eRF3/Sup35 and ERF1/Sup45. The involvement of the termination factors is independent of a termination event.

Protein synthesis in mammalian cells requires initiation factor eIF3, a ~750-kDa complex that controls assembly of 40S ribosomal subunits on messenger RNAs (mRNAs) being either a 5'-cap or an internal ribosome entry site (IRES) (Siridechadilok et al. 2005). eIF3 is a five-lobed particle, interacts with the hepatitis C virus (HCV) IRES RNA and the 5'-cap binding complex eIF4F via the same domain. Detailed modeling of eIF3 and eIF4 onto the 40S ribosomal subunit reveals that eIF3 uses eIF4 or the HCV IRES in structurally similar ways to position the mRNA strand near the exit site of 40S, promoting initiation complex assembly. Cryo-electron microscopic structures of (A) eIF3 and (B) eIF3-IRES are shown in Figure 12.31.

Translation of messenger RNA into protein is a complex and intricate process involving several steps and many step-specific protein factors. But one factor – eIF5A – seems to have a hand in every step (Merrick 2009). Evolutionary relationships between bacterial and eukaryotic translation factors are presented in Table 12.3.

Impaired Assembly of Ribosomes: Impaired assembly of cell's protein-synthesis factories, the ribosomes, can cause cell cycle arrest and disease (Ferreira-Cerca and Hurt 2009). This finding emphasizes the close link between cell proliferation and ribosome function. Ribosomal stress and p53 stabilization is shown in Figure 12.32 (Ferreira-Cerca and Hurt 2009). (A) In normal cells, after initial biogenesis of the 60S and 40S ribosomal subunits in the nucleolus, they are exported out of the nucleus, where the mature subunits come together to mediate mRNA translation. Under these conditions, MDM2 is free to mediate p53 degradation in the nucleus. (B) In mutant cells with impaired ribosome assembly, the L11 component of 60S can cause cell cycle arrest by binding to MDM2 and preventing it from degrading p53. Free L11 can be generated in two ways: through defective 60S biogenesis (b); or, (c) by increased translation of its mRNA, when 40S biogenesis is defective. The mRNA is scanned to locate the first AUG codon, which signals the beginning of the reading frame. Ineukaryotes, only AUG is recognized as initiation codon. Also, methionine is not formylated in eukaryotes. $tRNA^{Met}$ is present in eukaryotes compared to $tRNA_f^{Met}$ in prokaryotes. Newly synthesized bacterial proteins start with formyl-methionine and in eukaryotes proteins start with methionine.

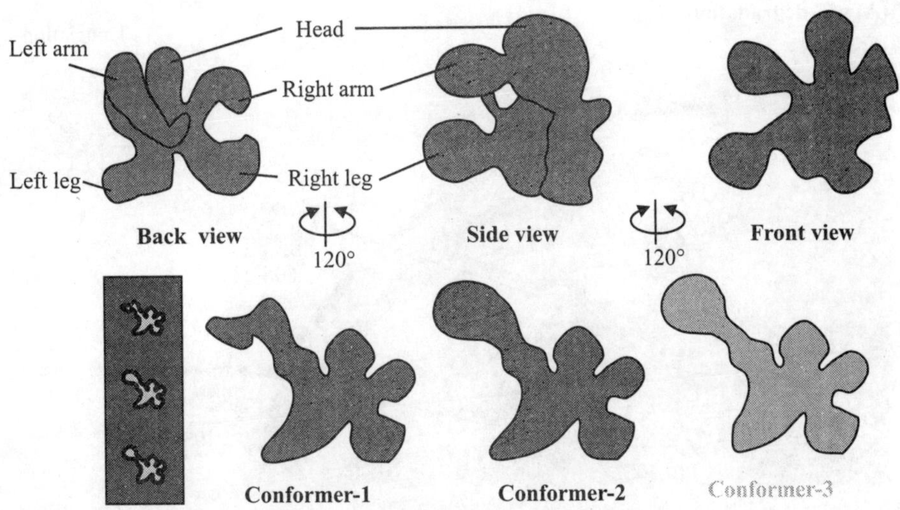

Figure 12.31 Cryo-electron microscopic structures of (A) eIF3 and (B) eIF3-IRES

Table 12.3 Evolutionary relationships between bacterial and eukaryotic translation factors

Bacterial factor	*Function*	*Eukaryotes*	*Function*
Hypothetical protein	Unknown	eIF1	Enhances AUG recognition
IF1	Enhances AUG recognition	eIF1A	Enhances AUG recognition
SelB	Binds GTP and selenocycteine-tRNA	eIF2γ	Binds GTP and met-tRNA
W2	A helicase function not defined	eIF4A	DEAD box helicase, subunit of eIF4F
EF-P	Elongation	eIF5A	Elongation (and more?)
IF2	Binds GTP and initiator tRNA	eIF5B	GTPase; subunit joining
EF1A	GTPase binds aminoacyl-tRNA to A site	eEF1A	GTPase binds aminoacyl-tRNA to A site
EF2	GTPase ribosome translocation	eEF2	GTPase, ribosome translocation
(EF1A)	GTPase binds aminoacyl-tRNA to A site	eRF3	GTPase termination

Regulation of Cap-Dependent Translation: Eukaryotic messenger RNAs contain a modified guanosine, termed as a cap, at their 5' ends. Translation of mRNAs requires the binding of an initiation factor, eIF4E, to the cap structure. Richster and Sonnenberg (2005) describe a family of proteins that through a shared sequence regulate cap-dependent translation. The biological importance of this translational regulation is immense, and affects such processes as cell growth, development oncogenic transformation and perhaps even axon path finding and memory consolidation. Translational control of eIF4E is shown in Figure 12.33. Translation initiation occurs when the 40S ribosomal subunit is recruited to the 5' ends of capped (that is, 7mG-containing) mRNAs through an eIF3–eIF4G–eIF4E interaction. Initiation is disrupted by 4E-BP, which binds and sequesters eIF4E by interacting with eIF4G; this process occurs on a number of mRNAs because 4E-BP is not tethered to any particular sequence. Two examples of tethered 4E-BPs are represented by Maskin (*Xenopus*) and Cup (*Drosophila*). Through its association with CPEB, Maskin interacts with the eIF4E only on RNAs that contain a cytoplasmic polyadenylation element (CPE); disruption of the eIF4E–eIF4G complex by this protein is therefore mRNA-specific. In a similar manner, Cup, through its association with Bruno,

(A) p53 degradation

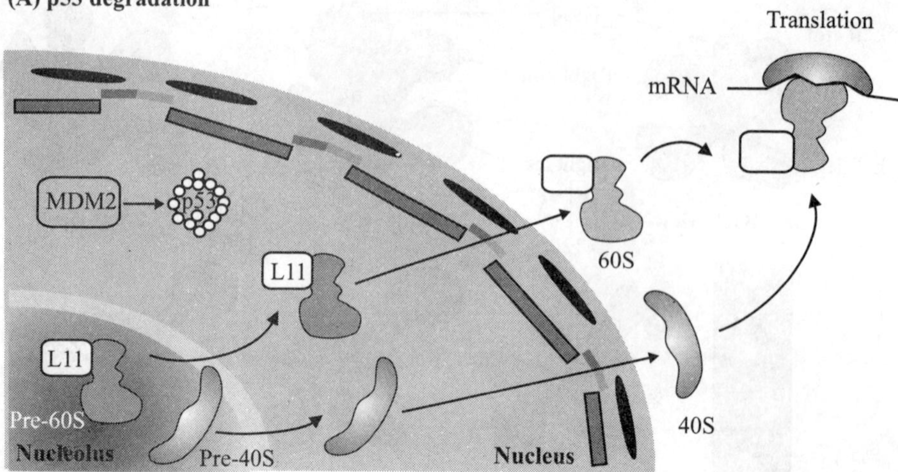

(B) b,c p53 stabilization

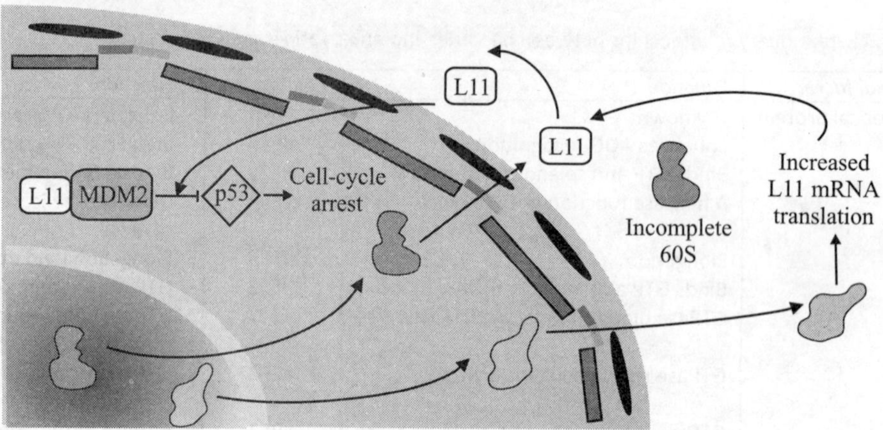

Figure 12.32 Ribosomal stress and p53 stabilization

binds and displaces the eIF4G from eIF4E only on mRNAs that contain a Bruno response element (BRE). Note, however, that Cup also binds Smaug, a protein that binds nanos mRNA; thus Cup associates with the eIF4E on this mRNA as well.

Regulation of eIF4E inhibitory proteins is shown, as suggested by Richster and Sonnenberg (2005), in Figure 12.34. The kinase FRAP/mTOR hyperphosphorylates 4E-BP on several sites; this causes the liberation of eIF4E from 4E-BP, and the association of eIF4E with both capped mRNA and eIF4G. The inhibition of FRAP/mTOR by rapamycin leads to the hypophosphorylation of 4E-BP and enhanced binding to eIF4E. Maskin binding to eIF4E excludes the eIF4G–eIF4E interaction on CPE-containing mRNAs. The inhibition of translation by Maskin is abrogated by cytoplasmic polyadenylation, which is induced by Aurora A-catalyzed CPEB phosphorylation. The newly elongated poly(A) tail is bound by poly(A) binding protein, whose association with eIF4G helps disrupt the Maskin–eIF4E complex and facilitate initiation. The compartmentalization of Maskin- (or Cup-) bound eIF4E probably makes it resistant to further regulation by 4E-BP.

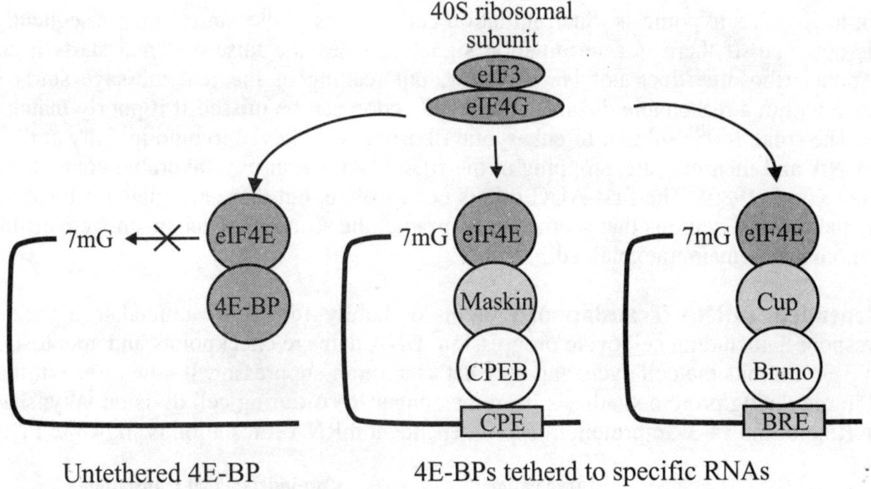

Figure 12.33 Translational control by eIF4E inhibitory proteins

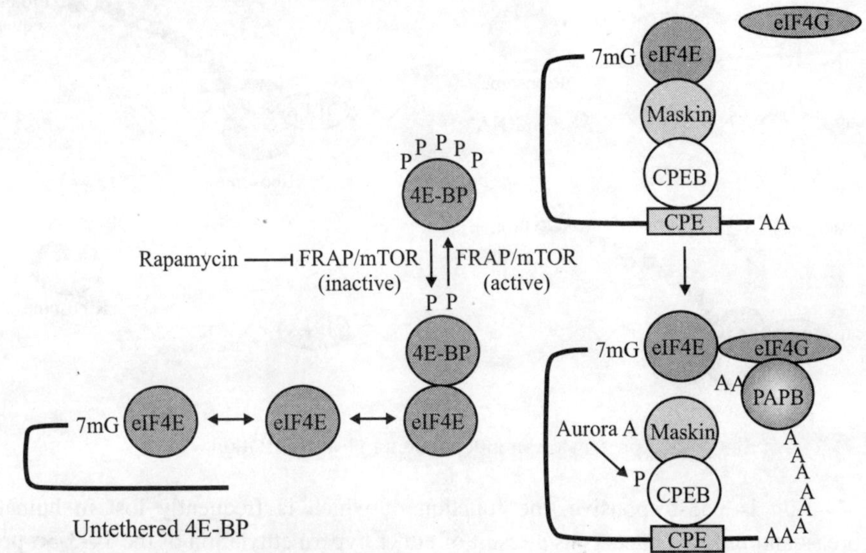

Figure 12.34 Regulation of eIF4E inhibitory proteins

Nuclear Import of tRNAs in Yeast: Previous evidence suggested that transfer RNAs (tRNAs) cross the nuclear envelope to the cytosol only once after maturing in the nucleus. Takano et al. (2005) present evidence for nuclear import of tRNAs in yeast. Several export mutants accumulate mature tRNAs in the nucleus even in the absence of transcription. Import requires energy but not the Ran cycle. These results indicate that tRNAs shuttle between the nucleus and cytosol.

Kozak's Scanning Hypothesis: The first initiation codon AUG in 90 per cent of the cases occurs in the form of consensus sequence PuNNAUGG. This is known as Kozak's sequence. In 5 per cent cases one or more AUG codons occur upstream to first AUG codon. These extra AUG codons are also read according to Kozak's hypothesis. Such initiation codons make false starts and consequently extended protein is obtained. If there is a termination signal between the false and real starts irrespective of reading frame, ribosome does not leave mRNA but reading of the real message starts if AUG is encountered within a reasonable distance. An AUG codon can be missed if it poorly matches Kozak's sequence. The small (40S) subunit of eukaryotic ribosome is believed to bind initially at the capped 5′-end of mRNA and then migrate, stopping at the first AUG codon in a favorable context for initiating translation (Kozak 1989). The first-AUG rule is not absolute, but there are rules for breaking the rule. Some anomalous observations that seemed to contradict the scanning mechanism were artifacts. A few genuine anomalies remain unexplained.

Cap-Independent mRNA Translation: Proteins of family 14-3-3 are crucial in a wide variety of cellular responses including cell cycle progression, DNA damage checkpoints and apoptosis. A protein called 14-3-3σ inhibits the cell cycle and may act as a tumor suppressor. It now turns out that it is also involved in regulating protein synthesis from messenger RNA during cell division (Wynshaw-Boris et al. 2007). Role of the 14-3-3σ protein in cap-independent mRNA translation is shown in Figure 12.35.

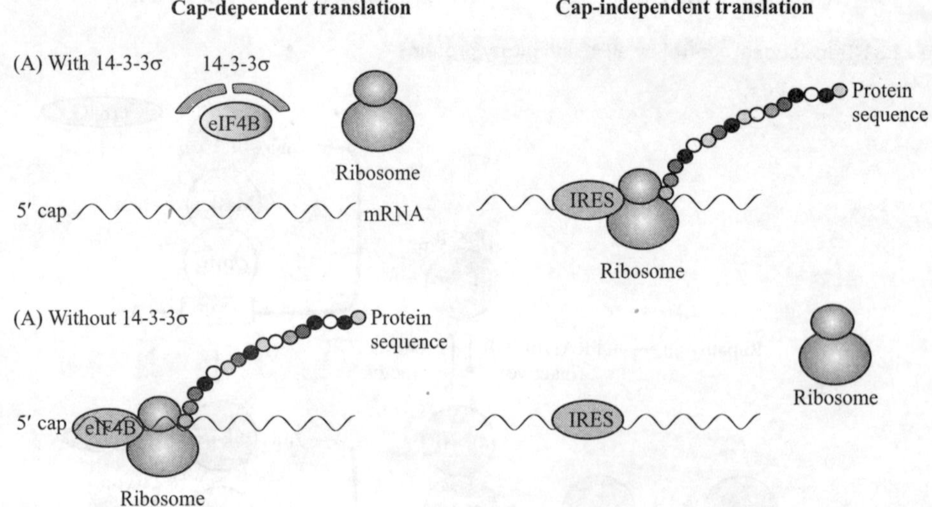

Figure 12.35 Role of the 14-3-3σ protein in cap-independent mRNA translation

Gene *14-3-3σ* is p53-responsive, the function of which is frequently lost in human tumors, including breast and prostate cancers as a result of either hypermethylation of the 14-3-3σ promoter or induction of an estrogen-responsive ubiquitin ligase that specifically targets 14-3-3σ for proteasomal degradation. Loss of 14-3-3σ protein occurs not only within the tumors themselves but also in the surrounding pre-dysplastic tissue (so-called field cancerization), indicating that 14-3-3σ might have an important tumor suppressor function that becomes lost early in the process of tumor evolution. The molecular basis for the tumor suppressor function of 14-3-3σ is unknown. Wilker et al. (2007) report a previously unknown function for 14-3-3σ as a regulator of mitotic translation through its direct mitosis-specific binding to a variety of translation/initiation factors, including eukaryotic initiation factor 4B in a stoichiometric manner. Cells lacking 14-3-3σ, in marked contrast to normal cells, cannot suppress cap-dependent translation and do not stimulate cap-independent translation during and immediately after mitosis. This defective switch in the mechanism of translation results in reduced mitotic-specific expression of the endogenous internal ribosomal entry site (IRES)-dependent form of the cyclin-dependent kinase Cdk11 (p58 PITSLRE), leading to impaired cytokinesis, loss of Polo-like kinase-1 at the mid-body, and the accumulation of binucleate cells. The aberrant mitotic phenotype of 14-3-3σ-depleted cells can be rescued by forced expression of p58 PITSLRE or by extinguishing cap-dependent translation and increasing cap-independent translation during mitosis by using rapamycin. These findings show how aberrant mitotic translation in the absence of 14-3-3σ impairs mitotic exit to generate binucleate cells and provides a potential explanation of how 14-3-3σ-deficient cells may progress on the path to aneuploidy and tumorigenesis.

Coordinated Ribosomal Biogenesis and Translation: Cell growth and proliferation require coordinated ribosomal biogenesis and translation. Eukaryotic initiation factors (eIFs) control translation at the rate-limiting step of initiation. So far, only two eIFs connect extracellular stimuli to global translation rates; eIF4E acts in the eIF4F complex and regulates binding of capped messenger RNA to 40S subunits, downstream of growth factors, and eIF2 controls loading of the ternary complex on the 40S subunit and is inhibited on stress stimuli. No eIFs have been found to link extracellular stimuli to the activity of the large 60S ribosomal subunit. eIF6 binds 60S ribosomes precluding ribosome joining in vitro. In yeast, eIF6 is required for ribosome biogenesis rather than translation. Gandin et al. (2008) show that mammalian eIF6 is required for efficient initiation of translation, in vivo. eIF6 is the first eIF associated with the large 60S subunit that regulates translation in response to extracellular signals.

Eukaryotic translation elongation

Elongation factor eEF3 ia an ATPase that, in addition to the two canonical factors eEF1A and eEF2, serves an essential function in the translation cycle of fungi. eEF3 is required for the binding of the aminoacy-tRNA-eEF1a-GTP ternary complex to the ribosomal A-site and has been suggested to facilitate the clearance of deacyl-tRNA from the E-site. Andersen et al. (2006) present the crystal structure of *Saccharomyces cerevisiae* eEF3, showing that it contains an amino-terminal HEAT repeat domain, followed by a four-helix bundle and two ABC-type ATPase domains, with a chromodomain inserted in ABC2. Moreover, they present the cryo-electron microscopy structure of the ATP-bound form of eEF3 in complex with the post-translation-state 80S ribosome from yeast. eEF3 uses an entirely new factor binding site near the ribosomal E-site, with the chromodomain likely to stabilize the ribosomal L1 stalk in an open conformation, thus allowing tRNA release. Figure 12.36 (Andersen et al. 2006) shows (a) schematic representation of the eEF3 sequence. Different domains indicated are Chromo (chromodomain) and C-term (carboxy-terminal domain).

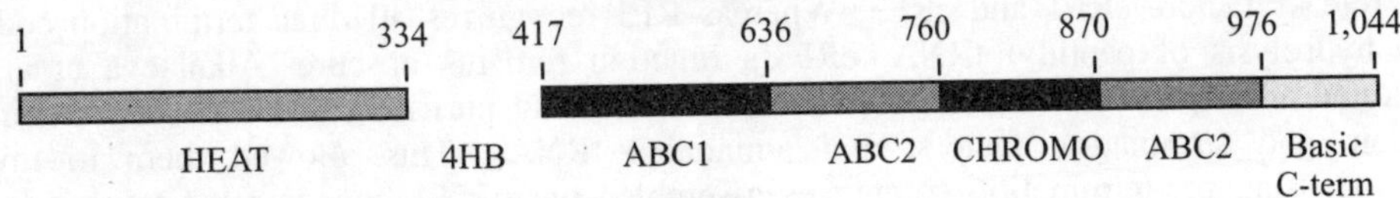

Figure 12.36 Structure of *Saccharomyces cerevisiae* eEF3

Model of the role of eEF3 in the fungi elongation cycle is given in Figure 12.37 (Andersen et al. 2006). (A) The post-state ribosome with a locked E-site tRNA owing to the L1 stalk in the 'in' position and the conformation of the 40S head (Post, locked E). (B) Hypothetical initial interaction of eEF3 in the open tandem or intermediate conformation (Post*, locked E). (C) Ribosome interaction triggers the ATP-dependent closed tandem formation and high-affinity ribosome binding by eEF3, as observed by

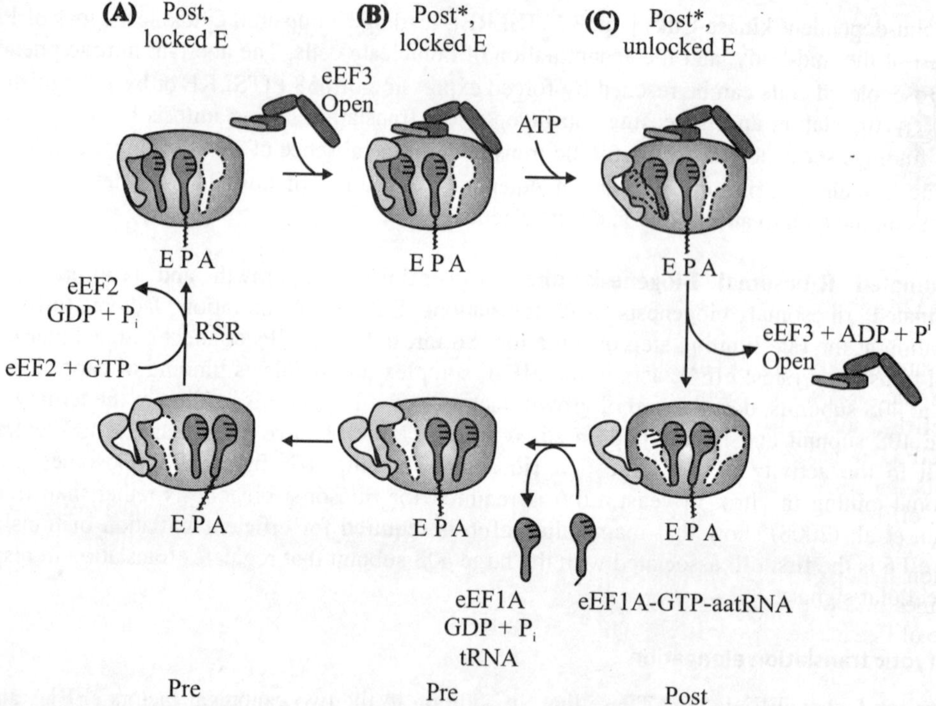

Figure 12.37 Model of the role of eEF3 in the fungi elongation cycle (aatRNA, aminoacyl-tRNA. RSR, ratchet-like subunit rearrangement)

cryo-EM (Post*). A conformational switch of the chromodomain stabilizes the L1 stalk in the 'out' position (unlocked E). ATP hydrolysis of the closed tandem results in the dissociation of eEF3, E-site opening, and unlocking of the 40S head (Post). Now, eEF1A–GTP–aminoacyl-tRNA can bind and the E-site deacyl-tRNA is released. ATP hydrolysis by eEF3, tRNA release, and A-site loading by eEF1A may take place as a joint event.

Eukaryotic translation termination

The released peptidyl-tRNA comes in the cytoplasm. Eukaryotic translation termination is triggered by peptide release factors eRF1 and eRF3. Whereas RF1 recognizes all three termination codons and induces hydrolysis of peptidyl tRNA, eRF3's function remains obscure. Alkalaeva et al. (2006) reconstituted all steps of eukaryotic translation in vitro using purified ribosomal subunits; initiation, elongation, and termination factors; and aminoacyl tRNAs. This allowed them to investigate termination using pre-termination complexes assembled on mRNA encoding a tetrapeptide. In this model, binding of eRF1, eRF3, and GTP to pre-termination complexes first induces a structural rearrangement that is manifested as a 2-nucleotide forward shift of the toeprint attributed to pre-termination complexes that leads to GTP hydrolysis followed by rapid hydrolysis of peptidyl tRNA. Cooperativity of between eRF1 and eRF3 required the eRF3 binding C-terminal domain of eRF1. A model for translation termination in eukaryotes , as proposed by Alkalaeva et al. (2006), is given in Figure 12.38. The process is shown in two steps: (1) Binding of eRF1, eRF3, and GTP to the pre-TC, (2) Pre-TC translocation/structural rearrangement, (3) GTP hydrolysis by eRF3, and (4) Peptidyl-tRNA hydrolysis and peptide release.

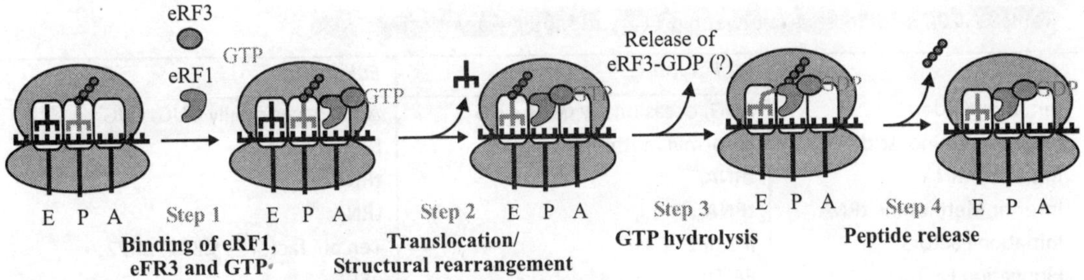

Figure 12.38 A model for translation termination in eukaryotes

Pisarev et al. (2007) investigated eukaryotic recycling using post-TCs assembled on a model mRNA encoding a tetrapeptide followed by a UAA stop codon and report that initiation factors eIF3, eIF1, eIF1A, and eIF3j, a loosely associated subunit of eIF3, can promote recycling of eukaryotic post-TCs. eIF3 is the principal factor that promotes splitting of post-termination ribosomes into 60S subunits and tRNA- and mRNA-bound 40S subunits. Its activity is enhanced by eIFs3j, eIF1 and eIF1A. eIF1 also mediates release of P site tRNA, whereas eIF3j ensures subsequent dissociation of mRNA. A model for eukaryotic ribosomal recycling is presented in Figure 12.39 (Pisarev et al. 2007). Initiation factor eIF3, in cooperation with its eIF3j subunit, eIF1 and eIF1A, split post-termination ribosomes into 60S subunits and tRNA- and mRNA-bound 40S subunits. eIF1 promotes subsequent release of P site deacylated tRNA, which is followed by dissociation of mRNA mediated by eIF3j.

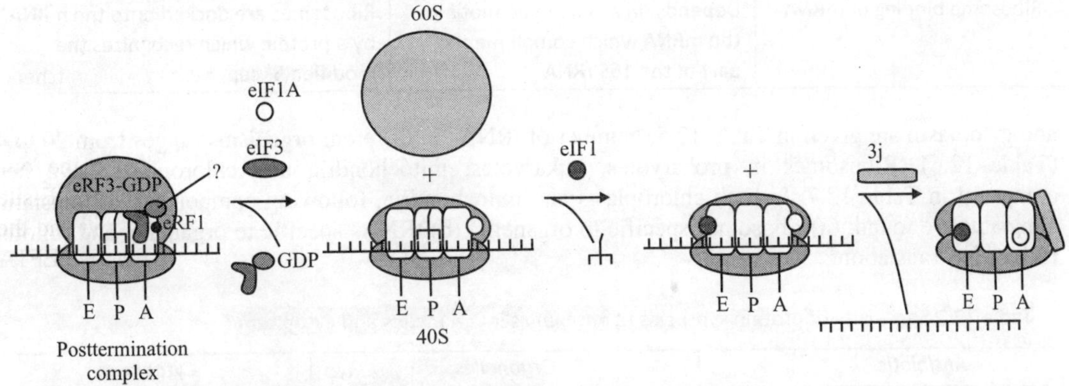

Figure 12.39 A model for eukaryotic ribosomal recycling

Some of the differences in translation machinery of prokaryotes and eukaryotes are listed in Table 12.4.

Protein Synthesis in Chloroplasts and Mitochondria

In mitochondria and chloroplasts, following components of protein synthesis are found: DNA, DNA polymerase, RNA polymerase, ribosomes, tRNAs, initiation factors, elongation factors and termination factors. All these components are specific to cell organelles. Some of these components of protein synthesis may be synthesized outside these organelles and then transported to the organelles. Protein synthesis apparatus in chloroplasts and mitochondria differ from that in cytoplasm in eukaryotes in various respects. Differences with respect to sensitivity of protein synthesis to antibiotics in organelles

Table 12.4 Some differences between prokaryotic and eukaryotic translation

	Prokaryotes	*Eukaryotes*
Initiation Codon	AUG, occasionally GUG, CUG	AUG, occasionally GUG, CUG
Initiation Amino Acid	N-formyl methionine	Methionine
Initiation tRNA	$tRNA_f^{Met}$	$tRNA^{Met}$
Interior Methionine tRNA	$tRNA_m^{Met}$	$tRNA_m^{Met}$
Initiation Factors	IF1, IF2, IF3	Ten eIF factors + CBP1, CBP2
Elongation Factor	EF-Tu	eEF1
Elongation Factor	EF-Ts	eEF1
Translocation Factor	EF-G	eEF2
Release Factors	RF1, RF2	eRF
Location of transcription and translation	Transcription and translation occur simultaneously in the same cellular compartment.	Transcription is restricted to the nucleus and RNA must be exported to the cytoplasm for translation
Half life of mRNA	mRNA has limited half-life (several minutes for the most stable transcripts	Some mRNAs are unstable while most are stable for hours or even days (e.g., in eggs)
Processing	Transcripts are used directly for translation	Transcripts are extensively processed and modified before they can be used for translation
Ribosome binding of mRNA	Depends on a conserved motif in the mRNA which complements part of the 16S rRNA	Ribosomes are docked onto the mRNA by a protein which recognizes the modified 5′ cap

and cytoplasm are given in Table 12.5. Number of tRNAs in different organisms ranges from 20 to 49 (Table 12.6). Ribosomes in prokaryotes, eukaryotes, mitochondria and chloroplasts have been compared in Table 12.7. In both chloroplasts and mitochondria, following components of translation apparatus are found: (a) ribosomes specific to organelle, (b) tRNAs specific to organelle, and (c) other factors for translation.

Table 12.5 Sensitivity of protein synthesis to antibiotics in organelles and cytoplasm

Antibiotic	*Organelles*	*Cytoplasm*
Chloramphenicol	Inhibitor	No effect
Cycloheximide	No effect	Inhibitor

It has been shown that the translation apparatus in chloroplasts and mitochondria differ from that in cytoplasm in eukaryotes in the following respects: (a) Ribosomes in these organelles are smaller in size (70S) than those in the cytoplasm (80S). (b) The tRNAs are specific and differ, the number of tRNAs; in mitochondria being 22 as against 55 in cytoplasm. (c) Initiation of translation takes place by formyl-methionyl tRNA both in chloroplasts and mitochondria, although no formylation takes place in cytoplasm. (d) Translation in chloroplasts and mitochondria can be inhibited by chloramphenicol, as in bacteria since the 70S ribosomes are sensitive to chloramphenicol and not to cycloheximide; on the other hand, the translation in cytoplasm is inhibited by cycloheximide, since 80S ribosomes are sensitive to cycloheximide. There are other antibiotics like spectinomycin, lincomycin and erythromycin

Table 12.6 Number of tRNAs in different organisms

Type of tRNA	*Group*	*Organism*	*No. of tRNAs*
Viral		T5 virus	>20
Archaebacterial		*Halobacterium volcanii*	49
Eubacterial		*Escherichia coli*	45
		Bacillus subtilis	31
Chloroplast		Tobacco, rice	30
		Liverwort *Merchantia polymorphia*	27
		Unicellular green algae *Euglena gracilis*	25
Mitochondrial	Ciliate protozoan	*Tetrahymena pyriformis*	36 (10 encoded by mtDNA)
	Fungi	*Saccharomyces cerevisiae/ Schizosaccharomyces pombe*	24
		Neurospora crassa	25
		Aspergilus nidulans	27 or 28
		Mycoplasma capricolum	29
		S. cerevisiae	42
	Plants	*Petunia hybrida*	17
	Animals		22
Cytoplasmic	Nematodes		20
	Single-celled organisms and fungi	*Dictyostelium discoideum*	20
		Saccharomyces cerevisiae	40

Table 12.7 Comparison of ribosomes in prokaryotes, eukaryotes, mitochondria and chloroplasts

	Prokaryotes	*Eukaryotes*	*Mitochondria*	*Chloroplasts*
Monomer unit	70S	80S	70-80S	70S
Molecular weight (Da)	2.7×10^6	4.5×10^6	30-40S	30S
Small subunit	30S	40S		
Molecular weight (Da)	0.9×106	1.5×106		
Ribosomal RNA	16S	18S	12-13S (animals) 14-18S (Others)	16S
Large subunit	50S	60S	40-45S	50S
Molecular weight (Da)	1.5×106	3.0×106		
Ribosomal RNA	5S, 23S	5S, 5,8S, 28S	5S (Higher plants) 16S-19S (Animals)	4.5S, 5S, 23S

which also inhibit translation in bacteria and in cell organelles of eukaryotes. These antibiotics help to stop the translation preferentially either in cytoplasm (by cycloheximide) thus permitting protein synthesis only in chloroplasts for mitochondria or in organelles (by chloramphenicol) thus permitting protein synthesis only in the cytoplasm. Mitochondrial apparatus for protein synthesis is assembled from RNA synthesized in the mitochondria and proteins imported from the cytoplasm in yeast.

Chlorophyll-binding proteins play a major role in the primary reactions of photosynthesis, translation and its regulation. In many plant species, in the absence of light, the mRNA of chlorophyll-binding proteins, including those encoded by chloroplast genome, remain associated with thylakoid bound polysomes, but are not translated. Transfer of seedlings to light induces synthesis of proteins, although no increase in transcription has been observed. This suggests that post-transcriptional processes play a key role in light- induced chloroplast gene expression. It has also been shown that the polycistronic mRNA encoded by the *psbB* operon in maize, need not be cleaved into tri-, di- or monocistronic forms for successful translation (*psbB* operon = *psbB* + *psbH* + *petB* + *petD*). This suggests that plastid ribosomes can bind directly to internal initiation regions for initiation of translation as in prokaryotes. However, RNA processing does take place in chloroplasts and mitochondria, and the transcript may differ in their translatability.

There are also nuclear genes, whose products are essential for translation of specific genes. Several of these nuclear encoded factors interact with the 5' untranslated region of chloroplast messages, but little is known about the identity and precise function of these factors. These products may be translational activators for prokaryotic and eukaryotic genes, respectively. Protein synthesis in isolated chloroplasts of pea was studied by Ellis (1975, 1981). The proteins which could be synthesized in isolated chloroplasts included (a) large subunit of Fraction 1 protein. (b) five unidentified proteins of the internal lamellar system, and (c) two or three unidentified polypeptides of the envelop. These are only a few of a large number of proteins found in chloroplasts. Proteins synthesized in mitochondria have also been identified by inhibiting protein synthesis in cytoplasm by cycloheximide.

Cooperation between nuclear and chloroplast DNA

A well known example of cooperation between nuclear and chloroplast DNA is synthesis of protein is mentioned here. Small subunit of fraction I protein is coded by nuclear DNA and is synthesized in the cytoplasm and is then transported to the chloroplast. The large subunit of fraction I protein is synthesized in the chloroplast. The small subunit is then transported to the chloroplast. The small and large subunits of fraction I protein join each other in the chloroplast (Ellis 1975) (Figure 12.40).

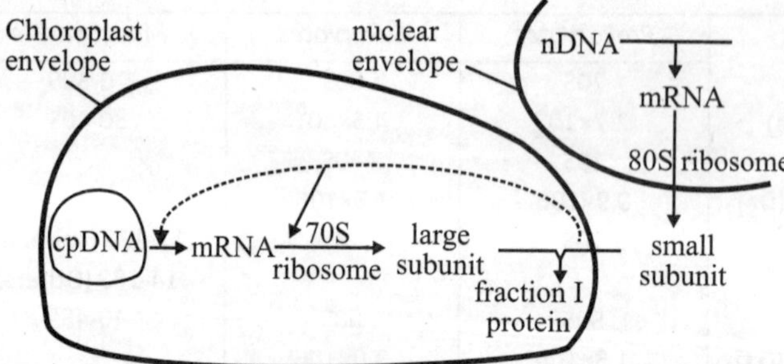

Figure 12.40 A model showing cooperation between nuclear DNA and chloroplast DNA in peas

Cooperation between nuclear and mitochondrial DNA

Although some proteins are synthesized in mitochondria, not all proteins present in mitochondria are synthesized there. There are some mitochondrial proteins which are synthesized in cytoplasm and then transported. There are still others having dual origin, some polypeptides having cytoplasmic origin and others having mitochondrial origin. Some proteins, which are known to be synthesized in mitochondria include the following: (a) Cytochrome oxidase contains seven different kinds of polypeptides, of which three are synthesized on mitochondrial ribosomes, the remaining four being synthesized on cytoplasmic ribosomes. (b) Adenosine triphosphatase (ATPase) is an important component of mitochondrial membrane, which plays an important role in coupling respiration to ATP formation.

There are other polypeptides present in this complex and only two are definitely known to be synthesized on mitochondrial ribosomes. (c) In mitochondrial ribosomes, although ribosomal proteins are known to be synthesized outside mitochondria and under the influence of nuclear DNA, but rRNA is transcribed from mtDNA, since both 21S rRNA and 155 rRNA (in yeast mitochondria, 21S rRNA and 15S rRNA occur instead of 23S and 16S rRNAs); hybridize with specific regions of mtDNA.

ROLE OF MicroRNAs IN PROTEIN SYNTHESIS

MicroRNAs (also known as miRNAs) are 21-23 nucleotide long RNA molecules that play role, in addition to many diverse biological functions of the cell, in protein synthesis also in several different ways

.

MicroRNAs Inhibit Protein Synthesis

To elucidate how microRNAs mediate their repressive effects, Chendrimada et al. (2007) identified new factors in miRNA pathway. They show that human RISC (RNA-induced silencing complex) associates with a mutiprotein complex containing MOV10 – which is the homolog of *Drosophila* translational repressor Armitage – and proteins of the 60S ribosome subunit. Notably, this complex contains the anti-association factor eIF6 (also called TGB4BP or P27BBP), a ribosome inhibitory protein known to prevent productive assembly of the 80S ribosome. Depletion of eIF6 in *C. elegans* diminishes lin-4 miRNA-mediated repression of the endogenous LIN-14 and LIN-28 target protein and mRNA levels. These results uncover an evolutionarily conserved function of the ribosome anti-association factor eIF6 in miRNA-mediated post-transcriptional silencing.

Thermann and Hentze (2007) developed a cell-free system from *Drosophila melanogaster* embryos that faithfully recapitulates miR2-mediated translational control by means of the 3′ untranslated region of the *D. melanogaster* reaper messenger RNA. They showed that miR2 inhibited translation initiation without affecting mRNA stability. Surprisingly, miR2 induces the formation of dense (heavier than 80S) microribonucleoproteins (miRNPs) (pseudopolysomes) even when polyribosome formation and 60S ribosomal subunit joining are blocked. An mRNA bearing an $A_{ppp}G$ instead of an $m^7G_{ppp}G$ cap structure escapes the miR2-mediated translational block. These results directly show the inhibition of $m^7G_{ppp}G$ cap-mediated translation initiation as the mechanism of miR2 function, and uncover pseudo-polysomal messenger ribonucleoprotein assemblies that may help to explain earlier findings.

MicroRNAs Influence Stability of mRNAs

In metazoa, microRNAs act by imperfectly base-pairing with the 3′ untranslated region of target messenger RNAs (mRNAs) and repressing protein accumulation by an unknown mechanism. Pillai et al. (2005) demonstrate that endogenous let-7 miRNPs or the tethering of Argonaute (Ago) proteins to reporter mRNAs in human cells inhibit translation initiation. M^7G-cap-independent translation is not subject to repression, suggesting that microRNPs interfere with recognition of the cap. Repressed mRNAs, Ago proteins and microRNAs were all bound to accumulate in processing bodies. They propose that localization of mRNAs to these structures is a consequence of translational repression.

MicroRNAs can pair to sites in messages. They are known to influence the evolution and stability of many mRNAs, but their global impact on protein output has not been examined. Baek et al. (2008) measure the response of thousands of proteins after introducing microRNAs into cultured cells and after deleting mir-223 in mouse neutrophils. The identities of the responsive proteins indicate that targeting is primarily through seed-matched sites located within favorable predicted contexts in the 3' untranslated region. Hundreds of genes are directly repressed, albeit each to a modest degree, by individual micoRNAs. Although some targets were repressed without detectable changes in mRNA levels, those translationally repressed by more than a third also displayed detectable mRNA destabilization, and, for the more highly repressed targets, mRNA destabilization usually comprised the major component of repression. The impact of microRNAs on the proteasome indicated that for most interactions microRNAs act as rheostats to make the fine-scale adjustments to protein output.

Single MicroRNA can Repress the Production of Hundreds of Proteins

Animal miRNAs regulate gene expression by inhibiting translation and/or by inducing degradation of target messenger RNAs. Selbach et al. (2008) used a new proteomic approach to measure changes in synthesis of several thousand proteins in response to miRNA transfection or endogenous miRNA knockdown. In parallel, they quantified mRNA levels using microarrays. They show that a single miRNA can repress the production of hundreds of proteins, but that this repression is typically relatively mild. A number of known features of the miRNA-binding site such as the seed sequence also govern repression of human protein synthesis, and they report additional target sequence characteristics. The results demonstrate that, in addition to downregulating mRNA levels, miRNAs also directly repress translation of hundreds of genes. Finally, these data suggest that a miRNA can, by direct or indirect effects, tune protein synthesis from thousands of genes.

HYBRID ARRESTED TRANSLATION

Hybrid arrested translation is based on the fact that an mRNA will not direct the synthesis of a protein in a cell-free system when it is in a hybrid from with its complementary DNA. Rabbit mRNA is a mixture of α-globin and β-globin mRNAs. It was mixed with denatured DNA of plasmid pBG1 which contains β-globin cDNA sequence from rabbit. DNA-RNA hybrid was obtained. Hybridized mRNA was not translated. When with brief heating mRNA was recovered, translational activity of β-globin mRNA was observed.

HYBRID RELEASED TRANSLATION

Hybrid Released Translation (HRT) enables a cloned DNA to be correlated with the protein(s) which it encodes. HRT is a direct method in which cloned DNA is bound to a cellulose nitrate filter and hybridized with an unfractionated preparation of mRNA or even total cellular RNA. The filter is washed and hybridized mRNA is eluted by heating in low salt buffer. Recovered mRNA is then translated in a cell-free translation system.

PROTEIN ENGINEERING

One serious limitation facing protein engineers is the availability of only twenty proteinogenic amino acids encoded by natural messenger RNA. Technology has been developed for incorporating non-standard amino acids into polypeptide by ribosome based translation. In this technology, the genetic code is expanded through creation of a 65th codon-anticodon pair from unnatural nucleoside bases having non-standard H-bonding patterns. This new codon anticodon pair efficiently supports translation in vitro to yield polypeptides containing a non-standard amino acid. The versatility of the ribosome as a synthetic tool offers new possibilities for protein engineering and compares with another approach in which the genetic code is simply rearranged to recruit stop codons to play a coding role.

In an experiment, Bain et al. (1992) prepared two messenger RNA molecules (Figure 12.41) (Bain et al. 1992) that encoded identical hexadecapeptides that were preceded by identical consensus 5′ untranslated sequences. One molecule has the UAG termination codon at position 9 as the signal for incorporation of the non-standard amino acid L-iodotyrosine whereas the other uses the novel non-standard (iso-C)AG codon at the 65th position for the same purpose (Figure 12.42). These mRNA

A

	1	2	3	4	5	6	7	8	9	10	11	12	13	14	15	16	17	18
	AUG	GGU	UUA	UAU	UUG	GGC	CUU	UUU	UAG	GGA	CUC	UAC	CUA	GGG	CUG	UUC	UAA	UGA
a	Met	Gly	Leu	Tyr	Leu	Gly	Leu	Phe	*End*									
b	Met	Gly	Leu	Tyr	Leu	Gly	Leu	Phe	iTyr	Gly	Leu	Tyr	Leu	Gly	Leu	Phe	*End*	
c	Met	Gly	Leu	Tyr	Leu	Gly	Leu	Phe	Arg	Asp	Cys	Tyr	*End*					

B

	1	2	3	4	5	6	7	8	9	10	11	12	13	14	15	16	17	18
	AUG	GGU	UUA	UAU	UUG	GGC	CUU	UUU	iCAG	GGA	CUC	UAC	CUA	GGG	CUG	UUC	UAA	UGA
a	Met	Gly	Leu	Tyr	Leu	Gly	Leu	Phe	*End*									
b	Met	Gly	Leu	Tyr	Leu	Gly	Leu	Phe	iTyr	Gly	Leu	Tyr	Leu	Gly	Leu	Phe	*End*	
c	Met	Gly	Leu	Tyr	Leu	Gly	Leu	Phe	Arg	Asp	Cys	Tyr	*End*					

Figure 12.41 The mRNA molecules used to compare rearranging (A) and expanding (B) the genetic code with the 65th codon as alternative strategies for incorporating non-strand amino acids into translated peptides. Different protein products are also shown

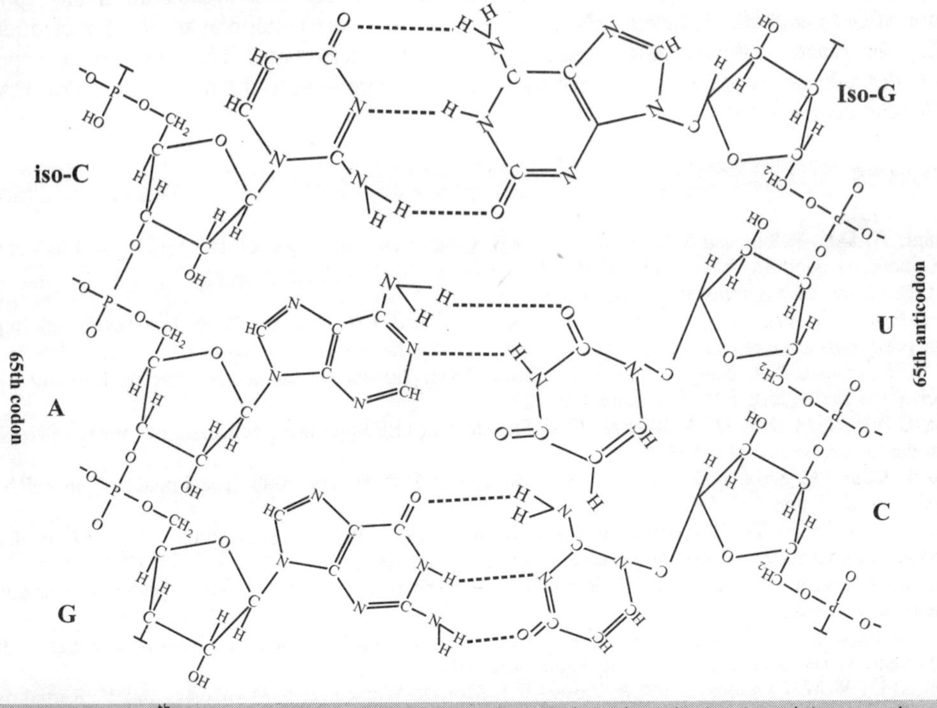

Figure 12.42 The 65th codon (incorporating the non-stranded nucleoside iso-C and its complementary anticodon incorporating iso-dG) allows the incorporation of non-stranded amino acids into proteins synthesized by translation

molecules were incubated separately with rabbit reticulocylate containing L-3H leucine and L-35S methionine. Ribosome-mediated peptide synthesis was evaluated in the presence and absence of the corresponding charged and uncharged transfer RNA molecules incorporating either CUA or CU(iso-dG) as the anticodon where iso-G is a non-standard purine complementary to iso-C. Translation products were isolated by precipitation or high performance liquid chromatography, A suppression level of 63 or 67 per cent was observed for read through of the UAG codon in the presence of a semi-synthetic suppressor tRNA incorporating the CUA anticodon and charged with iodotyrosine. In contrast, read through of the (iso-C)AG codon was 90 or 91 per cent. In the presence of corresponding non-standard $tRNA_{CU(iso\text{-}dG)}$ charged with iodotyrosine.

The specificity of translation of 65th codon is high. Neither semi-synthetic suppressor tRNA allowed reading of the (iso-C)AG nor any natural tRNA did, as shown by the absence of detectable full length product in translation mixtures containing the non-standard mRNA in the absence of charged $tRNA_{CU(iso\text{-}dG)}$. When the ribosome encountered the (iso-C)AG codon in the absence of charged $tRNA_{CU(iso\text{-}dG)}$, the primary outcome was continued translation following a frameshift that skipped the iso-C base. In contrast, attempted translation of a message containing the UAG stop codon in the absence of corresponding charged $tRNA_{CAU}$ resulted in termination at this position, yielding a truncated peptide; there was no detectable frameshifting.

Both genetic code rearrangement and expansion strategies suffer from an intrinsic disadvantage – the low yields of cell-free translation. The presence of release factors in translation mixtures and their requirement for correct termination create a further limitation. To obtain high yields of translation products, the release factors that bind UAG in competition with semi-synthetic tRNA must be removed or inactivated. However, this would require removal of UAG stop signals to avoid continued translation after frameshift. However, this technology promises to exploit non-standard nucleotides for expanding the genetic code and also suggest how translation terminates. The most simple hypothesis for this is that release factors present in the translation mixture bind to the non-sense codon UAG but not to the 65th codon (iso-C)AG.

REFERENCES

Adilakshmi, T., D.L. Bellur, and S.A. Woodson. 2008. Concurrent nucleation of 16S folding and induced fit in 30S ribosome assembly. Nature 455: 1268-72.

Ainsworth, C. 2005. Lost in translation. Nature 435: 556-8.

Alkalaeva, E.Z., A.V. Pisarev, L.Y. Frolova, L.L. Kisselev, and T.V. Pestova. 2006. In vitro reconstitution of eukaryotic translation reveals cooperativity between release factors eRF1 and eRF3. Cell 125: 1125-36.

Amrani, N., S. Ghosh, D.A. Nangus, and A. Jacobson. 2008. Translation factors promote the formation of two states of the closed-loop mRNP. Nature 453: 1276-80.

Anderson, C.B.F., T. Becker, M. Blau, et al. 2006. Structure of eEF3 and the mechanism of transfer RNA release from the E-site. Nature 443: 663-8.

Baek, D., J. Villén, C. Shin, F.D. Camargo, S.P. Gygi, and D.P. Bartel. 2008. The impact of microRNAs on protein output. Naure 455: 64-71.

Bain, J.D., C. Switzer, A.R. Chamberlin, and S.A. Brenner. 1992. Ribosome-mediated incorporation of a non-standard amino acid into a peptide through expansion of the genetic code. Nature 356: 537-9.

Beebe, K., M. Mock, E. Merriman, and P. Schimmer. 2008. Distinct domains of tRNA synthetase recognize the same base pair. Nature 451: 90-3.

Bingel-Erlenmeyer, R., R. Kohler, G. Kramer, et al. 2008. A peptide deformylase-ribosome complex reveals mechanism of nascent chain processing. Nature 452: 108-11.

Brodersen, D.E., W.M. Clemons, Jr., A.P. Carter, R.J. Morgan-Warren, B.T. Wimberly, and V. Ramakrishnan. 2000. The structural basis for the action of the antibiotics tetracycline, pactamycin, and hygromycin B on the 30S ribosomal subunit. Cell 103: 1143-54.

Carter, A.P., W.M. Clemons, D.E. Brodersen, R.J. Morgan-Warren, B.T. Wimberly, V. Ramakrishnan, 2000. Functional insights from the structure of the 30S ribosomal subunit and its interactions with antibiotics. Nature 407: 340-8.

Chendrimada, T.P., K.J. Finn, X. Ji, et al. 2007. MicroRNA silencing through RISC recruitment of eIF6. Nature 447: 823-8.

Cochella, L., and Green, R. 2005. An active role for tRNA in decoding beyond codon:anticodon pairing. Science 308: 1178-80.

Crick, F.H.C. 1966. Codon-anticodon pairing: The wobble hypothesis. J. Mol. Biol. 19: 548-55.

Dar, A.C., T.E. Dever and F. Sicheri. 2005. High-order substrate recognition of eIF2α by the RNA-dependent protein kinase PKR. Cell 122: 887-900.

Daviter, T., F.V. Murphy IV, and V. Ramakrishnan. 2005. A renewed focus on transfer RNA. Science 308: 1123-4.

Dirheimer, G., G. Keith, P. Dumas, and E. Westhof. 1995. Primary, secondary and tertiary structures of tRNAs. In: *tRNA: Structure, Biosynthesis, and Function.* Söll, D., and U. RajBhandary (Editors). Pp. 93-126. Wasington, D.C.: American Society for Microbiology.

Ellis, R.J. 1975. Chloroplast protein synthesis. Nature 254: 13.

Ellis, R.J. 1981. Chloroplast proteins: synthesis, transport, and assembly. Annu. Rev. Pl. Physiol. 32: 111-37.

Ferreira-Cerca, S., and E. Hurt. 2009. Arrest by ribosomes. Nature 459: 46-7.

Fredrick, K., and M. Ibba. 2009. Protein synthesis: Errors rectified in retrospect. Nature 457: 157-8.

Gandin, V., M. Miluzio, A.M. Barbieri, et al. 2008. Eukaryotic initiation factor 6 is rate-limiting in translation, growth and transformation. Nature 455: 684-8.

Gao, H., Z. Zhou, U. Rawat, et al. 2007. RF3 induces ribosomal conformational changes responsible for dissociation of class I release factors. Cell 129: 929-41.

Gross, J.D., N.J. Moerke, T. von der Haar, et al. 2003. Ribosome loading onto the mRNA cap is driven by conformational coupling between eIF4G and eIF4E. Cell 115: 739-50.

Halic, M., T. Becker, M.R. Pool, et al. 2004. Structure of the signal recognition particle interacting with the elongation-arrested ribosome. Nature 427: 808-14.

Ingolia, N.T., S. Ghaemmaghami, J.R.S. Newman, and J.S. Weissman. 2009. Genome-wide analysis in vivo of translation with nucleotide resolution using ribosome profiling. Science 324: 218-23.

Jenner, L., P. Romby, B. Rees, et al. 2005. Transcriptional operator of mRNA on the ribosome: how repressor proteins exclude ribosome binding. Science 308: 120-3.

Jergensen, R., A.R. Merril, S.P. Yates, et al. 2005. Exotoxin A-eEF2 complex structure indicates ADP ribosylation by ribosome mimicry. Nature 436: 979-84.

Khaholz, B.P., A.G. Myasnikov, and M.V. Heel. 2004. Visualization of release factor 3 on the ribosome during termination of protein synthesis. Nature 427: 862-5.

Komili, S., N.G. Farny, F.P. Roth, and P.A. Silver. 2007. Functional specificity among ribosomal proteins regulates gene expression. Cell 131: 557-71.

Korostelev, A., S. Trakhanov, M. Laurberg, and H.F. Noller. 2006. Crystal structure of a 70S ribosome-tRNA comlex reveals functional interactions and rearrangements. Cell 126: 1065-77.

Kozak, M. 1989. The scanning model for translation: an update. J. Cell Sci. 92: 325-8.

Lauberg, M., H. Asahara, A. Korostelev, J. Zhu, Z.S. Trakhanov, and H.F. Noller. 2008. Structural basis for translation termination on the 70S ribosome. Nature 454: 852-7.

Lazaris-Karatzas, A., K.S. Montine, and N. Sonenberg. 1990. Malignant transformation by a eukaryotic initiation factor subunit that binds to mRNA 5′ cap. Nature 345: 544-7.

Liljas, A. 2008. Getting close to termination. Science 322: 863-4.

Marzi, S., A.G. Myasnikov, A. Serganov, et al. 2007. Structured mRNAs regulate translation initiation by binding to platform of the ribosome. Cell 130: 1019-31.

McIntosh, K.B., and J.R. Warner. 2007. Yeast ribosomes: Variety is the spice of life. Cell 131: 450-1.

Merrick, W. 2009. Translation: Till termination do us part. Nature 459: 44-5.

Mitra, K., C. Schaffitzel, T. Shaikh, et al. 2005. Structure of *E. coli* protein-conducting channel bound to a translating ribosome. Nature 438: 318-24.

Mlot, C. 1989. On the trail of transfer RNA identity. Bioscience 39: 756-9.

Moazed, D., and H.F. Noller, 1989a. Interaction of tRNA with 23S rRNA in the ribosomal A, P, and E sites. Cell 57: 585-97.

Moazed, D., and H.F. Noller. 1989b. Intermediate states in the movement of tRNA in the ribosome. Nature 342: 142-8.

Moore, P.B. and Steitz, T.A. 2003. The structural basis of large ribosomal subunit function. Annu. Rev. Biochem. 72: 813-850.

Moore, P.B., and T.A. Steitz. 2002. The involvement of RNA in ribosome function. Nature 418: 229-34.

Moore, P.R. 1989. Protein synthesis, elongation remodelled. Nature 342: 127-8.

Moras, D. 1990. Synthetases gain recognition. Nature 344: 195-7.

Namy, O., S.J. Moran, D. Stuart, R.J.C. Gilbert, and I. Brierley. 2006. A mechanical explanation of RNA pseudoknot function in programmed ribosomal frameshifting. Nature 441: 244-7.

Nissen, P., N. Bau, J. Hensen, P.B. Moore, and J.A. Steitz. 2000. The structural basis of ribosome activity in peptide bound synhesis. Science 289: 920-30.

Petry, S., D.E. Brodersen, F.V. Murphy, et al. 2005. Crystal structure of the ribosome in complex with release factor RF1 and RF2 bound to a cognate stop codon. Cell 123: 1255-66.

Pillai, R.S., N. Bhattacharyya, C.G. Artus, et al. 2005. Inhibition of translational initiation by Let-7 microRNA in human cells. Science 309: 1573-6.

Pisarev, A.V., C.U.T. Hellen, and T.V. Pestova. 2007. Recycling of eukaryotic posttermination ribosomal complexes. Cell 131: 286-99.

Qin, Y., N. Polacek, O. Vesper, Staub, et al. 2006. The highly conserved LepA is a ribosomal elongation factor that back-translocates the ribosome. Cell 127: 721-33.

Richter, J.D., and N. Sonnenberg. 2005. Regulation of cap-dependent translation by eIF4E inhibitory protein. Nature 433: 477-80

Roy, H., and M. Ibba. 2006. Sticky ends in protein synthesis. Nature 443: 41-2.

Sampath, P., B. Mazumder, V. Seshadri, et al. 2004. Noncanonical function of glutamyl-prolyl-tRNA synthetase: Gene specific silencing of translation. Cell 119: 195-208.

Schmeing, T. M., K. S. Huang, D. E. Kitchen, S. A. Strobel, and T. A. Steitz. 2005a. Structural insights into the roles of water and the 2' hydroxyl of the P site tRNA in the peptidyl transferase reaction. Mol. Cell 20: 437-48.

Schmeing, T.M., K.S. Huang, S.A. Strobel, and T.A. Steitz. 2005b. An induced-fit mechanism to promote peptide bond formation and exclude hydrolysis of peptidyl-tRNA. Nature 438: 520-4.

Schuwirth, B.S., M.A. Borovinskaya, C.W. Hau, et al. 2005. Structures of the bacterial ribosome at 3.5Å resolution. Science 310: 827-34.

Selbach, M., B. Schwanhausser, N. Thierfelder, Z. Fang, R. Khanin, and N. Rajewsky. 2008. Widespread changes in protein synthesis induced by micro RNAs. Nature 455: 58-63.

Selmer, M., C.M. Dunham, M.V. Murphy IV, et al. 2006. Structure of the 70S ribosome complexed with mRNA and tRNA. Science 313: 1935-42.

Shine, J., and L. Dalgarno. 1974. The 3'-terminal sequence of *Escherichia coli* 16S ribosomal RNA complementarity to nonsense triplets and ribosome binding sites. Proc. Natl. Acad. Sci.USA 71: 1342-6.

Shine, J., and L. Dalgarno. 1975. Determinant of cistron specificity in bacterial ribosomes. Nature 254: 34-8.

Siridechadilok, B., C.S. Fraser, R.J. Hall, J.A. Doudna, and E. Nogales. 2005. Structural roles for human translation factor eIF3 in initiation of protein synthesis. Science 310: 1513-5.

Steitz, J.A., and K.T. Tycowski. 1995. Small RNA chaperones for ribosome biosynthesis. Science 270: 1626-7.

Syntichaki, P., K. Troulinaki, and N. Tavernarakis. 2007. eIF4E function in somatic cells modulates ageing in *Caenorhabditis elegans*. Nature 445: 922-6.

Takano, A., T. Endo, and T. Yoshihisa. 2005. tRNA actively shuttles between nucleus and cytosol in yeast. Science 309: 140-2.

Takyar, S., R.P. Hickerson, and H.F. Noller. 2005. mRNA helicase activity of the ribosome. Cell 120: 49-58.

Talkington, M.W., G. Siuzdak, and J.R. Williamson. 2005. An assembly landscape for the 30S ribosomal subunit. Nature 438: 628-32.

Thermann, R. and Hentze, M.W. 2007. *Drosophila* miR2 induces pseudo-polysomes and inhibits translation initiation. Nature 447: 875-8.

Valle, M., A. Zavialov, J. Sengupta, U. Rawat, M. Ehrenberg, and J. Frank. 2003. Locking and unlocking of ribosomal motions. Cell 114: 123-34.

Vasudevan, S., and J.A. Steitz, 2007. AU-rich-element-mediated upregulation of translation by FXR1 argonaute 2. Cell 128: 1105-18.

Watanabe, K., Y. Toh, K. Suto, et al. 2007. Protein-based peptide-bond formation by aminoacyl-tRNA protein transferase. Nature 449: 867-71.

Weixibaumer, A., H. Jin, C. Meubauer, et al. 2008. Insights into translational termination from the structure of RF2 bound to the ribosome. Science 322: 953-6.

Wen, J.-D., L. Lancaster, C. Hodges, et al. 2008. Following translation by single ribosomes one codon at a time. Nature 452: 598-603.

Wilker E.W., M.A. van Vugt, S.A. Artim, et al. 2007. 14-3-3 sigma controls mitotic translation to facilitate cytokinesis. Nature 446: 329-32.

Wilson, F.H., A. Hariri, A. Farhi, et al. 2004. A cluster of metabolic defects caused by mutation in a mitochondrial tRNA. Science 306: 1190-4.

Wu, X.-Q., and U.L. RajBhandary. 1997. Effect of the amino acid attached to Escherichia coli initiator tRNA on Its affinity for the initiation factor IF2 and on the IF2 dependence of its binding to the ribosome. J. Biol. Chem. 272: 1891-5.

Wynshaw-Boris, A. 2007. Cell biology: Lost in mitotic translation. Nature 446: 274-5.

Yonath, A. 2000. Decoding the genetic information on the ribosome at close to atomic resolution. Acta Cryst. A56: s1.

Yonath, A. 2005. Can structures lead to better drugs? Lessons from ribosomal antibiotics. Acta Cryst. A61: c118-119.

Yonath, A., K.S. Bartels, F. Frolow, et al. 1987. Crystallography of intact ribosomal particles. Acta Cryst. A43: C39.

Yonath, A.E., J. Muessig, B. Tesche, S. Lorenz, V.A. Erdmann, and H.G. Wittmann. 1980. Crystallization of the large ribosomal subunits from *Bacillus stearothermophilus*. Biochem. Int. J. 1(5): 428-35.

Youngman, E.M., J.L. Brunelle, A.B. Kochanaik, and R. Green. 2004. The active site of the ribosome is composed of two layers of conserved nucleotides with distinct roles in peptide bond formation and peptide release. Cell 117: 589-99.

Yusupova, G., L. Jenner, B. Rees, D. Moras, and M. Yusupova. 2006. Structural basis for messenger RNA movement on the ribosome. Nature 444: 391-4.

Zaher, H.S., and R. Green. 2009. Quality control by the ribosome following peptide bond formation. Nature 457: 161-5.

Zavialov, A.V., and M. Ehrenberg. 2003. Peptidyl-tRNA regulates the GTPase activity of transcription factors. Cell 114: 113-22.

13

Fate of Nascent Proteins, Ribosomes and Messenger RNAs

Nascent protein is a protein as it is being formed by a ribosome before it folds into its active shape. In this chapter we shall discuss the mechanism involved in deciding the fate of the finished/nascent proteins. The proteins synthesized in the cell are either transported to the site of their further modification or action or are secreted out. Carrier proteins are used to transport proteins across cellular membranes such as the plasma membrane, endoplasmic reticulum and nuclear envelope. Proteins are also trafficked between membrane-bound organelles inside membrane vesicles. The proteins synthesized on rough endoplasmic reticulum (RER) are transported while still being synthesized. Signal hypothesis has been proposed for this transport. According to this hypothesis, ribosomes synthesizing proteins are attached to the membrane via the leader sequence.

PROTEIN MODIFICATION

Before folding of the nascent protein, certain post-translational modifications take place. Post-translational modification (PTM) is the chemical modification of a protein after its translation. It is one of the later steps in protein biosynthesis, and thus gene expression, for many proteins. Protein post-translational modification (PTM) increases the functional diversity of the proteome by the covalent addition of functional groups or proteins, proteolytic cleavage of regulatory subunits or degradation of entire proteins. These modifications include phosphorylation, glycosylation, ubiquitination, nitrosylation, methylation, acetylation, lipidation and proteolysis and influence almost all aspects of normal cell biology and pathogenesis. Therefore, identifying and understanding PTMs is critical in the study of cell biology and disease treatment and prevention.

Protein targeting is discussed here, under two headings – protein secretion and protein translocation (sorting or trafficking), separately in prokaryotes and eukaryotes. In bacteria, the choice of destination is between the cytoplasm, the inner and outer cell membranes, and the periplasmic space between them. Proteins can also be secreted. In eukaryotes, synthesis of all polypeptides encoded by nuclear genes begins in the cytosol. The large and small ribosomal subunits associate with each other and with the 5′-end of an mRNA molecule, forming a functional ribosome that starts making the polypeptide. When the polypeptide is about 30 amino acids long, it enters one of two alternative pathways – co-translational import and post-translational import. Targeting in eukaryotes is necessarily more complex due to the multitude of internal compartments: nucleus, mitochondria, peroxisomes, chloroplasts, endoplasmic reticulum (ER), Golgi, lysosomes, and secretory granules. Nuclear targeting is unusual since it is a two-way traffic. Proteins are not transported through the nuclear membrane but rather through a complex pore called the nuclear pore. Nuclear pore comprises of (a) about 100 different proteins – proteins smaller than 20 kDa that move by diffusion, proteins larger than 20 kDa that move by selective transport (nuclear localization signal) and (b) cluster of 4-8 positively charged amino acids (example: PKKKRLV). Signal sequence binds to receptor on the pore called importin.

PROTEIN SECRETION IN PROKARYOTES

Protein Secretion in Gram-Negative Bacteria

Secretion is the process of elaborating, releasing, and oozing chemicals, or a secreted chemical substance from a cell or gland. In contrast to excretion, the substance may have a certain function, rather than being a waste product. The classical mechanism of cell secretion is via secretory portals at the cell plasma membrane called porosomes. Porosomes are permanent cup-shaped lipoprotein structure at the cell plasma membrane, where secretory vesicles transiently dock and fuse to release intra-vesicular contents from the cell. Secretion in bacterial species means the transport or translocation of effector molecules for example: proteins, enzymes or toxins (such as cholera toxin in pathogenic bacteria for example *Vibrio cholerae*) from across the interior (cytoplasm or cytosol) of a bacterial cell to its exterior. Secretion is a very important mechanism in bacterial functioning and operation in their natural surrounding environment for adaptation and survival. There are six secretion systems known in bacteria.

Type I Secretion System (T1SS): It is similar to the ATP binding cassette (ABC) transporter, however it has additional proteins that, together with the ABC protein, form a contiguous channel traversing the inner and outer membranes of Gram-negative bacteria. It is a simple system, which consists of only three protein subunits: the ABC protein, membrane fusion protein (MFP), and outer membrane protein (OMP). Type I secretion system transports various molecules, from ions, drugs, to proteins of various sizes (20-900 kDa). The molecules secreted vary in size from the small *Escherichia coli* peptide colicin V, (10 kDa) to the *Pseudomonas fluorescens* cell adhesion protein LapA of 900 kDa. The best characterized are the RTX toxins and the lipases. Type I secretion is also involved in export of non-proteinaceous substrates like cyclic β-glucans and polysaccharides. ABC type transporters are common to all the three domains of life.

Proteins targeted to the translocator carry an (uncleaved), poorly conserved secretion signal of approximately 50 residues (Holland et al. 2005). In *E. coli* the HlyA toxin interacts with both the MFP (HlyD) and the ABC protein HlyB, (a half transporter) triggering, via a conformational change in HlyD, recruitment of the third component, TolC, into the transenvelope complex. In vitro, HlyA, through its secretion signal, binds to the nucleotide binding domain (NBD or ABC-ATPase) of HlyB in a reaction reversible by ATP that may mimic initial movement of HlyA into the translocation channel. HlyA is then transported rapidly, apparently in an unfolded form, to the cell surface, where folding and release takes place. Whilst recent structural studies of TolC and MFP-like proteins are providing atomic detail of much of the transport path, structural analysis of the HlyB NBD and other ABC ATPases, have revealed details of the catalytic cycle within an NBD dimer and a glimpse of how the action of HlyB is coupled to the translocation of HlyA.

Type II Secretion System (T2SS): Proteins secreted through the type II system, or main terminal branch of the general secretory pathway, depend on the Sec or Tat system for initial transport into the periplasm. Once there, they pass through the outer membrane via a multimeric (12-14 subunits) complex of pore forming secretin proteins. In addition to the secretin proteins, 10-15 other inner and outer membrane proteins compose the full secretion apparatus, many with as yet unknown function. Gram-negative type IV pili use a modified version of the type II system for their biogenesis, and in some cases certain proteins are shared between a pilus complex and type II system within a single bacterial species.

In bacteria, proteins destined to be secreted are synthesized as pre-proteins with N-terminal signal sequences sometimes termed leader peptides. These signal sequences are short (25 residues) and comprise of a hydrophobic central core which can adopt an α-helical structure, flanked by region containing several charged residues. A number of chaperone proteins can bind to the nascent protein as the leader peptide emerges from the ribosome, to prevent mis-folding. One such protein, SecB, plays a predominant role in protein export because it binds another component of the secretory system, SecA, a chaperone associated with the translocation apparatus, SecA mediates translocation by feeding the substrate protein through the translocation apparatus (comprising of transmembrane proteins SecE and SecY) in a manner which is dependent on ATP hydrolysis. The leader peptide is then cleaved by an enzyme called leader peptidase. In the mature protein, amino acids with small residual groups are often found adjacent to the cleavage site (the –1 position) and the next but one upstream residue (the –3 position). This phenomenon is known as the –1 and –3 rule.

Zimmer et al. (2008) report the crystal structure of SecA bound to the SecY complex, with a maximum resolution of 4.5Å, obtained for components from *Thermotoga martima*. One copy of the SecA is an intermediate state of ATP hydrolysis is bound to one molecule of the SceY complex. Both partners undergo important conformational changes on interaction. The peptide-cross-linking domain of SecA makes a large conformational change that could capture the translocation substrate in a 'clamp'. Polypeptide movement through the SecY channel could be achieved by the motion of a 'two-helix finger' of SecA inside the cytoplasmic tunnel of SecY, and by the coordinated tightening and widening of SecA's clamp above the SecY pore. SecA binding generates a 'window at the lateral gate of the SecY channel and it displaces the plug domain, preparing the channel for signal sequence binding and channel opening. Model for SecA-mediated protein translocation is given in Figure 13.1 (Zimmer et al. 2008). (A) Proposed polypeptide translocation pathway through the SecA–SecYEG complex is shown. The signal sequence of a hypothetical polypeptide is intercalated into the lateral gate of SecY. (B) Upon ATP binding, the clamp of SecA would widen, allowing the two-helix finger to bind to the polypeptide and move it into the channel. After ATP hydrolysis, the clamp would tighten and the finger would reset. SecA probably adopts more than two conformational states during ATP hydrolysis. The structure might represent a situation in which the two-helix finger is in its 'down-state' and the clamp is already closed, although the conformation of the clamp may be different in the presence of a translocating polypeptide.

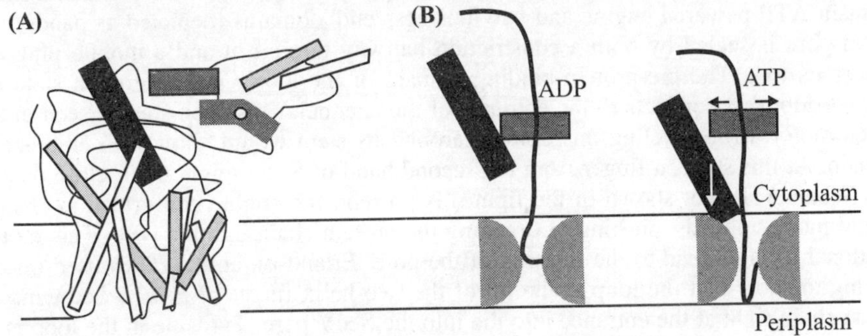

Figure 13.1 Model for SecA-mediated protein translocation

Over 30 per cent of proteins are secreted across or integrated into membranes. The only high-resolution structure of a translocon available is from the archaeon *Methanococcus jannaschi*, which lacks SecA. Tsukazaki et al. (2008) present a 3.2-Å resolution structure of the SecYE translocon from a SecA-containing organism, *Thermus thermophilus*. The structure, solved as a complex with an anti-SecY Fab fragment, revealed a 'pre-open' state of SecYE, in which several transmembrane helices are shifted, as compared to the previous SecYEβ structure, to create a hydrophobic crack open to the cytoplasm. Fab and SecA bind to a common site at the tip of the cytoplasmic domain of SecY. Molecular dynamics and disulfide mapping analyses suggest that the pre-open state might represent a SecYE conformational transition that is induced by SecA binding. Moreover, they identified a SecA-SecYE interface that comprises of SecA residues originally buried inside the protein, indicating that both the channel and motor components of the Sec machinery undergo cooperative conformational changes on formation of the functional complex. Multiple modes of SecA-SecY interactions are shown in Figure 13.2 (Tsukazaki et al. 2008). According to the dimer model, one copy of SecY serves as a SecA-docking site and the other functions as a translocation pore. The SecA–SecY interaction observed here should represent the one between the non-translocating copy of SecY and SecA, and is crucial for the SecA ATPase activation. Both the SecA and SecY components undergo conformational changes on their interaction, as shown by bidirectional arrows. The orientation of SecY protomers in the dimeric assembly is shown arbitrarily.

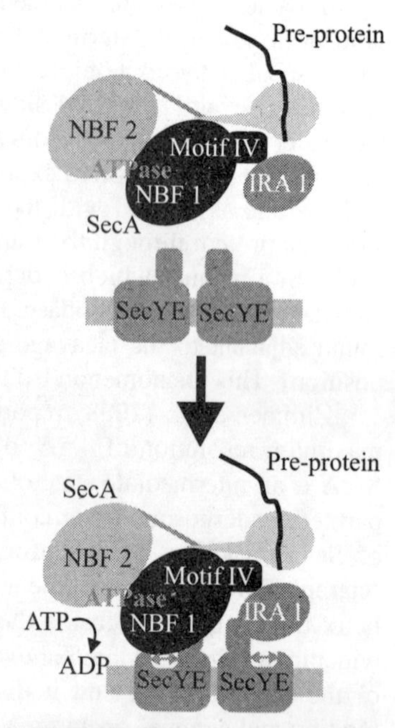

Figure 13.2 Multiple modes of SecA-SecY interactions

Core features of the major mechanism for the translocation of proteins across membranes are conserved throughout biology. The bacterium *E. coli*, like all other organisms, synthesizes secreted proteins with amino-terminal signal sequences, which direct such proteins to a translocation machinery located in the cytoplasmic membrane (Figure 13.3).

Precisely how proteins snake their way through channels in cell membranes is unclear. Complexes between the SecY channel and its motor protein, and the use of a 'molecular ensoscope', provide fascinating clues (Economou 2008). Mechanism of protein translocation in bacterial cell is depicted in Figure 13.4. The SecA motor lies flat against the cytoplasmic side of the SecY channel and consists of a two-domain ATP-powered engine and two 'business-end' domains (depicted as hands). (a) Initially, the channel pore is sealed by both a constriction halfway through it and a mobile plug domain (not shown) near its exit. The pre-protein-binding domain of the motor is in the open state exposing an elongated corridor that connects to the entrance of the channel. This open state is seen in structures of the isolated motor. (b) Swiveling this domain around its stem would allow it to embrace a secretory protein chain. At this stage, a finger from the second hand of SecA might be in close contact with the chain. (c) When ATP (not shown in the figure) is present, the engine conformation changes and the finger could move upwards, pushing or dragging the protein chain into the pore. This motion or other conformational changes lead to the opening of the pore. Erlandson et al. (2008) used disulfide-bridge cross-linking to show that the loop at the tip of the two-helix finger of *E. coli* SecA interacts with a polypeptide chain right at the entrance into the into the SecY pore. Tyrosine in the loop is particularly important for translocation.

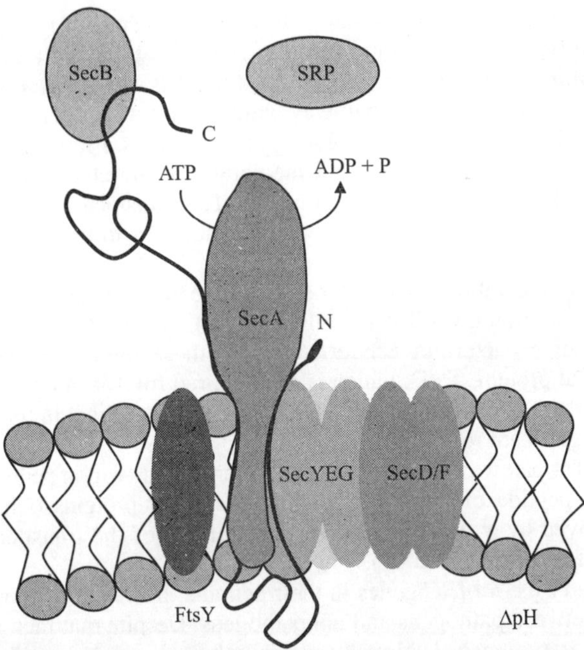

Figure 13.3 Protein secretion. Core features of the major mechanism for the translocation of proteins across membranes are conserved throughout biology. The bacterium *Escherichia coli*, like all other organisms, synthesizes secreted proteins with amino-terminal signal sequences, which direct such proteins to a translocation machinery located in the cytoplasmic membrane (Redrawn, with permission, from whttp://beck2.med.harvard.edu/secretion/secretion.htm)

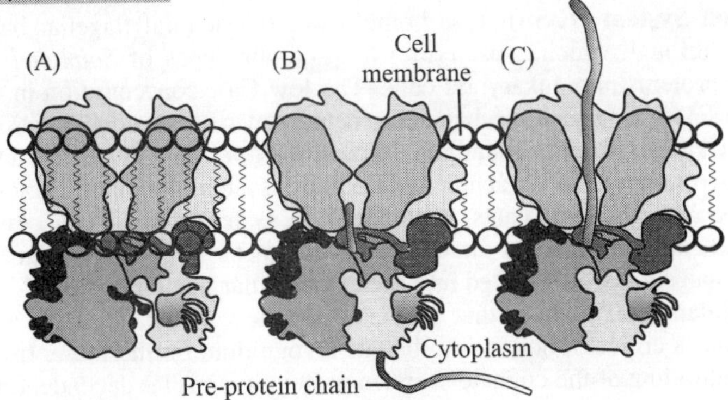

Figure 13.4 Mechanism of protein translocation in bacterial cell

Some secreted proteins are translocated across the cytoplasmic membrane by the Sec translocon, which requires the presence of an N-terminal signal peptide on the secreted protein. Others are translocated across the cytoplasmic membrane by the twin-arginine translocation pathway (Tat). The Tat pathway is a protein export, or secretion pathway found in plants, bacteria, and archaea. The Tat system operates in the chloroplast thylakoid and the plasma membranes of a wide range of bacteria (Robinson and Bolhuis 2004). It recognizes substrates bearing cleavable signal peptides in which a twin-arginine motif almost invariably plays a key role in recognition by the translocation machinery.

These signal peptides are surprisingly similar to those used to specify transport by Sec-type systems, but the Tat pathway differs in fundamental respects from Sec-type and other protein translocases. Its key attribute is its ability to translocate large, fully folded (even oligomeric) proteins across tightly sealed membranes. In contrast to the Sec pathway which transports proteins in an unfolded manner, the Tat pathway serves to actively translocate folded proteins across a lipid membrane bilayer. In plants, the Tat translocase is located in the thylakoid membrane of the chloroplast, where it acts to export proteins into the thylakoid lumen. In bacteria, the Tat translocase is found in the cytoplasmic membrane and serves to export proteins to the cell envelope, or to the extracellular space (Sargent et al. 2006).

In the plant thylakoid membrane and in Gram-negative bacteria the Tat translocase is composed of three essential membrane proteins – TatA, TatB, and TatC. In the most widely studied Tat pathway, that of the Gram-negative bacterium *Escherichia coli*, these three proteins are expressed from an operon with a fourth Tat protein, TatD, which is not required for Tat function. A fifth Tat protein TatE that is homologous to the TatA protein is present at a much lower level in the cell than TatA and is not believed to play any significant role in Tat function.

The name of the Tat pathway relates to a highly conserved twin-arginine motif which is found in the N terminal Signal peptide of the corresponding passenger proteins (Chaddock et al. 1995). The signal peptide is removed by a signal peptidase after release of the transported protein from the Tat complex (Frielingsdorf and Klösgen 2007).

Not all bacteria carry the *tatABC* genes in their genome; however, of those that do, there seems to be no discrimination between pathogens and nonpathogens. Despite that fact, some pathogenic bacteria such as *Pseudomonas aeruginosa*, *Legionella pneumophila*, *Yersinia pseudotuberculosis*, and *E. coli* O157:H7 rely on a functioning Tat pathway for full virulence in infection models. In addition, a number of exported virulence factors have been shown to rely on the Tat pathway. One such category of virulence factors are the phospholipase C enzymes, which have been shown to be Tat-exported in *Pseudomonas aeruginosa*, and thought to be Tat-exported in Mycobacterium tuberculosis.

Type III Secretion System (T3SS): It is homologous to bacterial flagellar basal body. It is like a molecular syringe through which a bacterium (e.g. certain types of *Salmonella*, *Shigella*, *Yersinia*, *Vibrio*) can inject proteins into eukaryotic cells. The low Ca^{2+} concentration in the cytosol opens the gate that regulates T3SS. One such mechanism to detect low calcium concentration has been illustrated by the lcrV (Low Calcium Response) antigen utilized by *Y. pestis*, which is used to detect low calcium concentrations and elicits T3SS attachment. The Hrp system in plant pathogens injects hairpins through similar mechanisms into plants. This secretion system was first discovered in *Y. pestis* and showed that toxins could be injected directly from the bacterial cytoplasm into the cytoplasm of its host's cells rather than simply be secreted into the extracellular medium.

Akeda and Galan (2005) show that InvC, an ATPase associated with a *S. enterica* type III secretion system, has a critical function in substrate recognition. Furthermore, InvC induces chaperone release from and unfolding of the cognate secreted protein in an ATP-dependent manner. These results show a similarly between the mechanism of substrate recognition by type III secretion systems and AAA+ATPase disassembly machines. A model for TTS is presented in Figure 13.5 (Akeda and Galan 2005). Substrate folding is required for TTS.

Type IV Secretion System (T4SS): It is homologous to conjugation machinery of bacteria (and archaeal flagella). It was discovered in *Agrobacterium tumefaciens*, which uses this system to introduce the T-DNA portion of the Ti plasmid into the plant host, which in turn causes the affected area to develop into a crown gall (tumor). *Helicobacter pylori* uses a type IV secretion system to deliver CagA into gastric epithelial cells, which is associated with gastric carcinogenesis. *Bordetella pertussis*, the causative agent of whooping cough, secretes the pertussis toxin partly through the type

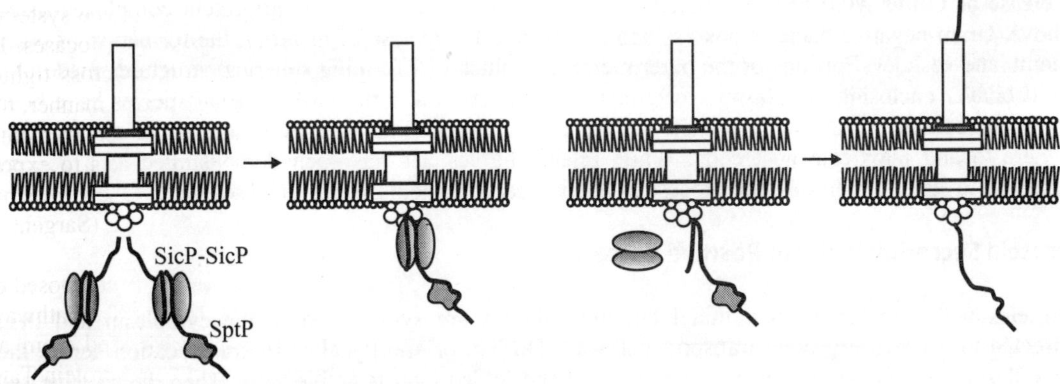

Figure 13.5 Model for type III protein secretion systems (TTS)

IV system. *Legionella pneumophila*, the causing agent of legionellosis (Legionnaires' disease) utilizes a type IVB secretion system, known as the *icm/dot* (intracellular multiplication/defect in organelle trafficking genes) system, to translocate numerous effector proteins into its eukaryotic host. The prototypic Type IVA secretion system is the VirB complex of *Agrobacterium tumefaciens*. T4SS is the general mechanism by which bacterial cells secrete or take up macromolecules. T4SS is encoded on Gram-negative conjugative elements in bacteria.T4SS are cell envelope-spanning complexes or in other words 11-13 core proteins that form a channel through which DNA and proteins can travel from the cytoplasm of the donor cell to the cytoplasm of the recipient cell. Additionally, T4SS also secrete virulence factor proteins directly into host cells as well as taking up DNA from the medium during natural transformation, which shows the versatility of this macromolecular secretion apparatus.

Type V Secretion System (T5SS): Also called the autotransporter system, type V secretion involves use of the Sec system for crossing the inner membrane. Proteins which use this pathway have the capability to form a beta-barrel with their C-terminus which inserts into the outer membrane, allowing the rest of the peptide (the passenger domain) to reach the outside of the cell. Often, autotransporters are cleaved, leaving the beta-barrel domain in the outer membrane and freeing the passenger domain. Some people believe remnants of the autotransporters gave rise to the porins which form similar beta-barrel structures. A common example of an autotransporter that uses this secretion system is Trimeric Autotransporter Adhesins (TAAs).

Type VI Secretion System (T6SS): Type VI secretion systems were identified in 2006 by the group of John Mekalanos at the Harvard Medical School (Boston, USA) in two bacterial pathogens, *Vibrio cholerae* and *Pseudomonas aeruginosa*. Since then, Type VI secretion systems have been found in most genomes of proteobacteria, including animal, plant, human pathogens, as well as soil, environmental or marine bacteria. While most of the early studies of Type VI secretion focus on its role in the pathogenesis of higher organisms, more recent studies suggest a broader physiological role in defense against simple eukaryotic predators and its role in inter-bacteria interactions. The Type VI secretion system gene clusters contain from 15 to more than 20 genes, two of which, *Hcp* and *VgrG*, have been shown to be nearly universally secreted substrates of the system. Structural analysis of these and other proteins in this system bear a striking resemblance to the tail spike of the T4 phage.

Release of Outer Membrane Vesicles: In addition to the use of the multiprotein complexes listed above, Gram-negative bacteria possess another method for release of material: the formation of outer membrane vesicles. Portions of the outer membrane pinch off, forming spherical structures made of a lipid bilayer enclosing periplasmic materials. Vesicles from a number of bacterial species have been found to contain virulence factors, some have immunomodulatory effects, and some can directly adhere to and intoxicate host cells. While release of vesicles has been demonstrated as a general response to stress conditions, the process of loading cargo proteins seems to be selective.

Protein Secretion in Gram Positive Bacteria

Proteins with appropriate N-terminal targeting signals are synthesized in the cytoplasm and then directed to a specific protein transport pathway. During, or shortly after its translocation across the cytoplasmic membrane, the protein is processed and folded into its active form. Then the translocated protein is either retained at the extracytoplasmic side of the cell or released into the environment. Since the signal peptides that target proteins to the membrane are key determinants for transport pathway specificity, these signal peptides are classified according to the transport pathway to which they direct proteins. Signal peptide classification is based on the type of signal peptidase (SPase) that is responsible for the removal of the signal peptide. The majority of exported proteins are exported from the cytoplasm via the general Secretory (Sec) pathway. Most well known virulence factors (e.g. exotoxins of Staphylococcus aureus, protective antigen of *Bacillus anthracis*, listeriolysin O of *Listeria monocytogenes*) that are secreted by Gram-positive pathogens have a typical N-terminal signal peptide that would lead them to the Sec-pathway. Proteins that are secreted via this pathway are translocated across the cytoplasmic membrane in an unfolded state. Subsequent processing and folding of these proteins takes place in the cell wall environment on the *trans*-side of the membrane. In some *Staphylococcus* and *Streptococcus* species, the accessory secretory system handles the export of highly repetitive adhesion glycoproteins.

In addition to the Sec and accessory-Sec systems, some Gram-positive bacteria contain the Tat-system that is able to translocate folded proteins across the membrane. This is especially appropriate for proteins that need co-factors, such as iron-sulfur clusters and molybdopterin, which are incorporated in the cytoplasm. Pathogenic bacteria may contain certain special purpose export systems that are specifically involved in the transport of only a few proteins. For example, several gene clusters have been identified in mycobacteria that encode proteins that are secreted into the environment via specific pathways (ESAT-6) and are important for mycobacterial pathogenesis. Specific ABC transporters direct the export and processing of small antibacterial peptides called bacteriocins. Genes for endolysins that are responsible for the onset of bacterial lysis are often located near genes that encode for holin-like proteins, suggesting that these holins are responsible for endolysin export to the cell wall. The Tat pathways of Gram-positive bacteria differ in that they do not have a TatB component. In these bacteria the Tat system is made up from a single TatA and TatC component, with the TatA protein being bifunctional and fulfilling the roles of both *E. coli* TatA and TatB (Barnett et al. 2008).

Gram-positive bacteria often secrete large amounts of proteins into the surrounding medium. This feature makes them attractive as hosts for the industrial production of extracellular enzymes. Compared to *Escherichia coli*, relatively little is known about the mechanism of protein secretion in these organisms. However, the recent identification of *Bacillus subtilis* genes whose gene products are highly homologous to some of the Sec (secretion) proteins of *E. coli* strongly suggests that important principles of protein translocation across the plasma membrane might be highly conserved. In contrast, the steps following the actual translocation event might be different in Gram-positive and Gram-negative bacteria.

PROTEIN TRANSLOCATION IN PROKARYOTES

The prokaryotic signal recognition particle (SRP) targets membrane proteins into the inner membrane. It binds translating ribosomes and screens the emerging nascent chain for a hydrophobic signal sequence, such as the transmembrane helix of inner membrane proteins. If such a sequence emerges the SRP binds tightly, allowing the SRP receptor to lock on. This assembly delivers the ribosome-nascent chain complex to the protein translocation machinery in the membrane. Schaffitzel et al. (2006) obtained a structure of the *Escherichia coli* complex with a translating ribosome containing a nascent chain with a transmembrane helix anchor. They also obtained structural information on the SRP bound to an empty *E. coli* ribosome. The structures reveal the regions that are involved in complex formation, provide insights into conformation of the SRP on the ribosome and indicate the conformational changes that accompany high-affinity SRP binding to ribosome nascent chain complexes upon recognition of the signal sequence. A model of signal anchor-dependent docking of SRP to the ribosome is given in Figure 13.6. In the absence of a nascent chain, the SRP binds via its N domain to L23, and the M domain scans for a nascent chain with a signal anchor. In this state, the M domain with Ffs is flexible. Upon recognition of a signal anchor, the SRP binds tightly to the ribosome and forms three additional contacts with 50S. The signal anchor is buried in the hydrophobic pocket of the M domain, and the NG domain has an increased affinity for GTP, which is a prerequisite for FtsY binding.

In bacteria, N-terminal signal sequences promote translocation across the cytoplasmic membrane, which surrounds the entire cell, but some proteins are nevertheless secreted in one part of the cell by poorly understood mechanisms. Carlsson et al. (2006) catalyze localized secretion in the Gram-positive pathogen *Streptococcus pyrogenes*, and show that the signal sequences of two surface proteins, M protein and F protein (PrtF) direct secretion to different subcellular regions. The signal sequence of M protein promotes secretion of the division septum, whereas that of PrtF preferentially promotes

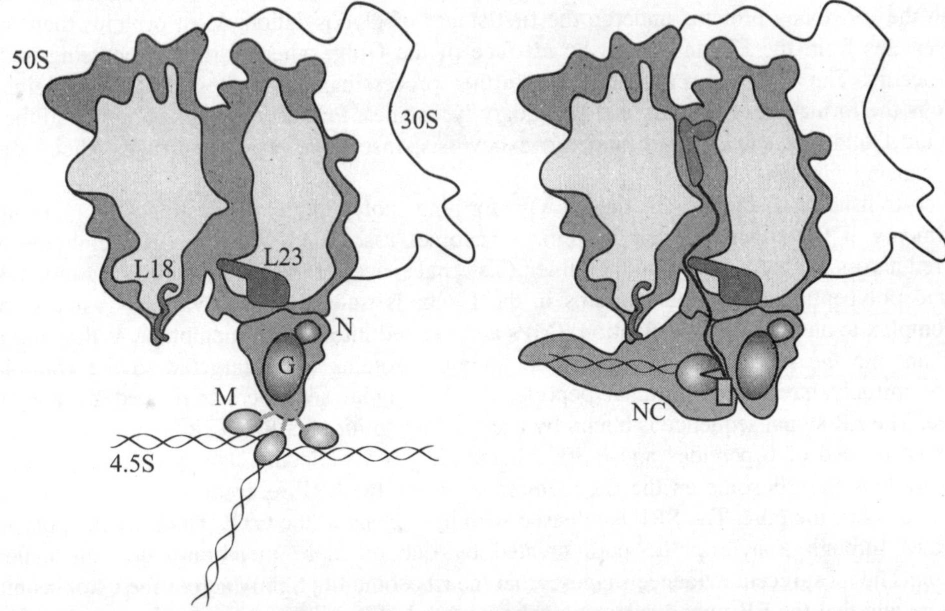

Figure 13.6 Model of signal anchor-dependent docking of SRP to the ribosome

secretion at the old pole. This work thus shows that a signal sequence may contain information that directs the secretion of a protein to one subcellular region, in addition to its classical role in promoting secretion. This finding identifies a new level of complexity in protein translocation and emphasizes the potential of bacterial systems for the analysis of fundamental cell-biological problems.

CO-TRANSLATIONAL TARGETING (SECRETORY PATHWAYS) IN EUKARYOTES

Proteins that are secreted from the cell must pass through the plasma membrane to the exterior. These proteins start their synthesis in the same way as membrane-associated proteins but pass entirely through the system instead of halting at some particular point within it. Interaction between the membrane systems of the cell and parts of the protein structure (sequences) or groups added covalently to it give signals. By means of such signals, a protein is recognized by receptors located within the membrane. A travelling protein may then be secreted through the membrane or may be passed on to another membrane. Free polysomes synthesize proteins that are released directly into the cytosol. Proteins synthesized on membrane-bound ribosomes pass into the endoplasmic reticulum, along to the Golgi, and then through the plasma membrane, unless they possess signals that cause retention at one of the steps on the pathway.

Co-translational targeting (secretory) pathways are present in endoplasmic reticulum (ER), Golgi, lysosomes, plasma membrane, and secreted proteins. Examples of such proteins are: translational machinery, metabolic enzymes, cytoskeletal proteins, and many signal transduction proteins. Co-translational insertion of protein occurs into or through ER membrane via attached ribosomes (rough ER). A signal sequence of 16-30 amino acids at N-terminus (hydrophobic) is the emerging signal sequence of nascent protein on free ribosome binds to signal recognition particle (SRP). The translation is arrested. SRPs consist of 6 proteins and one RNA molecule (7S RNA). The SRP-signal sequence-mRNA-ribosome complex docks with receptor on ER membrane. The signal sequence crosses ER membrane. Translation continues with polypeptide chain being pulled into the ER lumen. While in the ER, many proteins undergo the first stages of glycosylation. Most proteins then migrate inside vesicles from the ER and enter the *cis* face of the Golgi where further processing and final sorting occurs. The Golgi is responsible for further processing and final sorting of proteins. One example is the formation of primary and secondary lysosomes. Primary lysosomes bud from the *trans* face of the Golgi and subsequently undergo exocytosis, fuse with vesicles to digest their contents, rupture, causing autolysis.

In co-translational import, if the newly forming polypeptide is destined for any of the compartments of the endomembrane system, it becomes associated with the ER membrane and is transferred across the membrane into the lumen (cisternal space) of the ER as synthesis continues. The completed polypeptide then either remains in the ER or is transported via various vesicles and the Golgi complex to another final destination. IMPs are inserted into the ER membrane as they are made, rather than into the lumen. In co-translational import, proteins to be targeted to the endoplasmic reticulum initially have an N-terminal peptide, the ER signal sequence, translated by a cytosolic ribosome. The ER signal sequence is bound by a signal-recognition particle (SRP), a ribonucleoprotein complex composed of 6 peptides and a 300 nucleotide RNA molecule. The SRP binds to the SRP receptor to dock the ribosome on the ER membrane. When the SRP receptor binds GTP, the nascent polypeptide enters the pore. The SRP is released with hydrolysis of the GTP. The growing polypeptide translocates through a hydrophilic pore created by one or more membrane proteins called the translocon. The most recent evidence suggests that the ribosome fits tightly across the cytoplasmic side of the pore and that the ER-lumen side is somehow closed off until the polypeptide is about 70 amino acids long. When the polypepide is complete, the signal peptidase cleaves the signal to release the protein into the ER lumen while retaining the signal peptide, for a time, in the membrane. Afterwards the ribosome is released and the pore closes completely.

G. Palade was awarded Nobel Prize in 1974 for studying role of ribosomes, cytoplasmic endoreticulum and other subcellular components in the synthesis of secretory proteins of pancreatic exocrine cell and role played by proteins in secretory process (Palade 1955, 1958). The proteins destined for secretion are initially targeted to the ER. This requires an N-terminal hydrophobic signal peptide which is recognized by a cytoplasmic ribonucleoprotein complex termed the signal recognition particle (SRP). The SRP comprises six proteins and a small cytoplasmic RNA (7S RNA), which is homologous to the *Alu* element. It binds to the signal sequence as it emerges from the ribosome and stalls protein synthesis until the complex reaches the ER. Here, the SRP binds to its receptor, a dimeric docking protein located on the endoplasmic reticulum membrane. The SRP has three functional domains which sponsor signal binding, receptor binding and elongation arrest, respectively). The ribosome then associates with its own receptor, a trimeric transmembrane protein called Sec61, and protein synthesis recommences. Sec translocon is Sec 61, which comprise of α-, γ- and β-subunits in eukaryotes. The polypeptide is fed into the ER lumen, a process termed vectorial discharge or co-translational import, through the Sec61 complex (some proteins require an additional component called TRAM). After synthesis, STP exchange releases the ribosome from the SRP, and the SRP itself is released from its receptor back into the cytoplasm. In the lumen of ER, proteins are folded by molecular chaperones.

Signal sequence at the ribosomal tunnel exit

Membrane and secretory proteins can be co-translationally inserted into or translocated across the membrane. This process is dependent on signal recognition on the ribosome by SRP, which results in targeting of the ribosome-nascent chain complex to the protein-conducting channel at the membrane. Halic et al. (2006) present an ensemble of structure at subnanometer resolution, revealing the signal sequence both at the ribosomal tunnel exit and in the bacterial and eukaryotic ribosome-SRP complexes. Molecular details of signal sequence interaction in both prokaryotic and eukaryotic complexes were obtained. The signal sequence is presented at the ribosomal tunnel exit in an exposed position ready for accommodation in the hydrophobic groove of the rearranged SRP54 M domain. Upon ribosome binding, the SRP 54 NG domain also undergoes a conformational rearrangement, priming it for the subsequent docking reaction with the NG domain of the SRP receptor.

Signal sequences target proteins for secretion from cells or for integration into cell membranes. As nascent proteins emerge from the ribosome, signal sequences are recognized by the signal recognition particle (SRP), which subsequently associates with its receptor (SR). In this complex, the SRP and SR stimulate each other's GTPase activity, and GTP hydrolysis ensures unidirectional targeting of cargo through a translocation pore in the membrane. To define the mechanism of reciprocal activation, Egea et al. (2004) determine the 1.9-Å structure of the complex formed between these two GTPases. These two partners form a quasi-two-fold symmetrical heterodimer. Complex formation aligns the two GTP molecules in a symmetrical, composite active site, and the 3' groups are essential for association, reciprocal activation and catalysis. This unique circle of twinned interactions is served twice on hydrolysis, leading to complex dissociation after cargo delivery.

The signal for membrane insertion is coded into the first one to three dozen amino acids of membrane-bound proteins. This signal peptide takes part in a chain of events leading to membrane attachment by the ribosome and membrane insertion of the protein. The first step occurs when the signal peptide becomes accessible outside of the ribosome. A striking verification of this hypothesis came about through recombinant DNA techniques in which a signal sequence was placed in front of α-globin gene, whose protein product is normally not transported through a membrane. Translation of this gene resulted in the ribosome becoming membrane-bound and the protein passing through the membrane.

Since different proteins enter different membrane-bound compartments (e.g., Golgi apparatus), some mechanism must exist that directs a nascent protein to its proper membrane. This specificity seems to depend on the exact signal-sequence receptors. Apparently, after the ribosome binds to the docking protein, the signal peptide interacts with a signal-sequence receptor, which presumably determines whether that protein is specific for that membrane (Figure 13.7). If it is, the remaining processes continue. If not, the ribosome may be released from the membrane.

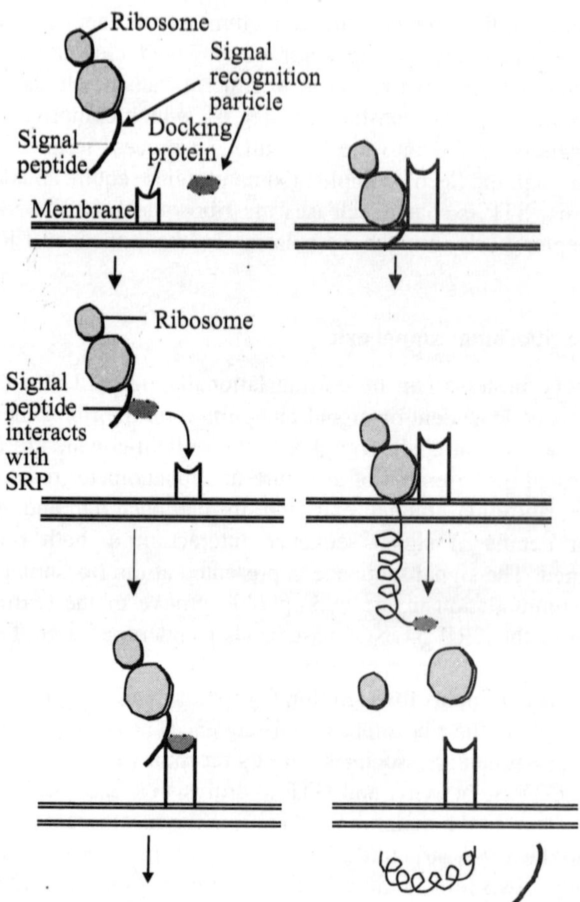

Figure 13.7 The signal hypothesis

The signal peptide does not seem to have a consensus sequence like the transcription or translation recognition boxes. Rather, similarities (at least for the ER and bacterial membrane-bound proteins) include a positively charged (basic) amino acid (commonly lysine or arginine) near the beginning (N-terminal end) followed by about a dozen hydrophobic (non-polar) amino acids, commonly alanine, isoleucine, leucine, phenylalanine, and valine. The signal peptide of the bovine prolactin protein reported by Sasavage et al. (1982) is: NH_2-Met Asp Ser Lys Gly Ser Ser Gln Lys Ser **Arg Leu Leu**

Leu Leu Leu Val Val Ser Asn Leu Leu Leu Cys Gln Gly Val Val Ser Thr Pro Val .... Asn Asn Cys-COOH. The amino acids represented in bold separate the signal peptide from the rest of the protein which consists of 199 amino acid residues.

Protein-conducting channel

Secreted and membrane proteins are translocated across or into cell membranes through a protein-conducting channel (PPC). Mitra et al. (2005) report a cryo-electron microscopy reconstruction of the *E. coli* PCC, SecYEG, complexed with the ribosome and a nascent chain containing a signal anchor. This reconstruction shows a messenger RNA, three transfer RNAs, the nascent chain, and detailed features of both a translocating PCC bound to mRNA hairpins. This translocating PCC forms connections with ribosomal RNA hairpins on two sides and ribosomal proteins at the back, leaving a frontal opening. Normal mode-based flexible fitting of the archaeal SecYEβ structure into the PCC electron microscopy densities favors a front-to-front arrangement of two SecYEG complexes in the PCC, and supports channel formation by the opening of two linked SecY halves during polypeptide translocation. On the basis of their observation in the translocating PCC of two segregated pores with different degrees of access to bulk lipid, they propose a model for co-translational protein translocation.

Co-translational translocation (of proteins across or into membranes is a vital process in all kingdoms of life. It requires that the translating ribosome be targeted to the membrane by signal recognition particle (SRP), an evolutionarily conserved ribonucleoprotein particle. SRP recognizes signal sequences of nascent protein chains emerging from the ribosome. Subsequent binding of SRP leads to a pause in peptide elongation and to the ribosome docking to the membrane-bound SRP receptor. Halic et al. (2004) present the structure of a targeting complex consisting of mammalian SRP bound to an active 80S ribosome carrying a signal sequence. This structure solved to 12-Å enables them to generate a molecular model of SRP in its functional conformation. The model shows how the S domain of SRP contacts the large ribosomal subunit at the nascent chain exit site to bind the signal sequence, and that of the Alu domain reaches into the elongation-factor-binding site of the ribosome, explaining its elongation arrest activity. Signal-sequence-dependent SRP-ribosome interaction is shown in Figure 13.8 (Halic et al. 2004). On the signal sequence binding by SRP54, a kinked

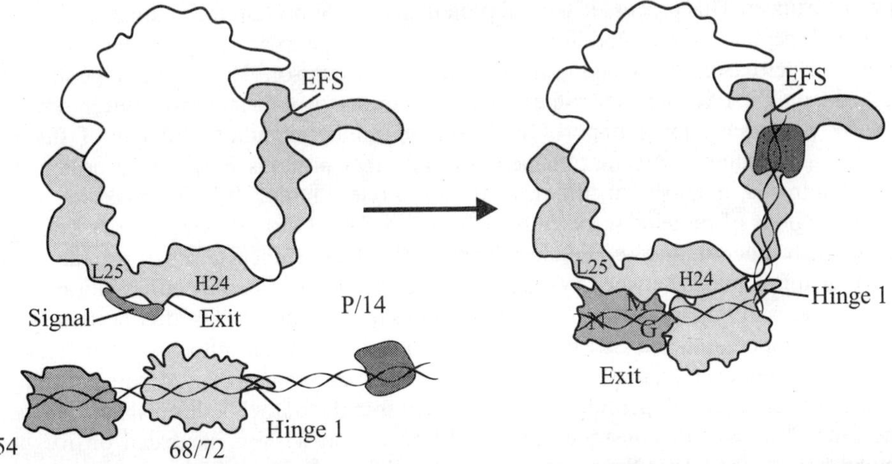

Figure 13.8 Signal-sequence-dependent SRP-ribosome interaction (Exit, peptide tunnel exit; EFS, elongation-factor-binding site)

conformation of SRP involving possibly SRP68/72 and a rotation around hinge 1 is stabilized. As a result, SRP interacts with the ribosome, stretching from the peptide exit (S domain) to the elongation-factor-binding site in the intersubunit space (Alu domain), where it causes elongation arrest by competition with elongation factors.

Secretory proteins are transported in coated vesicles. Pinching off from a membrane generates a vesicle. It fulfills its function by fusing with another membrane to release its contents. Some proteins have constitutive secretion while others are secreted through exocytosis. Proteins are imported into the cell by endocytosis. Exocytosis is the pathway for regulated secretion of proteins. The predominant protein in coat of vesicles is clathrin. The specificity of clathrin-coated vesicles presumably is conferred by other proteins located in coats.

POST-TRANSLATIONAL TARGETING IN EUKARYOTES

In eukaryotes, ribosomes may be free or may be attached to the endoplasmic reticulum (ER). Membrane-bound ribosomes synthesize proteins that are exported from the cell into the surrounding serum. 60S subunit of ribosome is attached to ER. The ER membrane is used to form a vesicle containing the protein. This vesicle then becomes secretary message whose membrane fuses with the cell membrane, emptying its contents outside the cell. On the other hand, free ribosomes synthesize specific non-serum liver proteins. It is possible that leader sequences on mRNAs determine which ribosomes shall translate them. Some short proteins are not synthesized on ribosomes. Such proteins are antibiotic polypeptides, such as gramicidin and tyrocidine.

Proteins synthesized on cytoplasmic ribosomes of eukaryotic cells can either be retained in the cytoplasm or transferred to subcellular organelles or inserted into membranes or secreted outside the cell (Verner et al. 1988). All living cells require specific mechanisms that target proteins to the cell surface. In eukaryotes, the first part of this process involves recognition in the endoplasmic reticulum of amino-terminal signal sequences and translocation through Sec translocons, whereas subsequent targeting to different surface locations is promoted by internal sorting signals. In eukaryotes, proteins can also be targeted to any one of several intracellular organelles, as well as to the nucleus. Since all bacterial proteins and most eukaryotic proteins are synthesized in the cytoplasm, targeted proteins must carry recognizable sequences or structures which allow them to be transported to the appropriate cellular compartment. This process is termed protein sorting or protein trafficking.

If the polypeptide is destined for the cytosol or for import into the nucleus, mitochondria, chloroplasts, or peroxisomes, its synthesis continues in the cytosol. When the polypeptide is complete, it is released from the ribosome and either remains in the cytosol or is transported into the appropriate organelle by post-translational import. In the endoplasmic reticulum, folding of the newly-made proteins may also require molecular chaperones and other proteins involved in protein folding. Bip (binding protein), a member of the Hsp70 chaperone family, briefly binds to and stabilizes hydrophobic regions of proteins (especially rich in Trp, Phe, Leu) allowing proper folding instead of aggregation with other immature proteins. Protein disulfide isomerase catalyzes the formation and breakage of disulfide bonds between cysteine residues to produce a stable conformation.

Post-translational import allows some polypeptides to enter organelles after protein synthesis. Like co-translational import into the ER, post-translational import into a mitochondrion (and chloroplast) involves a signal sequence (called a transit sequence), a membrane receptor, pore-forming membrane proteins, and a peptidase. Polypeptides being imported into the mitochondrion span both membranes at the same time. This was demonstrated in a cell-free import system incubated on ice in which the polypeptides begin to penetrate the mitochondrion but then stall. The transit sequence is cleaved by the transit peptidase present in the matrix, indicating that the N-terminus of the polypeptide is within the mitochondrion. At the same time, most of the polypeptide molecule is can be attacked by exogenously added proteolytic enzymes on the outside of the mitochondrion. Therefore, the polypeptide must span both membranes transiently during import at a contact site between the two membranes.

However, in the mitochondrion, the membrane receptor recognizes the signal sequence directly without the intervention of a cytosolic SRP. Furthermore, chaperone proteins play several crucial roles in the mitochondrial process: (1) Chaperones keep the polypeptide partially unfolded after synthesis in the cytosol so that binding of the transit sequence and translocation can occur. (2) Chaperones drive the translocation itself by binding to and releasing from the polypeptide within the matrix, an ATP-requiring process, and (3) Chaperones often help the polypeptide fold into its final conformation.

Polypeptides synthesized on cytosolic ribosomes but destined for either the intermembrane space or the inner membrane of the mitochondrion require two separate targeting sequences (both located at the N-terminus): (1) The polypeptide is directed to a contact (translocation) site on the mitochondrion by a positively charged or amphipathic transit sequence. (2) Cleavage of the transit sequence by a peptidase in the mitochondrial matrix uncovers a highly hydrophobic second signal sequence. This second signal sequence causes the polypeptide to be inserted into the inner membrane in the same way that mitochondrially-encoded polypeptides are targeted to this membrane.The remainder of the polypeptide is then moved across the membrane into the intermembrane space (or into the inner membrane for integral inner membrane proteins). Cleavage by a second peptidase can release the polypeptide into the intermembrane space leaving the signal sequence behind in the inner membrane.

Many different proteins are synthesized on the rough endoplasmic reticulum (RER) and transported to the Golgi apparatus in transport vesicles. From the Golgi apparatus these proteins are then sent to their proper locations in the cell. For example, lysosomal proteins are packaged into primary lysosomes, integral membrane proteins are sent to the plasma membrane or to particular regions of the plasma membrane, secretory proteins like peptide hormones or digestive enzymes must be packaged into secretory vesicles, and acrosomal proteins must be packaged into acrosomal vesicles. To insure that these different proteins are sent to their proper locations, they must be sorted, one from the other, in the Golgi apparatus. Thus one function of the Golgi apparatus is protein sorting.

Ribosomes are either free in the cytoplasm or associated with membranes. The choice is determined by the type of protein being synthesized. Membrane-bound ribosomes, polysomes, synthesize proteins that enter membrane or in eukaryotes are passed into membrane-bound organelles (e.g., the Golgi apparatus, mitochondria, chloroplasts, vacuoles) or transported outside the cell membrane. The mechanism for membrane attachment is explained by the signal hypothesis developed by Blobel and Dobberstein (1975) and Milstein et al. (1972). Signal hypothesis explains the process of secretion of proteins. According to this hypothesis, ribosomes synthesizing proteins are attached to the membrane via leader sequence. The mechanism is applicable for both prokaryotes and eukaryotes. G. Blobel was awarded Nobel Prize in 1999 for his discovery that proteins have intrinsic signals that govern their transport and location in the cell.

The signal sequence is cleaved as the polypeptide enters the ER lumen by a pentameric signal peptidase. Proteins with asparagine residues in the tripeptide motif An-Xaa-Ser/Thr (known as a sequon) are N-glycosylated with preformed oligosaccharide units. Not all sequons are glycosylated so other residues may be also be involved in the determination of a glycosylation site or the structure of the protein itself may inhibit the glycosylation of certain polypeptides.

Transmembrane proteins possess internal hydrophobic stop transfer sequence which lodges the protein in the ER membrane. The remainder proteins are packaged into vesicles and pass to the Golgi apparatus, although those proteins which are to be retained in the ER lumen are recognized and selectively returned to the ER. Such proteins often have a retention signal with the consensus sequence KDEL.

In the Golgi apparatus, N-linked glycan chains can be further modified and de novo O-glycosylation of serine, threonine and hydroxylysine residues may occur. In some cases, the glycosylation is required for correct protein folding and function whereas in others the glycosylation itself acts as a targeting signal. Proteins with a signal patch are modified by addition of mannose-6-phosphate, which targets them to the lysosomes. The remainder of the proteins is secreted, although some may possess sequences which cause them to be retained in the membrane of one of the organelles in the secretory pathway.

Halic et al. (2006) present the 8-Å structure of a "docking complex" consisting of a SRP-bound 80S ribosome and the SRP receptor. Interaction of the SRP receptor with both SRP and the ribosome rearranged the S domain of SRP such that a ribosomal binding site for the translocon, the L23e/L35 site, became exposed, whereas *Alu*-domain-mediated elongation arrest persisted. Dynamic behavior of SRP upon SR interaction is shown in Figure 13.9 (Halic et al. 2006). SR interaction with RNC and SRP induces rearrangement of the S domain but leaves the *Alu* domain unchanged: The NG domain of SRP54 is delocalized after NG-twin formation, resulting in exposure of the L23e/L35 universal ribosomal adaptor site and access of the translocon to its ribosomal binding site Ct4.

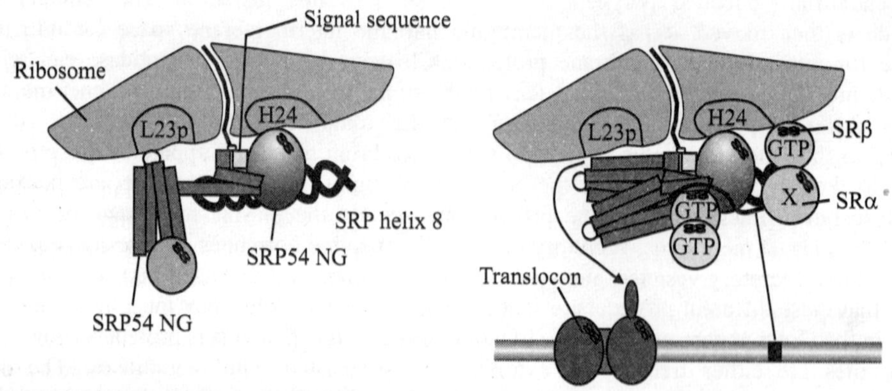

Figure 13.9 Dynamic behavior of SRP upon SR interaction

PROTEIN FOLDING

Deficiencies in the protein-folding capacity of the endoplasmic reticulum (ER) in all eukaryotic cells lead to ER stress and trigger the unfolded protein response (UPR). ER stress is sensed by Irel, a transmembrane kinase/endoribonuclease, which initiates the non-conventional splicing of the messenger RNA encoding a key transcriptional activator, Hacl in yeast to XBP1 in metazoans. In the absence of ER stress, ribosomes are stalled on unspliced *HAC1* mRNA. The translational control is by a base-pairing interaction between the *HAC1* intron and the *HAC1* 5′ untranslated region. After excision of the intron, transfer TNA ligase joins the several exons, lifting the translational block and allowing synthesis of Hacl from the spliced *HAC1* mRNA to ensue. Hacl in turn derives the UPR gene expression program comprising 7-8 per cent of the yeast genome to counteract ER stress. Aragon et al. (2009) show that on activation Irel molecules cluster in the ER membrane into discrete foci of high-order oligomers, to which unspliced *HAC1* mRNA is recruited by means of a conserved bipartite targeting element contained in the 3′ untranslated region. Disruption of either Irel clustering or *HAC1* RNA recruitment impairs UPR signaling. The *HAC1* 3′ untranslated region element is sufficient to target other mRNAs to Irel foci, as long as their translation is repressed. Translational repression afforded by the intron fulfils this requirement for *HAC1* mRNA. Recruitment of mRNA to signaling centers provides a new paradigm for the control of eukaryotic gene expression.

The contribution of co-translational chaperone functions to protein folding is poorly understood. Ribosome-associated trigger factor (TF) is the first molecular chaperone encountered by nascent polypeptides in bacteria. Kaiser et al. (2006) have shown that TF interacts with ribosomes and translating polypeptides in a dynamic reaction cycle. Ribosome binding stabilizes TF in an open, activated conformation. Activated TF departs from the ribosome after a mean residence time of ~10 s, but may remain associated with the elongating nascent chain for upto 35S, allowing entry of new TF molecule at the ribosome docking site. The duration of nascent-chain interaction correlates with the occurrence of hydrophobic motifs in translating polypeptides, reflecting a high aggregation propensity.

The ER was long considered to be the only compartment of the eukaryotic cell in which protein folding is accompanied by the enzyme-catalyzed disulfide bond formation. Cells harbor a second oxidizing compartment, the mitochondrial intermembrane space, where disulfide formation facilitates protein translocation from the cytosol. Protein oxidation has been implicated in many mitochondria-associated processes central for human health such as apoptosis, aging and regulation of the respiratory chain. Whereas the machineries of ER and mitochondria both form disulfides between cysteine residues, they do not share evolutionary origins and exhibit distinct mechanistic properties. Riemer et al. (2009) summarize the current knowledge of these oxidation systems and discuss their functional similarities and differences.

Unfolded Protein Response

Protein folding in endoplasmic reticulum is a complex process whose malfunction is implicated in aging. By use of cell's endogenous sensor (the unfolded protein response), Jonikas et al. (2009) identify several hundred yeast genes with roles in endoplasmic reticulum folding and systematically characterized their functional independencies by measuring unfolded protein response levels in double mutants. This strategy reveals multiple conserved factors critical for endoplasmic reticulum folding, including an intimate dependence on the later secretory pathway, a previously uncharacterized six-protein transmembrane complex, and a co-chaperone complex that delivers tail-anchored proteins to their membrane insertion machinery. The use of a quantitative reporter in a comprehensive screen followed by systematic analysis of genetic dependencies should be broadly applicable to functional dissection of complex cellular processes from yeast to man.

Mis-Folded Proteins

After insertion into endoplasmic reticulum (ER), proteins that fail to fold there are destroyed. Through a process termed dislocation such mis-folded proteins arrive in the cytosol, where ubiquitination, deglycosylation and finally proteasomal proteolysis dispense with the unwanted polypeptides. The human cytomegalovirus-encoded glycoproteins US2 and US11 catalyze the dislocation of class I major histocompatibility complex (MHC) products, resulting in their rapid degradation. Liley and Pleogh (2004) show that US11 uses its transmembrane domain to recruit class I MHC products to a human homolog of yeast Der1p, a protein essential for the degradation of class I MHC molecules catalyzed by US11, but not by US2. Derlin-1 is an important factor for the extraction of certain aberrantly folded proteins from the mammalian ER.

Mis-folding proteins of mis-charged tRNAs

Mis-folding proteins are associated with several pathological conditions, including neurodegenaration. Although some of these abnormally folded proteins result from mutations in genes encoding disease-associated proteins (for example, repeat-expansion diseases), more general mechanisms that lead to mis-folded proteins in neurons remain largely unknown. Lee et al. (2006) demonstrate that low levels of mis-charged tRNAs can lead to an intracellular accumulation of mis-folded proteins in neurons. These accumulations are accompanied by upregulation of cytoplasmic protein chaperones and by induction of the unfolded protein response. They report that the mouse sticky mutation which causes cerebellar Purkenje cell loss and ataxia, is a mis-sense mutation in the editing domain of the alanyl-tRNA synthetase gene that compromises the proofreading activity of this enzyme during aminoacylation of tRNAs. These findings demonstrate that disruption of translational fidelity in terminally differentiated neurons leads to the accumulation of mis-folded proteins and cell death, and provide a novel mechanism underlying neurodegradation.

Elimination of mis-folded proteins from the endoplasmic reticulum (ER) by retro-translocation is an important physiological adaptation to ER stress. The process requires recognition of a substrate in the ER lumen and its subsequent movement through the membrane by the cytosolic p97 ATPase. Ye et al. (2004) identify a p97-interacting membrane protein complex in the mammalian ER that links these two events. The central component of the complex, Derlin-1, is a homolog of Der1, a yeast protein whose inactivation prevents the elimination of mis-folded luminal ER proteins. Derlin-1 associates with different substrates as they move through the membrane, and inactivation of derlin-1 in *C. elegans* causes ER stress. Derlin-1 interacts with US11, a virally encoded ER protein that specifically targets MHC class I heavy chains for export from the ER, as well as with VIMP, a novel membrane protein that recruits the p97 ATPase and its cofactor. A model for US11-mediated retro-translocation of MHC class I heavy chains is given in Figure 13.10 (Ye et al. 2004). US11 recognizes HC in the ER lumen and targets it to Derlin-1, a proposed component of the retro-translocation channel. The p97 ATPase complex is recruited to Derlin-1 by VIMP. HC emerging into the cytosol is bound by p97. Poly-ubiquitin chains (Poly-Ub, red) are attached and recognized by both the N-domain (N) of p97 and the cofactor Ufd1/Npl4 (U/N). ATP hydrolysis by p97 moves HC into the cytosol. The retro-translocation of mis-folded ER proteins may occur similarly, with US11 being replaced by other targeting components.

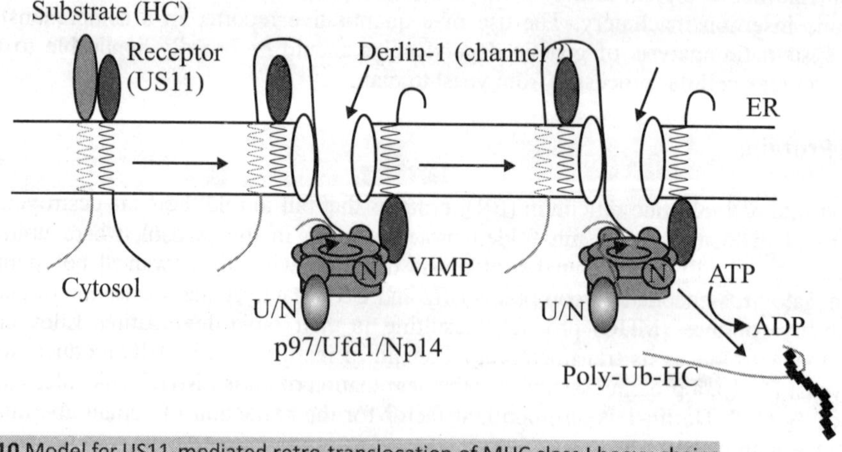

Figure 13.10 Model for US11-mediated retro-translocation of MHC class I heavy chains

Cell-to-cell protein transportation pathways in plants have been depicted in Figure 13.11. A model of the potential intracellular distribution mechanism of TMV-MP and KN1 via interaction with MPB2C is shown in Figure 13.12.

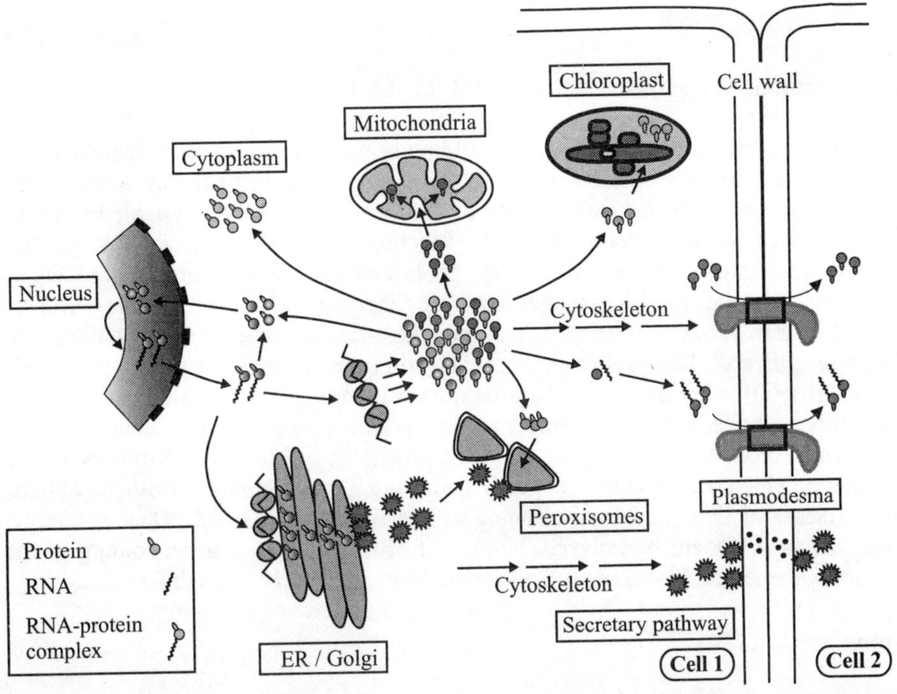

Figure 13.11 Cell-to-cell protein transportation pathways in plants (Reprinted, with permission, from http://www.univie.ac.at/ibmz/groups/kragov.htm)

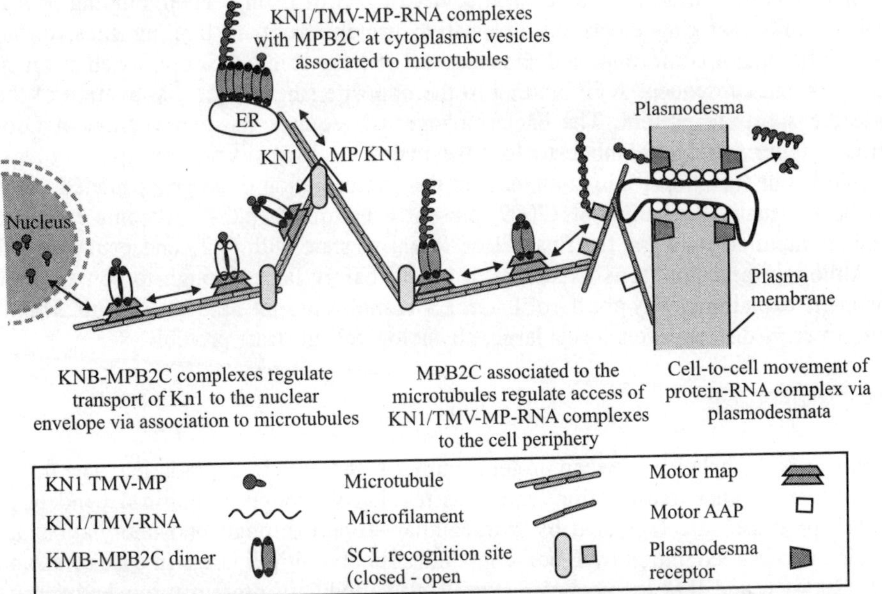

Figure 13.12 Model of the potential intracellular distribution mechanism of TMV-MP and KN1 via interaction with MPB2C (Reprinted, with permission, from http://www.univie.ac.at/ibmz/groups/kragov.htm)

MOLECULAR CHAPERONE

A molecular chaperone is a protein that is needed for the assembly or proper folding of some other protein but which itself is not a component of the target complex (Ellis 1991). The molecular chaperones provide a structure on which proper folding of proteins takes place. The chaperones do not appear to provide the three dimensional structure of the proteins they help, but rather the chaperones seem to bind to a protein in its early stages of folding and prevent unproductive folding. Thus molecular chaperones allow proteins to find a functional, stable state by allowing the proteins opportunities to fold into a thermodynamically stable and functional configuration. Each cycle of refolding requires ATP energy. Well-studied class of chaperones is known as chaperonins, or hsp60 proteins. They occur in bacteria, chloroplast and mitochondria. The best known of these chaperonins is the protein GroE of *E. coli.* This protein in its active form is composed of two components. GroEL and GroES. GroEL (hsp60) is composed of two disks, each composed of seven copies of a polypeptide. GroES (hsp10) is a smaller component composed of seven copies of a small subunit. GroEL forms a barrel within which protein folding takes place. As a protein begins to emerge from the ribosome in *E. coli*, two proteins which are products of genes DnaJ and DnaK,, complex with it. A third protein, GrpE, causes release of DnaJ and DnaK leading to interaction with GroEL. Once inside, GroES can cycle on and off as ATPs are hydrolyzed, helping GroEL chaperone correct folding of the protein. Finally, the protein emerges. If still not folded correctly, it can recycle through GroEL again.

Chaperonins

A subset of essential cellular proteins requires the assistance of chaperonins (in *E. coli*, GroEL and GroES), double-ring complexes in which the two rings act alternatively to bind, encapsulate and fold a wide range of nascent or stress-denatured proteins. The process starts by the trapping a substrate protein on hydrophobic surfaces in the central cavity of a GroEL ring. Then, binding of ATP and co-chapernonin GroES that ring ejects the non-native protein from its binding sites, through forced unfolding or other major conformational changes, and enclose it in a hydrophilic chamber for folding. ATP hydrolysis and subsequent ATP binding to the opposite ring trigger dissociation of the chamber and release the substrate protein. The bacteriophage T4 requires its own version of GroES, gp31, which forms a taller folding chamber, to fold the major viral capsid protein gp23. Polypeptides are known to fold inside the chaperonin complex, but the conformation of an encapsulated protein has not previously been visualized. Clare et al. (2009) present structure of gp23-chaperonin complex, showing both the initial captured state and the final, close-to-native state with gp23 encapsulated in the folding chamber. Although the chamber is expanded, it is still barely large enough to contain the elongated gp23 monomer, explaining why the GroEL-GroES complex is not able to fold gp23 and show the chaperonin structure distorts to enclose a large, physiological substrate protein.

PROTEIN TURNOVER

Protein turnover is the balance between protein synthesis and protein degradation. The turnover of Jun proteins, like that of other transcription factors, is regulated through ubiquitin-dependent proteolysis. Usually, such processes are regulated by extracellular stimuli through phosphorylation of the target protein, which allows recognition by F box-containing E3 ubiquitin ligases. In case of c-Jun and JunB, Gao et al. (2004) found that extracellular stimuli also modulate protein turnover by regulating the activity of an E3 ligase by means of its phosphorylation. Activation of the Jun amino-terminal kinase (JNK) mitogen activated protein kinase cascade after T cell stimulation accelerated degradation of c-

Jun and JunB through phosphorylation-dependent activation of the E3 ligase Itch. This pathway modulates cytokine production by effector T cells.

PROTEIN DEGRADATION

Lysosomes are organelles that store enzymes which rapidly degrade other proteins and nucleic acids. A famous target sequence is "KDEL". Initial targeting occurs via secretory pathway. In order to keep a cell working it needs to remove incorrectly synthesized proteins (with errors in amino acid sequence), damaged proteins (i.e. oxidative damage), cell-cycle specific proteins, the other signaling proteins which are no longer necessary. One mechanism of protein degradation is via lysosomes. Lysosomes are acidic vesicles that contain about 50 different enzymes involved in degradation. Examples of such enzymes are (a) proteases (cathepsins) that cleave peptide bonds, (b) phosphatases that remove covalently bound phosphates, (c) nucleases that cleave DNA/RNA, (d) lipases that cleave lipid molecules, and (e) carbohydrate-cleaving enzymes that remove covalently bound sugars from glycoproteins. Lysosomes often secrete their contents into the extracellular medium via exocytosis. Lysosomes can also target damaged organelles in a process called autophagy. Sometimes, lysosomes are triggered to rupture inside a cell, resulting in autolysis, also called apoptosis or programmed cell death.

Another major mechanism of protein degradation is via ubiquitin labeling of surplus proteins. Ubiquitin (a small 76-residue protein) is attached to the protein. First, an activating enzyme attaches itself to the carboxy terminus of free ubiquitin in an ATP-dependent process. Then, the activated ubiquitin is transferred onto a second enzyme which at the same time recognizes damaged proteins. The activated ubiquitin is then covalently linked to lysine residues on the surface of the damaged protein. These ubiquitin-tagged proteins are now recognized by specific proteases in the cytosol which in turn cleave and degrade the tagged protein. These proteases are combined in a very large protein complex called the proteasome. The proteasome (20S) is comprised of 28 subunits and has a molecular weight of 700 kDa.

Several models have been suggested for control of protein degradation. N-end rule and PEST hypothesis are well known models. These two models are explained here briefly.

N-end Rule

According to N-end rule the amino acid at the amino or N-terminal end of a protein is a signal to proteases that control the average length of life of a protein. In a recent experiment, almost complete predictability was achieved in determining the life of β-galactosidase protein based on N-terminal amino acid. Protein life spans range from 2 minutes for those with N-terminal arginine to greater than 20 h for those with N-terminal methionine (Table 13.1).

Table 13.1 N-terminal amino acid and half life of proteins chloroplasts (Reprinted, with permission, Miglani, G.S. 2006. *Developmental Genetics*. IKI International Publishing House, Pvt. Ltd., New Delhi)

N-terminal amino acid	*Half life*
Met, Ser, Ala, Thr, Val, Gly	20 h
Ile Glu	30 min
Tyr, Gln	10 min
Pro	7 min
Phe, Leu, Asp, Lys	3 min
Arg	2 min

The N-end rule states that the half-life of a protein is determined by the nature of its amino-terminal residue. Eukaryotes and prokaryotes use N-terminal destabilizing residues as a signal to target proteins for degradation by the N-end pathway. In eukaryotes, an E3 ligase, N-recognin recognizes N-end rule substrate and mediates their ubiquitination and degradation by the proteasome. In *E. coli*, N-end rule substrates are degraded by the AAA + chaperone ClpA in complex with the ClpPpeptidase (ClpP). Erbse et al. (2006) report the ClpP-specific adaptor, ClpS, is essential for degradation of N-end rule substrates by ClpAP in bacteria. ClpS binds directly to N-terminal destabilizing residues through its substrate binding site distal to the ClpS-ClpA interface, and targets these substrates to ClpAP for degradation. Degradation by the N-end rule pathway is more complex than anticipated and several other features are involved, including a net positive charge near the N terminus and an unstructured region between the N-terminal signal and the folded protein substrate. Through interaction with this signal, ClpS converts the ClpAP machine to a protease with exquisitely defined specificity, ideally suited to regulatory proteolysis.

PEST Hypothesis

According to PEST hypothesis, protein degradation is determined by one of the four amino acids – proline (P), glutamic acid (E), serine (S) and threonine (T); P, E, S and T are one-letter abbreviations for these four amino acids. The amino acid sequences of ten proteins with intracellular half-lives less than 2 h contain one or more regions rich in proline, glutamic acid, serine, and threonine (Rogers et al. 1986). Different proteins are programmed to survive different programs in the cell. But program based on N-terminal amino acid, which in PEST is amino acid within the protein. The proteins E1A, Cmyc, C-phos, α-casein and β-casein have been found to contain regions in one of these four amino acids (Table 13.2).

Table 13.2 Amino acid sequences showing extremes of PEST amino acid sequences

Protein	*Sequence*	*Half-life (h)*	*Sequence*
EIA	177-202	0.5	RTCCMFYYSPVSFPEPEPEPEPEPAR
c-myc	241-249	0.5	HEETPPTTSSDSEEEOEDEEEIDVVSVEK
c-fos	128-139	0.5	KVEQLSPEELEX
α-casein	54-79	2-5	KEMEAESISSSEEIVPNSVOEK
β-casein	1-25	2-5	RELEELNVPGEIVESLESSSEESITR

These PEST regions are generally, but not always, flanked by clusters containing several positively charged amino acids. Similar inspection of 35 proteins with intracellular half-lives between 20 and 220 h revealed that only three contain a PEST region. On the basis of this information, it was anticipated that caseins, which contain several PEST sequences, would be rapidly degraded within eukaryotic cells. This expectation was confirmed by red blood cell-mediated microinjection of ^{125}I-labeled caseins into HeLa cells where they exhibited half-lives of less than 2 h. The rapid degradation of injected α- and β-casein as well as the inverse correlation of PEST regions with intracellular stability indicates that the presence of these regions can result in the rapid intracellular degradation of the proteins containing them.

Insertion of integral membrane proteins

There are two possible mechanisms for the insertion of integral membrane proteins having a single transmembrane segment: (1) Type I: Insertion of a polypeptide with both a terminal ER signal sequence and an internal stop-transfer sequence. The terminal peptide is eventually cut off, leaving a transmembrane protein with its N-terminus in the ER lumen and its C-terminus in the cytosol. (2) Type II: Insertion of a polypeptide with only a single, internal start transfer sequence, which both starts polypeptide transfer and anchors itself permanently in the membrane. The amino-carboxyl orientation of the completed protein depends on the orientation of the start-transfer sequence when it first inserts into the translocation apparatus.

From the elephant to *E. coli*: SRP-dependent protein targeting

It is usually assumed that biological mechanisms revealed from the study of prokaryotic cells will be applicable to higher eukaryotes. This view was summed up by Jacques Monod's assertion that what is true for *Escherichia coli* is true for the elephant. Less frequently, information flows in the other direction, and experiments performed on complex organisms provide insights into cellular processes in prokaryotes. An example is provided by studies of the pathways used by eukaryotic and prokaryotic cells to export proteins (Wolins 1994). In higher eukaryotes, biochemical studies revealed that the targeting of nascent secretory proteins to the endoplasmic reticulum (ER) required a cytoplasmic ribonucleoprotein (RNP) particle, the signal recognition particle (SRP), and its membrane-bound receptor. Characterization of SRP and its receptor in mammalian organisms made it possible to identify related components in several prokaryotes. Recent experiments investigating the role of the SRP and SRP receptor homologs suggest that these exceedingly conserved components function in the targeting of exported proteins to the *E. coli* cytoplasmic membrane. This information adds a novel twist to our understanding of protein export in *E. coli*, a process that had already been well defined using a powerful combination of biochemical and genetic approaches.

FATE OF RIBOSOMES AFTER TRANSLATION

If the ribosome is free (in the cytoplasm), the subunits disassociate and the polypeptide and mRNA strand detaches. I the ribosome is attached to the endoplasmic reticulum, the ribosome may detach and the subunits disassociate. The polypeptide is ejected into the ER. In either case, the ribosomes may be re-used should they encounter another strand of mRNA.

During normal translation termination, a stop codon entering the ribosome A-site recruits eukaryotic release factor 1 (eRF1) and eRF3·GTP. GTP hydrolysis by eRF3 triggers eRF1 rearrangements that induce nascent protein release and eRF3·GDP dissociation (Graille and Séraphin 2012). Recruitment of Rli1 (RNase L inhibitor 1) ensues and triggers subunit dissociation after ATP hydrolysis. Some termination events are associated with activation of mRNA decay by nonsense-mediated decay. Release of a stalled ribosome at a mRNA 3′-end is an example of non-stop decay (NSD) of a poly(A)-carrying mRNAs. After translation of the poly(A) tail, attachment of a poly-Lys extension to the nascent polypeptide and recruitment of the super killer 7 (Ski7)–exosome complex lead to ribosome dissociation and exosome-mediated mRNA degradation. The Dom34–Hbs1 complex was also recently shown to be involved in this process. The released protein is degraded by the proteasome after ubiquitylation by E3 ligases (not shown). Degradation of mRNAs that induce translational stalls by the no-go decay (NGD) pathway. Ribosome stalling can result from faulty mRNAs that carry a very stable stem–loop near the stalling site, an in-frame poly(A) stretch (at least 18 adenine residues) or a damaged base within the ribosomal A-site. In yeast, such mRNAs are endonucleolytically cleaved by Dom34 and Hbs1 and degraded by the 5′ to 3′ exoribonuclease Xrn1. The nascent peptide is degraded by Ltn1–Not4-mediated proteasomal degradation (not shown). In the 18S-rRNA decay (18S-NRD) pathway, ribosomes unable to elongate owing to a defective 40S subunit recruit Dom34, Hbs1 and Ski7, leading to 18S rRNA degradation.

MESSENGER RNA DECAY

Messenger RNA (mRNA) is recognized as a major control point in regulation of gene expression. After messenger RNA has been used as a template for translation, it is protected till needed by the cell or degraded by the cellular machiney when not required. mRNA can last a long time. For example, mammalian red blood cells eject their nucleus but continue to synthesize hemoglobin for several months. This indicates that mRNA is available to produce the protein even though the DNA is gone. mRNA has non-coding nucleotides at either end of the molecule. These segments contain information about the number of times mRNA is transcribed before being destroyed by ribonucleases. Hormones stabilize certain mRNA transcripts. For example, prolactin is a hormone that promotes milk production because it affects the length of time the mRNA for casein (a major milk protein) is available.

Ribonucleases are enzymes that destroy mRNA. Special structure and 3′ termini of many mRNAs appear to provide protection against rapid exonucleolytic digestion but selectivity of the decay process appears to be determined to a large extent by interactions between endonucleases or other factors and internal mRNA structures (Braweraman 1989). An apparent linkage between translation and mRNA decay adds further complexity to the decay process. Site-specific endonucleases responsible for the rate-limiting step in the mRNA decay have been shown to function in bacteria. Such endonuclease has been identified in higher organisms; there it could be an exonuclease. The synthesis rate of many proteins is influenced more by how rapidly their mRNA templates decay, or turnover, than by how rapidly these RNAs are synthesized (Ross 1989). For example, histone mRNA turnover in a cell free system quadrupled when we added histones to polyribosomes isolated from cells. Added histones had no effect on other mRNA molecules. Tubulin is such other protein that autoregulates its mRNA.

Processing Bodies

Translation and mRNA degradation are affected by a key transition where eukaryotic mRNAs exit translation and assemble an mRNP state that accumulates into processing bodies (P bodies). Cytoplasmic sites of mRNA degradation contain non-translating mRNAs, and the mRNA degradation machinery. Coller and Parker (2005) identify the decapping activators (Dhh1p and Pat1p as functioning as translational repressors and facilitators of P body formation. Strains lacking both Dhh1p and Pat1p show strong defects in mRNA decapping and P body formation and are blocked in translational repression. Contrastingly, overexpression of Dhh1p or Pat1p causes translational repression, P body formation, and arrests cell growth. Dhh1p, and its human homolog, RCK/p54, repress translation in vitro, and Dhh1p function is bypassed in vivo by inhibition of translational initiation. These results identify a broadly acting mechanism of translational repression that targets mRNAs for decapping and functions in translational control. This mechanism is competitively balanced with translation, and shifting this balance is an important basis of translational control. A general active machinery exists which is in competition with translation (Coller and Parker 2005) (Figure 13.13). This competition creates a finely balanced system setting the relative translation rate for an mRNA. mRNAs can be driven into either translation or repression by tipping the balance of this competition via any number of events.

mRNA is decapped and degraded in P-bodies

Eukaryotic cells contain non-translating messenger RNA concentrated in P-bodies, which are sites where the mRNA can be decapped and degraded. Brengues et al. (2005) present evidence that mRNA molecules within yeast P-bodies can also return to translation. First, inhibiting delivery of new mRNAs to P-bodies leads to their disassembly independent of mRNA decay. Second, P-bodies decline in a

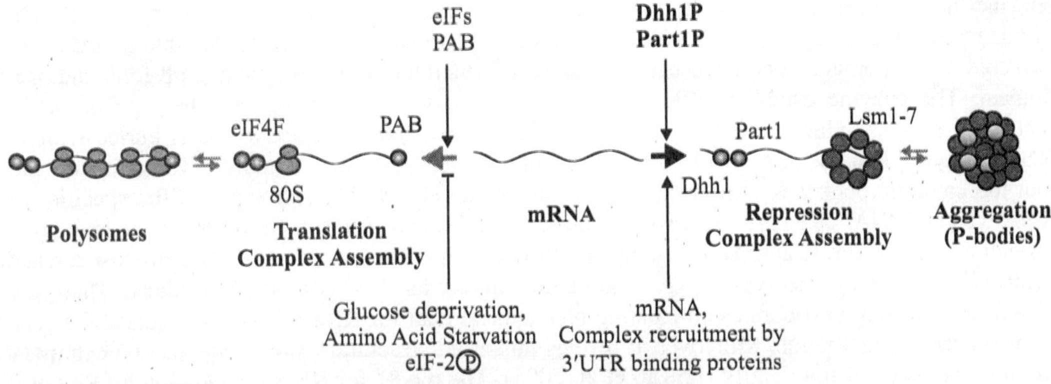

Figure 13.13 A general active machinery in competition with translation

translation-dependent manner during stress recovery. Third, reporter mRNAs concentrate in P-bodies when translation initiation is blocked and resume translation and exit P-bodies when translation is restored. Fourth, stationary phase yeast have large P-bodies containing mRNAs that reenter translation when growth resumes. The reciprocal movement of mRNAs between polysomes and P-bodies is likely to be important in the control of messenger RNA translation and degradation. Moreover, the presence of related proteins in P-bodies and maternal mRNA storage granules suggests this mechanism is widely adapted for mRNA storage.

EXOSOME

Exosomes are 50-90 nm vesicles secreted by a wide range of mammalian cell types (Keller et al. 2006). First discovered in maturing mammalian reticulocytes, they were shown to be a mechanism for selective removal of many plasma membrane proteins (van Niel et al. 2006. The exosome is a major eukaryotic nuclease located in both the nucleus and the cytoplasm that contributes to the processing, quality control and/or turnover of a large number of cellular RNAs. This large macromolecular assembly has been described as a 3'→5' exonuclease and shown to contain a nine-subunit ring structure evolutionarily related to archaeal exosome-like complexes and bacterial polynucleotide phosphorylases. Unlike its prokaryotic counterparts, the yeast and human ring structures are catalytically inactive. In contrast, the exonucleolytic activity of the yeast exosome core was shown to be mediated by the RNB domain of the eukaryotic-specific Dis3 subunit. Leberton et al. (2008) show that yeast Dis3 has an additional endoribonuclease activity mediated by the PIN domain located at the amino terminus of this multidomain protein. Simultaneous inactivation of the endonucleolytic and exonucleolytic activities of the exosome core generates a synthetic growth phenotype in vivo, supporting a physiological function for the PIN domain. This activity is responsible for the cleavage of some natural exosome substrates, independently of exonucleolytic degradation. These results show that eukaryotic exosome cores have both endonucleolytic and exonucleolytic activities, mediated by two distinct domains of the Dis3 subunit. The mode of action of eukaryotic exosome cores in RNA processing and degradation should be reconsidered, taking into account the cooperation between its multiple ribonucleolytic activities.

RNA degradation is a determining factor in the control of gene expression. The maturation, turnover and quality control of RNA is performed by many different classes of ribonucleases. RNaseII is a major exonuclease that intervenes in all of these fundamental processes; it can act independently or as a component of the exosome, an essential RNA-degrading multiprotein complex. RNase II-like enzymes are found in all three kingdoms of life, but there are no structural data for any of the proteins of this family. Frazao et al. (2006) have reported that *E. coli* RNase II is organized into four domains – two cold-shock domains, one RNB catalytic domain, which has an unprecedented αβ-fold, and one S1 domain. The enzyme establishes contact with RNA in two distinct regions, the 'anchor' and the 'catalytic' regions, which act synergistically to provide catalysis. The active site is buried within the RNB catalytic domain, in a pocket formed by four conserved sequence motifs. The structure shows that the catalytic pocket is only accessible to single-stranded RNA, and explains the specificity for RNA versus DNA cleavage. It also explains the dynamic mechanism of RNA degradation by providing the structural basis for RNA translocation and enzyme processivity. They propose a reaction mechanism for exonucleolytic RNA degradation involving key conserved residues. Their three-dimensional model corroborates all existing biochemical data for RNase II, and elucidates the general basis for RNA degradation. Moreover, it reveals important structural features that can be extrapolated to other members of this family (Frazao et al. 2006). The model for RNA degradation by RNase II is given in Figure 13.14. The ssRNA is threaded into the catalytic cavity and clamped between Tyr253 and Phe358. The additional stabilization of RNA inside the cavity drives the RNA translocation after each cleavage, upto a final four-nucleotide fragment.

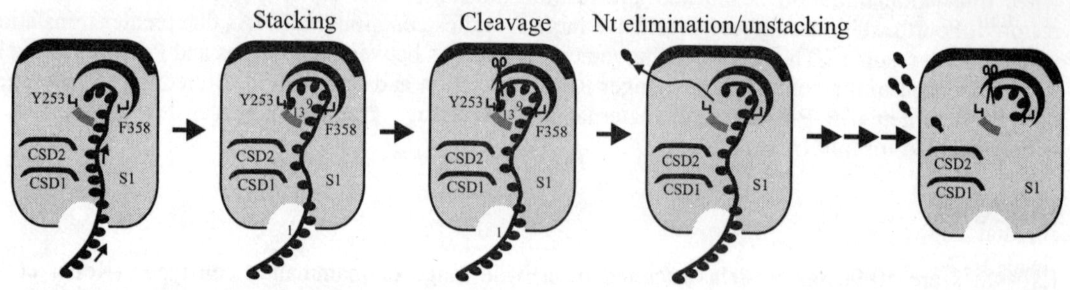

Figure 13.14 Model for RNA degradation by RNase II

Ribonuclease E has a key role in mRNA degradation and the processing of catalytic and structural RNAs in *E.coli*. Lee et al. (2003) report the discovery of an evolutionarily conserved 17.4 kDa protein, named RraA (regulator of ribonuclease activity) that binds to RNase E and inhibits RNase E endonucleolytic cleavages without altering cleavage site specificity or interacting detectably with substrate RNAs. Overexpression of RraA circumvents the effects of an autoregulatory mechanism that normally maintains the RNase E cellular level within a narrow range, resulting in the genome-wide accumulation of RNase E-targeted transcripts. While not required for RraA cation, the C-terminal RNase E region that serves as a scaffold for formation of a multiprotein degradasome complex modulates the inhibition of RNase E catalytic activity by RraA. These results reveal a possible mechanism for the dynamic regulation of RNA decay and processing by RNase binding proteins.

Although the primary mechanism of eukaryotic messenger RNA decay is exoribonucleolytic degradation in the 5'-to-3' orientation, it has been widely accepted that bacteria can only degrade RNAs with opposite polarity, i.e., 5' to 3'. Mathy et al. (2007) show that maturation of the 5' side of *Bacillus subtilis* 16S ribosomal RNA occurs via a 5'-to-3' exonucleolytic pathway, catalyzed by the widely distributed essential ribonuclease RNase J1. The presence of 5'-to-3' exoribonuclease activity in *B. subtilis* suggested an explanation for the phenomenon whereby mRNAs in this organism are stabilized for great distances downstream of "roadblocks" such as stalled ribosomes or stable secondary structures, whereas upstream sequences are never detected. They show that a 30S ribosomal subunit bound to a Shine Dalgarno-like element (Stab-SD) in the *cryIIIA* mRNA blocks exonucleolytic progression of RNase J1, accounting for the stabilizing effect of this element in vivo. Schematic view of *cryIIIA* mRNA degradation by RNase J1 is presented in Figure 13.15 (Mathy et al. 2007). 5' to 3' nucleolytic attack of the cryIIIA mRNA (wavy line) results in the production of mononucleotides (small black squares). The position of Stab-SD and site of translation initiation (SAD-AUG) are indicated.

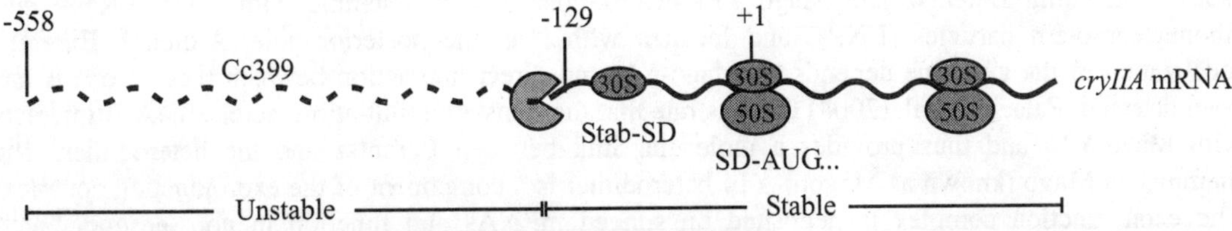

Figure 13.15 Schematic view of *cryIIIA* mRNA degradation by RNase J1

No-Go Devay Pathway

In many manufacturing processes, quality control is crucial, and gene expression is no exception. A new pathway monitors mRNAs – the intermediaries of gene expression – and destroys faulty molecules. No-go decay pathway, given by Tollervey (2006) and shown in Figure 13.16, results in rapid mRNA cleavage and degradation. This pathway is different from other mRNA degradation pathways in yeast, in that stalling of the ribosomes by a stem-loop structure leads to cleavage of the mRNA, a process that involves the Dom34 and Hbs1 proteins. This cleavage generates free ends that are subject to degradation: the enzyme Xrn1 chews up the mRNA from the 5′ end, whereas the multienzyme exosome complex degrades from the 3′ end. Dom34 and Hbs1 might interfere directly with the 'A site' of the stalled ribosome, possibly to promote release of the ribosome and allow subsequent mRNA cleavage, or they may directly stimulate cleavage of the mRNA (Doma and Parker 2006).

Non-sense-mediated mRNA Decay

Most eukaryotic genes are interrupted by non-coding introns that must be accurately removed from pre-messenger RNAs to produce translatable mRNAs. Splicing is guided locally by short conserved sequences, but genes typically contain many potential splice sites, and the mechanisms specifying the correct sites remain poorly understood. In most organisms, short introns recognized by the intron definition mechanism cannot be efficiently predicted on the basis of sequence motifs. In multicellular eukaryotes, long introns are recognized through exon definition and most genes produce

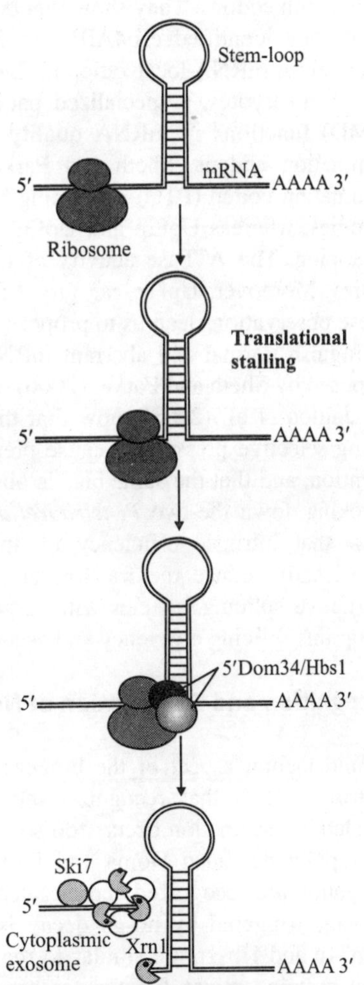

Figure 13.16 No-go decay pathway

multiple mRNA variants through alternative splicing. The non-sense-mediated mRNA decay (NMD) pathway may further shape the observed sets of variants by selectively degrading those containing premature termination codons, which are frequently produced in mammals.

The specification of both the germline and abdomen in *Drosophila* depends on the localization of *oskar* messenger RNA to the posterior of the oocyte. This localization requires several *trans*-acting factors, including Barentsz and Mago-Y14 heterodimer, which assemble with oskar mRNA into ribonucleoprotein particles (RNPs) and localize with it at the posterior pole. Although Barentsz localization in the germline depends on Mago-Y14, no direct interaction between these proteins has been detected. Palacios et al. (2004) demonstrate that the translation initiation factor eIF4AIII interacts with Mago-Y14 and thus provides a molecular link between Barentsz and the heterodimer. The mammalian Mago (known as Magoh)-Y14 heterodimer is a component of the exon junction complex. The exon junction complex is deposited on spliced mRNAs and function in non-sense-mediated mRNA decay (NMD), a surveillance mechanism that degrades mRNA with premature translation-termination codons. They show that both Barenstz and eIF4AIII are essential for NMD in human cells. Thus, they identified eIF4AIII and Barenstz as components of conserved protein complex that is essential for mRNA localization in flies and NMD in mammals.

In eukaryotes, a specialized pathway of mRNA degradation termed non-sense-mediated decay (NMD) functions in mRNA quality control by recognizing and degrading mRNAs with aberrant termination codons. Sheth and Parker (2006) demonstrate that NMD in yeast targets premature termination codon (PTC)-containing mRNA to P-bodies. Upf1p is sufficient for targeting mRNAs to P-bodies, whereas Upf2p and Upf3p act, at least in part, downstream of P-body targeting to trigger decapping. The ATPase activity of Upf1p is required for NMD after the targeting of mRNA to P-bodies. Moreover, Upf1p can target normal mRNAs to P-bodies but not promote their degradation. These observations lead us to propose a new model for NMD wherein two successive steps are used to distinguish normal and aberrant mRNAs. A model for the process of non-sense-mediated decay, as proposed by Sheth and Parker (2006), is given in Figure 13.17.

Jaillon et al. (2008) show that the tiny introns of the ciliate *Paramecium tetraurelia* are under strong selective pressure to cause premature termination of mRNA translation in the event of intron retention, and that the same bias is observed among the short introns of plants, fungi and animals. By knocking down the two *P. tetraurelia* genes encoding UPF1, a protein that is crucial in NMD, they show that intrinsic efficiency of splicing varies widely among introns that NMD activity can significantly ıeduce the fraction of unspliced mRNAs. The results suggest that, independently of alternative splicing, species with large intron numbers universally rely on NMD to compensate for suboptims splicing efficiency and accuracy.

Recognition and Degradation of Non-functional mRNAs

A fundamental aspect of the biogenesis and function of eukaryotic messenger RNA is the quality control systems that recognize and degrade non-functional mRNAs. Eukaryotic mRNAs where translation termination occurs too soon (non-sense-mediated decay) or fails to occur (non-stop decay) are rapidly degraded Doma and Parker (2006) show that yeast mRNAs with stalls in translation elongation are recognized and targeted for endonucleolytic cleavage, referred to as no-go decay. The cleavage triggered by no-go decay is dependent on translation and involves Dom34p and Hbs1p. Dom34p and Hbs1p are similar to the translation termination factors eRF1 and eRF3, indicating that these proteins might function in recognizing the stalled ribosome and triggering endonucleolytic

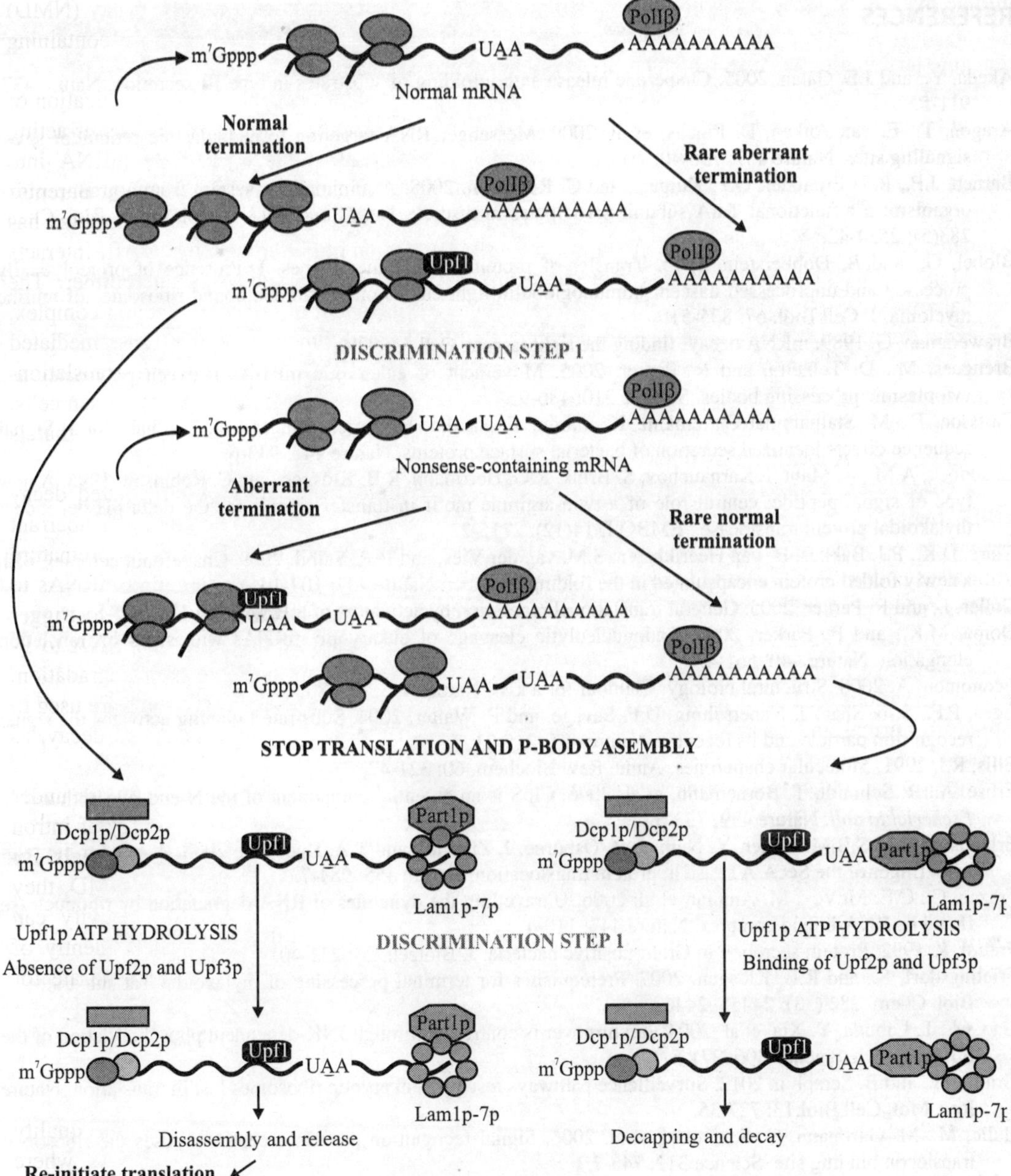

Figure 13.17 A model for the process of nonsense-mediated decay

cleavage. No-go decay provides a mechanism for clearing the cell of stalled translation elongation complexes, which could occur as a result of damaged mRNAs or ribosomes, or a mechanism of post-transcriptional control.

REFERENCES

Akeda, Y., and J.E. Galan. 2005. Chaperone release and unfolding of substrates in type III secretion. Nature 437: 911-5.

Aragon, T., E. van Anken, D. Pincus, et al. 2009. Messenger RNA targeting to endoplasmic reticulum stress signaling sites. Nature 457: 736-40.

Barnett, J.P., R.T. Eijlander, O.P. Kuipers, and C. Robinson. 2008. A minimal Tat system from a gram-positive organism: a bifunctional TatA subunit participates in discrete TatAC and TatA complexes. J. Biol. Chem. 283(5): 2534-42.

Blobel, G., and B. Dobberstein. 1975. Transfer of proteins across membranes. I. Presence of proteolytically processed and unprocessed nascent immunoglobulin light chains on membrane bound ribosomes of murine myeloma. J. Cell Biol. 67: 835-51.

Braweraman, G. 1989. mRNA decay: finding the right targets. Cell 57: 9-10.

Brengues, M., D. Telxeira, and R. Parker. 2005. Movement of eukaryotic mRNAs between polysomes and cytoplasmic processing bodies. Science 310: 486-9.

Carlsson, F., M. Stalhammar-Carlemalm, K. Flardh, C. Sandin, E. Carlemalm, and G. Lindhal. 2006. Signal sequence directs localized secretion of bacterial surface proteins. Nature 442: 943-6.

Chaddock, A.M., A. Mant, I. Karnauchov, S. Brink, R.G. Herrmann, R.B. Klösgen, and C. Robinson. 1995. A new type of signal peptide: central role of a twin-arginine motif in transfer signals for the delta pH-dependent thylakoidal protein translocase.. EMBO J. 14(12): 2715-22.

Clare, D.K., P.J. Bakkes, H. van Heerikhizen, S.M. van der Vies, and H.R. Saibil. 2009. Chaperonin complex with a newly folded protein encapsulated in the folding chamber. Nature 457: 107-10.

Coller, J., and R. Parker. 2005. General translational repression by activators of mRNA. Cell 122: 875-86.

Doma, M.K., and R. Parker, 2006. Endonucleolytic cleavage of eukaryotic mRNAs with stalls in translation elongation. Nature 440: 561-4.

Economou, A. 2008. Structural biology: Clamour for a kiss. Nature 455: 879-80.

Egea, P.F., S.-o. Shan, J. Naperschnig, D.F. Savage, and P. Walter. 2004. Substrate twinning activates the signal recognition particle and its receptor. Nature 427: 215-21.

Ellis, R.J. 1991. Molecular chaperones. Annu. Rev. Biochem. 60: 321-47.

Erbse, A., R. Schmidt, T. Bornemann, et al. 2006. ClpS is an essential component of the N-end rule pathway in *Escherichia coli*. Nature 439: 753-6.

Erlandson, K.J., S.B.M. Miller, Y. Nam, A.R. Osborne, J. Zimmer, and T.A. Rapoport. 2008. A role for the two-helix finger of the SecA ATPase in protein translocation. Nature 455: 984-7.

Frazao, C., C.E. McVey, M. Amblar, et al. 2006. Unravelling the dynamics of RNA degradation by ribonuclease II and its RNA-bound complex. Nature 443: 110-4.

Freudl, R. 1992. Protein secretion in Gram-positive bacteria. J. Biotech. 23: 231-40.

Frielingsdorf, S., and R.B. Klösgen, 2007. Prerequisites for terminal processing of thylakoidal Tat substrates. J. Biol. Chem. 282 (33): 24455–24462.

Gao, M., T. Labuda, Y. Xia, et al. 2004. Jun turnover is controlled through JNK-dependent phosphorylation of the E3 ligase Itch. Science 306: 271-5.

Graille M., and B. Séraphin. 2012. Surveillance pathways rescuing eukaryotic ribosomes lost in translation. Nature Rev. Mol. Cell Biol.13: 727-35.

Halic, M., M. Gartmann, O. Schienker, et al. 2006. Signal recognition particle receptor exposes the ribosomal translocon binding site. Science 312: 745-7.

Holland, L. I.B., Schmitt, and J. Young. 2005. Type 1 protein secretion in bacteria, the ABC-transporter dependent pathway. Mol. Memb. Biol. 22: 29-39.

Jaillon, O., K. Bouhouche, J,-F. Gout, et al. 2008. Translational control of intron splicing in eukaryotes. Nature 451: 359-62.

Jonikas, M.C., S.R. Collins, V. Denic, et al. 2009. Comprehensive characterization of genes required for protein folding in the endoplasmic reticulum. Science 323: 1693-7.

Kaiser, C.M., H.-C. Chang, V.R. et al. 2006. Real-time observation of trigger factor function on translating ribosomes. Nature 444: 455-60.

Kragler F., J. Monzer, K. Shash, B. Xoconostle-Cazares, and W.J. Lucas. 1998. Cell-to-cell transport of proteins: Requirement for unfolding and characterization of binding to a plasmodesmal receptor. Pl. Journal. 15: 367-81.

Lebreton, A., R. Tomecki, A. Dziembowski, and B. Seraphin. 2008. Endonucleotytic RNA cleavage by a eukaryotic exosome. Nature 456: 993-6.

Lee, J.W., K. Beebe, L.A. Nangle, et al. 2006. Editing-defective tRNA synthetase causes protein misfolding and neurodegeneration. Nature 443: 50-5.

Lee, K., J. Zhan, J. Gao, et al. 2003. RraA: A protein inhibitor of RNase E activity that globally modulates RNA abundance in *E. coli*. Cell 114: 623-34.

Liley, B.N., and H.I. Pleogh. 2004. A membrane protein required for dislocation of misfolded proteins from the ER. Nature 429: 834-40.

Mathy, N., L. Bernard, O. Pellegrini, R. Daou, T. Wen, and C. Condon. 2007. 5′-to-3′ exonuclease activity in bacteria: Role of RNase J1 in rRNA maturation and 5′ stability of mRNA. Cell 129: 681-92.

Milstein, C., G.G. Broenlee, T.M. Harrison, and M.B. Mathews. 1972. A possible precursor of immunoglobulin light chains. Nature New Biol. 239: 117-20.

Palacios, I.M., D. Gatfeld, D. St. Johnston, and E. Izaurralde. 2004. An eIF4AIII-containing complex required for mRNA localization and nonsense-mediated mRNA decay. Nature 427: 753-7.

Palade, G.E. 1955. A small particulate component of the cytoplasm. J Biophys Biochem Cytol. 1: 59-68.

Palade, G.E. 1958. Introduction. In: *Microsomal particles and Protein Synthesis*. Roberts, R.B., editor, New York: Pergamon Press, Inc.

Riemer, J., N. Bulleid, and J.M. Hermann. 2009. Disulphide formation in the ER and mitochondria: two solutions to a common process. Science 324: 1284-7.

Robinson, C., and A. Bolhuis. 2004. Tat-dependent protein targeting in prokaryotes and chloroplasts. Biochimica et Biophysica Acta 1694: 135-47.

Rogers, S., R. Wells, and M. Rechsteiner. 1986. Amino acid sequences common to rapidly degraded proteins: the PEST hypothesis. Science 234: 364-8.

Ross, J. 1989. The turnover of messenger RNA. Scient. Am. 260: 28-35.

Sargent, F., B.C. Berks, and T. Palmer. 2006. Pathfinders and trailblazers: a prokaryotic targeting system for transport of folded proteins. FEMS Microbiol. Lett 254(2): 198–207.

Sasavage, N., M. Smith, S. Gillam, R.P. Woychik, and F.M. Rottman. 1982. Variation in polyadenylation site of bovine prolactin messenger RNA. Proc. Natl. Acad. Sci. USA 79: 223-7.

Schaffitzel, C., M. Oswald, I. Berger, et al. 2006. Structure of the *E. coli* signal recognition particle bound to a translating ribosome. Nature 444: 503-6.

Tollervey, D. 2006. RNA lost in translocation. Nature 440: 425-6.

Tsukazaki, T., H. Mori, S. Fukai, et al. 2008. Conformational transition of Sec machinery inferred from bacterial SecYE structures. Nature 455: 988-91.

van Niel, G., Porto-Carreiro, I., Simoes, S. and Raposo, G. 2006. Exosomes: a common pathway for a specialized function. J. Biochem. 140: 13-21.

Verner, K., and G. Schate. 1988. Proteins translocation across membrane. Science 241: 1307-13.

Wolin, S.L. 1994. From the elephant to *E. coli*: SRP-dependent protein targeting. Cell 77: 787-90.

Ye, Y., Y. Shibata, D. Ron, and T.A. Rapoport. 2004. A membrane protein complex mediates retro-translocation from the ER lumen into the cytosol. Nature 429: 841-7.

Zimmer, J., Y. Nam, and T.A. Rapoport. 2008. Structure of a complex of the ATPase SecA and the protein – translocation channel. Nature 455: 936-47.

14

Gene Expression – An Overview

In genetics, gene expression is the most fundamental level at which the genotype gives rise to the phenotype. The genetic code stored in DNA is "interpreted" by gene expression, and the properties of the expression give rise to the organism's phenotype. Such phenotypes are often expressed by the synthesis of proteins that control the organism's shape, or that act as enzymes catalyzing specific metabolic pathways characterizing the organism.

GENE STRUCTURE

A gene is a stretch of DNA that encodes information. Genomic DNA consists of two antiparallel and reverse complementary strands, each having 5' and 3' ends. With respect to a gene, the two strands may be labeled the "template strand," which serves as a blueprint for the production of an RNA transcript, and the "coding strand," which includes the DNA version of the transcript sequence. With highly efficient nucleotide sequencing techniques, it has now been possible to understand gene structure much better.

GENE FUNCTION

What is function of a gene? Is only one gene responsible or sufficient for producing one functional product? With the coming up of concepts like one gene-many proteins, many genes-one protein, moveable genes, half genes, etc. it has still not been possible to have a unified view of gene function.

GENE EXPRESSION

Gene expression is the process by which information from a gene is used in the synthesis of a functional gene product. These products are often proteins, but in non-protein coding genes such as ribosomal RNA (rRNA), transfer RNA (tRNA) or small nuclear RNA (snRNA) genes, the product is a functional RNA. The process of gene expression is used by all known life - eukaryotes (including multicellular organisms), prokaryotes (bacteria and archaea), possibly induced by viruses - to generate the macromolecular machinery for life. Several steps in the gene expression process may be recognized, including the DNA rearrangement (in case of some genes only), transcription, RNA processing, RNA export, translation, post-translational modification, and translocation of a protein to its site of action.

DNA Rearrangement

This step of gene expression exists only in a few eukaryotic genes (e.g., immunoglobulin genes) where out of many different gene segments only one is picked up from each different region to have a template DNA strand available for transcription.

Transcription

The production of RNA copies of the DNA is called transcription, and is performed by RNA polymerase, which adds one RNA nucleotide at a time to a growing RNA strand. This RNA is complementary to the template 3' → 5' DNA strand, which is itself complementary to the coding 5' → 3' DNA strand (Brueckner et al. 2009). Therefore, the resulting 5' → 3' RNA strand is identical to the coding DNA strand with the exception that thymines (T) are replaced with uracils (U) in the RNA. A coding DNA strand reading "ATG" is indirectly transcribed through the non-coding strand as "AUG" in RNA. Transcription in prokaryotes is carried out by a single type of RNA polymerase (RNAP), which needs a DNA sequence called a Pribnow box as well as a sigma factor (σ factor) to start transcription. In eukaryotes, transcription is performed by three types of RNA polymerases, each of which needs a special DNA sequence called the promoter and a set of DNA-binding proteins – transcription factors – to initiate the process. RNA polymerase I (RNAPI) is responsible for transcription of ribosomal RNA (rRNA) genes. RNA polymerase II (RNAPII) transcribes all protein-coding genes but also some non-coding RNAs (e.g., snRNAs, snoRNAs or long non-coding RNAs). RNAPII includes a C-terminal domain (CTD) that is rich in serine residues. When these residues are phosphorylated, the CTD binds to various protein factors that promote transcript maturation and modification. RNA polymerase III (RNAPIII) transcribes 5S rRNA, transfer RNA (tRNA) genes, and some small non-coding RNAs (e.g., 7SK). Transcription ends when the polymerase encounters a sequence called the terminator.

RNA Processing

While transcription of prokaryotic protein-coding genes creates messenger RNA (mRNA) that is ready for translation into protein, transcription of eukaryotic genes leaves a primary transcript of RNA (pre-mRNA), which first has to undergo a series of modifications to become a mature mRNA. These include 5'-capping, which is set of enzymatic reactions that add 7-methylguanosine (m^7G) to the 5'-end of pre-mRNA and thus protect the RNA from degradation by exonucleases. The m^7G cap is then bound by cap binding complex heterodimer (CBC20/CBC80) which aids in mRNA export to cytoplasm and also protects the RNA from decapping.

Another modification is 3' cleavage and polyadenylation. They occur if polyadenylation signal sequence (5'- AAUAAA-3') is present in pre-mRNA, which is usually between protein-coding sequence and terminator. The pre-mRNA is first cleaved and then a series of ~200 adenines (A) are added to form poly(A) tail which protects the RNA from degradation. Poly(A) tail is bound by multiple poly(A)-binding proteins (PABP) necessary for mRNA export and translation re-initiation.

A very important modification of eukaryotic pre-mRNA is RNA splicing. The majority of eukaryotic pre-mRNAs consist of alternating segments called exons and introns. During the process of splicing, an RNA-protein catalytical complex known as spliceosome catalyzes two transesterification reactions, which remove an intron and release it in form of lariat structure, and then splice neighboring exons together. In certain cases, some introns or exons can be either removed or retained in mature mRNA. This so-called alternative splicing creates series of different transcripts originating from a single gene. Because these transcripts can be potentially translated into different proteins, splicing extends the complexity of eukaryotic gene expression.

Extensive RNA processing may be an evolutionary advantage made possible by the nucleus of eukaryotes. In prokaryotes transcription and translation happen together whilst in eukaryotes the nuclear membrane separates the two processes giving time for RNA processing to occur.

In most organisms non-coding genes (ncRNA) are transcribed as precursors which undergo further processing. In the case of ribosomal RNAs (rRNA), they are often transcribed as a pre-rRNA which contains one or more rRNAs. The pre-rRNA is cleaved and modified (2′-O-methylation and pseudouridine formation) at specific sites by approximately 150 different small nucleolus-restricted RNA species, called snoRNAs. SnoRNAs associate with proteins, forming snoRNPs. While snoRNA part base-pairs with the target RNA and thus positions the modification at a precise site, the protein part performs the catalytical reaction. In eukaryotes, in particular a snoRNP called RNase, MRP cleaves the 45S pre-rRNA into the 28S, 5.8S, and 18S rRNAs. The rRNA and RNA processing factors form large aggregates called the nucleolus (Sirri et al. 2008).

In the case of transfer RNA (tRNA), for example, the 5' sequence is removed by RNase P (Frank and Pace 1998), whereas the 3'-end is removed by the tRNase Z enzyme (Ceballos and Vioque 2007) and the non-templated 3' CCA tail is added by a nucleotidyl transferase (Weiner 2004). In the case of micro RNA (miRNA), miRNAs are first transcribed as primary transcripts or pri-miRNA with a cap and poly-A tail and processed to short, 70-nucleotide stem-loop structures known as pre-miRNA in the cell nucleus by the enzymes Drosha and Pasha. After being exported, it is then processed to mature miRNAs in the cytoplasm by interaction with the endonuclease Dicer, which also initiates the formation of the RNA-induced silencing complex (RISC), composed of the Argonaute protein.

Even snRNAs and snoRNAs themselves undergo series of modifications before they become part of functional RNP complex. This is done either in the nucleoplasm or in the specialized compartments called Cajal bodies. Their bases are methylated or pseudouridinilated by a group of small Cajal body-specific RNAs (scaRNAs) which are structurally similar to snoRNAs.

Alternative RNA splicing and RNA editing take place in many of the eukaryotic nuclear and organelle genes. Understanding these two strategies of RNA processing will help us understand gene expression better.

RNA Export

In eukaryotes most mature RNA must be exported to the cytoplasm from the nucleus. While some RNAs function in the nucleus, many RNAs are transported through the nuclear pores and into the cytosol. Notably this includes all RNA types involved in protein synthesis (Köhler and Hurt 2007). In some cases RNAs are additionally transported to a specific part of the cytoplasm, such as a synapse; they are then towed by motor proteins that bind through linker proteins to specific sequences (called "zipcodes") on the RNA (Jambhekar and Derisi 2007).

Translation

During the translation, tRNA charged with amino acid enters the ribosome and aligns with the correct mRNA triplet. Ribosome then adds amino acid to growing protein chain. For some RNA (non-coding RNA) the mature RNA is the final gene product (Amaral et al. 2008). In the case of messenger RNA (mRNA) the RNA is an information carrier coding for the synthesis of one or more proteins. mRNA carrying a single protein sequence (common in eukaryotes) is monocistronic whilst mRNA carrying multiple protein sequences (common in prokaryotes) is known as polycistronic. Every mRNA consists of three parts - 5' untranslated region (5'UTR), protein-coding region or open reading frame (ORF) and 3' untranslated region (3'UTR). The coding region carries information for protein synthesis encoded by the genetic code to form triplets. Each triplet of nucleotides of the coding region is called a codon and corresponds to a binding site complementary to an anticodon triplet in transfer RNA. Transfer RNAs with the same anticodon sequence always carry an identical type of amino acid. Amino acids are then chained together by the ribosome according to the order of triplets in the coding region. The ribosome helps anticodon in transfer RNA to temporarily pair with codon in messenger RNA and takes the amino acid from each transfer RNA and makes a structureless protein out of it (Hansen et al. 2003; Berk and Cate 2007). Each mRNA molecule is translated into many protein molecules, on average ~900 in mammals (Schwanhäusse et al. 2011).

In prokaryotes, translation generally occurs at the point of transcription (co-transcriptionally), often using a messenger RNA that is still in the process of being created. In eukaryotes translation can occur in a variety of regions of the cell depending on where the protein being written is supposed to be. Major locations are the cytoplasm for soluble cytoplasmic proteins and the membrane of the endoplasmic reticulum for proteins that are for export from the cell or insertion into a cell membrane. Proteins which are supposed to be expressed at the endoplasmic reticulum are recognized part-way through the translation process. This is governed by the signal recognition particle – a protein which binds to the ribosome and directs it to the endoplasmic reticulum when it finds a signal sequence on the growing (nascent) amino acid chain (Hegde and Kang 2008).

Protein Folding

The polypeptide folds into its characteristic and functional three-dimensional structure from a random coil (Alberts et al. 2002). Each protein exists as an unfolded polypeptide or random coil when translated from a sequence of mRNA into a linear chain of amino acids. This polypeptide lacks any developed three-dimensional structure (the left hand side of the neighboring figure). Amino acids interact with each other to produce a well-defined three-dimensional structure, the folded protein (the right hand side of the figure) known as the native state. The resulting three-dimensional structure is determined by the amino acid sequence (Anfinsen's dogma) (Anfinsen 1972). The correct three-dimensional structure is essential to function, although some parts of functional proteins may remain unfolded (Berg et al. 2002). Failure to fold into the intended shape usually produces inactive proteins with different properties including toxic prions. Several neurodegenerative and other diseases are believed to result from the accumulation of misfolded (incorrectly folded) proteins (Selkoe 2003). Many allergies are caused by the folding of the proteins, for the immune system does not produce antibodies for certain protein structures (Bruce 2010).

Enzymes called chaperones assist the newly formed protein to attain (fold into) the 3-dimensional structure it needs to function (Hebert and Molinari 2007). Similarly, RNA chaperones help RNAs attain their functional shapes (Russell 2008). Assisting protein folding is one of the main roles of the endoplasmic reticulum in eukaryotes.

Protein Transport

Many proteins are destined for other parts of the cell than the cytosol and a wide range of signaling sequences are used to direct proteins to where they are supposed to be. In prokaryotes this is normally a simple process due to limited compartmentalization of the cell. However, in eukaryotes there is a great variety of different targeting processes to ensure the protein arrives at the correct organelle. Not all proteins remain within the cell and many are exported, for example digestive enzymes, hormones and extracellular matrix proteins. In eukaryotes the export pathway is well developed and the main mechanism for the export of these proteins is translocation to the endoplasmic reticulum, followed by transport via the Golgi apparatus (Moreau et al. 2007; Prudovsky et al. 2008).

Protein degradation

Once protein synthesis is complete the level of expression of that protein can be reduced by protein degradation. There are major protein degradation pathways in all prokaryotes and eukaryotes of which the proteasome is a common component. An unneeded or damaged protein is often labeled for degradation by addition of ubiquitin.

GENE REGULATION

Gene regulation gives the cell control over structure and function, and is the basis for cellular differentiation, morphogenesis and the versatility and adaptability of any organism. Gene regulation may also serve as a substrate for evolutionary change, since control of the timing, location, and amount of gene expression can have a profound effect on the functions (actions) of the gene in a cell or in a multicellular organism. Same organism utilizes different mechanisms for regulating the expression of different genes. "Gene regulation" in fact is short term used for denoting "regulation of gene expression" and thus is important component of gene expression system. Regulatory aspect of gene expression deserves to be dealt with in its minutest details and is therefore not discussed in this book.

EXPRESSION SYSTEM

An expression system is a system specifically designed for the production of a gene product of choice. This is normally a protein although may also be RNA, such as tRNA or a ribozyme. An expression system consists of a gene, normally encoded by DNA, and the molecular machinery required to transcribe the DNA into mRNA and translate the mRNA into protein using the reagents provided. In the broadest sense this includes every living cell but the term is more normally used to refer to expression as a laboratory tool. An expression system is therefore often artificial in some manner. Expression systems are, however, a fundamentally natural process. Viruses are an excellent example where they replicate by using the host cell as an expression system for the viral proteins and genome.

Tetracycline-controlled transcriptional activation is a method of inducible gene expression where transcription is reversibly turned on or off in the presence of the antibiotic tetracycline or one of its derivatives (e.g. doxycycline). In nature, the P_{tet} promoter expresses TetR, the repressor, and TetA, the protein that pumps tetracycline antibiotic out of the cell. The difference between "Tet-on" and "Tet-off" is not whether the transactivator turns a gene on or off, as the name might suggest; rather, both proteins activate expression. The difference relates to their respective response to doxycycline; Tet-Off activates expression in the absence of Dox, whereas Tet-On activates in the presence of Dox. Doxycycline is also used in "Tet-on" and "Tet-off" tetracycline controlled transcriptional activation to regulate transgene expression in organisms and cell cultures

In addition to these biological tools, certain naturally observed configurations of DNA (genes, promoters, enhancers, repressors) and the associated machinery itself are referred to as an expression system. This term is normally used in the case where a gene or set of genes is switched on under well defined conditions. For example, the simple repressor-switch expression system in Lambda phage and the lac operator system in bacteria. Several natural expression systems are directly used or modified and used for artificial expression systems such as the Tet-on and Tet-off expression system.

GENE EXPRESSION NETWORKS

Genes have sometimes been regarded as nodes in a network, with inputs being proteins such as transcription factors, and outputs being the level of gene expression. The node itself performs a function, and the operation of these functions has been interpreted as performing a kind of information processing within cells and determines cellular behavior. Gene networks can also be constructed without formulating an explicit causal model. This is often the case when assembling networks from large expression data sets. Covariation and correlation of expression is computed across a large sample of cases and measurements (often transcriptome or proteome data). The source of variation can be either experimental or natural (observational). There are several ways to construct gene expression networks, but one common approach is to compute a matrix of all pair-wise correlations of expression across conditions, time points, or individuals and convert the matrix (after thresholding at some cut-off value) into a graphical representation in which nodes represent genes, transcripts, or proteins and edges connecting these nodes represent the strength of association (Brueckner et al. 2009; Chesler et al. 2004).

ANALYSIS OF GENE EXPRESSION

The mRNA and protein quantification assays help in analysis of gene expression. Many transcriptomic and proteomic tools are in used for this purpose.

mRNA Quantification

Levels of mRNA can be quantitatively measured by Northern blotting which gives size and sequence information about the mRNA molecules. A sample of RNA is separated on an agarose gel and hybridized to a radioactively labeled RNA probe that is complementary to the target sequence. The radiolabeled RNA is then detected by an autoradiograph. Because the use of radioactive reagents makes the procedure time consuming and potentially dangerous, alternative labeling and detection methods, such as digoxigenin and biotin chemistries, have been developed. Perceived disadvantages of Northern blotting are that large quantities of RNA are required and that quantification may not be completely accurate, as it involves measuring band strength in an image of a gel. On the other hand, the additional mRNA size information from the Northern blot allows the discrimination of alternately spliced transcripts.

Another approach for measuring mRNA abundance is RT-qPCR. In this technique, reverse transcription is followed by real-time quantitative PCR (qPCR). Reverse transcription first generates a DNA template from the mRNA; this single-stranded template is called cDNA. The cDNA template is then amplified in the quantitative step, during which the fluorescence emitted by labeled hybridization probes or intercalating dyes changes as the DNA amplification process progresses. With a carefully constructed standard curve, qPCR can produce an absolute measurement of the number of copies of original mRNA, typically in units of copies per nanoliter of homogenized tissue or copies per cell. qPCR is very sensitive (detection of a single mRNA molecule is theoretically possible), but can be expensive depending on the type of reporter used; fluorescently labeled oligonucleotide probes are more expensive than non-specific intercalating fluorescent dyes.

For expression profiling, or high-throughput analysis of many genes within a sample, RT-qPCR may be performed for hundreds of genes simultaneously in the case of low-density arrays. A second approach is the hybridization microarray. A single array or "chip" may contain probes to determine transcript levels for every known gene in the genome of one or more organisms. Alternatively, "tag-based" technologies like Serial Analysis of Gene Expression (SAGE) and RNA-Seq, which can provide a relative measure of the cellular concentration of different mRNAs, can be used. An advantage of tag-based methods is the "open architecture", allowing for the exact measurement of any transcript, with a known or unknown sequence. Next-generation sequencing (NGS) such as RNA-Seq is another approach, producing vast quantities of sequence data that can be matched to a reference genome. Although NGS is comparatively time-consuming, expensive, and resource-intensive, it can identify single-nucleotide polymorphisms, splice-variants, and novel genes, and can also be used to profile expression in organisms for which little or no sequence information is available.

Single-cell gene expression profiling

A key goal of biology is to relate the expression of specific genes to a particular cellular phenotype. However, current assays for gene expression destroy the structural context. By combining advances in computational fluorescence microscopy with multiplex probe design, Levsky et al. (2002) devised technology in which the expression of many genes can be visualized simultaneously inside single cells with high spatial and temporal resolution. Analysis of 11 genes in serum-stimulated cultured cells revealed unique patterns of gene expression within individual cells. Using the nucleus as the substrate for parallel gene analysis, they thus provided a platform for the fusion of genomics and cell biology: "cellular genomics."

Meiome

Meiome is the term used in functional genomics for meiotic transcriptome. Meiosis is a key feature for all sexual reproducing eukaryotes in which homologous chromosome pairing, synapse and recombination are unique. Since meiosis in most organisms occur in a short time period, study of meiotic transcript profiling is extremely hard due to the challenge of isolation of meiotic cells. There are a number of groups studying on meiotic transcriptome using high throughput techniques such as microarray and sequencing technologies. Meiome is the way to study transcript profiling using mRNA isolated from enriched meiotic cells (meiocytes). Currently there are two major approaches to understand the RNA accumulation in meiocytes: (1) RNA-seq and (2) Microarray. The RNA-seq tech is more powerful for whole transcriptome analysis, the available analyzers such as Illumina Genome Analyzer 2 and Roche 454 analyzer made this approach more powerful.

GeneCalling

GeneCalling is a mRNA profiling technology finding increased use in the field of genomics and invented by Dr. Jonathan M. Rothberg (Shimkets et al. 1999) and developed by CuraGen Corporation. This genomics technique rapidly identifies candidate genes for use in drug discovery and development. Differences between gene expression in healthy tissues and disease or drug responsive tissues are examined and compared in this technology. Known as well as novel genes in any organism - most commonly humans, animals, plants and pathogens - can be detected by this open expression system.

This mRNA profiling technique for determining differential gene expression utilizes, but does not require, prior knowledge of gene sequences. This method permits high-throughput reproducible detection of most expressed sequences with a sensitivity of greater than 1 part in 100,000. Gene identification by database query of a restriction endonuclease fingerprint, confirmed by competitive PCR using gene-specific oligonucleotides, facilitates gene discovery by minimizing isolation procedures. This process was validated by analysis of the gene expression profiles of normal and hypertrophic rat hearts following in vivo pressure overload.

The GeneCalling process is robust and efficient. It is well validated and requires only minimal sized samples. Furthermore, it is widespread in its scope, identifying nearly all genes in a given area of research. Beyond this, it identifies subtle changes in gene expression while maintaining reproducibility. The ability to share information obtained through databases allows for a complete analysis of gene and protein function.

Protein Quantification

For genes encoding proteins the expression level can be directly assessed by a number of means with some clear analogies to the techniques for mRNA quantification. The most commonly used method is to perform a Western blot against the protein of interest – this gives information on the size of the protein in addition to its identity. A sample (often cellular lysate) is separated on a polyacrylamide gel, transferred to a membrane and then probed with an antibody to the protein of interest. The antibody can either be conjugated to a fluorophore or to horseradish peroxidase for imaging and/or quantification. The gel-based nature of this assay makes quantification less accurate but it has the advantage of being able to identify later modifications to the protein, for example proteolysis or ubiquitination, from changes in size.

Western blotting is a technique used to identify and locate proteins based on their ability to bind to specific antibodies. Western blot analysis can detect your protein of interest from a mixture of a great number of proteins. Western blotting can give you information about the size of your protein (with comparison to a size marker or ladder in kDa), and also give you information on protein expression (with comparison to a control such as untreated sample or another cell type or tissue). Western blot is dependent on the quality of antibody use to probe for your protein of interest, and how specific it is for this protein.

The enzyme-linked immunosorbent assay (ELISA) works by using antibodies immobilized on a microtiter plate to capture proteins of interest from samples added to the well. Using a detection antibody conjugated to an enzyme or fluorophore the quantity of bound protein can be accurately measured by fluorometric or colourimetric detection. The detection process is very similar to that of a Western blot, but by avoiding the gel steps more accurate quantification can be achieved.

Proteomics, which focuses on gene products, is complementary to genomics and transcritomics and is a thrust area of molecular biotechnological research. Verification of a gene product by methods o proteome analysis serves a very useful purpose for annotation of the genome. Post-translational modification of proteins can be examined only by methods of proteome analysis. Sometimes results of transcriptome analysis do not correlate with those of proteome analysis, thus making it necessary to examine protein expression levels. Protein function is also regulated by proteolysis and recycling or sequestration of products in various cell compartments. Protein-protein interactions also determine the function of gene and gene product, the protein.

Various tools of proteome analysis are: Mass spectrometry (MS), two-dimensional gel electrophoresis 2D-GE), Isoelectric focusing IEF), polyacrylamide gel electrophoresis (PAGE), comparative 2-D gel approach, protein chip approach, ELISA, liquid chromatography coupled with electrospray-ionization tandem mass spectroscopy.

Post-translational modification of proteins involves phosphorylation, glycosylation, sulfation, etc., which are very important for protein function. Various variants of eastern blotting are used to analyze specific types of post-translational modifications.

Study of protein-protein interaction

A study of protein-protein interactions is an important component of proteomics research. These interactions can be exploited for biotechnological applications. Purification of entire multiprotein complexes can be done by affinity based method.

Phage Display: In this method, a vector (f1, M13 or λ phage) is designed, which gives a hybrid protein resulting from the fusion of coat protein with the protein produced due to the cloned gene, This allows display of foreign protein on phage coat, thus permitting its possible interaction with other proteins that phage encounters. Phage display thus allows the study of protein-protein interactions.

Yeast-Two Hybrid (Y2H) System: Yeast-two hybrid is an in vivo system for protein-protein interaction studies. A transcription factor generally consists of a DNA-binding domain (DBD) and an activation domain (AD), the former helping in DNA binding and the later facilitating the activation of a gene lying downstream. If one wants to study interaction between two proteins X and Y, two hybrid gene constructs are prepared, one having a hybrid gene coding for fusion protein DBD + X along with a reporter gene and the other coding for the fusion protein AD + Y. Yeast cells are co-transformed with both hybrid constructs, so that X and Y interact physically. This will bring DBD and AD in close proximity, AD will not be available at DBD site and reporter gene will not be expressed. Thus yeast-two hybrid system is used for study of protein-protein interaction.

Applications of Gene Expression Analysis

Class discovery

A popular approach to class discovery involves grouping similar genes or samples together. The simplest form of class discovery would be to list all the genes that changed by more than a certain amount between two experimental conditions. While the statistics may reliably identify which gene products change under experimental conditions, making biological sense of expression profiling rests on knowing which protein each gene product makes and what function this protein performs. Gene annotation provides functional and other information, for example the location of each gene within a particular chromosome. Some functional annotations are more reliable than others; some are absent. Gene annotation databases change regularly, and various databases refer to the same protein by different names, reflecting a changing understanding of protein function. Use of standardized gene nomenclature helps address the naming aspect of the problem, but exact matching of transcripts to genes remains an important consideration.

Categorizing Regulated Genes: Having identified some set of regulated genes, the next step in expression profiling involves looking for patterns within the regulated set. Do the proteins made from these genes perform similar functions? Are they chemically similar? Do they reside in similar parts of the cell? Gene ontology analysis provides a standard way to define these relationships. Gene ontologies start with very broad categories, e.g., "metabolic process" and break them down into smaller categories, e.g., "carbohydrate metabolic process" and finally into quite restrictive categories like "inositol and derivative phosphorylation". Genes have other attributes beside biological function, chemical properties and cellular location. One can compose sets of genes based on proximity to other genes, association with a disease, and relationships with drugs or toxins. The molecular signatures database and the comparative toxicogenomics database are examples of resources to categorize genes in numerous ways.

Finding Patterns and Links Among Regulated Genes: Regulated genes are categorized in terms of what they are and what they do, important relationships between genes may emerge. For example, we might see evidence that a certain gene creates a protein to make an enzyme that activates a protein to turn on a second gene. This second gene may be a transcription factor that regulates yet another gene. Observing these links we may begin to suspect that they represent much more than chance associations in the results, and that they are all on our list because of an underlying biological process.

Coverage of Genes Associated with Entire Pathways: The application of such technology to toxicology, toxicogenomics, promises substantial dividends in mechanistic toxicity research and also, possibly, the ability to predict adverse toxicity for novel or untested compounds. Pennie et al. (2001) have developed a custom approach to this technology, designing cDNA microarray platforms specifically for gene expression events of relevance to a large number of toxicological endpoints. Such arrays allow comprehensive coverage of genes associated with entire pathways (such as oxidative stress, signal transduction, stress response, epithelial biology) and enable simultaneous measurement of more than ten thousand gene expression events.

Large-scale gene function studies

Microarray transcript profiling and RNA interference are two new technologies crucial for large-scale gene function studies in multicellular eukaryotes. Both rely on sequence-specific hybridization between complementary nucleic acid strands, inciting us to create a collection of gene-specific sequence tags (GSTs) representing at least 21,500 *Arabidopsis* genes and which are compatible with both approaches. The GSTs were carefully selected to ensure that each of them shared no significant similarity with any other region in the Arabidopsis genome. They were synthesized by PCR amplification from genomic DNA. Spotted microarrays fabricated from the GSTs show good dynamic range, specificity, and sensitivity in transcript profiling experiments. The GSTs have also been transferred to bacterial plasmid vectors via recombinational cloning protocols. These cloned GSTs constitute the ideal starting point for a variety of functional approaches, including reverse genetics. Hilson et al. (2004) have subcloned GSTs on a large scale into vectors designed for gene silencing in plant cells. We show that in planta expression of GST hairpin RNA results in the expected phenotypes in silenced Arabidopsis lines. These versatile GST resources provide novel and powerful tools for functional genomics.

Limitations of Gene Expression Analysis

In general, expression profiling studies report those genes that show statistically significant differences under changed experimental conditions. This is typically a small fraction of the genome for several reasons. Different cells and tissues express a subset of genes as a direct consequence of cellular differentiation so many genes are turned off. Many of the genes code for proteins that are required for survival in very specific amounts so many genes do not change. Cells use many other mechanisms to regulate proteins in addition to altering the amount of mRNA, so these genes may stay consistently expressed even when protein concentrations are rising and falling. Financial constraints limit expression profiling experiments to a small number of observations of the same gene under identical conditions, reducing the statistical power of the experiment, making it impossible for the experiment to identify important but subtle changes. It takes a great amount of effort to discuss the biological significance of each regulated gene, so scientists often limit their discussion to a subset. Newer microarray analysis techniques automate certain aspects of attaching biological significance to expression profiling results, but this remains a very difficult problem.

SYSTEMS BIOLOGY APPROACH

Contrary to traditional central dogma of molecular biology (Gene → mRNA → protein), the new (contemporary) model explains that the genome (all the genes of an organism) gives rise to transcriptome (the complete set of mRNA), which is then translated to produce the proteome (complete set of proteins in a given cell under particular conditions). The new model is written as:

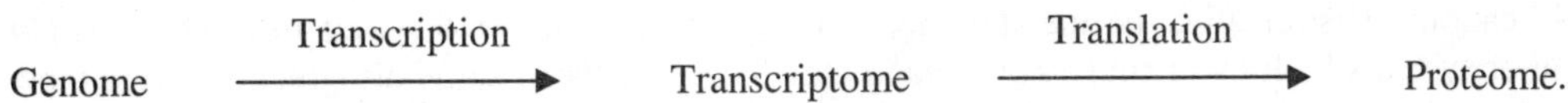

The genome is a static information source with a defined gene content that with a few exceptions remains the same regardless of the cell type or environmental conditions. Both transcriptome and proteome are dynamic entities, whose content can fluctuate dramatically under different conditions due to regulation of transcription, RNA processing, translation, and protein modification.

Which is More Complex – Transcriptome, Proteome or Genome?

The transcriptome and proteome are much more complex than genome. Why? A single gene can produce many different mRNAs, and hence proteins. Different transcripts from the same gene can be generated by (i) alternative splicing, (ii) alternative promoter usage, (iii) polyadenylation site usage, or special RNA processing strategies (like RNA editing). Different proteins can be generated from the same transcript by alternate use of translation start and stop codons. Proteins synthesized from one mRNA can be variously modified during or after translation.

By taking a system-biology approach, one aims to identify metabolic switches and transcriptional regulators and reveal the intrinsic biological networks that occur within cells. Mutants and metabolically-altered plants are also analyzed using transcriptomics (study of transcript profile), proteomics (study of protein profile) and metabolomics (study of metaboloite profile) to understand how different intrinsic networks are influenced.

Genomics studies DNA isolated from any part of the plant or animal cell, since DNA in different cells do not differ in time and space. Transcriptomics involves study of mRNA from each possible tissue, at each developmental stage and under each of a number of possible environments (stresses). Proteomics studies proteins from each possible tissue, at each developmental stage and under each of a number of possible environments (stresses).

Transcriptome and proteome are generally described for a specific tissue or an organ under specific conditions, the transcriptome of the plant as a whole would consist of mRNA available in all kinds of cells in a plant species in time and space. The same will be true for the study of proteome.

METABOLOMICS

Metabolomics is the scientific study of chemical processes involving metabolites. It is a study of metabolome in a cell or organism at different developmental stages, and under different conditions. Specifically, metabolomics is the "systematic study of the unique chemical fingerprints that specific cellular processes leave behind"; the study of their small-molecule metabolite profiles. The metabolome represents the collection of all metabolites in a biological cell, tissue, organ or organism, which are the end products of cellular processes. Thus, while mRNA gene expression data and proteomic analyses do not tell the whole story of what might be happening in a cell, metabolic profiling can give an instantaneous snapshot of the physiology of that cell. One of the challenges of systems biology and functional genomics is to integrate proteomic, transcriptomic, and metabolomic information to give a more complete picture of living organisms. The metabolome forms a large network of metabolic reactions, where outputs from one enzymatic chemical reaction are inputs to other chemical reactions. Such systems have been described as hypercycles.

Terms "metabolomics" and "metabonomics" are sometimes synonymously used. Some people define these two terms differently. is defined as "the quantitative measurement of the dynamic multiparametric metabolic response of living systems to pathophysiological stimuli or genetic modification". Historically, the metabonomics approach was one of the first methods to apply the scope of systems biology to studies of metabolism. There has been some disagreement over the exact differences between 'metabolomics' and 'metabonomics'. The difference between the two terms is not related to choice of analytical platform: although metabonomics is more associated with NMR spectroscopy and metabolomics with mass spectrometry-based techniques, this is simply because of usages amongst different groups that have popularized the different terms. While there is still no absolute agreement, there is a growing consensus that 'metabolomics' places a greater emphasis on metabolic profiling at a cellular or organ level and is primarily concerned with normal endogenous metabolism. 'Metabonomics' extends metabolic profiling to include information about perturbations of metabolism caused by environmental factors (including diet and toxins), disease processes, and the involvement of extragenomic influences, such as gut microflora. This is not a trivial difference; metabolomic studies should, by definition, exclude metabolic contributions from extragenomic sources, because these are external to the system being studied. However, in practice, within the field of human disease research there is still a large degree of overlap in the way both terms are used, and they are often in effect synonymous.

Metabolites

Intermediate products of metabolism are called metabolites. These are small molecules that exhibit enormous variation. A metabolite is usually defined as any molecule less than 1 kDa in size. However, there are exceptions to this depending on the sample and detection method. For example, macromolecules such as lipoproteins and albumin are reliably detected in NMR-based metabolomics studies of blood plasma. In plant-based metabolomics, it is common to refer to "primary" and "secondary" metabolites. A primary metabolite is directly involved in the normal growth, development, and reproduction. A secondary metabolite is not directly involved in those processes, but usually has important ecological function. Examples include antibiotics and pigments. By contrast, in human-based metabolomics, it is more common to describe metabolites as being either endogenous (produced by the host organism) or exogenous. Metabolites of foreign substances such as drugs are termed xenometabolites.

Metabolites are classified as primary metabolites and secondary metabolites. A primary metabolite is directly involved in the normal growth, development, and reproduction. A secondary metabolite has important ecological functions. Antibiotics, pigments and other metabolites of medicinal and aromatic value derived from plants are important secondary metabolites. Entire range of matabolites cannot be analyzed by a single analytical method.

Metabolome

All metabolites in complete set make metabolome (e.g., hormones and other signaling molecules, and secondary metabolites of medicinal or aromatic value). Although the metabolome can be defined readily enough, it is not currently possible to analyze the entire range of metabolites by a single analytical method. Over 50,000 metabolites have been characterized from the plant kingdom, and many thousands of metabolites have been identified and/or characterized from single plants. Analysis of metabolome involves (i) alignment of chromatographic and spectroscopic data using specific bioinformatics tools, and (ii) identification of different metabolites.

Metabolic Engineering

It is improvement of cellular activities by manipulation of enzymatic transport and regulatory functions of the cell with the use of recombinant DNA technology. Introduction of heterologous genes and regulatory elements made metabolic engineering very fascinating area of research. Cell function can be modified using targeted alterations in normal cellular activities. This may involve not only synthesis of a metabolite, but may sometimes also involve manipulation of processing pathways. Metabolic engineering has been successfully used for improving the nutritional value of crops like rice and maize. Different approaches used in metabolic engineering are:

Heterologous Genes for Metabolic Engineering: This approach has been used for production of new compounds. Completion of partial pathways to obtain new product, transfer of an entire biosynthetic pathway, heterologous host for producing novel products, altering nutrient uptake and metabolite flow are some of the applications.

Redirecting Metabolic Flow: In a branched biosynthetic pathway, there are several forks, where intermediates can enter alternative pathways. At such forks, resources (substrate, enzyme, transport system, ribosomes, etc.) are distributed equally or unequally to two or more metabolic pathways. When we are interested in one of the several possible products, at a fork the desired route can be made a priority, so that resources are minimized without decreasing cell viability. This can be achieved by (i) use of cloned genes suppressing feedback inhibition by the end products, and (ii) elevation of the activity of rate limiting enzyme.

Development of Novel Biosynthetic Pathways: Development of novel pathways can be achieved by matching and mixing genes from different sources (even from unrelated metabolic routes) and by creating new biosynthetic functions by (i) random mutagenesis, (ii) recombination, and (iii) selection. In this approach, no information on enzyme structure or catalytic mechanism involved in these pathways is needed.

Challenges in plant metabolomics

A typical plant metabolome exhibits enormous chemical diversity which needs multiple approaches for extraction. Individual metabolites exhibit a dynamic range of concentrations, so that special techniques are needed for separation of metabolites. There is a large variation in spatial/temporal distribution in metabolites, which means that metabolites differ in different organs and at different times (during life-span and during different seasons/climates). Genomic information cannot be used for study of metabolome. These limitations require different strategies for the extraction, identification and quantification of plant metabolites.

IONOME AND IONOMICS

The ionome is defined as the mineral nutrient and trace element composition of an organism and represents the inorganic component of cellular and organismal systems (Salt et al. 2008).

Ionomics is the study of elemental accumulation in living systems using high-throughput elemental profiling (Baxter 2010). This approach has been applied extensively in plants for forward and reverse genetics, screening diversity panels, and modeling of physiological states. It involves the quantitative and simultaneous measurement of the elemental composition of living organisms and changes in this composition in response to physiological stimuli, developmental state, and genetic modifications. Ionomics requires the application of high-throughput elemental analysis technologies and their integration with both bioinformatic and genetic tools. Ionomics has the ability to capture information about the functional state of an organism under different conditions, driven by genetic and developmental differences and by biotic and abiotic factors. The relatively high throughput and low cost of ionomic analysis means that it has the potential to provide a powerful approach to not only the functional analysis of the genes and gene networks that directly control the ionome, but also to the more extended gene networks that control developmental and physiological processes that affect the ionome indirectly.

REFERENCES

Amaral, P.P., M.E. Dinger, T.R. Mercer, and J.S. Mattick. 2008. The eukaryotic genome as an RNA machine. Science 319: 1787-9.

Anfinsen, C. 1972. The formation and stabilization of protein structure. Biochem. J. 128: 737-49.

Baxter, I. 2010. Ionomics: The functional genomics of elements. Brief. Funct. Genom. 9: 149-56.

Berg, J.M., J.L. Tymoczko, and L. Stryer. 2002. Protein structure and function. In: *Biochemistry*. San Francisco: W. H. Freeman.

Berk, V., and J.H. Cate. 2007. Insights into protein biosynthesis from structures of bacterial ribosomes. Curr. Opin. Struct. Biol. 17: 302-9.

Bruce, A., A. Johnson, J. Lewis, M. Raff, K. Roberts, and P. Walters. 2002. The shape and structure of proteins. In: *Molecular Biology of the Cell.* Fourth Edition. New York and London: Garland Science.

Bruce, A., D. Bray, K. Hopkin, A. Johnson, J. Lewis, M. Raff, K. Roberts, and P. Walter. 2010. Protein structure and function. Pg 120-170. In: *Essential Cell Biology*. Edition 3. New York: Garland Science, Taylor and Francis Group, LLC.

Brueckner, F., K.J. Armache, A. Cheung, et al. 2009. Structure–function studies of the RNA polymerase II elongation comple". Acta Crystallogr. D Biol. Crystallogr. 65 (Pt 2): 112-20.

Ceballos, M., and A. Vioque. 2007. tRNase Z. Protein Pept. Lett. 14(2): 137-45.

Chesler, E.J., L. Lu, J. Wang, R.W. Williams, and K.F. Manly. 2004. WebQTL: rapid exploratory analysis of gene expression and genetic networks for brain and behavior. Nat Neurosci 7(5): 485-6.

Frank, D.N., and N.R. Pace. 1998. Ribonuclease P: unity and diversity in a tRNA processing ribozyme. Annu. Rev. Biochem. 67: 153-80.

Hansen, T.M., P.V. Baranov, I.P. Ivanov, R.F. Gesteland, and J.F. Atkins. 2003. Maintenance of the correct open reading frame by the ribosome. EMBO Rep. 4(5): 499-504.

Hebert, D.N., and M. Molinari. 2007. In and out of the ER: protein folding, quality control, degradation, and related human diseases. Physiol. Rev. 87: 1377-408.

Hegde, R.S., and S.W. Kang. 2008. The concept of translocational regulation. J. Cell Biol. 182(2): 225-32.

Hilson, P., J. Allemeersch, T. Altmann, et al. 2004. Versatile gene-specific sequence tags for Arabidopsis functional genomics: transcript profiling and reverse genetics applications. Genome Res. 14: 2176-89.

Jambhekar, A., and J.L. Derisi. 2007. Cis-acting determinants of asymmetric, cytoplasmic RNA transport. RNA 13: 625-42.

Köhler, A, and E. Hurt. 2007. Exporting RNA from the nucleus to the cytoplasm. Nat. Rev. Mol. Cell Biol. 8(10): 761-73.

Levsky, J.M., S.M. Shenoy, R.C. Pezo, and R.H. Singer. 2002. Single-cell gene expression profiling. Science 297: 836-40.

Moreau, P., F. Brandizzi, and S. Hanton, et al. 2007. The plant ER-Golgi interface: a highly structured and dynamic membrane complex. J. Exp. Bot. 58: 49-64.

Pennie, W.D., N.J. Woodyatt, T.C. Aldridge, and G. Orphanide. 2001. Application of genomics to the definition of the molecular basis for toxicity. Toxicol. Lett. 31: 353-8.

Prudovsky, I., F. Tarantini, M. Landriscina, et al. 2008. Secretion without Golgi. J. Cell. Biochem. 103: 1327-43.

Russell, R. 2008. RNA misfolding and the action of chaperones. Front. Biosci. 13(13): 1-20.

Salt, D.E., I. Baxter, and B. Lahner. 2008. Ionomics and the study of the plant ionome. Annu. Rev. Plant Biol. 59: 709-33.

Schwanhäusser, B., D. Busse, G. Dittmar, J. Schuchhardt, J. Wolf, W. Chen, and M. Selbach. 2011. Global quantification of mammalian gene expression control. Nature 473: 337-42.

Selkoe, D.J. 2003. Folding proteins in fatal ways. Nature 426: 900-4.

Shimkets, R.A., D.G. Lowe, J.T. Tai, et al. 1999. Gene expression analysis by transcript profiling coupled to a gene database query. Nat. Biotechnol. 17:798-803.

Sirri, V., S. Urcuqui-Inchima, P. Roussel, and D. Hernandez-Verdun. 2008. Nucleolus: the fascinating nuclear body. Histochem. Cell Biol. 129: 13-31.

Weiner, A.M. 2004. tRNA maturation: RNA polymerization without a nucleic acid template. Curr. Biol. 14(20): R883-5.

Glossary

A

Abortive transcription initiation – During transcription initiation in vitro, prokaryotic and eukaryotic RNA polymerase can engage in abortive initiation – the synthesis and release of short (2-15 nucleotides) RNA transcripts – before productive initiation. Abortive initiation involves DNA "scrunching" in which RNA polymerase remains stationary and unwinds and pulls downstream DNA into itself. Scrunching requires RNA synthesis and depends on RNA length. Also see *DNA scrunching.*

Actin gene family – Actin genes in eukaryotes represent a multigene family, where the members are homologous but non-identical, giving rise to slightly different variants so that different members function either at different times or in different tissues at the same time.

Activation of tRNA – Before translation commences, activation of amino acids (formylated methionine in prokaryotes and methionine in eukaryotes) takes place in the cytosol at the expense of ATP. An amino acid reacts with ATP to become adenylated with concomitant release of pyrophosphate. As a result of adenylation, the amino acid is attached to adenylic acid via a high energy ester bond in which the carbonyl group of amino acid is joined to phosphoryl group of AMP.

Activators and inhibitors in alternative RNA splicing – Splicing of individual pre-mRNAs is frequently controlled by combinatorial or competitive effects of both activators and inhibitors.

Active site of spliceosome – Formation of catalytically active RNA structures within the spliceosome requires the assistance of proteins. Number and nature of proteins needed to establish and maintain the spliceosome's active site is not fully known.

Addition of cap at 5′-end of pre-mRNA – Eukaryotic mRNA always has a 5' cap composed of a 5'→5' triphosphate linkage between two modified nucleotides: a 7-methylguanosine and a 2' O-methyl purine. This cap is a modified guanine (m^7GpppN_1) which is added post-transcriptionally and serves to identify this RNA molecule as an mRNA to the translational machinery.

Addition of poly(A) tail at 3′-end – Polyadenylation signal in the train is AAUAAA in pre-mRNA. Polyadenylation involves two steps — cleavage and addition of poly(A) tail. Multiple copies of polyadenylation signal are of common occurrence.

Advancing transcription elongation complex – RNA polymerase combines two contradictory biochemical features: (a) exceptional stability for dissociation and (b) the ability to easily translocate along DNA. Thus, elongating RNA polymerase simultaneously behaves as a strong DNA binding protein and as a protein with no affinity for particular DNA sites. The combination of these features ensures processivity of RNA polymerase.

Allelic exclusion – In immunoglobulin-producing cells or plasma cells, only one member of each pair of alleles, is involved in the synthesis of immunoglobulin being expressed. Correct rearrangement of only one chromosome is needed for the generation of each expressed light or heavy chain genes. Since DNA rearrangement of one chromosome takes place, expression of the alternate allele is excluded.

Allelic mutations – Two mutations are alleleic if they do not show complementation.

Alteration in the meaning of code words – In almost all organisms, there seem to be available "recoding signals" in mRNAs leading to altered meaning of codons. These are exceptions to the universal character of genetic code. In atleast two bacterial genes and three mammalian genes, an internal UGA codon codes for 21st amino acid selenocysteine (SeCys) which has no unique codon in the genetic code dictionary. A hairpin structure called the 'selenocysteine insertion sequence' (SECIS) is present in the messenger RNA that corresponds to where selenocysteine (SeCys) is due to be inserted into the growing peptide chain. Pyrrolysine is the 22nd amino acid in proteins. UAG is normally the amber stop codon, but encodes pyrrolysine (Pyl) if a PYLIS element is present.

Alternative donor and acceptor splice sites – Presence of alternative donor and acceptor splice sites for a given exon results in the excision of intron of different lengths with complementary variation in exon size.

Alternative promoters and polyadenylation sites – Presence of alternative promoters and polyadenylation sites in the transcript may also yield different mRNAs molecules of different sizes by differential use of such sites.

Alternative RNA splicing – The process through which more than one functional mRNA molecules are produced from one and the same primary transcript by differential removal of introns. Alternative RNA splicing makes it possible for a single gene to produce more than one messenger RNA molecule.

Alternative RNA splicing factors – The factors that preferentially regulate removal of introns harboring (or closest to) certain clusters. Nova proteins are neuron-specific alternative RNA splicing factors.

Alternative RNA splicing, functional effects of – Functional effects of alternative RNA splicing are: (a) protein localization, (b) deletion of protein activity, (c) modification of protein activity, (d) novel protein activities, and (e) RNA stability and translational efficiency.

Alternative RNA splicing, patternss of – Six patterns of alternative RNA splicing are: combinatorial exons, mutually exclusive patterns, retained introns, alternative donor and acceptor splice sites, alternative promoters and polyadenylation sites, and splice site compatibility.

Alternative RNA splicing, regulation of – Splicing is regulated in a manner analogous to the transcriptional regulation, in that *trans*-acting proteins (repressors and activators) bind to *cis*-acting regulatory sites (silencers and enhancers) on the RNA.

Alternative RNA splicing, role of – Alternative RNA splicing (a) has role in developmental control by serving as on/off switch and through alternative RNA splicing of morphogenetic proteins, (b) allows a high degree of protein diversity at low genetic cost, (c) involved in localization of different proteins in different cells, deletion of function of gene product and production of isomeric proteins having different functions. (d) is the most efficient mechanism for generating protein diversity as it is reversible and allows genetic variation at a very low genetic cost, (e), genes encode proteins with myriad functions, and (f) is an important regulatory mechanism in metazoans and their viruses.

Ambiguous code – When one codon can code for more than one amino acid.

Amplification of ribosomal RNA gene repeats – Organisms maintain ribosomal RNA gene repeats (rDNA) at stable copy number by recombination; the loss of repeats results in gene amplification.

Ancient DNA – Refers to the organisms which lived long ago and some of which have become extinct. Such DNA is a genetic treasure, which is useful in studying the molecular changes that might have occurred in a species during evolution.

Ancillary Sites – Sequences located between –30 and –40 position. Also referred to as –35 sequence. In bacterial genes, these sequences are known as CAP (cAMP-activated protein) site. The term ancillary site applies to both prokaryotic and eukaryotic genes.

Aneuploidy and cancer – In contrast to normal cells, aneuploidy – alterations in the number of chromosomes – is consistently observed in virtually all cancers. A growing body of evidence suggests that aneuploidy is often caused by a particular type of genetic instability, called chromosomal instability, which may reflect defects in mitotic segregation in cancer cells.

Antibody diversity – Organisms have an ability to synthesize a large number (in millions) of different types of immunoglobulins to defend the body against a large number of foreign substances (antigens) which may otherwise harm the body. DNA rearrangements are responsible for antibody diversity.

Anticodon – Sequence of nucleotides on tRNA that complements with the codon on mRNA.

Antioncogenes – Antioncogenes are also known as *tumor-suppressor genes* (TSGs) or *recessive oncogenes*. Proteins encoded by antioncogenes can prevent tumorigenic transformation of cells. These genes act by suppressing malignant growth.

Antisense DNA strand – DNA template strand for a given mRNA. Also known as *non-coding strand, transcribing strand, template strand*, or *Watson strand.*

Antitermination – Process that causes the enzyme to continue transcription past the terminator sequence, an event called readthrough. Antitermination is used as a control mechanism in both bacterial operons and phage regulatory circuits.

Antiterminators – At some terminators, anti-termination causes the enzyme to continue transcription past the terminator sequence. There are certain specific ancillary factors that interact with RNA polymerase (RNAP or Pol). NusA protein controls the hairpin formation promoting termination, and stabilization of these contacts by phage lambda N protein leads to anti-termination. Also known as *antitermination factors*.

Associative introns – Intron that come into being when two unrelated exons were joined by an unequal recombination between two genes. Also known as *type A introns*. Type A intron would function evolutionarily by increasing the frequency of successful shifting of gene segments.

Attenuated (or killed) viruses – An attenuated virus is one whose virulence in human beings has been reduced by mutation in the course of passage through some different animal host, so that the virus becomes adapted to survival in that animal rather than in human beings. The injected attenuated virus is not itself likely to cause infection.

Attenuation – A mechanism that links the supply of an aminoacyl~tRNA to the ability of RNA polymerase to read through a termination site. The terminator is located at the beginning of the cluster of structural genes coding for the enzymes that synthesize amino acid carried by tRNA.

Attenuator – Aregulator site which serves as a barrier to transcription. The termination event at this site responds to the level of transcription.

Autotransporter system – Another name for Type V secretion system of Gram-negative bacteria.

B

Bacteria signal sequences – In bacteria, proteins destined to be secreted are synthesized as pre-proteins with N-terminal signal sequences sometimes termed leader peptides. Bacteria signal sequences are short (25 residues) and comprise of a hydrophobic central core which can adopt an α-helical structure, flanked by region containing several charged residues.

Bacterial promoters – Promoters lack any extensive conservation of sequence over the 60 bp associated with RNA polymerase but some short sequences within the promoters appear to be conserved, which could be referred to as signal sequences. A 6-bp sequence upstream from the start point is recognizable in almost all the promoters studied. This signal sequence is TATAAT, sometimes known as Pribnow box. Occurrence of bases in Pribnow box is $T_{89}T_{89}T_{50}A_{65}A_{65}T_{100}$. The actual location of this hexamer varies from –11 to –5 to –14 to –8. Pribnow box is 10 bp upstream the start point, so it is called –10 sequence. Centre of the Similarities of sequence also occur at other locations centered about –35 sequence, sometimes called recognition region. Consensus sequence of –35 region with commonly occurring bases is $T_{85}T_{83}G_{81}A_{61}C_{69}A_{52}$. The distance between –35 and –10 sites varies between 16 and 19 bp in known promoters. The –35 region is involved in the efficiency of the polymerase recognition.

Bacterial RNA polymerase – Bacterial RNA polymerase is composed of six polypeptides ($\alpha_2\beta\beta'\omega\sigma$) of five different kinds. This means five different genes are needed to make *Escherichia coli* RNAP. Active form of enzyme is called *holoenzyme*. Holoenzyme is made up of a core enzyme and a sigma (σ) factor. Core enzyme has five subunits ($\alpha_2\beta\beta'\omega$). No covalent bond runs between the various chains of RNAP.

Bacterial RNA polymerase inhibition – Myxopyronin inhibits bacterial RNAP. The antibiotic binds to a pocket deep inside the RNA polymerase clamp head domain, which interacts with the DNA template in the transcription bubble.

Blank codons – Amber non-sense codons and, to a lesser extent, ochre, opal, or frameshift codons have been used to specify unnatural amino acids (UAAs). A sense codon was used to encode an UAA in an auxotrophic strain by exploiting wobble position stability differences and limiting endogenous amino acid concentrations. A general solution that involves the conversion of sense codons into blank codons would be ideal.

C

CAAT box – A sequence present in promoters. This sequence GGT/CAATCT lies between –70 and –80.

Cancer cells – Cells that have a number of properties, viz., rapid division, invasion of new cellular territories, high metabolic rate, new membrane antigens, altered shapes, and so on. These cells obey no community rules and multiply without any regard to tissue organization. These calls are also termed as *anti-social cells* or *malignant cells.*

Cap-independent mRNA translation – Proteins of family 14-3-3 are crucial in a wide variety of cellular responses including cell cycle progression, DNA damage checkpoints and apoptosis. A protein called 14-3-3σ inhibits the cell cycle and may act as a tumor suppressor. It now turns out that it is also involved in regulating cap-independent protein synthesis from messenger RNA during cell division.

Cardinal nucleotide – The 3′ nucleotide of anticodon triplet is considered to be most important and is called cardinal nucleotide.

Cellular genomics – A platform for the fusion of genomics and cell biology.

Central dogma of molecular biology – Genetic information from DNA is transcribed to RNA and that information from RNA is translated to polypeptide chain. This concept about flow of biological information from DNA to RNA and then to protein was given by F.H.C. Crick in 1958. With the discovery of reverse transcription this concept was modified

Central dogma, new contemporary – This model explains that the genome (all the genes of an organism) gives rise to transcriptome (the complete set of mRNA), which is then translated to produce the proteome (complete set of proteins in a given cell under particular conditions). The genome is a static information source with a defined gene content that with a few exceptions remains the same regardless of the cell type or environmental conditions. Both transcriptome and proteome are dynamic entities, whose content can fluctuate dramatically under different conditions due to regulation of transcription, RNA processing, translation, and protein modification.

Chaperonins – A class of chaperones. Chaperonins occur in bacteria, chloroplast and mitochondria. The best known of these chaperonins is the protein GroE of *E. coli*. This protein in its active form is composed of two components. GroEL and GroES. GroEL (hsp60) is composed of two disks, each composed of seven copies of a polypeptide. GroES (hsp10) is a smaller component composed of seven copies of a small subunit. GroEL forms a barrel within which protein folding takes place. As a protein begins to emerge from the ribosome in *E. coli*, two proteins which are products of genes DnaJ and DnaK,, complex with it. A third protein, GrpE, causes release of DnaJ and DnaK leading to interaction with GroEL. Once inside, GroES can cycle on and off as ATPs are hydrolyzed, helping GroEL chaperone correct folding of the protein. Also see *chaperones*.

Charging of transfer RNAs – During this charging reaction, specific amino acid is covalently attached to a specific tRNA by enzyme called aminoacyl-tRNA synthetases. Each tRNA can be charged only with the amino acid for which its anticodon is appropriate. Charging reaction involves two steps, activation and transfer reactions, which are crucial for functioning of the tRNAs. During charging reaction, each of the 20 different amino acids gets attached to its corresponding tRNA by specific aminoacyl-tRNA synthetases. The amino acid is linked by an ester bond involving its carboxyl group to one of the last base of the tRNA.

Chloroplast DNA transcription – A template-binding polypeptide has been identified in pea chloroplast transcriptional complex. It has a molecular weight of 150 kDa and binds to both chloroplast ribosomal (16S rRNA) and messenger (psbA) promoters.

***Cis*-position** – When two alleles or elements are present in the same DNA strand.

Cistron – Operationally, a cistron is defined by *cis-trans* complementation tests. Two mutations belong to the same cistron if they do not show complementation in *cis-trans* test or in a simple *trans* complementation test.

Class I genes – Genes transcribed by RNA polymerase I. Pre-rRNA 45S (35S in yeast) genes

Class I oncogenes – The oncogenes (e.g., *src, yes, neu, abl, fps, fms, erb, ros, mos, fgr*) that are related to synthesis of cell surface receptor proteins. The cells surface receptor proteins are activated by growth factors produced by other cell types. An activated cell will proliferate as long as the second cell type produces its growth factor in response to some external stimulus. Class I oncogene products have kinase activity and are found in cytoplasm.

Class II genes – Genes transcribed by RNA polymerase II. Precursors of mRNAs and most snRNA and microRNAs. Many class II genes contain a region located between 19 and 27 bp upstream from the transcripttion start site whose consensus sequence TATANA is similar (except in location) to the prokaryotic TATA box, also known as Hogness box, where N means any nucleotide. Class II promoters contain yet another upstream element with consensus sequence GGGCGG. This sequence functions in either direction and is often present in multiple copies and binds the transcription factor SP1. Class II genes contain still other binding sites for transcription factors in a variety of locations.

Class II oncogenes – These oncogenes (e.g., *HA-ras*, *kl-ras*, *N-ras*) are commonly found in human tumors. Their protein products have a common characteristic of regulating cellular metabolism.

Class III oncogenes – These oncogenes (e.g., *myc*, *myb*, *fos*, *ski*, *p53*) regulate nuclear activities, possibly cell cycling.

Class IV oncogenes – These oncogenes (e.g., *cis*, *rel*, *B-lym*, *erb-A*, *ets*, *met*) are relatively unrelated. oncogenes, some of them produce cell growth factors (proteins).

Class discovery – This popular approach involves grouping similar genes or samples together. The simplest form of class discovery would be to list all the genes that changed by more than a certain amount between two experimental conditions. This approach helps in gene annotation.

Code letter – Nucleotides A, T, G, and C in DNA and A, U, G and C in mRNA. Codon Sequence of nucleotides specifying an amino acid.

Coding dictionary – A table of all code words that specify amino acids.

Codon length – When only as many amino acids are coded as there are code words in end-to-end sequence. (e.g., UUUCCC two code words; they code for only two amino acids). Code length is three code letters.

Cohesion complex – During cell division the cohesion complex (formed by cohesin molecules) mediates the pairing of sister chromatids. Cohesion also has roles in interphase cells. Cohesion is targeted to specific sites on chromosomes and implicate cohesion in the regulation of gene expression.

Colinearity – A point-for-point correspondence between gene and its protein product. Sequence of nucleotides in a prokaryotic gene determines sequence of amino acids in a protein/polypeptide. A prokaryotic gene and its protein product were colinear. Not all eukaryotic genes are colinear with their proteins. This is so because many eukaryotic genes contain interspersed blocks of sequences called intervening sequences or introns.

Combinatorial exons – In this type of alternative RNA splicing, inclusion or exclusion of a particular exon is independent of all other exons present in the transcript.

Commaless code – When there are no spacer nucleotides between code words.

Complementation – The ability of two recessive mutations to restore wild-type phenotype (partially or completely).

Concordant regulation – Perturbation of antisense RNA can alter the expression of sense messenger RNAs, suggesting that antisense transcription contributes to the control of transcriptional outputs. Gene expression profiling in mammals reveals frequent concordant regulation of sense/antisense pairs of transcripts.

Consensus sequence – A sequence found in majority of the sequences analyzed to find out structure of an element. Any consensus sequence is defined by aligning all known examples so as to maximize their homology.

Convergent transcription – Sister chromatids, the products of eukaryotic DNA replication, are held together by the chromosomal cohesion complex after their synthesis. This allows the spindle in mitosis to recognize pairs of replication products for segregation into opposite directions. Cohesin forms large protein rings that may bind DNA strands by encircling them. Cohesin localizes almost exclusively between genes that are transcribed in converging directions.

Cooperation between nuclear and chloroplast DNA – Small subunit of fraction I protein is coded by nuclear DNA and is synthesized in the cytoplasm and is then transported to the chloroplast. The large subunit of fraction I protein is synthesized in the chloroplast. The small subunit is then transported to the chloroplast. The small and large subunits of fraction I protein join each other in the chloroplast.

Cooperation between nuclear and mitochondrial DNA – There are many proteins in mitochondria that have dual origin, some polypeptides having cytoplasmic origin and others having mitochondrial origin.

Cooperation response genes – Transformation of normal cells into cancer cells entails concerted changes in the expression of many genes. Oncogenic mutations in the transcription factor p53 and in the small GTPase protein RAS – which individually have limited effects on promoting cancer – cooperate to transform normal cells into cancer cells. A large proportion of such genes are controlled synergistically by loss-of-function.

Coordinated regulation of gene expression – The coordinated regulation of gene expression is required for homeostasis, growth and development in all organisms. Such coordination may be partly achieved at the level of messenger RNA stability, in which the targeted destruction of subset of transcripts generates the potential for cross-regulating metabolic pathways.

Copia gene family – This gene family is 4-5 kb long and appears to consist of genes of very similar structure located at multiple sites that are characterized by different flanking sequences.

Co-transcriptional cleavage (CoTC) – This primary cleavage event within β-globin pre-messenger RNA, downstream of the poly(A) site is critical for efficient transcriptional termination by RNA polymerase II. CoTC process involves a self-cleaving activity. Autocatalytic core of the CoTC ribozyme has functional role in efficient termination in vivo. CoTC may be a general phenomenon and functionally it may provide an entry point for exonuclease involved in mRNA maturation, turnover and, in particular, transcriptional termination.

Co-translational protein translocation – Secreted and membrane proteins are translocated across or into cell membranes through a protein-conducting channel (PPC) when they are being synthesized. This translocating PCC forms connections with ribosomal RNA hairpins on two sides and ribosomal proteins at the back, leaving a frontal opening. For entry of these proteins into the ER, leader sequence is required at the N-terminal end of the protein. Also known as *co-translational transfer.*

D

Degenerate code – When there is more than one codon for a particular amino acid.

Divergent multigene family – Different members of the family have resulted from concerted evolution to the formation of divergent sets of duplicates. Examples of divergent multigene families are immunoglobulin genes, globin genes, actin genes, albumin genes, α-fetoprotein genes, serine protease genes, the interferon geness (for defense against viral infection) genes, and chorion protein-making genes

Divergent transcription – Transcription initiation by RNAP II is thought to occur unidirectionally from most genes. Transcription start site-associated RNAs (TSSa-RNAs) non-randomly flank active promoters, with peaks of antisense and sense short RNAs at 250 nucleotides upstream and 50 nucleotides downstream of TSSs, respectively. Divergent transcription over short distances is common for active promoters.

Divisive introns – Random insertions of DNA (bearing rudimentary splice signals) into previously intact genes. Also known as *type B introns*. Such type B introns could arise by aberrant recombination, retrovirus or transposon integration, or retrovirus-mediated insertion of cellular mRNA sequences.

DNA sequence divergence – Nucleotide sequencing techniques made it possible to know primary structure of many genes, which made it possible to compare pairs of homologous nucleotide sequences of a wide variety of genes in different organisms. This enabled us to evaluate the extent of sequence divergence in coding regions (exons) and non-coding regions (introns) of a gene. From the comparison of DNA sequences in coding regions it became possible to evaluate the extent of sequence divergence for two types of substitutions – (a) a nucleotide replacement leading to an amino acid change (mis-sense mutations) and (b) a nucleotide replacement leading to a synonymous codon change (same-sense mutations). In molecular terms, at DNA level, evolutionary rate is expressed as difference in nucleotide per site per year.

DNA supercoiling and transcription – Many prokaryotic genes are responsive to DNA supercoiling. Transcription itself can be major contribution to the level of supercoiling.

DNA topoisomerase type I – Type I DNA topoisomerase rectifies the situation behind by making a transient break in one strand of DNA.

DNA topoisomerase type II – Type II DNA topoisomerase relaxes negative supercoiling during transcription by introducing a transient double-stranded break in DNA.

DNA topoisomerases – RNA ppolymerase generates positive supercoiling ahead and leaves negative supercoiling behind. The enzymes that introduce or remove turns from the double helix by transient breakage of one or both polynucleotides. These enzymes are important during transcription. There exist two types of DNA topoisomerases – Type I and Type II.

Domains – Functional regions that form modular architecture in proteins, which is made up of discrete structural regions. Different exons code for the different domains of a protein. Each domain is also found in other proteins. Origin of genes coding for such proteins may have been by *exon shuffling*.

Dosage repetitions of DNA sequences – In case of dosage repetition there are many copies of a DNA sequence per nucleus. The rRNA, 5SRNA, tRNAs and histone genes show dosage repetition.

Down mutations – Mutations with lost or reduced transcription of the adjacent gene are known as down mutations. Down mutations could be due to deletion of an extragenic part of the promoter. Down mutations can be produced involving only a single base pair.

E

Early intron theory – Also known as exon theory of origin of introns, explains that exons are descents of ancient minigenes and the introns are descendents of the spacers between them. Genes large enough to encode proteins were first assembled from sets of exons. The machinery of splicing originated in an ancient RNA world. Introns were completely lost from both kingdoms of bacteria as well as several protist groups.

Editing reactions during translation – Synthesis of proteins containing errors (mis-translation) is prevented by aminoacyl tRNA synthetases through their accurate aminoacylation of cognate tRNAs and their ability to correct occasional errors of aminoacylation by editing reactions. A principle source of mis-translation comes from mistaking glycine or serine for alanine.

Effector gene – An effector gene is a gene that produces a regulatory molecule that drives the expression of another gene.

Effector plasmid – Effector plasmids carry a gene that expresses a regulatory protein which in turn regulates the expression of another gene carried in a reporter plasmid.

Encrypted genes – These genes are found as separate segments around the genome, so that, for example, all building blocks of a given mRNA molecule can be located, as modules, on separate chromosomes.

Engineered vaccines – Vaccines prepared by recombinant DNA technology.

Enhancers – A common feature of enhancersvis GGTGTGG AAAG. Enhancers have been detected in eukaryotic genes also. Enhancers mostly are *cis*-acting but *trans*-acting ones are also known. Enhancers are effective whether lying upstream or downstream from the promoter. They are active whether they lie in same or opposite polarity as the mature gene. They are equally effective regardless of the organism from which the gene is derived when attached to foreign DNA.

Enzyme-linked immunosorbent assay (ELISA) – This assay works by using antibodies immobilized on a microtiter plate to capture proteins of interest from samples added to the well. Using a detection antibody conjugated to an enzyme or fluorophore the quantity of bound protein can be accurately measured by fluorometric or colourimetric detection. The detection process is very similar to that of a Western blot, but by avoiding the gel steps more accurate quantification can be achieved.

Enhansons – The enhancer elements cooperate with one another or duplicates of themselves to enhance transcription. These elements are bipartite, being composed of subunits called enhansons, which can be duplicated or interchanged to create new enhancer elements. Enhansons differ from the enhancer elements because they are very sensitive to changes in spacing.

Epigenetic gate keepers – Epigenetic silencing genes *p16*, *SFRPs*, *GATA-4* and *GATA-5*, and *APC* in stem/precursor cells of adult cell-renewal systems that may serve to abnormally lock these cells into stem-like states that foster abnormal clonal expansion.

Eukaryotic ribosomes – Ribosomes of eukaryotes and prokaryotes differ in size and other details. The cytoplasmic ribosomes of eukaryotes are of 80S, size, contain 60 per cent rRNA and 40 per cent protein and dissociate into a smaller 40S subunit and a larger 60S subunit. 60S subunit has 5S, 5.8S and 28S rRNA and 40S subunit has 18S rRNA.

Eukaryotic RNA polymerases – Eukaryotic RNA polymerases are characterized by type of RNA they synthesize. Five types of eukaryotic RNA polymerases are well known – RNA polymerase I, RNA polymerase II, RNA polymerase III, RNA polymerase IV, and RNA polymerase V.

Eukaryotic translation elongation – Elongation factor eEF3 is an ATPase that, in addition to the two canonical factors eEF1A and eEF2, serves an essential function in the translation cycle of fungi. eEF3 is required for the binding of the aminoacy-tRNA-eEF1a-GTP ternary complex to the ribosomal A-site and has been suggested to facilitate the clearance of deacyl-tRNA from the E-site.

Eukaryotic translation initiation – In eukaryotes, there is existence of at least 11 eukaryotic initiation factors (eIF1, eIF2, eIF3, eIF4A, eIF4B, eIF4C, eIF4 D, eIF4E, eIF4F, eIF5, eIF6). There are two cap binding proteins (CBPI and CBP2)). CBP1 binds to the 5′ cap (m^7GpppX) of mRNA and facilitates formation of a complex between the mRNA and the 40S ribosomal subunit. CBP2 performs some yet unknown function.

The 5′ cap is required for efficient translation. The eIF1 assists mRNA binding, eIF2 binds met-tRNA. It has 3 subunits - α (binds ATP), β (may be recycling factor) and γ (binds met-tRNA$_f^{Met}$). The eIF3 binds mRNA. The initiation factor eIF4A assists mRNA binding and also binds ATP, eIF4B assists mRNA binding and unwinding, eIF4C binds 60S subunit. Function of eIF4D and eIF4E is unknown. Initiation factor eIF4F mediates the function of cap. Initiation factor eIF4E binds the cap. This step is thought to be regulated by phosphorylation. Initiation factor eIF5 releases eIF2 and eIF3 while eIF6 prevents 40S-60S joining. The eukaryotic initiation factor 4G (eIF4G) is the core of a multicomponent switch controlling gene expression at the level of translation initiation. It interacts with the small ribosomal subunit interacting protein, eIF3, and the eIF4E/cap-mRNA complex in order to load the ribosome onto mRNA during cap-dependent translation. Initiation factor eIF5A seems to have a hand in every step. mammalian eIF6 is required for efficient initiation of translation, in vivo.

Eukaryotic translation termination – In eukaryotic system, only one release factor, eRF, has been found. GTP seems to be necessary for activity of eRF. The released peptidyl-tRNA comes in the cytoplasm.

Exon – Sequences present in split (compound) genes whose complementary sequences are represented in messenger RNA.

Exon sharing – Likely mechanism for sharing of exons coding for different proteins is through the duplication and migration of exons during evolution.

Exon shuffling – Mixing up of exons during evolution. It can lead to significant genome rearrangement due to meiotic division. Exon shuffling is believed to be responsible for evolution of split genes in nuclear genes.

Exosome – Exosomes are 50-90 nm vesicles secreted by a wide range of mammalian cell types. Exosome constitute a mechanism for selective removal of many plasma membrane proteins. The exosome is a major eukaryotic nuclease located in both the nucleus and the cytoplasm that contributes to the processing, quality control and/or turnover of a large number of cellular RNAs.

Expansion of genetic alphabet – A new Watson-Crick base pair with a hydrogen pattern different from that in the A·T and G·C base pairs is incorporated into the duplex DNA and RNA by DNA and RNA polymerases, and expands the genetic alphabet from 4 to 6 letters. The genetic code can be extended artificially.

Expressed fraction of genome – Several supposedly silent intergenic regions in the genome of budding yeast are actually transcribed by RNA polymerase II, suggesting that the expressed fraction of the genome is higher than anticipated.

Expression system – A system specifically designed for the production of a gene product of choice. This is normally a protein although may also be RNA, such as tRNA or a ribozyme. An expression system consists of a gene, normally encoded by DNA, and the molecular machinery required to transcribe the DNA into mRNA and translate the mRNA into protein using the reagents provided. In the broadest sense this includes every living cell but the term is more normally used to refer to expression as a laboratory tool.

Extein – The remaining portins of the protein after inteins (protein introns) have been spliced out.

Extended anticodon hypothesis – The structure of anticodon loop and the proximal anticodon stem are related to the sequence of anitcodon. In other words, anticodon is extended into the nearby structure and consists of (a) two nucleosides at the 5′-end of the anticodon loop, (b) three nucleosides of anticodon, and (c) two nucleosides at the 3′-end of the anticodon loop (d) and five nucleoside pairs in the anticodon stem. Extended anticodons are involved in translation.

F

Fidelity of transcription – During transcription elongation, the hydrolytic reaction stimiulated by misincorporated nucleotides proofreads most of the misinccorporation events and thus serves as an intrinsic mechanism of transcriptional fidelity.

5S RNA gene family – These genes are present independently of the rDNA, which is localized in NOR in eukaryotes. In prokaryotes and yeast, *5S rRNA* genes are present in close vicinity of rDNA. The 5S rRNA genes are transcribed by RNA polymerase II.

Fold-back elements – Some inverted repeats may be immediately adjacent or separated by upto several thousand base pairs. During renaturation, as the reaction is intramolecular, the structures formed are called snap-back (SB) or fold-back (FB) structures. There are 2,000-4,000 pairs of inverted repeats or potential fold-back structures in *D. melanogaster* which comprise about 3 per cent of the total genome.

***412* gene family** – This gene family at each of the 30 or so sites contains a complete *412* gene. Each is 7.0 kb in length. It is terminally redundant.

Functional alleles – Alleles determined on the bases of complementation test.

G

Gene – (1) Mendel's "particles" are unit of heredity responsible for phenotype. (2) Morgan's "loci" are genes in a chromosome, i.e., it is a cellular entity that is part of chromosome and is mapable. (3) According to Watson and Crick, gene is a sequence of specific nucleotides along the length of a double helical DNA molecule. (4) From geneticist's point of view a gene is a discrete chromosomal region which is responsible for synthesis of a specific cellular product (mRNA, tRNA, rRNA), and it consists of a linear collection of potentially mutable sites each of which can exist in several alternative forms and between which crossing-over can occur. (5) Modern functional definition of gene is: a DNA sequence coding for a specific polypeptide but split gene DNA sequence must also include introns (non-coding segments) and exons (coding segments that make proteins) and. others DNA pieces. Any definition of gene must also include promoters, enhancers, regulator genes, operators, and also segments that code for rRNA, tRNA, and snRNP's. (6) A sequence in a nucleic acid (DNA, except in RNA viruses) that provides a code for a protein, polypeptide, or RNA of direct value to cellular metabolic processes, plus those parts, adjacent or internal, important to its being located, identified, transcribed, often translated, and processed. Internal transcribed regions, such as introns, are also a part of the gene.

Gene annotation – Gene annotation provides functional and other information, for example the location of each gene within a particular chromosome.

GeneCalling – A mRNA profiling technology finding increased use in the field of genomics and invented by Dr. Jonathan M. Rothberg and developed by CuraGen Corporation. This genomics technique rapidly identifies candidate genes for use in drug discovery and development.

Gene discoveries – Phase in understanding of gene function when many new different types of genes, viz., repeated genes, moveable genes (transposable genetic elements), pseudogenes, retrogenes, and related phenomenon such as multiple polyadenylation sites, protein *trans*-splicing, were discovered. These discoveries made the neoclassical concepts of gene structure and gene function obsolete.

Gene expression – Gene expression and transcription are considered synonymous. This is the first step in gene expression. The gene product is then translocated to the site of function. Used RNA and proteins molecules are then degraded. Ribosomes are reused. Amino acids and ribonulceitides go their respective pools.

Gene expression networks – Genes have sometimes been regarded as nodes in a network, with inputs being proteins such as transcription factors, and outputs being the level of gene expression. The node itself performs a function, and the operation of these functions has been interpreted as performing a kind of information processing within cells and determines cellular behavior.

Gene function – What does gene do? Answer to this question came by understanding relationship between gene and its product. First crucial step was to understand the genetic control of biochemical reactions. Relationship between genotype and phenotype was then understood. Final picture about gene function became clear as the fine structure of gene was known. Discovery of new types of genes changed our thinking about gene structure and gene function.

Gene fusion – (1) Mutations that cause portions of two genes to fuse together and form a hybrid gene are frequent in blood-related and lung cancers. (2)

Gene fusions – Certain exons can be members of at least two, and possibly even more, transcripts. These chimeric transcripts are formed by transcription of two consecutive genes into one RNA.

Gene structure concept, classical phase of – Gene was regarded as a hypothetical particle, performing a specific function and recombining with other genes.

Gene structure concept, modern phase of – Gene was a unit of function but not a unit of recombination or that of mutation. Fine structure of gene was defined in terms of cistron, recon and muton. Gene has well-defined internal structure.

Gene structure concept, transitional phase of – Some alleles may occupy complex locus which can be divided into subloci.

Gene targeting – A technique allowing scientists to block or alter a gene's function. It works by infusing strands of lab-made DNA into stem cells, where they latch on to their target genes. These cells are injected into a mouse embryo, spreading the new gene throughout. Embryonic stem cells are collected from fertilized eggs a few days after they have started to divide.

Gene, complex – Complex genes are so called because either there is rearrangement of DNA at different levels before 3'←5' strand of DNA is available for transcription (as is case of *immunoglobulin gene*) or pre-mRNA is cleaved through a series of steps before mRNA in finished form is available (as is the case of *dimorphic genes*, as exemplified by the kallikrein gene or these genes have a cryptic structure in that the ultimate active product or products are carried within the precursorial protein (as is the case of *cryptomorphic genes*, exemplified by yeast sex pheromone gene (*MF1*), mammalian pancreatic glucagon genes, which are released after enzymatic breakdown of the precursor and become functional following further processing.

Gene, compound – Genes in which coding sequences (exons) are separated by non-coding sequences (introns). Discovery of split genes was made independently by two groups led by Phillp A. Sharp and Richard J. Roberts in 1977. Also known as *split genes*.

Gene, regions of a eukaryotic – Ten regions: recognition region (~50 kb), transcription initiation site, 5' untranslated region, translation initiation, alternating exon/intron, splice donor and acceptor sites, translation stop site, 3' untranslated region, polyadenylation signal, and transcription stop site.

Gene, regions of a prokaryotic – Eight regions: recognition region (~50 bp), transcription initiation site, 5' untranslated region, translation initiation, coding region, translation stop site, 3' untranslated region, and translation stop site.

Gene, simple – A continuous sequence in a nucleic acid that specifies a particular polypeptide or functional RNA. This condition exists in the prokaryotes and viruses.

Gene-enzyme relationship – First clue to the nature of the primary gene function came from the studies on humans. Garrod in 1909 noted that several hereditary human defects are produced by recessive mutations. Such disorders were attributed to an absence or a defect in an enzyme. Thus, a relationship between genes and enzymes, which are responsible for conducting various biochemical reactions, was suggested.

Gene organization – Refers to the way genes are arranged in a cell. It may be considered from different points of view. One may consider whether individual genes are independently controlled or a group of them are under the control of one and the same operator, whether the message is contiguous or split, genes are overlapping or non-overlapping, genes are functional or non-functional, genes are repeated or not, whether the genes occupy a fixed position or they shift from one place to another.

General shutdown of protein synthesis – In response to binding viral double-stranded RNA by-products within a cell, the RNA-dependent protein kinase, PKR, phosphorylates α subunit of the translation initiation factor eIF2 on a regulatory site, Ser51. This triggers the general shutdown of protein synthesis and inhibition of viral propagation.

Genes, types of – From structure point of view genes are classified into three types — simple gene (most of the prokaryotic genes), compound gene (most of the eukaryotic genes), and complex gene.

Gene-specific translational silencing – Gene-specific translational (GST) silencing as a novel function of the fused glutamyl- and prolyl-tRNA synthetase (GluProRS). GluProRS is released from a multisynthetase translation complex in response to γ-interferon and forms a four-protein GAIT complex that silences translation of ceruloplasmin (Cp), a protein linked to the inflammatory response.

Genetic code – The processes involved in translating or decoding the information contained in the primary structure of DNA

Genetic code specificity – A particular base modification in a tRNA is important for reading the its specific codon for an amino acid for the specificity of its aminoacylation.

Genetic restoration – Organisms can sometimes rewrite their DNA on the basis of RNA messages inherited from generations past. Plants and animals can inherit allele-specific DNA sequence information that was not present in the genome of their parents but was present in previous generations. First found in *Arabidopsis thaliana*, later, the same phenomenon was also found in mammals, suggesting common occurrence.

Genome – Sum total of all the genes of an organism. Further classified according to location of the genome – nuclear genome, chloroplast genome, mitochondrial genome.

Genome evolution – Refers to evolution of a genome in viruses and bacteria to higher eukaryotes.

Genome interpretation – Refers to the mechanisms involved in negotiating the genome for transcription initiation. Spontaneous preferences of DNA binding proteins are a primary mechanism by which cells interpret the genome.

Genomics – Genomics studies DNA isolated from any part of the plant or animal cell, since DNA in different cells do not differ in time and space.

Genotype-phenotype relationship – Beadle and Ephrussi in 1937 conducted eye disc transplantation experiments in *Drosophila melanogaster*, which indicated a strong link between phenotype (eye color) and genotype (*v* and *cn*).

Germline theory of antibody diversity – The genome contains thousands of related but unique DNA sequences, each of which specifies one of the *V′* regions synthesized by an organism. Thus a large percentage of the mammalian genome is devoted to carrying variable sequence families and each person will inherit and transmit all *V* regions through his or her germline. Each lymphocyte will selectively express only one of these regions (genes) for the *V* portion of a light chain and only one of the *V* portion of a heavy chain. Because most lymphocytes will happen to select different combinations of sequences to express, the organism will come to express its vast diversity of antibody producing cells.

Global transcription machinery engineering (gTME) – An approach for reprogramming gene transcription to elicit cellular phenotypes important in technological applications.

Goldberg-Hogness box – Sequences discovered by Goldberg and Hogness in promoters of eukaryotic genes. Sequences similar to Pribnow box. Sequence of the Hogness box is $T_{82}A_{97}T_{93}A_{35}A_{43}/T_{27}A_{83}$ A_{50}/T_{37}. Hogness box is surrounded by G.C-rich sequence. Also called as *TATA box.*

Guide RNA – This RNA is involved in RNA editing. RNA interference and related RNA silencing phenomena use short antisense guide RNA molecules to repress the expression of target genes.

H

Half genes – Analysis of genome sequence of the small hyperthermophilic archaeal parasite *Nanoarchaeum equitans* has not revealed genes encoding the glutamate, histidine, tryptophan and intiator methionine transfer RNA species. A computational approach was used to search for widely separated genes encoding tRNA halves that, on the basis of structural prediction, could form intact tRNA molecules. The search revealed nine genes that encode tRNA halves; together they account for missing tRNA genes.

Haploinsufficiency – Reduced gene dosage. It may lead to some syndromes. For example, haploinsufficienvy may be one of the causes of Rubinstein-Tyabi syndrome (RTS). Greg syndrome also results from haplo-insufficiency.

Hidden genes – These are short open reading frames, i.e., sequences initiated by the alternative translation initiation codons ACG, AUA, and GUG in the 5′-terminal extra-cistronic region.

Hidden transcription – Cryptic unstable transcripts (CUTs) have been recently described as a principal class of RNAPII transcripts in budding yeast. These transcripts are targeted immediately for degradation immediately after synthesis by the action of the Nrd1-exosome-TRAMP complexes.

Highly unstable RNAs – The bulk of eukaryotic genomes is transcribed. A class of short, polyadenylated and highly unstable RNAs has been reported. These promoter upstream transcripts (PROMPTs) are produced ~0.5 to 2.5 kilobases upstream of active transcription start sites. PROMPT transcription occurs in both sense and antisense directions with respect to the downstream gene. It requires the presence of the gene promoter and is positively correlated with gene activity. PROMPT transcription is a common characteristic of RNA polymerase II transcribed genes with a possible regulatory potential.

Histocompatibility – Tissues can be transplanted between genetically identical animals without concern for immunologic rejection, whereas transplants between genetically non-identical animals are usually rejected with time. Also known as *tissue compatibility*.

Histocompatibility antigens – Those antigens that determine the acceptance or rejection of a tissue graft. These antigens are produced by histocompatibility genes.

Histocompatibility genes – The genes that code for histocompatibility antigens. These genes are involved in the production of cell surface antigens that are recognized by the rejection or tolerance of tissue transplants.

Histone gene family – There are five types of histones H1, H2a, H2b, H3 and H4. Genes synthesizing these histones are present in multiple copies.

HIV reverse transcriptase – The reverse transcriptase of human immunodeficiency virus (HIV) catalyzes a series of reactions to convert the single-stranded RNA genome of HIV into double-stranded DNA for host-cell integration. The active sites of the enzyme are correctly positioned to support one of three catalytic functions: RNA-directed DNA synthesis, DNA-directed DNA synthesis and DNA-directed RNA hydrolysis.

HLA genes, endogenous functions of – These genes may have played some role in our evolutionary history in response to some environmental selection pressure and/or may be intimately involved in cancer. Presumably, one of the normal functions of the HLA genes is to recognize cancer cells as a kind of "foreign" agent within the body. Some HLA alleles seem to confer resistance or susceptibility to certain specific diseases. The HLA genes also play an important role in normal immune response.

Homology of transcription mechanisms – TATA-binding protein (TBP) is a classless factor essential for all transcription in eukaryotic cell nuclei and Archaebacteria showing that transcription mechanisms of Archaebacteria and eukaryotes are fundamentally homologous.

Hormones and transcription – Study of sex differentiation and role of hormones shows that realization of genotypic capability depends upon and can be modified by hormones, while the production of hormones is itself guided by the genetic constitution of the individual.

Human leukocyte antigen (HLA) complex – A complex comprising of genes that are responsible for synthesis of antigens that determine acceptance or rejection of a tissue graft. HLA system is based on three classes of HLA genes on chromosome number 6 which determine immunological acceptability.

Hybrid arrested translation – Hybrid arrested translation is based on the fact that an mRNA will not direct the synthesis of a protein in a cell-free system when it is in a hybrid from with its complementary DNA.

Hybrid released translation – Hybrid released translation (HRT) enables a cloned DNA to be correlated with the protein(s) which it encodes. HRT is a direct method in which cloned DNA is bound to a cellulose nitrate filter and hybridized with an unfractionated preparation of mRNA or even total cellular RNA. The filter is washed and hybridized mRNA is eluted by heating in low salt buffer. Recovered mRNA is then translated in a cell-free translation system.

Hybridoma – The term hybridoma is applied to fused cells resulting due to fusion of cells from two different sources. The role of the myeloma cell in this process is to provide the normal B lymphocytes with the cancerous property of uncontrolled growth, thereby allowing hybridomas to grow easily both in tissue culture and after injection into mice. Normal non-cancerous B-cell would quickly die in either situation.

Hypercycles – The metabolome forms a large network of metabolic reactions, where outputs from one enzymatic chemical reaction are inputs to other chemical reactions. Such systems have been described as hypercycles.

I

Identical multigene families – Different members of the family are multiple copies of the same gene.

Immunoglobulin class-switch recombination (CSR) – Activation induced cytidine deaminase (AID) is required for the DNA cleavage step in immunoglobulin class switch recombination. AID is proposed to deaminate cytosine to generate uracil (U) in either mRNA or DNA. DNA cleavage depends on uracil DNA glycosylase (UNG) for removal of U. UNG is involved in repair step of CSR yet by an unknown mechanism. Uracil removal introduces immunoglobulin class-switch recombination.

Immunoglobulin genes – If convention that one polypeptide is coded for by one gene is accepted, each of the final DNA segment coding a complete heavy or light chain should be considered as a single gene. Various alternative sequences within each set (V, J, D or C) should according to this definition be termed simply as DNA sequences since they are all parts of a single gene coding for one polypeptide chain. With respect to immunoglobulin genes, the genome of somatic cells is not constant. They constitute a multigene system which comprises of three unlinked multigene families – λ, k, and H.

Immunoglobulin heavy chain genes – These genes code for light chains of Immunoglobulins. There are five types of heavy chains of constant region (C_H) known: μ, δ, γ, ε and α which define five types of immunoglobulins, IgM, IgD, IgG, IgE and IgA, respectively. Each of the five heavy chain classes may be associated with either k or λ light chains. L (leader), V (variable), J (joining), D (diversity) and C (constant) sequences are present in heavy chain genes.

Immunoglobulin light chain genes – These genes code for light chains of Immunoglobulins. Only two types of are known, k and λ, each differing in amino acid sequence of constant light (C_L) segment and includes light chains (L) with many different variable light chain (V_L) sequences. L (leader), V (variable), J (joining) and C (constant) sequences are present in light chain genes.

Immunoglobulins (Ig) – The antigenically-related proteins. They are antigen-binding proteins synthesized by B-lymphocytes.

Immunopurification – The process that involves separation of a specific antigen from a mixture of very similar antigens. This purified antigen can then be used for developing vaccine against a pathogen.

Importin – Receptor on the pore to which signal sequence binds.

Inchworming model of abortive transcription initiation – It invokes a flexible element in RNAP. In each cycle of abortive initiation, a module of RNAP containing the active center detaches from the remainder of RNA polymerase and translocated downstream; upon release of the abortive RNA, this module of RNA polymerase reverse translocates.

Informational gene family – It has individual members that can differ markedly in sequence from one another although all are homologous and obviously show an ancient ancestry.

Initiation codon – First codon in translation (usually AUG) is the initiation codon. GUG acts as initiation codon (although it has been assigned for valine) in vitro but with lesser efficiency.

Insertional editing – It involves editing of kinetoplastid mitochondrial mRNA. It involves base pairing to specific guide RNAs with poly(U) tails. Mechanism involves identification of editing sites in the mRNA by mis-matches with the guide RNA, cleavage of the mRNA and religation of mRNA halves with added or deleted U residues by TUTase.

Integral membrane protein – A protein that resides within a membrane.

Intein – An intein is a segment of a protein that is able to excise itself and rejoin the remaining portions (the exteins) with a peptide bond. Inteins have also been called protein introns.

Interchromosomal interactions in coordinated gene expression – Some eukaryotic genes (e.g., cytokine genes) located in separate chromosomes associating physically in the nucleus via interactions that may have a function in coordinating gene expression.

Intergenic spacers (IGS) – (1) The genomic regions that are not directly involved in the final product and are found between actual coding sectors of two adjacent genes. These spacers vary in length from a single base pair to many thousands base pairs. These are frequently referred to as *flanking regions*. Certain parts of these sequences exhibit specialization of function. The sector of the non-coding sequence preceding 5'-end of gene proper (mature gene, antisense strand) is called leader while that following the 3'-end is the train. (2) Each member of repeat unit in a ribosomal gene family has a coding region with genes that specify 18S, 5.8S and 28S rRNA molecules, a spacer region called intergenic spacers (IGS). The variation in number and size of the subrepeats in IGS is responsible for variation in length of the repeat unit (IGS + coding region).

Internal transcribed sequences – Each member of repeat unit in a ribosomal gene family has a coding region with genes that specify 18S, 5.8S and 28S rRNA molecules, a spacer region called intergenic spacers (IGS) and internal transcribed sequences (ITS).

Intron-exon junctions – Consensus sequences at the boundaries of intron-exon junctions have shown that GT was always found at the 5′-side of the intron and AG at 3′-side. This is known as Chambon's rule.

Introns – Sequences present in split (compound) genes whose complementary sequences are absent in messenger RNA. The sequences of non-coding DNA that interrupt the coding sequence of genes. Introns are excised from gene transcripts during RNA processing. Intron encoded proteins act as helpers in RNA splicing.

Introns, mitochondrial gene – Introns of split genes in fungal mitochondria are classified into two groups on the basis of their internal organization. Group I introns, are found in majority of the fungal mitochondrial split genes, do not carry any conserved sequence at intron-exon junctions, but carry internally a short conserved sequence called 'internal guide sequence'. Group II introns resemble nuclear genes and have consensus sequences (GT and APy) and a branch sequence that resembles the TACTAAC box.

Introns, chloroplasts gene – Split genes for ribosomal RNA (rRNA), transfer RNAs and some proteins have also been reported in the chloroplast genomes of several plants including *Chlamydomonas* and *Nicotiana.* Introns found in chloroplast genes can be classified into three groups on the basis of intron boundary sequences. Group I introns (e.g., in *trnL*) can be folded in a secondary structure similar to self-splicing rRNA procedure of *Tetrahymena.* These can be removed either by self-splicing or by a 'maturase' (as in cytochrome b and cytochrome oxidase mRNA precursors). Group II introns (e.g., majority of genes including *trnA* and *trnl*) can be folded into a complex secondary structure (as in introns of mitochondrial genes for cytochrome oxidase in maize and yeast). Group III introns (e.g., *trnG*, *trnK*, *trnV*, *rpl2*, *rps12*, *rps16*, etc.) have conserved sequence at their borders (GTGCGNY at 5'-end, and ATCHR YY(N)YYAY at 3'-end), similar to those in the eukaryotic nuclear genes.

Introns, types of – (1) Depending on mode of splicing, there are four different types of introns: Type 1 introns are self-splicing and are employed in nuclear, mitochondrial and chloroplast rRNA, tRNA and mRNA. In Type 1 introns, the 3′ OH group of a free guanosine acts a nucleophile, attacking the 5′ phosphate at the splice site, displacing it from the exon. The 3′ OH at the exon, usually a U, acts as a nucleophile, attacking the phosphodiester bond, removing the other end of the intron, and reforming the phosphodiester bond with itself. Type 2 introns are also self-splicing, and are used in genes in mitochondria and chloroplasts of fungi algae and plants. In Type 2 introns, a similar mechanism is used, except that an internal 2′ OH of an adenosine is used as the nucleophile, attacking the 5′ splice site of the exon to produce a lariat structure. In the final step, the 3′ OH of the 5′ exon attacks the phosphodiester bond of 3′ uracil, to remove the lariat. Type 3 introns require a spliceosome and small nuclear ribonucleoproteins (snRNPs) and are responsible for solely eukaryotic splicing. In type 3 introns, a lariat structure is formed, but the snRNPs are used to form the secondary structures and to mediate the reaction, though it still occurs without free energy input. There are 6 snRNP subunits that make up the splicesome. Type 4 introns require ATP and an endonuclease. (2) Depending on length, intron are of three types: (1) Short or very short introns are found in many tRNA genes. (2) Much longer introns are found in nuclear RNA and mitochondrial RNA genes and many mitochondrial mRNA genes. (3) The longest introns (a few hundred to tens of kb long) are found in genes specifying proteins.

Invariant DNA code – Variant code means that a mutation in the genetic code would place new amino acids in certain loci and entirely eliminate amino acids from the other loci of practically all proteins in an organism. But this actually is not the case as DNA code is invariant.

Ionome – The ionome is defined as the mineral nutrient and trace element composition of an organism and represents the inorganic component of cellular and organismal systems.

Ionomics – Ionomics is the study of elemental accumulation in living systems using high-throughput elemental profiling. This approach has been applied extensively in plants for forward and reverse genetics, screening diversity panels, and modeling of physiological states.

K

Kinetoplast DNA maxicircle – Maxicircle in Trypanosome kinetoplast DNA has 23 to 36 kbp. Maxicircle molecules are homologs of informational mitochondrial DNA molecules in animals and fungal cells and possess various genes. Several of maxicircle genes represent cryptogenes, i.e., incomplete genes whose transcripts are edited to yield translatable sequences.

Kinetoplast DNA minicircle – Minicircles in Trypanosome kinetoplast DNA has 465 to 2,500 bp. No obvious protein-encoding genes are known in minicircles.

Knockout' technique – O. Smithies in 1985 studied gene function by knocking off genes in mice. Knockout mice are genetically modified mice that have one or more genes silenced. Knockout mice are used to recreate human diseases in mice and to study the effect of individual genes on an organism's development.

Kozak's Scanning Hypothesis – The first initiation codon AUG in 90 per cent of the cases occurs in the form of consensus sequence PuNNAUGG. This is known as Kozak's sequence. In 5 per cent cases one or more AUG codons occur upstream to first AUG codon. These extra AUG codons are also read according to Kozak's hypothesis.

L

Large-scale gene function analysis – Microarray transcript profiling and RNA interference are two new technologies crucial for large-scale gene function studies in multicellular eukaryotes. Both rely on sequence-specific hybridization between complementary nucleic acid strands. For this, gene-specific sequence tags (GSTs) are created representing almost all the genes. GSTs resources provide novel and powerful tools for functional genomics.

Late intron view – Split genes arise from uninterrupted genes by insertion of introns. Genes large enough to encode proteins first arose (presumably from smaller genes) without participation of introns. The machinery of spliceosomal splicing arose from fragmented self-splicing introns. Spliceosomal introns were never observed in the ancestors of those organisms that now lack them.

Leader sequence – Sector of the non-coding sequence preceding 5'-end of gene proper (mature gene, antisense strand)

Leader peptides – In bacteria, proteins destined to be secreted are synthesized as pre-proteins with N-terminal signal sequences which are short (25 residues) and comprise of a hydrophobic central core which can adopt an α-helical structure, flanked by region containing several charged residues.

Leucine zippers – It is a family of eukaryotic transcriptional factors that are rich in leucine. It includes C/EBP, Fos, Jun and GCN4.

Live vaccines – Large DNA viruses such as Vaccinia virus, are used as a biological delivery causing agents. Vaccinia virus has a linear double-stranded DNA genome of approximately 1,85,000 bp and replicates in the cytoplasm of infected cells.

Localized protein secretion – A signal sequence may contain information that directs the secretion of a protein to one subcellular region, in addition to its classical role in promoting secretion. The signal sequences of two surface proteins, M (PrtM) and F (PrtF), direct secretion to different subcellular regions. The signal sequence of M protein promotes secretion of the division septum, whereas that of PrtF preferentially promotes secretion at the old pole.

Long non-coding (lnc) RNA – Many of the non-coding RNAs synthesized are longer than 200 nucleotides. These are called long non-coding (lnc) RNAs. Certain regulatory functions like controlling genome dynamics, cell biology and developmental programming have been associated with them.

M

Major splicing pathway – The major splicing pathway removes U2-type introns allowing the adjacent exons to be spliced together in the nucleus before export of the fully spliced transcripts to the cytoplasm. The transcripts that also contain U12-type introns are exported as partially spliced transcripts and their exons are spliced together by the U12 pathway in the cytoplasm. The mechanism by which they evade RNA surveillance in the nucleus is unknown. Although U2-type introns are found globally, U12-type introns may occur preferentially in particular sets of genes such as those involved in cellular proliferation.

Mammalian gene promoters for RNA polymerase II – They are typically organized in the following order: upstream sequence motif(s)/TATA box/initiation site. Polarity of transcription is primarily determined by the linear order of an upstream sequence relative to a TATA box, rather than by the individual orientations of either of these two elements.

Many genes-one polypeptide hypothesis – The immunoglobulin genes, which can be called assembled genes, do not fit any classical or neoclassical definition of the gene, since the genetic unit in the germline and in the mature immune cell is completely different. Functional genes of immunoglobulins mature by means of somatic recombination from a few units in the germline during the maturation of immune cells. The somatic recombination in immunoglobin genes led to many genes-one polypeptide hypothesis.

Mechanism of snRNP action – The term snRNP stands for small ribonucleoprotein. There are three possibilities of their mechanism of action of snRNPs: (a) they help to arrange the intron in a configuration that encourages self-splicing, (b) in some organism precursor mRNA forms lariat structure and splicing of intron occurs without the aid of protein or other RNA factors and (c) splicing may provide missing intron sequences necessary for spontaneous splicing.

Mechanisms changing the meaning of codon – The encoding of two non-universal amino acids involves dynamic redefinition of 'stop' signals in the genetic code. Mechanism changing the meaning of codon translation: (a) Stop codons can be reassigned independently of the context wherever they occur in any messenger RNA in a cell. Three example mRNA sequences are shown. This mode of changing codon meaning occurs in certain specialized genomes, such as those of many mitochondria. b and c, In context-dependent reassignment, either (b) site-specific codon redefinition can occur in response to a co-localized recoding signal (shown as a loop representing an RNA structure), or (c) mRNA-dependent redefinition of codons might happen in response to a recoding signal embedded outside the coding region.

Meiome – Meiome is the term used in functional genomics for meiotic transcriptome. It is the way to study transcript profiling using mRNA isolated from enriched meiotic cells (meiocytes).

Messenger RNA (mRNA) – One of the RNAs which contain codons and serve as a template during translation at the ribosome. mRNA is the only coding RNA. mRNA to carries the sequence information of DNA to ribosomes. Messenger RNA serves as a template on which a polypeptide is constructed and contains an initiation codon (AUG), at least one of the termination codons (UAA/UGA/UAG) and base sequence in form of triplet codons that dictate the order of amino acid in a polypeptide chain. mRNA also includes certain trailer and leader sequences that are not translated. Protein encoding regions of mRNA are composed of contiguous, non-overlapping string of codons called open reading frames (ORFs), each of which specifies single protein. Translation starts at 5′-end of open reading frame and proceeds one codon at a time to 3′-end. First and last codons of ORF are called start and stop codons, respectively.

Messenger RNA decay – A major control point in regulation of gene expression. After messenger RNA has been used as a template for translation, it is protected till needed by the cell or degraded by the cellular machiney when not required. mRNA has non-coding nucleotides at either end of the molecule. These segments contain information about the number of times mRNA is transcribed before being destroyed by ribonucleases. Hormones stabilize certain mRNA transcripts.

Messenger RNA quantification – Measurement of abundance of different transcripts in a transcript profile.

Metabolic engineering – It is improvement of cellular activities by manipulation of enzymatic transport and regulatory functions of the cell with the use of recombinant DNA technology. Introduction of heterologous genes and regulatory elements made metabolic engineering very fascinating area of research. Cell function can be modified using targeted alterations in normal cellular activities. This may involve not only synthesis of a metabolite, but may sometimes also involve manipulation of processing pathways. Metabolic engineering has been successfully used for improving the nutritional value of crops like rice and maize.

Metabolites – Intermediate products of metabolism are called metabolites. These are small molecules that exhibit enormous variation. A metabolite is usually defined as any molecule less than 1 kDa in size. In plant-based metabolomics, it is common to refer to "primary" and "secondary" metabolites.

Metabolome – Sum total of all the metabolites in the cell or tissue at different developmental stages, and under different conditions. Metabolites are unique chemical fingerprints that specific cellular processes leave behind.

Metabolome analysis – Metabolome anasysis involves (i) alignment of chromatographic and spectroscopic data using specific bioinformatics tools, and (ii) identification of different metabolites.

Metabolomics – The scientific study of chemical processes involving metabolites. It places a greater emphasis on metabolic profiling at a cellular or organ level and is primarily concerned with normal endogenous metabolism. Sometimes terms *metabolomics* and *metabonomics* are synonymously used.

Metabonomics – It extends metabolic profiling to include information about perturbations of metabolism caused by environmental factors (including diet and toxins), disease processes, and the involvement of extragenomic influences, such as gut microflora. Sometimes terms *metabolomics* and *metabonomics* are synonymously used.

Metastatic genes – Genes which specifically promote the dissemination of its cancerous cells to other organs.

MicroRNA – MicroRNAs (miRNAs) have emerged as key post-transcriptional regulators of gene expression, involved in diverse physiological and pathological processes. The precursor of a microRNA (pri-miRNA) is transcribed in the nucleus. It forms a stem-loop structure that is processed to form another precursor (pre-miRNA) before being exported to the cytoplasm. Further processing by the Dicer protein creates the mature miRNA, one strand of which is incorporated into the RNA-induced silencing complex (RISC). Base pairing between the miRNA and its target directs RISC to either destroy the mRNA or impede its translation into protein.

Minor splicing pathway – The minor spliceosome is ribonucleoprotein complex that catalyzes the removal of an atypical class of spliceosomal introns (U12-type) from eukaryotic messenger RNAs. First identified and characterized in animals, where it was found to contain several unique RNA constituents that share structural similarity with and seem to be functionally analogous to the small nuclear RNAs (snRNAs) contained in the major spliceosome. Subsequently, minor spliceosomal components and U12-type introns have been found in plants but not in fungi.

–1 and minus –3 rule – In the mature protein, amino acids with small residual groups are often found adjacent to the cleavage site (the –1 position) and the next but one upstream residue (the –3 position).

Mis-folded proteins – After insertion into endoplasmic reticulum (ER), proteins that fail to fold there are destroyed. Through a process termed dislocation such mis-folded proteins arrive in the cytosol, where ubiquitination, deglycosylation and finally proteasomal proteolysis dispense with the unwanted polypeptides.

Mis-folding proteins of mis-charged tRNAs – Some abnormally folded proteins result from mutations in genes encoding disease-associated proteins (for example, repeat-expansion diseases). Low levels of mis-charged tRNAs can lead to an intracellular accumulation of mis-folded proteins.

Models of protein degradation – Two well known models of protein degradation are: N-end rule and PEST hypothesis.

Modified central dogma of molecular biology – Crick's concept of central dogma was modified with the emerging knowledge that in some plant viruses RNA is the genetic material where genetic information flows from RNA to DNA with the help of an enzyme *RNA-dependent DNA polymerase*, also known as *reverse transcriptase*. Another modification in Crick's central dogma was that RNA can self replicate. Protein cannot self-replicate, nor can it use amino acid sequence information to reconstruct RNA or DNA. However, under laboratory conditions DNA can directly be used as a template for translation.

Molecular chaperone – A molecular chaperone is a protein that is needed for the assembly or proper folding of some other protein but which is not itself a component of the target complex. The molecular chaperones provide a structure on which proper folding of proteins takes place. Molecular chaperones allow proteins to find a functional, stable state by allowing the proteins opportunities to fold into a thermodynamically stable and functional configuration. Each cycle of refolding requires ATP energy.

Monoclonal antibody – The homogenous antibody derived from a single B-cell clone and therefore all bearing identical antigen binding sites and isotype.

Moveable genes – Movable genes are DNA elements that can move from one location to another in the genome of an organism. These genes were first discovered in maize by Barbara McClintock in 1950s. The existence of movable genes challenges the hypothesis of a fixed location of the gene in the chromosome.

mRNA helicase – Most mRNAs contain secondary structure, yet their codons must be in single-stranded form to be translated. Ribosome itself is an mRNA helicase. The mRNA helicase activity is localized to the middle of the downstream tunnel, between the head and shoulder of the 30S subunit.

mRNA reading frame maintenance – The triplet-based genetic code requires that translating ribosomes maintain the reading frame of a messenger RNA faithfully to ensure correct protein synthesis. However, in programmed –1 ribosomal frameshifting, a specific subversion of frame maintenance takes place, wherein the ribosome is forced to shift one nucleotide backwards into an overlapping reading frame and to translate an entirely new sequence of amino acids.

Multigene families – A multigene family is a group of nucleotide sequences or genes that exhibits four properties, namely, multiplicity, close linkage, sequence homology, and related or overlapping phenotypic functions.

Multiple allelism – The existence of a gene in various alternate forms.

Multiple polyadenylation sites – During the RNA processing in which the primary transcript ripens to form messenger RNA, about 200 adenosine nucleotides are added in a polyadenylation reaction at the 3′ end. These are not coded by the corresponding gene. In certain cases, there are multiple alternative polyadenylation sites in the primary transcript. This was first observed in adenoviruses.

Multiplicational gene family – It consists of 10^1 to 10^4 copies of a gene, 80-100 nucleotides long. Repeat units are essentially identical.

Multivalent vaccines – Multiple foreign genes are inserted into Vaccinia virus. Vaccinia virus strain VP168 carried three foreign genes – *InfHA*, *HSVgD* and *HBsAg*. Vaccinia virus recombinant VP168 responded by making antibodies to all three of the foreign antigens synthesized under Vaccinia virus regulation.

Muton – A unit of mutation defined by recombination test. Two mutations belong to the same muton if they do not show recombination. The smallest mutational unitin a double-stranded DNA molecule is a single nucleotide pair.

Mutually exclusive mRNA splicing – A eukaryotic gene contains a large number of introns; many of them are alternatively spliced. Four mechanisms of mutually exclusive splicing are: stearic interference, spliceosomal incompatibility, disposal by non-sense-mediated decay, and antagonism of repressor by Docker:selector pairing. Reverse splicing provides considerable insight into the catalytic flexibility of the spliceosome.

Mutually exclusive patterns – In this type of alternative RNA splicing, selection of exons of a pair is such that if one member of a pair is spliced in, other is necessarily to be excised out.

N

N-end rule – According to N-end rule the amino acid at the amino or N-terminal end of a protein is a signal to proteases that control the average length of life of a protein. Almost complete predictability was achieved in determining the life of β-glactosidase protein based on N-terminal amino acid. Protein life spans range from 2 minutes for those with N-terminal arginine to greater than 20 h for those with N-terminal methionine.

Nested gene – A single gene which produces two or more nested products by modulating the end point of protein synthesis. This can occur by leaky read through of a termination codon or by co-translational frameshifting. Similar strategies are used by eukaryotic RNA viruses and nested products can also be produced from eukaryotic genes by alternative RNA splicing or use of alternative polyadenylation sites. Nested genes serve as an example of a situation in which one gene resides within an intron of another gene

No-go decay pathway – This pathway results in rapid mRNA cleavage and degradation. This pathway is different from other mRNA degradation pathways in yeast, in that stalling of the ribosomes by a stem-loop structure leads to cleavage of the mRNA, a process that involves the Dom34 and Hbs1 proteins. This cleavage generates free ends that are subject to degradation: the enzyme Xrn1 chews up the mRNA from the 5′ end, whereas the multienzyme exosome complex degrades from the 3′ end. Dom34 and Hbs1 might interfere directly with the 'A site' of the stalled ribosome, possibly to promote release of the ribosome and allow subsequent mRNA cleavage, or they may directly stimulate cleavage of the mRNA

Non-ambiguous code – There is no ambiguity about a particular codon. A particular codon will always code for the same amino acid wherever it is found.

Non-coding RNA central dogma of molecular biology – mDNA is transcribed into pre-mRNA. uDNA is transcribed into uRNA which, as a part of spliceosome, is involved in the processing of pre-mRNA into mRNA. rDNA is transcribed into pre-RNA, which with the help of RNase MRP and snoRNP is processed into rRNAs, which are integral component of a ribosome. tDNA is transcribed into pre-rRNA, which with the help of RNase P matures into tRNA. SRP and tmRNA are involved in translocating nascent proteins to the site of action.

Non-degenerate code – When there is only one codon for each amino acid.

Non-overlapping code – When only as many amino acids are coded for as there are code words represented in end-to-end sequence (UUUCCC has two code and two codons actually present are UUU and CCC.

Non-sense codon – A codon that does not code for an amino acid. Also called a *termination codon* or *stop codon*.

Non-sense-mediated mRNA decay – In multicellular eukaryotes, long introns are recognized through exon definition and most genes produce multiple mRNA variants through alternative RNA splicing. The non-sense-mediated mRNA decay (NMD) pathway may further shape the observed sets of variants by selectively degrading those containing premature termination codons, which are frequently produced in mammals. Thus NMD functions in mRNA quality control. Independently of alternative RNA splicing, species with large intron numbers universally rely on NMD to compensate for suboptims splicing efficiency and accuracy.

Non-transcribed spacer (NTS) – These spacers in a ribosomal gene family makes only a part of the intergenic spacers (IGS)

Nuclear import of tRNAs in yeast – The transfer RNAs (tRNAs) were thought to cross the nuclear envelope to the cytosol only once after maturing in the nucleus. Several export mutants accumulate mature tRNAs in the nucleus even in the absence of transcription. Import requires energy but not the Ran cycle. The tRNAs shuttle between the nucleus and cytosol.

Nuclear organizer regions (NORs) – Regions where rRNA heteroclusters are present. For the production of about 10 million ribosomes per cell in eukaryotes, rRNA genes are present in multiple copies in tandem array at of specific satellite chromosomes. The number of these genes may vary from 50-30,000 in a cell and this number may be equally distributed on NORs. The DNA comprising these genes is called ribosomal DNA (rDNA), which is repetitive in nature.

Nuclear RNA editing in mammals – The post-transcriptional modification of a glutamine codon CAA to create a translational stop codon UAA has been reported in in apolipoprotien B (*apoB*) RNA in mammals.

Nucleic acid vaccines – When a plasmid DNA encoding an antigen of interest is injected, it results in sustained expression of the antigen and generation of an immune response.

Nucleosme binding proteins – The proteins that act to modulate the promoter chromatin architecture and transcription of target genes. For example, histone H1 and poly(ADP-ribose) polymerase-1 (PARP-1), exhibit a reciprocal pattern of chromatin binding at many RNA polymerase II-transcribed promoters.

Nucleotide sequence difference (K) – The number of nucleotide substitutions per site in a pair of aligned sequences. Using RNA code, one can calculate how the homologous sites differentiated from each other in the course of evolution starting from the common ancestor.

O

Oncogene, fine structure of an – In an attempt to locate the sequence responsible for oncogenicity, with restriction enzymes, the proto-oncogene and oncogene are cleaved at the same site, the fragments recombined, and the recombinants tested for transforming activity. In this way, exact position was narrowed to 350-bp region, which was sequenced. The cause of oncogenic activity was found to be a single base change (G→A) that results in a valine substitution for glycine.

Oncogenes – The genes that confer the ability to convert cells to a tumorigenic state. All human and subhuman species contain a complement of oncogenes that in their non-activated state produce substances necessary for normal cell proliferation and cell surface properties.

Oncogenes, classes of – There are four main classes of oncogenes: Class I-IV.

Oncogenes, origin of – Viral oncogenes and cellular oncogenes are formed from cellular proto-oncogenes.

One cistron-one polypeptide hypothesis – With the discovery of cistron by Benzer in 1962 and colinearity by Yanofsky in 1964, cistron was considerd that part of the DNA molecule which coded information for the synthesis of a single polypeptide. This lead to the proposal of one cistron-one polypeptide hypothesis. Some consider this hypothesis as an alternative name of *one gene-one poplypeptide hypothesis*.

One codon can code for two different amino acids – Strict one-to-one correspondence between codons and amino acids is thought to be an essential feature of the genetic code. However, it has been reported that one codon can code for two different amino acids with the choice of the inserted amino acid determined by a specific 3′ untranslated region structure and location of the dual-function codon within the messenger RNA (mRNA). For example, codon UGA specifies insertion of selenocysteine and cysteine in the ciliate *Euplotes crassus*, that the dual use of this codon can occur within the same gene.

One enzyme-two functions concept – One enzyme may have two functions. For example, of enzyme pon2 (human paraoxonase 2) has two functions, enzymatic lactonase activity and reduction of intracellular oxidative stress. This led to one enzyme-two functions concept.

One gene-many proteins hypothesis – Alternative RNA splicing is a mechanism by which more than one protein is made from a single gene. This observation gave rise to one gene-many proteins hypothesis. It means that the old paradigm that one gene makes one protein is clearly in need of revision. The discovery of polyproteins led to one gene-many protein hypothesis.

One gene-one antigen hypothesis – This hypothesis states that each of the genes in a tissue transplant governing the host's reaction to it is responsible for the manufacture of a particular transplant antigen alone.

One gene-one chromomere hypothesis – On the basis of observation that there is one essential function expressed by genetic information contained in each chromomere, this hypothesis was put forward. The one chromomere-one function relationship holds regardless of the amount of DNA the chromomere contains.

One gene-one enzyme hypothesis – This hypothesis states that each gene controls the reproduction, function and specificity of a particular enzyme. In this way, Beadle and Tatum in 1945 demonstrated relationship between gene and enzyme in *Neurospora*. Ample support to one gene-one enzyme hypothesis came from following studies on microorganisms like *E. coli*, *Salmonella*, and *Neurospora*.

One gene-one mRNA-one protein hypothesis – One gene-one ribosome-one protein hypothesis was rejected when it was reported that RNA is translated into protein by ribosome and this led to one gene-one mRNA-one protein hypothesis.

One gene-one polypeptide hypothesis – One gene-one enzyme hypothesis could not be universally applied. Studies on mutant hemoglobins, lactate dehydrogenase and tryptophan synthetase challenged one gene-one enzyme hypothesis and lead to one gene-one polypeptide relationship. . One gene-one polypeptide hypothesis states that one gene controls the synthesis of one polypeptide.

One gene-one primary cellular function hypothesis – One gene-one mRNA-one protein hypotheses put forward to explain gene function were based on the studies on protein-encoding genes. In addition to the structural genes, which code for polypeptides, there are other genes (tRNA genes and rRNA genes) which produce non-genetic RNAs required in protein biosynthesis. The major three types of non-genetic RNAs (mRNAs, tRNAs and rRNAs) are considered primary cellular products of genes. Thus one gene was noted to perform one primary cellular function. This led to one gene-one primary cellular function hypothesis.

One gene-one reaction hypothesis – This hypothesis states that every specific biochemical reaction is under the ultimate control of a different gene. It was later refined as one-gene.one-enzyme hypothesis.

One gene-one ribosome-one protein hypothesis – According to an early version of the theory of the information structure of protein synthesis, the RNA transcript was thought to provide the RNA moieties for newly formed ribosomes. Hence, each gene was imagined to give rise to the formation of one specialized kind of ribosome, which in turn would direct the synthesis of one and only one kind of protein - a scheme that Brenner and coworkers in 1961 epitomized as the one gene-one ribosome-one protein hypothesis. Now we know that synthesis of new kinds of ribosomes is not a precondition for the synthesis of new kinds of proteins.

Operator – A site to which repressor binds.

Overlapping code – When more amino acids are coded for than there are code words represented in end-to-end sequence. UUUCCC has two code words but in case of overlapping code, codons may be UUU, UUC, UCC, CCC. But this actually is not the case.

Overlapping genes – A single nucleotide sequence coding for more than one polypeptide. Overlapping genes share some of the same sequence. Such an arrangement of genes was first discovered by Barrell and coworkers in 1976. Later overlapping genes were also found in eukaryotic multicellular organisms.

P

Pall's model of carcinogenesis – Pall in 1981 proposed a model that induced carcinogenesis is due to the production of tandem duplications of a proto-oncogene in the cell genome, producing 5-10 fold increase of the proto-oncogene product. The initial event in carcinogenesis is proposed to be a mutation that produces a single tandem duplication of a proto-oncogene.

Paracodons – The existence of a second code which is imprinted in the structure of the AASs. It matches the amino acid with structural features of the tRNA, the phenomenon known as paracodons. The main features of this code are that it is non-degenerate, deterministic and older than the classical genetic code.

Parallelism between RNA editing and self-splicing – There is parallelism between RNA editing (done by gRNA) and self-splicing of introns in hnRNA (done by ribozymes) A parallelism between the function of IG-S in RNA splicing and the proposed role of guide RNAs in RNA editing in kinetoplastid mitochondria has been recognized. They identified chimeric intermediate with gRNA covalently joined to 3' portion of the mRNA in vitro suggesting an evolutionary analogy to the catalytic mechanism involved in RNA splicing. Like self-splicing of introns, RNA-editing also involves transesterification, but with the help of guide RNA. In the first step, the guide RNA aligns itself (by base pairing) with the unedited RNA, splits it into two, and makes a new bond between one of the broken ends and the uridine at the tip of the tail of gRNA at its 3'-end.

Peptide bond formation in eubacteria – Eubacterial leucyl/phenylalanyl-tRNA protein transferase (LF-transferase) catalyzes peptide-bond formation by using Leu-tRNA^{Leu} (Phe-tRNA^{Phe}) and an amino-terminal Arg (or Lys) of a protein, as donor and acceptor substrates, respectively. The protein-based mechanism for peptide-bond formation by LF-transferase is similar to the reverse reaction of the acylation step observed in the peptide hydrolysis reaction by serine proteases.

Peptidyl transferase – Peptide forming enzyme peptidyl transferase is required one copy for one 50S subunit.

Periodic introns – Many of the anomalous introns are periodic, that is, relics of internal sequence repetitions within the ancestral gene. Some of these periodic-intron patterns are shared between related genes.

Pervasive transcription – Certain transcripts either existed stably in cells (stable unannotated transcripts, SUTs) or are rapidly degraded by the RNA surveillance pathway (cryptic unstable transcripts, CUTs). One characteristic of pervasive transcription is the extensive overlap of SUTs and CUTs. Transcription of SUTs and CUTs can be functional, through mechanisms involving the generated RNAs or their generation itself.

PEST hypothesis – According to PEST hypothesis, protein degradation is determined by one of the four amino acids – proline, glutamic acid, serine and threonine; the one-letter abbreviations for these four amino acids, respectively, are P, E, S and T. Different proteins are programmed to survive different programs in the cell.

Phage display technology – This technology allows production of monoclonal antibodies for chosen target antigen, without the use of any animal or hybridoma. DNA sequences coding for antibody V region are fused with DNA sequence coding for amino-terminal of the minor coat protein pIII of the phage. This fused DNA sequence, inserted in phage genome, will produce a fused protein (carrying the antibody) that will display on the surface of the phage (due to coat protein sequence).

Polycistronic mRNA – mRNA containing multiple ORFs are called polycistronic mRNA which often encode proteins that perform related functions such as different steps in the biosynthesis of amino acids and nucleotides. Prokaryotic mRNA frequently contains two or more ORFs and hence can encode multiple polypeptide chains.

Polyploidy – Refers to multiple copies of a complement in a single cell. It is a common phenomenon in the evolution of most species of plants but it is much rarer phenomenon in animals. Polyploidy has played role in evolution by having several copies of the same genome but it does not seem to have played role in causing evolutionary changes in genome size.

Polyprotein genes – Many genes encode for one single large polypeptide which, however, after translation, is cleaved enzymatically into smaller subunits. Such polyprotein genes are also known in multicellular eukaryotes. Polyprotein genes thus contradict the hypothesis adopted by the neoclassical view of the gene that each gene encodes for a single polypeptide.

Polyribosome formation – If only one ribosome is attached to a single mRNA, protein synthesis will not be efficient. Instead, a single mRNA usually has several ribosomes attached to it, as many as 1 for every 90 nucleotides. This mRNA with its attached ribosomes is called a polyribosome or polysome.

Polyteny – It is the multiplication of the number of DNA strands within a chromosome. This occurs in certain animal tissues, e.g., the salivary glands, the gut and the Malphighian tubules of Diptera. Polyteny can increase amount of DNA manifold in somatic cells but not in germ cells. Thus polyteny is not a general mechanism for increasing amount of DNA in evolution.

Porosome – Permanent cup-shaped lipoprotein structure at the cell plasma membrane, where secretory vesicles transiently dock and fuse to release intra-vesicular contents from the cell.

Post-termination complexes in eukaryotes – Eukaryotes do not encode a ribosome recycling factor (RRF) homolog. In eukaryotes, a loosely associated subunit of initiation factors eIF3 promotes recycling of eukaryotic post-TCs. eIF3 is the principal factor that promotes splitting of post-termination ribosomes into 60S subunits and tRNA- and mRNA-bound 40S subunits. Its activity is enhanced by eIFs3j, eIF1 and eIF1A. eIF1 also mediates release of P site tRNA, whereas eIF3j ensures subsequent dissociation of mRNA. Initiation factor eIF3, in cooperation with its eIF3j subunit, eIF1 and eIF1A, split post-termination ribosomes into 60S subunits and tRNA- and mRNA-bound 40S subunits. eIF1 promotes subsequent release of P site deacylated tRNA, which is followed by dissociation of mRNA mediated by eIF3j.

Post-termination complexes in prokaryotes – After translational termination, mRNA and P site deacylated tRNA remain associated with ribosomes in post-termination complexes (post-TCs), which must therefore be recycled by releasing mRNA and deacylated tRNA by dissociating ribosomes into subunits. Recycling of bacterial post-TCs requires elongation factor EF-G and a ribosome recycling factor (RRF).

Post-translational modifications – A post-translational event that involves phosphorylation, glycosylation, sulfation, etc., which are very important for protein function. Various variants of eastern blotting are used to analyze specific types of post-translational modifications.

Post-translational steps – From newly synthesized polypeptide, N-formyl group is removed by enzyme deformylase. Then the removal of first amino acid takes place by methionine-specific amino-peptidase. After that terminal tRNA is removed and complete protein becomes available.

Pre-messenger RNA processing – Messenger RNA is not made directly in a eukaryotic cell. RNA transcribed from eukaryotic genes is called heterogeneous RNA (hnRNA) one of them is pre-mRNA which is a large primary transcript and is processed into mRNA. It is 5-10 per cent of total RNA in the cell. It codes for amino acid sequence. Therefore, it is also known as coding RNA. Various steps of pre-mRNA processing are: removal of parts of leader and train, addition of "Cap" at 5'-end, addition of poly(A) tail at 3'-end, splicing of non-coding sequences, called introns and retaining only some sequences, known as exons and methylation of some adenines present.

Pre-ribosomal RNA processing – Prokaryotic cells have three types of rRNA: 16S, 23S and 5S. Genomic arrangement of genes in the sequence 5'-16S, tRNAs, 23S, 5S, tRNA(s)-3' is (in a single operon. The entire operon is often transcribed as a unit. Cleavage enzymes are required at a number of points even to separate several classes of pre-rRNAs. Ribosomal RNA transcripts are processed both in prokaryotes and eukaryotes to generate the mature rRNA molecules with unique 3′ and 5′ ends. The rRNA transcript undergoes various changes either during or, less likely, after its synthesis but before it's processing: (a) To each transcript, ~110 methyl groups are added to the ribose moieties; these methyl groups are retained in the mature molecules. Thus ribose methylation seems to act as a marker for the cleavage sites. (b) The RNA becomes bound to proteins so that it is processed as ribonulceoprotein complex rather than as free RNA.

Pre-transfer (soluble) RNA processing – Different stages of tRNA maturation in fixed temporal order are: (a) Trimming of the 5' and 3' ends in which RNase P acts on the 5' terminus endonucleolytically and generates mature 5'-terminus by characterizing mature tRNA sequence. The 3'-end maturation takes place by exonuclease activity by RNase D and 3'-CCA sequences of tRNA C (as in prokaryotes). When -3'-CCA is not encoded in DNA, this triplet is enzymatically added. But this triplet has to be added before removal of intron. Second situation is present in eukaryotes. (b) Modification of the bases in mature coding region can occur on either the precursor molecular or cleavage products. (c) Removal of introns takes place when present. Introns are very short (13-60 nucleotides). The cleavage and splicing at intron junctions has been found to be highly dependent upon the presence of the -CCA and piece.

Pre-tRNA intron removal – Intervening sequences (IVS) are located in most of the tRNA precursors of eukaryotes and in advanced genera of prokaryotes, at a point, one site removed downstream from anticodon. However, in the gene for $tRNA^{Leu}$, the intron splits the anticodon sequence. For removal of intron from tRNA transcript, first step is cleavage of tRNA sections from the inserted sequence. These different tRNA precursors are separated by RNase E and III producing single tRNA precursors. The monomeric tRNA precursors interact with one or more RNases reducing the size of originally long tRNA precursor and permitting the mature tRNA precursor to assume its final secondary and tertiary structure. The introns of tRNA precursors are excised by precise endonucleolytic cleavage and ligation reactions catalyzed by special splicing endonuclease and ligase activities.

Pribnow box – Sequences discovered by Pribnow and coworkers in 1981 in promoters of prokaryotic genes. Also called as *TATA box*.

Primary metabolite – A primary metabolite is directly involved in the normal growth, development, and reproduction.

Processing (P) bodies – Translation and mRNA degradation are affected by a key transition where eukaryotic mRNAs exit translation and assemble an mRNP state that accumulates into processing bodies (P bodies). Cytoplasmic sites of mRNA degradation contain non-translating mRNAs, and the mRNA degradation machinery. Decapping activators (Dhh1p and Pat1p function as translational repressors and facilitators of P body formation. This competition creates a finely balanced system setting the relative translation rate for an mRNA. mRNAs can be driven into either translation or repression by tipping the balance of this competition via any number of events. Eukaryotic cells contain non-translating messenger RNA concentrated in P-bodies, which are sites where the mRNA can be decapped and degraded. mRNA molecules within yeast P-bodies can also return to translation.

Processing of 5S rRNA – The processing for maturation of 5S RNA takes place in two steps – the early steps and the late steps. In early steps, after transcription, the primary transcript is cleaved into two parts, one bears the genes for the minor rRNA and one or two tRNAs present on the intergenic spacer and other contains the major and the supplemental rRNAs, depending upon whether or not a tRNA occurs on the 3'-end. Neither RNase III nor E is involved in initial enzymatic reaction. In late step, the 10S RNA contains 6 stem-and-loop structures generated by RNase III. Then by action of RNase P, fragment 7S RNA is generated which contains 5S rRNA and a termination stem. Removal of this stem is accomplished by RNase E.

Processing of the 5.8S rRNA – Processing of the supplemental species of rRNA is confined to the nucleolus and involved small RNA U3 and another type of RNA, 7-1. During much of its maturation, it is hydrogen bonded to the other pre-rRNAs. First product of this species is 12S RNA particle, followed by one sedimenting at 8S. Finally, 5.8S rRNA is produced.

Processing the major species of rRNA – The primary transcript of shorter operons, 30S pre-rRNA is in the form of two loops containing the minor and major species, respectively, each closed at one side by a strongly base-paired stem. It is these stems that are cleaved by RNase III, the enzyme which specifically acts on double-stranded segments. The final trimming of 30S pre-rRNA occurs at the ribosome to produce 23S RNA.

Progenote stage – DNA sequences of the most primitive unicellular eukaryotes may have originated from primordial DNA sequences which had been randomly generated. Before a self-replicating cell could come into existence but after biological energy-generating system had evolved, DNA molecules were synthesized in the primordial soup by random addition of the four nucleotides without the help of templates. The nucleotide sequences that coded for proteins were selected from the randomly generated nucleotide sequences in the primordial soup by natural selection. This represents the "progenote" stage in the terminology of Woese and Fox given in 1977.

Prokaryotic ribosome – It has sedimentation coefficient 70S. It contains 65 per cent rRNA and 35 per cent protein. Prokaryotic ribosomes have two subunits. Larger subunit has sedimentation coefficient 50S. It contains one molecule of 5S rRNA, one molecule of 23S rRNA and 34 different proteins. The 50S subunit has a fairly compact body (~150×200×200Å) from which a central protuberance and a stalk, stick out; this subunit is relatively more spherical than the smaller one. The partition between the head and body of the small subunit fix into the large subunit. There could be a space or tunnel between the two subunits. Smaller subunit (30S) has one molecule of 16S rRNA and 21 different proteins. The 30S subunit has an elongated and asymmetrical shape (about 55×220×220Å). It has a constricted region and cleft that separates the head from the base, while its platform projects out the base. The 30S and the 50S subunits may be held together by an association between discrete areas of these subunits.

Promoter – Specific sequences of a given gene that serve in recognition of the gene and subsequent attachment with the transcribing enzyme. Promoter is part of the leader region. It contains Pribnow box (term used in prokaryotic genes) or Goldberg-Hogness box (term used for eukaryotic genes). Minimum size of promoter is 12 bp.

Promoter melting – This is crucial step in transcription initiation. The N terminus of the β' subunit (amino acids 1-314) and amino acids (94-507) of the σ subunit, together less than one-fifth of RNAP holoenzyme, were able to melt an extended –10 promoter in a reaction remarkably similar to that of authentic holoenzyme.

Promoter, consensus sequences in – The start point in greater than 90 per cent cases is a purine. Just upstream of the start point, a 6-bp region is recognized in almost all the promoters. The centre of the hexamer generally is close to 10 bp upstream of the start point. This hexamer is, therefore, often called the –10 sequence. This –10 sequence is TATAAT; conservation of these bases varies from 45 to 96 per cent (T_{80} A_{95} T_{45} A_{60} A_{50} T_{96}). At –35 sequence of promoter, consensus sequence is T_{82} T_{84} G_{78} A_{65} C_{54} A_{45}. Distance between –10 and –35 sequence in 90 per cent of cases is between 16 and 18 bp. This distance between two sites may be important in appropriate separation for the geometry of RNA polymerase. Thus an optimal promoter is defined as a sequence consisting of the –35 hexamer, separated by 17 bp from the –10 hexamer, lying 7 bp upstream of the start point.

Proofreading in translation – Protective role of proofreading in translation is evident from a case in which a small mistake has a catastrophic effect. Transfer of information from the DNA sequence of a gene to the corresponding amino-acid sequence of a protein requires transcription of the sequence into messenger RNA, then translation of that RNA into an amino-acid sequence at ribosomes, through the agency of amino-acid-specific transfer RNAs. These steps are all prone to error.

Proteasome – In protein degradation, ubiquitin-tagged proteins are recognized by specific proteases in the cytosol which in turn cleave and degrade the tagged protein. These proteases are combined in a very large protein complex called the proteasome. The proteasome (20S) is comprised of 28 subunits and has a molecular weight of 700 kDa.

Protein-encoding gene – The gene which on transcription yields messenger RNA (mRNA) or precursor of mRNA (pre-mRNA) which after various processing steps matures into functional mRNA. Also known as *structural gene.*

Protein engineering – Technology has been developed for incorporating non-standard amino acids into polypeptide by ribosome based translation. For this, the genetic code is expanded through creation of a 65th codon-anticodon pair from unnatural nucleoside bases having non-standard H-bonding patterns. This new codon anticodon pair efficiently supports translation in vitro to yield polypeptides containing a non-standard amino acid.

Protein folding – Co-translational chaperones function in protein folding. Ribosome-associated trigger factor (TF) is the first molecular chaperone encountered by nascent polypeptides in bacteria. TF interacts with ribosomes and translating polypeptides in a dynamic reaction cycle. Ribosome binding stabilizes TF in an open, activated conformation. Activated TF departs from the ribosome after a mean residence time of ~10 s, but may remain associated with the elongating nascent chain for upto 35S, allowing entry of new TF molecule at the ribosome docking site. The duration of nascent-chain interaction correlates with the occurrence of hydrophobic motifs in translating polypeptides, reflecting a high aggregation propensity. Protein mis-folding events during translation may provide a paradigm for the regulation of nucleotide-independent chaperones.

Protein quantification – Measurement of abundance of different proteins in a protein profile.

Protein secretion – Free ribosomes synthesize specific non-serum liver proteins. It is possible that leader sequences on mRNAs determine which ribosomes shall translate them. In eukaryotes, ribosomes may be free or may be attached to the endoplasmic reticulum (ER). Membrane-bound ribosomes synthesize proteins that are exported from the cell into the surrounding serum. 60S subunit of ribosome is attached to ER. The ER membrane is used to form a vesicle containing the protein. This vesicle then becomes secretary message whose membrane fuses with the cell membrane, emptying its contents outside the cell.

Protein sorting – Protein sorting involves interaction of cell membranes either with parts of newly synthesized protein or with a protein sequence added to this new protein. These protein sequences interacting with membranes are called signal sequences or transit peptides, or leader sequences, which are recognized by receptors located within the membrane. The transfer of proteins may be coupled with translation — co-translational transfer (proteins synthesized on ribosomes attached to endoplasmic reticulum) or may take place after the translation — post-translational transfer (proteins synthesized on free ribosomes).

Protein sorting signals – Amino acid sequences found in transported proteins that selectively guide the distribution of the proteins to specific cellular compartments.

Protein synthesis in cell organelles – In mitochondria and chloroplasts, following components of protein synthesis are found: DNA, DNA polymerase, RNA polymerase, ribosomes, tRNAs, initiation factors, elongation factors and termination factors. All these components are specific to cell organelles. Some of these components of protein synthesis may be synthesized outside these organelles and then transported to the organelles.

Protein synthesis in organelles versus cytoplasm – It has been shown that the translation apparatus in chloroplasts and mitochondria differ from that in cytoplasm in eukaryotes in the following respects: (a) Ribosomes in these organelles are smaller in size (70S) than those in the cytoplasm (80S). (b) The tRNAs are specific and differ, the number of tRNAs; in mitochondria being 22 as against 55 in cytoplasm. (c) Initiation of translation takes place by formyl-methionyl tRNA both in chloroplasts and mitochondria, although no formylation takes place in cytoplasm. (d) Translation in chloroplasts and mitochondria can be inhibited by chloramphenicol, as in bacteria since the 70S ribosomes are sensitive to chloramphenicol and not to cycloheximide; on the other hand, the translation in cytoplasm is inhibited by cycloheximide, since 80S ribosomes are sensitive to cycloheximide.

Protein-protein interaction – Proteins do not act in isolation to perform a biological function in a cell. Protein-protein interaction is an important step in many cellular functions. Thus a study of protein-protein interactions is an important component of proteomics research. These interactions can be exploited for biotechnological applications. Purification of entire multiprotein complexes can be done by affinity based method.

Protein synthesis inhibition by microRNAs – miR2 inhibited translation initiation without affecting mRNA stability. Inhibition of $m^7G_{ppp}G$ cap-mediated translation initiation is the mechanism of miR2 function.

Protein synthesizing machinery – In addition to mRNA, tRNA, rRNA and ribosome, other factors are also necessary for protein synthesis. Peptidyl transferase forms peptide bonds between amino acids. Protein factors catalyze partial reactions in the initiation, elongation and termination of peptides. Miscellaneous factors, such as ATP, GTP, Mg^{2+}, K^+, $NH4^+$ are important for various biochemical reactions. The enzymes required for protein synthesis are aminoacyl synthetases, phosphatase, transformylase, deformylase, methionine-specific peptidase, etc.

Protein targeting – The processes directing proteins to particular destinations in the cell. In bacteria, the choice of destination is between the cytoplasm, the inner and outer cell membranes, and the periplasmic space between them. Proteins can also be secreted. In eukaryotes, proteins can also be targeted to any one of several intracellular organelles, as well as to the nucleus. Since all bacterial proteins and most eukaryotic proteins are synthesized in the cytoplasm, targeted proteins must carry recognizable sequences or structures which allow them to be transported to the appropriate cellular compartment. This process is termed *protein sorting* or *protein trafficking*.

Protein targeting in eukaryotes – Proteins synthesized on cytoplasmic ribosomes of eukaryotic cells can either be retained in the cytoplasm or transferred to subcellular organelles or inserted into membranes or secreted outside the cell. All living cells require specific mechanisms that target proteins to the cell surface. In eukaryotes, the first part of this process involves recognition in the endoplasmic reticulum of amino-terminal signal sequences and translocation through Sec translocons, whereas subsequent targeting to different surface locations is promoted by internal sorting signals.

Protein targeting in prokaryotes – The prokaryotic signal recognition particle (SRP) targets membrane proteins into the inner membrane. It binds translating ribosomes and screens the emerging nascent chain for a hydrophobic signal sequence, such as the transmembrane helix of inner membrane proteins. If such a sequence emerges the SRP binds tightly, allowing the SRP receptor to lock on. This assembly delivers the ribosome-nascent chain complex to the protein translocation machinery in the membrane.

Protein translation without ribosomes – Some short proteins are not synthesized on ribosomes. Such proteins are antibiotic polypeptides, such as gramicidin and tyrocidine.

Protein trans-splicing – Intein-mediated protein splicing is a self-catalytic process in which the intervening intein sequence is removed from a precursor protein and the flanking extein segments are ligated with a native peptide bond. Splice junction proximal residues and internal residues within the intein direct these reactions. The protein *trans*-splicing poses problem for gene concept that start and end sites are not determined by genes.

Protein turnover – It is the balance between protein synthesis and protein degradation. The turnover of some transcription factors is regulated through ubiquitin-dependent proteolysis. Usually, such processes are regulated by extracellular stimuli through phosphorylation of the target protein, which allows recognition by F box-containing E3 ubiquitin ligases.

Proteome – Sum total of all the proteins in a cell at a particular developmental stage or environment.

Proteome analysis – Various tools of proteome analysis are: Mass spectrometry (MS), two-dimensional gel electrophoresis 2D-GE), Isoelectric focusing IEF), polyacrylamide gel electrophoresis (PAGE), comparative 2-D gel approach, protein chip approach, ELISA, liquid chromatography coupled with electrospray-ionization tandem mass spectroscopy.

Proteomics – A study focused on gene products (proteins) in a cell to conduct qualitative and quantitative measurements of different proteins.

Proto-oncogenes – Normal cellular genes from which the viral oncogenes are derived. Proto-oncogenes do not produce cancer cells under normal circumstances but can do so if they are modified or changed by their incorporation into viral genomes. Almost every viral oncogene tested so far has one or more closely related DNA sequences in uninfected host cells.

Proto-oncogenes, activation of – (1) A point mutation induced by a chemical or radiation carcinogen can cause a change in the protein, thereby initiating cancerous growth. (2) Induction of a chromosomal rearrangement that places in the proto-oncogene next to the regulatory region of an immunoglobulin gene. This will lead to the expression of the proto-oncogene inappropriately. (3) If the proto-oncogene is amplified either as repeat segments within the chromosome or extrachromosomally, the gene product will be over produced. (4) The retroviral transfer of a proto-oncogene picks up another animal cell. Thus, transcription of proviral regulatory signals, not by the signals associated with the proto-oncogenes in the cell's genome, abnormal regulation and/or abnormal product of v-onc expression yield(s) the tumor phenotype.

Pseudoalleles – Those alleles which are allelic in complementation test and non-allelic on recombination test. Large progeny studies showed that no such alleles exist. Now discarded term.

Pseudogenes – Defective copies of genes present in an organism's genome. Such genes may arise due to mutations in initiation codons. Pseudogenes seem to arise from reverse transcription of mRNA. Two important characteristics of pseudogenes are: (a) Most of the pseudogenes outnumber their genes and are therefore repetitive sequences. (b) They are flanked by 6-21 bases long direct repeats.Pseudogenes are a common feature of many multigene families in higher eukaryotes. In fact, pseudogenes are ancient genes which have lost their function, and pseudogenes which have in the past been protein coding genes usually contain many stop codons, i.e., their reading frame has been closed during the course of evolution.

R

Reading frame – The particular nucleotide sequence that starts at a specific translation initiation point and is then partitioned into codons until the final word (in fact a termination codon) of that sequence is reached.

Recoding – Certain instructions in mRNA may result in reprogramming of genetic information which mRNA receives from DNA. This reprogramming of mRNA is reflected in amino acid sequence of proteins. Most RNA editing systems are mechanistically diverse, informally restorative, and scattershot in eukaryotic lineages. In contrast, genetic recoding by adenosine-to-inosine RNA editing seems common in animals; usually, altering highly conserved or invariant coding positions in proteins.

Recoding signals – The set of instructions in mRNA which brings about this recoding are called recoding signals. Two mechanisms for recoding have been proposed: first, change in linear read out of nucleotide sequence and, second, change in meaning of the code.

Recoding, applications of – (1) One can expand the genetic code of an organism to include new amino acids by adding new components to this template-directed biosynthetic machinery. (2) Additional genetically encoded amino acids may offer evolutionary advantages to modern organisms and certainly extend our ability to manipulate the physicochemical and biological properties of proteins. (3) UAAs with chemically reactive groups can be used as bio-orthogonal handles for the site specific in vitro, and in some cases intracellular, modification of proteins; they can also be used to introduce a new or enhanced catalytic function into proteins. (4) Genetically encoded UAAs can also allow the study of cellular processes, protein structure and function. (5) UAA mutagenesis is beginning to find many applications in the generation of therapeutic proteins, where the production of large quantities of homogenously modified protein is desired.

Recognition and degradation of non-functional mRNAs – Eukaryotic mRNAs where translation termination occurs too soon (non-sense-mediated decay) or fails to occur (non-stop decay) are rapidly degraded. Yeast mRNAs with stalls in translation elongation are recognized and targeted for endonucleolytic cleavage, referred to as no-go decay. The cleavage triggered by no-go decay is dependent on translation and involves Dom34p and Hbs1p. Dom34p and Hbs1p are similar to the translation termination factors eRF1 and eRF3, indicating that these proteins might function in recognizing the stalled ribosome and triggering endonucleolytic cleavage.

Recon – A unit of recombination. Defined by recombination test. Two mutations belong to the same recon if they do not show recombination. The smallest recombinational unit probably is two neighboring nucleotides.

Redundant genes – Presence of several copies of a single gene in a cell. It allows the organism to obtain large amounts of the gene product in short time interval. These are known as redundant genes. Examples are: rRNA genes, tRNA genes, histone genes, antibody genes.

Regulation of alternative RNA splicing by histone modifications – Most human genes are alternatively spliced in a cell or tissue specific manner. Pre-mRNA splicing is influenced by RNA polymerase II elongation rate, chromatin remodelers and histone deacetylase inhibitors.

Regulation of Cap-dependent translation – Eukaryotic messenger RNAs contain a modified guanosine, termed as a cap, at their 5' ends. Translation of mRNAs requires the binding of an initiation factor, eIF4E, to the cap structure. family of proteins through a shared sequence regulate cap-dependent translation.

Removal of introns – Introns in many eukaryotic genes begin with dinucleotide 5′-GT-3′ and with dinucleotide 5′-AG-3′. Process of pre-mRNA splicing is a two-step pathway. In the first step, a precise cleavage occurs at the 5′-end of the intron, and a 2′-5′ phosphodiester linkage is formed between the 5′ position of the G at intron cleavage site and a conserved A residue located near the 3′-end of the intron. In the second step, the two exons are joined by a normal 3′-5′ phosphodiester bond, and the lariat-shaped intron is released. Small RNAs, known as U RNAs, are used in removal of introns.

Repeated gene family – It consists of many identical or nearly identical genes that cohabit a single haploid genome. In some cases the genes are contained within tandemly repeated units.

Repeated genes – Certain genes have many copies in the genome of a cell. The genes of ribosomal RNA are repeated in several tandem copies. Each one consists of one transcription unit, but the gene cluster is usually transmitted from one generation to the next as a single unit. Thus, the units of transmission and transcription are not always the same. The histone genes have been observed to be repeated in such tandem repeats in many higher eukaryotic organisms.

Repeats in promoters – The nucleosome-free region directly upstream of genes (the promoter region) is enriched in repeats. One-fourth of all gene promoters contain tandem repeat sequences. Genes driven by these repeat-containing promoters showed significantly higher rate of transcriptional divergence. Tandem repeats are variable elements in promoters that may facilitate evolutionary tuning of gene expression by affecting local chromatin structure.

Replisome-RNA polymerase collision – Replication forks are impeded by DNA damage and protein-nucleic acid complexes such as transcribing RNA polymerase. Head-on collision of the replisome with RNA polymerase results in replication fork arrest. However, co-directional collision of the replisome with RNA polymerase has little or no effect on fork progression.

Reporter gene – A reporter gene is a gene that researchers attach to a regulatory sequence of another gene of interest in bacteria, cell culture, animals, or cells.

Reporter plasmid – Reporter plasmid carries a gene whose expression is driven by product of a regulatory gene carried by a effector plasmid.

Represnting gene in literature – An mRNA sequence is complementary to sense strand while it is similar to antisense strand except that where there is thymine in DNA, there is uracil in RNA. Therefore, in literature a gene in represented through the sense DNA strand (in 5′→3′ direction).

Retained introns – Retained introns may also result in the production of different mRNAs with corresponding different translated products. Intron retention, one form of alternative RNA splicing, is common in plants but rare in higher eukaryotes, because messenger RNAs with retained introns are subject to cellular restriction at the level of cytoplasmic control and expression.

Retrogene – In certain genes there is reverse transcription of an mRNA and insertion of the product into genome. This is a case of RNA to DNA flow of information. Some examples of retrogenes are *RPL36AL*, *HSPA2*, *SPIN2B* and *FAM50B*.

Reverse RNA splcing – Introns are self-spliced in a very simple way. Catalytic steps of splicing can be efficiently reversed under appropriate conditions. Finding that reverse splicing occurs in vivo provides a piece of jigsaw that one day may allow us to reconstruct the paths that introns take when on the move.

Reverse transcription – The process where mRNA is used as a template for synthesis of DNA (called *complementary DNA, cDNA*).

Reverse transcription in bacteria – Reverse transcriptase encoded by cellular genes has also been described in bacteria – myxobacteria and *E. coli*. This enzyme is thus not unique to eukaryotes. This suggests that reverse transcriptase existed before retroviruses. Retrons with reverse transcriptase became retroposons, which with their ability to transpose became reterotransposons, which with long terminal repeats became retroviruses, which with the ability to form virus became pararetroviruses, which lost the ability to integrate and transpose, became DNA viruses.

Reverse transcription in DNA viruses – One expects reverse transcription in RNA viruses only. Reverse transcription takes place in DNA viruses also. Reverse transcription step has been proposed in replication of cauliflower mosaic virus (CaMV), a double-stranded DNA virus, which infects plants, on the basis of structural studies of encapsidated CaMV DNA and of intracellular DNA and RNA species judged to be intermediates in the replication of CaMV.

Rho-dependent terminators – In case of rho-dependent terminators, there is a need for addition of rho (ρ) factor for termination in vitro. Rho factor is required for termination in vivo also. Most of the known rho-dependent terminators are found in phage genomes.

Rho-independent terminators – Rho-independent or simple terminator core enzyme can terminate in vitro at certain sites in absence of any other factor. These terminators have two structural features — a hairpin in secondary structure and a run of 6 uracils (U's) at the very end of the unit. Both features are needed for termination. The hairpins contain a G-C rich region near base of the stem. Probably all the hairpins that form in the RNA product cause the polymerase to slow or pause in RNA synthesis. Pausing creates an opportunity for termination to occur.

Ribonuclease E – This enzyme has a key role in mRNA degradation and the processing of catalytic and structural RNAs in *E. coli.* An evolutionarily conserved 17.4 -kDa protein, named RraA (regulator of ribonuclease activity) that binds to RNase E and inhibits RNase E endonucleolytic cleavages without altering cleavage site specificity or interacting detectably with substrate RNAs. Overexpression of RraA circumvents the effects of an autoregulatory mechanism that normally maintains the RNase E cellular level within a narrow range, resulting in the genome-wide accumulation of RNase E-targeted transcripts.

Ribonuclease III family – Members of the ribonuclease III (RNase III) family are double-stranded RNA (dsRNA) specific endoribonucleases characterized by a signature motif in their active centers and a two-base 3' overhang in their products. Bacterial RNase III is structurally simpler.

Ribosomal elongation cycle – This cycle describes a series of reactions prolonging the nascent polypeptide chain by one amino acid and driven by two universal elongation factors termed EF-Tu and EF-G in bacteria. Extremely conserved LepA protein, present in all bacteria and mitochondria, is a third elongation factor required for accurate and efficient protein synthesis.

Ribosomal RNA genes – The genes which on transcription yields precursor of rRNAs (pre-rRNA) which after various processing steps mature into functional rRNA.

Ribosomal RNAs (rRNAs) – rRNAs are important constituent of ribosome, the site of protein synthesis. A large number of copies of 3 to 4 RNA molecules are present in each cell. Genes coding for rRNA are present in nucleolar organizing regions (NOR) of the chromosome.

Ribosomal RNA gene family – A family of non-coding RNAs responsible for principal functions of the ribosome. rRNAs contribute directly to the catalytic properties of protein synthesis. Ribosomes have two subunits – large subunit and small subunit. Large subunit has three rRNAs of size 28S, 5.8S and 5S. Small unit has 18S rRNA. There are 4 RNA genes. *18S rRNA*, *5.8S RNA* and *28S rRNA* genes are present in form of clusters, 18S rRNA-5.8S rRNA-28SrRNA, known as heteroclusters. In most eukaryotes, the rRNA genes are found in clusters composed of 100 to 5,000 repeats of these genes and the genes for different rRNA molecules are transcribed as a single precursor. Also known as *Class I genes*.

Ribosomal RNA transcription regulation – Ribosomal RNA (rRNA) transcription is regulated primarily at the level of initiation from rRNA promoters. The unusual kinetic properties of these promoters result in their specific regulation by two small molecule signals, ppGpp and the initiating NTP, that bind to RNA polymerase at all promoters.

Ribosomal translocation – During the ribosomal translocation, the binding of elongation factor G (EF-G) to the pre-translational ribosome leads to a ratchet-like rotation of the 30S subunit relative to the 50S subunit in the direction of the mRNA movement. This rotation is accompanied by a 20 Å movement of the L1 stalk of the 50S subunit, implying that this region is involved in the translocation of deacylated tRNAs from the P to the E site. These ribosomal motions can occur only when the P-site tRNA is deacylated. Prior to peptidyl-transfer to the A-site tRNA or peptide removal, the presence of the charged P-site tRNA locks the ribosome and prohibits both of these motions.

Ribosome – The ribosome translates the DNA code into life. A ribonucleoprotein structure that coordinates the correct recognition of mRNA by each tRNA and catalyzes the peptide bond formation between growing polypeptide chain and amino acid attached to selected tRNA. Ribosome is the site of protein synthesis. The intact ribosome contains three tRNA binding sites that reach between two subunits; an A- site where charged tRNA enters the ribosome, a P-site that contains peptidyl-tRNA and E-site, where deacylated tRNA exits the ribosome. The ribosome consists of a large submit which contains peptidyl transferase center and small subunit which contains decoding centre. Each subunit is composed of one or more rRNAs and multiple proteins. It is the proteins in the ribosome that perform a largely structural function, not the RNA.

Ribosome code – There is a possibility of ribosome code. In the budding yeast, 59 of the 79 cytoplasmic ribosomal proteins are encoded by two genes, stemming from an ancient genome duplication event. These paralogous genes are not functionally equivalent, suggesting the possible existence of a "ribosome code". Duplicated genes escape gene loss by conferring a dosage benefit or evolving diverged functions.

Ribosome profiling – A strategy that is based on the deep sequencing of ribosome-protected mRNA fragments and enables genome-wide investigation of translation with subcodon resolution.

Riboswitch-mediated base-pairing changes – Bacteria make extensive use of riboswitches to sense metabolites and control gene expression, and typically do so by modulating premature transcription termination or translation initiation. The most widespread riboswitch class known in bacteria responds to the coenzyme thiamine pyrophosphate (TPP), which is a derivative of vitamin B_1. Representatives of this class have also been identified in fungi and plants, where they are predicted to control messenger RNA splicing or processing.

RNA editing – Recent challenge to central dogma of molecular biology has been phenomenon of RNA editing. It is defined as a process resulting in difference in nucleotide sequence of RNA from its template DNA as a result of any process other than splicing. RNA editing is involved in the insertion, deletion or modification of nucleotides. RNA editing is a process that results in changes in the nucleotide sequence of mitochondrial transcripts such that the RNA sequence differs from the DNA template from which it is transcribed. RNA editing is thus a post-transcriptional process that results in radical changes in amino acid specified by a codon. Most of the editing events occur in the first two nucleotides of the coding sequences.

RNA editing adenosine-to-inosine – Primary transcripts of certain microRNA (miRNA) genes are subject to RNA editing that converts adenosine to inosine. A-to-I RNA editing leads to transcriptome diversity.

RNA editing cytosine-to-uracil – All canonical transfer RNAs (tRNAs) have a uridine at position 8, involved in maintaining tRNA tertiary structure. However, the hyperthermophilic archeon *Methanopyrus kandleri* harbors 30 (out of 34) tRNA genes with cytidine at position 8. Cytosine-to-uracil (C-to-U) editing at this location in the tRNA's tertiary core has the cytidine deaminase domain linked to a tRNA-binding THUMP domain. Presence of this C-to-U editing enzyme guarantees the proper folding and functionality of tRNAs. Three RNA editing mechanisms are potentially responsible for C-to-U conversion. Mechanism 1 involves deamination (or transamination) of the C-6 amide of cytosine to convert the base to uracil. Mechanism 2 involves transglycosylation of the ribosyl moiety to replace the base. Mechanism 3 involves deletion and insertion of a new nucleoside monophosphate.

RNA editing cytosine-to-uracil through base exchange – Exchange of base without cleavage of the sugar-phosphate backbone is another possible way of biochemical modification. Such transglycosylate reaction has been described in transfer RNAs. Pyrimidine replacement is also found in DNA repair metabolism. This process like deamination does not include any cut in the RNA chain.

RNA editing cytosine-to-uracil through deamination/Amination – The simplest mechanism for C-to-U conversion is deamination at position U of cytosine leading to a uracil residue. This could be achieved by cytidine deaminase, an enzyme of the nucleotide metabolism. The reverse U-to-C modification found in plant mitochondria could be carried out by a CTP synthetase in an ATP-dependent reaction. However, existence of reverse editing from U-to-C implies the presence of two different enzymatic activities and not reverse action of the same enzyme.

RNA editing cytosine-to-uracil through nucleotide exchange – Nucleotide deletions at RNA editing sites have been reported in wheat mitochondrial cDNA clones. These cDNAs were interpreted to represent single-nucleotide deleted intermediates of the editing process, and a deletion and insertion mechanism was proposed by these investigators. Nucleotide exchange involves cleavage of the RNA chain, deletion of a cytosine, addition of uracil, followed by religation. The nucleotide exchange involves several steps and is similar to the one described for U addition and deletion in editing of mRNA of trypanosomes. This mechanism is most complex and would probably need an apparatus composed of multiple factors.

RNA editing in chloroplasts – ACG codon appears at 5'-terminus of chloroplast genes where initiation codon ATG would be expected.

RNA editing in plant mitochondria – Mitochondria have adopted various modes of transcript maturation, such as RNA editing and *trans*-splicing. Transcripts of C-terminal gene modules undergo polyadenylation, and contiguous mRNAs are generated via cocatenation of separate module transcripts. Range of bases being edited varied from 0.8 to 5.8 per cent of the total nucleotides present in a transcript. The mechanism of RNA editing in plant mitochondria requires at least two steps: (a) the specific identification of the cytosines to be edited and (b) the biochemical modification. Several conclusions on the editing mechanism in plant mitochondria can be drawn from the information available to date.

RNA editing mechanisms – Various hypotheses proposed for mechanisms of RNA editing are: pseudogene hypothesis, variant of pseudogene hypothesis, edited template refractive to hybridization, cryptic signal sequences, self-replication edited RNA, alternative RNA/RNA heteroduplex model, and multienzyme editing complex.

RNA editing, characteristics of – General characteristics of RNA editing are: (a) Guide RNA editing is a post-transcriptions process. (b) Editing frequently results in radical changes in amino acid(s) specified by a codon.

(c) Most editing events occur in the first two nucleotides of the codon. (d) RNA editing efficiently modifies the transcripts such that a relatively homogenous population of polypeptide is produced. (e) RNA editing generally occurs in the coding sequences.

RNA editing, classes of – RNA editing has been divided into two classes – insertional editing and substitutional editing.

RNA editing, consequences of – RNA editing shows that genetic information resides not only in the genes per se but also in the apparatus that edits the transcript. RNA editing could be a relic of a RNA sequence correlation system that was utilized be for polymerase-dependent nucleic acid synthesis evolved. RNA editing could regulate gene expression, perhaps, in a tissue-specific manner.

RNA editing, functional significance of – (a) RNA editing, which serves as a translational control mechanism, adds ribonucleotides for N-terminal amino acids in type II editing and created the whole protein-encoding region in type III editing. (b) It also regulates maxicircle gene expression at developmental level.

RNA editing, patterns of – Various patterns of RNA editing are: (a) Insertion or deletion of non-templated uridines, as in Trypanosoma mitochondrial transcripts. (b) Insertion of non-temlplated guanosines, as in paramyxoviral transcripts. (c) Creation of a stop codon by conversion of a specific cytosine to a uridine, as in apolipoprotein B transcripts in mammalian intestinal tissue. (d) Conversion of multiple cytosines to uridines or uridines to cytosines, as in higher plant mitochondrial transcripts.

RNA molecules, types of – In vivo, one might think of two classes of RNA molecules: (1) Unstable molecules such as mRNAs, whose function is to present a linear sequence of bases, are prevented from folding into compact internally base-paired structures by specific proteins whose function is to do just this. (2) Stable molecules, such as rRNAs and tRNAs, fold into very characteristic secondary and tertiary structures that are an indispensable for their particular functions.

RNA polymerase I – Synthesizes pre-rRNA 45S (35S in yeast), which matures into 28S, 18S and 5.8S rRNAs which form the major RNA sections of the ribosome.

RNA polymerase I-specific promoters – Promoter sequences for RNAPI are located upstream (before start point of transcription).

RNA polymerase II – Synthesizes precursors of mRNAs and most snRNA and microRNAs.

RNA polymerase II in "backtracked" state – Transcribing RNA polymerases oscillate between three stable states: pre-translocated state, post-translocated state, and reverse translocated or "backtracked" state. The defining feature of the backtracked structure is a binding site for the first backtracked nucleotide. This binding site is occupied in case of nucleotide misincorporation in the RNA or damage to the DNA, and is termed the "P" site because it supports proofreading. The predominant mechanism of proofreading is the excision of a dinucleotide in the presence of the elongation factor SII (TFIIS).

RNA polymerase II-specific promoters – Promoters for RNAPII are present upstream and showed three regions at start point. These three regions are TATA or 'Hogness box' (7-bp long) which is located at position –20, CAAT box, which is located between –70 to –80, and GC box, which is located between –60 to –100.

RNA polymerase II transcription preinitiation complex (PIC) – In PICs, the TFIIB linker and core domains are positioned over the central cleft and wall of the RNAPII. This positioning is not observed in the smaller RNAPII-TFIIB complex.

RNA polymerase II ubiquitylation sites – Transcriptional arrest triggers ubiquitylation of RNA polymerase II. The yeast RNAPII ubiquitylation sites have been mapped. They play an important role in transcription elongation and the DNA damage response.

RNA Polymerase III – It carries out about 10 per cent of cellular transcription in eukaryotes. The genes transcribed by this enzyme are those for small non-coding RNA molecules such as 5S rRNA, tRNAs, the U6 snRNA found in spliceosmes and the 7S RNA component of signal recognition particle.

RNA polymerase III-specific promoters – Promoter sequences for RNAPIII they are located downstream with exception of 7S genes of signal recognition particles where they are present both upstream and downstream.

RNA Polymerase IV – Found in plants. Presumed maize RNAPIV is involved in paramutation, an inherited epigenetic change facilitated by an interaction between two alleles, as well as normal maize development. It does not engage in the efficient RNA synthesis typical of the three major eukaryotic DNA-dependent RNA polymerases. It employs abnormal RNAP activities to achieve genome-wide silencing and that its absence affects both maize development and heritable epigenetic changes. Also Synthesizes siRNA in plants microRNAs.

RNA polymerase V – Synthesizes RNAs involved in siRNA-directed heterochromatin formation in plants.

RNA polymerase recognition site – A site on template strand which RNA polymerase initially recognizes and binds. The bases of this site are not transcribed.

RNA polymerase secondary channel – High-resolution crystal structures have highlighted functionally important regions in multisubunit RNAPs, including the secondary channel, or pore, which is postulated to allow the diffusion of small molecules both into and out of the active center of the enzyme. Regulatory factors and small molecules can exploit the secondary channel to gain access to the active site and modify the transcription properties of RNA polymerase.

RNA polymerase DNA-binding site – After initial binding to recognition site, RNA polymerase diffuses to RNAP binding site which is A=T rich region. This binding region is called TATA box. RNA polymerase tracks along DNA as template rotation generates twin domains of supercoiling. The binding sites at which RNAP forms a stable initiation complex with DNA lie within promoter.

RNA polymerases and cancer – Oncogenic stimuli and tumor suppressors control RNA polymerase III-dependent transcription. Overexpression of a transcription factor of the RNA polymerase III apparatus transforms cells. Elevated tRNA synthesis can promote cellular transformation.

RNA processing – (1) Refers to various post-transcriptional steps that a transcript undergoes before it becomes a functional RNA molecule. (2) RNA processing may be defined as modification, mainly through cleavage and/or splicing, of RNA transcripts so as to release functional transfer RNA (tRNA), ribosomal RNA (rRNA) and messenger RNA (mRNA) and other species of RNA molecules from them.

RNA quality-control systems – Eukaryotic cells contain numerous RNA quality-control systems that are important for shaping the transcriptome of eukaryotic cells. These systems not only prevent accumulation of non-functional RNAs, but also regulate normal mRNAs, repress viral and parasitic RNAs, and potentially contribute to the evolution of new RNAs and hence proteins. RNA quality-control circuits depend on specific adaptor proteins that target aberrant RNAs for degradation as well as coupling of individual steps in mRNA biogenesis and function.

RNA transport – How mRNA gets to the cytoplasm? The successive steps of mRNA maturation need to be emphasized. Studies on regulation of RNA transport by HIV-1 Rev protein have provided only a glimpse into the complex interplay between nuclear RNA splicing and RNA transport. RNA processing and assembly require many factors but the nucleus apparently lacks any active transport system to deliver these to the RNA. spliced mRNA is exported to the cytoplasm for translation. Mechanism for recruitment of the conserved mRNA export machinery (TREX complex) to mRNA has been investigated. Human TREX complex is recruited to a region near the 5'-end of mRNA, with the TREX component Aly bound closest to the 5' cap. Both TREX recruitment and mRNA export require the cap, and these roles for the cap are splicing-dependent. CBP80, which is bound to the cap, associates efficiently with TREX to the 5'-end of mRNA, where it functions in mRNA export. As a consequence, the mRNA would be exported in a 5' to 3' direction through the nuclear pore.

RNA world hypothesis – This hypothesis holds that during evolution the structural and enzymatic functions initially served by RNA were assumed by proteins, leading to the latter's domination of biologicl catalysis. This progression can still be seen in modern biology, where ribozymes, such as ribosome and RNase P, have evolved into protein-dependent RNA catalysis (RNPzymes). Similarly, group I introns use RNA-catalyzed splicing reactions, but many function as RNPzymes bound to proteins that stabilize their catalytically active RNA structure.

RNA-dependent RNA polymerase – RNA polymerase II (RNAPII), in addition to DNA-dependent RNA polymerase activity, also possesses RNA-dependent RNA polymerase (RdRNAP) activity. RNAPII can use a homopolymeric RNA template, can extend RNA by several nucleotides in the absence of DNA, and has been implicated in the replication of RNA genomes of hepatitis delta virus (HDV) and plant viriods.

RNA-directed RNA synthesis – Non-coding small RNAs regulate gene expression in all organisms, in some cases through direct association with RNA polymerase. For example, the DNA-dependent RNA polymerase uses bound 6S RNA as a template for RNA synthesis, producing 14- to 20-nucleotide RNA product (pRNA).

RNA-DNA hybrid in biological functions – Formation of RNA-DNA hybrids is important in transcription of DNA, reverse transcription of viral RNA and DNA replication.

S

Scrunching model of abortive transcription initiation – It invokes a flexible element in DNA. In each cycle of abortive initiation, RNA polymerase unwinds downstream DNA and pulls it into itself, accommodating the accumulated DNA as single-stranded bulges in the unwound region; upon release of the abortive RNA, RNA polymerase extrudes the internalized DNA.

Second code – Correct recognition of transfer RNAs by aminoacyl-tRNA synthetases is central to the maintenance of translational fidelity. The anticodon is indeed important for some of the 20 *E. coli* isoaccepting groups. For many of the isoaccepting groups, the acceptor stem or position 73 (or both) is important as well. For accurate translation of genetic messages the precision of two matchings is very important. Firstly that of amino acids with transfer RNA and secondly that of tRNA with mRNA. The later matching is a strait forward interaction of the codon of mRNA with anticodon in tRNA but the first matching is indirect and is mediated by specific enzymes, the aminoacyl synthetases (AAS). These enzymes are specific for each amino acid, there being 20 AASs for 20 essential amino acids.

Secondary metabolites – A secondary metabolite is not directly involved in those processes, but usually has important ecological function. Antibiotics, pigments and other metabolites of medicinal and aromatic value derived from plants are important secondary metabolites.

Self-splicing of rRNA – Group II introns are type of RNA enzyme (ribozyme) that catalyzes their own excision from RNA transcripts and insertion into new genetic locations. Structural and functional analogies support the hypothesis that group II introns and the spliceosome share a common ancestor. The introns of *Tetrahymena* rRNA precursors are removed autocatalytically in a unique reaction mediated by the RNA molecule itself. Self-splicing of intron in 28S rRNA precursor occurs through transesterification mechanism. In this process, a guanosine nucleotide acts as nucleophile and gets attached to the 5'-end of the existing intervening sequence via a normal 3'-5' phosphodiester bond and then -OH of the 5' exon would cause cleavage to produce 5'-phosphate and 3'-OH terminus. The ligation occurs without hydrolysis of ATP with the removal of covalent circular intron. GMP initiates the process. A circular intron is created. 5′ splice site selection by the self-splicing *Tetrahymena* ribosomal RNA precursor requires the formation of a particular RNA helix by base pairing between an intron ‘guide sequence’ and bases surrounding the 5′ splice junction. Part of the guide sequence subsequently holds the positions the 3′-end of the cut 5′ exon to attack the 3′ splice junction.

Sense DNA strand – Strand complementary to the antisense strand. Also known as *coding strand, non-transcribing strand, non-template strand*, or *Crick strand.*

Sense word – A codon that specifies an amino acid normally present in that position in a protein.

Sequon – Proteins with asparagine residues in the tripeptide motif An-Xaa-Ser/Thr that are N-glycosylated with preformed oligosaccharide units, or other residues may be also be involved in the determination of a glycosylation site, or the structure of the protein itself may inhibit the glycosylation of certain polypeptides.

Shine-Dalgarno sequence – A short sequence many prokaryotic ORFs contain upstream on the 5′-end of start codon called ribosome binding site that facilitates the binding by a ribosome. This sequence typically located 3-9 bp on the 5′ side of start codon is complementary to the sequence located near 3′-end of one of RNA components (16S rRNA). Ribosome binding site base-pairs with 16S rRNA component thereby aligning ribosome with beginning of ORF.

Sigma (σ) factors – Bacterial transcription factors that bind core RNA polymerase and direct transcription initiation at cognate promoter sites.

Signal hypothesis – The mechanism for membrane attachment is explained by the signal hypothesis. Signal hypothesis explains the process of secretion of proteins. According to this hypothesis, ribosomes synthesizing proteins are attached to the membrane via leader sequence. The mechanism is applicable for both prokaryotes and eukaryotes. Proteins have intrinsic signals that govern their transport and location in the cell.

Signal peptide – Signal peptide takes part in a chain of events leading to membrane attachment by the ribosome and membrane insertion of the protein. The signal peptide does not seem to have a consensus sequence like the transcription or translation recognition boxes. Rather, similarities (at least for the endoplasmic reticulum and bacterial membrane-bound proteins) include a positively charged (basic) amino acid (commonly lysine or arginine) near the beginning (N-terminal end) followed by about a dozen hydrophobic (non-polar) amino acids, commonly alanine, isoleucine, leucine, phenylalanine, and valine. The signal peptide of the bovine prolactin protein reported by Sasavage et al. (1982) is: NH2-Met Asp Ser Lys Gly Ser Ser Gln Lys Ser **Arg Leu Leu Leu Leu Leu Val Val Ser Asn Leu Leu Leu Cys Gln Gly Val Val Ser** Thr Pro Val .... Asn Asn Cys-COOH. The amino acids represented in bold separate the signal peptide from the rest of the protein which consists of 199 amino acid residues.

Signal sequences – The amino acid sequences that target proteins for secretion from cells or for integration into cell membranes. As nascent proteins emerge from the ribosome, signal sequences are recognized by the signal recognition particle (SRP), which subsequently associates with its receptor (SR). In this complex, the SRP and SR stimulate each other's GTPase activity, and GTP hydrolysis ensures unidirectional targeting of cargo through a translocation pore in the membrane. The signal sequence is presented at the ribosomal tunnel exit in an exposed position ready for accommodation in the hydrophobic groove of the rearranged SRP54 M domain. Upon ribosome binding, the SRP 54 NG domain also undergoes a conforma-tional rearrangement, priming it for the subsequent docking reaction with the NG domain of the SRP receptor. Also known as *leader sequences*.

Signal transduction – Signal transduction pathways precede most of the mechanisms of regulation gene expression. In eukaryotes, a cascade of molecules leading to the activation of one or more specific transcription factors is involved. Among oncoproteins, G proteins are represented by Ras proteins. Since G-proteins transduce signals, they are also described as transducers.

Simple-sequence gene family – The simple sequence family encompasses segments of DNA derived from 10^2-10^7 repetitions of a short fundamental sequence, generally 6-15 nucleotides in length. Degree of homology amongst the repeat units is often 80-100 per cent.

Single-cell gene expression profiling – Simultaneous visualization of gene expression profiling inside single cell with high spatial and temporal resolution.

Single-polypeptide nuclear RNA polymerase – Transcription of some mRNAs in humans and rodents is mediated by previously unknown single-polypeptide nuclear RNA polymerase (spRNAPIV). It is expressed from an alternative transcript of the mitochondrial RNA polymerase gene *POLRMT*. The spRNAP-IV lacks 262 amino-terminal amino acids of mitochondrial RNA polymerase, including the mitochondrial-targeting signal, and localizes into the nucleus.

Small nuclear RNA (snRNA) gene family – snRNA genes are of 11-13 types. Major 5-6 types of snRNA genes (U1, U2, U3, U4, U5 and U6) are present in many copies and are transcribed frequently. These small nuclear RNAs, called U RNAs, constitute spliceosome formed during splicing of introns from precursors of messenger RNA. The rest small nuclear RNA genes are less frequent. snRNAs stands for small nuclear RNAs. No intervening sequences are found in U1, U2, and U3 genes. These genes are dispersed throughout the genome. Putative adenylation signal (AATAAA) and the termination signal (TTT, at the 3'-end of the RNA) are not found for the U1 gene. U1 RNA is not polyadenylated. Suggested number of genes for each snRNA is 2,000. They are also called *U RNAs*.

Smart transcription factor – Smart transcription factor first gets phosphorylated and then it binds to co-activator binding protein. After that this complex binds to DNA element and this leads to initiation of transcription. For example, smart transcription factor cAMP response element binding (CREB) protein first gets phosphorylated and then it binds to co-activator CREB binding protein (CBP). After that this complex binds to DNA element known as cAMP-regulated enhancer and this leads to initiation of transcription.

snRNP – snRNAs are complexed with proteins to form small nuclear ribonucleoproteins (snRNPs).

Snurps in pre-mRNA splicing – The introns of nuclear mRNA precursor transcripts are spliced out in a two-step reaction carried out by complex ribonucleoprotein particles called spliceosomes. They also attach new non-coding segments to the leading and trailing ends of the mRNA. This process is called splicing. Snurps stands for small nuclear ribonucleoprotein particles that help to remove meaningless introns from the messages issues by a cell's genes. Without snurps cellular activity would come to a halt.

Somatic diversity theory – Only a small number of V sequences are transmitted from parents to offspring, perhaps as few as 3 V_k sequences, 5 V_γ sequences and a few V_H sequences. These sequences then experience thousands of different base-pair changes in individual lymphocyte lines creating the observed diversity. One lymphocyte clone will, for example, carry sequences $V_{\delta 1a}$ and $V_{\gamma 11c}$, another $V_{\delta 1b}$ and $V_{\delta 11e}$, and so on. Two of these sequences are then selected for expression in a given lymphocyte clone.

Spickles in RNA splicing – Pre-mRNA splicing is a predominantly co-transcriptional event, which involves a large number of essential splicing factors within the mammalian cell nucleus. Most splicing factors are concentrated in 30-40 distinct domains called specklets. Speckles are highly dynamic structures that respond specifically to activation of nearby genes. These dynamic events are dependent on RNA polymerase II transcription and are sensitive to inhibitors of protein kinases and Ser/Thr phosphatases. When single genes were transcriptionally activated in living cells splicing factors left speckles in peripheral extensions and accumulated at new sites of transcription. One function of speckles is to supply splicing factors to active genes.

Splice site compatibility – This phenomenon leads to differential use of splice sites as certain splice site combinations are specifically favored over the others and different mRNAs are formed. Differential transfactors also affect splice site compatibility as they have ability to regulate the selection of differential splice site utilization in a positive or negative manner.

Splice site recognition – SR proteins have important roles in facilitating splice site recognition. For example, they recruit the U1 snRNP to the 5′ splice site and the U2AF complex and U2 snRNP to the 3′ splice site by binding to an ESE and directly interacting with protein targets. Inhibition of splice site recognition can be achieved in many ways. First, when splicing silencers are located close to splice sites or to splicing enhancers, inhibition can occur by sterically blocking the access of snRNPs or of positive regulatory factors. Some silencers can be over 100-200 bp away from enhancers. Such splicing inhibitors function by masking splice site recognition through multimerization along the RNA.

Spliceosome – A dynamic structure assembled on pre-mRNA templates in a stepwise fashion and then dissociates as the splicing reaction is completed. Even a small error in splicing will be intolerable resulting into non-functional polypeptide. Thus RNA processing involves nigh degree of specificity. Specialized nuceoriboprotein structures (Snurps)-like particles identified in cytoplasm and nucleolus are, respectively, named as "scRNPs" (or "scyrps") and "snorps". The multiple snRNP nature of spliceosome also raises the possibility that different introns may be recognized by different combinations of snRNPs.

Spliceosome assembly – Exon-intron 5′ splice site is recognized by U1 snRNP. The precursor RNA is probably also associated with hnRNP core proteins. U5 snRNP binds to the 3′ splice site at this stage. Incubation of substrate RNA in the presence of ATP results in the rapid formation of a U2 snRNP complex. Further incubation yields assembly of the spliceosome containing at least U2, U4, U5 and U6 snRNAs. The U4 and U6 snRNAs can be associated in one snRNP particle. The two RNAs characteristics of the intermediate in splicing, the 5′ exon and the lariat intervening sequence terminating in the 3′ exon RNAs, are found exclusively in the spliceosome.

Splicing and polyadenylation – Pattern of alternative RNA splicing and alternative cleavage and polyadenylation were strongly correlated across tissues, suggesting coordinated regulation of both splicing and polyadenylation.

Splicing in mRNA cytoplasmic localization – An important newly discovered function for splicing is regulation of messenger ribonucleoprotein complex assembly and organization for mRNA cytoplasmic localization.

Splicing mechanisms – Two types of splicing mechanisms, known as major and minor splicing mechanisms, have been identified. Accordingly, major and minor spliceosomes have been identified. These two splicing pathways are spatially separated in the cell and may have distinct functions.

Startise – Start point of transcription. A site on template strand required for initiation of transcription. Start point is the base pair on DNA that corresponds to the first nucleotide incorporated into transcript by RNA polymerase.

Startise – The site at which the first nucleotide is added during transcription. Also known as *startpoint*.

Storage protein gene family – Storage proteins in crop plants are encoded by multiple gene families and are represented mainly as prolamins or globulins. Homologies in genes coding for prolamins and globulins in different crop plants has also been shown. Their multiplicity meets the demand for rapid synthesis of storage proteins in the developing seeds.

Structural alleles – Alleles determined on the bases of recombination test.

Substitutional editing – No guide RNA is implicated. The phosphodiester backbone of the RNA is not cut and the new base is synthesized in situ. It blurs the distinction between editing and RNA modification.

Synergistic transcription – Gene expression is said to be synergistic when combined quantitative effect of two transcriptional activation elements is more than the sum of their individual expression.

Synonymous codons – Different codons that specify the same amino acid in a degenerate code.

Synonymous Mutations – Synonymous mutations do not alter the encoded proteins, but they can influence gene expression.

Synthesis of 28S and 18S rRNA – At the nucleolar organizer region, 45S RNA is synthesized which acts as a precursor molecule for the synthesis of 28S and 18S rRNAs. An endonulease cleaves 45S RNA into two short precursor molecules of size 41S and 20S, which are cleaved into 28S and 18S rRNAs, respectively. The 41S RNA under goes three sequential cleavage steps by exonuclease to yield 28S rRNA whereas 20S RNA undergoes only one cleavage step by exonuclease to produce 18S rRNA.

Synthetic vaccines – These vaccines do not contain intact viruses or complete polypeptides but merely small peptides that have been synthesized in laboratory to mimic a very small region of the outer coat of the virus. These peptides elicit antibodies capable of neutralizing the virus. Synthetic vaccines prepared through short synthetic peptide chains. Also known as *synthetic peptide vaccines*.

System-biology approach – This approach aims to identify metabolic switches and transcriptional regulators and reveal the intrinsic biological networks that occur within cells. Mutants and metabolically-altered plants are analyzed using transcriptomics (study of transcript profile), proteomics (study of protein profile) and metabolomics (study of metaboloite profile) to understand how different intrinsic networks are influenced.

T

T-cell receptor genes – T-cell receptor genes synthesize T-cell receptors and are more complex in structure than immunoglobulin genes.

T-cell receptors (TCRs) – Proteins present on cell membrane. They bind to special molecules to facilitate their entry into the cell.

Telomerase RNA processing – The spliceosome is the best known for shepherding primary messenger RNA transcripts to maturity. This enzyme complex also contributes to the synthesis of an enzyme that maintains chromosome ends. Chromosomal ends contain telomeric repeats, which are added to the 3'-end of DNA molecules by the telomerase enzyme complex. Telomerase consists of a reverse transcriptase enzyme (T), which uses another component of this complex, the telomerase RNA, as a template for telomere synthesis.

Temporal sequence of gene expression – Regulation of gene expression during the life cycles of virulent bacteriophages is quite different from the reversible on-off switches characteristic of bacterial operons. In phage-infected bacteria, the viral genes are expressed in genetically preprogrammed sequences or cascades.

Tet-on and Tet-off – Tet-off activates expression in the absence of Dox, whereas Tet-on activates in the presence of Dox. Doxycycline is also used in "Tet-on" and "Tet-off" tetracycline controlled transcriptional activation to regulate transgene expression in organisms and cell cultures.

Tetracycline-controlled transcriptional activation – A method of inducible gene expression where transcription is reversibly turned on or off in the presence of the antibiotic tetracycline or one of its derivatives (e.g. doxycycline). In nature, the Ptet promoter expresses TetR, the repressor, and TetA, the protein that pumps tetracycline antibiotic out of the cell. The difference between "Tet-on" and "Tet-off" is not whether the transactivator turns a gene on or off, as the name might suggest; rather, both proteins activate expression. The difference relates to their respective response to doxycycline; Tet-off activates expression in the absence of Dox, whereas Tet-on activates in the presence of Dox. Doxycycline is also used in "Tet-on" and "Tet-off" tetracycline controlled transcriptional activation to regulate transgene expression in organisms and cell cultures.

3'-end processing of pre-mRNAs – During 3'-end processing, histone pre-mRNAs are cleaved 5 nucleotides after a conserved stem loop by an endonuclease dependent on the U7 small nuclear ribonucleoprotein (snRNP). The upstream cleavage product is degraded by a 5'-3' exonuclease, also dependent on the U7 snRNP.

Torpedo model of transcriptional termination – 'Torpedo' model postulates that poly(A) site cleavage provides an unprotected RNA 5'-end that is degraded by 5'→3' exonuclease activities (torpedoes) and so induces dissociation of RNAPII from the DNA template. The 3'-end of the cleavage RNA is likely to be degraded by the exosome (Exo).

Train sequence – Sector of the non-coding sequence following the 3'-end of gene proper (mature gene, antisense strand). Train contains signals concerned with termination of transcription. Typically, there exists a series of T-A bp, in prokaryotes usually six or more in number, while in metazoans four times seems to suffice. A stem-and-loop structure could conceivably form in the transcript.

Transcription – Synthesis of RNA using DNA as a template. Transcription is a vital process in gene expression . It is through this process that the biological road map encoded in a strand of DNA is used to produce a complementary RNA copy. During transcription, only one of the two DNA strands is transcribed into RNA. *DNA-dependent RNA polymerases*, or simply called *RNA polymerases* (RNAPs) are the transcribing enzymes.

Transcription and recombination – Effect of RNA polymerase II-dependent transcription on recombination has been reported in *S. cerevisiae.* Stimulation in recombination is mediated by events initiating within the integrated plasmid sequences.

Transcription coexpression – Transcription coexpression of interacting gene products is required for complex molecular processes. C*is*-regulatory elements orchestrate coexpression. Syntactical rules governing this regulatory function are flexible but become highly constrained evolutionarily once they are established in a particular element.

Transcription elongation – Elongation of transcription proceeds in alternating laps of monotonous and inchworm-like movement with the flexible DNA polymerase configuration being subject to direct sequence control. During this step, RNAP catalyzes nucleophilic attack by the 3′-OH group of the growing RNA chain on α phosphorous of the ribonucleoside triphosphate, resulting in formation of a new phosphodiester linkage and release of pyrophosphate. 12 nucleotides of the template strand base-pair with newly synthesized RNA to form a hybrid RNA:DNA helix. When a new nucleotide joins the 3′-end of the growing RNA chain and lengthens the RNA: DNA hybrid by following steps: (a) The length of the RNA:DNA hybrid must be reduced by melting one bp at the upstream end of the transcription bubble. (b) The DNA bp must be melted at the downstream end of bubble. (c) The DNA bp must be reformed at the upstream end of bubble. (d) RNAP must advance along the DNA by one nucleotide, positioning its active site over the next unpaired nucleotide on the template strand. Elongation continues till a termination complex is formed.

Transcription elongation complex – The mechanism of substrate loading in multisubunit RNA polymerase is crucial for understanding the general principles of transcription.

Transcription elongation fators – These factors enhance overall activity of RNA polymerase II leading to increase in elongation rate. Two such factors are TFIIF and TFIIS. TFIIF accelerates RNA chain growth uniformly and TFIIS Helps in elongation by relieving obstacles. TFIIS causes hydrolytic cleavage at 3′-end of RNA chains which are blocked. Its action is similar to cleavage by GreA and GreB in *E. coli.* Rpb9 is small subunit of yeast RNAPII participating in elongation.

Transcription elongation tuning – RNA chain elongation is a highly processive and accurate process that is finely regulated by numerous intrinsic and extrinsic signals. Bar-Nahum and coworkers in 2005) describe a general mechanism that that governs RNAP movement and response to regulatory inputs such as pause, terminators, and elongation factors. *E. coli* RNA polymerase moves by a complex Brownian ratchet mechanism, which acts prior to phosphodiester bond formation. The incoming substrate and flexible F bridge domain of the catalytic center serve as two separate ratchet devices that function in concert to drive forward translocation. The adjacent G loop domain controls F bridge motion, thus keeping the proper balance between productive and inactive states of the elongation complex. This balance is critical for cell viability since it determines the rate, processivity, and fidelity of transcription.

Transcription factor CCAAT-enhancer-binding proteins (C/EBPs) – Transcription factors of family C/EBP use a bipartite structural motif to bind DNA. The C/EBP family consists of six transcription factors (α, β, γ, δ, ε and ζ). Two protein chains dimerize through a set of amphipathic α-helices termed the leucine zippers. "Scissors grip" model has been given to explain how C/EBP binds DNA in the major groove.

Transcription factors – Those proteins that are required for initiation of eukaryotic transcription.

Transcription in bacteriophages – When a temperate bacteriophage such as lambda (λ) infects a bacterium, it can follow either of two pathways of development. A temperate phage can either enter the lytic cycle, during which it reproduces and lyses the host cell, just like a virulent phage or enter the lysogenic pathway, during which its chromosome is inserted into the chromosome of the host and thereafter is replicated like any other segment of that chromosome. When integrated in the chromosome of the host cell, the temperate phage chromosome is called a prophage. In a lysogenic bacterium, the genes of the prophage that encode products involved in the lytic pathway must not be expressed.

Transcription in non-bacterial viruses – In cells infected by viruses that encapsidate DNA, synthesis of viral m-RNA must precede production of proteins. The majority of DNA including the parvo-, papova-, hepadna-, adeno- and herpesviruses, depend on cellular RNA polymerase II, the enzyme that produces cellular mRNA. The retroviruses, which establish themselves as DNA proviruses early in their infections cycles, also use the cellular RNA polymerase II system.

Transcription in viruses – The viral polymerases are diverse, and include some forms which can use RNA as a template instead of DNA. This occurs in negative-strand RNA viruses and dsRNA viruses, both of which exist for a portion of their life cycle as double-stranded RNA. However, some positive-strand RNA viruses, such as polio, also contain these RNA-dependent RNA polymerases.

Transcription Initiation – During this step of transcription RNA polymerase initially recognizes and binds to RNA polymerase recognition site, then it diffuses to RNA binding site

Transcription of centromeric repeats – Heterochromatin in eukaryotic genomes regulates diverse chromosomal processes including transcriptional silencing. However, in fission yeast RNA polymerase II transcription of centromeric repeats is essential for RNA interference-mediated heterochromatin assembly. RNA polymerase II transcription during S phase is linked to the loading of RNA interference and heterochromatin factors such as the Ago1 subunit of the RITS complex and the Clr4 methyltransferse complex subunit Rik1.

Transcription of Line-1 elements – LINE-1 (L1) elements are retrotransposons that comprise large fractions of mammalian genomes. Transcription through L1 open reading frames is inefficient owing to an elongation defect, inhibiting the robust expression of L1 RNA and proteins, the substrate and enzyme(s) for retrotransposition. This elongation defect probably controls L1 transposition frequency in mammalian cells. This transcriptional defect has been bypassed by synthesizing the open reading frames of L1 from synthetic oligonucleotides, altering 24 per cent of the nucleic acid sequence without changing the amino acid sequence. Such resynthesis led to greatly enhanced steady-state L1 RNA and protein levels.

Transcription preinitiation complex (PIC) – No eukaryotic RNA polymerase appears to bind specifically and directly to DNA. Instead, interactions between polymerase and DNA require the presence of sequence-specific accessory proteins called transcription factors. PIC includes a set of general transcription factors (TFIIA B, D, E, F and H) along with RNA polymerase II.

Transcription proofreading mechanism – Mistakes can occur as RNA polymerase copies DNA into transcripts. A proofreading mechanism that removes incorrect RNA is triggered by the erroneous RNA itself. Correction of misincorporated errors at the growing end of the transcribed RNA is stimulated by the misincorporated nucleotide.

Transcription termination – Termination of transcription is an important regulatory step closely related to transcriptional interference, and even transcriptional initiation. Once RNA polymerase starts transcription, the enzyme continues to move along the template synthesizing RNA until it meets a signal to stop this synthesis. At this stage, enzyme stops adding nucleotide to the RNA chain, releases the completed product and dissociates from the DNA template. Termination process in prokaryotes involves formation of 'hairpin structure'.

Transcription terminator – DNA sequence that provides the signal to cease or stop transcription. Terminators require that all the hydrogen bonds holding RNA-DNA together must be broken. Termination of transcription occurs because the elongation complex is less stable when transcribing certain specific DNA sequences. Terminators may be rho-dependent or rho-independent. RNA polymerase continues transcription until it meets a terminator sequence. At this point, the enzyme stops adding nucleotides to the growing RNA chain, releases the transcript and dissociates from the DNA template. Identification of terminators is provided by systems in which RNA polymerase terminates in vitro. At some terminators, termination event can be prevented by specific ancillary sequences (factors) that interact with RNA polymerase. Responsibility for termination lies with the sequences already transcribed by RNA polymerase.

Transcriptional apparatus in Archaea – The transcriptional apparatus in Archaea can be described as a simplified version of its eukaryotic RNA polymerase II counterpart, comprising of an RNAPII-like enzyme as well as two general transcription factors, the TATA-binding protein (TBP) and the eukaryotic TFIIB ortholog TFB.

Transcriptional complexity – Transcribed portions of several organisms are larger and more complex than expected, and many functional properties of transcripts are based not on coding sequences but on regulatory sequences in untranslated regions or non-coding RNAs. Alternative start and polyadenylation sites and regulation of intron splicing add additional dimensions to the rich transcriptional output.

Transcriptional elements – Transcriptional elements have a consensus sequence. They play crucial role in gene expression. For example, maximal transcription of heat shock protein *hsp70* gene requires two copies of a 14-bp heat shock element (HSE) with consensus sequence CNNGAANNTTCNNG.

Transcriptional factors – Sequence-specific binding proteins that control transcription by RNA polymerase II. They interpret the genetic regulatory information, such as in transcriptional enhancers and promoters, Salient properties of transcriptional factors are: (1) They are composed of functional modules; they regulate transcription via recruitment of coactivators and corepressors. (3) They can be regulated by post-translational modifications. (4) They are often members of mutiprotein family. (5) Chromatin is an integral component in the function of transcriptional factors. (3) Recognition sites for transcriptional factors tend to be located in clusters.

Transcriptional forests – Genomic mapping of the transcriptome reveals transcriptional forests, with overlapping transcription on both strands, separated by deserts in which few transcripts are observed. The data provide a comprehensive platform for the comparative analysis of mammalian transcriptional regulation in differentiation and development.

Transcriptional hub – Why would two distinct genes on separate chromosomes and from different nuclear locations unite in response to signals for gene expression? They might be seeds for formation of transcriptional hubs. In the absence of estrogen, two genes (TEF1 and GREB1) that are activated by this hormone and found on different chromosomes, reside in different locations within the nucleus.

Transcriptional pausing – Transcriptional pausing by RNA ploymerase plays an important role in the regulation of gene expression. Defined, sequence-specific pause sites have been identified biochemically. Single-molecule studies have also shown that bacterial RNA polymerase pauses frequently during transcription elongation. Elongation may be impeded by pause sites which induces temporary reversible block to nucleotide addition or by arrest or dead ends which stops the transcription which could only be resumed by factors like Gre A and Gre B. Gre A and Gre B are elongation factors involved in transcription elongation including suppression of transcription arrest, enhancement of transcription fidelity and facilitating transcription from abortive initiation to productive elongation. During transcription, RNAP moves processively along a DNA template, creating a complementary RNA. This stage in early elongation appears to be an important broadly used target of gene regulation. Also known as *transcriptional stuttering*,

Transcriptional processivity – The affinity of RNA polymerase for template DNA is referred to as transcriptional processivity. The interactions that occur during elongation process come under this phenomenon. Repressors bind to selected sets of genes at sites called silencers. They interfere with functioning of activators and slow down transcription. Three types of factors play role in transcriptional processivity. (a) Basal factors respond to activators and position RNA polymerase at the start of the protein-encoding region of a gene and send the enzyme on its way. (b) Coactivators are the adaptor molecules that integrate signals from activators and repressors and relay the result to basal factors. (c)Activators bind to genes at sites called enhancers. They help to determine which genes will be switched on, and they speed up rate of transcription.

Transcriptional unit – A sequence of DNA transcribed into a single RNA, starting at the promoter and ending at the terminator

Transcription-coupled alternative RNA splicing – Two models have been proposed to explain the role of RNA polymerase II in the regulation of alternative RNA splicing. The first model is known as the *recruitment model*, in which RNA polymerase II and transcription factors interact, directly or indirectly, with splicing factors, thereby increasing or decreasing the efficiency of splicing. A second, *kinetic model* proposes that the rate of transcription elongation influences the inclusion of alternative exons by affecting whether the splicing machinery is recruited sufficiently quickly for spliceosome assembly and splicing to occur.

Transcriptome – Total RNA complement synthesize by a cell. It is specific for organism, cell type, developmental stage and environment. The transcritomes of eukaryotic cells are incredibly complex.

Transcriptomics – It involves study of mRNA from each possible tissue, at each developmental stage and under each of a number of possible environments (stresses).

Transcriptome analysis – Analysis transcriptional profile of a cell. These transcriptome analyses indicate that many of the non-coding regions, previously thought to be functionally inert, are actually transcriptionally active regions with various features.

Transfer of amino acid to tRNA – Adenylated amino acid, which remains tightly bound to synthetase, reacts with tRNA. Each amino acid thus gets bound to accepter stem of tRNA by a specific aminoacyl-tRNA synthetase. These enzymes face two important challenges, they must recognize the correct set of tRNAs for a particular amino acid and they must charge all of these isoaccepting tRNAs with the correct amino acid. Some features of tRNA, called identity elements, aminoacyl-tRNA synthetases to discriminate isoaccepting tRNA from rest of 19 amino acids. The accepter stem is an especially important determinant for the specificity of tRNA synthetase recognition.

Transfer RNA (tRNA) – A non-coding RNA to which amino acids must attach prior to their incorporation into polypeptides. The secondary structure of tRNA has a characteristic cloverleaf configuration. It is further folded through alternative hydrogen bonding geometries (including Hoogsteen base pairs) into an L-shaped tertiary structure. tRNA contains several unusual (modified) bases. The principal features of tRNA clover leaf are an accepter stem, three stem loops which are referred to as ψ loop, D loop and the anticodon loop, fourth variable loop. Transfer RNA is the direct interface between amino-acid sequence of a protein and the information in DNA. Also known an *adaptor RNA*.

Transfer RNA (tRNA) genes – The genes which on transcription yields precursor of tRNAs (pre-tRNA) which after various processing steps mature into functional tRNA.

Transfer RNA, types of – There are two types of tRNAs based on size of extra or variable loop - Class I tRNAs and Class II tRNAs.

Transformation – In oncogenesis, transformation means cancerous growth of cells.

Transition from transcription initiation to elongation mode – Once the primer reaches 10-nucleotide length, the RNA holoenzyme undergoes a transition from the initiation to the elongation mode and the transcription complex moves away from the promoter along the DNA template. Transition to elongation stage of transcription is associated with a conformational change in holoenzyme that releases the σ factor. An accessory protein NusA binds to the core enzyme and prevents σ factor from reassociating during elongation. NusA-core enzyme elongates the RNA chain.

Translation – The process in which genetic information contained within the order of nucleotides is used to generate the linear sequence of amino acids in proteins. Four major steps involved in the process of translation are: charging of tRNA, initiation , elongation and termination The machinery responsible for translating the language of mRNA to language of proteins is composed of four primary components - mRNA, ribosome, rRNA and aminoacyl-tRNA synthetase.

Translation elongation – Polypeptide chain is lengthened by covalent attachment of successive amino acid units each carried to ribosomes by tRNAs, which base pairs with corresponding codon in mRNA at A-site. Elongation is promoted by elongation factors. Elongation requires (a) the initiation complex, (b) the next aminoacyl-tRNA, specified by the next codon in the mRNA, (c) a set of three soluble cytosolic proteins called elongation factors (EF-Tu, EF-Ts, and EF-G), and (d) GTP. Three steps are recognized in translation elongation. In the first step, the next aminoacyl-tRNA is first bound to a complex of a complex of EF-Tu containing a molecule of bound GTP. The resulting aminoacyl-tRNA-EF-Tu. GTP complex is then bound to the A site of the 70S initiation complex. The GTP is hydrolyzed, an EF-Tu. GDP complex is regenerated. In the second step, a new peptide bond is formed between the amino acids bound by their tRNAs to the A and P sites on the ribosome. The amino acid-tRNA that binds to the (A) site is specifically determined by the triplet codon of the mRNA that occupies the (A) site on the ribosome.

Translation initiation – Three steps are involved in this process. In the first step, the 30S initiation complex is formed. It requires mRNA, 30S subunit of ribosome, a special initiating species of aminoacyl~tRNA (that apparently starts all polypeptide chain, and three initiation protein factors — IF1, IF2, IF3. The initiation factor 3 (IF-3) prevents the 30S and 50S subunits from combining prematurely. Binding of the mRNA to the 30S subunit then takes place in such a way that the initiation codon (AUG) binds to a precise location on the 30S subunit. The initiating AUG is guided to the correct position on the 30S subunit by an initiating signal called the Shine-Dalgarno sequence in the mRNA. The first amino acid incorporated in prokaryotes is formylated N-methionine. At this stage, formation of 30 initiation complex is complete. In the second step of the initiation process, the complex consisting of the 30S subunit, IF-3, and mRNA now forms a still larger complex by binding IF-2, which already is bound to GTP and the initiating fMet-$tRNA_f^{Met}$. The anticodon of this tRNA pairs correctly with the initiation codon in this step. In the third step, this large complex combines with the 50S ribosomal subunit; simultaneously, the GTP molecule bound to IF-2 is hydrolyzed to GDP and

Pi (which are released). IF-3 and IF-2 also depart from the ribosome. The 70S initiation complex is thus formed. During initiation, the mRNA bearing the code for polypeptide to be formed binds to smaller subunit of ribosome, this is followed by binding of initiating amino acyl-tRNA and large subunit to form initiation complex. The initiating aminoacyl-tRNA base-pairs with mRNA codon AUG at P-site that signals the beginning of polypeptide chain.

Translation initiation complex – This step involves reactions prior to forming the peptide bond between the first two amino acids of the protein which is a relatively slow step in protein synthesis. The initiation of polypeptide synthesis in prokaryotes requires (a) the 30S ribosomal subunit, which contains 16S rRNA (b) the mRNA coding for the polypeptide to be made, (c) the initiating fMet-$tRNA_f^{Met}$, (d) a set of three proteins called initiation factors (IF-1, IF-2 and IF-3), (e) GTP, (f) the 50S ribosomal subunit and (g) Mg^{2+}.

Translation preinitiation – Activation and charging of transfer RNAs, binding of charged tRNA to the ribosome, loading of mRNA to the ribosome are important translation preinitiation steps.

Translation termination – The completion of the polypeptide chain is signaled by a termination codon in the mRNA. The polypeptide chain is then released aided by proteins called release factors. Once a termination codon occupies the ribosomal A site three termination or release factors, the proteins RF1, RF2 and RF3 contribute to (a) the hydrolysis of the terminal peptidyl-tRNA bond, (b) release of the free polypeptide and the last tRNA, now uncharged, from the P site, and (c) the dissociation of the 70S ribosome into its 30S and 50S subunits, ready to start a new cycle of polypeptide synthesis. RF1 recognizes the termination codons UAG and UAA, and RF2 recognizes UGA and UAA.The class II RF3 is a GTPase that removes class I RFs (RF1 and RF2) from ribosome after release of nascent polypeptide. Termination of polypeptide synthesis, is signaled by one of three termination codons in the mRNA (UAA, UAG, UGA), immediately following the last amino acid codon. RF1 is required for UAG and RF2 for UGA. UAA can accept either RF1 or RF2. RF3 activates RF1 and RF2.

Translocation-and-pause cycles – Translation occurs through successive translocation-and-pause cycles. The distribution of pause lengths, with a median of 2.8S, indicates that at least two rate-limiting processes control each pause. Each translocation step measures three bases – one codon – and occurs in less than 0.1 s. Translocation and RNA unwinding are strictly coupled ribosomal functions.

Transmembrane proteins – These proteins possess internal hydrophobic stop transfer sequence which lodges the protein in the ER membrane.

***Trans*-position** – When two alleles or elements are present in different DNA strands

Transposons – Also known as *transposable genetic elements* or *insertion sequences*. These elements are identified in a group of many prokaryotes (bacteria) and eukaryotes *Drosophila*, maize, etc.), which have the ability to get inserted at various different places of the genome. These insertion sequences, if carrying any other gene(s), have the ability to insert them also in the genome alongwith itself. This system is thus capable of incorporating genes from unrelated species into another species. This leads to an increase in the genome size. This is also a mode for generating genetic variability.

Triplet code – Three consecutive nucleotides in DNA or RNA make one code.

tRNA acceptance – During transfer RNA (tRNA) selection, a cognate codon-anticodon interaction triggers a series of events that ultimately results in the acceptance of that tRNA into the ribosome for peptide-bond formation. High-fidelity discrimination between the cognate tRNA and near- and cognate ones depends on their differential dissociation rates from the ribosome and on specific acceleration of forward rate constants by cognate species.

tRNA CCA-adding polymerase – CCA-adding polymerases mature the essential 3'-CCA terminus of transfer RNA without any nucleic-acid template. Transfer RNA nucleotidyltransferases (CCA-adding enzymes) are responsible for the maturation or repair of the functional 3'-end of tRNAs by means of the addition of the essential nucleotides CCA.

tRNA defect, diseases owing to – A single change in a person's tRNA gene can contribute to a range of life-shortening risk factors, including high blood pressure, high cholesterol, and other metabolic disorders.

tRNA gene family – Each type of tRNA gene has 10 to several hundred copies per haploid genome. The tRNA genes are also present in form of heteroclusters. These heteroclusters may be tandemly repeated or 50-60 sites in different chromosomes. Ten to several hundred genes for each tRNA are present in each haploid genome.

tRNAs, Class I – Have only 3-5 bases in their extra loop; they represent ~75 per cent of all tRNAs.

tRNAs, Class II – Have larger loop having 13-21 bases in their extra loop and 5 base pairs in the stem.

Tumor-progression genes – Genes involved in the development of new tumors.

Tumors – Aggregates of cells derived from an initial aberrant founder cell that although surrounded by the normal tissue is no longer integrated into the environment.

Tumor-suppressive mechanisms – Evolution has installed in the proliferative programs of mammalian cells a variety of innate tumor-suppressive mechanisms that trigger apoptosis or senescence, should proliferation become aberrant. These contingent processes rely on a series of sensors and transducers that act in a coordinated network to target the machinery responsible for apoptosis and cell cycle arrest of different points.

Tumor-suppressor protein – Principal tumor-suppressor protein is nuclear transcription factor p53, which accumulates in cells in response to DNA damage, oncogene activation and other stresses. It acts as a nuclear transcription factor that transactivates genes involved in apoptosis, cell cycle regulation and numerous other processes.

Two genes-one polypeptide hypothesis – The lambda chain seems to be encoded by two separate germline genes (a specificity region gene and a common region gene) which are expressed as a single, continuous polypeptide chain. Antibody light chains appear to be an exception to the rule of one gene-one polypeptide chain hypothesis. There is still a debate whether two genes-one polypeptide chain is fact or fiction.

Type I secretion system (T1SS) – It consists of only three protein subunits: the ABC protein, membrane fusion protein (MFP), and outer membrane protein (OMP). Type I secretion system transports various molecules, from ions, drugs, to proteins of various sizes (20-900 kDa). The molecules secreted vary in size from the small *Escherichia coli* peptide colicin V, (10 kDa) to the *Pseudomonas fluorescens* cell adhesion protein LapA of 900 kDa.

Type II secretion system (T2SS) – It depends on the Sec or Tat system for initial transport into the periplasm. Once there, they pass through the outer membrane via a multimeric (12-14 subunits) complex of pore forming secretin proteins. In addition to the secretin protein, 10-15 other inner and outer membrane proteins compose the full secretion apparatus.

Type III secretion system (T3SS) – It is like a molecular syringe through which a bacterium can inject proteins into eukaryotic cells. The low Ca^{2+} concentration in the cytosol opens the gate that regulates T3SS.

Type IV secretion system (T4SS) – It is homologous to conjugation machinery of bacteria. It was discovered in *Agrobacterium tumefaciens*, which uses this system to introduce the T-DNA portion of the Ti plasmid into the plant host, which in turn causes the affected area to develop into a crown gall (tumor).

Type V secretion system (T5SS) – Also called the autotransporter system, type V secretion involves use of the Sec system for crossing the inner membrane. Proteins which use this pathway have the capability to form a beta-barrel with their C-terminus which inserts into the outer membrane, allowing the rest of the peptide (the passenger domain) to reach the outside of the cell.

Type VI secretion system (T6SS) – The Type VI secretion system gene clusters contain from 15 to more than 20 genes, two of which, *Hcp* and *VgrG*, have been shown to be nearly universally secreted substrates of the system. Structural analysis of these and other proteins in this system bear a striking resemblance to the tail spike of the T4 phage.

U

Unfolded protein response – Protein folding in endoplasmic reticulum is a complex process whose malfunction is implicated in disease aging. By use of cell's endogenous sensor (the unfolded protein response), several hundred yeast genes have been identified with roles in endoplasmic reticulum folding and systematically characterized their functional independencies by measuring unfolded protein response levels in double mutants.

Universality of code – Utilization of the same genetic code in all organisms. A codon codes for same amino acid in all the organisms. Mitochondrial genetic systems tolerate relatively frequent changes in codon assignment. This tolerance of mitochondrial genomes for changes in codon assignment is explained by RNA editing.

Up mutations – Mutations with increased transcription level are known as up mutations. Up mutations can be produced involving only a single base pair, which shows that RNA polymerase reaction with these sites must be very much specific.

V

V(D)J recombination – Developing B lymphocytes assemble immunoglobulin genes from widely scattered gene segments, using a somatic DNA rearrangement process known as V(D)J recombination. The V(D)J recombination is catalyzed by a "recombinase" enzymatic machinery. An immunoglobulin heavy chain variable region gene is generated from three sequences of DNA, V_H, D and J_H. Conserved hepta- and decanucleotides in vicinity of *V*, *D*, and *J* genes form recognition system required for recombination. These conserved sequences are separated by a 10-11 or 21-23 bp non-homologous stretch of nucleotides. Presumably similar sequences are found in the vicinity of *D* regions of the heavy chain genes. Histone H3 trimethylation at lysine 4 is involved in V(D)J recombination. V(D)J recombination assembles antigen receptor genes from component gene segments.

Vaccination – Any preparation which is used to confer immunity to a disease by inoculation. Vaccination is one of man's most significant inventions. It has eliminated smallpox and brought under control such diseases as diphtheria, poliomyelitis and measle.

Vaccines – The antigens that do not reproduce but elicit production of antibodies.

Variant repetitions of DNA sequences – Some repetitive DNA sequences differ in the number of repeats in different individuals. The different members of a variant repetitive gene family evolve by gene duplication and divergence of an ancestral gene.

Viral oncogenes – Many of the oncogenic retroviral genomes carry coding segments other than the usual viral genes. These segments are modified cellular genes that account for oncogenicity. Viral oncogenes are altered normal cellular gene.

VSG switching – *Trypanosome brucei* is the causative agent of African sleeping sickness in humans and one of the causes of nagana in cattle. This protozoan parasite evades the host immune system by antigenic variation, a periodic switching of its variant surface glycoprotein (VSG) coat. Introduction of a DNA double-strand break (DSB) adjacent to the ~70 bp repeats upstream of the transcribed *VSG* gene increases switching in vitro. A DSB is a natural intermediate of VSG gene conversion and that *VSG* switching is the result of this DSB by break-induced replication. Antigenic switching is induced by a single I-sceI-generated DSB.

W

Western blotting – A technique used to identify and locate proteins based on their ability to bind to specific antibodies. Western blotting can give information about the size of your protein (with comparison to a size marker or ladder in kDa), and also give you information on protein expression (with comparison to a control such as untreated sample or another cell type or tissue).

Wobble rules – The base at 5′-end of the anticodon is not as spatially confined as the other two, allowing it to form hydrogen bonds with any of several bases located at 3′-end of codon. For example, U at the wobble position can pair with either adenine or guanine, while I can pair with U, C or A.

X

Xenometabolites – Metabolites of foreign substances such as drugs are termed xenometabolites.

Y

Yeast protein RAP1 – The yeast protein RAP1, initially describes as a transcriptional regulator, binds in vitro to sequences found in a number of seemingly unrelated genomic loci. These include the silencers at the transcriptionally repressed mating-type genes, the promoters of many genes important for cell growth and the poly[(cytosine)1-3adenine] repeats of telomeres. RAP1 may be involved in telomere formation in vivo.

Yeast transcription activator GCN4 – "bZIP", a DNA-binding protein, consists of a basic region that contacts DNA and an adjacent "leucine zipper" that mediates protein dimerization. A peptide model for basic regions of yeast transcription activator GCN4 has been developed in which leucine zipper has been replaced by disulfide bond.

Yeast-two hybrid (Y2H) system – Yeast-two hybrid is an in vivo system for protein-protein interaction studies. A transcription factor generally consists of a DNA-binding domain (DBD) and an activation domain (AD), the former helping in DNA binding and the later facilitating the activation of a gene lying downstream. If one wants to study interaction between two proteins X and Y, two hybrid gene constructs are prepared, one having a hybrid gene coding for fusion protein DBD+X along with a reporter gene and the other coding for the fusion protein AD+Y. Yeast cells are co-transformed with both hybrid constructs, so that X and Y interact physically. This will bring DBD and AD in close proximity, AD will not be available at DBD site and reporter gene will not be expressed. Thus yeast-two hybrid system is used for study of protein-protein interaction.

Z

Zinc fingers – Zinc fingers constitute important eukaryotic DNA binding domains, being present in many transcription factors.

Subject Index

About the Author

Gurbachan S. Miglani. Retired as Professor of Genetics from Department of Plant Breeding, Genetics and Biotechnology, Punjab Agricultural University (PAU), Ludhiana, India after 35 years of service in the profession. Taught general genetics, advanced genetics, biotechnology, biochemical genetics, molecular genetics, immunogenetics, developmental genetics, and evolution to undergraduate and graduate students of constituent colleges of the PAU. Also taught genetics to undergraduates at Howard University, Washington, D.C., USA, as a Graduate Teaching Assistant.

School of Agricultural Biotechnology, PAU invited him in March 2010 as Adjunct Professor, to teach biotechnology to B.Sc. (Biotechnology), M.Sc. (Biotechnology) and Ph.D. (Biotechnology) students, which he did till December 2010. He was rehired in January 2011 as Visiting Professor.

Authored 136 publications (including research papers, review papers, books, book chapters, short notes/communications, abstracts, popular science articles) in Indian and foreign journals/magazines, and guided eleven M.Sc. and one Ph.D. students. Working for popularization of science by way of radio talks and writing popular articles for magazines and newspapers. Completed two prestigious research projects funded by the University Grants Commission, New Delhi, India. Authored six laboratory manuals for different genetics courses. which were published by the Punjab Agricultural University, Ludhiana, India. Authored *Dictionary of Plant Genetics and Molecular Biology* (1998), *Basic Genetics* (2000); *Advanced Genetics* ş First edition (2002), *Developmental Genetics* (2006), *Advanced Genetics* ş Second Edition (2007), *Fundamentals of Genetics* (2008), and *Genetic Material* (2013). Contributed two book chapters for books edited by foreign authors.

Associated with *The Journal of Plant Science Research* published by the Society for the Promotion of Plant Science Research, Jaipur (India), for last more than 12 years in different capacities, including editor. Recipient of Meritorious Teachers Award of the Punjab Agricultural University, Ludhiana (for the year 1997-98) and Sneh Prabha Shukla Memorial Award of Honor by Punjab Sahitya Kala Manch (Regd.), Ludhiana, Punjab, India for the year 2001.